T0224357

Communications in Computer and Information Science 623

Commenced Publication in 2007
Founding and Former Series Editors:
Alfredo Cuzzocrea, Dominik Ślęzak, and Xiaokang Yang

More information about this series at http://www.springer.com/series/7899

Wanxiang Che · Qilong Han
Hongzhi Wang · Weipeng Jing
Shaoliang Peng · Junyu Lin
Guanglu Sun · Xianhua Song
Hongtao Song · Zeguang Lu (Eds.)

Social Computing

Second International Conference
of Young Computer Scientists,
Engineers and Educators, ICYCSEE 2016
Harbin, China, August 20–22, 2016
Proceedings, Part I

Springer

Editors
Wanxiang Che
Harbin Institute of Technology
Harbin
China

Qilong Han
Harbin Engineering University
Harbin
China

Hongzhi Wang
Harbin Institute of Technology
Harbin
China

Weipeng Jing
Northeast Forestry University
Harbin
China

Shaoliang Peng
National University of Defense Technology
Changsha
China

Junyu Lin
Harbin Engineering University
Harbin
China

Guanglu Sun
Harbin University of Science
and Technology
Harbin
China

Xianhua Song
Harbin University of Science
and Technology
Harbin
China

Hongtao Song
Harbin Engineering University
Harbin
China

Zeguang Lu
Harbin Sea of Clouds and Computer
Technology
Harbin
China

ISSN 1865-0929 ISSN 1865-0937 (electronic)
Communications in Computer and Information Science
ISBN 978-981-10-2052-0 ISBN 978-981-10-2053-7 (eBook)
DOI 10.1007/978-981-10-2053-7

Library of Congress Control Number: 2016945792

Printed on acid-free paper

This Springer imprint is published by Springer Nature
The registered company is Springer Science+Business Media Singapore Pte Ltd.

Preface

As the general and program co-chairs of the Second International Conference of Young Computer Scientists, Engineers and Educators 2016 (ICYCSEE 2016), it is our great pleasure to welcome you to the proceedings of the conference, which was held in Harbin, China, during August 20–22, 2016, hosted by Harbin Engineering University. The goal of this conference is to provide a forum for young computer scientists, engineers, and educators.

The call for papers of this year's conference attracted 338 paper submissions. After the hard work of the Program Committee, 91 papers were accepted to appear in the conference proceedings, with an acceptance rate of 27 %. The main theme of this conference was "Social Computing." The accepted papers cover a wide range of areas related to social computing such as: science and foundations for social computing, computation infrastructure for social computing, big data management analysis for social computing, evaluation methodologies for social computing and social media, intelligent computation for social computing, natural language processing techniques and culture analysis in social computing and social media, mobile social computing and social media, privacy and security in social computing and social media, public opinion analysis for social media, social modeling, social network analysis, user-generated content (wikis, blogs), and visualizing social interaction.

We would like to thank all the Program Committee members – 178 members from 84 institutes – for their hard work in completing the review tasks. Their collective efforts made it possible to attain quality reviews for all the submissions within a few weeks. Their diverse expertise in each individual research area helped us to create an exciting program for the conference. Their comments and advice helped the authors to improve the quality of their papers and gain deeper insights.

Our thanks also go to the authors and participants for their tremendous support in making the conference a success. Moreover, we thank Dr. Lanlan Chang and Jian Li from Springer, whose professional assistance was invaluable in the production of the proceedings.

Besides the technical program, this year ICYCSEE offered different experiences to the participants. We hope you enjoy the conference proceedings.

June 2016

Qilong Han
Wanxiang Che
Hongzhi Wang
Shoaling Peng
Junyu Lin

Organization

The Second International Conference of Young Computer Scientists, Engineers and Educators (ICYCSEE) 2016 (http://2016.icycsee.org) took place in Harbin, China, during August 2016 20–22, hosted by Harbin Engineering University.

ICYCSEE 2016 Steering Committee

Jianzhong, Li	Harbin Institute of Technology, China
Ting, Liu	Harbin Institute of Technology, China
Zhongbin, Su	Northeast Agricultural University, China
Guisheng, Yin	Harbin Engineering University, China

General Chairs

Qilong, Han	Harbin Engineering University, China
Wanxiang, Che	Harbin Institute of Technology, China

Program Chairs

Hongzhi, Wang	Harbin Institute of Technology, China
Shaoliang, Peng	National University of Defense Technology, China
Junyu, Lin	Harbin Engineering University, China

Organization Chairs

Hongtao, Song	Harbin Engineering University, China
Zeguang, Lu	Sea of Clouds and Computer Technology Services Ltd., China

Publication Chairs

Guanglu, Sun	Harbin University of Science and Technology, China
Zhaowen, Qiu	Northeast Forestry University, China

Publication Co-chairs

Weipeng, Jing	Northeast Forestry University, China
Xianhua, Song	Harbin University of Science and Technology, China

Education Chairs

Yingtao, Zhang	Harbin Institute of Technology, China
Zhongyang, Han	Heilongjiang Institute of Technology, China

Industrial Chair

Jiquan, Ma   Heilongjiang University, China

Demo Chairs

Changjian, Zhou   Northeast Agricultural University, China
Qi, Han   Harbin Institute of Technology, China

Panel Chairs

Haiwei, Pan   Harbin Engineering University, China
Hui, Gao   Harbin Huade University, China

Registration/Financial Chairs

Yong, Wang   Harbin Engineering University, China
Fa, Yue   Sea of Clouds and Computer Technology Services Ltd.,
                                    China

Post/Expo Chair

Tingting, Chen   SuperMap Software Co., Ltd

ICYCSEE Steering Committee

Guanglu, Sun   Harbin University of Science and Technology, China
Hai, Jin   Huazhong University of Science and Technology,
                                    China
Haoliang, Qi   Heilongjiang Institute of Technology, China
Hongzhi, Wang   Harbin Institute of Technology, China
Jiajun, Bu   Zhejiang University, China
Jian, Chen   PARATERA
Junyu, Lin   Harbin Engineering University, China
Liehuang, Zhu   Beijing Institute of Technology, China
Min, Zhu   Sichuan University, China
Qilong, Han   Harbin Engineering University, China
Shaoliang, Peng   National University of Defense Technology, China
Tao, Wang   Peking University, China
Tian, Feng   Institute of Software Chinese Academy of Sciences,
                                    China
Wanqing, He   Qihoo360 Cloud Company
Wanxiang, Che   Harbin Institute of Technology, China
Weipeng, Jing   Northeast Forestry University, China
Xiaohui, Wei   Jilin University, China
Xiaoru, Yuan   Peking University, China

Xuebin, Chen	North China University of Science and Technology, China
Yanjuan, Sang	Beijing Gooagoo Technology Service Co., Ltd., China
Yiliang, Han	Engineering University of CAPF
Yingao, Li	Neuedu
Yinhe, Han	Institute of Computing Technology, Chinese Academy of Sciences, China
Yu, Yao	Northeastern University, China
Yunquan, Zhang	Institute of Computing Technology, Chinese Academy of Sciences, China
Zeguang, Lu	Harbin Sea of Clouds and Computer Technology Services Ltd., China
Zhaowen, Qiu	Northeast Forestry University, China
Zheng, Shan	The PLA Information Engineering University

Program Committee

Tian, Bai	Jilin University, China
Zhifeng, Bao	University of Tasmania, Australia
Jiajun, Bu	Zhejiang University, China
Zhipeng, Cai	Georgia State University, USA
Wanxiang, Che	Harbin Institute of Technology, China
Xuebin, Chen	Hebei United University, China
Wenliang, Chen	Soochow University, China
Siyao, Cheng	Harbin Institute of Technology, China
Dansong, Cheng	Harbin Institute of Technology, China
Yuan, Cheng	Harbin University of Science and Technology, China
Yan, Chu	Harbin Engineering University, China
Lei, Cui	Microsoft Research
Beiliang, Cui	Nanjing Tech University, China
Bin, Cui	Peking University, China
Jianrui, Ding	Harbin Institute of Technology, China
Minghui, Dong	Institute for Infocomm Research, Singapore
Xunli, Fan	Northwest University, China
Chunxiang, Fan	University of Ulm, Germany
Guangsheng, Feng	Harbin Engineering University, China
Yansong, Feng	University of Edinburgh, UK
Guohong, Fu	Heilongjiang University, China
Hui, Gao	Harbin Huade University, China
Shang, Gao	Jilin University, China
Jing, Gao	University at Buffalo, USA
Dianxuan, Gong	North China University of Science and Technology, China
Yi, Guan	Harbin Institute of Technology, China
Quanlong, Guan	Jinan University, China
Yuhang, Guo	Beijing Institute of Technology, China

Ma, Han	Georgia State University, USA
Qilong, Han	Harbin Engineering University, China
Zhongyuan, Han	Harbin Institute of Technology, China
Qi, Han	Harbin Institute of Technology, China
Xianpei, Han	Institute of Software, Chinese Academy of Sciences, China
Zhongjun, He	Baidu Inc.
Zhenying, He	Fudan University, China
Yu, Hong	Soochow University, China
Zhengang, Jiang	Changchun University of Science and Technology, China
Cheqing, Jin	East China Normal University, China
Peng, Jin	Peking University, China
Weipeng, Jing	Northeast Forestry University, China
Leilei, Kong	Heilongjiang Institute of Technology, China
Dapeng, Lang	Harbin Engineering University, China
Dan, Le	Harbin Institute of Technology, China
Mei, Li	China University of Geosciences (Beijing), China
Shuaicheng, Li	City University of Hong Kong, SAR China
Jie, Li	Harbin Institute of Technology, China
Zhixun, Li	Harbin Institute of Technology, China
Junbao, Li	Harbin Institute of Technology, China
Ao, Li	Harbin University of Science and Technology, China
Peng, Li	Institute of Information Engineering, CAS, China
Maoxi, Li	Jiangxi Normal University, China
Chenliang, Li	Wuhan University, China
Junyu, Lin	Harbin Engineering University, China
Xianmin, Liu	Harbin Institute of Technology, China
Shaohui, Liu	Harbin Institute of Technology, China
Ming, Liu	Harbin Institute of Technology, China
Xiaoguang, Liu	Nankai University, China
Hailong, Liu	Northwestern Polytechnical University, China
Chenguang, Liu	Samsung
Zhiyuan, Liu	Tsinghua University, China
Zeguang, Lu	Harbin Sea of Clouds and Computer Technology Services Ltd., China
Nan, Lu	Shenzhen University, China
Jizhou, Luo	Harbin Institute of Technology, China
Zhiyong, Luo	Harbin University of Science and Technology, China
Chengguo, Lv	Heilongjiang University, China
Yanjun, Ma	Baidu Inc.
Shuai, Ma	Beihang University, China
Jiquan, Ma	Heilongjiang University, China
Dapeng, Man	Harbin Engineering University, China
Tiezheng, Nie	Northeastern University, China
Haiwei, Pan	Harbin Engineering University, China

Liqiang, Pan	Harbin Institute of Technology, China
Wei, Pan	Northwestern Polytechnical University, China
Zhijuan, Peng	Nantong University, China
Shaoliang, Peng	National University of Defense Technology, China
Jian, Peng	Sichuan University, China
Yuwei, Peng	Wuhan University, China
Shaojie, Qiao	Southwest Jiaotong University, China
Zhijing, Qin	University of California, Irvine, USA
Xipeng, Qiu	Fudan University, China
Zhaowen, Qiu	Northeast Forestry University, China
Ying, Shan	Harbin Guangsha University, China
Juan, Shan	Pace University, USA
Bin, Shao	Microsoft Research Asia
Shengfei, Shi	Harbin Institute of Technology, China
Xianhua, Song	Harbin University of Science and Technology, China
Yangqiu, Song	West Virginia University, USA
Jie, Su	Harbin University of Science and Technology, China
Jinsong, Su	Xiamen University, China
Hailong, Sun	Beihang University, China
Xiaoling, Sun	Dalian University of Technology, China
Weiwei, Sun	Fudan University, China
Jianguo, Sun	Harbin Engineering University, China
Chengjie, Sun	Harbin Institute of Technology, China
Guanglu, Sun	Harbin University of Science and Technology, China
Dongpu, Sun	Harbin University of Science and Technology, China
Xu, Sun	Peking University, China
Buzhou, Tang	Harbin Institute of Technology of Shenzhen Graduate School, China
Jintao, Tang	National University of Defense Technology, China
Zhanyong, Tang	Northwest University, China
Jianhua, Tao	Chinese Academy of Sciences, China
Yongxin, Tong	Beihang University, China
Xifeng, Tong	Northeast Petroleum University
Zhiying, Tu	Harbin Institute of Technology, China
Zumin, Wang	Dalian University, China
Hongya, Wang	Donghua University, China
Haofen, Wang	East China University of Science and Technology, China
Xingmei, Wang	Harbin Engineering University, China
Hongzhi, Wang	Harbin Institute of Technology, China
Jinbao, Wang	Harbin Institute of Technology, China
Tiantian, Wang	Harbin Institute of Technology, China
Kechao, Wang	Harbin Institute of Technology, China
Wei, Wang	Institute of Software, Chinese Academy of Sciences, China
Botao, Wang	Northeastern University, China

Chaokun, Wang	Tsinghua University, China
Yanjie, Wei	Shenzhen Institutes of Advanced Technology, Chinese Academy of Sciences, China
Wei, Wei	Xi'an University of Technology, China
Xiuxiu, Wen	Harbin Engineering University, China
Xiaojun, Wen	Shenzhen Polytechnic, China
Xiangqian, Wu	Harbin Institute of Technology, China
Sai, Wu	Zhejiang University, China
Rui, Xia	Nanjing University of Science and Technology, China
Min, Xian	Utah State University, USA
Tong, Xiao	Northeastern University, China
Hui, Xie	Harbin Institute of Technology, China
Yu, Xin	Harbin Engineering University, China
Junchang, Xin	Northeastern University, China
Zheng, Xu	Harbin Institute of Technology, China
Jianliang, Xu	Hong Kong Baptist University, SAR China
Ying, Xu	Hunan University, China
Yongzeng, Xue	Harbin Institute of Technology, China
Ziye, Yan	Beijing Institute of Technology, China
Shaohong, Yan	North China University of Science and Technology, China
Hailu, Yang	Harbin University of Science and Technology, China
Xiaochun, Yang	Northeastern University, China
Yajun, Yang	Tianjin University, China
Xiaoyan, Yin	Northwest University, China
Shouyi, Yin	Tsinghua University, China
Xz, Yu	Harbin Institute of Technology, China
Haining, Yu	Harbin Institute of Technology, China
Zhengtao, Yu	Kunming University of Science and Technology, China
Xiaohui, Yu	Shandong University, China
Ye, Yuan	Northeastern University, China
Yingjun, Zhang	Beijing Jiaotong University, China
Rong, Zhang	East China Normal University, China
Liguo, Zhang	Harbin Engineering University, China
Zhiqiang, Zhang	Harbin Engineering University, China
Yingtao, Zhang	Harbin Institute of Technology, China
Delong, Zhang	Harbin Institute of Technology, China
Yu, Zhang	Harbin Institute of Technology, China
Ru, Zhang	Harbin University of Commerce, China
Meishan, Zhang	Heilongjiang University, China
Jiajun, Zhang	Institute of Automation Chinese Academy of Sciences, China
Rui, Zhang	Jilin University, China
Jian, Zhang	Northeast Forestry University
Xiao, Zhang	Renmin University of China, China
Hu, Zhang	Shanxi University, China

Contents – Part I

Contents – Part II

Industry Track

Demo Track

Research Track

A Context-Aware Model Using Distributed Representations for Chinese Zero Pronoun Resolution

Bingbing Wu[✉] and Tiejun Zhao

School of Computer Science and Technology,
Harbin Institute of Technology, Harbin, China
{wbb,tjzhao}@mtlab.hit.edu.cn

Abstract. Previous approaches to Chinese zero pronoun resolution mainly use syntactic information and probabilistic methods, but the context information is ignored. To make full use of the context and semantic information, we build a context-aware model. We propose a key words extraction strategy and design a classification model by using distributed representations as context feature. To our best knowledge, this is the first work using distributed representations in Chinese zero pronoun resolution. Experimental results show that our approach achieves a better performance than previous supervised methods.

Keywords: Chinese zero pronoun · Context-awareness · Distributed representations · SVM

1 Introduction

Coreference resolution has been an important technique in Natural Language Processing (NLP) for decades, such as discourse analysis, information extraction and question answering tasks. Coreference resolution has also been used in social computing field. The goal of coreference resolution is to decide whether some noun phrases refer to one real entity. Zero pronoun (ZP) is a special case which is different from overt pronouns. A zero pronoun is a gap which is a null form used to refer to a real-world entity that supplies information about the gap in a sentence [1]. Because lack of grammatical attributes essential for overt pronoun resolution such as Number and Gender, ZP resolution is more difficult and challenging than overt pronoun resolution.

A zero anaphor can be classified into either anaphoric or non-anaphoric, depending on whether it has an antecedent in the discourse. If a ZP corefers with one or more certain noun phrases (NPs) in the preceding text, we call this type of zero pronoun as the anaphoric zero pronoun (AZP). Compared with English and some other languages, zero pronoun is more likely to occur in Chinese. Here is an example of anaphoric zero pronoun from OntoNotes 5.0 corpus [2]:

[中国 机电 产品 进出口 贸易]$_{NP1}$ 继续 增加, *pro* 占 总 进出口 的 比重 继续 上升。

*[China electronic products import and export trade]$_{NP1}$ continues increasing, *pro* represents total import and export's ratio continues increasing.*

© Springer Science+Business Media Singapore 2016
W. Che et al. (Eds.): ICYCSEE 2016, Part I, CCIS 623, pp. 3–11, 2016.
DOI: 10.1007/978-981-10-2053-7_1

The anaphoric zero pronoun *pro* is coreferring to noun phrase NP1. Chinese zero pronouns have been studied in syntactic level and features are extracted from syntactic trees [1]. In addition, tree kernel method [3] is proposed as well. These methods may not work well when the syntactic tree is simple such as conversation text. And using syntactic features tends to generate inconsistent samples because two sub-trees with same structure may have opposite labels. Furthermore, the lexical and semantic information are ignored. To overcome these shortcomings, we use distributed representations to build a context-aware model. Distributed representations are called word embeddings. Each dimension of the embedding represents a latent feature of the word. Distributed representations are used in many NLP tasks and are proved to be good at capturing syntactic and semantic regularities in language [4].

Our motivation is to search for the relation of distributed representations between AZP's antecedent and context below the AZP position. We propose a strategy to extract key words from the context and candidate antecedents. By using key words' distributed representations, we train a classifier to identify whether a candidate is the real antecedent of a certain AZP. The experiment results show our model outperforms previous supervised methods.

The rest of this paper is organized as follows. In the Sect. 2, we introduce previous works. In Sect. 3 we describe how to use distributed representations to build our context-aware model. In Sect. 4, we introduce the baseline methods and report the experiment results. Finally, we conclude our work and forecast future work.

2 Related Work

Chinese Zero Pronoun. Previous methods about resolution of zero pronoun are rule-based. Converse employ Hobb's algorithm [6] and select antecedent by syntactic structure in the Chinese Tree Bank documents. Maximum entropy models are also used to find the antecedents for overt, third-person pronouns [5]. Yeh and Chen [7] employ Centering Theory (Grosz et al. [8]) and constraint rules to identify the antecedents of zero anaphors by using shallow parsing.

Zhao and Ng [1] propose a feature-based method which is the first supervised machine learning approach. They extract a set of syntactic and positional features in zero pronoun identification and resolution tasks. The two tasks are regarded as binary classification problems respectively. Kong and Zhao [3] propose a tree-kernel method to resolve Chinese zero pronoun with appropriate syntactic parse tree structures. They build a unified framework dealing with zero pronoun identification, anaphoricity determination and antecedent selection by using tree-kernel method. Chen and Ng [9] extend Zhao and Ng [1] feature set and exploit the coreference links between zero pronouns. Chen and Ng [10] propose an unsupervised approach by using ranking model and Integer Linear Programming. They assume that zero pronouns and overt pronouns have similar distributions and train an unsupervised model to resolve Chinese zero pronoun. Rao et al. [11] builds a novel model that tracks the flow of focus in a discourse. Chen and Ng [12] propose an unsupervised probabilistic model for zero pronoun resolution.

Distributed Representations. A word representation is regarded as a vector which each dimension's value corresponds to a feature. A variety of researchers (Miller et al. [13]; Huang and Yates [14]) use clustering to induce unsupervised word representations. With the development of neural network, neural language models (Bengio et al. [15]; Mnih and Hinton [16]; Collobert and Weston [17]) is proposed and is used to induce dense real-valued low-dimensional distributed representations. Mikolov et al. [18] find vector-space word representations are good at capturing syntactic and semantic regularities in language. For example, with the induced vector representations, "King - Man + Woman" results in a vector very close to "Queen". Distributed representations have been used in a variety of NLP tasks and have demonstrated good performance. Tools like Word2vec [19] and Glove [20] are designed to generate distributed representations.

3 Context-Aware Model via Distributed Representations

In this section, we show how to build a context-aware model to resolve Chinese zero pronoun. We define the task of antecedent selection as a binary classification problem. We employ a strategy to generate antecedent candidates and create positive or negative samples. A heuristic method of extracting key words from context is proposed and an instance form is designed in order to transform key words to representation features. Finally, we build a binary classifier to identify candidate antecedents.

3.1 Overview of Our Approach

ZP resolution is composed of AZP identification and AZP resolution. In this paper we concentrate on AZP resolution task. In other words, we focus on determining whether an AZP and a candidate antecedent are coreferent.

We employ Zhao and Ng [1] strategy to generate candidate antecedents and add a restriction. If there is an AZP z in a sentence, we choose noun phrases (NPs) that either is maximal NPs or modifier NPs preceding z as candidate antecedents. Figure 1 shows an example that NP_1, NP_2 and NP_3 are all NP candidates of AZP's antecedent. NP_1 is a maximal noun phrase and NP_2, NP_3 are modifier noun phrase.

In order to use context information after an AZP's position, we design an instance form that contains key words from the context and candidate antecedents. It is obvious that in a sentence, parts of words following an AZP have semantic regularities with AZP's antecedent. We propose a set of rules to select key words that may have relation with antecedent. When key words are extracted, we combine key words and the real antecedent as positive instance as well as combine key words and other candidates as negative instances. Words in all instances are transformed to distributed representations and we train a classification model using these distributed representations.

3.2 Rules of Key Words Extraction

The principle of extracting key words is that what we extract should have relation with the antecedent as much as possible. However, it is difficult to make some rules that can

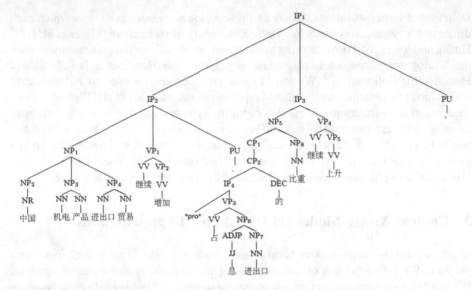

Fig. 1. The parse tree that corresponds to the AZP example in Sect. 3.2

fit all sentence patterns. According to our data analysis, we find that the verbs and nouns close to AZP tend to have semantic links with the antecedent more than other words. We define m as the maximal number of key words extracted from context after AZP's position in a sentence. The extraction rules are as follows:

(1) Extract verbs and nouns from AZP's following context sequentially.
(2) If the total number of key words not exceeds m, fill symbol '*' until the total number equals m.
(3) If the total number of key words exceeds m, delete words back to front until the total number equals m, nouns have higher priority to be deleted than verbs, modifier nouns have higher priority to be deleted than terminal nouns.

Additionally, because of the word numbers of candidate antecedents are different we define the maximal number of candidate's words as c. We extract c words from every candidate. If the total number of candidate is over c, keep the last c words; else fill symbol '*' in front of the candidate until the word number is c.

In OntoNotes 5.0 corpus, AZP has annotation with its antecedent. So we can get tag t of every sample easily. In a supervised task, we define the form of every sample as [t, $a_1,\ldots, a_c, w_1,\ldots, w_m$]. In this form, t is the tag of sample, a_i present the key word in AZP's candidate antecedent, w_i is the key word from AZP's following context. For example, in Fig. 1, let $m = 6$, $c = 3$, we get the forms:

[+1 产品 进出口 贸易 占 进出口 比重 继续 上升 *]
[-1 * * 中国 占 进出口 比重 继续 上升 *]
[-1 * 机电 产品 占 进出口 比重 继续 上升 *]

From example above, the two words 进出口 (import and export) and 贸易 (trade) from antecedent have closely relation with the word 进出口 (import and export) from context in semantic level. On the contrary, the words 中国 (China) and 产品 (products) have little relation with key words in AZP's following context like words 占 (represents) and 上升 (increasing).

3.3 Context-Aware Classification Model

There are CBOW model and Skip-gram model [23] in word2vec. CBOW model predict the current word based on the context while Skip-gram model predict the context based on the current word. We use word2vec to train a Skip-gram model and get the distributed representations dictionary. The skip-gram model tries to maximize classification of a word based on another word in the same sentence. In the Skip-gram model, every word w corresponds to two parameter vectors u_w and v_w. u_w is the input vector of w and v_w is the output vector of w. Given w_i, the prediction probability of w_j is as follows:

$$P(w_i|w_j) = \frac{exp(u_{w_j}^T v_{w_j})}{\sum_{w=1}^{W} exp(u_{w_j}^T v_{w_j})} \qquad (1)$$

W is the size of vocabulary. The structure of Skip-gram model is shown in Fig. 2. In action, when the training data is not large, Skip-gram model gives better distributed representations than CBOW model.

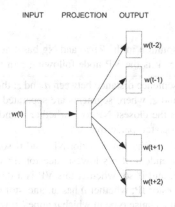

Fig. 2. Skip model structure

We define the dimension of every word embedding as d. All the words in the samples are replaced by its distributed representations and every sample is transformed as a feature vector with $(c + m) * d$ dimensions. The symbol '*' in samples should be replaced by zero vector (with d dimensions). We use SVM model to train a binary classifier. This classifier is used to identify whether a candidate is the AZP's antecedent.

4   Experiment

4.1   Dataset

We employ Chinese portion of the OntoNotes 5.0 corpus which was used in the official CoNLL-2012 [21] task. Because only the training set and development set contain ZP coreferential annotation in CoNLL-2012 shared dataset, we employ training set for model training and development set for model evaluation. In OntoNotes 5.0 corpus, a ZP is tagged as *pro* and all ZPs that have explicit coreferential annotation are regarded as anaphoric ZPs. The Chinese portion of OntoNotes 5.0 contains six types of source: Broadcast News (BN), Newswire (NW), Broadcast Conversation (BC), Magazine (MZ), Telephone Conversation (TC), Web Blog (WB).

4.2   Experimental Setup

Baseline Systems. We introduce three baseline systems which are all supervised machine learning models. (1) Zhao and Ng [1]. They propose a method of generating candidate antecedents and design 26 features for the zero pronoun and candidate. Soon et al.'s method [22] is used to create positive and negative instances. The description of these features is listed in Table 1 [10]. (2) Kong and Zhao [3]. They use tree-kernel method to resolve Chinese zero pronoun. They use Z&N's method to generate candidate antecedents and use Soon's method to create instances. (3) Chen and Ng [9]. They extend Z&N's method and add features about the contextual information between candidate antecedents and ZPs. They also create coreference links to resolve far-away antecedents for ZPs.

Table 1. Features for AZP resolution in the Zhao and Ng baseline system. z is a zero pronoun. a is a candidate antecedent of z. V is the VP node following z in the parse tree.

Features between a and z (4)	The sentence distance between a and z; the segment distance between a and z, where segments are separated by punctuations; whether a is the closest NP to z; whether a and z are siblings in the associated parse tree
Features on a (12)	Whether a has an ancestor NP, and if so, whether this NP is a descendent of a's lowest ancestor IP; whether a has an ancestor VP, and if so, whether this VP is a descendent of a's lowest ancestor IP; whether a has an ancestor CP; the grammatical role of a; the clause type in which a appears; whether a is an adverbial NP, a temporal NP, a pronoun or a named entity; whether a is in the headline text
Features on z (10)	Whether V has an ancestor NP, and if so, whether this NP node is a descendent of V's lowest ancestor IP; whether V has an ancestor VP, and if so, whether this VP is a descendent of V's lowest ancestor IP; whether V has an ancestor CP; the grammatical role of z; the type of the clause in which V appears; whether z is the first or last ZP of the sentence; whether z is in the headline of the text

The baseline systems above have the same strategy of generating candidate antecedents and instances, but use different ways to classify whether a candidate antecedent have coreference with an AZP. The three methods all use SVM model to train the binary classifier while Z&N and C&N using SVM^{light} tool and K&Z using SVM$^{light-TK}$. Our linear model uses the same data and same strategies of generating candidate antecedents and instances with the baseline methods. Therefore, we use Chen and Ng [10] experimental results of baselines as the comparative data in Table 2.

Table 2. Resolution results on the test set.

	Our Approach			Baselines								
				Chen and Ng			Zhao and Ng			Kong and Zhou		
	P	R	F	P	R	F	P	R	F	P	R	F
All	47.3	51.0	**49.0**	47.7	47.7	47.7	41.5	41.5	41.5	44.9	44.9	44.9
BC	55.9	66.0	**60.5**	52.7	52.7	52.7	44.7	44.7	44.7	43.5	43.5	43.5
BN	39.2	47.7	43.0	47.2	47.2	**47.2**	43.1	43.1	43.1	51.0	51.0	51.0
MZ	36.3	25.2	29.7	34.6	34.6	**34.6**	28.4	28.4	28.4	32.7	32.7	32.7
NW	33.8	43.1	37.9	38.1	38.1	38.1	40.5	40.5	**40.5**	34.5	34.5	34.5
TC	70.0	77.7	**73.7**	51.2	51.2	51.2	42.8	42.8	42.8	48.4	48.4	48.4
WB	45.3	41.6	43.3	46.1	46.1	**46.1**	40.1	40.1	40.1	45.4	45.4	45.4

Context-Aware Classification Model. We implement the aforementioned idea about building a linear model by using distributed representations. After tuning, we set the parameters $c = 3$, $m = 7$ and the dimension of distributed representations is 200 and then get the best performance.

4.3 Results and Analysis

In Table 2, we can see that our method has a better performance in the whole data than baseline systems and exceeds C&N's method 2.3 % in F-score value. Particularly in the sources BC (Broadcast Conversation) and TC (Telephone Conversation), our method has a significant improvement than other baseline methods. Through data analysis, we find that in sources BC and TC, AZP's antecedents are more likely pronouns such as 他 (he) and 我们 (we). Our model can identify the relation between AZP and key words from context better when finding these distributed representations. Some sentences in dataset can be very short such as conversation data. The baseline methods that using syntactic tree information may no longer work in this case because the syntactic trees contain very little information. Furthermore, features from syntactic tree just describe AZP and its antecedent's structural characteristics, structural features may cause ambiguity when two instances of AZP and candidate antecedent have the same structure but different tags. On the contrary, using distributed representations can easily avoid this because two instances with same structure having different distributed representations features.

5 Conclusions and Future Work

In this paper, we propose a context-aware model using distributed representations to resolve Chinese zero pronoun. We design a set of rules to extract key words in context and a key words form in order to combine antecedent and these key words. We use lexical and semantic information to build a classification by transforming context key words to distributed representations. The experimental results indicate the effectiveness of using distributed representations in this task. To the best of our knowledge, this is the first work in Chinese zero pronoun resolution by using distributed representations.

In the future work, we will redesign the rules of key words extraction in syntactic level and explore different application methods to improve the accuracy by using distributed representations. Furthermore, deep neural networks are suitable for this task such as RNN, LSTM model. On the other hand, it is foreseeable that combining distributed representations features and syntactic features is another way to resolve Chinese zero pronoun.

References

1. Zhao, S., Ng, H.T.: Identification and resolution of Chinese zero pronouns: a machine learning approach. In: EMNLP-CoNLL, vol. 2007, pp. 541–550 (2007)
2. Weischedel, R., Palmer, M., Marcus, M., Hovy, E., Pradhan, S., Ramshaw, L., Xue, N., Taylor, A., Kaufman, J., El-Bachouti, M.: Ontonotes release 5.0 LDC2013T19. Linguistic Data Consortium, Philadelphia (2013)
3. Kong, F., Zhou, G.: A tree kernel-based unified framework for Chinese zero anaphora resolution. In: Proceedings of the 2010 Conference on Empirical Methods in Natural Language Processing, pp. 882–891 (2010)
4. Mikolov, T., Yih, W.T., Zweig, G.: Linguistic regularities in continuous space word representations. In: HLT-NAACL, pp. 746–751 (2013)
5. Converse, S. P.: Pronominal anaphora resolution in Chinese. Doctoral Dissertation, University of Pennsylvania (2006)
6. Hobbs, J.R.: Resolving pronoun references. Lingua **44**(4), 311–338 (1978)
7. Yeh, C.L., Chen, Y.C.: Zero anaphora resolution in Chinese with shallow parsing. J. Chin. Lang. Comput. **17**(1), 41–56 (2007)
8. Grosz, B.J., Weinstein, S., Joshi, A.K.: Centering: a framework for modeling the local coherence of discourse. Comput. Linguist. **21**(2), 203–225 (1995)
9. Chen, C., Ng, V.: Chinese zero pronoun resolution: some recent advances. In: EMNLP, pp. 1360–1365 (2013)
10. Chen, C., Ng, V.: Chinese zero pronoun resolution: an unsupervised approach combining ranking and integer linear programming. In: AAAI, pp. 1622–1628 (2014)
11. Rao, S., Ettinger, A., Hal Daumé, I.I.I., Resnik, P.: Dialogue focus tracking for zero pronoun resolution. In: Proceedings of the Annual Conference of the North American Chapter of the Association for Computational Linguistics (NAACL), pp. 494–502 (2015)
12. Chen, C., Ng, V.: Chinese zero pronoun resolution: a joint unsupervised discourse-aware model rivaling state-of-the-art resolvers, vol. 2, Short Papers, p. 320 (2015)
13. Miller, S., Guinness, J., Zamanian, A.: Name tagging with word clusters and discriminative training. In: HLT-NAACL, vol. 4, pp. 337–342 (2004)

14. Huang, F., Yates, A.: Distributional representations for handling sparsity in supervised sequence-labeling. In: Proceedings of the Joint Conference of the 47th Annual Meeting of the ACL and the 4th International Joint Conference on Natural Language Processing of the AFNLP, pp. 495–503 (2009)
15. Bengio, Y., Schwenk, H., Senécal, J.S., Morin, F., Gauvain, J.L.: Neural probabilistic language models. In: Innovations in Machine Learning, pp. 137–186 (2006)
16. Mnih, A., Hinton, G.: Three new graphical models for statistical language modeling. In: Proceedings of the 24th International Conference on Machine Learning, pp. 641–648 (2007)
17. Collobert, R., Weston, J.: A unified architecture for natural language processing: deep neural networks with multitask learning. In: Proceedings of the 25th International Conference on Machine Learning, pp. 160–167 (2008)
18. Mikolov, T., Yih, W.T., Zweig, G.: Linguistic regularities in continuous space word representations. In: HLT-NAACL, pp. 746–751 (2013)
19. Mikolov, T., Chen, K., Corrado, G., Dean, J.: word2vec (2014)
20. Pennington, J., Socher, R., Manning, C.D.: Glove: global vectors for word representation. In: EMNLP, vol. 14, pp. 1532–1543 (2014)
21. Pradhan, S., Moschitti, A., Xue, N., et al.: CoNLL-2012 shared task: modeling multilingual unrestricted coreference in OntoNotes. In: Joint Conference on EMNLP and CoNLL-Shared Task, pp. 1–40. Association for Computational Linguistics (2012)
22. Soon, W.M., Ng, H.T., Lim, D.C.Y.: A machine learning approach to coreference resolution of noun phrases. In: Computational Linguistics, pp. 521–544 (2001)
23. Mikolov, T., Chen, K., Corrado, G., et al.: Efficient estimation of word representations in vector space (2013). arXiv preprint arXiv:1301.3781

A Hierarchical Learning Framework
for Steganalysis of JPEG Images

Baojun Qi[✉]

Luoyang University of Foreign Languages, Luoyang 471003, Henan, China
bjqi.ae@gmail.com

Abstract. JPEG Steganalysis is an important technique for forensic analysis of images on online social networks. This paper proposes a novel hierarchical learning framework for JPEG steganalysis. It is based on the observation that both regions of an image with different textural complexity and regions of different images with similar textural complexity tend to have different embedding probabilities. In the training stage of our framework, images are firstly clustered into a number of categories using Gaussian Mixture Model (GMM). Then, images in each category are decomposed into smaller blocks, and these blocks are also clustered into limited classes. Finally, a classifier is trained for each class of blocks. In the testing stage, an image and its blocks are also classified using trained GMM, and each block is tested on corresponding classifiers to make the final decision by weighed sum of individual results. Extensive experimental results show a better performance of our framework compared with some other previous learning framework.

Keywords: Steganography · Steganalysis · Ensemble · Framework · Wavelet · GMM

1 Introduction

For online social network users, steganography is a widely used technique to share secret messages, which hides secret data in innocent-looking cover data such that no one can detect the existence of the hidden data. To avert illegal usage of steganography by criminals or terrorist, steganalysis is proposed to discover the presence of hidden message in social media data, and it has been extensively studied and has made great progresses over recent years [16]. Blind steganalysis, designed for analyzing any steganographic schemes using proper training datasets, is capable of detecting previous unseen stego methods in realistic scenarios, and thus has attracted more attention than targeted steganalysis [5].

A blind steganalyzer mainly consists of two parts: an image model for describing embedding changes of steganographic algorithms, and a machine learning tool for exploring classifiers of cover and stego images. The image model, aiming at extracting steganalytic features sensitive to embedding changes while insensitive to image contents, tends to adopt high dimensional features for modern

© Springer Science+Business Media Singapore 2016
W. Che et al. (Eds.): ICYCSEE 2016, Part I, CCIS 623, pp. 12–23, 2016.
DOI: 10.1007/978-981-10-2053-7_2

steganlytic algorithms, such as SRM [4], J+SRM [14], PSRM [6], PHARM [8], DCTR [7] and so on. The machine learning tool widely used in steganalytic community is the ensemble learning, due to its low computational complexity and satisfactory performance in high dimensional feature spaces for large training datasets [4].

In the literature, there are two approaches to improve the performance of blind steganalysis algorithms. The first is developing more effective steganalytic features to detect a wide spectrum of steganographic algorithms [19]. To this end, the steganalytic features are usually with high dimensionality to capture as many statistical dependencies among neighboring pixels as possible [4,6–8,14,19]. However, high dimensionality always leads to other problems in steganalysis, such as the lack of large scale training datasets, the degradation of the generalization abilities and the complexity of classifier training [13].

The second approach is adjusting the learning scheme of steganlysis algorithms [3,11,12]. To alleviate the cover source mismatch problem of supervised learning based methods, Ker and Pevný proposed a novel clustering based paradigm for steganalysis in Ref. [11] and later a local outlier factor based steganalysis in Ref. [12], both making good performance in realistic scenarios. Hou et al. adopt Gaussian Mixture Model(GMM) to improve blind steganalysis by firstly clustering datasets into limited categories and then training corresponding classifiers for each category [10]. In addition to these datasets clustering based methods, Cho et al. describe another framework for constructing content-adaptive steganalysis algorithms [3], which firstly decomposes images into small blocks and then trains a classifier for each class of blocks. Similarly, Wang et al. believe that the embedding probability of steganographic algorithms varies across different textural regions in each image [20,21], and propose an image segmentation based steganalysis algorithm to improve the performance of Ref. [3]. All these algorithms have shown a better performance compared with previous methods. However, the datasets clustering based methods learn classifiers based on entire images and fail to make use of content diversity [1,10], while the image decomposition based methods don't take into account of the embedding probability of regions with similar texture in different classes of images [3,20,21].

In this paper, a novel hierarchical learning framework is proposed to improve the performance of JPEG image steganalysis methods described above. It is based on the observation that both regions of an image with different textural complexity and regions of different images with similar textural complexity tend to have different embedding probabilities. This phenomenon can be explained in Fig. 1, where the surface regions with similar textural complexity from different images have different embedding probability for J-UNIWARD algorithm [9] with payload value 0.5bpnzAC(bit per non-zero AC DCT coefficients).

In the training stage, there are three steps to construct an effective steganalysis detector. Firstly, the training dataset is clustered into M categories according to the Wavelet Packet Decomposition(WPD) based texture features via GMM. Secondly, images of each subdataset are decomposed into smaller image blocks, and these blocks are clustered into N classes for each subdataset, obtaining MN

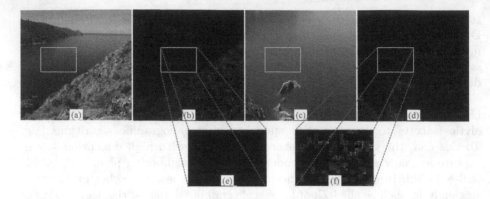

Fig. 1. Cover images((a) and (c)) and corresponding embedding changes((b) and (d)) in spatial domain for J-UNIWARD algorithm with a payload 0.5bpnzAC [9], (e) and (f) are the zoom in of (b) and (d) respectively. As shown in (a)–(d), regions with similar textural complexity from different images tend to have different embedding probabilities.

classes of image blocks. Finally, the hierarchical learning framework based steganalyzer is obtained by weighted sum of MN ensemble classifiers, each of whom is trained from a class of image blocks, with a corresponding weight derived from the proportion of the subdataset and the precision of trained classifiers. In the testing stage, a test image and its decomposed image blocks are firstly classified using trained GMM respectively, and then the final decision is made by fusing classifying results of decomposed image blocks. Extensive experiments of the proposed steganalysis framework have shown a better performance compared to some previous described frameworks in Refs. [3,10].

The rest of the paper is organized as follows. Section 2 introduces the proposed hierarchical learning framework. The performance of our framework is evaluated by extensive experiments in Sect. 3, and Sect. 4 concludes the paper.

2   Hierarchical Learning Framework

2.1   Framework Overview

For content adaptive steganographic algorithms [9,18], extensive experiments have show that, not only regions of an image with different texture complexity are usually with different embedding probability, but also regions from different images with similar texture complexity tend to have various embedding probability. Therefore, this paper proposes a novel hierarchical learning framework to improve the performance of previous frameworks [3,10]. The training procedure is shown in Fig. 2.

There are three stages in the training procedure: namely, dataset segmentation, block pools generation & classification, and classifiers training. Detailed descriptions of these stages are as follows. First, we compute the WPD based

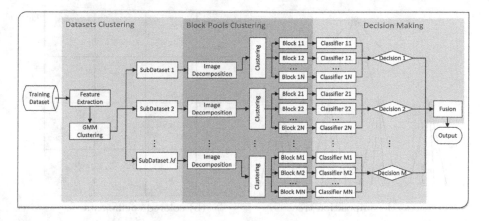

Fig. 2. Training procedure of the proposed framework.

texture features of all training images, and use the GMM clustering algorithm to segment the training dataset into M subdatasets with regard to the texture complexity. Second, to better utilize the content characteristics of images, the training images in each subdataset are decomposed into smaller blocks to generate block pools using the method described in Ref. [3], and then each block pool is segmented into N class of blocks via GMM clustering. Third, for each class of blocks, we train an ensemble classifier and record the corresponding precision value. The final decision is made by weighted sum of the outputs of ensemble classifiers, where the weights are computed based on the proportions of the trained images.

In the testing/application stage, a test image is firstly classified into some subdataset by analyzing its posterior probability belonging to each category using the WPD based texture features. Then, the test image is decomposed into smaller image blocks, and each block is classified using the trained GMM. Finally, these classified image blocks are tested on corresponding ensemble classifiers, and the outputs are integrated by weighted sum to make the final decision whether the tested image is innocent.

In the following subsections, we will give detailed descriptions of the WPD based texture features and the GMM clustering method. Additionally, a brief instruction of adopting ensemble classifiers is also presented.

2.2   Feature Extraction

The feature sets used in this paper are texture features and steganalysis features. The former is dedicated to dataset segmentation for designing content adaptive steganalytic algorithms, while the latter is used for image blocks steganalysis. The computation procedure and the adoption reasons of these features are described as follows.

For daset segmentation applications in this paper, the WPD based texture features are adopted. Unlike standard wavelet transforms which apply filtering

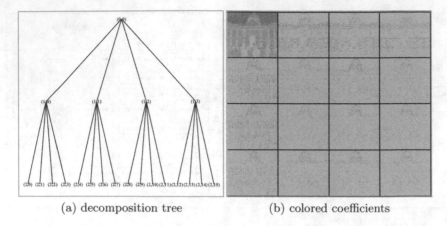

(a) decomposition tree   (b) colored coefficients

Fig. 3. Two levels wavelet packet decomposition (WPD) of an image. (a) decomposition tree of the WPD, (b) colored wavelet coefficients. (Color figure online)

only iteratively on approximate coefficients, the WPD applies high and low-pass filtering both in high and low frequency components at each iteration, and thus can provide a richer representation of textures from various scales and orientations, which is shown to be more effective for texture analysis [17]. Assume h and g to be low- and high-pass filters of a wavelet respectively, and the approximate coefficients of a image at level j is $a_j(m, n)$, the WPD at level j for $a_j(m, n)$ is given by

$$a_{j-1}(m, n) = \sum_{l \in \mathbb{Z}} \sum_{k \in \mathbb{Z}} h_{l-2m} h_{k-2n} a_j(l, k)$$

$$d_{j-1}^1(m, n) = \sum_{l \in \mathbb{Z}} \sum_{k \in \mathbb{Z}} h_{l-2m} g_{k-2n} a_j(l, k)$$

$$d_{j-1}^2(m, n) = \sum_{l \in \mathbb{Z}} \sum_{k \in \mathbb{Z}} g_{l-2m} h_{k-2n} a_j(l, k) \qquad (1)$$

$$d_{j-1}^3(m, n) = \sum_{l \in \mathbb{Z}} \sum_{k \in \mathbb{Z}} g_{l-2m} g_{k-2n} a_j(l, k)$$

where a_{j-1} is the approximate coefficient, $d_{j-1}^i, i = 1, 2, 3$ are detailed coefficients at horizontal, vertical and diagonal directions respectively. Similar decompositions are also carried out for $d_j^i, i = 1, 2, 3$ respectively. The procedure is iterated at each scale, and an instruction of two levels WPD of an image is shown as Fig. 3.

To compute the WPD based texture features, we apply three levels WPD on images, and compute four types of features for each node except for nodes $(i, 0), i = 0, 1, 2, 3$, which are corresponding to all approximate coefficients of the decomposed images. These used features are the average feature, the standard deviate feature, the entropy feature and the energy feature, which can be obtained respectively by

$$f_1 = \frac{1}{Z} \sum_m \sum_n a^j(m,n) \tag{2}$$

$$f_3 = \sqrt{\frac{1}{Z} \sum_m \sum_n (a^j(m,n) - f_1)^2} \tag{3}$$

$$f_2 = -\frac{1}{Z} \sum_m \sum_n a^j(m,n) \log_2 a^j(m,n) \tag{4}$$

$$f_4 = \frac{1}{Z} \sum_m \sum_n (a^j(m,n))^2 \tag{5}$$

where Z is the normalized factor, equaling to the number of elements of a^j. Thus, the dimension of the texture features for training images clustering is $(4 + 4^2 + 4^3 - 3) * 4 = 324D$.

For steganalysis applications of image blocks, the DCTR(Discrete Cosine Transform Residual) feature set is adopted, due to its competitive performance with a lower dimensionality compared with other rich models [4,6,8,14], which is highly required for classifier training. Additionally, DCTR is an effective feature set with low computational complexity based on the undecimated DCT [7], which is regarded as having much more information for steganalysis and can be efficiently obtained. Detailed descriptions of the DCTR feature set and its codes are provided in Ref. [7]. In this paper, DCTR is used in the applications of image blocks clustering and image blocks steganalysis.

2.3   Learning Methods

To fulfil the hierarchical learning framework, the GMM and the ensemble method are adopted for dataset clustering and image blocks classification respectively. The GMM equipped with Expectation-Maximization (EM) algorithm can make soft assignments of samples in clustering applications, which makes it more suitable for our framework.

The GMM clustering consists of parameters learning and samples classification, and is based on the assumption that any distribution of samples can be modeled using the Gaussian mixture distribution, which is defined as [2]:

$$p(x) = \sum_{k=1}^{K} \pi_k \mathcal{N}(x|\mu_k, \Sigma_k) \tag{6}$$

where π_k are the mixing coefficients, satisfying $\sum_k \pi_k = 1$. $\mathcal{N}(x|\mu_k, \Sigma_k)$ is the kth Gaussian distribution with mean μ_k and covariance Σ_k. Parameters learning of π_k, μ_k and Σ_k is carried out by EM algorithm through maximizing the log likelihood function:

$$\ln p(X|\pi, \mu, \Sigma) = \sum_{n=1}^{N} \ln\{\sum_{k=1}^{K} \pi_k \mathcal{N}(x_n|\mu_k, \Sigma_k)\} \tag{7}$$

The EM for GMM training consists of four steps:

(1) Initialize π_k, μ_k and Σ_k.

(2) E-step: evaluate the responsibility of the nth sample x_n belonging to the kth Gaussian by

$$\gamma(z_{nk}) = \frac{\pi_k \mathcal{N}(x_n|\mu_k, \Sigma_k)}{\sum_{j=1}^{K} \pi_j \mathcal{N}(x_n|\mu_j, \Sigma_j)} \tag{8}$$

(3) M-step: update the parameters using current $\gamma(z_{nk})$ by

$$\mu_k = \frac{1}{N_k} \sum_{n=1}^{N} \gamma(z_{nk}) x_n$$

$$\Sigma_k = \frac{1}{N_k} \sum_{n=1}^{N} \gamma(z_{nk})(x_n - \mu_k)(x_n - \mu_k)^T \tag{9}$$

$$\pi_k = N_k/N$$

where $N_k = \sum_{n=1}^{N} \gamma(z_{nk})$.

(4) Evaluate the log likelihood function (Eq. 7) and check for convergence conditions to decide whether repeating the above procedure.

In the application stage, the GMM uses trained parameters to classify a new sample x through computing its posterior probability by

$$\gamma(z_k) = \frac{\pi_k \mathcal{N}(x|\mu_k, \Sigma_k)}{\sum_{j=1}^{K} \pi_j \mathcal{N}(x|\mu_j, \Sigma_j)} \tag{10}$$

Equations 8–10 constitute the GMM clustering method we used. In the training stage of our framework, parameters of the GMM are learned by training samples. In the testing stage, samples are classified into corresponding categories by analyzing their posterior probabilities belonging to each cluster using Eq. 10.

Additionally, the ensemble method described in Ref. [13] is adopted as a tool for our classifier training due to the following reasons. First, it is a simple classification methodology allowing fast design of steganalysis detectors, and it scales well both with high dimensionality of features and large number of images, which is regarded as a tough problem for Gaussian SVMs. Second, it can make a comparative or even better performance in comparison to the Gaussian SVMs [13]. In our framework, a classifier is trained for each class of blocks, and the Out-Of-Bag (OOB) error of each ensemble classifier is saved during training, which will be used as weights for integration of individual classification results to form the final decision in the testing stage.

3 Experiments

3.1 Dataset and Experimental Setup

To evaluate our proposed learning framework, two datasets are adopted as sources to generate cover images. The first is the BOSSbase1.01 database, and the second is the BOWS2 database, both consist of $10,000$ 8-bit grayscale images with a size of 512×512, captured by different cameras in variant scenes. The JPEG cover images were created by Matlab's function *imwrite* with a fixed quality factor of 75. The content-adaptive JPEG steganographic algorithm J-UNIWARD was tested in our experiments [9], and we make stego images with a range of different payloads 0.05,0.1,0.2,...,0.5 bpnzAC(bit per non-zero AC DCT coefficients). These cover/stego images constitute the training/testing datasets in our experiments.

To generate smaller image blocks for hierarchical learning shown as in Fig. 2, the image decomposition method described in Ref. [3] was carried out due to its simplicity and effectiveness. Based on the conclusion that large block sizes and large block numbers benefit the performance of block-based steganalysis [3], the block size and the step size used in this paper were 256×256 and $512/8 = 64$ respectively, generating 16 blocks per image. Additionally, the haar wavelet filters were used in our experiments to compute the WPD based texture features.

In the testing stage, the image blocks of a test image were classified into corresponding clusters, and were analyzed by trained ensemble classifiers with results of $c_{i,j}$, where i was the image class label and j was its blocks class label. The final decision was made by:

$$P = \sum_{j=1}^{N} \omega_{i,j} c_{i,j} \tag{11}$$

here, $\omega_{i,j}$ was the weight derived from the OOB error of the trained ensemble classifiers by

$$\omega_{i,j} = \frac{1/OOB_{i,j}}{\sum_j 1/OOB_{i,j}} \tag{12}$$

According to Eq. 12, a classifier with a smaller OOB error provides a larger weight value for the final decision.

3.2 Experimental Results and Discussions

The experiments in this subsection consisted of two parts. Firstly, To evaluate the performance of the proposed framework on adaptive steganography detection, we tested it on BOSSbase 1.01 dataset, and the steganographic algorithm was J-UNIWARD [9]. In our experiments, 80 % of the images in BOSSbase 1.01 were taken for steganalyzer training, while the others were used for testing. Performance was evaluated by the average error probability defined as

$$P_{err} = (P_{FP} + P_{FN})/2 \tag{13}$$

where P_{FP} and P_{FN} are the probabilities of false positives and false negatives respectively.

Figure 4 showed a comparison of our proposed method over other previous framework, namely the dataset-clustering method and the image-segmentation

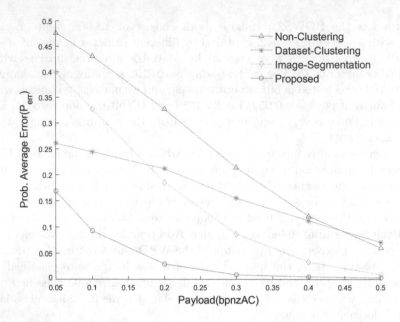

Fig. 4. Performance evaluation of the proposed framework on BOSSbase1.01 dataset, the dataset-clustering method and the image-segmentation method are the framework described in Refs. [10] and [3] respectively.

method described in Refs. [10] and [3] respectively. In the experiments, we implemented the framework of Refs. [10] and [3] using DCTR features, due to the fact that the purpose of the comparisons was to evaluate the performance of learning framework, and the steganalysis feature used in Refs. [10] and [3] were not so effective as DCTR [7] for adaptive steganography. As shown in Fig. 4, due to the refined learning in classifier training, the dataset-clustering based method, the image-segmentation based method and our proposed method all performed better than non-clustering method.

Secondly, to illustrate the ability of our framework to handle the cover-source mismatch problem, we applied the BOWS2 dataset to evaluate the steganalysis detectors, which was trained on BOSSbase 1.01 dataset. Note that, testing trained classifiers on another dataset was a widely used method for evaluating the ability of a classifier in cover-source mismatch problem [15]. Figure 5 showed the experimental results of variance methods on the BOWS2 dataset with classifiers trained on the BOSSbase 1.01 dataset. As shown in Fig. 5, our proposed framework obtained better performance than other frameworks, which demonstrated a better ability to handle cover-source mismatch problems.

According to the above experimental results, the proposed hierarchical learning framework tends to perform better than previous framework [3,10]. The reasons are that the proposed framework has taken into account both the statistical properties of images and the diversity of image regions in applications, and thus giving a more refined steganalysis than the framework described in Refs. [3,10].

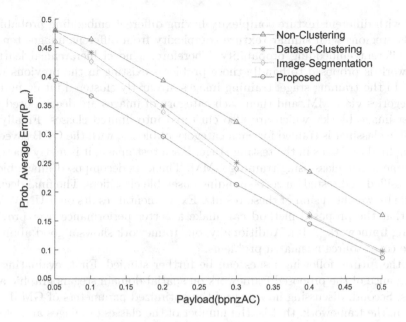

Fig. 5. Performance evaluation of various methods on the BOWS2 dataset with classifiers trained on the BOSSbase1.01 dataset. Lower curves correspond to lower average error probabilities.

Although the experiments are carried out for steganographic algorithms in JPEG domain, our framework can also be used for steganographic algorithms in spatial domain by updating the steganalysis features, which will be discussed in the future work.

The computational complexity of the proposed learning framework can be explained as follows. Because both the WPD based features and the DCTR features having efficient implementations [7], the computational complexity in the training stage mainly depends on the ensemble classifiers training, and the number of classifiers is affected by the number of categories of the image blocks. In our experiments, fifteen ensemble classifiers should be trained in the training stage according to the parameters setting methods described in Subsect. 3.1, which requires about 10 h to train these classifiers on our computer with a processor of Intel Core i5-4200M CPU and 2.50 GHZ ROM. However, in the application stage, a test image can be determined whether being innocent in less than 2 s.

4   Conclusions

As an important forensic technique for analyzing images on social networks, JPEG steganalysis has received much attention in recent years. In this paper, we have proposed a novel hierarchical learning framework for steganalysis of JPEG images, and it is based on the observation that not only regions of an image with different texture complexity having different embedding probability, but also regions with similar texture complexity from different images tend to have different embedding probability. Therefore, a novel hierarchical learning framework is proposed to handle those problems existing in the previous work [3,10]. In the training stage, training images are firstly clustered into a number of categories via GMM, and then each category of images are decomposed into smaller image blocks, which are also clustered into limited classes. Finally, an ensemble classifier is trained for each category of blocks, with the OOB saved for making final decisions in the testing stage. For a test image, it is firstly classified into some image class using training GMM. Then, its decomposed image blocks are classified and tested on corresponding ensemble classifiers. The final decision is made by weighted sum of these results. Experimental results on J-UNIWARD show that the proposed method can make a better performance over previous learning framework [3,10]. Additionally, our framework shows a good ability to handle cover-source mismatch problems.

In the future, following issues can be further studied. First, evaluating the performance of the proposed framework on spatial domain steganographic algorithms; Second, discussing how to find the optimized parameters of GMM clustering in the framework, that is, the number of the classes of images and blocks; Third, exploring more effective features for texture classification and image steganalysis.

References

1. Amirkhani, H., Rahmati, M.: New framework for using image contents in blind steganalysis systems. J. Electron. Imaging **20**, 013016:1–013016:14 (2011)
2. Bishop, C.M.: Pattern Recognition and Machine Learning. Springer, New York (2006)
3. Cho, S., Cha, B.H., Gawecki, M., Kuo, C.C.J.: Block-based image steganalysis: algorithm and performance evaluation. J. Vis. Commun. Image R. **24**(7), 846–856 (2013)
4. Fridrich, J., Kodovský, J.: Rich models for steganalysis of digital images. IEEE Trans. Inf. Forensics Secur. **7**(3), 868–882 (2012)
5. Fridrich, J., Goljan, M.: Practical steganalysis of digital images: state of the art. In: Proceedings of SPIE, vol. 4675, pp. 1–13 (2002)
6. Holub, V., Fridrich, J.: Random projections of residuals for digital image steganalysis. IEEE Trans. Inf. Forensics Secur. **8**(12), 1996–2006 (2013)
7. Holub, V., Fridrich, J.: Low-complexity features for JPEG steganalysis using undecimated DCT. IEEE Trans. Inf. Forensics Secur. **10**(2), 219–228 (2015)
8. Holub, V., Fridrich, J.: Phase-aware projection model for steganalysis of JPEG images. In: Proceedings of SPIE, vol. 9409. pp. 94090T1–94090T11 (2015)
9. Holub, V., Fridrich, J.: Digital image steganography using universal distortion. In: Proceedings of the First ACM Workshop on Information Hiding and Multimedia Security, pp. 59–68 (2013)
10. Hou, X., Zhang, T., Xiong, G., Lu, Z., Xie, K.: A novel steganalysis framework of heterogeneous images based on GMM clustering. Sig. Process. Image Commun. **29**(3), 385–399 (2014)

11. Ker, A.D., Pevný, T.: A new paradigm for steganalysis via clustering. In: Proceedings of SPIE, vol. 7880, pp. 78800U1–78800U13 (2011)
12. Ker, A.D., Pevný, T.: Identifying a steganographer in realistic and heterogeneous data sets. In: Proceedings of SPIE, vol. 8303, pp. 83030N1–83030N13 (2012)
13. Kodovský, J., Fridrich, J.: Steganalysis in high dimensions: fusing classifiers built on random subspaces. In: Proceedings of SPIE, vol. 7880, pp. 78800L1–78800L13 (2011)
14. Kodovský, J., Fridrich, J.: Steganalysis of JPEG images using rich models. In: Proceedings of SPIE, vol. 8303, p. 83030A (2012)
15. Kodovsky, J., Sedighi, V., Fridrich, J.: Study of cover source mismatch in steganalysis and ways to mitigate its impact. In: IS&T/SPIE Electronic Imaging, pp. 90280J–90280J (2014)
16. Li, B., He, J., Huang, J., Shi, Y.Q.: A survey on image steganography and steganalysis. J. Inf. Hiding Multimedia Signal Process. **2**(2), 142–172 (2011)
17. Ma, W.Y., Manjunath, B.S.: A comparison of wavelet transform features for texture image annotation. In: Proceedings International Conference on Image Processing, vol. 2, pp. 256–259 (1995)
18. Pevný, T., Filler, T., Bas, P.: Using high-dimensional image models to perform highly undetectable steganography. In: Böhme, R., Fong, P.W.L., Safavi-Naini, R. (eds.) IH 2010. LNCS, vol. 6387, pp. 161–177. Springer, Heidelberg (2010)
19. Song, X., Liu, F., et al.: Steganalysis of adaptive JPEG steganography using 2D Gabor filters. In: IH&MMSec, pp. 15–23 (2015)
20. Wang, R., Ping, X., Xu, M., Zhang, T.: Steganalysis of JPEG images using block texture-based rich models. J. Electron. Imaging **22**(4), 043033 (2013)
21. Wang, R., Xu, M., Ping, X., Zhang, T.: Steganalysis of JPEG images by block texture based segmentation. Multimedia Tools Appl. **74**, 1–22 (2015)

A Multi-agent Organization Approach for Developing Social-Technical Software of Autonomous Robots

Sen Yang[✉], Xinjun Mao, Yin Chen, and Shuo Yang

College of Computer, National University of Defense Technology, Changsha 410073, China
yangsen.nudt@hotmail.com

Abstract. Software of autonomous robot is a complex physical and social technical system that is context-aware, autonomous and capable of self-management to achieve tasks. It typically consists of a large amount of autonomous entities and interactions. To develop such system needs high-level metaphors and effective mechanisms independent of physical and technical details of various robots. The paper presents a multi-agent organization approach to developing autonomous robot software that is modelled as social organization, in which each agent is bound to specific roles with specified responsibilities that are tightly related with robot's characteristics and tasks. These agents form diverse organization structure and patterns to achieve flexible cooperation in order to achieve assigned tasks. The paper details multi-agent organization model of autonomous robot software and various roles in the model. We have implemented a framework called *AutoRobot* that realizes the approach and supports the development and running of autonomous robot software. A case is studied by using NAO robot to show the effectiveness of our proposed approach.

Keywords: Autonomous robot · Multi-agent organization · Environment

1 Introduction

Robots have succeeded in traditional industry manufacture fields and are expected to gain success in several emerging application domains (e.g. hospital, family, military, space exploration). These applications are characterized with open environment, autonomous behavior, and friendly human-robot interaction [1, 2]. Autonomous robot performs computational and physical behaviors in an autonomous way, according to the assigned tasks. They often operate in such situations where human control is either infeasible or not cost-effective 5, e.g., radiation field, battle space, deep sea, outer space, etc., and accomplish tasks without the intervention of human beings or other systems.

Development of autonomous robot involves with the design of various systems like hardware and software. In the past years, many efforts have been made on the researches of the theories and practices of autonomous robots by intersecting multiple disciplines, such as biology, nature system, mechanics, materials science, artificial intelligence, computer science [3]. As the continuous maturity and standardization in robot hardware technology and the decreases in the cost of robot equipment (like sensor, motor, battery) [2], the quick assembling and massive autonomous robots applications in various

© Springer Science+Business Media Singapore 2016
W. Che et al. (Eds.): ICYCSEE 2016, Part I, CCIS 623, pp. 24–38, 2016.
DOI: 10.1007/978-981-10-2053-7_3

domains become possible. As a result, software for autonomous robot becomes critical for satisfying various requirements.

Developing software for an autonomous robot is quite a challenging and complex task [5]. Firstly, autonomous robot represents specific applications in which software are expected to be context-aware, autonomous and capable of self-management to perceive various changes and control robot behaviors. The existing software engineering technologies for developing autonomous robot software are mostly based on object-orientation paradigm, from robot programming languages to development framework such as Robot Operating System (ROS) [16]. However, the software model of object-oriented paradigm implies that the elementary components implemented as objects cannot actively perceive their own changes as well as those occurring in the context. They perform methods with the same name and parameters as the received messages and therefore have no autonomous behaviors. The structure and behavior of objects will not change after they are instantiated and therefore they lack flexibility and variation to deal with various changes. Secondly, the development of robotic system for a specific application still remains difficult, time-consuming and expensive [4]. Developers have to learn many details related to the robots and develop software from low-level physical action such as execution components and image recognition components to high-level application-dependent behavior control components such as task accomplishment components. Current development tools for autonomous robot software either depend on general-purpose OOP languages (like C++, Python) that lack of enough supports for autonomy and robotic properties, or specific programming frameworks or languages that highly rely on specified robot hardware or type [13]. Thirdly, autonomous robot is typically composed of a number of heterogeneous software and hardware components that adopt different technical solutions and should inter-operate with each other. More-over, many of these components have high degree of autonomy. For example, when exceptions or errors occur in sonar sensor, the software component related with the sensor should support the self-repair of the sensor, which has little relevance with other components like motors and vision sensors. Therefore, self-management is expected for such system to simplify the maintenance and control complexity.

Recently in the academic and industry fields, there are many efforts put on the development technologies for autonomous robot software, from programming languages [5, 10], middleware [7], framework [6], software architecture [20] and design pattern, etc. Many mainstream software engineering technologies are borrowed and integrated together to support the development, running and evolution of autonomous robot software, e.g. service-oriented computing [8, 9], component-based approach [12, 14]. An important work is ROS which is actually a software framework emphasizing on large-scale integration, inter-operation and reuse of robotics researches and practices [16]. ROS provides a peer-to-peer, tools-based, multi-lingual and thin supports for robot software developers to meet a specific challenges encountered when developing large-scale service robots, especially when robotic systems grow ever more complex. However, ROS does not deal with the issues of autonomy that are specific for autonomous robots. Another important trend on autonomous robot research is to adopt Multi-Agent System (MAS) technology to support the development of autonomous robot software. There are several researches in the literature of robotics in recent years, ranging from agent-oriented programming language [15, 20], software architecture [17, 18, 22], development framework [19], etc. In this approach, autonomous robot software are modeled as autonomous agent that interacts with the environment and takes autonomous behaviors. Most of works take autonomous robot software as single software agent that is coarse granularity, which result in difficult reuse and self-management, and the proposed technologies focus on the agenthood and lack of the consideration on the integration of robot technologies with mainstream software engineering technologies.

This paper presents a multi-agent organization model and corresponding development platform for autonomous robot software. The remaining sections are organized as follows. Section 2 discusses the characteristics and technical demands of autonomous robot software. Section 3 presents the proposed multi-agent organization model of autonomous robot software. Section 4 introduces a development framework *AutoRobot*. Section 5 illustrates our approach and framework with a case study. Conclusions are made in Sect. 6.

2   Autonomous Robot Software as Social and Physical Technical System

This section details characteristics of autonomous robot software with some discussions on the challenges with a sample of autonomous robot.

2.1   Sample of Autonomous Robot

First let us consider a sample where an autonomous robot operates in an open environment (e.g., house, hospital) and intends to accomplish the task named "Follow Me". The case requires robot to follow a target and keep a proper distance in an autonomous way. When the robot starts to execute the task, it will turn on the sensors and search the target. If it finds the target, the robot will follow it and keep specified distance, otherwise it will continue to search it or interact with other robots to find it. Controls by human beings over robot are not permitted at runtime.

Though the sample seems simple, there are several points deserve to be considered for the robot to accomplish the above task. (1) The environment of the robot is open and dynamic. Objects in this environment may be unknown and unexpectedly appear. For example, new obstacle may occur, the target may be moved to a new position, etc. The autonomous robot should perceive the environment in terms of various sensors to obtain the information of environment's and itself changes. (2) The robot need to decide how to take appropriate behaviors to fulfill the task based on the environment percepts, e.g. finding the target, moving one step, or avoiding obstacle. (3) Interaction and coordination is necessary. It is impossible for the robot to obtain all of knowledge and information that it needs to fulfill the task. Therefore, the robot should interact with other robots or access data and services deployed in the Internet to satisfy the requirements of task accomplishment. For example, if the robot loses the target, it can coordinate with other robots in the environment to gain the current position of the tracing object. Moreover, the robot should adapt to various changes occurring in the environment or itself, in order to improve the performance, efficiency or simplify management and maintenance. For example, when battery is low, autonomous robot will change its motion patterns and adjust moving parameters. In the situation, the robot may also turn off some unnecessary sensors or optimize path to save energy. Besides, if the velocity of the target that autonomous robot follows have changed, the autonomous robot should adjust moving velocity by itself in order to keep a proper distance with the target. Some devices may break down or be out of order, it needs to recover its functionalities in term of self-repair.

2.2 Responsibilities of Autonomous Robot Software

Typically autonomous robots are designed to perform tasks independently, or with very limited external control [5]. They are required to take actions autonomously without human intervention. Different from traditional industry robots or other cyber-physical systems, autonomous robots have a number of distinct characteristics.

1. Interaction with open and diverse environment.
   Autonomous robots are normally situated in open environments that involve physical, social and cyber factors. The changes of environment are often dynamic, evolving, unpredictable, non-deterministic, and will significantly affect autonomous robots and their behaviors. Therefore autonomous robots are typically equipped with various sensors (e.g. sonar, camera) to perceive the environment information. Moreover, autonomous robots can influence its situated environment by taking actions in term of its physical devices.
2. Task-driven.
   Autonomous robots are essentially task-driven systems. They receive orders and generate tasks in terms of interactions with human beings and other robots, and are required to decide behaviors and take actions autonomously to accomplish tasks without human beings' interventions. In some cases, autonomous robots are even expected to proactively generate tasks according to the perceived changes.
3. Social-Awareness.
   In many cases, autonomous robots are required to interact with human beings, other robots or software systems in the environment to receive orders, obtain information and services in order to collectively accomplish tasks. They may form organization in which they behave, according to the roles they play, the protocol they adopt, and the regulations they respect.
4. Adaptability.
   Autonomous robots typically have limited resource (e.g. energy, computation, storage). They are usually operated in open environment that human beings are unable or unwilling to reach, and various expected or unexpected internal or external events may occur during operations. For example, they may encounter emerging obstacles in their planed paths, some sensors in robot fail to perceive environment information, etc. Moreover, the behaviors of autonomous robots may have safety issues, e.g. the move of arm may damage some facilities in the house or even endanger human beings. Therefore, autonomous robots are required to flexibly adapt

to various changes and effective manage various resources by themselves (e.g. self-configuration, self-optimization, self-healing) so that they can operate in an effective, efficient and safe way.

Software is extremely critical in autonomous robot because the robot's behaviors are essentially controlled and driven by software, the perceived environment information is accurately analyzed and processed by software, the limited resources and situations is effectively managed by software. Generally, autonomous robot software takes the following responsibilities.

1. Perceive and process environment information.
   Autonomous robot software should interact with the situated environment (including human beings, physical elements, other robots and Internet etc.) to perceive the environment information in terms of various equipped sensors. Typically as the openness and dynamics of the environment, autonomous robot software should effectively identify the environment elements to decide what to be sensed, continuously obtain a variety of perceived data in a timely-fashion way, accurately and correctly process these data to gain valuable environment information, efficiently solve the inconsistent and conflict problems during the operation of robot. The scope and size of environment is typically large and the concerned changes of environment may vary depending on the tasks to be accomplished and the behaviors to be taken, therefore autonomous robot software should flexibly control sensors to collect required and concerned data in order to reflect the environment status and satisfy the requirements of behavior decisions. The sensor data should be processed in term of various artificial intelligence technologies and algorithms such as image recognition, natural language understanding.

2. Controlling robot's behaviors.
   The autonomy of autonomous robots are actually achieved in terms of software. According to the interactions with the environment, software for autonomous robots should deliberate what tasks should be accomplished and what behaviors should be taken without any interventions from external objects like human beings. Moreover, safety and resources restrictions issues should also be considered in the process of behavior decisions. Autonomous robot software are also responsible for performing the planned behaviors and evaluating whether they are helpful to accomplish tasks. As the continuous and unpredictable changes of the environments, together with the evolution of the assigned tasks, the behavior decisions and executions should be flexible and adaptive, which means the planned behaviors are expected to be continuously adjusted or optimized to respond to various changes and improve the effectiveness, efficiency and rationality of autonomous robot behaviors.

3. Managing robot resources and situations.
   Software for autonomous robot should effectively manage the devices, computational resources and energy to satisfy the requirements of task. Furthermore, various expected events (e.g. the energy is used up) and unexpected events (e.g. sonar sensor cannot work) may occur during the operations of autonomous robots and environment may change greatly, therefore parameters for autonomous robots should be configured dynamically. Errors or failures should be eliminated or solved

autonomously, the plan and behaviors should be optimized continuously. Such management should be performed in an autonomous and adaptive way so as to promptly respond to the changes, improve efficiency and effectiveness, and decrease management complexity and maintenance costs.

Autonomous robot software is actually kind of complex social and physical technical systems (see Fig. 1). In such system, software should interact with physical elements to monitor, manage and control various physical devices, e.g., motor, sensor, leg, etc. Autonomous robot software should manage these physical components in an autonomic way (e.g., self-configuration, self-organization,self-optimization, etc.) in order to decrease maintenance complexity and cost, improve operation flexibility and efficiency. Moreover, autonomous robot software is also a social technical system, in which software should interact with the users and other robots. In multiple-robot case, they form complex social system, in which software plays various roles depending on the tasks of robots, the responsibilities and positions of software in the robot. The interactions between autonomous robot software relies on the social structure of whole system and relationship between these software.

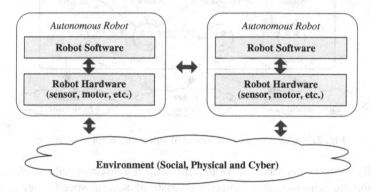

Fig. 1. Autonomous robot software as social and physical technical system

3 A Multi-Agent System Organization Model for Autonomous Robot Software

Undoubtedly software are playing a significant role in autonomous robot system and it shows some new characteristics like context-awareness, autonomy, self-adaptation and self-management, etc. This section discusses our considerations on the development of autonomous robot software and presents a multi-agent organization model for constructing and implementing such kind of software.

MAS implies novel epistemology and methodology to analyze, design and implement complex software systems that typically consist of independent and autonomous elements, constantly interact with environment, and cooperatively solve problems. In essence, autonomous robot software are such a kind of complex software system. Several researches argue that robot software should be modelled and designed as intelligent agent [25], and multi-robot system can be implemented as MAS.

However in this paper we model and design autonomous robot software as MAS organization rather than single agent. Each agent in autonomous robot software plays different roles and takes distinct behaviors, and cooperates with each other to support the operation of autonomous robot. The whole software system is modelled as social organization, in which each agent operates and interacts depending on its playing roles and responsibilities (see Fig. 2). Different robot applications require diverse multi-agent organization structure and pattern to satisfy task requirements. Moreover, the organization structure and interaction pattern of whole software and the roles that the agents play are flexible and can be dynamically adjusted in order to adapt to various changes occurring in the robot or the situated environment.

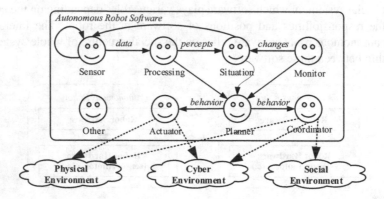

Fig. 2. MAS organization model of autonomous robot software

Actually autonomous robots software represents a domain-specific application and problem-solving approach that typically consists of several specific agents playing various roles like sensors, actuators, planners, coordinator, etc. These agents operate autonomously and coordinate with each other to exchange information, access situation and provide service, etc., and ultimately achieve the tasks. The variety of autonomous robot applications result in the difference of the roles, social relationship of agents and the organization structure of the whole system. However, several roles and agents can be identified according to the general characteristics and requirements of autonomous robot.

– *Sensor agents*, they are responsible for controlling and interacting with physical sensors, and obtain raw sensor data of the situated environment. For autonomous robot, sensor agents should also be responsible for managing the physical sensors in an autonomic and adaptive way. When the environment changes or the internal events (e.g. some errors in the physical sensor) arise, sensors should be self-configured, self-optimized and even self-healed. Besides, multiple sensor agents are required to cooperate with each other to respond to some changes. For example, when sonar sensor finds some obstacle, camera sensor is required to get the obstacle's image information.

- *Sensor Information Processing agents*, such agents are responsible for processing sensor raw data collected by sensor agents, obtaining valuable percepts about the environment. Typically, these agents process data by means of technologies and algorithms from artificial intelligence, Internet of the Thing, e.g. image and voice recognition. Such processing is normally performed in a reactive way, which means the processing is performed when sensors' data arrives.
- *Monitor agents*, they are responsible for monitoring the internal changes of robot. Various software probes should be developed and integrated into the monitor agents to obtain robot's state. The monitored internal changes are sent to situation agents for forming situation model or to planner agents for responding in a time-fashion way.
- *Situation agents*, such kind of agent is responsible for managing the percepts produced by processing agents and events monitored by monitor agents, generating the situation model of the observed external environment and the internal robot, and maintaining the consistency and completeness of the situation model in term of the cooperation with sensor agents and monitor agents. Situation agent is not necessary for robots in which there is requisite to establish the explicit model of the world.
- *Planner agents*, they are responsible for responding to the perceived information collected by sensor agents, the monitored events collected by monitor agents or the situation managed by situation agents. Planner agents output tasks or behaviors (e.g. new task is created, and behaviors for current task are produced). There are several ways for planner agents to generate tasks or behaviors either in a reactive way or in a deliberative way, which mainly depends on the robot applications.
- *Actuator agents*, they are responsible to execute the generated behaviors that consist of physical ones and computational ones. For physical behaviors, actuator agent should interact with the physical devices to take actions (e.g. move one step, grasp the object). Computational behaviors (e.g. method, service) that may be deployed in different computing nodes should be accessed and executed. As the independence and continuous changes of robot's physical devices, actuator agents typically take the responsibility to self-manage the robot physical body, like self-configure parameters, self-recover physical system.
- *Coordinator agents*, as a cyber-physical and social ecosystem, autonomous robot should interact and coordinate with other robots in the situated environment or software agents in the Internet environment, or the human beings in the social environment. Coordinator agents are responsible for cooperating with other agents in social, physical and cyber environment in term of various manners like language and protocol.

The above MAS model of autonomous robot software extensively exploits a number of fundamental tools to manage complexity, and improve the reusability and flexibility of software system. These tools are proved to be effective in many software engineering practices.

4 *AutoRobot*: Development Framework for Autonomous Robot Software

4.1 An Overview of *AutoRobot*

We have developed a software development framework *AutoRobot* for autonomous robot software based on the MAS organization model. There are several important design decisions behind the framework.

- *Separate different software requirements of autonomous robot software.* In autonomous robot, different kinds of software are needed, from low-level device control and data processing software to high-level behavior planning and self-management software. We believe existing programming technologies and languages like OOP can satisfy the development requirements of low level software for autonomous robot. Therefore, our framework focuses on the development and running supports for high-level autonomous robot software, which demands the capability of autonomous decision on tasks and behaviors, self-adaptation and self-management of resources and devices, and the interactions and coordination of autonomous components.
- *ROS as fundamental infrastructure for autonomous robot software.* As an open-source development framework, ROS provides a great number of reusable resources for developing robot software. In current software engineering practices, open source software and crowdsourcing have become important way to improve development efficiency and quality. Therefore, AutoRobot is to be built on the ROS and our framework is responsible for establishing the connection between the autonomous robot software developed by AutoRobot and the ones developed by ROS.
- *Integration with mainstream development technologies in current robot software development practices.* The framework should be developed and used in terms of mainstream and general-purpose programming tools so that the reusable packages can be reused and integrated into a wide class of autonomous robot software. Such choice is important for the framework to be adopted and accepted by the developers of autonomous robot software.
- *Simplifying the development of autonomous robot software in term of encapsulation and reuse.* The specific functional requirements and basic capabilities of autonomous robot software should be encapsulated as reusable packages in form of various software agents or as some design patterns for particular autonomous robot applications.

As the Fig. 3 shows, the architecture of *AutoRobot* consists of three levels: infrastructure level, running support level and development level. The infrastructure level is totally based on ROS, while development support level provides reusable software package and software development tools for developer to implement autonomous robot software. Running engine level supports the execution of autonomous robot software.

Software agents in the same autonomous robot interact with each other in term of events, messages and ACL languages, which depends on whether they are located in the same computing devices. When a software agent for autonomous robot intends to invoke and perform an action deployed in ROS, the running engine will establish the connection

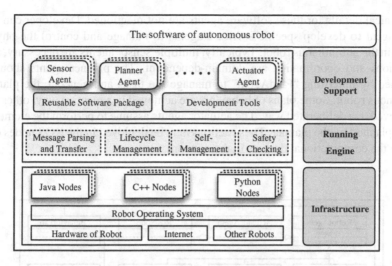

Fig. 3. The Architecture of *AutoRobot*

between the software agent and the ROS. If software agents in different robots interact with each other, the connection is to be established through platform.

AutoRobot separates the control and execution of autonomous robot software. Each level of *AutoRobot* has its responsibility in the development and running of autonomous robot software. Development support level mainly supports the developers to construct autonomous robot software. Running support level supports the execution and management of autonomous robot software, including the decision-making, self-adaptation and self-management. It also provide several tools to monitor and analyze the running of autonomous robot software for managers of autonomous robot software. Infrastructure however provides the connections between the high-level software and low-level software of autonomous robot. Especially, various low-level software for performing robot's actions can be developed or reused at this level.

4.2   Development Supports

Various software agents should be developed with the aid of development supports. Each agent has its role and responsibility in multi-agent system of autonomous robot software, and they interact with each other to achieve the robot's tasks. The supports are achieved in term of reusable software package that *AutoRobot* provides and encapsulates the fundamental functionalities of software agents for autonomous agents, and some software tools to simplify the development activities like designing the software agents for autonomous robots and automatically generating program codes.

Figure 4 shows part of reusable software package provided by *AutoRobot* for developing autonomous robot software. The software agents in package are autonomous and self-adaptive in behaviors and are implemented as diverse software architectures such as reactive or Belief-Desire-Intension (BDI) architecture. Due to the space restriction, the technical details for these software agents are not mentioned. Developer can reuse sensor agent to develop specific software agents to manage and control the physical sensors in the autonomous robot. Typically, multiple sensor agents are necessary, while interactions and coordination are domain-dependent. The planner agents should be developed by inheriting "PlanAgent" to manage the tasks and generate the plans for autonomous robot. Some plans of behaviors are achieved in a reactive way, others may be achieved in a deliberated way. The actuator agents assume to perform the elementary actions planned by the planner. These actions may be deployed in the ROS nodes or the robot or the computational devices in the network.

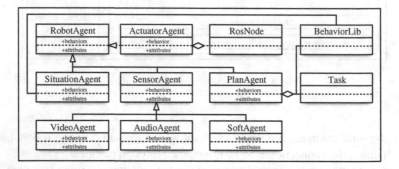

Fig. 4. The software development package of *AutoRobot*

4.3   Running Engine

Running engine provides several supports for autonomous robot software at runtime. Firstly, it acts as the medium between development support and infrastructure. The sequence actions that are generated by planner agents and executed by actuator agents will be requested and transferred to the infrastructure through running engine. The feedback of action execution will be sent back to actuator agents through running engine. Running engine actually provides functionalities to parse and transfer the messages between various agents and the infrastructure. Secondly, running engine is also responsible for checking the safety of autonomous robot's behaviors that are decided by the planner and executed by actuator. Different from other cyber-physical systems, the behaviors of autonomous robot may be precarious and harmful to the environment (e.g. human beings or other robots) or itself. For example, some actions of robot may hurt the human beings or cause damage to robot's physical devices. Thirdly, running engine also provides basic services to implement the self-adaptation and self-management of autonomous robot software so that when the specified events occur and conditions are satisfied, the adaptation and self-configuration, self-optimization, etc., can be accomplished. Lastly, running engine is also designed to manage the software agents in the autonomous robot system.

4.4   Infrastructure and Interaction with ROS

Infrastructure can be seen as the fundamental execution environment for low-level robot software, which is tightly relevant with the robot. In *AutoRobot*, the infrastructure is based on ROS, as we convince that the programs of autonomous robot software should be developed and reused in an open-source manners, and interacted with each other through common technical standards.

The interior communication of infrastructure relies on the mechanism of ROS that contains a set of nodes and topics. ROS is a distributed system and the inner nodes communicate through topics. When a node subscribes one topic, it will start to execute actions when other nodes publish messages on this topic. Every node in ROS is an independent program and is able to response to multiple requests. The encapsulation of behavior or action into nodes make the system more clear and loosely-coupled.

5   Case Study

This section introduces a case study for the sample described in Sect. 2. This case is developed based on our *AutoRobot* and NAO robot. According to the requirements description, a red ball is taken as the target that NAO will follow with. We develop software for NAO to accomplish the task. Four kinds of software agents should be developed, including NAOSonar, NAOActuator, NAOPlanner and NAOVideo. NAOSonar is responsible for controlling and interacting with sonar of NAO, and obtain the sonar information. NAOActuator is responsible for executing the behaviors sequence, and interacting with the running engine to take actions in ROS. NAOPlanner is responsible for responding to the perceived information collected by NAOSonar and producing behavior sequence. NAOVideo is responsible for controlling and interacting with camera sensor of NAO, and obtaining the vision information.

The software agents are launched once the system gets start. NAOPlanner generates the behavior that searches the red ball while the NAOVideo requests the information of camera and detects whether there are any red balls in the vision horizon of NAO. Once the red ball is found, the position will be sent to NAOPlanner and new behaviors that follow the red ball and detect obstacles are generated. At the same time, NAOVideo stops and NAO begins to follow the red ball. NAOSonar requests the information of sonar and detects whether there are any obstacles in front of NAO. If there are obstacles that are close enough to NAO, NAOPlanner will stop the current behavior and generate the new behavior that avoids the obstacle. After avoiding the obstacle, next round of searching and following will start. Figure 5 depicts several snapshots of the case implemented based on above description. Figure 5(1) shows the initial state of robot which verify the position of red ball by image recognition. In this phase, sonar is turned on to detect the distance of the obstacles. In Fig. 5(2), the robot finds the red ball and moves to it. During the process, the robot detects an obstacle that is located at the right-front and is close enough to it. Therefore, the robot stops the current behavior, plans and executes another behavior to avoid obstacle. Figure 5(3) and (4) illustrate that the robot has successfully avoided obstacle. Figure 5(5) and (6) show that the robot have accomplished the assigned task.

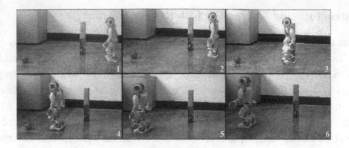

Fig. 5. The snapshots of the case

6   Conclusions and Future Works

Autonomous robot is essentially a complex social and physical technical system that performs various computational and physical behaviors by interacting with the open environment in an autonomous way. As an important constituent, software plays an important role to support the operation, management and maintenance of autonomous robot. The responsibilities of autonomous robot software are various, ranging from low-level hardware control and sensor data processing to high-level social interaction, behavior decision and self-management. As the distinct characteristics of autonomous robot (e.g. open and diverse environment, autonomous and adaptive behaviors, social coordination), autonomous robot software are expected to be context-aware, autonomous and capable of self-management. As the increasing demands on autonomous robots in various domains, it will be a great challenge to develop such kind of software in an efficient way.

This paper presents a MAS organization approach to understanding and modeling the formation of robot software, the individuals in the system and their interactions, and the dynamic adjustment and flexible management of the whole system. Different existing methods, the autonomous entity in robot software is modelled as agent, the interactions among agents depends on their social relationship, and the whole system is a social organization. From software engineering perspective, multi-agent system provides alternative paradigm to understand how to solve the problem, and model to develop software in terms of autonomous agent and their interaction. Essentially, autonomous robot software can be modelled as an autonomous system that perceives environment information, plans robot's behaviors and takes actions to achieve the assigned task. According to several important observations to autonomous robot, we can find that the software for autonomous robot are multiple-level, decentralized to control robot, and cooperative to accomplish tasks. Therefore, different from existing works, autonomous robot software in this paper are modelled and constructed as a multi-agent system. Each agent takes different responsibilities, interacts with each other to control robot's behavior in an autonomous way, and manage robot's resource in an adaptive way. Autonomous robot software are actually a domain-dependent MAS application with various agents like sensor agent, processing agent, planner agent, actuator agent, coordinator agent, etc. Various design patterns can be defined depending on the identified agents and their interactions.

Such model of autonomous robot software represents a number of important software engineering principles to tackle the complexity of software system, e.g. decomposition, organization, abstraction, reuse, which are believed to greatly improve the development efficient and quality of autonomous robot software. Moreover, MAS-based mode also provides method to construct software that is context-aware, autonomous, adaptive and flexible in behaviors, and cooperative with other software and resources in Internet, e.g. cloud service and robot, big data.

Based on the above approach and model, we have developed a framework called *AutoRobot* to support the development and running of autonomous robot software. The framework is established on ROS so that developers can effectively reuse great amount of robot software resources. The running environment of *AutoRobot* supports the integration between autonomous robot software and ROS code, management and communication of robot agents, etc. A reusable software package and some development tools are provided for developers to construct autonomous robot software. We have successfully developed several cases of autonomous robot applications with NAO robot, e.g. "Following Me". The result proves that our model and framework is effective to develop autonomous robot software, and can significantly improve the development efficiency.

References

1. Seto, M.L., Paull, L., Saeedi, S.: Introduction to autonomy for marine robots. In: Seto, M.L. (ed.) Marine Robot Autonomy, 1–46. Springer, New York (2013)
2. Gates, B.: Robot in every home. Sci. Am. **296**, 58–65 (2007)
3. Bekey, G.A.: Autonomous robots: from biological inspiration to implementation and control. J. Artif. Life **13**(4), 419–421 (2007)
4. Pan, Z., Polden, J., Larkin, N., Duin, S.V., Norrish, J.: Review: recent progress on programming methods for industrial robots. Robot. Comput. Integr. Manuf. **28**(2), 1–8 (2012)
5. Ziafati, P.: Programming autonomous robots using agent programming languages. In: Proceedings of AAMAS 2013, pp. 1463–1464 (2013)
6. Iñigo-Blasco, P., Diaz-Del-Rio, F., Romero-Ternero, M.C., Cagigas-Muñiz, D., Vicente-Diaz, S.: Robotics software frameworks for multi-agent robotic systems development. Robot. Auton. Syst. **60**(6), 803–821 (2012)
7. Chen, P., Cao, Q.: A middleware-based simulation and control framework for mobile service robots. J. Intell. Robot. Syst. **76**(3–4), 489–504 (2014)
8. Brugali, D., Fonseca, A.F.D., Luzzana, A., Maccarana, Y.: Developing service oriented robot control system. In: Proceedings of the 2014 IEEE 8th International Symposium on Service Oriented System Engineering, pp. 237–242. IEEE Computer Society (2014)
9. Insaurralde, C.C.: Service-oriented agent architecture for unmanned air vehicles. In: 2014 IEEE/AIAA 33rd Digital Avionics Systems Conference (DASC), pp. 8B1–1–8B1–14. IEEE (2014)
10. Pineda, L.A., Salinas, L., Meza, I.V., Rascon, C., Fuentes, G.: SitLog: A Programming Language for Service Robot Tasks, pp. 358–369 (2013)
11. Dragone, M., Abdel-Naby, S., Swords, D., Broxvall, M.: A programming framework for multi-agent coordination of robotic ecologies. In: Proceedings Of ProMAS, pp. 69–84 (2013)
12. Burke, T.M., Chung, C.-J.: Autonomous robot software development using simple software components, In: Proceedings of Intelligent Robots and Computer Vision XXII: Algorithms, Techniques, and Active Vision, p. 107 (2004)

13. Nordmann, A., Hochgeschwender, N., Wrede, S.: A survey on domain-specific languages in robotics. In: Brugali, D., Broenink, J.F., Kroeger, T., MacDonald, B.A. (eds.) SIMPAR 2014. LNCS, vol. 8810, pp. 195–206. Springer, Heidelberg (2014)

14. Abdellatif, T., Bensalem, S., Combaz, J., de Silva, L., Ingrand, F.: Rigorous design of robot software: a formal component-based approach. Robot. Auton. Syst. **60**(12), 1563–1578 (2012)

15. Ziafati, P., Dastani, M., Meyer, J.-J., van der Torre, L.: Agent programming languages requirements for programming autonomous robots. In: Dastani, M., Hübner, J.F., Logan, B. (eds.) ProMAS 2012. LNCS, vol. 7837, pp. 35–53. Springer, Heidelberg (2013)

16. Quigley, M., Conley, K., Gerkey, B., Faust, J., Foote, T., Leibs, J., Berger, E., Wheeler, R.: ROS: an open-source Robot Operating System (2009)

17. Skrzypczyński, P.: Multi-agent software architecture for autonomous robots: a practical approach. Manage. Prod. Eng. Rev. **1**, 55 (2010)

18. Benaissa, S., Moutaouakkil, F., Medromi, H.: New multi-agent's control architecture for the autonomous mobile robots. Int. Rev. Comput. Softw. **6**, 477 (2011)

19. Lacouture, J., Noël, V., Arcangeli, J.P.: Engineering agent frameworks: an application in multi-robot systems. In: Demazeau, Y., Pěchouček, M., Corchado, J.M., Pérez, J.B. (eds.) Advances on Practical Applications of Agents and Multiagent Systems. AISC, vol. 88, pp. 79–85. Springer, Heidelberg (2011)

20. Chella, A., Cossentino, M., Gaglio, S., Sabatucci, L., Seidita, V.: Agent-oriented software patterns for rapid and affordable robot programming. J. Syst. Softw. **83**(4), 557–573 (2010)

21. Choi, H.-T., Sur, J.: Issues in software architectures for intelligent underwater robots. In: Kim, J.-H., Matson, E., Myung, H., Xu, P. (eds.) Robot Intelligence Technology and Applications 2. AISC, vol. 274, pp. 831–839. Springer, Heidelberg (2014)

22. Koubâa, A., Sriti, M.F., Bennaceur, H., Amaar, A., Javed, Y., Alajlan, M., AI-Elaiwi, N., Tounsi, M., Shakshuki, E.: COROS: a multi-agent software architecture for cooperative and autonomous service robots. Stud. Comput. Intell. **604**, 3–30 (2015)

23. Jennings, N.R.: An agent-based approach for building complex software systems. Commun. ACM **44**(4), 35–41 (2001)

24. van Riemsdijk, M.B.: 20 years of agent-oriented programming in distributed AI: history and outlook. In: Proceedings of AGERE! pp. 7–10 (2013)

25. Brooks, R.A.: A robust layered control system for a mobile robot. IEEE J. Robot. Autom. **2**(1), 14–23 (1986)

A Novel Approach for the Identification of Morphological Features from Low Quality Images

Weiqiang Xia[1(✉)], Zhijun Hu[1], Huijuan Zhai[1], Jian Kang[1],
Jingqun Song[1], and Guanglu Sun[2]

[1] Beijing Institute of Astronautical Systems Engineering, Beijing, China
xiaweiqiang@126.com
[2] School of Computer Science and Technology, Harbin University of Science
and Technology, Harbin, Heilongjiang, China

Abstract. In this paper, a novel mathematical morphological approach is proposed, which is combined with an active threshold-based method for the identification of morphological features from images with poor qualities. The algorithm is very fast and needs low computing power. First, a mixed smooth filtering is designed to remove background noises. Second, an active threshold-based method is discussed to create a binary image to achieve rough segmentation. Third, some simple morphological operations, such as opening, closing, filling, and so on, are designed and applied to get the final result of segmentation. After morphological analysis, morphological features, such as contours, areas, numbers, locations, and so on, are obtained. Finally, the comparisons with other conventional methods validate the effectiveness, and an additional experimental result proves the repeatability of the proposed method.

Keywords: Mathematical morphology · Automatic image segmentation and analysis · Active threshold · Morphological operator

1 Introduction

In the middle of the 19th century, the invention of microscope made people can study on human fundamental biological processes at the cellular or organism level. Nowadays, computer-assisted automatic morphological analysis of cells and tissues from microscopy images, such as cell size, cell morphology, density of staining, and so on, are useful methods to reduce hand data recording and user's errors for the detection and diagnosis of some diseases [1, 2] or some special vivo cell phenomenon [3]. But for cell microscopy images, problems remain for image segmentation, shape description and analysis, especially the automatic border detection of cell object boundaries as well

---

This research is partly supported by the innovative research fund of aerospace, research fund for the program of new century excellent talents in Heilongjiang provincial university No. 1155-ncet-008 and the Natural Science Foundation of Heilongjiang Province under grant No. QC2015084, F201132.

© Springer Science+Business Media Singapore 2016
W. Che et al. (Eds.): ICYCSEE 2016, Part I, CCIS 623, pp. 39–47, 2016.
DOI: 10.1007/978-981-10-2053-7_4

as surface contours is still a difficult problem [4, 5]. All difficulties are due to the distinguishing qualities of microscopy imaging. It is known that:

- There is a very low contrast in microscopy image, so that objectives are hard to be separated from background.
- Cell edge in microscopy images is fuzzy and smooth, and moreover, cells may be contiguous and overlapped.
- Noises such as salt and pepper noise, Gaussian white noise and remarkable electrical noise are worse than any other digital system.
- Dust, little bubble, blink point, and blur may be zoomed in by microscopy views, so that cell images may be confused in microscopy images.

For the characters of microscopy images, numerous segmentation methods have, up to now, been proposed; their performances depend largely on the type of images to be processed and on the a priori knowledge relative to the object features. These methods can be roughly classified into several categories, according to the nature of handled features: threshold-based methods, contour-based methods, region-based methods, and methods based on the integration of these techniques.

The threshold-based methods [6, 7] try to separate pixels within or without cell bodies base on the histogram threshold of cell gray-scale. For example, double threshold way, minimum-error way, adaptive threshold way, Otsu way, and so on. Calculations are simple but poor quality images and complex backgrounds may affect the separation result.

The contour-based methods generally deal with the detection and localization of edge points by differential operators and associated regularization processes [8–10], Such as Laplacian operator, Sobel operator, LOG operator, Canny operator, and so on. But they often result in many unclosed frontiers, when they are applied to highly textured objects or weakly contrasted images. Recently, wavelets provide a powerful tool to let microscopy images be studied in nonlinear multiresolution [11, 12]. But, poor quality images and complex algorithms are still problems.

The region-based methods try to detect the areas sharing one or many homogeneity criteria [13, 14], for example, fuzzy C-means, K-means, Mean-Shift, and so on. But they often suffer from their poor localization of boundaries and from over-segmentation. Another interesting region-based technique is based on mathematical morphology [15–17], which is a branch of image analysis based on algebraic, set-theoretic and geometric principles. It is particularly useful in providing basic building blocks to more sophisticated imaging applications. Using mathematical morphology, image is filtered to either preserve or remove features of interest, sizing transformations can be constructed, and information relating to shape, form and size can be easily applied.

Among the mixed methods, one has to mention the integration of two or more methods to conquer the shortage oneself. Integration of threshold and region growing algorithm [18], integration of active contours and mathematical morphology tools [19], integration of wavelet and mathematical morphology algorithm [20, 21], and so on, can perform well.

Our approach in this paper is based on the integration of a region-based method (mathematical morphology) and an active threshold-based method for low quality microscopy images (non-uniform illumination, low contrast, blur or faint image). We based our work on this technique because it is very fast and requires lower computing power. So the final system can run on a very poor computer system, even on an embedded computer-based (such as FPGA, DSP) mini system.

2 Basic Operations in Mathematical Morphology

Morphology processing is a nonlinear filter, its basic concept is to use construction element called as probe to search image information and do morphological transform according to requirement, and so it can be called as morphological filtering. The basic morphological operations are erosion and dilation. Based on them, some important combined operations, such as opening, closing, opening residue and closing residue are defined. They are described as following [15, 16].

A binary image A (i, j) consists of a set of points with the value 1 or 0 at the point (i, j). Usually, let A be an input image, and B a structuring element.

$$\text{Dilation}: A \rightarrow A \oplus B = \cup \{A + b : b \in B\} \tag{1}$$

$$\text{Erosion}: A \rightarrow A \Theta B = \cup \{A - b : b \in B\} \tag{2}$$

Where the symbols $\oplus$ and Θ are the dilation and erosion operators. The erosion is also expressed by the dilation as:

$$\text{Erosion}: A \rightarrow A \Theta B = (A^C \oplus B)^C \tag{3}$$

The superscript C means the complement of the set. The important applications of the basic operations, the opening and the closing of a set A by a set B, are the following:

$$\text{Opening}: A \rightarrow AOB = (A \Theta B) \oplus B = (A^C \oplus B)^C \oplus B \tag{4}$$

$$\text{Closing}: A \rightarrow A \bullet B = (A \oplus B) \Theta B = ((A \oplus B)^c \oplus B)^c \tag{5}$$

Where the symbols and are the opening and the closing operators, respectively. It is clear that an opening operation is anti-extensive, which has the ability to remove positive impulses and peaks smaller than the diameter of structuring element, and a closing operation can remove negative impulses and valleys.

3 Relevant Algorithms and Application

The overall procedure of identification of cell morphological features is demonstrated in Fig. 1.

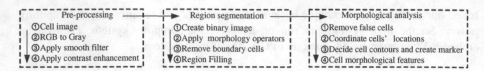

Fig. 1. Procedure of the identification of cell morphological features

3.1 Pre-processing

The red, green and blue (RGB) space of the original image is usually got, so images must be transformed to gray-scale first. The gray image (blood cell) is obtained and shown in Fig. 2(a). High noises and very low contrast usually exist, as discussed in Sect. 1. So, it is necessary to pre-process cell images to get rid of the drawbacks exposed above.

A mixed smooth filter is designed to remove background noises. Linear smooth filter can eliminate Gaussian noise and uniformly distributed noises effectively, but may import image blurring. And, median filter can suppress high-frequency noises, but not suit for eliminating Gaussian noise[22]. So, a combined filter mentioning the integration of a linear filter and a median filter is designed not only for eliminating the aforementioned noises and some kinds of additive noises, but also for reducing calculation volume. For the filter, a 5*5 model is designed, as shown in Eq. (6), and the combined filter can be expressed in Eq. (7). In equations, $a_{i,j}$ is the gray value at (i, j) in image; $g(x, y)$ is the medium value of 5 elements; and MED is a sampling medium value operator.

$$\begin{bmatrix} & & a_{i-2,j} & & \\ & & a_{i-1,j} & & \\ a_{i,j-2} & a_{i,j-1} & a_{i,j} & a_{i,j+1} & a_{i,j+2} \\ & & a_{i+1,j} & & \\ & & a_{i+2,j} & & \end{bmatrix} \tag{6}$$

$$g(x, y) = MED\left\{ \frac{1}{2}(a_{i,j-2} + a_{i,j-1}), \frac{1}{2}(a_{i,j+1} + a_{i,j+2}), a_{i,j}, \frac{1}{2}(a_{i-2,j} + a_{i-1,j}), \frac{1}{2}(a_{i+1,j} + a_{i+2,j}) \right\}. \tag{7}$$

In the pre-processing, smooth filtering is along the cross directions as shown in Eq. (6). This filter model can keep the details in horizontal and vertical directions and eliminate the details in diagonal line, which corresponds with man's visual sense. After smooth filtering, $g(x, y)$ is set as the new gray value at (i, j). Repeatedly, filtering is carried out in whole images. Base on this method, both effective noise suppression and simplifying calculation are satisfying. After smooth filtering, operate histogram equalization to increases the contrast of the image. The pre-processing result is as shown in Fig. 2(b).

3.2   Segmentation

The segmentation is in two steps, in the context of morphological segmentation, the initialization is achieved by means of active threshold-based method first to get the binary image by a primary segmenting, and the second step relates to the localization of boundaries by applying the morphological operators.

After applying pre-processing, the threshold value is calculated by a iteration process, as following:

① Getting initial threshold T_0: $T_0 = \frac{1}{2}(Z_{max} + Z_{min})$. Z_{max} and Z_{min} are maximum gray value and minimum gray value of images individually.

② Dividing image into background and foreground based on T_k, and calculating the mean gray values of background and foreground, E_a and E_b, individually.

③ Getting next new threshold T_{k+1}: $T_{k+1} = (E_a + E_b)/2$.

④ If $T_{k+1} = T_k$, the best threshold is obtained, else turn to ② to continue the iterative loop. End the iterative operation, best threshold T_n is obtained:

$$T_n = (\frac{\sum\limits_{i<T_{n-1}} P_i \times i}{\sum\limits_{i<T_{n-1}} P_i} + \frac{\sum\limits_{i>T_{n-1}} P_i \times i}{\sum\limits_{i>T_{n-1}} P_i})/2 \qquad (8)$$

In Eq. (8), i (i = 0, 1..., 255) is the gray value of a certain pixel; is the amount of the pixels, where the gray value is i; and is the n time and n-1 time iterative results individually.

Base on the best threshold, rough cells segmentation from background is done, and a binary image is shown in Fig. 2(c). The process can be expressed in Eq. (9):

$$BW(i,j) = \begin{cases} 1, & T(i,j) \geq T_n \\ 0, & T(i,j) < T_n \end{cases} \qquad (9)$$

$T(i,j)$ is the gray value at (i,j) in the image; $BW(i,j)$ is the result of binary operation at this point.

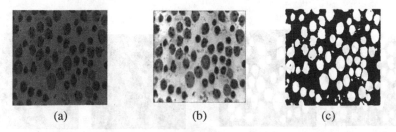

Fig. 2. The procedure of rough segmentation from cell microscopy image

The second step of segmentation is mathematical morphological operation. In Fig. 2(c), it is clearly to see that some discontinuous concave convenes exist at the edge of cells, conglutination or overlap phenomenon appears among the cells, and fracture frontiers are still at cell edges. Moreover, holes in cells and specks in background are in wide distribution. So obeying Eqs. (4) and (5), an opening operation of basic morphological operations is applied (Structuring element is "disk, 4*4"), and a closing operation is executed subsequently ("disk, 2*2"). The result is shown in Fig. 3(a). But still, it is found that bigger holes in cells and bigger specks still exist. So, Morphological reconstructions are operated to solve these problems. Border objects are removed firstly. Next, 8-connected domain is selected to search the boundary of cells and the background, then set background as "0" and the other as "1" to finish the segmentation, as shown in Fig. 3(b).

3.3 Morphological Analysis

Obeying the 4-neighbourhood rule to explore the whole binary image to identify the separate connected regions (set cell regions to "1", set background to "0"), numbered $1 \sim N$, N is the total amount of cell regions. Counting the pixels amount of "1" (named n) to calculate the average area of cells m, $m = n/N$. If the area of a separate connected region is smaller than $1/3$ m (Depend upon the given cell species), this region should be regarded as a false cell and set to "0" to be removed. After removing false cells, new real amount and average area of all cell regions are computed, marked N1, $\overline{m}$. The following step is to get morphological features of cells. Gravity model approach is adopted to get the center point of each cell, which is significant for position finding and tracing of cells (Fig. 4(a)). For example, if one cell region numbered n, center point (x_n, y_n) is got from Eq. (10).

$$x_n = \sum x_i \Big/ S_n, y_n = \sum y_i \Big/ S_n \qquad (10)$$

Next, counting the areas of all close regions, if one is bigger than $5/3\overline{m}$ (Depend upon the given cell species), which is regarded as a remarkable cell. After that, cell information such as serial number, location of the center point, and area is recorded. This information may indicate abnormal cells or cell proliferation. Finally, 8-connected domain is selected to extract the cell contour, as shown in Fig. 4.

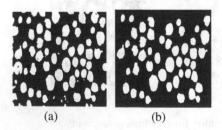
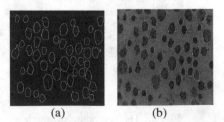

(a)   (b)   (a)   (b)

Fig. 3. Morphological operation results   **Fig. 4.** Identification of cell contours

4 Results and Comparisons

Cell morphological features are obtained after the proposed mathematical morphological operation in Sect. 3. For the experimental cell image, information is got as following: Pixels: 512*512; Amount: 47 cells; Average area of cells: 1554; Biggest area: 3469, No. 20, (207, 384). Abnormal cells: No. 20, 24, 38 and 46. Moreover, base on the algorithm, lots of other cell morphological features can be learned easily, such as cell tracing, cell growth analysis and so on.

In this paper, it is confirmed that identification of cell f morphological features from microscopy images applying proposed mathematical morphological algorithm is better than of the conventional methods. The comparison is show in Fig. 5.

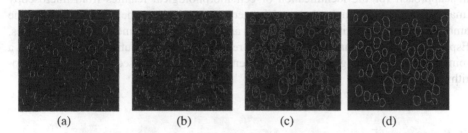

| (a) | (b) | (c) | (d) |

Fig. 5. Comparison between proposed algorithm and some conventional methods

In Fig. 6, all results are processed from Fig. 2. Hereinto, (a) is got by applying the Sobel operation, (b) is got by applying the Laplacian operation, (c) is got by applying the Canny operation, and (d) is got by applying the proposed method. It is easy to see that (a) and (b) do not got the cell edges for the potential reasons of low contrast and smooth edges in microscopy images. And, (c) represents too many edges and unclosed frontiers for the potential reasons of terrible noises and interferences such as dust, little bubble, blink point, and blur in microscopy images. In (d), by applying the proposed method, clear cell edges are detected and cell morphological features are identified. It is indicated that the proposed method can perform much better for the identification of cell morphological features from microscopy images than the others do obviously.

To prove the universal property of method, another experiment is done for the identification of one endothelial cell image. The result is shown in Fig. 6. Figure 6(a) is

| (a) | (b) | (c) |

Fig. 6. Identification of endothelial cells based on the same method

the original image of endothelial cells, and (b), (c) are the results by applying the same proposed method. It is clear to see that the proposed method performs well too.

5   Conclusions and Expectations

For the identification cell morphological features from microscopy images with poor qualities, we proposed an attractive mathematical morphological approach combining with an active threshold-based method. This method can be taken into both account some shortages of microscopy image and morphological characters of cells. By applying some simple morphological operations, such as opening, closing and filling, the cell morphological features are identified. The results not only demonstrate the validity of this approach for the identification of cell morphological features from microscopy images, but also highlight the need for techniques such as these by illustrating the ambiguity associated with the conventional and manual measurement of such properties. Especially, the algorithm is very fast and requires lower computing power. So that, a mini automatic imaging and analysis system (FPGA&DSP-based) based on the algorithm is under design.

References

1. Leland, D.S., Ginocchio, C.C.: Role of cell culture for virus detection in the age of technology. Clin. Microbiol. Rev. **20**, 49–78 (2007)
2. Hall, K.K., Lyman, J.A.: Updated review of blood culture contamination. Clin. Microbiol. Rev. **19**, 788–802 (2006)
3. Wingate, R., Kwint, M.: Imagining the brain cell: the neuron in visual culture. Nat. Rev. Neurosci. **7**, 745–752 (2006)
4. Stephens, D.J., Allan, V.J.: Light microscopy techniques for live cell imaging. Science **300**, 82–87 (2003)
5. Germain, R.N., Miller, M.J., Dustin, M.L., et al.: Dynamic imaging of the immune system: progress, pitfalls and promise. Nat. Rev. Immunol. **6**, 497–507 (2006)
6. Sahoo, P.K., Soltam, S., Wong, A.K., et al.: A survey of thresholding techniques. Comput. Vis. Graph. Image Process. **41**, 233–260 (1988)
7. Otsu, N.: A threshold selection method from gray-level histogram. IEEE Trans. Syst. Man Cybern. **9**, 62–66 (1979)
8. Canny, J.: A computational approach to edge detection. IEEE Trans. Pattern Anal. Mach. Intell. **8**, 679–698 (1986)
9. Witkin, A.P.: Scaling-space filtering. In: International Joint Conferences on Artificial Intelligence, vol. 2, pp. 1019–1022 (1983)
10. Marr, D., Hildreth, E.: Theory of edge detection. Proc. R. Soc. Lond. **B207**, 187–217 (1980)
11. Mallat, S.: multifrequency channel decompositions of images and wavelet models. IEEE Trans. Acoust. Speech Signal Process. **37**, 2091–2110 (1989)
12. Mattlat, S.: A theory for multiresolution signal decomposition: the wavelet representation. IEEE Trans. Pattern Anal. Mach. Intell. **7**, 674–693 (1989)
13. Cheng, Y.: Mean shift, mode seeking, and clustering. IEEE Trans. Pattern Anal. Mach. Intell. **17**, 790–799 (1995)

14. Comanieiu, D., Meer, P.: Mean Shift analysis and applications. In: Proceedings of the Seventh IEEE International Conference on Computer Vision, vol. 2, pp. 1197–1203 (1999)
15. Serra, J.: Introduction to Mathematical Morphology. Comput. Vis. Graph. Image Process. **35**, 283–305 (1986)
16. Breen, E.J., Jones, R., Talbot, H.: Mathematical morphology: a useful set of tools for image analysis. Stat. Comput. **10**, 105–120 (2000)
17. Lee, J.R., Smith, M.L., Smith, L.N., et al.: A mathematical morphology approach to image based 3D particle shape analysis. Mach. Vis. Appl. **16**, 282–288 (2005)
18. Pavlidis, T., Liow, Y.T.: Integrating region growing and edge detection. IEEE Trans. Pattern Anal. Mach. Intell. **12**, 225–233 (1990)
19. Elmoataz, A., Schupp, S., Clouard, R., et al.: Using active contours and mathematical morphology tools for quantification of immunohistochemical images. Signal Process. **71**, 215–226 (1998)
20. Goutsias, J., Heijmans, H.: Nonlinear multiresolution signal decomposition scheme-part I: morphological pyramids. IEEE Trans. Image Process. **11**, 1862–1876 (2000)
21. Heijmans, H., Goutsias, J.: Nonlinear multiresolution signal decomposition scheme-part II: morphological wavelets. IEEE Trans. Image Process. **11**, 1897–1913 (2000)

A Novel Filtering Method for Infrared Image

Jian Kang[1](✉), Chunxiao Zhou[1], Weiqiang Xia[1], Chaopeng Shen[1],
and Guanglu Sun[2]

[1] Beijing Institute of Astronautical Systems Engineering, Beijing, China
Kangjian004@163.com
[2] School of Computer Science and Technology, Harbin University of Science
and Technology, Harbin, Heilongjiang, China

Abstract. The image filtering technology is widely used in many fields, such as
environmental monitoring and assessment, space remote sensing, recognition
and tracking of infrared target. In this paper, aiming at the problem of poor
generalization capability and over-fitting with artificial neutral network for
infrared image filtering, a new filtering method is presented. The structure
elements are used to set up the training samples. And then, based on support
vector machine theory, it builds the learning ma-chine with proper model and
trains the samples. The result can be used to suppress the background SNR of
following image. The experimental result with infrared image shows that the
method can obtain higher SNR than conventional neutral network and fixed
Top-Hat operator method, especially in low SNR.

Keywords: Image filtering · Neutral network · Support vector machine ·
Top-Hat operator

1 Introduction

The technology of image filtering is widely used in many fields, such as environmental
monitoring and assessment, space remote sensing, recognition and tracking of infrared
target, and so on. With the development of mathematical morphology, it has become a
important method for image filtering and many researchers use it to filter the background noise. But, the traditional morphology filter all depends on the experience of the
designer, which is difficult to select the optimal structural elements [1]. So, a morphology weighted neural network is used for target recognition by Won [2], and the
learning ability of morphology neural network is researched by Ritter [3]. The application of the morphology neural network is summed up by Grana [4]. But, in practice,
the neural network has the problem of over-fitting and poor generalization capability.
In the paper. the training algorithm based on Support Vector Machine (SVM) is presented. SVM [5] realizes the structural risk minimum, and minimums the experience

This research is partly supported by the innovative research fund of aerospace, research fund for the
program of new century excellent talents in Heilongjiang provincial university No. 1155-ncet-008
and the National Natural Science Foundation of China under grant No. 60903083, 61502123.

© Springer Science+Business Media Singapore 2016
W. Che et al. (Eds.): ICYCSEE 2016, Part I, CCIS 623, pp. 48–55, 2016.
DOI: 10.1007/978-981-10-2053-7_5

risk and the bound of VC dimension. SVM can get less practice risk and better generalization. In this paper, it is used for the filtering of infrared image.

2 Target Observation Models

For the infrared image, the targets are close to a point source (several pixels), because of aerooptic disturbance and air turbulence. The model of targets can be substitute by a 2D IR optical blur function, which is shown as follow:

$$ f_T(x,y) = \tau * \exp\left\{ -\frac{1}{2}\left(\frac{x}{\delta_x}\right)^2 + \left(\frac{y}{\delta_y}\right)^2 \right\} \tag{1} $$

Where, $f_T(x,y)$ is the targets intensity; τ is the targets intensity amplitude, δ_x and δ_y are horizontal and vertical extent parameter, x and y are the coordinates of targets.

The observation sequence of a random infrared image embedded with moving targets can be modeled as follow

$$ f(x,y,k) = f_T(x,y,k) + B(x,y,k) + N(x,y,k) $$

Where, $B(x, y, k)$ is the exterior clutter background; $N(x, y, k)$ is the noise of the image; k is the sampling time.

Figure 1 is the real infrared image, which is get from thermal infrared imager, which is made by Sofradir company, France. It is 320×240, 8 bit infrared target image sequence. According to Fig. 1, the local properties of infrared targets are convex.

3 Grayscale Morphology Filter Algorithm

The basic operation of grayscale morphology is erosion, dilation, open and close, which are defined as follow.
Defining one:
Structural element B erodes image f:

$$ (f\Theta B)(x) = \min_{t \in B} f(x+t) $$

Structure element B dilates image f:

$$ (f \oplus B)(x) = \min_{t \in B} f(x-t) $$

Defining two:
Opening

$$ f \circ B(x) = (f\Theta B)(x) \oplus B $$

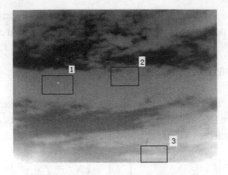

(a) Original infrared image with three targets

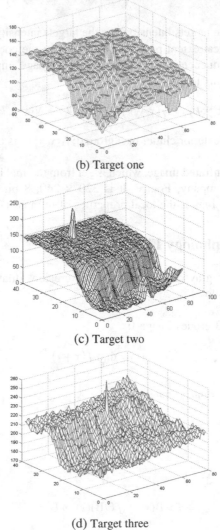

(b) Target one

(c) Target two

(d) Target three

Fig. 1. The local property of infrared target.

Closing

$$f \circ B(x) = (f \oplus B)(x) \ominus B$$

Open operation likes a non-linear low-pass filter, but it is different from the frequency domain low-pass filter. When the big and small structures both have high frequencies in one image, the open operation only allows the big structures passing and prevents the small structures passing.

As discussing above, the target is the "convex" structure in infrared image. After filtering, original image subtracting the background can immediately get the candidate targets.

Open Top-Hat operation and Close Top-hat operation is defined as follow and Fig. 2.

$$\mathrm{OTH}_{f,B}(x) = (f - f \circ B)(x)$$

$$\mathrm{CTH}_{f,B}(x) = (f \circ B - f)(x)$$

Original image

Morphology opening result   Subtracted image

4 An Illustration of Mathematical Morphological Top-Hat Transform

Open Top-Hat operation can find the peak of one image. Close Top-Hat operation can find the valley of one image. Morphological Top-Hat operation can effectively identify spot targets in a complicated background. But for spot targets in high noise image, the traditional Top-Hat Morphology operation is helpless. So, the learning algorithm based on Support vector machine (SVM) is provided.

5 Training Method Based on SVM

The size of structural elements decides the input vector dimension of learning machine. For example, if the length of structural elements is $n \times n$, the input vector dimension must be n^2.

Suppose the set of training samples is

$$(x_1, y_1), (x_2, y_2), \ldots, (x_l, y_l) \in \mathbf{R}^l \times \{-1, 1\}$$

Where, x_i is a n^2 dimension vector and $y_i \in \{-1, 1\}$. The standard format of SVM is shown as follows:

$$\min_{\omega,b,\varepsilon}(\frac{1}{2}\omega^T\omega + C\sum_{i=1}^{l}\xi_i), y_i[\omega^T\phi(x_i) + b] \geq 1 - \xi_i \tag{2}$$

Where, ξ_i usually is called punishing coefficient, which satisfies $\xi_i \geq 0, i = 1, 2, \ldots, l$. ω is called weighted vector. When $\phi(x_i) = x_i$, formula (2) is called linear kernel under linear separable conditions. For the non-linear inseparable problem, x_i is usually mapped to a high dimension space, then it becomes a SVM of non-linear kernel. In order to solve the objective function under restricting condition, it must solve the following dual problem.

$$\min_{\alpha}(\frac{1}{2}\alpha^T Q\alpha - e^T\alpha), y^T\alpha = 0 \tag{3}$$

Where, $0 \leq \alpha_i \leq C_i = 1, \ldots, l$, α_i is the undetermined coefficient, Q is $l \times l$ positive semidefinite matrix. $Q_{ij} = y_i y_j \phi(x_i)^T\phi(x_j)$, e is the matrix whose elements all are one. $K(x_i, x_j) = \phi(x_i)^T\phi(x_i)$ is called kernel function. It can be seen from the solution $\omega = \sum_{i=1}^{l} y_i \cdot \phi(x_i) \cdot \phi(x_j)$. The only input vector, which satisfies $\alpha_i > 0$, is called support. According to the training sample, the decision-making function is shown as follow.

$$f(x) = \text{sgn}(\sum_{i=1}^{ns} \alpha_i y_i K(x_i, x) + b) \tag{4}$$

Where, ns is the number of support vectors, b is a constant number. For a testing vector x, if $\sum_{i=1}^{ns} \alpha_i y_i K(x_i, x) + b > 0$, it will be classified as the class of "1", otherwise classified as the class of "−1", the learning machine can be generalized by radius basis function (RBF), which is shown as follows:

$$K(x_i, x) = \exp\left(\frac{\|x - x_i\|^2}{-2\sigma^2}\right) \tag{5}$$

Where, σ is the parameter of kernel function.

Synthesize this two kernel function; a mixed kernel is shown as follows:

$$K_{mix} = \rho K_{poly} + (1 - \rho)K_{rbf} \tag{6}$$

According to the decision-making function, for a image of $R \times L$, the operation times is $ns \times R \times L$, the complicated extent of the calculation is depend on the size of the image $O(R \times L)$, the support vector number of the learning machine $O(ns)$ and the $n \times n$ structural element. So, the operation times is shown as follow

$$Operations \approx ns \times R \times L \times n^2$$

(a)   Original image

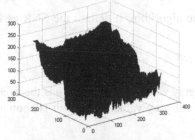

(b)   3D image of the original image

(c)   Filtering image

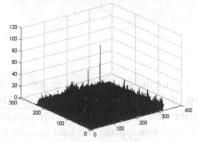

(d)   3D image of filtering image

Fig. 2. The filtering result

6 Experiment Result

In this experiment, it gets a real infrared target image sequence, which is used to verify the ability of the algorithm, which is presented in this paper. Figure 2 shows the filtering result. Figure 2(a) is the original image. Figure 2(b) is the 3D image of the original image. Figure 2(c) is the image, which is filtered using SVM. Figure 2(d) is the 3D image after filtering. The size of the target is only several pixels, and because of the sunshine's irradiation, the gray degree value at the right bottom is much higher than another place. Figure 2(b) shows an obvious staircase, the target is hard to distinguish. According to Fig. 2(c) and (d), the background is flat after filtering and the SNR increases.

In order to compare the algorithm, it defines the SNR as follow:

$$SNR = \frac{f_{Tm} - u}{\sigma} \qquad (7)$$

Where, f_{Tm} is the minimal gray value of the targets; u is the average value of gray value; σ is the standard deviation of gray value. The result is shown in Table 1.

Table 1. SNR after filtering

SNR	SNR after filtering		
	Method one	Method two	Method three
1.7	8.37	13.41	19.70
2.1	15.22	20.18	25.37
4	37.56	38.46	40.15

Method one is using SVM. Method two is using neural network. Method three is using Top-Hat operator, which is fixed. As shown above, three methods are both performing well in high SNR, but in low SNR, the SVM method has the best filtering result.

7 Conclusion

In this paper, a new filtering method of infrared image is presented. The structure elements is taken to set up the training samples, and then, based on support vector machine theory, it builds the learning machine with proper model and trains the samples. The results can be used to suppress the background SNR of following image. The experiment result with infrared image shows that the method can obtain higher SNR than traditional neutral network and fixed Top-Hat operator method, especially in low SNR.

References

1. XIE, K., XU, G., LIU, X.: Optimized design for mathematics morphology filter. Comput. Eng. Appl. **35**, 103–105 (2003)
2. Won, Y.G., Gader, P.G., Coffield, P.D.: Morphological shared-weight networks with applications to automatic target recognition. Trans. Neural Netw. **8**(5), 1195–1203 (1997)
3. Ritter, G.X., Sussner, P., Diza-de-Leon J.L. Morphological neural networks. In: Neural Networks Proceedings IJCNN 2001 International Joint Conference, vol. 4(15), pp. 2518–2523 (2001)
4. Grana, M., Raducanu, B.: Some applications of morphological neural networks. In: Neural Networks Proceedings IJCNN 2001 International Joint Conference, vol. 4(15), pp. 2518–2523 (2001)
5. Zheng, S., LIU, J., Tian, J.: Research of SVM-based infrared small object segmentation and clustering method. Sig. Proc. **21**(5), 515–519 (2005)

A Novel Quantum Noise Image Preparation Method

Xianhua Song[✉]

School of Applied Science, Harbin University of Science and Technology,
Harbin 150080, China
showshinesong@163.com

Abstract. Quantum noise image has important role in evaluating quantum image quality and testing processing algorithms. A novel preparation method of quantum Gauss noise image is proposed. Furthermore, the experimental simulation proves the efficiency of the method. The research about quantum noise image is of great significance to evaluate and test the schemes for quantum image authentication and secure communication.

Keywords: Quantum computation · Quantum image representation · Quantum noise image · Quantum Gauss noise

1 Introduction

Quantum computer is an inevitable stage in the development process of computer. Quantum information, especially quantum media information, will have important researching value in the era of quantum computer. Quantum image processing, an area focusing on extending conventional image processing tasks and operations to the quantum computing framework, will be a key issue in quantum information processing field [1].

At present, the quantum image representation methods can be divided into the following categories: lattice method based [2–4], entanglement based [5], vector based [6,7], FRQI (Flexible Representation of Quantum Images) method [8] and NEQR (Novel Enhanced Quantum Representation) method [9]. The two methods of FRQI and NEQR are widely used in the research of quantum image processing and security. In FRQI, the gray level and the position of the image are expressed as a normalized quantum superposition state. NEQR quantum image indicates an increase in the number of quantum bits required for representing the gray level of image. Therefore, FRQI representation method is taken into consideration to design quantum noise image preparation.

Noise comes mainly from the acquisition, transmission and processing of images. The noise usually used in image processing is Gauss noise (i.e. white noise), uniform noise and impulse noise (i.e. salt and pepper noise). A suitable quantum noise image is helpful to test the effectiveness of the quantum image processing algorithms [10,11] and/or quantum image security schemes [12–14].

© Springer Science+Business Media Singapore 2016
W. Che et al. (Eds.): ICYCSEE 2016, Part I, CCIS 623, pp. 56–62, 2016.
DOI: 10.1007/978-981-10-2053-7_6

Therefore, producing a quantum noise image whose possibility density function is consistent with a given formulation is a necessary stage to test or verify other quantum image processing methods. Although producing a classical noise image is very simple, its quantum version isn't so easy because quantum image production need to satisfy the principle of quantum computation. In this paper, a quantum Gauss noise image is prepared using basic quantum transforms based on FRQI quantum image. The research about quantum noise image provides convenience for testing or verifying the effectiveness of the quantum image processing algorithms [15].

The rest of the paper is organized as follows. Section 2 gives the related works on quantum image representations. In Sect. 3, the quantum noise image and its preparation method is proposed. Section 4 is devoted to the simulation results and result analyses. Finally, Sect. 5 concludes the paper.

2   Flexible Representation of Quantum Images

Inspired by the pixel representation for images in classical computers, a flexible representation for quantum images is proposed in [8]. The quantum image corresponding to a classical image sized $2^n \times 2^n$ is defined by a quantum encoding state, i.e.,

$$|I(\theta)\rangle = \frac{1}{2^n} \sum_{i=0}^{2^{2n}-1} (\cos \theta_i |0\rangle + \sin \theta_i |1\rangle) \otimes |i\rangle,$$

$$\theta_i \in [0, \frac{\pi}{2}], i = 0, 1, ..., 2^{2n} - 1,$$

(1)

wherein $\cos \theta_i |0\rangle + \sin \theta_i |1\rangle$ encodes the color information and $|i\rangle$ encodes the corresponding position of the quantum image. The position information includes two parts: vertical and horizontal coordinates. Concretely, in a $2n$-qubits system,

$$|i\rangle = |y\rangle|x\rangle = |y_{n-1}y_{n-2} \cdots y_0\rangle|x_{n-1}x_{n-2} \cdots x_0\rangle,$$
$$x, y \in \{0, 1, \cdots, 2^n - 1\},$$
$$|y_j\rangle, |x_j\rangle \in \{|0\rangle, |1\rangle\}, j = 0, 1, \cdots, n - 1,$$

(2)

where $|y\rangle = |y_{n-1}y_{n-2} \cdots y_0\rangle$ encodes the first n-qubits along the vertical location and $|x\rangle = |x_{n-1}x_{n-2} \cdots x_0\rangle$ encodes the second n-qubits along the horizontal axis.

3   Quantum Noise Image and its Preparation

The noise of quantum image mainly comes from the transmission and processing of the image, which is mainly based on the additive noise. The noise superimposition process of quantum images can be represented by a degenerate function:

$$|G\rangle = |F\rangle + |\eta\rangle$$

(3)

where $|F\rangle$ denotes the input quantum image, $|\eta\rangle$ represents the additive noise and $|G\rangle$ is the degenerated image after superimposing noise.

Gauss noise and salt and pepper noise are two most common additive noise. Through the degenerate function, it is easy to conclude that the superposition process of variable intensity noises can be equivalent to the quantum noise image preparation process. The feasibility study of quantum preparation of Gauss noise image will be proposed as follows.

Suppose the quantum Gauss noise has the following possibility distribution:

$$p(\varphi) = \frac{1}{\sqrt{2\pi}\sigma} e^{-\frac{(\varphi-\mu)^2}{2\sigma^2}} \tag{4}$$

wherein, Gauss random variable φ represents gray value coding, μ denotes the expectation value and σ expresses the standard deviation. When φ satisfy Eq. (4), ninety-five percent its values fall into the range of $[\mu - 2\sigma, \mu + 2\sigma]$. In many practical situations, noise can be approximated by the Gauss noise. Specially, the preparation process of the FRQI quantum Gauss noise image can be described as follows:

Step 1: Suppose the size of the FRQI image is $N = 2^n \times 2^n$ and the number of Gauss distribution noise pixels is $M, M \leq N$. The initial quantum image state without noise is:

$$|I(\theta)\rangle = \frac{1}{2^n} \sum_{i=0}^{2^{2n}-1} (\cos\theta_i|0\rangle + \sin\theta_i|1\rangle) \otimes |i\rangle \tag{5}$$

Step 2: Sampling the range of $[\mu-2\sigma, \mu+2\sigma]$ to S bins denoting $\{\varphi_1, \varphi_2, \ldots, \varphi_S\}$ which is displayed in Fig. 1.

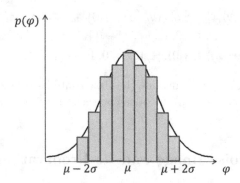

Fig. 1. Sampling bins of Gauss noise probability density function.

It is significant that

$$\sum_{i=1}^{S} p(\varphi_i) \cong 95\%$$

Step 3: Selecting $[M \cdot p(\varphi_1)]$ pixels randomly from the image $|I(\theta)\rangle$ to construct a set P_1, where signal "$[\cdot]$" denotes the rounding operation.

Step 4: Applying unitary transform R_1 to quantum image $|I(\theta)\rangle$, wherein

$$R_1 = I \otimes \sum_{j=0, j \notin P_1}^{2^{2n}-1} |j\rangle \langle j| + R(2\varphi_1) \otimes \sum_{j \in P_1} |j\rangle \langle j|$$

$$R(2\varphi_1) = \begin{pmatrix} \cos \varphi_1 & -\sin \varphi_1 \\ \sin \varphi_1 & \cos \varphi_1 \end{pmatrix}$$

If denoting the result as

$$R_1(|I(\theta)\rangle) \triangleq |I(\theta)\rangle_1$$

then this step realize the aim of adding noise whose gray coding is φ_1 to to the quantum image $|I(\theta)\rangle$ randomly.

Step 5: For

$$k = 2, 3, \dots, S$$

selecting randomly $[M \cdot p(\varphi_k)]$ pixels from the image $|I(\theta)\rangle_{k-1}$ to construct a set P_k, here

$$P_k \cap P_{k-1} \cap \cdots \cap P_1 = \varPhi$$

Then implementing the following operations:

$$R(|I(\theta)\rangle_1) = \left(\prod_{k=2}^{S} R_k \right) (|I(\theta)\rangle_1) \triangleq |I(\theta)\rangle_n$$

where

$$R_k = I \otimes \sum_{j=0, j \notin P_k}^{2^{2n}-1} |j\rangle \langle j| + R(2\varphi_k) \otimes \sum_{j \in P_k} |j\rangle \langle j|$$

$$R(2\varphi_k) = \begin{pmatrix} \cos \varphi_k & -\sin \varphi_k \\ \sin \varphi_k & \cos \varphi_k \end{pmatrix}$$

Thus, the final quantum state $|I(\theta)\rangle_n$ is the quantum Gauss noise image satisfying Eq. (4).

From above *steps* 1 to 5, the preparation of quantum Gauss noise image is finished.

The whole preparation of quantum noise image costs no more than $O(2^{4n})$ for a $2^n \times 2^n$ image. This indicates the efficiency of the preparation process.

4 Experiments

The simulations are based on linear algebra with complex vectors as quantum states and unitary matrices as unitary transforms on the software of Matlab 7.12.0(R2011a). The final step in quantum computation is the measurement, which converts the quantum information into the classical form as probability distributions. Experiments are performed on desktop computer with Intel(R) Xeon(R) CPU E5620 2.40 GHz, 2.39 GHz(4 processors) 8 GB RAM and 64-bit operating system. Figure 2 illustrates the simulation results of quantum original image and quantum Gauss noise image. (a1) is the original image Lena and (a2) is the version after added Gauss noise. (b1) is the original image Cameraman and (b2) is the quantum noise image added Gauss noise. Through simulation, it is verified that preparing a quantum noise image satisfying some probability density is totally feasible.

Fig. 2. Quantum noise image illustration.

5 Conclusions

In this paper, a novel quantum noise image preparation scheme based on FRQI image has been proposed. The proposed strategy comprises of five steps which include only elementary unitary transforms. Quantum noise image provides convenience for testing or verifying the effectiveness of the quantum image processing algorithms. Experimental simulation results show the preparation algorithm is effective. Exploring quantum algorithm for other noises methods will be our future work.

Acknowledgments. This work is supported by the National Natural Science Foundation of China(61501148).

References

1. Yan, F., Iliyasu, A.M., Venegas-Andraca, S.E.: A survey of quantum image representations. Quantum Inf. Process. **15**, 1–35 (2016)
2. Venegas-Andraca, S.E., Bose, S.: Storing, processing, and retrieving an image using quantum mechanics. In: Proceedings of SPIE, Quantum Information and Computation, Orlando, FL, United States: International Society for Optics and Photonics, vol. 5105, pp. 137–147 (2003)
3. Li, H.S., Qing, X.Z., Lan, S., et al.: Image storage, retrieval, compression and segmentation in a quantum system. Quantum Inform. Process. **12**(6), 2269–2290 (2013)
4. Yuan, S., Mao, X., Xue, Y., et al.: SQR: a simple quantum representation of infrared images. Quantum Inform. Process. **13**(6), 1353–1379 (2014)
5. Venegas, A.S., Ball, J.: Processing images in entangled quantum systems. Quantum Inform. Process. **9**(1), 1–11 (2010)
6. Latorre, J.I.: Image compression and entanglement. arXiv preprint quantph/0510031 (2005)
7. Hu, B.Q., Huang, X.D., Zhou, R.G., et al.: A theoretical framework for quantum image representation and data loading scheme. Sci. China Inform. Sci. **57**(032108), 1–11 (2014)
8. Le, P.Q., Dong, F., Hirota, K.: A flexible representation of quantum images for polynomial preparation, image compression and processing operations. Quantum Inf. Process. **10**(1), 63–84 (2010)
9. Zhang, Y., Lu, K., Gao, Y.H., Wang, M.: NEQR: a novel enhanced quantum representation of digital images. Quantum Inf. Process **12**(8), 2833–2860 (2013)
10. Le, P.Q., Iliyasu, A.M., Dong, F., Hirota, K.: Fast geometric transformations on quantum images. IAENG Int. J. Appl. Math. **40**(3), 113–123 (2010)
11. Caraiman, S., Manta, V.I.: Histogram-based segmentation of quantum images. Theor. Comput. Sci. **529**, 46–60 (2014)
12. Zhou, R.G., Wu, Q., Zhang, M.Q., Shen, C.Y.: Quantum image encryption and decryption algorithm based on quantum image geometric transformations. Int. J. Theor. Phy. **52**(6), 1802–1817 (2012)
13. Yang, Y.G., Xia, J., Jia, X., Zhang, H.: Novel image encryption/decryption based on quantum fourier transform and double phase encoding. Quantum Inf. Process. **12**(11), 3477–3493 (2013)

14. Song, X.H., Wang, S., Ei-Latif, A.A.A., Niu, X.M.: Quantum image encryption based on restricted geometric and color transformations. Quantum Inf. Process. **13**(8), 1765–1787 (2014)
15. Mastriani, M.: Quantum Boolean image denoising. Quantum Inf. Process. **14**(5), 1647–1673 (2015)

A Personalized Recommendation Algorithm
with User Trust in Social Network

Yuxin Dong[1($\boxtimes$)], Chunhui Zhao[2], Weijie Cheng[1], Liang Li[1],
and Lin Liu[1]

[1] College of Computer Science and Technology, Harbin Engineering University,
Harbin 150001, Heilongjiang, China
yuxinhrbeu@126.com
[2] College of Information and Communication Engineering,
Harbin Engineering University, Harbin 150001, Heilongjiang, China

Abstract. In the era of big data, personalized recommendation has become an important research issue in social networks as it can find and match user's preference. In this paper, the user trust is integrated into the recommendation algorithm, by dividing the user trust into 2 parts: user score trust and user preference trust. In view of the common items in user item score matrix, the algorithm combines the number of items with the score similarity between users, and establishes an asymmetric trust relationship matrix so as to calculate the user's score trust. For the non-common score items, we use the attribute information of items and the scoring weight to calculate the user's preference trust. Based on the user trust in social network, a new collaborative filtering recommendation algorithm is proposed. Besides, a new matrix factorization recommendation algorithm is proposed by combining the user trust with matrix factorization. We did the experiments comparing with the related algorithms on the real data sets of social network. The results show that the proposed algorithms can effectively improve the accuracy of recommendation.

Keywords: Recommendation system · Collaborative filtering · Matrix factorization · User trust · Social network

1 Introduction

Recommendation system can help not only find commodities we are interested in, but also find friends that have been lost in touch for long time [1]. At present, recommendation system technologies have been widely exploited in many fields of social network [2], and most focus on movies [3], music, TV shows [4], friends of microblog and so on. The data used in recommendation algorithm mainly comes from the previous history of customers, for example, from customers' past ratings, their tags. User-based collaborative filtering algorithm is a prevailing algorithm in recommender system, with the core concept: regarding the user program scoring matrix and the target user as the input; calculating the similarities between users by the score information. The algorithm needs to identify similar users of the target user accurately [5]. The number of neighbor users can be manually designed. The objects which have never

© Springer Science+Business Media Singapore 2016
W. Che et al. (Eds.): ICYCSEE 2016, Part I, CCIS 623, pp. 63–76, 2016.
DOI: 10.1007/978-981-10-2053-7_7

been predicted by the target use, can get their score by the predicted score of the first K nearby users.

Although the thought of collaborative filtering algorithm is simple, there are still some problems to be solved. One of the typical problems is that the accuracy and efficiency of recommendation is relatively low due to sparse data [6]. This paper can efficiently solve the problem by introducing user trust to the algorithm. Actually, user trust relationship has already become the focus of many research fields. The structure of the paper will be: Sect. 2 is the introduction of recent related work. Section 3 introduces the algorithm to get user trust and show the concrete process of recommendation. In Sect. 4, we show experimental results and analysis. Section 5 concludes the paper and presents future work.

2 Related Work

User trust mainly comes from explicit trust and implicit trust of social network [7]. Explicit trust means the directly statement trust to other users, which can be excavated from the basic information of the user history. Implicit trust means the potential trust to other users. At present, many researches focus on explicit trust between users, but in real social network, many users may use false registration information in order to protect their privacy, which makes it impossible to exploit explicit trust directly. For implicit trust, present algorithms mainly consider the relationship between the user scores. However, it is difficult when the data is sparse and even rare to apply implicit trust to recommendation in social network [8].

For the explicit trust between users, Massa proposes a trust network model when designing the recommendation algorithm. The potential desirable commodities can be predicted by the history score of trust neighborhood, which can greatly enhance the prediction accuracy [9]. Guo et al. make estimated score by trust of neighborhood who has similar interests in social network. However, with the increase of the number of the recommended list, the data sparsity has a significant effect on the algorithm [10]. Although explicit trust between users can improve the efficiency of recommendation. But in real social network, many users may use false registration information in order to protect their privacy, which makes it impossible to exploit explicit trust directly. In most of the recommendation systems, explicit trust is not easy to collect, so it is necessary to make recommendation for users through the implicit trust. For implicit trust, Tang et al. considered global trust and local trust comprehensively, built a user trust model and proposed the algorithm to calculate global trust [11]. Liu et al. combined user rating credibility and score similarity to improve the accuracy and robustness of the recommendation algorithm [12]. Lu et al. calculated the local trust degree and global trust degree of the users respectively. It improves the accuracy of the recommendion by mining the potential trust relationship from the massive user history data [13].

Trust-based recommendation system has greatly improve the quality of recommendation by combining user trust with CF algorithm. And matrix decomposition recommendation algorithm can achieve a higher accuracy in the prediction score. Yang et al. believed that the factors influencing user prediction score lie in not only the people whom the user trust, but also the ones who trust the user. So Yang combined matrix decomposition algorithm to propose a hybrid recommendation(TrustMF)which considered trusted users and the ones who trust them [14]. A new method named trustsvd is proposed according to the rating similarity calculation of user trust between users, and users will score the trust relationship into the matrix decomposition technique, and achieved good results in predicting the score [15].

In this paper, we put forward a recommendation algorithm which is based on the users trust, and the algorithm comprehensively considers both common score of users and non-common score items and propose the algorithm to calculate user score trust degree. The algorithm takes into account the user's all scoring information, to improve the accuracy of the recommendation.

3 A Recommendation Algorithm Based on User Trust

3.1 User Trust

The idea of collaborative filtering algorithm is to let the user's "friends" to help users filter out the items he may be interested in, so how to find the user's "friends" is of great importance. Trust relationships between users can be a good solution to this problem in the social networking sites. However, unreasonable recommendations make it hard to satisfy users. Traditional trust relationship is mainly displayed to allow users to specify who he trusts, but it can only help users get friends who they already knows through this direct way. There are some limitations and problems in the accuracy of recommendation. The trust information can be obtained from the users' own information, but since many users may use false registration information in order to protect their privacy, this method may have some disadvantages. This section excavates the trust relationships between users by defining the user trust containing the score trust and the preference trust.

3.1.1 Score Trust of User (STUser)

Traditional score trust relationship is symmetrical but the trust relationship matrix may be asymmetric, that is to say, the trust degree of A to B may not be equal to that of B to A. The reason of considering different trust degree is that the number of common scored items is different. Obviously, if two users' scored items are the same, then the size of trust between them is the same.

For user A and B, the trust measurement of A to B is designed as the proportion of the number of common scoring items among them in all the rating items of A, which is called the trust factor, designed as follow:

$$\theta_{AB} = \frac{|I_A \cap I_B|}{|I_A|} * \frac{|I_A \cap I_B|}{|I_A| \cup |I_B|} \tag{1}$$

In the formula, θ_{AB} is the level of trust of A to B, I_A and I_B are two sets which includes items scored by user A and B.

In order to calculate score trust similarities between users more accurately, it needs to consider the relationship between common scoring items. The two common similarity algorithms are Cosine Similarity and Pearson Correlation Coefficient.

Cosine similarity regards the user score of items as the vector of 1*n dimension. The interest similarity between two users can be calculated by using the cosine of the angle between users on the items of the score vector. The cosine similarity calculation formula is shown:

$$sim_{uv} = \cos(u, v) = \frac{\sum_{i \in I_{uv}} r_{ui} \times r_{vi}}{\sqrt{\sum_{i \in I_u} r_{ui}^2} \sqrt{\sum_{i \in I_v} r_{vi}^2}} \tag{2}$$

The correlation coefficient of Pearson is used to consider the linear correlation degree between two user scores and the similarity ranges from −1 to 1, in which positive represents users interest are in positive correlation. The closer to 1, the greater the similarity is. Shown as follow:

$$sim_{uv} = \frac{\sum_{i \in I_{uv}} (r_{ui} - \bar{r}_u) \times (r_{vi} - \bar{r}_v)}{\sqrt{\sum_{i \in I_u} (r_{ui} - \bar{r}_u)^2} \sqrt{\sum_{i \in I_v} (r_{vi} - \bar{r}_v)^2}} \tag{3}$$

But the accuracy of those methods is not high. In this paper, we consider each user have their own interests, also have their score standard, some users score more strictly, so their scores are relatively low. Some users are more relatively light, and their scores are also higher. In order to reflect the degree of easing user score better, we should consider the gap between the user ratings and also should consider the average score. The user rating similarity formula is calculated as follows:

$$sim_{AB} = \frac{\sum_{l=1}^{N} \left(1 - \frac{|\bar{R}_A - \bar{R}_B + R_{Bl} - R_{Al}|}{2(Max - Min)}\right)}{|I_A \cap I_B|} \tag{4}$$

In the formula, Max is the highest score, and Min is the lowest score. The trust factor should be taken into consideration, the final degree of the score trust of A to B is:

$$sim_{st}(A, B) = sim_{AB} * \theta_{AB} \tag{5}$$

3.1.2   Preference Trust of User (PTUser)

Data sparsity means that most users are only interested in a small number of items. That is to say, the number of common score items between users is very small, so the algorithm have disadvantages in solely relying on common item rating to compute trust value. At present, researches of most recommendation systems have ignored or reduced the influence of non-common score items on user trust degree [16]. For non-common score items between users, the paper uses the information of the own label of items to calculate the trust of their preferences.

In recommendation system, users can label the project that they scored. Each item has 1 or more tags, these tags can represent their content information. For example, a movie has a labels like comedy, romance, horror label and star. Investigating labels help to improve the quality of the recommendation system. This paper uses scores made by users to items. After calculating the user preference on each label, items of high scores should be strengthened to the corresponding label preference and items of low scores should be reduced. Finally, the preference with positive labels represents user interest degree, the greater the more interested.

For example: Table 1 shows the ratings of 8 movies which User1 scored. In the table, 1 represents the movie including this label and the average score is 3.0. Then, we consider users have interest in those movies with more than 3.0 score and have less interest in those movies with less than 3.0 score. The positive preference denoted as Pd and negative preference is denoted as Nd. Calculation method is as follows: first, calculate the user preference on each label, such as calculating of label 1, as shown in Fig. 1:

Table 1. User tag rating matrix

Movies	Ratings	T1	T2	T3	T4	T5	T6
M1	5	0	0	1	1	0	0
M2	3	1	1	0	0	0	0
M3	3	1	0	0	1	0	1
M4	1	1	1	0	1	0	0
M5	4	0	0	0	0	0	1
M6	5	0	0	1	0	0	1
M7	1	1	1	0	0	1	0

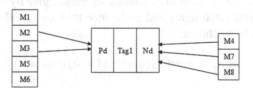

Fig. 1. User preference of tag1

The user's positive and negative preference to the label is calculated as follows:

$$sim_{u,t}^{pd} = \frac{1}{\left|I_u^{pd}(t)\right|} * \sum_{i \in I_u^{pd}(t)} \frac{R_{ui} - \overline{R}}{\sqrt{\sum_{j=1}^{I} R_{uj}^2}} \tag{6}$$

$$sim_{u,t}^{nd} = \frac{1}{\left|I_u^{nd}(t)\right|} * \sum_{i \in I_u^{nd}(t)} \frac{\overline{R} - R_{ui}}{\sqrt{\sum_{j=1}^{I} R_{uj}^2}} \tag{7}$$

The degree of preference of u for the tag t is calculated by the end user:

$$sim_u(t) = sim_{u,t}^{pd} - sim_{u,t}^{nd} \tag{8}$$

If the user is interested in labels, and vice versa. Similarity is calculated as follows for the user u's and v's preference trusts:

$$sim_{pt}(u, v) = \frac{\sum_{t \in (T_u \cap T_v)} sim_u(t) * sim_v(t)}{\sqrt{\sum_{t \in T_u} sim_u^2(t)} * \sqrt{\sum_{t \in T_v} sim_v^2(t)}} \tag{9}$$

T_u is the tag set of the user u' rating items, and T_v the tag set of the user v' rating items.

3.2   Collaborative Filtering Recommendation Algorithm Based on User Trust (SPTUserCF)

STUserCF is based on UserCF, and considers the trust factor and user rating similarity on the algorithm, which improve the recommended results. PTUserCF algorithm is to map users' preferences on tag attributes of non-common rated items to the user preference trust matrix, as a supplement to the STUserCF. In this section, we propose SPTUserCF, which is considered together with PTUserCF comprehensively making full use of all the items that have been scored by users to calculate their trust relationship. Because the algorithm not only takes into account the impact of the common behavior of the user on the trust, but also consider the impact of non-common behavior on trust.

Next part will give the method to calculate user trust degree by combining the score trust on users' common rated items and preference trust on users' non-common rated items comprehensively as shown:

$$sim(u, v) = asim_{st}(u, v) + (1 - a)sim_{pt}(u, v) \tag{10}$$

In this formula, a is an adjusting parameter, whose values is between 0 and 1. By adjusting the value it can make the recommendation algorithm to achieve the best. Predict the user's rating for the non-rated item as shown:

$$r_{ui} = \bar{r}_u + \frac{\sum_{v \in S(u,k) \cap N(i)} sim_{uv}(r_{vi} - \bar{r}_v)}{\sum_{v \in S(u,k) \cap N(i)} |sim_{uv}|} \tag{11}$$

In this formula, S(u,k) is a collection of user u's most trusted k uses.

3.3 A Matrix Decomposition Recommendation Algorithm Based on User Trust (SPTSVD)

Collaborative filtering recommendation algorithm has some defects in recommendation accuracy and system scalability, while the matrix decomposition considering various factors can improve these deficiencies in a great extent.

3.3.1 Latent Factor Models

The idea of singular value decomposition (SVD) is: for a m × n user item rating matrix R, we need to complete the missing values of R by using the average value or a fixed value. The formula is showed as follow:

$$R = U^T S V \qquad (12)$$

In order to achieve the purpose of reducing the dimension of the matrix, only the largest k singular value of the singular value matrix can be chosen, k < min(m,n), so we get the formula as follow:

$$R' = U^T S V \qquad (13)$$

In the formula, U stands for the k × m orthogonal matrix, and S is a k × k singular value matrix. Besides, V is a k × n orthogonal matrix.

SVD is a commonly used method in recommender systems, but it is very complex and not suitable for practical applications. Latent factor model (LFM) which improves the traditional SVD decomposition, is a good matrix decomposition algorithm in the recommendation system and has a good predictive ability [17]. Latent factor model is decomposed into two low dimensional matrix product so as to predict the score as follows:

$$R' = P^T Q \qquad (14)$$

P represents user's characteristic matrix, Q represents items' characteristic matrix, and then the user's prediction score for the project can be obtained. The prediction formula is as follows:

$$\hat{r}_{ui} = p_u^T q_i \qquad (15)$$

The main purpose of the recommendation is to make the $\hat{r}_{ui}$ as close to the true score as possible. So we use the following loss function to learn the user characteristic matrix and the item characteristic matrix with the purpose to make the loss function value to achieve the minimum. Due to over fitting problem in the model, it is also required to avoid over fitting by adding the regularization parameter:

$$C(p,q) = \sum_{u} \sum_{i \in I_u} (r_{ui} - p_u^T q_i)^2 + \lambda(\|p_u\|^2 + \|q_i\|^2) \qquad (16)$$

Then the gradient descent learning method is used to optimize the function of p_u and q_i. For each parameter in the derivative the steepest descent direction is calculated, and then through iterative method the loss function reaches the optimal value. Once p_u and q_i is set, the final forecasting result can obtained.

As the potential factor model can be easily integrated into other related features, it can get a better result. The predictive score of the underlying factor model is shown in the following:

$$\hat{r}_{ui} = b_{ui} + p_u^T q_i \tag{17}$$

$$b_{ui} = \mu + b_u + b_i \tag{18}$$

In the formula, μ is the overall average rating, b_u and b_i indicate the observed deviations of user u and item i, respectively. bi and bu are also obtained by machine learning training, and the initial value is 0, which is obtained by the gradient descent learning method.

3.3.2 Neighborhood Models

Although matrix decomposition latent factor models can predict a better score, it does not consider the impact of user's historical rating information on rating prediction, while Koren improved the latent factor models by adding implicit feedback. A neighborhood model [18] was proposed to predict score as follows:

$$\hat{r}_{ui} = \mu + b_u + b_i + q_i^T \left(p_u + \frac{1}{\sqrt{|I_u|}} \sum_{j \in I_u} y_j \right) \tag{19}$$

I_u is the collection of items which is similar to the user u, and y_j is the user u's rating. The essence of the neighborhood model is to improve the traditional collaborative filtering algorithm by adding implicit feedback to the traditional collaborative filtering algorithm.

3.3.3 A Matrix Decomposition Recommendation Algorithm Based on User Trust

Matrix decomposition achieves a high accuracy, whereas its interpretation is not enough. In the recommendation system, user trust can be a good solution to this problem. This section combines the implicit rating trust and preference trust degree into the neighborhood model, and proposes a matrix decomposition (SPTSVD) recommendation algorithm, which is based on the fusion of user's trust:

$$\hat{r}_{ui} = b_{ui} + q_i^T \left(p_u + \frac{1}{\sqrt{|I_u|}} \sum_{j \in I_u} y_j + \frac{1}{\sqrt{|PT_u|}} \sum_{a \in PT_u} w1_a + \frac{1}{\sqrt{|ST_u|}} \sum_{b \in ST_u} w2_b \right) \tag{20}$$

PT_u represents the set of user u's trusted users based on preference trust, and ST_u is the set of user u's trusted users based on score trust. $w1_a$ is the characteristic matrix for user u's trusted users based on preference trust. $w2_b$ is feature matrix for user u's trusted users based on scoring trust. Then we get the formula showed below:

$$L = (r_{ui} - \hat{r}_{ui}) + \lambda_1 b_u^2 + \lambda_2 b_i^2 + \lambda_3 \|p_u\|^2 + \lambda_4 \|q_i\|^2$$
$$+ \lambda_5 \sum_{j \in I_u} \|y_j\|^2 + \lambda_6 \sum_{a \in PT_u} \|w1_a\|^2 + \lambda_7 \sum_{b \in ST_u} \|w2_b\|^2 \tag{21}$$

In order to prevent over fitting, the steepest descent learning method to achieve the minimum value of L is used.

SPTSVD considers the proposed rating trust and preference trust, and takes full account of the trust relationship between users in order to provide users with the better recommendation service.

4 Experiment Results and Analysis

This paper uses the dataset CIAODVD to validate the proposed algorithm as it is mostly cited in the research of recommendation system in social network. The rated items involve 17 types of information which have comedy, action, adventure, record, cartoon, romance, fiction, crime etc. In the experiment, 80 % ranking scores are randomly chosen as the training set, and the rest are chosen as the testing set.

The metrics to evaluate the predicted score are the mean absolute error (MAE) and root mean square error (RMSE) as defined formula below:

$$MAE = \frac{\sum_{u,i \in T} |R_{ui} - \hat{R}_{ui}|}{|T|} \tag{22}$$

$$RMSE = \sqrt{\frac{\sum_{u,i \in T} (R_{ui} - \hat{R}_{ui})^2}{|T|}} \tag{23}$$

The goal is to find the items that users are most interested in, rather than just to predict what users will rate a score for one item. In order to provide service in the social networking sites, the general idea is to use the top-N method to give users an N-item list of recommendations. Top-N recommendation accuracy is generally measured by precision which is defined by the following formula:

$$precision = \frac{\sum_{u \in U} |R(u) \cap T(u)|}{\sum_{u \in U} |R(u)|} \tag{24}$$

In the formula, $R(u)$ represents a list of recommended items in training set, and $T(u)$ represents actual items in the test set.

The comparative algorithms include collaborative filtering recommendation algorithm based on users (UserCF), collaborative filtering recommendation algorithm based on items (ItemCF) [19], collaborative filtering recommendation algorithm based on the tag trust (TagCF) [20], collaborative filtering recommendation model algorithm in the trust-based (TrustCF) [21] and the SVD ++ algorithm [18] and the TrustSVD algorithm [15].

4.1 STUserCF Contrast with PTUserCF

The comparison on MAE and RMSE is between STUserCF and PTUserCF. The experimental data is obtained by changing the neighbor number of k and the results are shown in Table 2.

Table 2. Comparison of MAE and RMSE

k	MAE(PT)	RMSE(PT)	MAE(ST)	RMSE(ST)
10	0.869	1.073	0.788	0.996
20	0.833	1.012	0.782	0.983
30	0.812	0.998	0.779	0.961
40	0.799	0.984	0.774	0.970
50	0.792	0.976	0.757	0.938
60	0.785	0.971	0.761	0.941
80	0.802	0.981	0.777	0.955

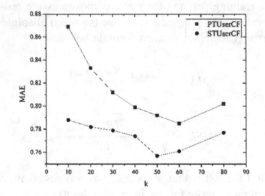

Fig. 2. MAE at different values of k

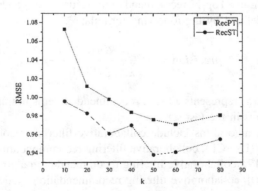

Fig. 3. RMSE at different values of k

Figures 2 and 3 are the comparison of the two proposed algorithms on MAE and RMSE:

It can be seen: PTUerCF's performance on the MAE and RMSE is lower than STUserCF. When the number of nearest neighbors is between 50 and 60, the result is the most accurate.

Figure 4 shows a comparison of the precision of several algorithms. STUserCF takes full account of the relationship between the rating and co-rating numbers, and its precision is higher than other algorithms, and PTUserCF do not consider the number of users so its precision is slightly lower than other algorithms, but it is better than algorithms not considering the weight threshold (TagCF).

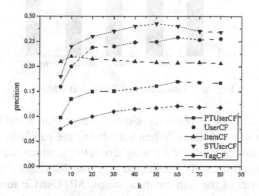

Fig. 4. Comparison with other algorithms on precision

4.2 SPTUserCF Comparison Experiment

For co-rated items STUserCF is used, and for items not rated commonly by two users PTUserCF is used. When a is set between 0.1 and 0.9, the results is shown in Fig. 5:

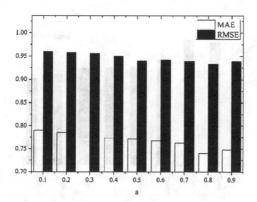

Fig. 5. Influence of parameter a on algorithm

Figure 5 shows that the two metrics decrease with the increase of a. When a is around 0.8, the algorithm works best. In order to verify the effectiveness of SPTUserCF with a as 0.8, RMSE and MAE of comparative results between algorithms in this paper and other algorithms are shown in Fig. 6:

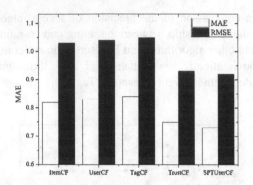

Fig. 6. Comparison of MAE and RMSE

As seen in Fig. 6, ItemCF performs slightly better than UserCF. TagCF did not consider the number of rated items. When calculating the similarity of user preference on item labels, they did not consider the weight of ratings, which reduced the prediction accuracy. TrustCF algorithm fully considered user trust for rating similarities, but it did not consider the impact of the non-common items. SPTUserCF took advantage of all users' ratings to calculate the user trust, which made the predicted ratings more accurate.

4.3   SPTSVD Algorithm Comparison Experiment

Figure 7 shows the comparative results between SPTSVD and other algorithms on MAE and RMSE. Matrix factorization algorithm have better performance in terms of prediction rankings than other algorithms. TrustSVD considered the trust level of a user on the basis of the neighborhood model. The prediction results are better than the SVD

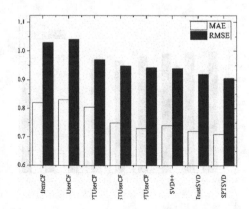

Fig. 7. Comparative results of CIAODVD Dataset

++ algorithm, which did not consider the user trust. SPTSVD considered user score trust and user preference trust. So it had the lowest RMSE. It can be seen, when two types of trust is combined with the matrix factorization recommendation algorithm, the accuracy of predicted ratings can be improved significantly.

5 Conclusion

This paper integrated users' trust relationship in online social networks into the recommendation algorithm. Current recommendation systems are not accurate enough and have difficulties to handle sparse data. We proposed two recommendation algorithms based on the trust of users' ratings and their interest, respectively. Then collaborative filtering recommendation algorithm integrating users' trust was proposed. The proposed algorithm makes full use of users' ratings to items. Then a new matrix factorization recommendation algorithm is proposed by combining the user trust with matrix factorization technology. Finally, the experiment results indicate that the proposed algorithm has better performance in both accuracy of predicting ratings and precision of recommending items. The future work is to take the time into consideration to increase the performance.

Acknowledgments. This work is supported by the National Natural Science Foundation of China under Grants No. 61272186 and the Foundation of Heilongjiang Postdoctoral under Grant No. LBH-Z12068.

References

1. Guy, I., Ronen, I., Wilcox, E.: Do you know?: Recommending people to invite into your social network. In: Proceedings of the 14th International Conference on Intelligent User Interfaces, vol. 3, pp. 77–86 (2009)
2. Park, D.H., Kim, H.K., Choi, I.Y., et al.: A literature review and classification of recommender systems research. Expert Syst. Appl. **39**(11), 10059–10072 (2012)
3. Carrer-Neto, W., Valencia-García, R.: Social knowledge-based recommender system. Application to the movies domain. Expert Syst. Appl. **39**(12), 10990–11000 (2012)
4. Su, J.H., Yeh, H.H., Yu, P.S., et al.: Music recommendation using content and context information mining. Intell. Syst. IEEE **25**(1), 16–26 (2010)
5. Choi, K., Suh, Y.: A new similarity function for selecting neighbors for each target item in collaborative filtering. Knowl. Based Syst. **37**(1), 146–153 (2013)
6. Huang, Z., Chen, H., Zeng, D.: Applying associative retrieval techniques to alleviate the sparsity problem in collaborative filtering. ACM Trans. Inf. Syst. **22**(1), 116–142 (2004)
7. Jamali, M., Ester, M.: TrustWalker: a random walk model for combining trust-based and item-based recommendation. In: ACM SIGKDD International Conference on Knowledge Discovery & Data Mining, pp. 397–406. ACM (2009)
8. Lathia, N., Hailes, S., Capra, L.: Trust-based collaborative filtering. In: Karabulut, Y., Mitchell, J., Herrmann, P., Jensen, C.D. (eds.) Trust Management II. IFIP – International Federation for Information Processing, vol. 263, pp. 119–134. Springer, New York (2008)

9. Massa, P., Avesani, P.: Trust metrics on controversial users: balancing between tyranny of the majority. Int. J. Seman. Web Inf. Syst. **3**(1), 39–64 (2007)
10. Guo, G., Zhang, J., Thalmann, D.: Merging trust in collaborative filtering to alleviate data sparsity and cold start. Knowl. Based Syst. **57**(2), 57–68 (2013)
11. Tang, J., Hu, X., Gao, H., et al.: Exploiting local and global social context for recommendation. In: Proceedings of the 23rd International Joint Conference on Artificial Intelligence, pp. 2712–2718 (2013)
12. Liu, S.-Z., Liao, Z.-F., Wu, Y.-F., et al.: A collaborative filtering algorithm combined with user rating credibility and similarity. J. Chin. Comput. Syst. **35**(5), 973–977 (2014)
13. Lu, K., Xie, L., Li, M.-C.: Research on implied-trust aware collaborative filtering recommendation algorithm. J. Chin. Comput. Syst. **37**(2), 241–245 (2016)
14. Yang, B., Lei, Y., Liu, D., et al.: Social collaborative filtering by trust. In: Proceedings of the Twenty-Third International Joint Conference on Artificial Intelligence, pp. 2747–2753. AAAI Press (2013)
15. Guo, G., Zhang, J., Yorke-Smith, N.: TrustSVD: Collaborative Filtering with both the explicit and implicit influence of user trust and item ratings. In: Proceedings of the 29 AAAI Conference on Artificial Intelligence, pp. 123–130 (2015)
16. Ji, K., Shen, H.: Making recommendations from top-N user-item subgroups. Neurocomputing **165**(C), 228–237 (2015)
17. Lu, Z., Dou, Z., Lian, J., et al.: Content-based collaborative filtering for news topic recommendation. In: Proceedings of the Twenty-Ninth AAAI Conference on Artificial Intelligence, pp. 217–223 (2015)
18. Koren, Y.: Factor in the neighbors: scalable and accurate collaborative filtering. ACM Trans. Knowl. Discovery Data **4**(1), 2010 (2010)
19. Sarwar, B., Karypis, G., Konstan, J., et al.: Item-based collaborative filtering recommendation algorithms. In: Proceedings of the 10th International Conference on World Wide Web, pp. 285–295. ACM (2001)
20. Su, J.H., Chang, W.Y., Tseng, V.S.: Personalized music recommendation by mining social media tags. Procedia Comput. Sci. **22**(10), 303–312 (2013)
21. Lai, C.H., Liu, D.R., Lin, C.S.: Novel personal and group-based trust models in collaborative filtering for document recommendation. Inf. Sci. **239**(4), 31–49 (2013)

A Preprocessing Method for Gait Recognition

Hong Shao, Yiyun Wang[✉], Yang Wang, and Weihao Hu

School of Information Science and Engineering, Shenyang University
of Technology, Shenyang 118070, Liaoning, China
296884226@qq.com

Abstract. The results of image preprocessing may directly affect gait feature extraction in the area of gait recognition. Due to the influence of light, shelter and other external factors of gait image, some problems such as loss of information, image shadows, and improper threshold of image preprocessing may occur. In order to solve these problems, an image preprocessing method of gait recognition is proposed. Firstly, background image is extracted by background modeling, secondly, the target profile is extracted by the direct difference method; thirdly, the shadow elimination based on the HSV color model is carried out on the target profile map; Finally, the complete target profile is obtained by threshold segmentation. Experimental results on CASIA_A database demonstrate that this proposed method is quite effective on both target profile extraction and proportion comparison with the real area.

Keywords: Gait recognition · Image pre-processing · Shadow elimination

1 Introduction

With the development of computer science and technology, information security issues have become increasingly prominent, the identification of the identity of the technical requirements are getting higher and higher. Gait recognition has attracted more and more researchers because of its advantages of non contact, non aggression and long distance recognition [1, 2].

Gait identification is not considering clothing, camera angle, and background under people walk for identification. It is mainly divided into three steps: gaits detection, feature extraction and classification. Thus, gait detection plays a key role in gait recognition, and will directly affect the subsequent steps, and also determine the In gait image detection, there may be dynamic background changes, dark background light, low contrast of background and objectives (even the human eyes is difficult to distinguish), the shadows are shown by the sun or by the light, all these factors are making the target result not ideal, the real outline of the objective cannot be reflected even through the repair operation accuracy of recognition.

Wang Liang has proposed LMEDS (Least Median of Squares LMEDS) method for background modeling in gait preprocessing, and indirect difference determination

---

Supported by Program for Liaoning Excellent Talents in University.

W. Che et al. (Eds.): ICYCSEE 2016, Part I, CCIS 623, pp. 77–86, 2016.
DOI: 10.1007/978-981-10-2053-7_8

threshold is used for binarization processing, morphology and the registration method based on contour edge correlation is used to further track the foreground region [3]. Although this method can get full target profile, but relevant registration method based on contour edge algorithm is much more complicated. Reference [4] uses a simple threshold segmentation and morphological processing to extract object contour map, but the distortion of contour is more serious. Reference [5] obtains the background model by subtraction method and gets the target image by the maximum entropy threshold segmentation. Although maximum entropy threshold segmentation can get a clear outline, but shadow exist between the two legs. Yuan-yuan Zhang selects the threshold by iterative threshold method, and repair the incomplete silhouette graph by the gait silhouette algorithm based on probability model [6]. Reference [7] directly constructs new gait characteristics from the original video sequence, does not do the pretreatment operation but marks interest points. However, this method is only effective for the extraction of specific features, not a common solution. The preprocessing algorithm in a certain extent to the extracted silhouette graph of morphological processing and repair, this will lead to distortion of target image. Therefore, a gait preprocessing method is put forward, which is to remove the image shadow at first, then extract the silhouette by the method of threshold segmentation and finally denoise and get the target contour map.

2   Preprocessing Methods for Gait Recognition

The proposed approaches mainly include the gait image background modeling, gait image background deduction, RGB conversion HSV color model, shadow elimination of HSV color model, image binarization and median method denoising, image normalization and other steps.

2.1   Gait Image Background Modeling

Background modeling is used to extract the background of gait image sequences, the purpose of which is to prepare for extracting the target area. Common used background modeling method includes optical flow method, frame difference method and background subtraction method. Optical flow method is used to detect moving regions in image sequences with the change of time of moving objects, it will have a good performance if there is special hardware support, but its computational complexity and ability to resist the noise is poor. Inter frame difference method uses a temporal sequence of images in the adjacent two or three frames image to do subtraction to get difference image, and then selects target motion information through threshold. This method is simple and easy to implement, but it is difficult to obtain the complete outline of the moving object, which is easy to produce 'double shadow' and 'empty' in the target, the information cannot be detected accurately [8]. The basic idea of background subtraction method is to construct a background firstly, and then with the subtraction of current frame and the background, detects the moving target according to the difference image. This method is simple and easy to implement and can be used to extract the characteristic data of the target. Thus this method is chosen for background modeling of gait image sequences.

The essence of image background subtraction method is to make a difference between the sequence of target image and the background [9]. Typical background subtraction method includes Mean filtering method, Median method, Adjacent frame difference method, Gaussian mixture background model and so on. Mean filtering method is the average of N frame image in the pixel gray value of each pixel, so that the impact of the pixel value of the mutation on the background image is larger; The adjacent frame difference method is too complex to be too high; Gaussian mixture background model is suitable for the detection of small size and fast moving objects in outdoor light and weather changes; Median method is to strike a multi-frame image of a point in the median as the background pixel value of the point, it can avoid the impact of abnormal pixel values of the background image. Considering the background of the selected database is relatively fixed, the median method is chosen to build the background. The intermediate value of the pixel value of the continuous N image is used as the pixel value of the background image. Make $\{I_j, j = 1, 2, 3,... N\}$, j represents a sequence of N frame images, (x, y) for pixel coordinates, the background image B can be expressed as a formula 1:

$$B(x, y) = median \ (I_j(x, y)) \tag{1}$$

Figure 1 for N frames image sequence, Fig. 2 is obtained by median method for background.

Fig. 1. N frames image sequence

Fig. 2. Median background images

2.2 The Background Subtraction of Gait Image

Background subtraction is to measure the image sequence and the background difference image to get the foreground. To make A(x, y), B(x, y) are the original image and the background image, then the foreground map C(x, y) can be expressed by the formula (2):

$$C(x,y) = |A(x,y) - B(x,y)| \tag{2}$$

In the formula (x, y) represents the pixel coordinate. Figure 3 for the original image, Fig. 4 as the background, Fig. 5 is the original image and the background image difference image.

Fig. 3. Original image

Fig. 4. Background image

Fig. 5. Background subtraction image

2.3   Shadow Elimination Based on HSV Color Model

After the observation of the image, it is found that there is a shadow cast between two legs in the human body. In order to remove the influence of the cast shadow, the color space model based on HSV is chosen to eliminate the shadow.

(1) Shadow characteristics

Shadow is a darker area, which is formed by the direct illumination of the target surface, which is divided into its own shadow and cast shadow. When the target position changes, the shadow is the moving shadow. It is mainly analyzed in this paper. The luminance value of the moving shadow is lower than that of the non-shadow region, and the color and saturation changes slightly. In order to remove the moving shadows from moving objects, a method is needed to describe the projection shadows of the moving regions without considering the shadow of other static objects. Assuming a point on the surface of the target object is illuminated by white light, supposing this point is a moving shadow then it is modeled as formula (3):

$$S_k(x, y) = E_k(x, y)P_k(x, y) \tag{3}$$

Among them, S is expressed in the K moment, the coordinates (x, y) point of the brightness, $P_k(x, y)$ for the point of reflection coefficient, $E_k(x, y)$ for the object surface unit area received light intensity. Assuming that the target is far from the light source, the distance between the object surface and the light source is constant, the light source emits parallel light and the distance between the light source and the observation point is a fixed value. $E_k(x, y)$ can be approximated as a formula (4):

$$E_k(x, y) = \begin{cases} C_A & \text{if covered by a shadow} \\ C_A + C_p \cos \partial & \text{if the illuminated area} \end{cases} \tag{4}$$

C_A and C_p are ambient light and light source intensity, which is the angle between the light source direction and the normal direction of the target surface. Assuming that the reflection coefficient $P_k(x, y)$ does not change with time when the background is fixed, the reflection coefficient of the background is equal to that of the detected point, that is, the formula (5) and (6):

$$P(x, y) = p_K(x, y) \tag{5}$$

$$L_k(x, y) = P(x, y)[E_k(x, y), E(x, y)] \tag{6}$$

If the front frame of k–1 is bright, and the K frame is the shadow covered then $L_k(x, y)$ value should be relatively large. The difference between foreground and background detection method leads to different shadow elimination methods. The commonly used method is to calculate the ratio of the intensity of the pixel and the background intensity, whether it is the shadow of the motion, the use of the previous formula, there are (7):

$$R_k(x, y) = \frac{E_k(x, y)}{E(x, y)} \qquad (7)$$

(2) Shadow elimination based on HSV color model

HSV (Hue, Saturation and Brightness) color model is one of the color systems used to choose the color from the color palette or the color wheel. This color system is closer than the RGB system in people's experience and perception of color, especially for very bright, very dark objects that can reflect the corresponding information well. When processing the image in HSV color space, the three parameters H, S, and V are not in judgment; in normal light condition, usually V (image brightness) is selected as the parameter because it can reflect the useful information of image no matter in color or non-color conditions; While the H and S components in certain circumstances is different; the tone S falls below a certain threshold, the image is color, which determines the saturation of the H component is relatively less important.

When a pixel is covered with shadows, the human eye can perceive its brightness to be compared to that of uncovered areas dark, the saturation value changes small; when the moving object is occluded, the brightness changes may become larger or smaller, and the saturation value is larger. Shadow detection method based on HSV color space can effectively detect the shadow, and the specific algorithm is shown in the formula (8):

$$S(x, y) = \begin{cases} 1, & \alpha \le \frac{V_I(x,y)}{V_B(x,y)} \le \beta \cap [S_I(x, y) - S_B(x, y)] \le T_S \cap |H_I(x, y) - H_S(x, y)| \le T_H \\ 0 & others \end{cases}$$

$$(8)$$

$V_I(x, y)$ values for the brightness of the current frame, $V_B(x, y)$ is the brightness of the background frame value, $S_I(x, y)$ values for the color of the current frame, $S_B(x, y)$ values for the color background frame, $H_I(x, y)$ value for the saturation of the current frame, $H_S(x, y)$ background frame of the saturation value; The parameters of $0 < \alpha < \beta \le 1$, the parameter α value should be considered shadow strength When the background is stronger, the smaller the α, β is to enhance the robustness of the noise, that is, the current frame of the brightness cannot be too close to the background brightness. T_S and T_H are hue and saturation threshold, which is selected according to a large number of experiments and empirical values determined, respectively 4 and 1; due to changes in the brightness of each frame has the fluctuation range of α it is 0.15 – 0.3, β is 1.

2.4   Image Binarization

In order to eliminate the potential impact of clothing changes brought on gait feature extraction, the foreground image of binarization processing is introduced. There are two main modes of gray level in the background pixel and the target pixel, a suitable threshold value T is selected to separate the two modes. Meet any $F(x, y) \ge T(x, y)$ as

the target, the other is the background point. The image G(x, y) is defined as the
threshold value:

$$G(x,y) = \begin{cases} 1 & if\ F(x,y) \geq T \\ 0 & if\ F(x,y) \leq T \end{cases}$$

OTSU method is chosen to determine the threshold for eliminating the fixed
threshold for binarization effect.

2.5   Median Method Noise Removal and Image Normalization

In the process of segmentation threshold around the human body and background
region will be the presence of a number of small noise, the method of median filter is
used to remove the noise and redundant information, and the body silhouette size is not
unified in the normalization process; Firstly, calculates the target minimum bounding
rectangle; secondly, calculates the midpoint coordinates of the rectangular wide
through the external rectangular coordinate of the upper left corner and the width of the
rectangle; finally calculates the size of each target image in the database, obtains the
height and the width through experience, and normalizes the image to 160 * 90 sizes
and get the images as shown in Fig. 6.

Fig. 6. Target contour map

3   Experimental Results and Performance Analysis

In order to verify the validity of the method, respectively, through the following two
methods: (1) Experiment was compared using this method in the reference [5, 10]
extraction of object contour map. (2) Experiment was compared target contour area
reference [11] extraction by morphological processing contour map area, by the
proposed method and the actual area.

3.1   Area of Profile

Area of profile refers to the area of the foreground in the two value image, which is a measure of the size of the foreground area in the image. Roughly speaking, the area is not the number of pixels in the image, the area calculation of this paper is not a simple calculation of the number of non-zero pixels; different pixels are weighted according to different pixels. Weighted data compensates for the inherent distortion caused by the continuous image using discrete image.

3.2   Comparison of Target Contour Effect

Used in this paper is provided by Institute of automation, Chinese Academy of Sciences the CASIA database a, Fig. 7 (a), (b), (c) and (d) shown respectively to get the target image, [5, 10] and the method in this paper the contour contrast effect.

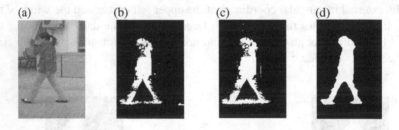

Fig. 7. (a) Original image (b) Contour map (c) Contour map (d) Contour map

Threshold segmentation by maximum entropy method is used in Reference [5]. The target outline is clear in Fig. 7, but the human body between the legs has cast shadows obviously, it may influence the extraction of gait features. The method of maximum variance ratio used in the reference [10] also has the same projection shadow. Two methods described in reference [5, 10] were used to remove the cast shadows by using label and morphological image processing methods, but the effect is not optimistic, not only because of the greatly increased amount of computations but also the distortion of shape of the contour brought up by morphological process.

With comparing the above methods, in this paper, firstly separate brightness, saturation and hue segmentation through HSV color model method, and then use Otsu threshold segmentation method, thus will avoid the morphological distortion, edge sawtooth profile, keyhole and part of the missing information.

3.3   Morphological Processing and Comparison of Area

In the Institute of Automation, Chinese Academy of Sciences, the gait database provided by the contour of the object silhouette graph selected three gait angles respectively 0 degrees, 45 degrees and 90 degrees in one frame. Method of reference [11]

target silhouette contour map information has been lost. In this paper, on the basis of the selected disc shape of the structure, and through twice image expansion and corrosion in Table 1 as shown in the image. On contrast, this method can obtain the contour area similar to the actual area, clear and complete target, not only retains the shape of the human body, but also avoid the loss of image distortion and the important information.

Table 1. Morphology and method comparison chart

Experiment object	Original image	Reference [11]	Morphological processing	Paper method
Ljg-0degree				
zjg45degree				
Zl-90degree				

The morphological processing method for missing information used by reference [11] which is described in Table 1 cannot make up the deficiency, it also led to distortion and leg important information description properly and does not get ideal effect; the expansion and corrosion morphology processing cause the contour area has increased, which is also an important reason for contour distortion. Table 2 shows the comparison of the area and the actual area of the target contour map obtained by the morphological processing of [11].

Table 2. Area Comparison Chart

Experiment object	Actual area	Reference [11]	Morphological processing	Paper method
Ljg-0degree	3.54	3.10	4.10	3.38
Zjg-45degree	3.19	2.50	3.63	3.28
Zl-90degree	9.26	6.08	7.70	9.21

Table 2 indicates that the areas computed by the method proposed by this page are much closer to the actual area of the object contour map than the other two methods. The area in reference [11] losses more that will cause incomplete information and the area processed by morphological method increases more will lead to image distortion.

4 Conclusion

Gait recognition has recently become a new research direction in the field of computer vision. This paper presents an effective method for preprocessing of gait. On the one hand, the method of background modeling and the background subtraction method are proposed to extract the target image to avoid the sudden changes for image brightness impact on the background. On the other hand, the shadow elimination method based on HSV color model is proposed to remove the influence of the cast shadow. The method can accurately determine the optimal threshold segmentation.

References

1. Boulgouris, N.V., Chi, Z.X.: Gait recognition using radon transform and linear discriminant analysis. IEEE Trans. Image Process. **16**, 731–740 (2007)
2. Tao, D., Li, X., Wu, X.: General tensor discriminant analysis and gabor features for gait recognition. IEEE Trans. Pattern Anal. Mach. Intell. **29**, 1700–1715 (2007)
3. Wang, L., Hu, W., Tan, T.: Identity recognition based on gait. J. Comput. Sci. **26**(3), 354–360 (2003)
4. Zhang, S., Bei, S.: Gait identification method based on closed area feature of lower limbs. In: Proceedings of International Conference on Electronic and Mechanical Engineering and Information Technology, pp. 2056–2059 (2011)
5. Hou, B.: Research on Gait Recognition Based on Gait. Harbin Engineering University, Heilongjiang (2007)
6. Zhang, Y.: Research on Gait Recognition Algorithm Based on Sequence Statistics. Shandong University, Shandong (2010)
7. Kusakunniran, W.: Recognizing gaits on spatio-temporal feature domain. IEEE Trans. Inf. Forensics Secur. **9**, 1416–1422 (2014)
8. Wang, K., Xian, Y.B., Zhao, Y.: Gait recognition in gait detection and sequence preprocessing automation technology and its application. **28**(8), 69–72 (2009)
9. Hong, S., Lee, H., Toh, K.-A., Kim, E.: Gait recognition using multi-bipolarized contour vector. Int. J. Control Autom. Syst. **7**, 799–808 (2009)
10. Ye, B., Wen, Y.: Gait recognition algorithm based on wavelet transform and support vector machine. J. Image Graph. **12**, 1055–1063 (2007). Chinese
11. Wang, L., Ning, H., Hu, W., Tan, T.: A new attempt to gait-based human identification. In: International Conference on Pattern Recognition (ICPR), Quebec, Canada, vol. 1, pp. 115–118, 11–15 August 2002

A Real-Time Fraud Detection Algorithm
Based on Usage Amount Forecast

Kun Niu[1], Zhipeng Gao[2(✉)], Kaile Xiao[1], Nanjie Deng[2],
and Haizhen Jiao[1]

[1] School of Software Engineering,
Beijing University of Posts and Telecommunications, Beijing 100876, China
[2] State Key Laboratory of Networking and Switching Technology, Beijing
University of Posts and Telecommunications, Beijing 100876, China
gaozhipeng@bupt.edu.cn

Abstract. Real-time Fraud Detection has always been a challenging task, especially in financial, insurance, and telecom industries. There are mainly three methods, which are rule set, outlier detection and classification to solve the problem. But those methods have some drawbacks respectively. To overcome these limitations, we propose a new algorithm UAF (Usage Amount Forecast). Firstly, Manhattan distance is used to measure the similarity between fraudulent instances and normal ones. Secondly, UAF gives real-time score which detects the fraud early and reduces as much economic loss as possible. Experiments on various real-world datasets demonstrate the high potential of UAF for processing real-time data and predicting fraudulent users.

Keywords: Real-time Fraud Detection · Usage Amount Forecast · Telecom industry

1 Introduction

With development of society and evolution of technology, economic fraud which is less in the past has gradually risen [1, 2], resulting in heavy loss of many enterprises and organizations. Therefore, from theoretical research to practical application, identification and monitoring fraud [3, 4] have caught more attention than before.

1.1 Related Work

Sieve method based on rule set used historical data related to fraud users' behavior feature to define a series of rules [4–6]. If users break pre-defined rules, system will warn administrators by reporting an emergency. For example, a mobile phone user is presumed to be fraud if his 'monthly cumulative charge exceeds 1,000 USD.

Outlier detection uses intelligent model to detect special samples in total, then system submits the outliers to administrators [7]. For example, by using density-based algorithm DBOM [8], abnormal degree of each instance in feature space is measured by LOF (local outlier factor).

© Springer Science+Business Media Singapore 2016
W. Che et al. (Eds.): ICYCSEE 2016, Part I, CCIS 623, pp. 87–95, 2016.
DOI: 10.1007/978-981-10-2053-7_9

Another solution is category discrimination [9]. It uses classification methods in data mining, such as decision tree [10], support vector machine [11], neural network [12–14], to classify and evaluate new samples. According to such IF-THEN rules, a person whose monthly outbound times are more than 6,000 may be regarded as a fraud user.

However, those methods are not good at processing stream data. Among those methods, some are not easy to set up parameters, and some others cannot teach themselves to fit variable data. In addition, those methods limit the capacity of application system for their high calculation complexity [15].

1.2 Our Contributions

To overcome these limitations, we present a new algorithm UAF (Usage Amount Forecast). We analyze variables independent of total amount to predict whether a user is fraudulent. The experiment shows that UAF is superior over existing relative methods in terms of runtime, accuracy, and robustness.

Overall, the contributions of our work on real-time fraud detection are as follows:

1. UAF does not need cumulative variables, which makes it has low computational cost.
2. UAF only computes variables which are independent of total amount, so it is able to catch fraud timely.
3. UAF can be used on real-time scenarios. The scores update synchronously while bills are inputted continuously.

The rest parts are organized as follows. In Sect. 2, we demonstrate the main idea of UAF, and give notions and definitions. The complete process of UAF with pseudo-code is showed in Sect. 3. Experiment results are presented in Sect. 4 and we conclude our work in Sect. 5.

2 Preparation of Your Paper

2.1 Notions

Assume that dataset D is the feature space to be studied. It contains n instances and m attributes. It is represented as $D = \{x_1, \ldots, x_n\}$ and the matrix form is $D = \{x_1^T, \ldots, x_n^T\} \in Z^{n*m}$. For any instance x_i of D, we have $x_i = \{x_{i1}, \ldots, x_{im}\}^T$. Here x_{ik} is the discretized result of the ith instance on k^{th} attribute. For each fraudulent sample, we also have $y_j = \{y_{j1}, \ldots, y_{jm}\}^T$. Here y_{jk} is the discretized result of the jth sample on k^{th} attribute.

2.2 Real-Time Fraud Detection

To find out whether a user is fraudulent, we have to know how to accurately divide the target user sets into two subsets, fraud and normal.

A. Usage Amount Forecast

In telecom industry, users pay bills periodically. The billing cycle is usually a month, users randomly generate call records, surf the internet and purchase value-added services. Data scientists working for operators collect and analyze these consuming data with big data techniques. The attributes used to describe users can be divided into two types, cumulative attributes and feature ones.

As shown in Fig. 1, the cumulative attributes are increasing monotonously when consuming records generate, but the feature attributes are stable throughout the whole billing cycle. The feature attributes are independent of usage amount and almost constant for a single user. That is why they are called feature attributes.

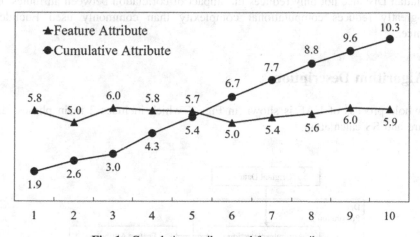

Fig. 1. Cumulative attribute and feature attribute

With large sample analysis, we find out that the two types have some specific correlations. The cumulative attributes can be predicted by feature attributes. When we detect fraud, the cumulative attributes are useless because they need long enough time to increase and warn, which is really belated. So when we use feature attributes only, as shown in Fig. 2, we may estimate the potential risk of total usage and locate fraud timely.

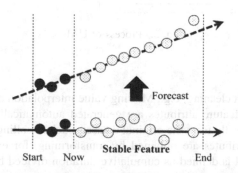

Fig. 2. The time window of usage amount forecast

B. Similarity Evaluation

Although we know feature attributes are more useful in fraud detection, a mechanism of scoring is still needed. Generally speaking, close objects have similar patterns, such as K-NN (K-Nearest Neighbors) algorithm [17]. The user who shows similar features to given fraudulent samples has a higher risk of fraud.

Therefore, we give the definition of Similarity Score (SS).

$$\text{Definition}: \forall i = 1, \cdots, n; j = 1, \cdots, n', SS(x_i) = min_j \left(\sum_{k=1}^{m} |x_{ik} - y_{jk}| \right) \quad (1)$$

$\sum_{k=1}^{m} |x_{ik} - y_{jk}|$ is the Manhattan Distance between user x_i and fraudulent user y_j. Manhattan Distance not only reduces the impact of correlation between attributes, but also greatly reduces computational complexity than commonly used Euclidean Distance.

3 Algorithm Description

The whole process of UAF is shown in Fig. 3, which includes 2 main phases, data prepare and SS calculation.

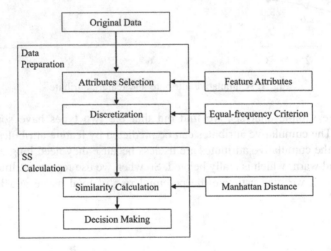

Fig. 3. Process of UAF

3.1 Data Prepare

First of all, we do data cleaning, e.g., Missing value interpolation and outlier detection. Then the predefined feature attributes are generated automatically. Some basic attributes are obtained directly from the original datasets such as calling duration and times. The other feature attributes are obtained by transforming, for example, the average duration of single call is defined as cumulative duration divided by cumulative times.

The third step is discretization. Because of frequent left avertence of normal distribution in telecom industry, equal-frequency criterion is more suitable than the common equal-width criterion. For example, assuming L is range of an attribute, K is number of segments, N is number of instances, the critical values of equal-width method are $\left\{0, \frac{L}{K}, \frac{2*L}{K}, \ldots, \frac{K*L}{K}\right\}$, the critical values of equal-frequency method are $\left\{x_1, x_{\left[\frac{N}{K}\right]}, x_{\left[\frac{2*N}{K}\right]}, \ldots, x_{\left[\frac{K*N}{K}\right]}\right\}$.

3.2 SS Calculation

Firstly, SS is calculated by (1). After that, SS has to be normalized and reversed for displaying. Scoring range is from 0 to 100. So we have (2).

$$SS(x_i) = 100 - \frac{100 \times (SS(x_i) - SS_{min})}{SS_{max} - SS_{min}} \tag{2}$$

The last step is decision process. When SS is higher than decision threshold, the user will be assumed as fraud, and the system triggers alarm to administrators, otherwise updates the user SS score. The decision threshold is an important parameter which can adjust and optimize by actual results.

4 Experiments and Results

4.1 Empirical Evaluation

A. Datasets
In this work, we use nine datasets to evaluate the performance of UAF. Description of the datasets is given in Table 1, for example, the date set A-1 means the data is from city A, which has 1,715,459 bills and 177,761 users in the first month. Additionally, a library which includes 6 international roaming fraudulent users is used as reference.

Table 1. Description of used datasets

Datasets	#Bills	#Users
A-1	1,715,459	177,761
B-1	224,697	43,461
C-1	72,191	7,246
B-2	465,266	41,720
B-3	199,459	38,708
B-4	578,252	51,631
B-5	292,244	56,165
B-6	468,491	45,382
B-7	463,587	46,854

All feature attributes are divided into boxes of total number n with equal-frequency discretization. Improper n may result in failure or over-fitting. The following results are the best performance of different n.

B. Attributes
Considering usage amount forecast mechanism, we select attributes which are dependent of total amount, such as average call duration and average times of each number. Description of attributes is showed in Table 2.

Table 2. Description of used attributes

Name	Method of Calculation
Avg_call_dur	$= \frac{Total\ call\ duration}{Total\ call\ numbers}$
Fluc_call_dur	$= \frac{Stdev(total\ call\ duration)}{avg\_call\_dur}$
Avg_call_invl	$= \frac{First\ call\ start\ time - last\ call\ finish\ time}{total\ call\ numbers}$
Cnt_call_num	$= Count(non - repeated\ call\ number)$
Avg_times_num	$= \frac{Total\ call\ times}{total\ call\ numbers}$
Cnt_ctry	$= Count(non - repeated\ roaming\ countries)$
High_fee_share	$= \frac{Total\ high\ settlement\ call\ fee}{total\ call\ fee}$
3rd_ctry_share	$= \frac{Total\ third\ country\ call\ fee}{total\ call\ fee}$

C. Decision Threshold
After finishing tests and adjustments, decision threshold may be 90 % of minimum score of all fraudulent users in the last month. If the system has a higher false rate compared with missing rate, the decision threshold should be increased, otherwise, it should be reduced.

D. Evaluation Criteria
We designed two ways to evaluate the effectiveness and robustness of UAF: post-testing, and pre-testing.

Post-Testing: Examine whether the given fraudulent users get a higher score than normal users. Conduct Experiments on different cities and different months to ensure that UAF is applicable for different situations.

Pre-Testing: With continuous input of bills, users' real-time scores can be calculated simultaneously. Pre-testing focuses on the proportion of bills occupied when a fraud user is caught. The lower the rate is, the more effective UAF is.

4.2 Results and Analysis

A. Post-Testing
To illustrate the performance of UAF, both normal and fraudulent users' scores of nine datasets are calculated, as shown in Table 3.

Table 3. Post-testing results

Datasets		A-1	B-1	C-1	B-2	B-3	B-4	B-5	B-6	B-7
Best n		20	20	20	20	20	20	20	20	20
Ordinaries	max	93	98	94	95	89	92	99	96	94
Fraudulent Samples	No.1	152	121	154	196	189	186	176	169	153
	No.2	148	**93**	153	169	152	143	151	141	127
	No.3	113	116	146	139	121	124	136	119	136
	No.4	123	129	148	141	135	133	140	131	143
	No.5	112	112	122	129	103	122	126	110	129
	No.6	116	118	149	140	129	101	137	121	139

(i) Different Cities in the Same Month

Obviously, comparing the result of A-1, B-1, and C-1, all fraudulent users get scores higher than 100 except No.2 in B-2. Since normalizing is based on normal users, the fraudulent users have specific features. That is why the fraudulent users get much higher scores.

However, there is a score lower than 100 in dataset B-2, and by sorting all scores and studying the bills, we are convinced that there is really a fraud user in B-2.

(ii) Different Months at the Same City

Analyzing 7 months' results of B city, the fraudulent users' scores are always higher than the normal users' scores, which proves that UAF works steadily through a long time.

B. Pre-Testing
In this part, the program of UAF reads in bills continuously simulating a data stream. Then, it calculates the usage rate of bills until fraud is detected, as show in Table 4.

Table 4. Pre-testing results

Datasets	A-1	B-1	C-1	B-2	B-3	B-4	B-5	B-6	B-7
Best n	20	20	20	20	20	20	20	20	20
No.1	1.74 %	2.61 %	1.94 %	2.63 %	2.58 %	2.34 %	2.43 %	2.21 %	2.69 %
No.2	63.19 %	59.36 %	52.12 %	58.26 %	26.38 %	31.64 %	28.65 %	20.87 %	26.31 %
No.3	1.82 %	1.82 %	1.98 %	1.32 %	1.74 %	1.95 %	1.65 %	1.98 %	1.52 %
No.4	13.02 %	12.50 %	13.26 %	10.89 %	11.32 %	10.98 %	11.32 %	11.65 %	10.63 %
No.5	5.54 %	8.86 %	7.65 %	8.12 %	8.09 %	8.35 %	7.86 %	8.32 %	8.09 %
No.6	0.01 %	0.01 %	0.01 %	0.01 %	0.01 %	0.01 %	0.01 %	0.01 %	0.01 %

(i) Different Cities in the Same Month

Obviously, the fraudulent users can be detected when only 0.01 % bills produced under the best condition, and for the worst, it needs 63.19 %. In this table, the usage rate will be 10.75 % on average, which means the model is robust and performs steadily for each dataset.

(ii) Different Months at the Same City

From Table 4, each fraudulent user in the 7 datasets of city B can be detected timely.

C. Parameter n
Due to different sizes of datasets, the parameter n may affect the performance remarkably. For example, datasets A-1 has 177,761 users, where n = 10 is not big enough for distinguishing each attribute. As shown in Table 5, the No.3 fraudulent user only gets 98. When n increases to 20, the scores become more reasonable.

Table 5. Contrast experiment on A-1

A-1		n = 10	n = 20
Ordinaries	max	93	98
Fraudulent Samples	No. 1	101	152
	No. 2	112	168
	No. 3	**98**	113
	No. 4	103	123
	No. 5	101	112
	No. 6	101	116

But n is not the larger the better. To illustrate this puzzle, experiment results are shown in Table 6. There are 3 different n on B-2: 10, 20 and 30. When n is10, the minimum is 110. It rises to 131when n increases to 20, while it drops to 126 when n is 30. That is a typical example of overfitting.

Table 6. Contrast experiment on B-2

B-2		t = 10	t = 20	t = 30
Ordinaries	max	100	95	93
Fraudulent Samples	No. 1	132	196	181
	No. 2	122	169	153
	No. 3	118	139	128
	No. 4	129	151	142
	No. 5	**110**	**131**	**126**
	No. 6	118	147	132

5 Conclusion

In this paper, we provide a new algorithm UAF to tackle the problem of real-time fraud detection. UAF selects feature attributes which are independent of total amount and uses equal-frequency criterion for discretization. After that, similarity calculation is proceeded by computing and comparing Manhattan distance between users. The experiments demonstrate that UAF is more precise than the state-of-the-art techniques in this domain and also has more effectiveness and scalability. In future studies, we will extend our algorithm to handle more complicated data types.

Acknowledgements. This paper is supported by the National Natural Science Foundation of China (61272515), and National Science & Technology Pillar Program (2015BAH03F02).

References

1. Hoath, P.: What's new in telecoms fraud. Comput. Fraud Secur. **2**, 13–19 (1999)
2. Hoath, P.: Telecoms fraud, the gory details. Comput. Fraud Secur. **20**, 10–14 (1998)
3. Ghosh, M.: Telecoms fraud. Comput. Fraud Secur. **2010**, 14–17 (2010)
4. Rosset, S., Murad, U., Neumann, E., Idan, Y., Pinkas, G.: Discovery of fraud rules for telecommunications - challenges and solutions. In: International Conference on Management of Data and Symposium on Principles of Database Systems, pp. 409–413. ACM, New York (1999)
5. Estévez, P.A., Held, C.M., Perez, C.A.: Subscription fraud prevention in telecommunications using fuzzy rules and neural networks. Expert Syst. Appl. **31**, 337–344 (2006)
6. Panigrahi, S., Kundu, A., Sural, S., Majumdar, A.: Use of dempster-shafer theory and bayesian inferencing for fraud detection in mobile communication networks. In: Pieprzyk, J., Ghodosi, H., Dawson, E. (eds.) ACISP 2007. LNCS, vol. 4586, pp. 446–460. Springer, Heidelberg (2007)
7. Gupta, D., et al.: An analysis of telecommunication fraud using outlier detection model based on similar coefficient sum. Int. J. Soft Comput. Eng. (IJSCE) **4**, 2231–2307 (2014)
8. Cárdenas-Montes, M.: Depth-based outlier detection algorithm. In: Polycarpou, M., de Carvalho, A.C., Pan, J.-S., Woźniak, M., Quintian, H., Corchado, E. (eds.) HAIS 2014. LNCS, vol. 8480, pp. 122–132. Springer, Heidelberg (2014)
9. Hollmén, J.: User profiling and classification for fraud detection in mobile communications networks. Helsinki Univ. Technol. **29**, 31–42 (2000)
10. Zou, K., Sun, W., Hongzhi, Y.: ID3 decision tree in fraud detection application. In: 2012 International Conference on Computer Science and Electronics Engineering, pp. 399–402. IEEE Press, Hangzhou (2012)
11. Subudhia, S., Panigrahib, S.: Quarter-sphere support vector machine for fraud detection in mobile telecommunication networks. Procedia Comput. Sci. **48**, 353–359 (2015)
12. Moreau, Y., Verrelst, H., Vandewalle, J.: Detection of mobile phone fraud using supervised neural networks: a first prototype. In: Gerstner, W., Hasler, M., Germond, A., Nicoud, J.-D. (eds.) ICANN 1997. LNCS, vol. 1327, pp. 1065–1070. Springer, Heidelberg (1997)
13. Burge, P., Shawe-Taylor, J.: An unsupervised neural network approach to profiling the behavior of mobile phone users for use in fraud detection. J. Parallel Distrib. Comput. **61**, 915–925 (2001)
14. Hilas, C.S., Mastorocostas, P.A.: An application of supervised and unsupervised learning approaches to telecommunications fraud detection. Knowl. Based Syst. **21**, 721–726 (2008)
15. Yufeng, K., Lu, C.-T., Sirwongwattana, S., Huang, Y.-P.: Survey of fraud detection techniques. In: 2004 IEEE International Conference on Networking. Sensing and Control, pp. 749–754. IEEE Press, Taiwan (2004)

A Self-determined Evaluation Method
for Science Popularization Based on IOWA
Operator and Particle Swarm Optimization

Tianlei Zang[✉], Yan Wang, Zhengyou He, and Qingquan Qian

School of Electrical Engineering, Southwest Jiaotong University,
Chengdu 610031, Sichuan, China
zangtianlei@126.com

Abstract. With the increase of science popularization, evaluation of science popularization has become an urgent demand. Considering science popularization bases as independent agents, a self-determined evaluation approach for science popularization using induced ordered weighted averaging (IOWA) operator and particle swarm optimization (PSO) is proposed in this paper. Firstly, six factors including science popularization personnel, space, fund, media, activity and influence are selected to construct an index system for science popularization evaluation. On this basis, the absolute dominance and relative dominance of evaluation indexes are used as induced components, and the prior order of the evaluation indexes is determined. Besides, the optimization model of index weighted vectors is established by IOWA operator, index weighted vectors are calculated by particle swarm optimization algorithm, and index weighted vectors and evaluation value vectors are obtain. Finally, the optimal evaluation vectors and evaluation results are given according to the Perron-Frobenius decision eigenvalve theorem .

Keywords: Science popularization · Self-determined evaluation · Induced Ordered Weighted Averaging operator · Particle swarm optimization · Perron-Frobenius decision eigenvalve theorem

1 Introduction

With the rapid development of science and technology, the industry deeply aware that to promote the marketization process of the science and technology, they need to further strengthen the science popularization work and improve public response speed for new technology products. With the increasing of the popularization, it has become an urgent demand to carry out evaluation work to measure the effect and to promote the quality of the science popularization work. The evaluation method for science popularization has gradually become one of the research hotspots in social computing.

In the world, the study on science popularization evaluation is mostly a summary of the practice method. The evaluation indexes mainly consider the universal surface, influence on the attitude and behavior of the public. Information collection is the first step, which mainly include the field interviews, questionnaires survey and Internet feedbacks [1–3]. In China, science popularization evaluation has made some achievements.

© Springer Science+Business Media Singapore 2016
W. Che et al. (Eds.): ICYCSEE 2016, Part I, CCIS 623, pp. 96–108, 2016.
DOI: 10.1007/978-981-10-2053-7_10

Li Zhaohui et al. evaluated the science popularization infrastructure development from the scale, structure, and effect 3 aspects, employed the Delphi method to determine the weights of the index [4]. Ren Rongrong et al. established the evaluation index system of regional science popularization ability in 5 aspects including investment, facilities, personnel, creation and activity organization. They also combined the entropy weight method with GEM to determine the index weight [5]. Zuo Qingfu et al. established the effect evaluation index system of the project on science popularization benefiting peasants and prospering the rural, employed analytic hierarchy process method to calculate index weight [6]. Wu Huagang established the evaluation index system of science popularization resources construction, employed the global principal component analysis method to evaluate the science popularization resource construction level in 31 provinces / autonomous regions / municipalities in China [7].

It can be seen from the research status on science popularization evaluation that the existing science popularization evaluation methods are all considered evaluation target as passive "evaluation object", which has no "discourse right" in the evaluation. While the views of the evaluation target should be adopted comprehensively in the actual evaluation, carry out self-determined evaluation with "fair competition" concept [8]. In view of this, this paper give full independence to the science popularization base and introduce the Induced Ordered Weighted Averaging (IOWA) operator into the science popularization level evaluation to construct the self-determined evaluation model. The model is solved by particle swarm optimization algorithm, and then the self-determined evaluation results of science popularization are given.

2 The Evaluation Index System of Science Popularization

The primary task of science popularization evaluation is to establish a scientific evaluation index system. This paper refer to science popularization statistic survey which made by the Ministry of Science and Technology in the People's Republic of China. The science popularization personnel, space, fund, media, activity and influence six first level indexes are selected, and be subdivided into twenty-two second level indexes, establishing a science popularization evaluation index system, as shown in Table 1.

3 A Self-determined Evaluation Model for Science Popularization Based on IOWA

3.1 Formation of Decision Matrix for Science Popularization Evaluation

The n science popularization base to be evaluated is $E_i(i = 1, 2, \cdots, n \in N)$, where the j-th evaluation index is $x_j(j = 1, 2, \cdots, m \in M)$, then the indexes set of science popularization base is $I_i = \{x_{i1}, x_{i2}, \cdots, x_{ij}, \cdots, x_{im}\}$, where x_{ij} represents the value of the j-th evaluation index for the i-th science popularization base. All x_{ij} consist of the evaluation index matrix.

Table 1. Evaluation index system for science popularization

First-level index	Second-level index
Science popularization personnel	Full-time personnel I_1
	Part-time personnel I_2
	Volunteers I_3
Science popularization space	Area of exhibition hall I_4
	The number of exhibits I_5
	Equipment I_6
Science popularization fund	Annual funds raised I_7
	Annual amount used I_8
	Special funds I_9
Science popularization media	Annual number of science popularization books published I_{10}
	Annual number of science popularization periodicals published I_{11}
	Annual total page view of science popularization website I_{12}
	Other science popularization readings I_{13}
Science popularization activity	Person-time of participants in science popularization lectures and training I_{14}
	Person-time of visitors for science popularization exhibitions I_{15}
	Person-time of participants in science popularization competitions I_{16}
	Person-time of participants in international communication of science popularization I_{17}
	Person-time of young participants in science popularization I_{18}
	Person-time of participants in science and technology activity week I_{19}
Science popularization influence	Affirmation and naming I_{20}
	Science popularization award I_{21}
	Media coverage I_{22}

$$\mathbf{X} = (x_{ij})_{n \times m} = \begin{bmatrix} x_{11} & x_{12} & \cdots & x_{1j} & \cdots & x_{1m} \\ x_{21} & x_{22} & & x_{2j} & \cdots & x_{2m} \\ \vdots & \vdots & \ddots & \vdots & \ddots & \vdots \\ x_{n1} & x_{n2} & \cdots & x_{nj} & \cdots & x_{nm} \end{bmatrix}. \tag{1}$$

Since the dimension of the evaluation index is not consistent, X should be normalized before the evaluation, and the normalized matrix is called the evaluation decision matrix $X' = \left(x'_{ij} \right)_{n \times m}$. Because all the evaluation indexes selected in this paper are the benefit indexes, so they can be standardized as follows:

$$x'_{ij} = \frac{x_{ij} - \min_i \{x_{ij}\}}{\max_i \{x_{ij}\} - \min_i \{x_{ij}\}}. \tag{2}$$

3.2   Absolute and Relative Dominance Calculation of Evaluation Index

In the self-determined evaluation, the competition field of the evaluation subject should be determined at first. For science popularization base E_i, the determination method of its competition field as follows:

(1)   For $\forall j \in M$, if $x_{ij} \geq x_{kj}(k \in N, k \neq i)$, then science popularization base E_i is better than E_k, not form a competitive relationship.
(2)   For $\forall j \in M$, if $x_{ij} \leq x_{kj}(k \in N, k \neq i)$, then science popularization base E_i is worse than E_k, not form a competitive relationship.
(3)   For $\forall j \in M$, if both $x_{ij} \geq x_{kj}$ and $x_{ij} \leq x_{kj}(k \in N, k \neq i)$ existing, then science popularization base E_i and E_k form a competitive relationship.

For science popularization base E_i, the set of the science popularization bases which form the competitive relationship with it is called competition field of E_i, as $H_i = \left\{ E_1^{(i)}, E_2^{(i)}, \cdots, E_{n_i}^{(i)} \right\}$, where n_i represents the number of the science popularization bases which form the competitive relationship with E_i.

Set the competition intensity of the j-th evaluation index x_i of popularization science base E_i relative to E_k in the competition field [8].

$$d_{ik}^{(j)} = x_{ij} - x_{kj}(i, k \in N; j \in M). \tag{3}$$

For science popularization base E_i, the absolute dominance and relative dominance of the evaluation indexes are:

$$\lambda_i^{(j)} = \frac{\sum\limits_{m=1}^{P_i} d_{im}^{(j)}}{\sum\limits_{m=1}^{n_i} \left| d_{ik}^{(j)} \right|}. \tag{4}$$

$$\lambda_i'^{(j)} = \frac{\exp(\sum\limits_{k=1}^{n_i} d_{ik}^{(j)})}{\sum\limits_{j=1}^{m} \exp(\sum\limits_{k=1}^{n_i} d_{ik}^{(j)})}. \tag{5}$$

Where, P_i is the number of the non-negative value of the competition strength of science popularization base E_i relative to all the science popularization bases in the competition field.

3.3   Determination of Position Weight Vector

For science popularization base E_i, the position weight vector of the evaluation index is $w = (w_1, w_2, \cdots w_j, \cdots, w_m)^T$, where

$$w_j = \frac{q^{\sum\limits_{l=1}^{j} \eta_l}}{\sum\limits_{j=1}^{m} q^{\sum\limits_{l=1}^{j} \eta_l}}. \tag{6}$$

In Eq. (6), $\eta_1 = 1 - \left(\alpha\lambda_i^{(l)} + \beta\lambda_i^{'(l)}\right), (l \in M), 0 < q < 1, \alpha\lambda_i^{(l)} + \beta\lambda_i^{'(l)}$, is the overall competitive advantage of the l-th index, α and β are the preference of the absolute and relative dominance of the evaluation expert respectively, $\alpha + \beta = 1, \alpha, \beta \in [0, 1]$, this paper set $\alpha = \beta = 0.5$.

Take the absolute and relative preference dominance as the induced components, the evaluation value of the science popularization base E_k with E_i as the evaluation subject is [9]

$$Y_k[(\lambda_1^{(1)}, \lambda_1^{'(1)}, x_{1m}), (\lambda_2^{(2)}, \lambda_2^{'(2)}, x_{2m}), \cdots, (\lambda_i^{(k)}, \lambda_i^{'(k)}, x_{km})]$$
$$= \sum_{j=1}^{m} w_j a_{kj} (k \in N). \tag{7}$$

Where, a_{kj} is the value of the j-th ranked evaluation index of the science popularization base E_k.

According to the ordered weight averaging operator, the following optimization model can be used to solve the position weight vector:

$$\max \quad orness(w) = \frac{1}{m-1} \sum_{j=1}^{m} [(m-j)w_j]$$
$$s.t. \begin{cases} w_j = \dfrac{q^{\sum\limits_{l=1}^{j} \eta_l}}{\sum\limits_{j=1}^{m} q^{\sum\limits_{l=1}^{j} \eta_l}}, j \in M \\ 0 < q < 1, 0 < w_1 \le 0.5. \end{cases} \tag{8}$$

3.4   Aggregation of Self-determined Evaluation Values

If science popularization base E_i is the evaluation subject, whose given evaluation values of all the science popularization bases are $y^{(i)} = \left(y_1^{(i)}, y_2^{(i)}, \cdots, y_n^{(i)}\right)^T$, the evaluation value vector of all the science popularization bases is recorded as $Y = \left(y^{(1)}, y^{(2)}, \cdots, y^{(n)}\right)$. The sum of the angle between the optimal self-determined evaluation vector $y^* = \left(y_1^*, y_2^*, \cdots, y_n^*\right)^T$ and the vector $y_1^{(i)}, y_2^{(i)}, \cdots, y_n^{(i)}$ should be the smallest, so that y^* can be obtained according to the Perron-Frobenius decision eigenvalve theorem [10].

Theorem 1. For $\forall y \in R^n$, $\max\limits_{\|y\|_2} \sum\limits_{i=1}^{n} \left(y^T, y^{(i)}\right)^2 = \sum\limits_{i=1}^{n} \left((y^*)^T, y^{(i)}\right)^2 = \lambda_{max}$, where λ_{max} is the largest eigenvalve of YY^T, $Y = \left(y^{(1)}, y^{(2)}, \cdots, y^{(n)}\right)$ is the positive eigenvectors of λ_{max} correspond to YY^T, and $\|y^*\|_2 = 1$.

4   Optimal Evaluation Value Solve Based on Particle Swarm Optimization Algorithm

4.1   Particle Swarm Optimization Algorithm

Particle swarm optimization algorithm is an intelligent iterative optimization algorithm based on group oriented search, which is easy to implement, and has good robustness and fast convergence speed [11–14]. In the previous iteration of the standard particle swarm optimization algorithm, the particle i is updated according to the velocity v and position x:

$$v_i^{k+1} = \omega v_i^k + c_1 r_1 (p_i^k - x_i^k) + c_2 r_2 (p_g^k - x_i^k). \tag{9}$$

$$x_i^{k+1} = x_i^k + v_i^{k+1}. \tag{10}$$

Where, k is the number of iterations, p_i^k is the history optimal position of the i-th individual particle in the k th iteration; p_g^k is the history optimal position of group in the k-th iteration; ω is inertial weight; c_1 and c_2 is the learning factor which controlling the maximum step size; r_1 and r_2 are the random number between [0,1]; to ensure the search efficiency of the particle, usually limited velocity $v_i = [-v_{min}, v_{max}]$.

4.2   Solving Process of Optimal Evaluation Value

For the optimization model of Eq. (8), the PSO algorithm is employed to solve the optimal position weight vector, the optimal evaluation value vector, and then obtain the evaluation result. The calculation flow is shown in Fig. 1.

5   Case Analysis

Through the issuance of science popularization survey, collected the operational data of annual Guangzhou new energy and renewable energy science popularization base, Beijing new energy automobile exhibition center, smart grid demonstration hall of Sichuan electric power company of state grid, Yancheng new energy automobile industrial park and the Xin'ao group energy innovation experience center 5 science popularization bases ($E_1 \sim E_5$), as shown in Table 2, where p-t means person-times.

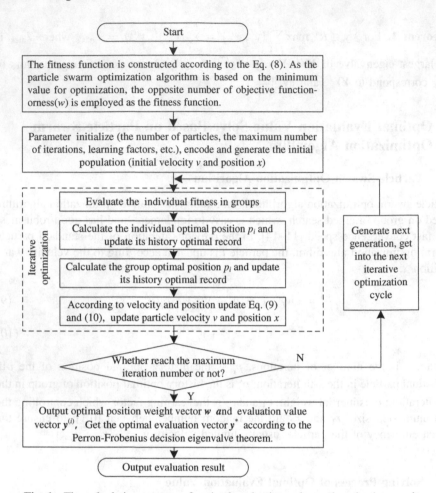

Fig. 1. The calculation process of optimal evaluation value and evaluation result

Table 2. Operation data from science popularization bases

Index	E_1	E_2	E_3	E_4	E_5
I_1	2	1	5	2	2
I_2	16	15	7	45	0
I_3	0	80	28	60	0
I_4	450 m^2	1100 m^2	900 m^2	600 m^2	660 m^2
I_5	17 pcs.	6 pcs.	20 pcs.	200 pcs.	4 pcs.
I_6	10 sets	51 sets	20 sets	15 sets	30 sets
I_7	280000	215000	800000	200000	8000000
I_8	260000	215000	500000	200000	8000000
I_9	0	0	200000	100000	2000000

(*Continued*)

Table 2. (*Continued*)

Index	E_1	E_2	E_3	E_4	E_5
I_{10}	0	0	600	0	10000
I_{11}	0	0	0	0	12000
I_{12}	8000	0	0	4000	0
I_{13}	1000	5847	500	5000	0
I_{14}	1000 p-t	0	400 p-t	200 p-t	6000 p-t
I_{15}	5000 p-t	5847 p-t	1200 p-t	5000 p-t	3099 p-t
I_{16}	200 p-t	0	0	0	300 p-t
I_{17}	0	503 p-t	0	20 p-t	576 p-t
I_{18}	2500 p-t	1671 p-t	400 p-t	300 p-t	500 p-t
I_{19}	2000 p-t	0	200 p-t	100 p-t	0
I_{20}	3	6	2	1	2
I_{21}	0	0	0	3	0
I_{22}	3 times	23 times	18 times	27 times	100 times

The evaluation process of this method is as follows:

STEP1. Form the science popularization evaluation decision matrix.

According to Table 2 and Eq. (2), normalized the index data of the science popularization evaluation, obtain the decision matrix X.

STEP2. Calculate the absolute and relative dominance of the index.

According to Eq. (3)–(5), ensure the evaluation index priority order of the 5 science popularization base, and calculate the absolute and relative dominance, as shown in Table 3.

STEP3. Solve evaluation value vector.

In this paper, the PSO algorithm is employed to solve the optimization model of position weight vector. In order to find a PSO algorithm which has more advantage for solving, the accuracy, speed and stability of the standard particle swarm optimization (SPSO) [11], modified particle swarm optimization (MPSO) [12], improved particle swarm optimization algorithm (IPSO) [13] and time-varying acceleration coefficients particle swarm optimization algorithm (TACPSO) [14] are tested respectively. The object E_5 is chosen as an example to be evaluated, set the number of particles 10, 20 and 50 respectively, statistical data of fitness values and convergence iteration numbers are given of the four PSO algorithms for the 20 times continuous operation. The results are shown in Tables 4 and 5; when the number of particles is 50 and the maximum iteration number is 100, the convergence curve of the average fitness changing with the iteration number is shown in Fig. 2.

It can be seen from the figure and tables, the accuracy of the four PSO algorithms are very high, and the convergence speeds are very fast. In general, the performance of the four algorithms ranked as: TACPSO>IPSO>MPSO>SPSO.

The optimal q value and the minimum fitness are obtained by particle swarm optimization, and the corresponding evaluation value vector $y^{(i)}$ is shown in Table 6.

Table 3. The evaluation index priority order and dominance of the science popularization base

E_1			E_2			E_3			E_4			E_5		
x_j	$\lambda_1^{(j)}$	$\lambda_1'^{(j)}$	x_j	$\lambda_2^{(j)}$	$\lambda_2'^{(j)}$	x_j	$\lambda_3^{(j)}$	$\lambda_3'^{(j)}$	x_j	$\lambda_4^{(j)}$	$\lambda_4'^{(j)}$	x_j	$\lambda_5^{(j)}$	$\lambda_5'^{(j)}$
x_{19}	0.2500	0.0050	x_{20}	0.0000	0.0014	x_1	1.0000	0.5337	x_{21}	0.2500	0.0040	x_{11}	0.2500	0.0017
x_{12}	0.4727	0.0078	x_6	0.4259	0.0066	x_9	0.1129	0.0071	x_5	1.0000	0.1565	x_8	0.0000	0.0005
x_{18}	0.0000	0.0010	x_3	1.0000	0.1431	x_4	0.4000	0.0146	x_2	0.8837	0.0347	x_{10}	0.0000	0.0004
x_{16}	0.0000	0.0009	x_{13}	1.0000	0.1236	x_3	0.8319	0.0698	x_{13}	0.1485	0.0022	x_7	0.2842	0.0015
x_{15}	0.1143	0.0036	x_4	0.0090	0.0026	x_6	0.1549	0.0098	x_3	1.0000	0.3109	x_{14}	0.0000	0.0009
x_{20}	0.0000	0.0013	x_{15}	1.0000	0.1830	x_{19}	0.2679	0.0110	x_{15}	0.0820	0.0019	x_{22}	0.6818	0.0052
x_2	0.0173	0.0029	x_{17}	0.0018	0.0027	x_7	0.1915	0.0102	x_{12}	0.0000	0.0022	x_{16}	1.0000	0.1438
x_{14}	0.0130	0.0030	x_{18}	0.0018	0.0028	x_{20}	0.0991	0.0088	x_9	0.0000	0.0023	x_{17}	1.0000	0.1498
x_1	0.0000	0.0007	x_2	0.0000	0.0006	x_{10}	1.0000	0.2521	x_{22}	0.5000	0.0067	x_9	1.0000	0.0351
x_{13}	0.0000	0.0029	x_{22}	0.0000	0.0027	x_5	0.1607	0.0097	x_1	0.0000	0.0023	x_6	1.0000	0.1480
x_5	0.0000	0.0031	x_5	0.0000	0.0029	x_{22}	0.0000	0.0076	x_4	0.0000	0.0025	x_4	1.0000	0.1572
x_7	1.0000	0.2747	x_8	0.0000	0.0018	x_2	0.0000	0.0046	x_{19}	0.7500	0.0181	x_1	0.0000	0.0006
x_8	0.1450	0.0024	x_7	1.0000	0.1414	x_8	0.0461	0.0038	x_6	0.9410	0.0581	x_{15}	0.0000	0.0003
x_{11}	0.3243	0.0054	x_{11}	0.0000	0.0022	x_{14}	0.0882	0.0081	x_{17}	0.0286	0.0022	x_{20}	1.0000	0.1204
x_{21}	0.8706	0.0236	x_{21}	1.0000	0.0557	x_{13}	0.0000	0.0010	x_{14}	0.8706	0.0190	x_{18}	0.2248	0.0011
x_{10}	0.8571	0.0439	x_{10}	0.0000	0.0015	x_{18}	0.0000	0.0039	x_{11}	0.0000	0.0013	x_{21}	1.0000	0.0807
x_{22}	0.0000	0.0012	x_{19}	0.9533	0.0920	x_{11}	0.0000	0.0031	x_8	0.0371	0.0012	x_{19}	1.0000	0.0634
x_6	1.0000	0.2119	x_{14}	0.8214	0.0306	x_{21}	0.0280	0.0045	x_{10}	0.0000	0.0011	x_5	0.0864	0.0008
x_{17}	1.0000	0.3899	x_{12}	0.0000	0.0025	x_{12}	0.2174	0.0108	x_7	0.0909	0.0027	x_{12}	0.0000	0.0009
x_3	0.5714	0.0101	x_{16}	1.0000	0.1931	x_{16}	0.1667	0.0093	x_{16}	0.0000	0.0011	x_2	0.1667	0.0013
x_4	0.0000	0.0031	x_1	0.0000	0.0029	x_{17}	0.0000	0.0076	x_{18}	1.0000	0.3642	x_3	0.0000	0.0011
x_9	0.0000	0.0017	x_9	0.2358	0.0044	x_{15}	0.1351	0.0090	x_{20}	0.3364	0.0046	x_{13}	1.0000	0.0855

Table 4. The statistical data of fitness values for convergence

| Number of particles | Optimization algorithm | Fitness values for convergence | | | |
		Optimal value	Worst value	Mean value	Standard deviation
10	SPSO	−0.9534409174	−0.9534373473	−0.9534403495	8.3018E-07
	MPSO	−0.9534409298	−0.9534409257	−0.9534409293	9.0273E-10
	IPSO	**−0.9534409298**	−0.9534409288	−0.9534409296	2.3167E-10
	TACPSO	**−0.9534409298**	**−0.9534409295**	**−0.9534409297**	**8.7945E-11**
20	SPSO	−0.9534409295	−0.9534386345	−0.9534404503	6.0019E-07
	MPSO	**−0.9534409298**	−0.9534409288	−0.9534409296	2.1754E-10
	IPSO	**−0.9534409298**	−0.9534409291	−0.9534409296	1.9105E-10
	TACPSO	**−0.9534409298**	−0.9534409296	**−0.9534409297**	**3.9564E-11**
50	SPSO	−0.9534409293	−0.9534404423	−0.9534408261	1.2766E-07
	MPSO	**−0.9534409298**	**−0.9534409297**	**−0.9534409297**	1.2102E-11
	IPSO	**−0.9534409298**	**−0.9534409297**	**−0.9534409297**	3.3404E-11
	TACPSO	**−0.9534409298**	**−0.9534409297**	**−0.9534409297**	**1.0032E-11**

Table 5. The statistical data of iteration numbers for convergence

| Number of particles | Optimization algorithm | Iteration numbers for convergence | | | |
		Maximum value	Minimum value	Mean value	Standard deviation
10	SPSO	96	47	75.00	13.9812
	MPSO	67	**11**	51.10	14.4801
	IPSO	68	37	54.45	**8.0163**
	TACPSO	**65**	21	**49.85**	12.7497
20	SPSO	85	37	67.25	13.0015
	MPSO	65	**20**	46.35	11.5953
	IPSO	**57**	23	**46.00**	**8.9736**
	TACPSO	66	22	46.70	9.7877
50	SPSO	77	10	50.15	19.4970
	MPSO	55	8	39.10	**10.9299**
	IPSO	**51**	7	36.75	12.8345
	TACPSO	53	**1**	**32.30**	12.7159

STEP4. Solve the self-determined evaluation value, give the evaluation results

According to the Table 6, $y^{(i)}$ is constructed to evaluate the matrix Y, get the characteristic root diagonal matrix D and its corresponding eigenvectors matrix V.

$$D = \begin{bmatrix} 0.4316 & & & & \\ & 0.6358 & & & \\ & & 0.7586 & & \\ & & & 1.0772 & \\ & & & & 2.5229 \end{bmatrix}. \tag{11}$$

$$\mathbf{V} = \begin{bmatrix} 0.1258 & 0.4023 & 0.7196 & -0.4176 & 0.3608 \\ -0.2434 & -0.8126 & 0.1601 & -0.1695 & 0.4755 \\ 0.8412 & -0.0708 & -0.2310 & 0.1763 & 0.4504 \\ -0.2957 & 0.3288 & -0.6242 & -0.4505 & 0.4602 \\ -0.3602 & 0.2544 & 0.1161 & 0.7502 & 0.4787 \end{bmatrix}. \tag{12}$$

The largest eigenvalve of YY^T is $\lambda_{max} = 2.5229$, by the Perron-Frobenius decision eigenvalve theorem, its corresponding eigenvalve column vector is the optimal

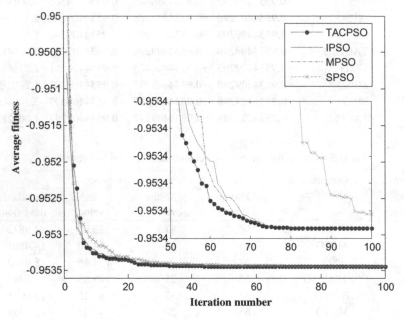

Fig. 2. The convergence curve of the average fitness (the number of particles is 50) (Color figure online)

Table 6. The evaluation value vector of science popularization bases

Science popularization base	q value	Minimum fitness	Evaluation value vector $\mathbf{y}^{(i)}$
E_1	0.1869	−0.9590986825	$(0.9716, 0.1171, 0.0582, 0.1811, 0.0796)^T$
E_2	0.1941	−0.9545581273	$(0.2250, 0.9976, 0.2336, 0.1995, 0.2473)^T$
E_3	0.2480	−0.9621242911	$(0.1291, 0.1979, 0.9124, 0.3415, 0.4745)^T$
E_4	0.1781	−0.9587023359	$(0.0852, 0.1341, 0.0567, 0.9827, 0.0045)^T$
E_5	0.1988	−0.9534409298	$(0.0125, 0.0072, 0.0283, 0.0057, 0.9996)^T$

self-determined evaluation value $y^* = (0.3608, 0.4755, 0.4504, 0.4602, 0.4787)^T$. So far, the science popularization level of the 5 science popularization bases are ranked as $E_5 > E_2 > E_4 > E_3 > E_1$.

6 Conclusion

(1) Overall considered science popularization personnel, space, fund, media, activity and influence these 6 aspects, established a science popularization evaluation index system, which ensure the systematicness of the evaluation.
(2) Combined the IOWA operator and PSO algorithm, studied the science popularization level self-determined method, fully considered the index value differences of the science popularization bases, and provided a new scientific quantitative analysis method for the science popularization evaluation.
(3) The PSO algorithm was employed to solve the position weight vector, which having the advantages of fast computation speed, high accuracy and good stability, could ensure the accuracy of the evaluation value vector.

Acknowledgment. The authors acknowledge the support by Graduate Science Popularization Ability Promotion Project of Chinese Association for Science and Technology (No. 2014KPYJD41) and Science and Technology Support Program of Sichuan Province (No. 2016RZ0079).

References

1. Wang, B.H.: Trans: 2005 Einstein Annual Reports. Science Press, Beijing (2008)
2. Koolstra, C.M.: An example of a science communication evaluation study: Discovery07, a Dutch science party. J. Sci. Commun. **6**(1), 1–8 (2008)
3. European Science Events Association: Science communication events in Europe. EUSCEA (2005)
4. Li, Z.H., Ren, F.J.: An evaluation of China's infrastructure development for popularization of science and technology. Sci. Technol. Rev. **29**(4), 64–68 (2011)
5. Ren, R.R., Zhen, N., Zhao, M.: Evaluation on regional science popularization capacity: based on entropy-GEM. Technol. Econ. **32**(2), 59–64 (2013)
6. Zuo, Q.F., Li, H., Wang, Y.N., et al.: Effect evaluation of 'the project on science popularization benefiting peasants and prospering the rural'. Agric. Eng. **4**(4), 173–177 (2014)
7. Wu, H.G.: Construction and evaluation research on the construction level of provincial science popularization resources in China. Sci. Technol. Manage. Res. (18), 66–69 (2014)
8. Yi, P.T., Guo, Y.J.: Multi-attribute decision-making method based on competitive view optimization under condition of weights nondictatorship. Control Decis. **22**(11), 1259–1263 (2007)
9. Yao, S., Guo, Y.J., Yi, P.T.: Multi-variable induced ordered weighted averaging operator and its application. J. Northeastern Univ. Nat. Sci. **30**(2), 298–301 (2009)

10. Zang, T.L., He, Z.Y., Qian, Q.Q.: Distribution network service restoration multiple attribute group decision-making using entropy weight and group eigenvalue. In: Proceedings of the 3rd International Conference on Intelligent System Design and Engineering Applications, Hong Kong, China, pp. 602–606 (2013)
11. Eberhart, R.C., Kennedy, J.: Particle swarm optimization. In: Proceeding of IEEE International Conference on Neural Network, Perth, Australia, pp. 1942–1948 (1995)
12. Mao, K.F., Bao, G.Q., Xu, C.: Particle swarm optimization algorithm based on non-symmetric learning factor adjusting. Comput. Eng. **36**(19), 182–184 (2010)
13. Cui, Z.H., Zeng, J.C., Yin, Y.F.: An improved PSO with time-varying accelerator coefficients. In: Proceeding of the Eighth International Conference on Intelligent Systems Design And Applications, Kaohsiung, China, pp. 638–643 (2008)
14. Tang, Z.Y., Zhang, D.X.: A modified particle swarm optimization with an adaptive acceleration coefficients. In: Proceeding of Asia-Pacific Conference on Information Processing, Shenzhen, China, pp. 330–332 (2009)

A Strategy for Small Files Processing in HDFS

Zhenshan Bao[✉], Shikun Xu, Wenbo Zhang, Juncheng Chen, and Jianli Liu

College of Computer Science, Beijing University of Technology, Beijing 100124, China
baozhenshan@bjut.edu.cn

Abstract. Hadoop distributed file system (HDFS) as a popular cloud storage platform, benefiting from its scalable, reliable and low-cost storage capability. However it is mainly designed for batch processing of large files, it's mean that small files cannot be efficiently handled by HDFS. In this paper, we propose a mechanism to store small files in HDFS. In our approach, file size need to be judged before uploading to HDFS. If the file size is less than the size of the block, all correlated small files will be merged into one single file and we will build index for each small file. Furthermore, prefetching and caching mechanism are used to improve the reading efficiency of small files. Meanwhile, for the new small files, we can execute appending operation on the basis of merged file. Contrasting to original HDFS, experimental results show that the storage efficiency of small files is improved.

Keywords: Hadoop · HDFS · Small file · File merging · Prefetching and caching · Appending operation

1   Introduction

Data gradually become the most valuable information in this age, and every year to grow exponentially. Many applications in the area of education, e-business, Biology consist of mass data, and every day each industry produces large amounts of data. According to the study of relevant authorities, the amount of data in 2013 reached 4.4 zettabytes, and is forecasting a tenfold growth by 2020 to 44 zettabytes [1]. In addition, type and size of the data are also varied. Today, there are many large E- commerce companies such as TaoBao, JD and Amazon. These sites store vast amounts of small files. We need to store and manage these huge amounts of small files more effectively, and a better using experience can be provided for users. Today single machine processes massive small files seem to be difficult, the expansion of the longitudinal computer performance, not only cost too much, but also meet bottleneck sooner or later. Thus distributed processing mode has become the key to solve those problems.

Hadoop as an open source distributed framework, because of reliability, high performance, high scalability, thus it attracts more and more individuals and organizations to use [2]. HDFS is one of the core components of Hadoop. It is a storage component which can be deployed in cheap hardware. Meanwhile, as a distributed file system, it's responsible for the distributed data storage and data management. When we store the large files in HDFS, they are divided into several blocks, and the blocks are stored

© Springer Science+Business Media Singapore 2016
W. Che et al. (Eds.): ICYCSEE 2016, Part I, CCIS 623, pp. 109–119, 2016.
DOI: 10.1007/978-981-10-2053-7_11

in the DataNode. Furthermore, blocks fit well with replication for providing fault tolerance and availability. To insure against corrupted blocks and disk and machine failure,
each block is replicated to small number of physically separate machines (typically
three) [3]. Each file block will generates metadata in the NameNode memory when the
machines start up. Metadata records a series of information. When the client wants to
read a file, first of all, it will read the metadata information in the NameNode, through
metadata information to find the corresponding DataNode, and then get target files. As
shown in Fig. 1 is the basic storage framework of HDFS.

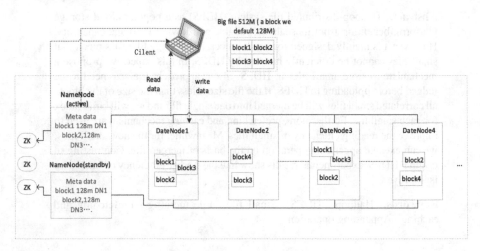

Fig. 1. Basic storage framework of HDFS

In Fig. 1, it is not hard to find that big file is divided into four blocks. Each block
size is 128 Mbytes (the new version of Hadoop default size). Meanwhile in NameNode
will generate corresponding metadata about blocks. Under the circumstances, HDFS for
small file management will be a problem. Because namenode is responsible for storing
file metadata in memory, the limit to the number of files in a filesystem is governed by
a mount of memory on the NameNode. In file management strategy of HDFS, if the file
size is less than a block size, this file will occupy a block alone, and in the NameNode
memory it will generate metadata. If there are 10 million small files, there will be 10
million blocks. Those small files will use NameNode about 3 Gigabyte memories [4].
If we have massive small files, the memory pressure of NameNode will be huge, maybe
beyond the capacity of current hardware. If you need to read a lot of small files from
clusters, frequently switching between nodes also bring a huge network load.

In this paper, we mainly to solve the massive small file problem in HDFS. To reduce
the NameNode memory usage, we use small files merging. On read and write we also
do the corresponding processing, this way to store small files in the HDFS has certain
performance improvements.

The rest of our paper is organized as follows: Sect. 2 we introduce the related
works; Sect. 3 explains the proposed approach for handing small files in HDFS.

Experimental results and analysis are presented in Sect. 4; Sect. 5 presents conclusions and provides future directions.

2 Related Works

Solutions for dealing with small files can be divided into two categories: Hadoop own solutions and current academic solutions. For Hadoop Archive (HAR) is mainly used to archive small files in HDFS, the purpose is that reducing memory usage of NameNode [5]. But for HAR low speed in terms of small files retrieving. Once the package is completed, if there are some new small files, HAR need all the small files repackaged, and we can't perform appending operations on the basis of merged file. Figure 2 is a HAR file structure. SequenceFile seems to play the role of a small file container, which uses <key, value> structure. However, this structure seems not optimistic on retrieval efficiency, because it is similar to link storage structure. If you want to search a certain key and get value, you need to traverse the entire SequenceFile file, which greatly reduces the retrieval efficiency. And converted into Sequencefile it takes a very long time [6]. CombineFileInputFormat can combine multiple files into a split, but we need to design Class to achieve, and it is not easy to implement [7]. Currently there are many different ways in academic dealing with small files. In [8] presented an approach for small files in the application in WEBGIS, although processing performance for small files has improved in some extent. However, this approach applies only to geographic information data. In [9] is mainly aimed at PPT files, this article described the operation of the prefetching mechanism, index files and the correlated files, but this is a specific application mode. In [10] mainly discuss MP3 files, a new storage method by use of the rich description of MP3 files, but it is only based on MP3 files storage model. In [11] authors have proposed a new architecture of HAR and we referred to as New Hadoop Archive(NHAR). This method can improve efficiency of accessing small files, but we can't do appending operation when we have some new small files. Meanwhile, files are not considered the correlations when archiving. In [12] proposed a method for merging files, reducing memory consumption, but did not highlight the file read and write efficiency results. In [13] describes the general framework for handling small files, and introduces the several components of the framework, but for files merging algorithm and file appending operation didn't make a specific description.

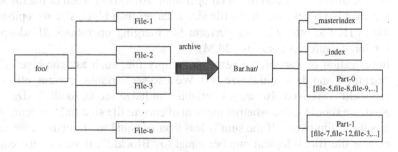

Fig. 2. Archive small files in HAR

3   Proposed Approach

In this part we will introduce our approach from 5 respects as follows:

3.1   File Merging

File Merging is the fundamental to solve the small file problems in HDFS. It can reduce the numbers of metadata files in NameNode memory. We merge massive correlated files into a single file, so the NameNode can just maintains the metadata of the single file. The main purpose of this approach is to reduce the memory load of NameNode. Neither the traditional processing way of Hadoop nor other storage strategies, the basic idea is file merging. But we must ensure that no small file is spitted across two blocks. This means that we must ensure the integrity of the file.

Figure 3 shows the basic idea of merging algorithm.

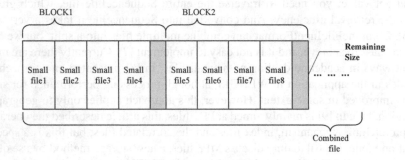

Fig. 3. Basic idea of merging algorithm

Remaining size in the Fig. 3 we can explain that we can't guarantee the merged file is just equal to the size of a block, when remaining size of current block can't suitable for the size of next small file, we should put this small file into next block.

Algorithm of small file merging

Step1:   Get all correlated files in a directory

Step2:   For file directory, we do traversal operation, Analyzing whether the file size is less than the block size. If the file size is larger than block size we upload this file to HDFS, otherwise we perform file merging operations. (Hadoop new version default block size is 128 M)

Step3:   Initialization of variables. Create two empty file, such as "combinefile" and "mapfile", and their initial size is 0. We also set variable "current_offset" and "BlockId" is 0. And also we set variable "limitsize" equal to block size

Step4:   Next, we should judge whether the sum of current file size and "current_offset" less than "limitsize", If the sum is less than or equal to "limitsize", we can set current file Block logical number equal to "BlockId", if sum is just equal to "limitsize", then we execute "BlockId++" operation, at this time "current_offset" equal to "limitsize + 1", and "limitsize" equal to "BlockId" multiplied by block size. Otherwise, "current_offset" equal to the sum of offset and file size of this file

Step5: if the sum of current file size and "current_offset" larger than "limitsize", then we should execute "BlockId ++" operation, block logical number of current file equal to "BlockId", and "current_offset" equal to "limitsize + 1", and "current_offset" equal to the sum of offset and file size of this file, and "limitsize" equal to "BlockId" multiplied by block size. Update the current file information to the mapping file (mapping file will introduce in next part)

Step6: Continue to repeat the *Step4* and the *Step5* until no small files. Then upload merged file to HDFS, and store mapping file in NameNode

3.2 Mapping File

Build the mapping file, the purpose is to find the target file information, through relevant information, we can quickly locate to block address about small files, and then extracted from the content of the block. Mapping file structure is shown in Fig. 4:

Small file name	Block logical number	Offset	Length

Fig. 4. Mapping file structure

Small File Name: This is the small file name under the directory, that's can serve as a unique identifier for a file, regardless of the Linux system, or under the Windows system, it is not allowed to small file name repetition.

Block Logical Number: A merge file contains to a lot of blocks, each small file should be in a certain block. So if we want to get a small file, we must get the block logical number of small file.

Offset: The starting address of each small file in merged file.

Length: That's mean small file size, by the offset and length we can get the small file content from merged file.

The following Fig. 5, it is a mapping file contents demo.

```
6472  15824.txt 03 71125504 157254
6473  15825.txt 03 71282758 157254
6474  15826.txt 03 71440012 20531
6475  15827.txt 03 71460543 20531
6476  15828.txt 03 71481074 157254
6477  15829.txt 03 71638328 157254
6478  1583.txt 03 71795582 85440
```

Fig. 5. Mapping file contents demo

3.3 Prefetching and Caching

In the HDFS, when a file is read, firstly the client must request the NameNode to get metadata information of merged file, and then we will get the block information about merged file. The NameNode also provides a mapping record of small file. So the question is here. When we visit small files frequently, NameNode will get a heavy load and access speed will be slow. Thus we must reduce load on NameNode and improve access speed. So we used prefetching and caching techniques, the more details we can refer to paper [14, 15].

3.3.1 The Metadata Information Caching

When we access to a small file, Firstly we need to get the metadata information about merged file from NameNode. If the client cache has file metadata information, then we can get those metadata in client cache directly.

3.3.2 Prefetching Mapping File Information

According to the merged file metadata information, the client decided to which block we should read, if the small file mapping records are perfected from the mapping file in advance, then we can read small file directly.

Metadata information caching and prefetching mapping file information these two mechanisms can accelerate I/O access speed, thus it can be improve the efficiency of file reading.

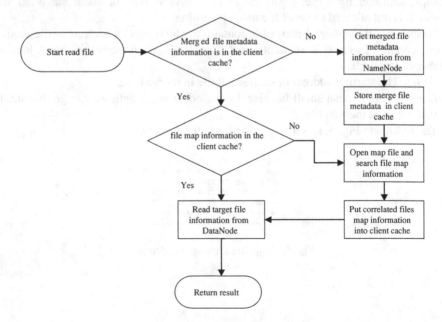

Fig. 6. File reading process

3.4 File Access Operation

When reading the files, at first we should determine whether the client cache has merged file metadata information, If the client cache has metadata information, then continue to determine whether client cache has file mapping information, if there is, we can read small file content from DataNode directly. If we can't find metadata information in client cache, we need to read it from NameNode, and store merged file metadata in client cache, then open mapping file and search file mapping information and put correlated files mapping information into client cache. At last, read target file information from DataNode. File reading process is shown in Fig. 6.

3.5 File Append Operation

In order to make full use of every piece of space, so we execute file append operations. When we merge files, we cannot guarantee the size of each block is equal to the sum of the file size. File append operation can also improve the efficiency of file merging. File appending process as shown in Fig. 7.

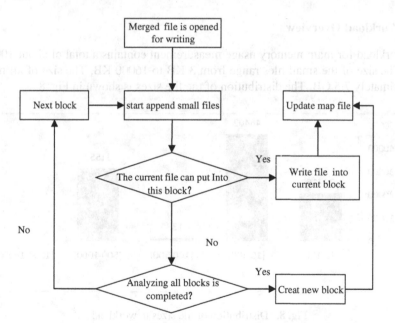

Fig. 7. File appending process

4   Evaluation and Results

4.1   Experimental Environment

Test platform includes a total of three machines. The number of copies of file blocks is set to 2 and the default block size is 64 M, The following we can see relevant configuration information.

Experimental Environment	Related parameters
JDK	1.7.0_79
OS	Ubuntu 12.04 64-bit
Hadoop	2.6.0
Memory	2G
Hard disk	100G
CPU	Intel core 2 /2.4GHz
NameNode	1
DateNode	2

4.2   Workload Overview

The workload for main memory usage measurement contains a total of about 100,000 files. The size of the small files range from 3 KB to 16000 KB, The size of all files is approximately 7.5 GB. The distribution of the file sizes is shown in Fig. 8.

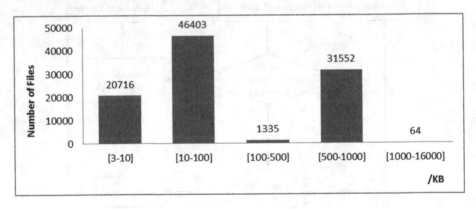

Fig. 8.   Distribution of file sizes in workload

4.3   Memory Usage Measurement

The different number of files stored in HDFS. Random read files after each startup NameNode, analyzing the NameNode Memory. The memory usage of NameNode for original HDFS and the proposed approach is shown in Fig. 9.

We can find that in Fig. 9 the memory used by proposed Approach is less than Original HDFS. However, when the number of small files is 10000, in this paper, the

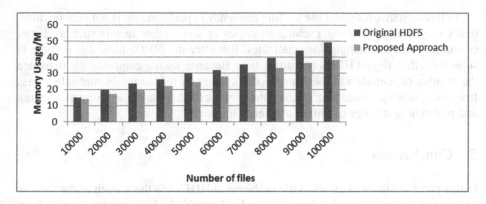

Fig. 9. Memory usages of HDFS and proposed approach

optimization of design effect is not obvious, with the increase of the number of files advantage gradually. The main reason proposed approach can save memory, because we use File merging strategy. The block metadata is stored by NameNode for single combined file and not for every single small file. So this can reduce memory usage.

4.4 Time Taken for File Access

File access time, group by the number of the small files in 10000, 20000, 30000, 40000, 50000, 60000, 70000, 80000, 90000 and 100000. The access time of every group is recorded. Each group is test 3 times. Ignoring the impact of network latency and the average of the residual values is obtains as the access time of the each group. The time taken for read operation in HDFS and proposed approach is depicted in the graph shown in Fig. 10.

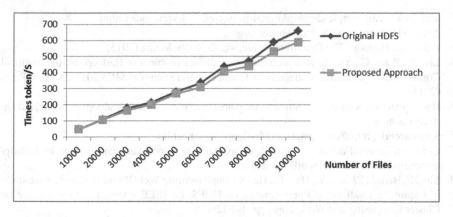

Fig. 10. Time taken for read operation

In this scenario, for small file reading efficiency in performance is not very obvious, that's maybe caused by our caching strategies or some other factors such as Time-consuming when searching for small files. But after the 50000 files, this scheme is superior to the original HDFS in reading time. Because we use combined file to reduce the number of metadata in the NameNode, it also can reduce the NameNode access frequency, in some extend that's reduce the network load. Meanwhile Using the caching and prefetching strategy can make efficiency improve.

5   Conclusions

In this paper we focused on small file problems in HDFS. As the growth of the number of small files, it will gradually bring the load to NameNode. In this paper, starting from this issue, we merged a large number of small file into single file to reduce the memory usage of NameNode. When reading the file we use caching and prefetching mechanism. We confirmed our method by experiments, this scheme can reduce the NameNode memory consumption, and improve the reading efficiency, but the effect is not too prominent. In future work, we will do more research in file reading, such as caching and prefetching strategy.

Acknowledgement. This research supported by Beijing Key Laboratory on Integration and Analysis of Large Scale Stream Data (ID: PXM2015_014204_500221) and the significant special project for Core electronic devices, high-end general chips and basic software products. (2012ZX01039-004).

References

1. http://www.emc.com/leadership/digital-universe/2014iview/index.html
2. Apache Hadoop. http://hadoop.apache.org/
3. White, T.: Hadoop: The Definitive Guide, 4E. O'Reilly Media (2015)
4. Liu, X., Peng, C., Yu, Z.: Research on the small files problem of Hadoop. In: International Conference on Education, Management, Commerce and Society (EMCS 2015). Atlantis Press (2015)
5. HadoopArchivesGuide.   http://hadoop.apache.org/docs/stable/hadoop-archives/Hadoop Archives.html
6. SequenceFile. http://wiki.apache.org/hadoop/SequenceFile
7. CombineFileInputFormat.   http://hadoop.apache.org/docs/stable/api/org/apache/hadoop/mapred/lib/CombineFileInputFormat.html
8. Liu, X., Han, J., Zhong, Y., Han, C., He, X.: Implementing WebGIS on Hadoop: a case study of improving small file I/O performance on HDFS. In: IEEE International Conference on Cluster Computing and Workshops, pp. 1–8 (2009)
9. Dong, B., Qiu, J., Zheng, Q., Zhong, X., Li, J., Li, Y.: A novel approach to improving the efficiency of storing and accessing small files on Hadoop: a case study by PowerPoint files. In: IEEE International Conference on Services Computing (SCC), pp. 65–72 (2010)

10. Zhao, X., Yang, Y., Sun, L.-L., et al.: Based on the Hadoop mass MP3 file storage structure. J. Comput. Appl. **32**(6), 1724–1726 (2012)
11. Vorapongkitipun, C., Nupairoj, N.: Improving performance of small-file accessing in Hadoop. In: 11th International Joint Conference on Computer Science and Software Engineering (JCSSE), pp. 200–205 (2014)
12. Patel, A., Mehta, M.A.: A novel approach for efficient handling of small files in HDFS. In: 2015 IEEE International Advance Computing Conference (IACC), pp. 1258–1262 (2015)
13. Changtong, L.: An improved HDFS for small file. In: 2016 18th International Conference on Advanced Communication Technology (ICACT) (2016). doi:10.1109/ICACT.2016.7423438
14. Peng, X., Feng, D., Jiang, H., Wang, F.: FARMER: a novel approach to file access correlation mining and evaluation reference model for optimizing peta-scale filesystem performance. In: Proceedings of the 17th International Symposium on High Performance Distributed Computing, pp. 185–196 (2008)
15. Dong, B., Zhong, X., Zheng, Q., Jian, L., Liu, J., Qiu, J., Li, Y.: Correlation based file prefetching approach for Hadoop. In: IEEE Second International Conference on Cloud Computing Technology and Science (CloudCom), pp. 41–48 (2010). [14]

A SVM-Based Feature Extraction
for Face Recognition

Peng Cui[⊠] and Tian-tian Yan

Harbin University of Science and Technology, Harbin 150080, China
cuipeng83@163.com

Abstract. Social computing, a cross science of computational science and social science, is affecting people's learning, work and life recently. Face recognition is going deep into every field of social life, and the feature extraction is particularly important. Linear Discriminant Analysis (LDA) is an effective feature extraction method. However, the traditional LDA cannot solve the nonlinear problem and small sample problem existing in high dimensional space. In this paper, the method of the Support Vector-based Direct Discriminant Analysis (SVDDA) is proposed. It incorporates SVM algorithm into LDA, extends SVM to nonlinear eigenspace, and optimizes eigenvalue to improve performance. Moreover, this paper combines SVDDA with the social computing theory. The experiments were tested on different face datasets. Compared with other existing methods, SVDDA has higher robustness and optimal performance.

Keywords: Discriminant analysis · Face recognition · Support vector machine · Feature extraction

1 Introduction

In recent years, on the social computing research has become an increasingly active area, related research and practice has been growing rapidly, and its appearance is closely related to the use of the Internet, Web2.0 technology and social software. It has a profound impact on all aspects of people's lives [1]. Social computing is a cross discipline combining computing technology with social science. It uses computational techniques to help us understand the social rules, mutual communication and cooperation, and uses the principle and method of swarm intelligence to solve the problem. However, there is no clear and universally accepted definition. On social computing environment, vast social softwares have been produced and used widely. This phenomenon promotes social computing to penetrate into every field of society, and achieves the digital dynamic blend in the field of science, technology, humanities in the fields of science, technology, humanities, etc.

The face recognition products have been widely used in the field of financial, legal, military, public security, frontier defense, government, aerospace, power, factories, education, health care, and many enterprises and institutions, etc. With the further development of technology and the improvement of social recognition, face

© Springer Science+Business Media Singapore 2016
W. Che et al. (Eds.): ICYCSEE 2016, Part I, CCIS 623, pp. 120–126, 2016.
DOI: 10.1007/978-981-10-2053-7_12

recognition technology will be applied in more fields. It is closely related to people's life. Moreover, feature extraction is a key step in face recognition.

In the past decades, subspace analysis methods are the most efficient method for feature extraction on face recognition. Among these methods, LDA and the improved algorithm based LDA are popular, and obtain good recognition performance [2, 3]. Yu and Yang proposed Direct Linear Discriminant Analysis (D-LDA) to find most important eigenvector projection in the eigenspace (nonzero eigenvalues subspace) between-class scatter or (zero eigenvalues subspace) within-class scatter matrix [4]. For improving performance, LDA is extended to nonlinear subspace by kernel trick, including Kernel Fisher Discriminant (KFDA) [5], Generative Discriminant Analysis (GDA) [6], Kernel Direct Discriminant Analysis (KDDA) [7] and Kernel Discriminant Analysis (KDA) [8]. Kim proposed Support Vector Machine Discriminant Analysis SVM-DA [9], which incorporated SVM into LDA to solve dimension disaster and small sample problem. However, the algorithm is same as traditional LDA, which ignored eigenvector set of within-class scatter matrix zero eigenvalues space. Inspired by SVM-DA, KDDA and DLDA, we propose a Support Vector–based Direct Discriminant Analysis (SVDDA), which extends SVM to nonlinear eigenspace, and optimizes eigenvalue to improve performance.

2 Support Vector-Based Direct Discriminant Analysis

The null space of within class scatter (S_W) is given up firstly in LDA or SVM-DA, which can include discriminant information. Secondly, the null space of between-class scatter S_B is given up. In the feature selection process, SVM-DA gives up some small eigenvalues. Though it is effective to eliminate noise, the null space of S_W is given up. We believe that the null space which can contain some important class information that should be retained. Firstly, diagonalize S_B, and solve eigenvalues; in terms of eigenvalues, give up useless null space. Then diagonalize S_W, but retain those small eigenvalues which can contain important information. If S_W is singular, let S_W equal to $S_W + S_B$ for solving small samples problem. We extend LDA to SVM nonlinear space, and propose SVDDA.

Let $\tilde{\Omega}_{ij}$ represents parameter of SVM hyperplane, which can separate, class i and class j into generative space H. The expression is

$$\tilde{\Omega}_{ij} = \sum_{\forall x_k \in S_{ij}} \alpha_{ij}^k y_{ij}^k \phi(x_k) = \Phi \alpha_{ij} \tag{1}$$

Where S_{ij} is set of support vectors, α_{ij}^k and y_{ij}^k are respectively weight and label corresponding to support vector x_k, $\Phi = [\varphi(x_1) \ldots \varphi(x_N)]$ is training set which is mapped to kernel feature space. $\alpha_{ij}^k y_{ij}^k$ denotes corresponding to support vectors in Φ. SVM-based between-class scatter matrix [9] in kernel feature space can be represented as

$$\tilde{S}_B = \sum_{i=1}^{c-1} \sum_{j=i+1}^{c} p_i p_j \frac{\tilde{\Omega}_{ij} \tilde{\Omega}_{ij}^T}{\left\|\tilde{\Omega}_{ij}\right\|^2} = \frac{1}{N^2} \sum_{i=1}^{c-1} \sum_{j=i+1}^{c} N_i N_j \frac{\tilde{\Omega}_{ij} \tilde{\Omega}_{ij}^T}{\left\|\tilde{\Omega}_{ij}\right\|^2} \tag{2}$$

In terms of kernel function, Eq. (2) can be converted into form of dot product:

$$\tilde{S}_B = \tilde{\Omega}A\tilde{\Omega}^T = \Phi\alpha A\alpha^T\Phi^T \tag{3}$$

Where $\tilde{\Omega} = [\Omega_{1,2},\ldots,\Omega_{1,c},\ldots,\Omega_{(c-1),c}]$, $A = diag[(N_1 \times N_2/N^2),\ldots,(N_{c-1} \times N_c/N^2)]$, and $\alpha = [\alpha_{1,2},\ldots,\alpha_{(c-1),c}]$. The within-class scatter matrix is

$$S_W = \sum_{i=1}^{c} \sum_{\forall x_i \in C_i} (\phi(x_j) - \overline{\phi_i})(\phi(x_j) - \overline{\phi_i})^T = \Phi(I - B - B^T + BB^T)\Phi^T \tag{4}$$

Where $\overline{\phi_i}$ denotes mean vector of samples in class i, $\overline{\phi_i} = \frac{1}{N_i}\sum_{\forall x_i \in C_i}^{c} \phi(x_j)$, B is $N \times N$ block diagonal matrix given by $(B_i)_{i=1,\ldots,c}$, B_i is $N_i \times N$ matrix of all terms which equal to $1/N_i$, $B_i = 1/N_i$, i = 1,2,…,c. In terms of Eqs. (3) and (4), the new Fisher criterion is

$$V = \arg\max_v \frac{|V^T\Phi\alpha A\alpha^T\Phi^T V|}{|V^T\Phi(I - 2B + BB^T)\Phi^T V|} = \arg\max_v \frac{|V^T\tilde{S}_B V|}{|V^T S_W V|} \tag{5}$$

As basis vector V corresponding to nonzero eigenvalues should locate in generative space of training vector, it exists efficient E satisfy

$$V = \sum_{i=1}^{N} e_i\phi(x_k) = \Phi E \tag{6}$$

Where $E = [e_1,\ldots,e_n]^T$ N-dimensional vector. Equation (6) is substituted into Eq. (7),

$$E = mag\max_E \frac{E^T\Phi^T\Phi\alpha A\alpha^T\Phi^T\Phi E}{E^T\Phi^T\Phi(I - 2B + BB^T)\Phi^T\Phi E} \tag{7}$$

As $\Phi^T\Phi$ only need to compute dot product in generative space H, it can be replaced with a kernel function. The function should be same as the kernel used in SVM hyperplane. Let K be $N \times N$ matrix, which be made up of dot product of training samples in H.

Let K replace $\Phi^T\Phi$, then Eq. (7) is represented as

$$E = m\arg\max_E \frac{E^T K\alpha A\alpha^T KE}{E^T K(I - 2B + BB^T)KE} = \frac{E^T\tilde{S}_B^K E}{E^T S_W^K E} \tag{8}$$

$\tilde{S}_B^K$ and S_W^K are new defined scatter, and E which can be solved by eigenvalues problem.

3   Feature Extraction of SVDDA

As the dimension of the feature space F is expressed as possible arbitrarily large, or may be infinitely, it is difficult to directly calculate the matrix $\tilde{S}_B^K$ and S_W^K. Eigenvalue problem of $\tilde{S}_B^K$ can be obtained indirectly by $\tilde{S}_B$.

(1) Feature analysis of $\tilde{S}_B$

First diagonalize $\tilde{S}_B$ and find $\tilde{\Lambda}_B$, $\tilde{\Lambda}_B$ is a diagonal matrix in descending order. If $\tilde{S}_B$ is singular, it is necessary to give up those eigenvalues and eigenvectors. $V_m = [v_1, \ldots, v_m] = \Phi E_m$, where $E_m = [e_1, \ldots, e_m]$ is corresponding eigenvector with eigenvalues greater than 0. $\tilde{\tilde{\Lambda}}_B\tilde{\tilde{\Lambda}}_B$ is main sub-array of $\tilde{\Lambda}_B$, which is $m \times m$ diagonal matrix with eigenvalues greater than 0.

(2) Feature analysis of S_W^K

Let $Z = V_m\tilde{\tilde{\Lambda}}_B^{-1/2}$, then $Z^T\tilde{S}_B Z = I$. We project S_W to S_W^K of generative space by Z

$$Z^T S_W Z = (E_m\Lambda_b^{-1/2})^T(\Phi^T S_W\Phi)(E_m\Lambda_b^{-1/2}) \tag{9}$$

Where $\Phi_b^T S_W\Phi_b$ equals to S_W^K. By diagonalization of $Z^T S_W Z$, we obtain Λ_W

$$GZ^T S_W ZG = \Lambda_W \tag{10}$$

Where G is eigenvector of $Z^T S_W Z$. The eigenvectors with maximum eigenvalues are given up, a M selected eigenvectors are represented as $G_W = [g_1, \ldots, g_M]$. Let $P = ZG_M$, then $P^T S_W P = \tilde{\Lambda}_W$, where $\tilde{\Lambda}_W = diag[\lambda_1', \ldots, \lambda_M']$ is $M \times M$ main sub-matrix of Λ_W,

$$(E_m\tilde{\tilde{\Lambda}}_B^{-1/2}G_M)^T(\Phi^T S_W\Phi)(E_m\tilde{\tilde{\Lambda}}_B^{-1/2}G_M) = \tilde{\Lambda}_W \tag{11}$$

we obtain

$$\tilde{\Lambda}_W^{-1/2}P^T S_W P\tilde{\Lambda}_W^{-1/2} = I_W = (E_m\tilde{\tilde{\Lambda}}_B^{-1/2}G_M\tilde{\Lambda}_W^{-1/2})^T(\Phi^T S_W\Phi)(E_m\tilde{\tilde{\Lambda}}_B^{-1/2}G_M\tilde{\Lambda}_W^{-1/2})$$
$$= E_{opt}^T S_W^K E_{opt} \tag{12}$$

Where E_{opt} is optimal eigenvector in generative space, and $E_{opt} = E_m\tilde{\tilde{\Lambda}}_B^{-1/2}G_M\tilde{\Lambda}_W^{-1/2}$. To make the method robust, modified criterion is used. When $Z^T S_W Z$ is singular, then

$$V = mag \max_V \frac{|(VS_B V)|}{|(V^T\tilde{S}_B V) + (VS_W V)|} \tag{13}$$

The modified criterion is equivalent to standard Fisher criterion. Hence $Z^T S_W Z$ in Eq. (10) is replaced with $Z^T(\tilde{S}_B + S_W)Z$.

(3) Dimension reduction and feature extraction.

Any input pattern z, eigenvector projection can be represented as

$$y = E_{opt}^T \Phi^T \phi(z) = (E_m \tilde{\tilde{\Lambda}}^{-1/2} G_M \tilde{\Lambda}_W^{-1/2})(\Phi^T \phi(z)) = E_{opt}^T \cdot \gamma(z) \qquad (14)$$

where $\gamma(z) = [k(x_1, z), \ldots, k(x_N, z)]^T$ is $N \times 1$ kernel vector.

4 Experiment and Result Analysis

4.1 Dataset and Experimental Method

To assess the proposed method, experiments are conducted on FERET, AR, PIE datasets. FERET dataset consists of 1400 images from 200 subjects, and image resolution is 80×80. AR dataset consists of 1200 images from 120 subjects, and image resolution is 40×50. PIE dataset consists of 1360 images from 68 subjects, and image resolution is 32×32. Figure 1 shows sample images from three datasets. The images in first row are randomly selected 10 images from FERET dataset. The images in second row are randomly selected 10 images from AR dataset. The images in third row are are randomly selected 10 images from PIE dataset.

Fig. 1. Some sample images from different datasets

To evaluate the proposed method, we compare SVDDA with LDA, DLDA, KDA, GDA, KDDA and SVM-DA. 4, 5, 5 images per person from FERET, AR and PIE datasets are randomly selected for training and the rest for testing. After 7 runs, results are averaged.

4.2 Experimental Results

Figure 2 shows the recognition rate from 7 different methods on FERET dataset. It can be seen that recognition rate of SVDDA is obviously better than other methods. The recognition rates of LDA, DLDA and SVM-DA are close. However, the PCA reduction dimension is all used by LDA and DLDA, so their dimensions are limited. KDDA and GDA achieve low recognition rate. Recognition rate of KDA is better than KDDA and GDA.

As seen in the three experiments, SVDDA receives the highest average recognition rate; DLDA is relatively close to the SVM-DA, they are better than other nonlinear

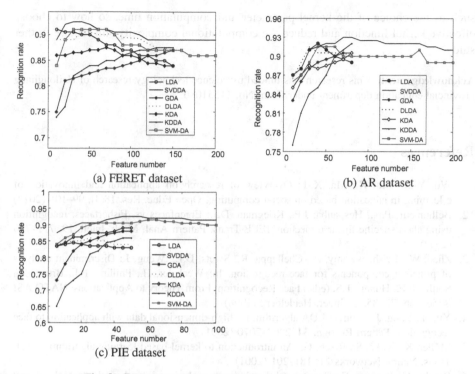

Fig. 2. Recognition rate of different methods on three datasets

method. Though KDDA receives lowest average recognition rate, the recognition rate is better in high dimensions. This shows that the non-linear methods don't obtain superior performance in all case. As the dimensionality of Fisher criterion-based subspace increases, there are more and more useless information, so discriminant information of low dimensions is particularly important. In comparison with other methods, computation time of SVDDA is slightly higher than others, but considering its high recognition rate, and its robustness that can be applied to all feature dimensions, thus its performance is optimal.

5 Conclusion

The emergence of social computing theory provides modern methods and means for effectively dealing with the complex and dynamic changing emerging social and the engineering problems. It promotes the formation and development of computational sociology. Based on the theory of social computing, we propose SVDDA. It is used to extracting nonlinear feature. This method uses SVM projection to replace the traditional mean shift, and modifies discriminant criterion to solve small sample problem existing in high dimensional space, and changes computation order of eigenvalues of scatter to retain important information by giving up null space of within-class scatter and retaining null space of within-class scatter. However there are still some problems, such as the choice of the kernel parameter and computation time, so how to choose effective kernel function and reduce the computational complexity still require further study.

Acknowledgement. This research was funded by science technology research of Heilongjiang provincial education department under Grant No. 11551086.

References

1. Wu, Y.-h., Liu, X., Ma, X.-l.: Overview of research on application and innovation of e-learning in education based on social computing. Open Educ. Res. **18**(4), 99–105 (2012)
2. Belhumeur, P.N., Hespanha, J.P., Kriegman, D.J.: Eigenfaces vs. Fisherfaces: recognition using class specific linear projection. IEEE Trans. Pattern Anal. Mach. Intell. **19**, 711–720 (1997)
3. Zhao, W., Krishnaswamy, A., Chellappa, R., Swets, D.L., Weng, J.: Discriminant analysis of principal components for face recognition. In: Wechsler, H., Phillips, P.J., Bruce, V., Soulié, F.F., Huang, T.S. (eds.) Face Recognition: From Theory to Applications. NATO ASI Series, pp. 73–85. Springer, Heidelberg (1998)
4. Yu, H., Yang, J.: A direct LDA algorithm for high-dimensional data with application to face recognition. Pattern Recogn. **34**, 2067–2070 (2001)
5. Müller, K., Mika, S., Rtsch, G.: An introduction to kernel-based learning algorithms. IEEE Trans. Neural Networks **21**, 181–201 (2001)
6. Baudat, G., Anouar, F.: Generalized discriminant analysis using a kernel approach. Neural Comput. **12**, 2385–2404 (2000)
7. Lu, J., Plataniotis, K.N., Venetsanopoulos, A.N.: Face recognition using kernel direct discriminant analysis algorithms. IEEE Trans. Neural Netw. **14**, 117–126 (2003)
8. Devi, H.S., Laishram, R., Thounaojam, D.M.: Face recognition using R-KDA with non-linear SVM for multi-view database. Procedia Comput. Sci. **54**, 532–541 (2015)
9. Kim, S.K., Park, Y.J., Toh, K.A.: SVM-based feature extraction for face recognition. Pattern Recogn. **43**, 2871–2881 (2010)
10. Li, F.-g., Liang, Y., Gao, X.-z.: Research on text categorization based on LDA-wSVM model. Appl. Res. Comput. **32**(1), 21–25 (2015)
11. Yang, L.-y., Qin, A.: Face recognition algorithms based on uncorrelated multilinear PCA. Radio Commun. Technol. **42**(1), 73–75 (2016)
12. Dong, Y., Xu, Y.-b., Zhuo, L., et al.: Research on spam identification based on social computing and machine learning. J. Shandong Univ. (Nat. Sci.) **48**(7), 72–78 (2013)
13. Meng, X.-f., Li, Y., Zhu, J.J.H.: Social computing in the era of big data: opportunities and challenges. J. Comput. Res. Develop. **50**(12), 2483–2491 (2013)

A Transductive Support Vector Machine Algorithm Based on Ant Colony Optimization

Xu Yu[1(✉)], Chun-nian Ren[1], Yan-ping Zhou[1], and Yong Wang[2]

[1] School of Information Science and Technology, Qingdao University
of Science and Technology, Qingdao 266061, China
yuxu0532@163.com
[2] College of Computer Science and Technology, Harbin Engineering University,
Harbin 150001, China

Abstract. Transductive support vector machine optimization problem is a NP problem, in the case of larger number of labeled samples, it is often difficult to obtain a global optimal solution, thereby the good generalization ability of transductive learning has been affected. Previous methods can not give consideration to both running efficiency and classification precision. In this paper, a transductive support vector machine algorithm based on ant colony optimization is proposed to overcome the drawbacks of the previous methods. The proposed algorithm approaches the approximate optimal solution of Transductive support vector machine optimization problem by ant colony optimization algorithm, and the advantage of transductive learning can be fully demonstrated. Experiments on several UCI standard datasets and the newsgroups 20 dataset showed that, with respect to running time and classification precision, the proposed algorithm has obvious advantage over the previous algorithms.

Keywords: Transductive inference · Support vector machine · NP problem · Ant colony optimization

1 Introduction

Statistic learning theory [1], proposed by Vapnik, is a theory on small sample problem. Support Vector Machines (SVMs) [2] is a new generation of learning algorithm developed on the basis of statistical learning theory. Since appeared, it has got a wide range of applications in various fields, such as face recognition [3–5], intrusion detection [6] and speech recognition [7]. Similar to the neural network classifier [8] and the decision tree classifier [9, 10], the traditional SVMs use inductive learning strategies, which firstly determine a decision function, and then classify the testing samples via this function. However, the traditional SVMs obeyed the following basic principle in machine learning fields. It is better to avoid solving a more general problem as an intermediate step when solved a given problem.

Unlike the idea of inductive SVMs learning, Vapnik et al. proposed the idea of transduction inference SVMs to improve the performance of SVMs [11]. This idea does not compute the decision function which can classify any unknown samples, but

W. Che et al. (Eds.): ICYCSEE 2016, Part I, CCIS 623, pp. 127–135, 2016.
DOI: 10.1007/978-981-10-2053-7_13

merely calculate the labels of testing samples. Transduction inference SVMs will perform better than inductive SVMS if we just want to label the testing samples.

Given iid labeled training data $(x_1, y_1), \cdots, (x_l, y_l), y \in \{-1, 1\}$, and unlabeled data $x_1^*, \cdots, x_k^*$ from the same distribution. Under the non-linear separable conditions, the transduction inference SVMs models can be described as the follows optimizing problem, denoted as OP 1.

$$\min_{y_1^*, \cdots, y_k^*, w, b, \xi_1, \cdots, \xi_n, \xi_1^*, \cdots, \xi_k^*} \frac{1}{2}\|w\| + C\sum_{i=1}^{n} \xi_i + C^*\sum_{j=1}^{k} \xi_j^*$$

$$y_i((w \cdot x_i) + b) \geq 1 - \xi_i \quad i = 1, \ldots, n$$
$$y_j((w \cdot x_j^*) + b) \geq 1 - \xi_j^* \quad j = 1, \ldots, k \tag{1}$$
$$\xi_i \geq 0, \quad i = 1, \ldots, n$$
$$\xi_j^* \geq 0, \quad j = 1, \ldots, k$$

Generally, the exact solution of this problem need to search all 2^k classification results in the sample set, which is a NP problem. As it is often difficult to obtain the appropriate approximate solution of OP 1, the advantages of transduction inference SVMs models are not obvious. In recent years, some domestic and foreign scholars are using heuristic algorithms to approximately compute the optimal solution of OP 1, and some classic algorithms are reviewed as follows:

(1) Joachims [12] proposes a feasible transduction inference algorithm, denoted as J_TSVM, in the field of text classification. The training process of this algorithm can be divided into the following steps:

Step 1:   Specify the parameters C and C^*, and use inductive inference SVMs on the training data to obtain an initial classifier. Set the number N of positive label samples according to a certain rule.

Step 2:   Compute the decision function values of all the unlabeled examples with the initial classifier. Label N examples with the largest decision function values as positive, and the others as negative. Set a temporary effect factor C_{tmp}^*.

Step 3:   Retrain the support vector machine over all the examples. Switch labels of one pair of different-labeled samples from the unlabeled sample set using a certain rule, and make the value of the objective function of OP 1 decrease as much as possible. This step is repeated until no pair of samples satisfying the switching condition is found.

Step 4:   Uniformly increase the value of the temporary factor C_{tmp}^* and return to step 3. If $C_{tmp}^* > C^*$, the algorithm is finished, and the result is outputted.

Unfortunately, J_TSVM algorithm also has some shortcomings, J_TSVM algorithm needs to specify in advance the number N of positive samples of the unlabeled sample sets. An easy method about computing N used in the J_TSVM algorithm is to assume the ratio of sample numbers between the positive class and the negative class in the labeled samples set equal to the ratio in the unlabeled samples set. Thus the number of N can be computed. However, when labeled samples are few, the result can be quite imprecise. This will cause a low and unstable classification performance of the J_TSVM algorithm.

(2) Chen et al. proposed a progressive transductive support vector machine [13], denoted as PTSVM. Unlike Joachims' J_TSVM algorithm in which the number of positive samples was estimated in advance, PTSVM does not estimate the number of positive samples, and the unlabeled examples will not be labeled altogether at the same time.

PTSVM is an iterated algorithm, and its main idea is by the following. In each iteration, one or two unlabeled examples are chosen and labeled. The selected samples can be labeled with the strongest confidence. The next iteration will be influenced by the newly labeled samples and the optimal hyper-plane will be changed a small shift. Therefore some earlier labeling samples may be improper and all the improper samples should be restored as unlabeled. The progressive labeling and dynamical adjusting is carefully arranged, and the separating hyperplane will approach the optimal solution of the optimizing problem.

As the PTSVM algorithm does not need to specify the number of positive samples, it can overcome the drawback of the J_TSVM algorithm. However, the PTSVM algorithm is with a high time complexity. Thus it is not an appropriate algorithm if the unlabeled samples are too many.

(3) Zhao et al. proposed the Semi-supervised Least Square Support Vector Machine algorithm [14], denoted by the SLS-SVM algorithm. Its basic idea is to train an initial classifier with SVM algorithm on the labeled training set. Then classify the unlabeled samples with the initial classifier. Finally, train a SVM classifier on all the samples and label the unlabeled samples with the obtained classifier. This method adjusts the labels of samples dynamically.

(4) Wang et al. proposed the K Means Clustering based Transductive Support Vector Machine algorithm [15], denoted by KMCTSVM algorithm. Firstly, the KMCTSVM algorithm clustered the unlabeled samples with the k-means algorithm due to its good clustering effect, and assigned samples in each cluster with the same label. Then train a classifier with all the samples by solving OP 1. As the KMCTSVM algorithm used clustering algorithms to decrease the scale of the test sample set, its running efficiency is high. However, its classification performance is unsatisfactory if k is assigned a bad number.

In this paper, a transductive support vector machine algorithm based on ant colony optimization is proposed to overcome the drawbacks of the above methods. The proposed algorithm approaches the approximate optimal solution of the Optimization Problem 1 by ant colony optimization algorithm, and the advantage of transductive learning can be fully demonstrated. Experiments on several UCI standard datasets and the newsgroups 20 dataset showed that, with respect to running time and classification precision, the proposed algorithm has obvious advantage over the previous algorithms.

This paper is organized as follows. In Sect. 2, the ant colony optimization algorithm is reviewed. In Sect. 3, the ant colony optimization algorithm is used to solve Optimization Problem 1, the trail intensity update equation is given, and meanwhile the transductive support vector machine algorithm based on ant colony optimization is proposed. In Sect. 4, several UCI data sets and the newsgroups 20 data set are used to conduct the experiments and the experimental results and a detailed analysis are also given in this part. Section 5 concludes the whole paper.

2 Review on the Ant Colony Optimization Algorithm

In 1992, Dorigo proposed the Ant Colony Optimization algorithm for the first time in his Ph.D. thesis [16]. Then ACO is studied by more and more scientists. An ACO is essentially a system based on agents which simulate the natural behavior of ants, including mechanisms of cooperation and adaptation.

In Dorigo's ACO algorithm, the constructed artificial ants are similar to real ants in the following aspects.

(1) Both artificial ants and real ants will give priority to paths with a larger amount of pheromone.
(2) The pheromone of shorter paths will increase faster.
(3) Both artificial ants and real ants use the pheromone deposited on each path to communicate with each other.

ACO algorithms was originally proposed in order to solve combinatorial optimization problems, and it has been shown to be both robust and versatile. Currently, too many combinatorial optimization problems in different research fields have been solved successfully by ACO.

ACO algorithms was inspired by the behavior of real ants. Ethologists have studied how blind animals, such as ants, could establish shortest paths from their nest to food sources. The medium that is used to communicate information among individual ants regarding paths is pheromone. A moving ant lays some pheromone on the ground, thus marking the path. The pheromone, while gradually dissipating over time, is reinforced as other ants use the same trail. Therefore, efficient trails increase their pheromone level over time while poor ones reduce to nil [17].

The basic idea of ACO algorithms are by the following:

(1) Each path in an ACO algorithm represents a candidate solution of the combinatorial optimization problems.
(2) When an ant follows a path, the amount of pheromone deposited on that path is proportional to the quality of the corresponding candidate solution for the target problem.
(3) When an ant has to choose between two or more paths, the path(s) with a larger amount of pheromone have a greater probability of being chosen by the ant [18].

As a result, all the artificial ants will choose a shorter path which represents the optimum or a near-optimum solution of the combinatorial optimization problems.

3 Algorithm Design

The ACO algorithm is reviewed in Sect. 2. In order to use ACO algorithm to solve OP 1, we need to convert the transduction inference SVMs problem into an ACO problem. Moreover, we should present the pheromone update method with respect to the ACO algorithm. Thus we can use the ACO algorithm to solve OP 1.

Let the feasible solution space of OP 1 is $X = \{y_1, y_2 \ldots, y_k\}$, where $y_j \in \{0, 1\}$, $j = 1, \ldots, k$. The feasible solution space is shown in Fig. 1,

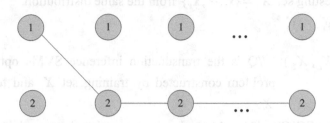

Fig. 1. Illustration of the solution space for transduction inference SVMs problem

where in each column of Fig. 1, there are two types of nodes labeled 1 and 2 respectively. Each y_j can only take one node, either the above one (denoted as the first node) or the below one (denoted as the second node).

If y_j takes the first node, its real value is 0. Otherwise its real value is 1. Thus if y_j takes the m_j-th node, where $j = 1, \ldots, k$, $m_i = 1, 2$, the corresponding solution of is by the following.

$$\{y_1, y_2, \ldots, y_k\} = \{m_1 - 1, m_2 - 1, \ldots, m_k - 1\} \tag{2}$$

The selection of the k variable of $y_1, y_2, \ldots, y_k$ is a k-grade decision problem. Firstly, m artificial ants are deployed in the first grade. The probability p_{ij} for selecting the i-th node can be computed by the following formula.

$$p_{ij} = \frac{\tau_{ij}}{\tau_{1j} + \tau_{2j}} \tag{3}$$

where τ_{ij} can be explained as the attractive intensity with respective of the i-th node in the j-th grade. The trail intensity update equation is as follows.

$$\tau_{ij}^{new} = \rho \tau_{ij}^{old} + \frac{Q}{f} \tag{4}$$

where ρ is a parameter that controls the pheromone persistence and is usually assigned in the range of 0.5–0.9. Q is a positive constant, and $f = \frac{1}{2} \|\mathbf{w}\| + C \sum_{i=1}^{n} \xi_i + C^* \sum_{j=1}^{k} \xi_j^*$ denotes the objective function value.

The trail intensity update equation is proposed above and therefore we can use ACO algorithm to solve OP 1. Below, we will propose A Transductive Support Vector Machine algorithm based on Ant Colony Optimization, denoted as the TSVMACO algorithm. Algorithm 1 shows the detailed process.

Algorithm 1 The TSVMACO algorithm

Input: the training set $X_1 = \{(\mathbf{x}_1, y_1), \cdots, (\mathbf{x}_l, y_l)\}$, where $y_i \in \{-1, 1\}$. The Unlabeled testing set $X_2 = \{\mathbf{x}_1^*, \cdots, \mathbf{x}_k^*\}$ from the same distribution.

Output: $y_1^*, \cdots, y_k^*$

method:

$Q = TSVM(X_1, X_2)$;   //Q is the transduction inference SVMs optimization problem constructed by training set X_1 and testing set X_2

$\{y_1^*, \cdots, y_k^*\} = ACO(Q)$;   //Use ACO algorithm to solve the transduction inference SVMs optimization problem and obtain the labels of samples

$ACO(Q)$

(1) Parameter initialization

   $nc = 0$;   // nc denotes the iterative steps or the search times

   $\tau_{ij} = 1$;   // *pheromone* initialization

   $m = 32$;   // the number of artificial ants is specified as 32

(2) Deploy the m artificial ants to the first grade

(3) Select one node for each ant according to the transition probability p_{ij}, and each ant goes through the k nodes

(4) Compute the objective function value of $f = \frac{1}{2}\|\mathbf{w}\| + C\sum_{i=1}^{n}\xi_i + C^*\sum_{j=1}^{k}\xi_j^*$, and record the best solution so far.

(5) Update the trail intensity: $\tau_{ij}^{new} = \rho\tau_{ij}^{old} + \dfrac{Q}{f}$

(6) $\Delta\tau_{ij} = 0$; $nc = nc + 1$;

(6) if $nc < l$, goto step (2);   // l is the predefined iterative steps

(7) Output the best solution

4   Experiments

In this paper, we perform experiments on the UCI standard data sets and the newsgroups 20 data set to test the performance of different algorithms. All the algorithms are implemented with Matlab 2015 and libsvm-mat-2.83. All experiments are run on 2.00 GHz, Intel (R) Core (TM) 2 CPU with 2 GB main memory under window 7.

4.1   An Experiment on the UCI Data Sets

Firstly, we test the performances of TSVMACO, J_TSVM, PTSVM and SLS-SVM on 3 different UCI data sets, namely *Australian*, *balance* and *iris*. In this experiment, we use the following Gaussian function as the kernel function,

$$K(\mathbf{x}, \mathbf{y}) = \exp(\frac{-\|\mathbf{x} - \mathbf{y}\|^2}{2\sigma^2}) \qquad (5)$$

where σ is a width parameter, x and y denote two n-dimensional vectors in the original sample space. In this experiment, we use 10-folds cross-validation [19] to compute the best values of parameters σ and C. For the multi-classification problem, we choose the one against all (1-v-r) approach [20], which is to transform a k-class classification problem into k two-class classification problems. Five runs of 10-fold cross-validation are performed for each algorithm, and the average result is reported in Tables 1 and 2.

Table 1. Running time comparison among different algorithms

Data sets	Running time/s				
	TSVMACO	J_TSVM	PTSVM	SLS-SVM	KMCTSVM
Australian	10.2	11.5	11.2	10.6	10.1
Balance	5.6	6.1	6.3	5.9	5.5
Iris	2.6	3.2	3.5	2.9	2.3

Table 2. Classification precision comparison among different algorithms

Data sets	Precision/%				
	TSVMACO	J_TSVM	PTSVM	SLS-SVM	KMCTSVM
Australian	82.1	81.1	81.3	80.9	80.7
Balance	83.2	81.2	81.9	82.1	80.8
Iris	92.5	90.3	91.5	91.2	90.6

4.2 A Text Classification Experiment

In this section, we perform an experiment on the newsgroups 20 data set. We use the mac and windows subset in the newsgroups 20 sample set for binary classification. The two subsets contains 1946 samples and each sample consists of 7511 dimensions. 500 samples are randomly chosen as training samples and the rest are chosen as testing samples. We also test the algorithms appeared in Experiment 1, and five runs of 10-fold cross-validation are performed, and Table 3 reports the corresponding results.

Table 3. Experimental results comparison on the newsgroup 20 dataset

Algorithms	Running time/s	Precision/%
TSVMACO	799	91.5
J_TSVM	1055	90.1
PTSVM	1021	89.2
SLS-SVM	803	90.3
KMCTSVM	775	88.6

4.3   Experimental Result Analysis

As shown in Tables 1, 2 and 3, although the KMCTSVM algorithm runs fast than the other algorithms, its classification precision is not very good. However the proposed TSVMACO algorithm performs better than the other transductive support vector machine algorithms considering both classification precision and running efficiency indexes. The main reason is as follows. The ACO designed in the proposed TSVMACO algorithm can compute the solution of OP 1 with a high efficiency and a good approximation. Thus the proposed TSVMACO algorithm shows a good classification performance.

5   Conclusion

As shown in Tables 1, 2 and 3, although the KMCTSVM algorithm runs fast than the other algorithms, its classification precision is not very good. However the proposed TSVMACO algorithm performs better than the other transductive support vector machine algorithms considering both classification precision and running efficiency indexes. The main reason is as follows. The ACO designed in the proposed TSVMACO algorithm can compute the solution of OP 1 with a high efficiency and a good approximation. Thus the proposed TSVMACO algorithm shows a good classification performance.

Acknowledgments. This work is sponsored by the National Natural Science Foundation of China (Nos. 61402246, 61402126, 61370083, 61370086, 61303193, and 61572268), a Project of Shandong Province Higher Educational Science and Technology Program (No. J15LN38, J14LN31), Qingdao indigenous innovation program (No. 15-9-1-47-jch), the Project of Shandong Provincial Natural Science Foundation of China (No. ZR2014FL019), the Open Project of Collaborative Innovation Center of Green Tyres & Rubber (No. 2014GTR0020), the National Research Foundation for the Doctoral Program of Higher Education of China (No. 20122304110012), the Science and Technology Research Project Foundation of Heilongjiang Province Education Department (No. 12531105), Heilongjiang Province Postdoctoral Research Start Foundation (No. LBH-Q13092), and the National Key Technology R&D Program of the Ministry of Science and Technology under Grant No. 2012BAH81F02.

References

1. Vapnik, V.: The Nature of Statistical Learning Theory. Springer, New York (1995)
2. Cortes, C., Vapnik, V.: Support vector networks. Mach. Learn. **20**, 273–297 (1995)
3. Wagner, A., Wright, J., Ganesh, A., Zhou, Z., Mobahi, H., Ma, Y.: Towards a practical face recognition system: robust registration and illumination by sparse representation. In: IEEE Conference on Computer Vision & Pattern Recognition, vol. 34, pp. 597–604. IEEE Computer Society (2009)
4. Cai, Z., Ducatez, M.F., Yang, J., Zhang, T., Long, L.-P., Boon, A.C., Webby, R.J., Wan, X.-F.: Identifying antigenicity associated sites in highly pathogenic H5N1 influenza virus hemagglutinin by using sparse learning. J. Mol. Biol. **422**(1), 145–155 (2012)

5. Cai, Z., Goebel, R., Salavatipour, M., Lin, G.: Selecting genes with dissimilar discrimination strength for class prediction. BMC Bioinform. **8**, 206 (2007)
6. Hofmeyr, S.A., Forrest, S., Somayaji, A.: Intrusion detection using sequences of system calls. J. Comput. Secur. **6**, 151–180 (1999)
7. Dennis, N., Mcqueen, J.M.: Shortlist B: a Bayesian model of continuous speech recognition. Psychol. Rev. **115**(2), 357–395 (2008)
8. Boland, M., Murphy, R.: A neural network classifier capable of recognizing the patterns of all major subcellular structures in fluorescence microscope images of HeLa cells. Bioinformatics **17**(12), 1213–1223 (2001)
9. Polat, K., Güneş, S.: A novel hybrid intelligent method based on C4.5 decision tree classifier and one-against-all approach for multi-class classification problems. Expert Syst. Appl. **36**(2), 1587–1592 (2009)
10. Yang, K., Cai, Z., Li, J., Lin, G.: A stable model-free gene selection in microarray data analysis. BMC Bioinform. **7**, 228 (2006)
11. Gammerman, A., Vapnik, V., Vowk, V.: Learning by transduction. In: Proceedings of the 14th Conference on Uncertainty in Artificial Intelligence, Wisconsin, pp. 148–156 (1998)
12. Joachims, T.: Transductive inference for text classification using support vector machines. In: ICML, vol. 99, pp. 200–209 (1999)
13. Chen, Y., Wang, G., Dong, S.: A progressive transductive inference algorithm based on support vector machine. J. Softw. **14**(3), 451–460 (2003)
14. Zhang, J., Zhao, Y., Yang, J.: Semi-supervised learning algorithm with a least square support vector machine. J. Harbin Eng. Univ. **29**(10), 1088–1092 (2008)
15. Wang, L., Li, J., Yue, Q.: K means clustering based transductive support vector machine algorithm. Comput. Eng. Appl. **49**(14), 144–146 (2013)
16. Dorigo, M., Birattari, M., Stutzle, T.: Ant colony optimization Computational. Intell. Mag. **1**(4), 28–39 (2006)
17. Liang, Y.C., Smith, A.E.: An ant colony optimization algorithm for the Redundancy Allocation Problem (RAP). IEEE Trans. on Reliab. **3**(3), 417–423 (2004)
18. Parpinelli, R.S., Lopes, H.S., Freitas, A.A.: Data mining with an ant colony optimization algorithm. IEEE Trans. Evol. Comput. **6**(4), 321–332 (2002)
19. Kohavi, R.: A study of cross-validation and bootstrap for accuracy estimation and model selection. In: Wermter, S., Riloff, E., Scheler, G. (eds.) Proceedings of the 14th Joint International Conference Artificial Intelligence, pp. 1137–1145. Morgan Kaufmann, San Mateo (1995)
20. Hsu, C.W., Lin, C.J.: A comparison on methods for multi-class support vector machines. IEEE Trans. Neural Netw. **13**(2), 415–425 (2001)

ABR: An Optimized Buffer Replacement Algorithm for Flash Storage Devices

Xian Tang[1](✉), Na Li[2], and Qiang Ma[2]

[1] School of Economics and Management,
Yanshan University, Qinhuangdao 066004, China
txianz@163.com
[2] School of Information Science and Engineering,
Yanshan University, Qinhuangdao 066004, China

Abstract. Flash disks are being widely used as an important alternative to conventional magnetic disks, although accessed through the same interface by applications, their distinguished feature, i.e., different read and write cost makes it necessary to reconsider the design of existing replacement algorithms to leverage their performance potential.

We propose an adaptive cost-aware replacement policy based on average hit distance (AHD) to control the movement of buffer pages when hits occur, thus pages that are re-visited within AHD will stay still. Such a mechanism makes our method adaptive to workloads of different access patterns. The experimental results show that our method not only adaptively tunes itself to workloads of different access patterns, but also works well for different kind of flash disks compared with existing methods.

1 Introduction

Social computing related applications bring great impacts to our daily lives, while produce large volume of data that need to be processed more efficiently. To this end, researchers developed flash-based storage devices to accelerate the computation. Flash-based storage devices have been steadily expanded into personal computer and enterprise server markets with ever increasing capacity of their storage and dropping of their price. A flash disk usually demonstrates extremely fast random read speeds, but slow random write speeds, and the best attainable performance can hardly be obtained from database servers without elaborate flash-aware data structures and algorithms [1], which makes it necessary to reconsider the design of IO-intensive and performance-critical software to achieve maximized performance.

During the past years, researchers proposed many buffer replacement algorithms [2–11] addressing the asymmetric read and write operation of flash disks, these methods, however, cannot work well for workloads suffering from "temporal locality". Here "temporal locality" means that in practice, a workload of databases, Web servers and operating systems, usually contains some pages that are requested with high frequency within a short time, but will not be requested in the future. We call such pages *once-frequently-requested* pages and the remaining pages except *once-requested* pages *frequently-requested* pages.

© Springer Science+Business Media Singapore 2016
W. Che et al. (Eds.): ICYCSEE 2016, Part I, CCIS 623, pp. 136–150, 2016.
DOI: 10.1007/978-981-10-2053-7_14

As *once-frequently-requested* pages will not be requested again in the future, keeping them in the buffer for a long time does not facilitate the improvement of system performance. Although existing methods [2–11] can tell the difference between pages that are requested only once and pages that are requested multiple times, they cannot tell the difference between *once-frequently-requested* pages and *frequently-requested* pages.

We propose an optimized buffer replacement strategy, namely ABR, which uses average hit distance (*AHD*) to control the movement of buffer pages when hits occur, thus pages that are re-visited within a distance less than AHD will stay still. As a result, *once-frequently-requested* pages can also be flushed out quickly. This mechanism makes ABR adaptive to workloads of different access patterns and can really improve the hit ratio of *frequently-requested* pages. Moreover, ABR maintains a buffer directory, namely, ghost buffer, to remember recently evicted buffer pages. The reference count of each entry in the ghost buffer is used to adaptively determine the length of the buffer list and the insertion position when it is re-visited, such that ABR can adaptively decide how many pages each list should maintain in response to an evolving workload.

2 Background and Related Work

2.1 Flash Memory

Flash disks usually consist of NAND flash chips. The three basic operations are read, write, and erase. Read and write operations are performed in unit of a page. Erase operations are performed in unit of a block, which is much larger than a page, usually contains 64 pages. NAND flash memory does not support in-place update, the write to the same page cannot be done before the page is erased. To overcome the physical limitation of flash memory, flash disks employ an intermediate software layer called Flash Translation Layer (FTL) to emulate the functionality of block device and hide the latency of erase operation as much as possible.

2.2 Buffer Replacement Policies

Assuming that the secondary storage consists of magnetic disks, the goal of existing buffer replacement policies is to minimize the buffer miss ratio for a given buffer size. Existing studies on magnetic disks, such as 2Q [12], ARC [13], LIRS [14], CLOCK [15], LRU-K [16], FBR [17] and LRFU [18] aim at improving the traditional LRU heuristic, which are not efficient when applied to flash disks due to the asymmetric access times.

The flash aware buffer policy (FAB) [3] maintains a block-level LRU list, of which pages of the same erasable block are grouped together. FAB is mainly used in portable media player applications where most write requests are sequential.

BPLRU [4] also maintains an block-level LRU list. Different from FAB, BPLRU uses an internal RAM of SSD as a buffer to change random write to sequential write to improve the write efficiency and reduce the number of erase operation. However, this method cannot really reduce the number of write requests from main memory buffer.

Clean first LRU (CFLRU) [2] is a flash aware buffer replacement algorithm for operating systems. It was designed to exploit the asymmetric performance of flash IO by first paging out clean pages arbitrarily based on the assumption that writing cost is much more expensive. The LRU list is divided into two regions: the working region and the clean-first region. Each time a miss occurs, if there are clean pages in the clean-first region, CFLRU will select the least recent referenced clean page in the clean-first region as a victim. Compared with LRU, CFLRU reduces the write operations significantly.

Based on the same idea, [5] makes improvements over CFLRU by organizing clean pages and dirty pages into different LRU lists to achieve constant complexity per request. In CFDC [6], dirty pages that are close to each other in the dirty queue are grouped into different clusters. Compared with CFLRU, CFDC improves the write efficiency.

Different from the above methods, ACR [7] addresses the problem that the ratio of write to read of different flash disks may vary significantly, and is adaptive to different types of flash disks.

3 The ABR Policy

3.1 Data Structures

As shown in Fig. 1, ABR splits the LRU list into two LRU lists, i.e., L_C and L_D. L_C keeps *clean* pages and L_D *dirty* pages. Assume that the buffer contains s pages when it is full, then $|L_C \cup L_D| = s \wedge L_C \cap L_D = \emptyset$. Further, $L_C(L_D)$ is divided into $L_{CT}(L_{DT})$ and $L_{CB}(L_{DB})$, and $L_{CT} \wedge L_{CB} = \emptyset(L_{DT} \wedge L_{DB} = \emptyset)$, $L_{CT}(L_{DT})$ contains *frequently-requested* clean (dirty) pages while $L_{CB}(L_{DB})$ contains *once-requested* and *once-frequently-requested* clean (dirty) pages, and *frequently-requested* clean (dirty) pages that are *not* referenced for a long time.

The sizes of L_{CB} and L_{DB} will be dynamically adjusted with the change of access patterns, which are controlled by δ_C and δ_D, respectively.

The difference of data structure between ACR [7] and ABR lies in that we use a ghost buffer L_H in ABR to trace the past references by recording the page id of those pages that are paged out from L_C or L_D. Fixing this parameter is potentially a tuning question, in our experiment, $|L_H| = s/2$. The notions used in this paper are shown in Table 1.

3.2 Cost-Based Eviction

If the buffer is full and the currently requested page p is in the buffer, it is served without accessing the auxiliary storage, otherwise, we will select from L_C or L_D a page x for replacement according to the metrics of "*cost*", not clean or dirty. The *cost* associated to L_C (L_D), say C_{L_C} (C_{L_D}), is a weighted value denoting the overall replacing cost.

Table 1. Notations used in this paper

Notation	Description
L_C	the clean list
L_{CT}	the top portion of L_C
L_{CB}	the bottom portion of L_C
δ_C	the number of clean pages contained in L_{CB}
L_D	the dirty list
L_{DT}	the top portion of L_D
L_{DB}	the bottom portion of L_D
δ_D	the number of dirty pages contained in L_{DB}
L_H	the ghost buffer containing page id of evicted pages
C_r	the cost of reading a page from a flash disk
C_w	the cost of writing a dirty page to a flash disk
s	the size of the buffer in pages
ρ_{rd}	the number of physical read operations of L_D
ρ_{wd}	the number of physical write operations of L_D
ρ_{rc}	the number of physical read operations of L_C
ι_d	the number of logical operations of L_D
ι_c	the number of logical operations of L_C

The *basic* idea is that the length of L_C (L_D) should be proportional to the ratio of the replacement cost of L_C (L_D) to that of all buffer pages according to recent m requests, in our experiment, $m = s/2$. This ratio can be formally represented as Formula 1:

$$\beta = C_{L_C}/(C_{L_C} + C_{L_D}) \tag{1}$$

The policy of selecting a victim page can be stated as: If $|L_C| < \beta \cdot s$, it means L_D is too long, and the LRU page in L_D should be paged out, otherwise the LRU page of L_C should be paged out.

Hereafter, we call the read and write operations that are served in buffer are logical, and ones that reach the disk are referred to as physical. Logical and physical operations are two different operations, and they all affect the overall performance. Assume that n is the number of pages in a file and s the number of pages allocated to the file in the buffer. The probability that a logical operation will be served in the buffer is s/n, and the probability that a logical operation will be translated to a physical one is $(1 - s/n)$. The probability is used to compute the values of C_{L_C} and C_{L_D}, as shown by Formulas 2 and 3, for which the meaning of each notion is shown in Table 1.

$$C_{L_C} = (\iota_c \cdot (1 - s/n) + \rho_{rc}) \cdot C_r \tag{2}$$

$$C_{L_D} = \iota_d \cdot (1 - s/n) \cdot (C_w + C_r) + \rho_{rd} \cdot C_r + \rho_{wd} \cdot C_w \tag{3}$$

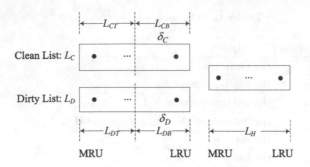

Fig. 1. Data organization of ABR

3.3 The ABR Algorithm

Simply using the above eviction strategy cannot solve the "temporal locality" problem. We introduce to use average hit distance to tackle this problem.

Average Hit Distance. For a sequence of requests $R = "r_1, r_2, \cdots, r_n"$, where each r_i refers to a data page, assume that r_i and $r_j(i < j)$ refer to the same page p and no other request $r_k(i < k < j)$ refer to p, we define the *Hit Distance d* of p between r_i and r_j is $p.d = j - i$. $p.d = 0$, if the number of requests ≤ 1 on p.

For example, for $"r_1(p_1), r_2(p_2), r_3(p_3), r_4(p_1), r_5(p_4), r_6(p_1)"$, if the current request is r_4, we have $p_1.d = 3$ and $p_2.d = p_3.d = 0$. If the current request in r_6, then $p_1.d = 2$. To differentiate the two Hit Distances of p_1, we denote them as $r_4.d = 3$ and $r_6.d = 2$, similarly, we have $r_1.d = r_2.d = r_3.d = r_5.d = 0$.

Definition 1 *(Average Hit Distance(AHD, ξ))*. *For the recent n requests* $R = "r_1, r_2, \cdots, r_n", R^+ = "r_1', r_2', \cdots, r_m'"$ *is the set of m requests of R (m $\leq$ n), where each request r of R^+ satisfies r.d > 0. The* Average Hit Distance ξ *of R is the average number of the Hit Distance of requests in R^+, as shown in Formula 4.*

$$\xi = \frac{\sum_{i=1}^{m} r_i.d}{m} \tag{4}$$

For example, assume the recent 8 requests are $R = "r_1(p_1), r_2(p_2), r_3(p_2), r_4(p_3), r_5(p_2), r_6(p_4), r_7(p_1), r_8(p_2)"$, we have $r_1.d = 0, r_2.d = 0, r_3.d = 1, r_4.d = 0, r_5.d = 2, r_6.d = 0, r_7.d = 6, r_8.d = 3$. According to Definition 1, the AHD for R is $\xi = (r_3.d + r_5.d + r_7.d + r_8.d)/4 = 3$.

Intuitively, if the Hit Distance of a request r on p is greater than or equal to AHD, i.e. $r.d \geq \xi$, it means that the interval between two continuously access to p is relatively large, thus p should be considered as a *frequently-requested* page, otherwise, if p will not be re-visited in the future, even if p is visited frequently within a short time, as its Hit Distance is less than AHD, p should be considered as a *once-frequently-requested* page instead of a *frequently-requested* page.

We use AHD to determine whether a hit on page p should be taken into account. Let r be the request corresponding to the hit on p, the idea of changing $p.hit$ can thus be stated as Formula 5.

$$p.hit = \begin{cases} p.hit, & r.d < \xi \\ p.hit + 1, & otherwise \end{cases} \tag{5}$$

However, there are still two problems that are needed to be solved before using the above method, otherwise, some frequently-requested pages with Hit Distance *less* than AHD will be wrongly moved out from the buffer, as shown in below.

$P1$: *where is the start position to compute Hit Distance?*

$P2$: *what is the upper bound of AHD?*

Example and Solution to $P1$. As shown in Fig. 2 (A), the dashed line represents a sequence of requests, where each large red dot denotes a request on page p, while each small black dot denotes a request on other page. d_1 to d_4 represent the Hit Distance of r_2 to r_5, respectively. If the current request is r_2, we have $r_2.d = d_1$. Since $d_1 < AHD$, $p.hit$ will not increase. Similarly, $p.hit$ will not increase when processing r_3, r_4 and r_5. Such case usually occurs if p is a page containing meta data or the root node of a B-tree index.

In our method, $p.hit$ will not increase when processing r_2, but will increase by 1 when processing r_3. The Hit Distance of r_3 is the distance from r_1 to r_3, instead of that between r_2 and r_3. That is, we use a flag to denote whether a certain request is a valid one. When processing r_2, since $d_1 < AHD$, r_2 is marked as an invalid request. When computing the Hit Distance of r_3, we will check the previous valid request of p, then we have $r_3.d=d'_1=d_1+d_2>AHD$. Similarly, $p.hit$ will not increase when processing r_4, but will increase by 1 when processing r_5, because $r_4.d=d_3< AHD$, while $r_5.d=d'_2=d_3+d_4>AHD$.

Example and Solution to $P2$. As shown in Fig. 2 (B), the thick line represents pages in the buffer. The circled blue numbers ① and ② represent two cases of AHD, the red dots denote the requests on page p. δ_C and δ_D are the size of L_{CB} and L_{DB}, respectively. Notice that the newly entered pages that have no page id in the ghost buffer are always inserted at the MRU position of either L_{CB} or L_{DB}.

Case ① $(AHD > \frac{\delta_C+\delta_D}{2})$: In this case, there may exist some request (e.g., r_2 on p) such that $\frac{\delta_C+\delta_D}{2} < r_2.d < AHD \wedge 2 \cdot r_2.d > (\delta_C + \delta_D)$. Thus r_2 is an invalid request and $p.hit$ doesn't increase by 1, thus p will not be moved to L_{CT} or L_{DT} when processing r_2. As a result, even if p is frequently revisited with a Hit Distance relatively large, p may be flushed out when processing r_3 in the future.

Case ② $(AHD \leq \frac{\delta_C+\delta_D}{2})$: In this case, for any request r_2 on p satisfying $r_2.d \leq AHD$, we have $2 \cdot r_2.d \leq (\delta_C + \delta_D)$. Thus in the worst case, p still can be moved to L_{CT} or L_{DT} when processing r_3.

Based on the above discussion, we use the following Formula to get AHD, not Formula 4.

$$\xi = \begin{cases} \frac{\sum_{i=1}^m r_i.d}{m}, & \xi \leq \delta_C + \delta_D \\ (\delta_C + \delta_D)/2, & otherwise \end{cases} \tag{6}$$

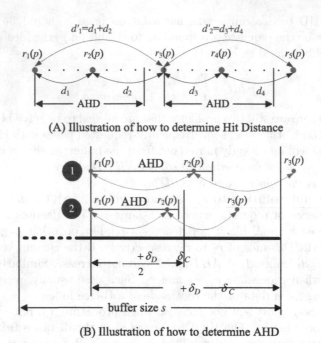

(A) Illustration of how to determine Hit Distance

(B) Illustration of how to determine AHD

Fig. 2. Problems about Hit Distance and AHD. (Color figure online)

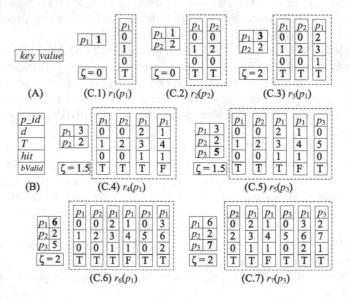

Fig. 3. Illustration of computing AHD. (A) is the structure of an item of a hash table H, (B) is the structure of an item of a priority queue Q, (C.1) to (C.7) show the process of a reference sequence.

AHD Computing. To compute the AHD of the past n requests, we use two data structures, a hash table H and a priority queue Q. For a reference sequence R, H is used to maintain the position of the most recent valid request on each page. Q is used to maintain the recent n requests. Figure 3 (A) shows the structure of an item of H, the *key* is a page id, the *value* is the position of an entry in Q, which denotes the start position to compute Hit Distance. Figure 3 (B) shows the structure of an item of Q, which consists of five member variables, p_{id} is the page id of p, d is the Hit Distance of r, T is the order of r in a reference sequence, *hit* is the reference count of p after processing r, *bValid* is a flag denoting whether the current request r on p is a valid request. If $bValid = $ T, it means the position of p in H should be changed according to r.

Let $R=$"$r_1(p_1),r_2(p_2),r_3(p_1),r_4(p_1),r_5(p_3),r_6(p_1),r_7(p_3)$" be a reference sequence. Assume $\delta_C + \delta_D = 4$ and $n = 6$ ($Q's$ length) in this example, i.e. we compute the AHD of the recent 6 requests. Initially, H and Q are both empty. The process of R is as follows.

(1) process $r_1(p_1)$. The status is shown in Fig. 3 (C.1).

(2) process $r_2(p_2)$. The status is as Fig. 3 (C.2).

(3) process $r_3(p_1)$. As shown in Fig. 3 (C.3), p_1 is hit and $r_3.T = 3$. From H in Fig. 3 (C.2) we know p_1's valid position is 1 and $\xi = 0$, then we have $r_3.d = 2$ and $r_3.hit = 1$ as $r_3.d \geq \xi = 0$. Thus $r_3.bValid =$T and we change p_1's valid position in H to 3. Finally, we set the value of ξ as $\xi = r_3.d/1 = 2$.

(4) process $r_4(p_1)$. As shown in Fig. 3 (C.4), p_1 is hit and $r_4.T = 4$. From H in Fig. 3 (C.3) we know that p_1's valid position is 3 and $\xi = 2$, then we have $r_4.d = 1$ and $r_4.hit = 1$ as $r_4.d \leq \xi = 2$. Thus $r_4.bValid =$F and p_1's valid position is still equal to 3. Finally, $\xi = \frac{r_3.d+r_4.d}{2} = 1.5$.

(5) process $r_5(p_3)$. It's the first time p_3 enters into Q, the value of its member variables is shown in Fig. 3 (C.5), p_3's valid position in H is set to 5.

(6) process $r_6(p_1)$. As shown in Fig. 3 (C.6), p_1 is hit and $r_6.T = 6$. From H in Fig. 3 (C.5) we know that p_1's valid position is 3 and $\xi = 1.5$, then we have $r_6.d = 3$ and $r_6.hit = 2$ since $r_6.d \geq \xi = 1.5$. Thus $r_6.bValid =$T and we change p_1's valid position in H to 6. Finally, $\xi = \frac{r_3.d+r_4.d+r_6.d}{3} = 2$.

(7) process $r_7(p_3)$. As shown in Fig. 3 (C.7), p_3 is hit and $r_7.T = 7$. As the size of Q is 6, before r_7 is added to the tail of Q, r_1 is firstly removed from the head of Q. From H in Fig. 3 (C.6) we know that p_3's valid position is 5 and $\xi = 2$, then we have $r_7.d = 2$ and $r_7.hit = 1$ since $r_7.d \geq \xi = 2$. Thus $r_7.bValid =$T and we change p_3's valid position in H to 7. Finally, as r_1 is not a hit request, we just need to add 1 and $r_7.d$ to the denominator and numerator of Formula 4 respectively, i.e. $\xi = \frac{r_3.d+r_4.d+r_6.d+r_7.d}{4} = 2$, otherwise, 1 and $r_1.d$ are firstly subtracted from the denominator and numerator of Formula 4, respectively.

As illustrated in this example, for each request, the computing of AHD can be done with time complexity of $O(1)$. In this example, we assume $\delta_C + \delta_D = 4$, thus ξ is directly computed by $\frac{\sum_{i=1}^{m} r_i.d}{m}$ according to Formula 6. In Algorithm 1, we use updateAHD(ξ) to denote the procedure of computing AHD, the pseudo-code is omitted for limited space.

The Algorithm. As shown in Algorithm 1, in the beginning stage before the buffer is full, i.e., $|L_C \cup L_D| < s \wedge |L_H| = 0$, if the request on p is a *miss-request* and $p's$ page id is not in L_H, Algorithm 1 will execute the code in Case III. Since $|L_C \cup L_D| < s$, we will call Procedure ReadIn() in line 16 to fetch p from the disk. After that updateAHD() is called in line 17. At last, δ_C or δ_D will increase by 1 by calling Procedure AdjustBottomProtionList(). If the current request on p is a *hit-request*, that is, $p \in L_C \cup L_D$, we will execute the code in Case I. Specifically, if $p \in L_{CB}$ (L_{DB}), it means that p should not stay anymore in L_{CB} (L_{DB}), since L_{CB} (L_{DB}) is used to maintain once-requested clean (dirty) pages. Then we move p to the MRU position of L_{CT} or L_{DT} and adjust the size of L_{CB} and L_{DB}, respectively.

If the buffer is full, for a *hit-request* corresponding to Case I and discussed already. If the current request is a *miss-request*, then we will check whether $p's$ id is contained in L_H. If $p's$ id is contained in L_H, which corresponds to Case II. In this case, we will firstly call evictPage() to select a victim page to make room for p. If $p.hit = 0$, it means that the size of L_{CB} or L_{DB} is too small, then δ_C or δ_D will increase by 1 (lines 10-11). After that, we will fetch p from disk by calling Procedure ReadIn(), then call updateAHD() in line 13, and adjust the length of L_{CB} and L_{DB} in line 14. If the current request is a *miss-request* and $p's$ id is not in L_H, which corresponds to Case III.

Based on Algorithm 1, we can effectively identify *once-frequently-requested* pages from pages that are requested multiple times to improve the overall performance. By using a hash table to maintain the pointers to each page in the buffer, the complexity serving each request is $O(1)$.

Analysis Adaptivity. The adaptivity means that our method continually revises the parameter δ_C and δ_D that are used to control the size of L_{CB} and L_{DB}. Moreover, the adaptivity means that if the workload is to change from one access pattern to another one or vice versa, we will track such change and adapt itself to exploit the new opportunity.

Scan-Resistant. When serving a long sequence of one-time-only requests, ABR will only evict pages in $L_{CB} \cup L_{DB}$. This is because, when requesting a new page p, i.e., $p \notin L_C \cup L_D \cup L_H$, p is always put at the MRU position of L_{CB} or L_{DB}. It will not impose any affect on pages in $L_{CT} \cup L_{DT}$ unless it is requested again before it is paged out from L_H. For this reason, we say ABR is scan-resistant. Furthermore, a buffer is usually used by several processes or threads concurrently, when a scan of a process or thread begins, less hits will be encountered in $L_{CB} \cup L_{DB}$ compared to $L_{CT} \cup L_{DT}$, and hence, according to Algorithm 1, the size of L_{CT} and L_{DT} will grow gradually, and the resistance of ABR to scans is strengthened again.

Loop-Resistant. We say that ABR is loop-resistant means that when the size of the loop is larger than the buffer size, ABR will keep partial pages of the loop sequence in the buffer, and hence, achieve higher performance. We explain this point from three aspects. (1) The loop requests only pages in L_C. In the first

Algorithm 1: ABR(page p, type T)

Case I: $p \in L_C \cup L_D$, a buffer hit has occurred.

```
1       if (p ∈ L_C) then  {ι_c ← ι_c + 1; if (p ∈ L_CB) then  {δ_C ← max{λ · s, δ_C − 1};}};
2       else {ι_d ← ι_d + 1; if (p ∈ L_DB) then  {δ_D ← max{λ · s, δ_D − 1};}}
3       if (T = read ∧ p ∈ L_C) then  {if (p.d ≥ ξ) then  {move p to MRU of L_CT;} }
4       else if (p.d ≥ ξ) then  {move p to the MRU position of L_DT;}
5       else if (T = write ∧ p ∈ L_C) then  {move p to the MRU position of L_DB;}
6       if (p.d ≥ ξ) then  {p.hit ← p.hit + 1; p.bHasHit = TRUE;}
7       updateAHD(ξ);
8       AdjustBottomPortionList();
```

Case II: $p \in L_H$, a buffer miss has occurred.

```
9       evictPage();
10      if (T = read ∧ p.bHasHit = FALSE) then  {δ_C ← min{|L_C|, δ_C + 1};}
11      if (T = write ∧ p.bHasHit = FALSE) then  {δ_D ← min{|L_D|, δ_D + 1};}
12      ReadIn(p, T, true);
13      updateAHD(ξ);
14      AdjustBottomPortionList();
```

Case III: $p \notin L_C \cup L_D \cup L_H$, a buffer miss has occurred.

```
15      if (|L_C ∪ L_D| = s) then  {evictPage();}
16      ReadIn(p, T, false);
17      updateAHD(ξ);
18      AdjustBottomPortionList();
```

Procedure evictPage()

```
1       β ← C_LC/(C_LC + C_LD);   /*β is computed based on the recent s/2 requests*/
```

Case I: $|L_C| < \beta \cdot s$ /* L_D is longer than expected*/.

```
2       ρ_wd ← ρ_wd + 1;
3       q ← FindEvictedPage(L_DB); write the content of q to disk;
4       if (|L_H| = s/2) then  {delete the item in the LRU position of L_H;}
5       delete q from L_DB and insert its page id as a new item in the MRU position of L_H;
```

Case II: $|L_C| \geq \beta \cdot s$ /* L_C is longer than expected*/.

```
6       if (|L_H| = s/2) then  {delete the item in the LRU position of L_H;}
7       q ← FindEvictedPage(L_CB); delete q and insert its page id to MRU of L_H;
```

Procedure ReadIn(page p, type T, bool $bTop$)

```
1       if (T = read) then  increase ρ_rc and ι_c by 1
else increase ρ_rd and ι_d by 1
2       fetch p from the disk; p.hit ← 0; p.bHasHit = FALSE;
3       if (T = read ∧ bTop = TRUE) then  insert it to the MRU of L_CT;
4       if (T = read ∧ bTop = FALSE) then  insert it to the MRU of L_CB;
5       if (T = write ∧ bTop = TRUE) then  insert it to the MRU of L_DT;
6       if (T = write ∧ bTop = FALSE) then  insert it to the MRU of L_DB;
```

Function FindEvictedPage(list L)

```
1       q ← L.Tail;
2       while (q.hit > 0) do  {q.hit ← q.hit/2; move q to MRU of L; q ← L.Tail;}
3       return q;
```

Procedure AdjustBottomPortionList()

```
1       if (|L_C ∪ L_D| = s) then
2          Move the MRU (or LRU) page of L_CB and L_DB (or L_CT and L_DT) to LRU
           (MRU) of L_CT and L_DT (or L_CB and L_DB) to make |L_CB| = δ_C ∧ |L_DB| = δ_D;
3       else  {δ_C ← |L_CB|; δ_D ← |L_DB|;}
```

cycle of the loop request, all pages are fetched into the buffer and inserted at the MRU position of L_{CB} sequentially. Before each insertion, ABR will select a victim page q. If q is the LRU page of L_{DB}, then after the insertion of p in the MRU position of L_{CB}, ABR will adjust the size of L_{CB} and p will be adjusted to the LRU position of L_{CT}; otherwise p is still at the MRU position of L_{CB}. With the processing of the loop requests, more pages of the loop sequence will be moved to L_{CT} and these pages are thus kept in buffer, therefore the hit ratio will not be zero anymore. (2) The loop requests only pages in L_D. This is same to (1). (3) The loop contains pages in both L_C and L_D. In this case, obviously, dirty pages will stay in buffer longer than clean pages and the order of the pages eviction is not same as they entered into the buffer, and hence, ABR can process them elegantly to achieve higher hit ratio.

4   Experiments

4.1   Experimental Setup

Due to the implementation of FTL is device-related and supplied by the disk manufacturer, and there is no interface supplied for users to trace the number of write and read, we choose to use the simulator of [19] to count the numbers of read and write operations.

Methods for Comparison. We implemented LRU, CFLRU [2] and CFDC [6]. Further, we implemented ACR [7] and ABR. All these methods were implemented on the simulator using Visual C++ 6.0. For CFLRU, we set the "window size" of "clean-first region" to 75 % of the buffer size, for CFDC, the "window size" of "clean-first region" is 50 % of the buffer size, and the "cluster size" of CFDC is 64, the value of these parameters are suggested by the papers.

Traces for Experiment. We simulated a database file of 64MB, which corresponds to 32 K physical pages and each page is 2KB, the buffer size is 4 K pages. We generated four synthetical traces which satisfy Zipf distributions. The statistics of the four traces are shown in Table 2, where $x\%/y\%$ in column "Read/Write Ratio" means that for a certain trace, $x\%$ of total requests are about read operations and $y\%$ about write operations; while $x\%/y\%$ in column "Locality" means that for a certain trace, $x\%$ of total operations are performed in a certain $y\%$ of the total pages.

Storage Mediums. We select two flash chips for our experiment, Samsung MCAQE32G5APP and Samsung MCAQE32G8APP-0XA [20]. The ratio of the cost of random read to that of random write is 1:118 and 1:2, respectively. The reason for the huge discrepancy of the two flash disks lies in that the first flash disk is based on MLC NAND chip, while the second flash disk is based on SLC NAND chip. Both type of flash disks are already adopted as auxiliary storage in many applications.

Metrics. We choose the following metrics to evaluate the nine buffer replacement policies: (1) number of physical read operations; (2) number of physical write operations, which includes the write operations caused by the erase

Table 2. The statistics of the traces

Trace	Total requests	Read/Write ratio	Locality
T1	3,000,000	90 % / 10 %	60 % / 40 %
T2	3,000,000	80 % / 20 %	50 % / 50 %
T3	3,000,000	60 % / 40 %	60 % / 40 %
T4	3,000,000	80 % / 20 %	80 % / 20 %

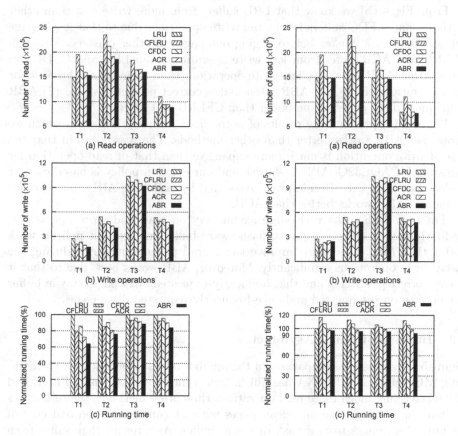

Fig. 4. The comparison of random read, random write and normalized running time on trace T1 to T4 for Samsung MCAQE32G5APP flash disk.

Fig. 5. The comparison of random read, random write and normalized running time on trace T1 to T4 for Samsung MCAQE32G8APP-0XA flash disk.

operations of flash disks; and (3) running time. Though there may exist some differences compared with the results tested on a real platform, they reflect the overall performance of different replacement policies by and large with neglectable tolerance.

4.2 Impacts of Large Asymmetry

Figure 4 (a) shows the comparison of the number of random read operations on trace T1 to T4 w.r.t. Samsung MCAQE32G5APP flash disk, from which we know that LRU has least read operations, the reason lies in that LRU does not differentiate read and write operations, thus it will not delay the paging out of dirty pages in the buffer. On the contrary, CFLRU firstly pages out clean pages, thus it needs to read in more pages than other methods.

From Fig. 4 (b) we know that LRU suffers from more write cost than other methods, and CFDC suffers from more write operations due to that it will page out all pages in a cluster before paging out pages in other clusters. Although CFLRU and ACR suffer from less write operations than LRU and CFDC, we can see that ABR consumes less write operations than them, this is because for the cost ratio of 1:118, (1) ABR often makes correct predictions, and (2) ABR keeps more dirty pages in the buffer than CFLRU and ACR.

Figure 4 (c) presents the results of normalized running time, from which we know that ABR works better than other methods. The reason lies in that the cost of write operation is much more expansive than that of read operation for Samsung MCAQE32G5APP flash disk and our eviction policy is based on cost of clean and dirty lists. Besides, we can see that by exploiting AHD and reference frequency, ABR works better than ACR.

Thus for flash disks with large asymmetry on read and write operations, by firstly paging out clean pages, flash-aware buffer replacement policies work better than LRU since they improve the overall performance by reducing the costly write operations significantly. Moreover, ABR works best, due to that it makes correct prediction and that frequently-requested dirty pages stay in buffer longer than once-requested and once-frequently-requested dirty pages.

4.3 Impacts of Small Asymmetry

Figure 5 (a) shows the comparison of the number of read operations w.r.t. Samsung MCAQE32G8APP-0XA flash disk, from which we know that CFLRU and CFDC consume much more read operations than other methods, the reason lies in that they firstly page out clean pages without considering the real cost of fetching clean pages from a flash disk into buffer. As a result, they suffer from large read cost. Although our methods keeps more dirty pages in buffer than clean pages since the cost of read operation is still cheaper than write operation, ABR achieves competing performance to LRU for read operation by improving the hit ratio of *frequently-requested* clean pages.

From Fig. 4 (b) we know that the number of write operations of ABR becomes larger than that in Fig. 4 (b), this is because the ratio of read and write becomes smaller than before, and our policy will pay more attention to clean pages. Though CFLRU and CFDC have less write operations than LRU, they waste many more read operations, which makes them achieving worse performance than LRU, as shown in Fig. 5 (c). Again, we can see that ABR works best by exploiting AHD and reference frequency.

Therefore, for flash disks with small asymmetry on read and write operations, ABR is better than LRU, CFLRU, CFDC and ACR, because ABR only consumes the same or less read operations than LRU, which is much less than that consumed by CFLRU and CFDC; though still need to consume more write operations than CFLRU and CFDC, the saved cost of read operation is far more than that wasted by write operations.

5 Conclusions

Considering that existing buffer replacement methods cannot process workload with temporal locality, we propose an adaptive cost-based replacement policy, namely ABR. ABR organizes buffer pages into clean list and dirty list, and the newly entered pages will not be inserted at the MRU position of either list, but at some position in middle. By exploiting average hit distance to control the movement of buffer pages, ABR is wise in identifying once-frequently-requested pages and the frequently-requested pages can stay in the buffer for a longer time. Besides, ABR considers reference count and is more adaptive than existing works to workloads of different access patterns, and thus, achieves better performance. The experimental results on different traces and flash disks show that ABR not only adaptively tunes itself to workloads of different access patterns, but also works well for different kind of flash disks compared with existing methods.

We plan to make further improvement on ABR by considering changing the write operations from random write to sequential write and implement ABR in a real platform to evaluate it with various real workloads for flash-based applications.

References

1. Lee, S.-W., Moon, B.: Design of flash-based DBMS: an in-page logging approach. In: SIGMOD, pp. 55–66 (2007)
2. Park, S.-Y., Jung, D., Kang, J.-U., Kim, J., Lee, J.: CFLRU: a replacement algorithm for flash memory. In: CASES, pp. 234–241 (2006)
3. Jo, H., Kang, J.-U., Park, S.-Y., Kim, J.-S., Lee, J.: FAB: flash-aware buffer management policy for portable media players. IEEE Trans. Consum. Electron. **52**(2), 485–493 (2006)
4. Kim, H., Ahn, S.: BPLRU: a buffer management scheme for improving random writes in flash storage. In: FAST, pp. 239–252 (2008)
5. Koltsidas, I., Viglas, S.: Flashing up the storage layer. PVLDB **1**(1), 514–525 (2008)
6. Yi, O., Harder, T., Jin, P.: CFDC: a flash-aware replacement policy for database buffer management. In: DaMoN, pp. 15–20 (2009)
7. Tang, X., Meng, X.: ACR: an adaptive cost-aware buffer replacement algorithm for flash storage devices. In: MDM, pp. 33–42 (2010)
8. Kim, B.-K., Lee, D.-H.: LSF: a new buffer replacement scheme for flash memory-based portable media players. IEEE Trans. Consum. Electron. **59**(1), 130–135 (2013)
9. Jin, R., Cho, H.-J., Chung, T.-S.: LS-LRU: a lazy-split LRU buffer replacement policy for flash-based B+-tree index. J. Inf. Sci. Eng. **31**(3), 1113–1132 (2015)

10. Jin, P., Yi, O., Härder, T., Li, Z.: AD-LRU: an efficient buffer replacement algorithm for flash-based databases. Data Knowl. Eng. **72**, 83–102 (2012)
11. On, S.T., Gao, S., He, B., Wu, M., Luo, Q., Xu, J.: FD-Buffer: a cost-based adaptive buffer replacement algorithm for flashmemory devices. IEEE Trans. Comput. **63**(9), 2288–2301 (2014)
12. Johnson, T., Shasha, D.: 2Q: a low overhead high performance buffer management replacement algorithm. In: VLDB, pp. 439–450 (1994)
13. Megiddo, N., Modha, D.S.: ARC: a self-tuning. low overhead replacement cache. In: FAST (2003)
14. Jiang, S., Zhang, X.: Making LRU friendly to weak locality workloads: a novel replacement algorithm to improve buffer cache performance. IEEE Trans. Comput. (TC) **54**(8), 939–952 (2005)
15. Babaoglu, O., Joy, W.N.: Converting a swap-based system to do paging in an architecture lacking page-reference bits. In: SOSP, pp. 78–86 (1981)
16. O'Neil, E.J., O'Neil, P.E., Weikum, G.: The LRU-K page replacement algorithm for database disk buffering. In: SIGMOD, pp. 297–306 (1993)
17. John, T., Robinson, M.V.: Data cache management using frequency-based replacement. In: SIGMETRICS, Devarakonda, pp. 134–142 (1990)
18. Lee, D., Choi, J., Kim, J.-H., Noh, S.H., Min, S.L., Cho, Y., Kim, C.-S.: LRFU: a spectrum of policies that subsumes the least recently used and least frequently used policies. IEEE Trans. Comput. (TC) **50**(12), 1352–1361 (2001)
19. Jin, P., Su, X., Li, Z.: A flexible simulation environment for flash-aware algorithms. In: CIKM, Lihua Yue, pp. 2093–2094 (2009)
20. http://www.datasheetcatalog.net

An Approach for Automatically Generating R2RML-Based Direct Mapping from Relational Databases

Mohamed A.G. Hazber[1], Ruixuan Li[1(✉)], Guandong Xu[2],
and Khaled M. Alalayah[3,4]

[1] School of Computer Science and Technology,
Huazhong University of Science and Technology, Wuhan, China
moh_hazbar@yahoo.co.uk, rxli@hust.edu.cn
[2] Faculty of Engineering and IT,
University of Technology Sydney, Sydney, Australia
guandong.xu@uts.edu.au
[3] Computer Science, IBB University, Ibb, Yemen
kh101ed2005@yahoo.com
[4] Computer Science, Najran University-Sharurah, Najran, Saudi Arabia

Abstract. For integrating relational databases (RDBs) into semantic web applications, the W3C RDB2RDF Working Group recommended two approaches, Direct Mapping (DM) and R2RML. The DM provides a set of mapping rules according to RDB schema, while the R2RML allows users to manually define mappings according to existing target ontology. The major problem to use R2RML is the effort for creating R2RML mapping documents manually. This may lead to appearance of many mistakes in the R2RML documents and requires domain experts. In this paper, we propose and implement an approach to generate an R2RML mapping documents automatically from RDB schema. The R2RML mapping reflects the behavior of the DM specification and allows any R2RML parser to generate a set of RDF triples from relational data. The input of generating approach is DBsInfo class that automatically generated from relational schema. An experimental prototype is developed and shows the effectiveness of our approach algorithms.

Keywords: Relational Database to Resource Description Framework (RDB2RDF) · Direct Mapping · R2RML · Relational database · Resource Description Framework (RDF)

1 Introduction

The continuous explosion of resource description framework (RDF) data opens door for new innovations in big data, social network analysis, and semantic web initiatives, which can be shared and reused through the application, enterprise and community boundaries. The semantic web [1] is one of the most important research fields that aim to construct a web of data based on the RDF [2]

© Springer Science+Business Media Singapore 2016
W. Che et al. (Eds.): ICYCSEE 2016, Part I, CCIS 623, pp. 151–169, 2016.
DOI: 10.1007/978-981-10-2053-7_15

data model. It allows data to be shared and reused through applications, enterprise and community boundaries. Relational databases (RDBs) are the primary sources of web data, "deep web" [3]. The main reason is one of the studies [3] showed that internet accessible databases contained up to 500 times more data compared to the static web, and roughly 70 % of websites are backed by RDBs. The W3C RDB2RDF Working Group recently recommended a specification for languages to map RDB (data and schemas) to RDF and OWL, tentatively called Direct Mapping (DM) [4] and R2RML (Relational Database to RDF Mapping Language) [5]. However, the W3C Working Group does not recommend any implementation for DM and R2RML. The DM provides a set of automatic mapping rules to construct an ontology schema (RDF(S) and OWL) from RDB schema and convert relational data to RDF graphs according to that schema [6]. The ontology constructed reflects the structure and content of the relational database. Nevertheless, the DM method may not be constantly sufficient or optimum, especially when mapping a relational database to an existing ontology. R2RML is a customized mapping language, which allows users to define mappings manually. In this approach, the expert user expresses the RDB schema using an existing target ontology in order to convert the relational data into RDF datasets.

The R2RML specification is accompanied by the DM specification [4], representing a standard approach for converting an RDB into RDF without the use of a customized mapping definition. Thus, the RDF generated using DM can be represented in R2RML. R2RML provides more flexibility than DM specification. Meanwhile, creating R2RML rules by domain experts manually is complex, time consuming process, cumbersome, mistakable, high cost process, and requires the supports of domain experts in knowledge acquisition. Moreover, the users who are interested to apply the R2RML for RDF generating from RDB are requested to learn how to create an R2RML mapping document, in addition to a significant gap between the structure of RDB and the R2RML mappings specifications. One of the ways to solve those problems and ease-of-loading for creating an R2RML document from users efficiently is to generate an initial R2RML mapping document automatically from RDB schema that reflects the conduct of the DM specification. Afterward users will be able to modify that document into a text editor or user interface (display screen). Thus, making the process engine of generating RDF triples (such as morph-RDB [7], nknos [8], RDF-RDB2RDF[1], etc.) takes the R2RML mapping document and RDB data as an input, and then provides an output corresponding RDF dataset (triples). This is done by automatically mapping RDB concepts to an ontology vocabulary, which could be used as a base to support generating RDF triples from RDB data. Recently, the two reports presented by a survey report [9] and W3C's RDB2RDF Implementation [10] are discussed and listed a few existing tools or ongoing projects that have been made available to support the task of mapping generation. However, some of those tools either create mappings in RDB2RDF languages such as ODEMapster GUI (creates R2O mappings) or only give

---

[1] https://metacpan.org/release/RDF-RDB2RDF.

syntactic sugar (form-based tools) to users, who still require a good knowledge of R2RML, which makes them not usable enough.

In this paper, we design and implement algorithms to automatically generate R2RML mapping documents that reflect the behavior of the Direct Mapping specification, which will be applicable as a base support generating the RDF triples from RDB data. Firstly, we design and implement an algorithm that takes an RDB schema as an input and extracts a DBsInfo class (has all the information about RDB schema) as an output. Secondly, we present an algorithm design approach to automatically generate an R2RML mapping document based on a DBsInfo class. Subsequently, generating an RDF dataset by integrating our work with the R2RML processor, which takes an R2RML mapping document and RDB data as inputs and generates the RDF triples as an output. The experimental results show important factors for the building R2RML mappings and their influence on the mapping generation time and size of R2RML and RDF file. These results together reflect the effectiveness of our algorithm and its implementation in Java with Jena API.

The rest of the paper is organized as follows. Section 2 provides an overview of the related works. Basic concepts which give a brief overview of the R2RML and DM with the relationship between them are described in Sect. 3. Section 4 proposes the approach and the algorithm. A prototype implementation of the architecture of our processor prototype and experimental results with discussion regarding the effectiveness and the run-time efficiency test on the proposed algorithms are presented in Sect. 5. Finally, Sect. 6 concludes this paper with the future work.

2   Related Work

Several approaches (auto or manual) have been proposed in the integrating RDB and semantic web, mainly concerning the creation and maintenance of mappings between RDF and RDB. Mapping RDB to RDF is a domain where quite a few works have been proposed over the last years [11]. Generally, the objective is to express the RDB contents using ontology (RDF graph) in a way that allows queries submitted to the RDF schema to be answered with data stored in the RDB. Also, for bringing data residing in RDB into the semantic web, several automated or semi-automated methods for ontology schemes representation have been created [12–14].

Currently, there are two main approaches recommended by W3C RDB2RDF Working Group for mapping RDB into RDF that we have mentioned previously: DM [4] and R2RML [5]. In the DM approach the ontology model is constructed from RDB model, and the contents of the RDB are transformed to generate ontology instances [6,12,15,16]. The approach [6,12] proposed (automatic-direct mapping rules) by investigating several cases of RDB schema to be directly mapped into ontology represented in RDF(S)-OWL and transformed RDB data to ontological instances (represented in RDF triples) based on the structure of the database schema. While in approach [16] a tool RDB2OWL language for mapping a database into an ontology in a compact notation within the ontology class and property annotations was presented. This tool was implemented by converting the RDB2OWL mappings into executable D2RQ mappings to produce the RDF dump of the source RDB, or to turn it into an SPARQL endpoint.

On the other side, the customized mapping approach such as ODEMapster [17], Triplify [18], D2R Server [19], and OpenLink Virtuoso [20] lets a domain expert to create a mapping between the relational schema and an existing target ontology, which is used to convert RDB content to RDF. However, early surveys of RDB-to-RDF tools [21] revealed that the tools typically adopt proprietary mapping languages. Triplify [18] offers a Linked Data publishing interface and provides a simplistic approach to publish RDF from RDB. D2R Server [19] is an engine that directly maps the RDB into RDF and uses D2RQ mappings to translate requests from external applications to SQL queries on the RDB. This implementation was first available for the D2R language and later for R2RML. Moreover, there are some tools for implantation DM and R2RML such as r2rml4net[2] and db2triples[3]. The r2rml4net is a library for processing the R2RML mapping documents, which provides functions to load R2RML mapping document and functions to convert relational data to the RDF dump. The db2triples-software tool is an RDB2RDF Antidot[4] Java implementation of the DM specifications and the R2RML for extracting data from RDBs and loading data into an RDF triple store. Recent efforts offered MIRROR system [22] for produce mappings in the R2RML language and an RML mapping language [23], an extension of R2RML, for non-relational sources and the integration of heterogeneous data formats to support XML and JSON data sources expressions in the mappings. In this work, we focus on the RDB schema. Meanwhile, other researches introduced a semi-automatic mapping approach for generating R2RML mappings based on a set of correspondence assertions (mapping between relational metadata and the vocabulary of a domain ontology) defined by domain experts [24,25]. Therefore, the user still needs to draw correspondence assertions (CAs) from the input system (source RDB schema and target ontology/RDF schema) to specify the mapping between them.

Based on the previous literature, mapping generation remains far from well understood and need to be further explored. Therefore, generation of R2RML mapping documents automatically from RDBs becomes an important challenge to avoid appearance of mistakes in R2RML mappings in addition, it reduces the generation time and no need for domain experts.

3   Basic Concepts

This section gives a brief overview of the R2RML and DM with the relationship between them. The W3C has recently standardized the RDB-to-RDF

---

[2] https://bitbucket.org/r2rml4net/core/wiki/Home.

[3] https://github.com/antidot/db2triples.

[4] http://www.antidot.net/.

(RDB2RDF) mapping mechanism and language to bridge the gap between RDBs and the semantic web. These standardized namely Direct Mapping (DM) of relational data to RDF [4] and R2RML: RDB to RDF mapping language [5]. The mapping engine of approaches/tools generates RDF dataset from RDB schema and its instances. The main step in this engine is to decide how to represent RDB schema concepts in terms of RDF classes and properties from tables and columns. This is done by mapping RDB concepts to an ontology vocabulary, to be used as the base to generate a set of RDF triples from relational data.

3.1   R2RML Standard

R2RML is a language for describing customized mappings from a relational database to RDF dataset. The input of an R2RML mapping is an RDB schema and its instance. The output is an RDF graph. This mapping definition is represented as an RDF graph using the R2RML vocabulary and serialized in the RDF Turtle syntax (RDF triple Language) [26] which is the recommended syntax to write R2RML mapping documents. The structure of an R2RML mapping document consists of one or more triples maps, which contains a logical table, a subject map, and a number of predicate-object maps. The logical table can either be an SQL table, an SQL view, or an SQL query statement. The triples map specifies a rule for mapping each row of a logical table to a set of RDF triples. The subject map contains the rules for generating the subject for each row, often represented as an IRI. While the predicate-object map contains the rules for generating a predicate maps and object maps (or referencing object maps) from the values in the table row. The referencing object map allows using the subjects of another triples map as an object. Since both triples maps may be based on different logical tables, it may require a correlation between the logical tables.

Furthermore, a triples map specifies RDF triples corresponding to a logical table while the subject map and the number of predicate-object maps used to specify how the triples should be. So, RDF triples are created by combining the subject map with a predicate map and a (referencing) object map, and applying these three to each logical table row.

3.2   Direct Mapping (DM)

The DM is a notable one as the W3C candidate recommendation [4]. It is default method to translate a relational database (schema and data) to an ontology (OWL/RDF(S) and RDF triples) automatically through directly mapping without user interaction. The ontology represented in OWL/RDF(S) format. The RDF will reflect the exact data model of the relational data, rather than the domain of the data. A direct mapping is typically working by transform each table to a class, column to property, and relationship to an object property. Each row in the table will be transformed to an individual that will be a member of the table's class. The foreign key transformed with a property that links one individual to another. The range of other properties will be literals.

Furthermore, generating IRI (prefix-name space) for the triples of the RDB schema and data (tables, columns in a table, and each row in a table) during the mapping process produced by combining base IRI and table name for each table, and base IRI, table name and column(s) name for each column in table. While the IRI for each row in table produced by combining base IRI and primary key column(s) of the table.

Therefore, a DM is the default and automatic way to translate RDBs into RDF without any input from the user, while R2RML is a mapping language, which allows users to manually define mappings. Thus, the DM can be represented in R2RML, which is accompanied with the DM specification, describing a standard method for generating RDF from RDB without using a customized mapping definition.

4   Approach and Algorithm

In this section, we introduced an approach that provides (RML-BDM) R2RML based on direct mapping rules from RDB to RDF(S)-OWL for automatically generating an R2RML mapping document from an RDB schema. Then any R2RML engine (e.g. nknos-r2rml parser) can be used to create the RDF dataset following the DM specification.

4.1   RDB Metadata Generation (DBsInfo Class)

In this section, we introduce a DBsInfo class, as a representation of an RDB's metadata, to be used as a source of information for RML-BDM generation. Basic information needed to proceed includes table names, view names and columns' properties that include column names, data types, size, and whether the column is nullable, index, primary key (PK), unique key (UK), and/or foreign key (FK). Moreover, the most important information needed when the attribute is FK are Ref_to_Table (reference to table) and Ref_to_Column (reference to column). The processor for producing a DBsInfo class, which contains all the information about the database, is shown in the Fig. 1. This processor has three levels, which contain forth algorithms (classes) to extract all the information about database tables, views, columns, datatype, columns properties, PKs, FKs, UKs, columns index, and relationships between tables through the foreign keys, etc. These algorithms are FillDBsInfo, FillTablesInfo, FillColumnsInfo, and FillTableRelationships.

Briefly, FillDBsInfo is the main algorithm that extracts the general important information about the database and invoking the FillTablesInfo algorithm to represent the functionality of tables and views. Algorithm FillTablesInfo extracts all the information about table and view, in addition to the information of table columns and table relationships with other tables by invoking algorithm FillColumnsInfo and algorithm FillTableRelationships, respectively.

A DBsInfo provides an image of metadata obtained from an existing RDB. The main purpose behind constructing a DBsIno class is to read essential metadata into memory outside the database's secondary storage. In this study, the DBsInfo class is designed to upgrade the semantic level of RDB and to play the role of an intermediate stage for database migration from RDB to RDF acting on both levels: schema translation and data conversion.

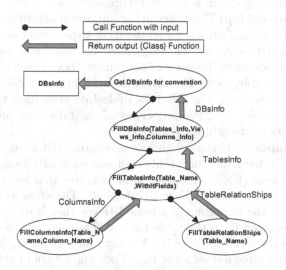

Fig. 1. Processor of algorithms for extracting RDB metadata (DBsInfo).

4.2 Rules of Approach: R2RML-Based Direct Mapping

This section defines the algorithms, which the mapping RDB schema to R2RML file is based on DM approach (RML-BDM). The RDB2RDF algorithm is the core of an R2RML engine. According to the algorithm ideas proposed in W3C recommendation R2RML [5], DM [4] and our previous works [6,12], we have designed a group of mapping algorithms, **R2RML Generator** (Algorithm 1), **GreateMapClass** (Algorithm 2), **GenerateLogicalTable** (Algorithm 3), **GenerateSubjectMap** (Algorithm 4), **GenerateTempate** (Algorithm 5), **GeneratepredicateObject** (Algorithm 6), and **GenerateRefObjMap** (Algorithm 7), to achieve the R2RML-BDC mapping file. Concisely, **R2RML Generator** is the main algorithm invoking the other algorithms to implement the functionality of R2RML triples map.

- **R2RML Generator:** This algorithm is the main algorithm to generate the R2RML mapping file based on the direct approach mapping.
- **GreateMapClass:** This algorithm creates map class name from the table/view name. The map class name is a triples map that used to translate each row in the logical table to number of RDF triples and link-connect between triples map classes (classes corresponding to tables that have the

relationships with each other). The output is a map class name corresponding table/view name.

- **GenerateLogicalTable:** This algorithm maps the table/view into logical table using the r2rml format depending on DM method, where the table name of RDB is rr:tableName in the logical table. This reflects which table/view is mapped to generate triples from its rows by using an R2RML parser. Then the return of this algorithm is the RDF triples.

- **GenerateSubjectMap:** This is one of the important algorithms. It generates the unique IRI used as a subject for all the RDF triples generated from the row of the table in rr:tableName. This algorithm invokes the GenerateTemplate algorithm to the identification form of the primary key of each triple generated from the row of table/view. The algorithm's input is a table/view name of DBsInfo.TablesInfo class. The DBsInfo.TablesInfo class used to store all the columns of the table/view, the properties of the table and properties of its columns, and its relationships to other tables.

- **GenerateTemplate:** This algorithm of generating IRI format for all triples from base IRI, table name, and table columns especially from primary keys (PKs), unique keys (UKs), or collect some columns that are not-null (when the table does not have any PK or UK defined). Therefore, this algorithm is characterized by the formation of a basic IRI key that is unrepeatable for all the triples generated from the table rows, where each row maps to a set of triples refer to the same subject (IRI key row) and all the triples of the table rows refer to the table name in rr:Class. Then the return of this algorithm is the RDF triples format in r2rml.

- **GeneratepredicateObject:** This algorithm maps the table/view column to PredicateObjectMap that includes a pair of predicate and object map. It generates the RDF terms for the predicate and object of a triple respectively. The value of rr:predicate is IRI consists of base IRI, table name and the columns name which are the algorithm inputs. The rr:object is a column name. Then the return of this algorithm is the RDF triples format in r2rml that will be associated with a subject (generated by the GenerateSubjectMap algorithm).

- **GenerateRefObjMap:** This algorithm maps all the table relationships to the reference triples, which are generated for *referencing object maps*, through a rr:joinCondition to another table similar to local triples. The *referencing object map* allows using the subjects of another triples map as an object (produced by a predicate-object map). Since both triples maps may base on different logical tables, it may require a link between the logical tables. All relationships of the table are stored in the DBsInfo.TableRelationShips, including FK_Table_Name, FK_Column_Name, Ref_To_TableName, and Ref_To_Column_Name, which are the inputs of algorithm. Then the return of this algorithm is the RDF triples format in r2rml corresponding to relationship of table with other tables. The output of the algorithm is associated with the objectMap generated by a PredicateObjectMap.

Algorithm 1. R2RML Generator (DBsInfo, NS Prefix)

```
1: Input: DBsInfo Has all information of database schema, NS Prefix;
2: Output: R2RML Mapping File: RDF Dataset;
3: Var
4: TBs: TableInfo[] //Class as List to save list information of tables in DBsInfo;
5: Col: ColumnsInfo[] //Class as List of table columns information;
6: TBR: TableRelationShipInfo[]//Class as List of table relationships;
7: Triples:String //to save triple generated;
8: Begin
9:   Triples ← θ;
10:  TBs ← DBsInfo.Schema_TableInfo;
11:  for each table T in TBs do
12:      Triples ← CreateMapClass (T.Table_Name) ;
13:      Triples ← Triples U GenerateLogicalTable (T.Table_Name, T.Table_Type) ;
14:      Triples ← Triples U GenerateSubjectMap (T.Table_Name, T) ;
15:      Col ← T.Table_ColumnInfo;
16:      for each Column C in Col do
17:          Triples ← Triples U GeneratepredicateObjectMap (T.Table_Name, C.Column_Name) ;
18:      end for
19:      TBR ← T.Table_RelationShipInfo;
20:      for each TableRealtionShip TR in TBR do
21:          Triples ← Triples U GenerateRefObjectMap(TR.getForeignKeyTable_Name (),
22:                  TR.getForeignKeyColumn_Name (), TR.getReferenceToPKTable_Name (),
23:                      TR.getReferenceToPKColumn_Name ());
24:      end for
25:  end for
26:  Triples ← Triples U "." + "\n\n";
27:  R2RMLMapping File ← Triples;
28:  Return R2RML Mapping File // to Save it in RDF File Format Or TTL
29: End
```

Algorithm 2. CreateMapClass(Table_Name)

```
1: Input: Tabl_Name: Name of table or view that will be mapped;
2: Output: MapClassName : Triples //ex map:Students;
3: Begin
4:   MapClassName ← "map : " + Table_Name + "s" + "\n";
5:   return MapClassName;
6: End
```

5   Prototype Implementation

5.1   Architecture

Figure 2 shows the architecture of our RML-BDM processor prototype. Depending on the proposed algorithms, we have implemented an R2RML-BDB processor prototype and have been integrated with nknos-r2rml parser[5] [8]. The processor takes system configuration, a DB connection to the relational database and a base IRI as inputs and produces automatically the R2RML mapping document and resulting RDF dataset as outputs shown by screen display.

The architecture and process flow of the R2RML-BDB processor prototype is illustrated in Fig. 2, where the functional modules are briefly described as follows.

---

[5] https://github.com/nkons/r2rml-parser.

Algorithm 3. GenerateLogicalTable(Table_Name,Table_Type)

1: **Input**: Tabl_Name: Name of table or view that will be mapped;
2:              Table_Type: table or view;
3: **Output**: Triples :Map table or view to Triple of LogicalTable;
4: **Begin**
5: $Triples \leftarrow \theta$;
6: $Triples \leftarrow Triples\ U\ "rr : logicalTable[rr : tableName'" + "\"" + Table\_Name + "\"" + "';];"$;
7: $Triples \leftarrow Triples\ U\ "\mathbf{\backslash n}"$;
8: **return** Triples;
9: **End**

Algorithm 4. GenerateSubjectMap(Table_Name,T)

1: **Input**: Tabl_Name: Name of table or view;
2:              T : has all table Information (DBsInfo.Schema_TableInfo);
3: **Output**: Triples :Map Row ID in table to subjectMap;
4: **Begin**
5: $Triples \leftarrow \theta$;
6: $Triples \leftarrow Triples\ U\ "rr : subjectMap[" + "\mathbf{\backslash n}"$;
7: $Triples \leftarrow Triples\ U\ GenerateTemplate(Table\_Name, T)$;
8: $Triples \leftarrow Triples\ U\ "rr : class\ \ NS : " + Table\_Name + "; \mathbf{\backslash n}"$;
9: $Triples \leftarrow Triples\ U\ "];"$;
10: $Triples \leftarrow Triples\ U\ "\mathbf{\backslash n}"$;
11: **return** Triples;
12: **End**

- *System config module:* This module configures the execution environment for the R2RML-BDB processor according to the user-specified settings, including:
    1. DB config: This is used to specify all parameters for connection to any database.
    2. R2RML file input type: It is used to specify the file name and type of R2RML document that will be used for storing an R2RML schema generated from RDB schema and then used it with any R2RML Parser to produce RDF triples from RDB data.
    3. RDF triples output type: It is used to specify the name file and type (format) of RDF graph to be used for storing RDB data as RDF graph format.
    4. Base IRI (NS Prefix): This NS-IRI is used to specify the namespace prefix of IRI for all the RDF triples.
- *DB connection:* This module uses to connect with the database (using JDBC driver engine in Java) and make it ready for reading. The input is DB config parameters and the output is DB-connection class.
- *DB analysis processor:* This module implements the algorithms of extracting RDB metadata (DBsInfo) (Fig. 1) from RDB. Metadata is extracted from RDB using JDBC driver engine in Java. The output is DBsInfo class that has many classes to store all the information about RDB schema such as

Algorithm 5. GenerateTemplate(Table_Name,T)

1: **Input**: Tabl_Name, T;
2: **Output**: Triples: Map Row ID in table to subjectMap;
3: **Var**
4: **Cols_Template** ← θ; //Array to save List of columns to generate Row Id as SubjectMap;
5: **NotNulls** ← θ; // Array to save List of columns are not null;
6: **PKs_Count** ← T.Table_ColumnPKs.length;
7: **UNQs_Count** ← T.Table_ColumnUniques.length;
8: **Begin**
9: Triples ← θ;
10: **if** (PKs_Count == 0 && UNQs_Count == 0) **then**
11:      NotNulls=getNotNullColumns(idx_TBs);
12: **end if**
13: **if** (PKs_Count > 0) **then**
14:      Cols_Template=DBs_Info.Schema_TableInfo[idx_TBs].Table_ColumnPKs;
15: **else if** (UNQs_Count > 0) **then**
16:      Cols_Template=DBs_Info.Schema_TableInfo[idx_TBs].Table_ColumnUniques;
17: **else if** (NotNulls.length > 0) **then**
18:      Cols_Template=NotNulls;
19:      **if** (Cols_Template.length > 2) **then**
20:          Cols_Template=getPartoFColumnsLeftPictureFileds(idx_TBs,Cols_Template,3);
21:      **end if**
22: **else**
23:      Cols_Template=ConvertArrayListToStringArray(DBs_Info.Schema_TableInfo[idx_TBs].Columns_Name);
24:      **if** (Cols_Template.length > 2) **then**
25:          Cols_Template=getPartoFColumnsLeftPictureFileds(idx_TBs,Cols_Template,3);
26:      **end if**
27: **end if**
28: Triples ← Triples U "rr : template'" + pfx.NS + Table_Name + "_";
29: **for** each column Col in Cols_Template **do**
30:      Triples ← Triples U "{\"" + Col + "\"}";
31: **end for**
32: Triples ← Triples U "';";
33: Triples ← Triples U "\n";
34: **return** Triples;
35: **End**

Algorithm 6. GeneratepredicateObjectMap(Table_Name,Column_Name)

1: **Input**: Tabl_Name, Column_Name;
2: **Output**: Triples :Map Column to Triple of PredicateObjectMap;
3: **Begin**
4: Triples ← θ;
5: Triples ← Triples U""rr : predicateObjectMap[" + "\n";
6: Triples ← Triples U "rr : predicate   NS : " + Table_Name + "." + Column_Name + "; \n";
7: Triples ← Triples U "rr : objectMap[rr : column\"" + Column_Name + "\"]" + "\n";
8: Triples ← Triples U "];";
9: Triples ← Triples U "\n";
10: **return** Triples;
11: **End**

Algorithm 7. GenerateRefObjectMap(FK_Table_Name, FK_Column_Name, Ref_To_TableName, Ref_To_Column_Name)

1: **Input**: FK_Table_Name: Name of table that has FK;
2:         FK_Column_Name : Name of column that is FK;
3:         Ref_To_TableName: Name of table referred to it;
4:         Ref_To_Column_Name: Name of column referred to it;
5: **Output**: Triples :Map relationships to Triples of RefObjectMap;
6: **Begin**
7: $Triples \leftarrow \theta$;
8: $Triples \leftarrow Triples\ U\ "rr : predicateObjectMap[" + "\backslash\mathbf{n}";$
9: $Triples \leftarrow Triples\ U\ "rr : predicate\ \ NS : " + FK\_Table\_Name + "." + FK\_$
    $Column\_Name + ";" + "\backslash\mathbf{n}";$
10: $Triples \leftarrow Triples\ U\ "rr : objectMap[" + "\backslash\mathbf{n}";$
11: $Triples \leftarrow Triples\ U\ "a\ \ \ rr : RefObjectMap;" + "\backslash\mathbf{n}";$
12: $Triples \leftarrow Triples\ U\ "rr : parentTriplesMap" + CreateMapClass(Ref\_To\_$
    $TableName) + ";" + "\backslash\mathbf{n}";$
13: $Triples \leftarrow Triples\ U\ "rr : joinCondition[" + "\backslash\mathbf{n}";$
14: $Triples \leftarrow Triples\ U\ "rr : child\backslash"" + FK\_Column\_Name + "\backslash";" + "\backslash\mathbf{n}";$
15: $Triples \leftarrow Triples\ U\ "rr : parent\backslash"" + Ref\_To\_Column\_Name + "\backslash";]" + "\backslash\mathbf{n}";$
16: $Triples \leftarrow Triples\ U\ "];" + "\backslash\mathbf{n}";$
17: $Triples \leftarrow Triples\ U\ "];";$
18: $Triples \leftarrow Triples\ U\ "\backslash\mathbf{n}";$
19: **return** Triples;
20: **End**

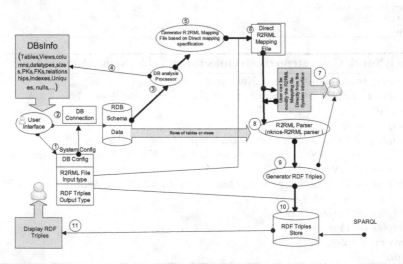

Fig. 2. A general overview of the R2RML mapping generation process in RML-BDM system.

tables, views, columns, data types, sizes, constraints, PKs, FKs, relationships, indexes, unique, and nulls, etc.

- *Generator R2RML Mapping File:* This module is used to automatically generate an R2RML mapping file based on the behavior of the DM specification from the DBsInfo class, according to our approach algorithms. The input of this processor is DBsInfo (contains all information about RDB schema) the output is R2RML mapping file formatted as rdf format (or TTL). The output file can encapsulate all mapping results into a standard input for any R2RML processor later to produce a set of RDF triples that is similar to those resulting from DM.

- *R2RML Parser (RDF processor):* It is used to generate a real RDF triples file from RDB data depending on R2RML mapping file to make it accessible to RDF store. Moreover, this stage is to satisfy our approach for generating R2RML mapping file. We used the open source tool nkons-r2rml-parser (or use any other R2RML processor) which is integrated with our RML-BDM system to generate a set of triples that correspond to the ones generated by DM approach.

- *Screen display (user interface):* The user can specify the configuration setting for execution environments (system config module), display the database information-schema (DBsInfo class), R2RML mapping file, and the resulting RDF triples on the tool screen through the user interface.

5.2   Implementation

The algorithms described in this paper have been implemented in RML-BDM processor prototype and integrated with nkons-r2rml-parser [8]. This prototype has been implemented on Netbeans IDE 7.3.1 (J2SE, JDK 1.7) platform. Thus, the inputs of processor are user-specific configuration system, a SQL connection to the relational database, and a base IRI. Meanwhile, the outputs are produced automatically, including the DBsInfo class, an R2RML mapping document and resulting RDF dataset. These outputs are shown in screen display, and the R2RML mapping document and RDF triples can be saved in an RDF file in different syntax formats (RDF/XML, N-TRIPLES, TURTLE (or TTL), and N3). Moreover, the RDF mapping file can be used with any R2RML parser for converting relational data to RDF triples. Current system prototype supports SQL connections to MySQL Server and already included drivers for major commercial and open source databases, including Postgres, SQL Server and Oracle.

5.3   Experimental Results and Discussion

We carried out RDB metadata and R2RML mapping extraction experiments with our R2RML-BDC tool on a Laptop with configurations as Windows 7 (32-bit), CPU Intel(R) Core i5-2410M 2.30 GHz, RAM 6 GB. A prototype for this experiment is implemented using MYSQL, Java programming language, Netbeans IDE 7.3.1 and Apache Jena tools. Experimental tests on the effectiveness and validity of our RDB2RDF mapping algorithms were conducted with the

164   M.A.G. Hazber et al.

Table 1. A list of RDB schema and data sizes

RDB	SizeDB (kb)	SizeDB Schema(kb)	View	Table	Column	FKs	PKS	Rows
rdblab	100035.2	160	0	6	16	5	7	100200
Iswc	20176	256	0	9	46	11	13	90000
Tracker	20000	1056	0	25	162	38	25	95003
Sakila	9132.26	625	7	16	131	22	18	92227
Norhwind	14048	576	16	13	191	13	16	120943

Table 2. A list of results for our approach

RDB	Time of extract metadata + GenerateR2RMLFile(ms)	SizeR2ML (kb)	R2RML Tuples	SizeRDF (mb)	RDF triples
rdblab	295	6.45	181	58.9	752916
Iswc	685	14.8	387	74.7	756094
Tracker	3551	50	1276	103	1083491
Sakila	3697	37.8	956	88	953581
Norhwind	4069	50.1	1141	133	1278842

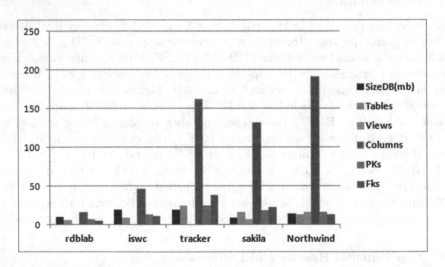

Fig. 3. Important factors of RDBs to build R2RML mapping file. (Color figure online)

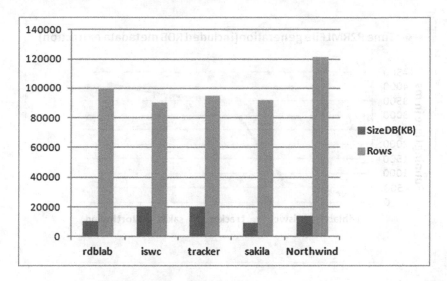

Fig. 4. Dataset sizes in RDBs (schema and data). (Color figure online)

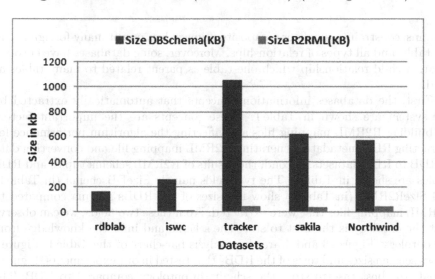

Fig. 5. Schema sizes in RDBs and R2RML mapping files. (Color figure online)

schema sizes of the five RDBs created with MYSQL Server and tested in our experiments. These RDBs are rdblab [15], iswc[6], tracker[7], sakila[8], and Northwind[9], which covering various of important RDB concepts such as tables, views,

---

[6] http://d2rq.org/example/iswc-mysql.sql.
[7] http://www.artfulsoftware.com/mysqlbook/sampler/mysqled1_appe.html#5-1.
[8] http://mysql-tools.com/en/downloads/mysql-databases/4-sakila-db.html.
[9] https://code.google.com/p/northwindextended/downloads/detail?
name=Northwind.MySQL5.sql.

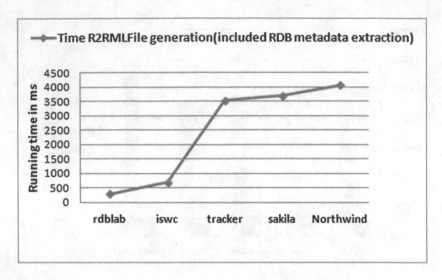

Fig. 6. The running time of algorithmic routines in RML-BDM.

columns, constraints, single or composite primary key, one or many foreign key in one table, and all types of relationships. Moreover, some databases have encoded a parent-child relationship which one table as parent related to many tables as child.

First, the databases information concepts that automatically extracted by our system are shown in Table 1. These concepts are the important factors for building R2RML mapping files and affecting the algorithm performance for extracting RDB metadata, generating R2RML mapping file and converting data of RDB to RDF datasets. Second, the results of R2RML schema tuples and RDF triples are shown in Table 2. The two fields namely SizeDBschema (in Table 1) and SizeR2RML (in Table 2) show the sizes of the RDBs schema compared to R2RML mapping files that were extracted. From these two fields, we can observe that the ontology is the best to store the schema and infer the knowledge from the ontology. Figures 3 and 4 are the analysis bar-chart of the Table 1. Figure 4 shows schema sizes and rows of the RDBs that tested in our experiments. Figure 5 reveals the best case to store the schema in ontology compared to RDB. The X-axis used to show the different domain datasets, whilst Y-axis shows the size of the database in kilobytes (kb). Moreover, the performance analysis of the different databases is shown in the Fig. 6. The execution time of creating R2RML mapping file included the running time for extracting RDB metadata-DBsInfo (RDB have data) and generating R2RML mapping file in algorithm **R2RML Generator**. Therefore, from the tables and figures analyses, we can conclude what are the most important factors for extracting R2RML mapping document from RDB and affecting the time of extraction. Although the size of tracker database greater than the size of sakila and northwind database, but the execution time of **R2RML Generator** algorithm is smaller, because there are other factors have influenced the execution time an R2RML mapping file generator with the size of the database.

6   Conclusion and Future Work

We introduced an approach tool for the automatically generating an R2RML mapping document from an RDB schema, and any R2RML engine can be use this mapping document to create a set of RDF dataset that following the DM specification. RML-BDM is a tool that enables domain experts and non-experts to automatically create an R2RML mapping files from RDBs by using the R2RML format. RML-BDM has been integrated with nknos-r2rml parser that can export RDB contents as RDF graphs, based on an R2RML mapping document.

The process was tested using five RDBs having different sizes of schema and covering several RDB concepts and types of relationships between tables. In future works, we will add other tools to address an expressing customized mappings from various types of data sources such as XML, NoSQL, and object-oriented to an RDF triples.

Acknowledgments. This work is supported by National Natural Science Foundation of China under grants 61572221, 61173170, 61300222, 61370230, 61433006 and U1401258, Innovation Fund of Huazhong University of Science and Technology under grants 2015TS069 and 2015TS071, Science and Technology Support Program of Hubei Province under grant 2014BCH270 and 2015AAA013, and Science and Technology Program of Guangdong Province under grant 2014B010111007.

References

1. Shadbolt, N., Hall, W., Berners-Lee, T.: The semantic web revisited. IEEE Intell. Syst. **21**(3), 96–101 (2006)
2. Manola, F., Miller, E., McBride, B.: RDF 1.1 Primer. W3C Working Group Note, 24 June 2014 (2014)
3. He, B., Patel, M., Zhang, Z., Chang, K.C.-C.: Accessing the deep web. Commun. ACM **50**(5), 94–101 (2007)
4. Arenas, M., Bertails, A., Prud, E., Sequeda, J.: A direct mapping of relational data to RDF. W3C Recommendation, 27 September 2012. http://www.w3.org/TR/rdb-direct-mapping/
5. Souripriya Das, O., Seema Sundara, O., Cyganiak, R.: R2RML: RDB to RDF mapping language. W3C. http://www.w3.org/TR/r2rml/
6. Hazber, M.A.G., Li, R., Zhang, Y., Xu, G.: An approach for mapping relational database into ontology. In: Proceedings of the 12th Web Information System and Application Conference (WISA 2015), Jinan, Shangdong, China, pp. 120–125 (2015)
7. Priyatna, F., Corcho, O., Sequeda, J.: Formalisation and experiences of R2RML-based SPARQL to SQL query translation using morph. In: Proceedings of the 23rd International Conference on World Wide Web (WWW 2014), New York, USA, pp. 479–490 (2014)

8. Konstantinou, N., Kouis, D., Mitrou, N.: Incremental export of relational database contents into RDF graphs. In: Proceedings of the 4th International Conference on Web Intelligence, Mining and Semantics (WIMS 2014), Thessaloniki, Greece, p. 33 (2014)
9. Michel, F., Montagnat, J., Faron-Zucker, C.: A survey of RDB to RDF translation approaches and tools. Research Report. I3S (2014). https://hal.inria.fr/hal-00903568
10. Villazn-Terrazas, B., Hausenblas, M.: RDB2RDF Implementation Report. W3C Working Group Note, 14 August 2012. http://www.w3.org/TR/rdb2rdf-implementations/
11. Sahoo, S.S., Halb, W., Hellmann, S., Idehen, K., Thibodeau Jr., T., Auer, S., Sequeda, J., Ezzat, A.: A survey of current approaches for mapping of relational databases to rdf. W3C RDB2RDF XG Incubator Report W3C (2009)
12. Mohamed, H., Jincai, Y., Qian, J.: Towards integration rules of mapping from relational databases to semantic web ontology. In: 2010 International Conference on Web Information Systems and Mining (WISM), Sanya, China, pp. 335–339 (2010)
13. Vavliakis, K.N., Grollios, T.K., Mitkas, P.A.: RDOTE publishing relational databases into the semantic web. J. Syst. Softw. **86**(1), 89–99 (2013)
14. Sequeda, J.F., Arenas, M., Miranker, D.P.: On directly mapping relational databases to RDF and OWL. In: Proceedings of the 21st international conference on World Wide Web (WWW 2012), Lyon, France, pp. 649–658 (2012)
15. Hazber, M.A.G., Li, R., Gu, X., Xu, G., Li, Y.: Semantic SPARQL query in a relational database based on ontology construction. In: Proceedings of the 11th International Conference on Semantics, Knowledge and Grids (SKG 2015). Beijing, China (2015)
16. Čerāns, K., Būmans, G.: RDB2OWL: a language and tool for database to ontology mapping. In: Proceedings of the CAiSE 2015 Forum at the 27th International Conference on Advanced Information Systems Engineering (CAiSE 2015), Kista, Sweden, pp. 81–88 (2015)
17. Barrasa, J., Corcho, Ó., Gómez-Pérez, A.: R2O, an extensible and semantically based database-to-ontology mapping language. In: Second Workshop on Semantic Web and Databases(SWDB 2004), pp. 1069–1070 (2004)
18. Auer, S., Dietzold, S., Lehmann, J., Hellmann, S., Aumueller, D.: Triplify: lightweight linked data publication from relational databases. In: : Proceedings of the 18th International Conference on World Wide Web (WWW 2009), pp. 621–630 (2009)
19. Bizer, C., Cyganiak, R.: D2R server-publishing relational databases on the semantic web. In: Presented at the 5th International Semantic Web Conference (ISWC 2006), Athens, GA, USA (2006)
20. OpenLink Virtuoso Universal Server. http://virtuoso.openlinksw.com
21. Sequeda, J.F., Tirmizi, S.H., Corcho, O., Miranker, D.P.: Survey of directly mapping sql databases to the semantic web. Knowl. Eng. Rev. **26**(4), 445–486 (2011)
22. de Medeiros, L.F., Priyatna, F., Corcho, O.: MIRROR: automatic R2RML mapping generation from relational databases. In: Cimiano, P., Frasincar, F., Houben, G.-J., Schwabe, D. (eds.) ICWE 2015. LNCS, vol. 9114, pp. 326–343. Springer, Heidelberg (2015)
23. Dimou, A., Vander Sande, M., Colpaert, P., Verborgh, R., Mannens, E., Van de Walle, R.: RML: a generic language for integrated RDF mappings of heterogeneous data. In: Proceedings of the 7th Workshop on Linked Data on the Web (LDOW 2014), Seoul, Korea (2014)

24. Neto, L.E.T., Vidal, V.M.P., Casanova, M.A., Monteiro, J.M.: *R2RML by assertion*: a semi-automatic tool for generating customised R2RML mappings. In: Cimiano, P., Fernández, M., Lopez, V., Schlobach, S., Völker, J. (eds.) ESWC 2013. LNCS, vol. 7955, pp. 248–252. Springer, Heidelberg (2013)
25. Pequeno, V.M., Vidal, V.M., Casanova, M.A., Neto, L.E.T., Galhardas, H.: Specifying complex correspondences between relational schemas and RDF models for generating customized R2RML mappings. In: Proceedings of the 18th International Database Engineering & Applications Symposium, Porto, Portugal, pp. 96–104 (2014)
26. Prud'hommeaux, E., Carothers, G., Beckett, D., Berners-Lee, T.: RDF 1.1 Turtle: terse RDF triple language. World Wide Web Consortium. http://www.w3.org/TR/turtle/. Accessed 24 Dec 2014

An Improved Asymmetric Bagging Relevance Feedback Strategy for Medical Image Retrieval

Sheng-sheng Wang[1,2(✉)] and Yan-ning Shao[1]

[1] College of Computer Science and Technology, Jilin University, Changchun, China
wss@jlu.edu.cn, shaoyanning@outlook.com
[2] College of Software, Jilin University, Changchun, China

Abstract. Much attention has been paid to relevant feedback in intelligent computation for social computing, especially in content-based image retrieval which based on WeChat platform for the medical auxiliary. It has a good effect on reducing the semantic gap between high semantics and low semantics of images. There are many kinds of support vector machines (SVM) based relevance feedback methods in image retrieval, but all of them may encounter some problems, such as a small size of sample, an asymmetric positive sample and negative sample as well as a long feedback cycle. To deal with these problems, an improved asymmetric bagging (IAB) relevance feedback algorithm is proposed. Furthermore, we apply a new fuzzy support machine (FSVM) to cooperate with IAB. To solve the over-fitting and real-time problems, we use modified local binary patterns (MLBP) as image features. Finally, experimental results demonstrate that our method performs other methods in terms of improving retrieval precision as well as retrieval efficiency.

Keywords: Social computing · Content-based image retrieval · Fuzzy support vector machine · Relevance feedback · Improved asymmetric bagging

1 Introduction

With the rapid development of digital technology, a large amount of data has been produced every day, most of which is multimedia data that is related to images and videos. This deluge of data is often referred as big-data, we need technical methods to help patients getting the image analysis from WeChat which is based on intelligent computation social computing. At the same much more methods need to be discovered for processing these massive images. However, due to the strong subjectivity and high time complexity of human annotation, traditional image retrieval technology cannot work well. In view of these reasons, we proposed our methods. In contrast, content based image retrieval (CBIR) [1] has become a leading image retrieval method considering its advantage in processing and analyzing images automatically from low-level to high-level. Soon after, a new framework including interaction module and user feedback module, related feedback [2], is put forward. Relevance feedback technique means a user can be introduced into the information retrieval process, which makes retrieval model interactive as many times as possible. It has further become the effective method

© Springer Science+Business Media Singapore 2016
W. Che et al. (Eds.): ICYCSEE 2016, Part I, CCIS 623, pp. 170–181, 2016.
DOI: 10.1007/978-981-10-2053-7_16

to improve the retrieval performance. In the process of relevance feedback, users just need to judge whether the retrieved result is relevant or not to the query. Then the system is based on the feedback from the users to learn and give better results.

In the past decade, research on relevance feedback technology has become an active academic domain, many kinds of relevance feedback algorithms are constantly developing. AULRFS [3] utilize the distance of sample to adapt to the user's aim. FW [4, 5] by adjust feature weight to improve the effect of retrieval. These methods play a limited role to enhance the feedback. From another angle, Researchers have put forward many algorithms by regarding the relevance samples and the irrelevance samples as two difference sets and aiming at finding a classifier to identify these two sets from each other. In other words, it becomes a classification problem. Relevance feedback based on SVM has become a hotspot in CBIR; because of the SVM based relevance feedback (RF) has shown hopeful results just like that generalization ability. SVM is good at pattern classification issues by minimizing the Vapnik-Chervonenkis dimension and achieving a minimal structural risk [9]. Recently, SVM-based relevance feedback [6, 7] has overcome the problem of small sample. Wu [10] proposed to use clustering to process the feedback samples but neglect the imbalance of positive samples and negative samples. Also many methods were proposed to keep balance of feedback samples [8].

Among these methods, the SVM [6–8] based relevance feedback achieve good effect owing to its outstanding generalization ability. However, in the field of medical auxiliary image retrieval the performance of these methods which based on SVM may become poor as refer to the retrieval of medical image, and all these methods will not effective when dataset too big, even result in misclassification. The number of negative feedback samples is too big. This is mainly due to the following reasons: (1) Problem of over-fitting, in relevance feedback samples, the dimension of feature vectors maybe much bigger than the size of training samples which may cause an over-fitting problem. (2) Optimal hyperplane of SVM may be biased when the negative feedback samples are much more than the positive feedback samples [11]. As in the process of relevance feedback, there are always fewer positive feedback samples than negative ones. On account of the imbalance of the training samples for the two classes, optimal hyperplane of SVM will be biased toward the negative feedback samples. Accordingly, SVM RF may mistake many query irrelevant images as relevant ones. (3) Feature problem, feature is necessary procedure for CBIR, yet there is no effective feature extraction algorithm for the CBIR of medical domain. (4) Problem of real-time, because the limited amount of feedback samples and there is no taking full advantage of feedback information to mining unmarked image information. There is no effective algorithm to add samples each time results in the problem of real-time and accuracy in the medical image sample field.

Even though, [8] using bagging and random subspace method to improve SVM whereas, the serious slow training problem exists and not stable for medical image retrieval according to our tests. So, in order to solve the above problems, we present an improved asymmetric bagging (IAB) relevance feedback approach for solving the small sample problem, the over-fitting problem and the biased hyper-plane problem. Our methods by using the density of feedback samples, and combine KNN to increase the feedback samples. The approach makes full use of positive and negative samples to mining more training samples from unmarked image [12] sets to make sure that the number of positive samples equal to the number of negative samples.

The remaining part of this paper is organized as follows: In Sect. 2, the architecture of our image retrieval system is introduced. We propose our algorithms base on SVM in Sect. 3. In Sect. 4, we introduce the process of feature extraction that paper can use. Section 5, By means of experiment demonstrate the effectiveness of the novel method that we proposed, with using the MIVIA dataset.

2   System Framework

The new framework of the system combines improved relevance feedback methods with FSVM for medical image retrieval, the system is designed for the user to obtain feedback from samples, and users should be familiar with basic knowledge of image. The details description of the develop method framework is shown in Fig. 1 and overall framework of the system is described as follows:

1. Users should submit query image for which they want to search, the system employs, by using Euclidean distance to retrieve similar images. First, the retrieved image is shown on the user interface ranked from high to low similarity, this phase is called the initial training phase that is executed only once. The Top N = 24 interfaces is shown in Fig. 5 and the interaction process is shown in Fig. 1.
2. The system returns N images and then they are evaluated by the users who have some knowledge about medical cells. Certainly, this system just needs a relevance tag and the unmarked images are assigned as irrelevant. According to the tag, the IAB is used to process these images. The flow chart is shown in Fig. 1.

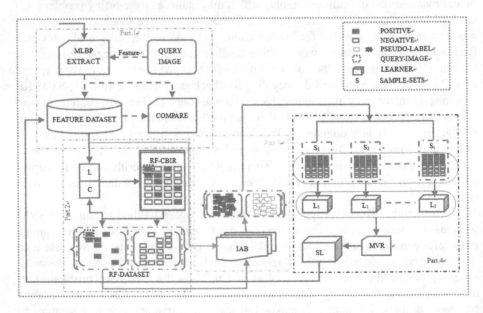

Fig. 1. The Framework of the System

3. Using the previous feedback samples to train a SVM, at the same time, the system make use of the marked image from the previous feedback to increase the next training set of sample by using the IAB method. Then a strong classier is trained based on a hybrid training samples by using MVR 13.
4. Repeat steps 2 and 3 until a stable retrieved result.

3   Algorithm Description

In this section, we give a specific description to the algorithm. The feedback process will be iterated until we get the satisfactory results. The system flow chart is shown in Fig. 1.

Due to asymmetric bagging [8] method ignores the differences in the sets of feedback samples and fails to aim at these issues, we employ the IAB based on an evaluated function to filter the negative feedback that has large correlation and then we obtain the evaluated samples. By eliminating the redundant negative feedback samples effectively, the method can solve the instability problems of the classification process. This method is proposed for SVM and the purpose is to obtain a large amount of information from the negative feedback sample. In the negative feedback sample set, select the minimum negative feedback and the average minimum positive feedback samples, then we obtain the average minimum distance between the negative and positive feedback samples. The evaluation function gives the number of positive and negative feedback samples. We define the evaluation function as follows:

$$R_i = \frac{S_{ni} - S_{pc}}{\max(S_{ni} - S_{pc})} \quad (i \in n) \tag{1}$$

Where, S_{ni} denotes ith negative feedback sample, S_{pc} denotes the center of positive feedback sample, n denotes the number of positive feedback sample. The expression of $max(S_{ni}\text{-}S_{pc})$ indicate to get the maximum that is the far distance from negative sample to positive center. A large dataset will lead to the same problems in [13], such as instable classifier, low classification accuracy and iterations. This is not good for the use of integrated learning device and cannot help get good results. To keep IAB on solving the problem of small sample, then we proposed to use the density of feedback samples to increase training samples. So IAB can reduce the number of user feedback and also can deal with negative sample asymmetry. Therefore, we propose cluster to calculate density of sample, and according to the result we select the unmarked image to add in training set. The input sample data is obtained by the above algorithm. S_p and S_n are obtained from the hybrid training sets. The defined functions get evaluated feedback samples from the collection of samples that is the feedback from the user. I is a base classifier, t is the number of classifiers, $R_i(S_n)$ is negative feedback that is filtered from the original collection of the negative collections. Then those samples are used in subsequent processing, the last step is to achieve classifier C^*. The detail of the algorithm is shown as follows:

```
Input: Sn ,Sp
Output: C*
1. Sn=Sun+Spln ; Sp=Sup+Splp
2. Init use Euclidean get Top N
3. Evaluate related or negative & learner trained
4. Learner engine select image for evaluate
5. Perform Clustering
```

6. $S_{ni}, S_p \in \{S_{un}, S_{up}, S_{pln}, S_{plp}\}$ hybrid training samples formed, where S_{un}, S_{up} are got from user and S_{pln}, S_{plp} are got from S_{un}, S_{up} .

```
7.  from i=1 to t
    {
      Sni=get subset from Ri (Sn)
      Term |Sni|=|Sp| & R(Sn) ⊂ Sn
      Ci=I(Sni ,Sn)
    }
```

8. $C^*(x) = C_t (x, S_{ni}, S_p)$ where $i \in [1, t]$

In order to use the extension of SVM, for user feedback and learning sample, we use different fuzzy membership functions. Specific new fuzzy function (2) and (3) for a feedback sample is defined as follow:

$$W(x_p) = \frac{n}{\sum\limits_{i=1}^{n} d(x_p - x_i)} \qquad (2)$$

Where, x_p is the relevant samples, x_i denotes arbitrarily samples, n is the number of feedback sample and $d(x_p-x_i)$ represents the distance between the sample and the average value of the sample. The sign $w(x_p)$ is used to measure the correlation degree between the feedback sample and the increased sample that according to intensity degree of feedback sample. The expression $w_2(x_p)$ in [9] and the quantity $w_3(x_p)$ indicates the distance from the center to the classification surface of the particle size. Eq. (3) is the membership function of the integrated samples and $G(x_p)$ is the membership degree.

$$G(x_p) = w(x_p) * w_1(x_p) * w_2(x_p) \qquad (3)$$

This paper puts forward IAB to solve the problem of negative feedback sample and small sample. The new sample center selection method can increase the diversity of the training sample set to enhance the differences between meta classifiers. This section describes the process of combining IAB with the new fuzzy support vector machine (FSVM). To this end, we need a new similarity measurement method, based on the fuzzy membership function $G(x_p)$, a new framework of IAB-FSVM retrieval system is formed, and the effect is detailed in the experimental section.

After the IAB, we obtain the hybrid training sets that by K-NN with using density degree of marked samples. These samples are trained to get the base classifier, then, the MVR algorithm is adopted to form a strong classifier. The samples are extracted by bootstrapping and MVR [13], MVR is the most simple and effective method to integrate multiple classifiers. For a set of weak classifiers $\{C_i(x), i \in [1,t]\}$, a new classifier is defined as follow:

$$C^*(x) = \text{sgn}\left\{ \sum_i C_i(x) - \frac{t-1}{2} \right\} \tag{4}$$

Where, C^* is the combination of the weak classifiers and, t is the number of classifiers, the classifier is remark ith. MVR will not consider any specific behavior differences of weak classifiers, but integrate all the classifiers that have the same classification results.

Due to clustering can easily obtain the accumulation degree of negative sample groups, we present a new sample selection algorithm based on the defined evaluation formula. In this way, high-quality and non-redundant negative samples are selected, thus, we establish a mcta-classifier which can obtain higher classification effectiveness and accuracy. IAB can solve the sample offset problems that is caused by imbalanced samples and also solve the real-time problem with fewer interactions.

4 Image Feature Extraction

LBP (local binary pattern) is firstly proposed by Ojala T and Pietika L Inenm in [14] to extract the texture features. The extracted features are local texture features which are used in many fields due to its good rotation, gray scale invariance and low computational complexity. Such as texture classification [14], face recognition [15], fingerprint recognition [16] image classification [17], etc.

The original LBP is defined in the 3*3 box and the gray value of the central pixel is a threshold. As shown in Fig. 2, the left matrix threshold value is 30 and the value of the 8 pixels in the center pixel is compared with that of the central pixel. If the pixel value is less than the gray value of the central pixel, the pixel position is marked as 0, otherwise the marker is 1, then we get the binary matrix, as show in Fig. 2, after the threshold value (i.e., 0 or 1), respectively, the corresponding position of the pixel weight is multiplied by 8 and that is the LBP value of the neighborhood, the calculation principle is shown in Fig. 2.

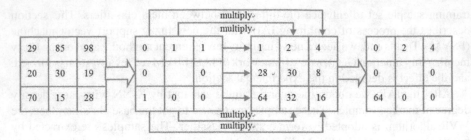

Fig. 2. The definition of original LBP value

The original LBP in the local texture feature extraction has good performance, but texture features extraction has certain limitations. For an example of Fig. 3, in a gray image, define a radius of a circular neighborhood R (r > 0) and P (P > 0) neighborhood pixels are evenly distributed on the circumference. The local texture feature of the local neighborhood is T, then, T can be defined as a function of $P + 1$ pixel in the neighborhood:

$$T = t(g_c, g_i, \cdots g_{p-1}) \tag{5}$$

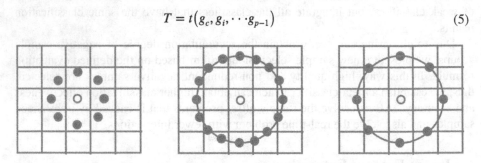

Fig. 3. The circular neighborhood for different P and R

Among them, the gray value of the central pixel of the neighborhood is g_c; the gray value of P neighborhood pixels is g_i.

Although Ojala circular neighborhood improvement can achieve rotation invariance operator, but it ignores the isotropy, in order to take advantage of this important structural information, in [18], elliptical near point definition method named the EBP is proposed.

In [19], a new algorithm LTP is proposed to solve the problem of the sensitivity to noise in near-uniform image regions and a model called modified binary patterns is proposed in [20] to obtain a texture descriptor that can overcome the noise. Modified LBP which is locally adaptive in nature and capable of discriminating the structural patterns is based on local statistical measures. Considering texture characters of the cell images, we adopt the MLBP method in this work. Indeed, other works in this area are in [20, 21], which use the preprocessing method to improve the classification results. For example, LBP descriptors in [21] are extracted and processed by the Gabor filter graph.

5   Experiments on Medical Image Database

For the experiment validation, we select the Hep-2 image dataset of MIVIA set[1], version 1.0. In this dataset, 28 pieces of it are images, which can be recorded as 28 classes, as is shown in Fig. 4. However ordinary person cannot find the subtle differences between images, we preprocess the images by do a number sign in each image. The resolution of the images in the dataset is different. Due to the texture feature, we also need to pre-process the image from color to gray and neglect the size of the images. There are 28 different categories of images in the database, each of which has less than 100 images. Feature extraction algorithm is used to extract the feature vector, which is stored in the image feature database. We use the feature extraction algorithm as is introduced in the fourth part of this paper.

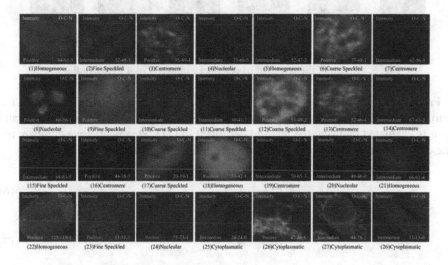

Fig. 4. Classes of the image in the database

To demonstrate the performance of our proposed algorithm, experiments are performed by comparing the results with that of AB-SVM and ABRS-SVM in the medical image databases. The experimental results are shown in Fig. 5 which gained from dataset by mapping feature vector to original image. The red box in Fig. 5 means the query image and the blue box means the negative sample. There is a checkbox at the bottom of each picture; the image is a positive image if it is marked true.

---

[1] For more information, please refer to: http://mivia.unisa.it/datasets/biomedical.

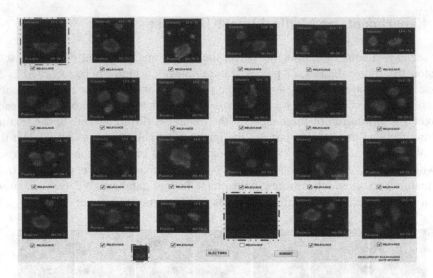

Fig. 5. Result of the retrieval of one query image (Color figure online)

From the result, we can find that the method we proposed is effective. Certainly, the performance is also validated by the proposed PVR curve, which is defined as in (6) and (7), it is evaluated by using the method described in [21]. Meanwhile, we compare the time of feedback with that of other alternatives under an identical iteration times.

$$P(S_q) = \frac{S_{qr}}{S_{qT}} \tag{6}$$

$$R(S_q) = \frac{S_{qr}}{S_{qt}} \tag{7}$$

In formula (6), the precision remark $P(S_q)$ and S_q indicates the query sample, S_{qr} indicates the number of samples associated with the query, S_{qT} indicates the number of feedback sample shown interface. In expression (7), $P(S_q)$ *express recall* and S_{qt} indicates the number of samples belongs to the same kind.

This paper designs two sets of experiments using Top = 24 and 12 to evaluate the performance which is performed on 1456 images; the results are shown in Fig. 6.

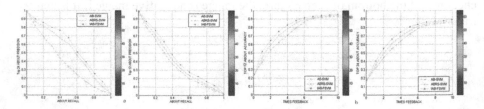

Fig. 6. Accuracy & Precision-versus-recall

It can be seen from Fig. 6-a:

1. IAB-FSVM achieves a rapid improvement when Top increased from 12 to 24 in terms of recall-precision;
2. The recall-precision of IAB-FSVM has a rapid improvement, especially at the scale position of 40 %;
3. IAB has a better performance compared with AB-SVM and ABRS-SVM all the time.

Due to our method is based on IAB, we redefine the fuzzy function of FSVM, and it makes full use of the samples clustering information to get better classification results. When the recall is relatively small, our method has much better recall-precision compared with AB-SVM and ABRS-SVM, as the recall increases by, these methods have fewer differences from each other.

We use a new measurement named *TRA* to evaluate the proposed method, as is defined in (8):

$$TRA(S_q) = \frac{F(S_{pt} * C_I)}{S_{qT}} \tag{8}$$

The new definition can be used to evaluate the effectiveness of the system and it can also achieve the purpose of the evaluation of RA in [16, 17]. Where, C_I refers to the number of iteration, S_{pt} means the number of feedback positive samples and S_{qT} are the same with that in (6). F filters the redundant images of feedback.

In Fig. 6-b, we can find that IAB has achieved better performance compared with AB-SVM and ABRS-SVM when Top is set as 24 and 12. It also has better convergence rate and retrieval precision. For the identical times of feedback, IAB performs better; the better the performance gets the smaller the times of feedback turns. In our experiments, the times of feedback is set as 10. Due to our method makes full use of the samples clustering information; it can achieve better performance compared with AB-SVM and ABRS-SVM.

6   Conclusions

Image retrieval is a hot research in social applications which is based on social computing and inteligent computing and it is being integrated into the field of information retrieval gradually. However, due to the semantic gap between high-level and low-level, the introduction of relevance feedback can be a good solution to these problems. In view of the use of SVM for small sample classification, this method is better than the other classification methods. The study on SVM also has a breakthrough. Yet small sample, asymmetric, over-fitting and real-time problems are still the main problems. In this paper, we find a way to solve the asymmetric problem, small sample problem, real-time problem and over fitting problem by the IAB method we proposed. The experimental results show that the proposed IAB has better performance than AB-FSVM and ABRS-SVM clearly. This kind of research which is based on social computing is very useful for medical auxiliary and more research should be concern on social computing.

Acknowledgment. This work is supported by the National Natural Science Foundation of China (No. 61472161, 61133011, 61402195, 61502198, 61303132, 61202308), Science & Technology Development Project of Jilin Province (No. 20140101201JC).

References

1. Smeulders, A.W.M., Worring, M., Santini, S., Gupta, A., Jain, R.: Content-based image retrieval at the end of the early years. IEEE Trans. Pattern Anal. Mach. Intell. **22**(12), 1349–1380 (2000)
2. Rui, Y., Huang, T.S., Mehrotra, S.: Content-based image retrieval with relevance feedback in MARS. In: Proceedings of the IEEE International Conference on Image Processing, vol. 2, pp. 815–818 (1997)
3. Revalillo-Herraez, M.A., Ferri, F.J.: An improved distance-based relevanced feedback strategey for image retrieval. Image Vis. Comput. **31**, 704–713 (2013)
4. Wei, C.L., Zong, Y.C., Ke, S.-W., Tsai, C.-F.: The effect of low-level image features on pseudo relevance feedback. Neurocomputing **166**, 26–37 (2015)
5. Wu, H., Fang, H.: An incremental approach to efficient pseudo-relevance feedback. In: ACM SIGIR Conference on Research and Development information Retrieval, pp. 553–562 (2013)
6. Xiang, Y.W., Hong, Y.Y., Yong, W.L.: A new SVM-based active feedback scheme for image retrieval. Eng. Appl. Artif. Intell. **37**, 43–53 (2015)
7. Wang, X.-Y., Chen, J.-W., Yang, H.-Y.: A new integrated SVM classifiers for relevance feedback content-based image retrieval uing EM parameter estimation. Appl. Soft Comput. **11**, 2287–2804 (2011)
8. Tao, D., Tang, X., Li, X., Wu, X.: Asymmetric bagging and random subspace for support vector machines-based relevance feedback in image retrieval. IEEE Trans. Pattern Anal. Mach. Intell. **28**, 1088–1099 (2006)
9. Vapnik, V.: The Nature of Statistical Learning Theory. Springer, New York (1995)
10. Kui, W., Kim, H.Y.: Fuzzy SVM for content-based image retrieval: a pseudo-label support vector machine framework. IEEE Comput. Intell. Mag. **1**(2), 10–16 (2006)
11. Wu, H., Lu, H., Ma, S.D.: A practical SVM based algorithm for ordinal regression in image retrieval. In: Proceedings of ACM Multimedia, pp. 612–621. Berkeley, USA (2003)
12. Zhou, Z.-H., Chen, K.-J., Jiang, Y.: Exploiting unlabeled data in content-based image retrieval. In: Boulicaut, J.-F., Esposito, F., Giannotti, F., Pedreschi, D. (eds.) ECML 2004. LNCS (LNAI), vol. 3201, pp. 525–536. Springer, Heidelberg (2004)
13. Bahler, D., Navarro, L.: Methods for combining heterogeneous sets of classifiers. In: Proceedings of the 17th National Conference American Association for Artificial Intelligence (2000)
14. Ojala, T., Pietikainen, M., Harwood, D.: A comparative study of texture measures with classification based on featured distribution. Pattern Recogn. **29**, 51–59 (1996)
15. Nanni, L., Lumini, A.: Region boost learning for 2D + 3D based face recognition. Pattern Recogn. Lett. **28**(15), 2063–2070 (2007)
16. Nanni, L., Lumini, A.: A reliable method for cell phenotype image classification. Artif. Intell. Med. **43**(2), 87–97 (2008)
17. Nanni, L., Lumini, A.: Local binary patterns for a hybrid fingerprint matcher. Pattern Recogn. **41**(11), 3461–3466 (2008)
18. Liao, S., Chung, A.C.S.: Face recognition by using elongated local binary patterns with average maximum distance gradient magnitude. In: Yagi, Y., Kang, S.B., Kweon, I.S., Zha, H. (eds.) ACCV 2007, Part II. LNCS, vol. 4844, pp. 672–679. Springer, Heidelberg (2007)

19. Hafiane, A., Seetharaman, G., Palaniappan, K., Zavidovique, B.: Rotationally invariant hashing of median binary patterns for texture classification. In: Campilho, A., Kamel, M. (eds.) ICIAR 2008. LNCS, vol. 5112, pp. 619–629. Springer, Heidelberg (2008)
20. Naresh, Y., Nagendraswamy, H.: Classification of medicinal plants: an approach using modified LBP with symbolic representation. Neurocomputing **173**, 1789–1797 (2015)
21. Zhang, W., Shan, H., Chen, X., Gao, W.: Local gabor binary patterns based on mutual information for face recognition. Int. J. Image Graph. **7**(4), 777–793 (2007)

An Incremental Graph Pattern Matching Based Dynamic Cold-Start Recommendation Method

Yanan Zhang[1(✉)], Guisheng Yin[2], and Qiushi Zhao[1]

[1] Software School, Harbin University of Science and Technology,
Harbin 150040, China
ynzhang_1981@163.com
[2] College of Computer Science and Technology,
Harbin Engineering University,
Harbin 150001, China

Abstract. In order to give accurate recommendations for cold-start user, researchers use social network to find similar users. These efforts assume that cold-start user's social relationships are static. However social relationships of cold-start user may change as time pass by. In order to give accurate and timely in manner recommendations for cold-start user, it is need to update social relationship continuously. In this paper, we proposed an incremental graph pattern matching based dynamic cold-start recommendation method (IGPMDCR), which updates similar users for cold-start user based on topology of social network, and gives recommendations based on the latest similar users' records. The experimental results show that, IGPMDCR could give accurate and timely in manner recommendations for cold-start user.

Keywords: Dynamic cold-start recommendation · Social network · Incremental graph pattern matching · Topology of social network

1 Introduction

Newly registered users usually have no or only a few purchase and comment records, this kind of user is known as cold-start user. To our knowledge, social relationships of users are not static all the time, as time pass by, cold-start user could have new friends, or might lose friends. It is not accurate to give recommendations for cold-start users based on former social relationship. In order to give accurate cold-start recommendation in a timely manner, many researchers have done a lot of efforts on real-time recommendation. Such as, researchers use time function to describe negative correlation attribute between time and trend of user interest. However, these efforts are based on cold-start user's own state without considering other users could also change their social relationships and properties. Another issue is the scale of social network is large that could lead to high time-complexity of updating similar users for cold-start user.

W. Che et al. (Eds.): ICYCSEE 2016, Part I, CCIS 623, pp. 182–195, 2016.
DOI: 10.1007/978-981-10-2053-7_17

Cold-start recommendation is one of the core technologies for many network applications. Researchers have done a lot of efforts on research of cold-start recommendation. For cold-start users have few historical information, research on cold-start recommendation mainly focus on mining, expanding information of cold-start users and finding similar users of cold-start user [1, 2]. The research on cold-start recommendation could be categorized into rule-based recommendation, content-based recommendation, collaborative filtering based recommendation, and trust based recommendation.

(1) Rule-based recommendation methods extract user characteristics to make rules and give recommendations follow the rules. Effectiveness of rule-based recommendation method depends on accurate characteristics extraction and appropriate rules. Such as Ren select suitable candidate and user consumption pattern to build rule for recommendation [3]. Ling extracted characteristics of cold-start users, grouped users by characteristics, and gave recommendations based on cluster [4]. Rule-based recommendation method could give clear recommendations, however rule-based recommendation method requires a lot of user information to make appropriate rules, which is not suitable for cold-start recommendation.

(2) Content-based recommendation methods extract keywords from commodity and users' historical information and give recommendations according to similarity of their keywords [5]. The main steps of content-based recommendation method includes: keywords extraction; find similar users, and give recommendations according to similar user's records. The effectiveness of content-based recommendation methods depend on accurate keywords extraction. Such as Cui proposes a latent class statistical mixed model TCAM, which could give recommendations in real-time [6]. Content based recommendation methods also need a lot of historical records, and its recommendations are usually duplicate with users' historical records.

(3) Collaborative filtering based recommendation methods give recommendations according to similarity degree of users or similarity degree between users and items [7]. Collaborative filtering based recommendation method could be categorized into user-based collaborative filtering recommendation method and item-based collaborative filtering recommendation method. User-based collaborative filtering recommendation methods normalize and compare similarity against comments of same items given by users. Item-based collaborative filtering recommendation methods calculate similarity of items and predicted ratings of item by user-item matrix. The effectiveness of collaborative filtering recommendation depends on degree of data sparseness of user-item matrix. It is difficult to determine similar users for cold-start user by sparse user-item matrix. To solve this issue, Jamali proposed a method using Non-Negative Matrix Factorization (NMF) which divided original user-item matrix into two low rank matrixes. The product of low rank matrix equals the original matrix [8]. Ma proposed a recommendation method based on probability matrix factorization (PMF) which introduce a condition that deviation between actual ratings and estimated ratings

follow normal distribution [9]. Wu proposed a recommendation method which involved user preferences factor to improve accuracy of recommendation [10]. In order to handle user interest drift issue, Koren proposed a time-sensitive collaborative filtering recommendation method to distinguish transient effects and long-term patterns of user interest [11]. Gu proposed a collaborative filtering recommendation method based on balance score prediction mechanism, which could adjust dynamic right and overall weight [12]. Collaborative filtering based recommendation methods could avoid adverse effectiveness of incomplete or inaccurate characteristics extraction, but still need plenty of users' historical information for recommendation.

(4) Trust based recommendation methods assume trust relationship is a stable social relationship. Researchers have done a lot of work on build trust relationship between users, including trust propagation strategy [13, 14], verifying small world feature of trust network [15], building trust network based on small world feature and weak relationship among users [16]. Wang proposed a trust based recommendation method which use gaming between relative feature and absolute feature in process of trust propagation [17]. Guha proposed a trust based recommendation method which could build trust relationship among users by a few trust or distrust [18]. Sun proposed a trust based recommendation method which amends deviation of recommendations according to Bayesian decision theory [19]. Kang proposed a trust based recommendation method which analyzed excessive trust propagation by dynamic social network model [20]. However existing researches ignore that trust relationship is only one kind of relationship in social network, and trust relationship could not comprehensively describe users' characteristics for recommendation.

As the increasing popularity of internet and social network service, there are various types of social relationships among users. Each kind of social relationship could reflect a characteristic of user. So we could predict users' preference by these social relationships. Cold-start users might have social relationship. It is feasible to find similar users for cold-start user according to social relationship, and give recommendations for cold-start user by similar users' records of purchase and comments. However, cold-start user and other users' social relationship may change as time pass by, in order to give accurate and timely in manner recommendations for cold-start users, we proposed an incremental graph pattern matching based dynamic cold-start recommendation method (IGPMDCR), which updates similar users for cold-start user based on topology of social network, and give recommendations for cold-start users by latest similar user's records.

2 Dynamic Cold-Start Recommendation Using Incremental Graph Pattern Matching

In order to give accurate and timely in manner recommendations for cold-start user, we proposed an incremental graph pattern matching based dynamic cold-start recommendation method (IGPMDCR) to update similar users for cold-start user continuously, and give recommendations based on latest updated similar users' records. The main steps of IGPMDCR are: (1) define topology of social network, (2) find similar users for cold-start user based on topology of social network, (3) update similar users by incremental graph pattern matching, (4) give recommendations according to latest similar users' records.

2.1 Definite Topology of Social Network

IGPMDCR updates similar users for cold-start user based on topology of social network, in which topology of social network is a directed graph $G = (V, E)$, where (1) V is a finite set of nodes, each node denotes a user; (2) $E \subseteq V \times V$, in which (u, u') denotes an edge from node u to u' Topology of social relationship is composed of nodes and edges, where nodes denote users and edges denote relationship among users. In topology of social relationship, edges have two attributes, one is type of social relationship, and the other denotes degree of tightness between users. $f_e(u, u')$ denotes distance between u and u' in a social relationship. Shorter distance means more tightness relationship, longer distance means less tightness relationship. $f_v(u, u')$ denotes type of social relationship. We regard topology of cold-start user's social relationship as pattern, and use graph pattern matching to find similar users with the pattern in directed graph G. We argue that the similar users have similar topology of social relationship with that of cold-start user.

2.2 Find Similar Users Based on Topology of Social Network

Similar users of cold-start users are found according to topology of social relationship. We proposed bounded graph simulation match algorithm to find similar users for cold-start user.

In bounded graph simulation match algorithm, cold-start user's topology of social relationship is regard as pattern. We use this pattern to match similar users in topology of social network. Before illustrating the algorithm, we first present notations it uses. Let u denote cold-start user. u' denotes user who have relationship with u, x, y, z, v, v_1 and v' denote nodes in G. $f_v(u, u')$ denotes type of social relationship between user u and u'. $f_e(u, u')$ denotes distance between user u and u'. The value of $f_e(u, u')$ could be integer k or ∞. V_C denotes nodes in pattern, and E_C denotes edges in pattern. V denotes all nodes in social network. Matching pattern is defined as $Q = (V_C, E_C, f_v, f_e)$. Let $S \subseteq V_C \times V$, the match condition of any node v is, (a) attribute of v is similar with attribute of u. (b) for (u, u'), there exist a non empty path $v/.../v'$ whose length is no more than $f_e(u, u')$, and $(u, v) \in S$. (c) for (u, u'), there exists $v/.../v'$ make $f_v(v, v_1)...$ $f_v(v_n, v')$ satisfy $f_v(u, u')$ on the path of (u, u'), where node order of $v/.../v'$ is $(v, v_1,...,v_n, v')$. Bounded graph simulation match algorithm is shown as Algorithm 1.

```
Algorithm 1 Bounded graph simulation match algorithm
INPUT: pattern P = (V_C,E_C,f_v,f_e) ,user data graph G=(V,E).
OUTPUT: Maximum match set S
1. Calculate distance Matrix M of G;
2. for each (u', u) ∈ E_C ,x ∈ V do
3. Calculate anc(f_e(u', u), f_v(u'), x), desc(f_e(u', u),
f_v(u'), x);
4. for each u∈ V_C do
5. mat(u) := {x|x∈ V, f_A(x) match f_v(u),and if out-
degree(a) ≠0 ,out-degree(x) ≠ 0 };
6. demat(u) := {x|x∈ V, out-degree(x) if out-degree(a)
≠0, and ∃(u', u) ∈ E_C (x'∈ mat(u), f_A(x) satisfy
f_v(u'),and len(x/.../x') ≤ f_e(u', u))};
7. while (u∈ V_C with demat(u) ≠∅) do
8. for each (u', u) ∈ E_C ,z ∈ demat(u) ∩ mat(u')) do
9. mat(u') := mat(u')\{z};
10. if (mat(u') = ∅) then return ∅;
11. for each u'' with (u'', u') ∈ E_C do
12. for (z'∈anc(f_e(u'', u'), f_v(u''), z) ∩ z'
/∈demat(u')) do
13. if (desc(f_e(u'', u'), f_v(u'), z') ∩ mat(u')= ∅)
14. then demat(u'):= demat(u')∪{z'};
15. demat(u) := ∅;
16. S := ∅;
17. for (u ∈ V_C and x ∈ mat(u)) do S := S ∪{(u, x)};
18. return S
```

The bounded graph simulation match algorithm, referred to as BGSM, Given P and G, BGSM returns a maximum match S for P or it returns empty set empty otherwise. Before illustrating process of the BGSM algorithm, we present notations in the algorithm, (1) distance matrix Matr maintains distances between all pairs of nodes in G. (2) For each node u in pattern P, we use a set mat(u) to record nodes in G that may match u, and a set demat(u) for those nodes that cannot match any parent of u. (3) For each node, we use anc() finds ancestor of a node by depth-first algorithm; desc() is used to find descendent of a node. (4) For each node $x\in V$ and edge $(u',u) \in E_C$, anc($f_e(u', u)$, $f_v(u')$, x) records nodes x' in the graph G such that (i) the distance from x' to x is within the bound imposed by $f_e()$, i.e., len($x'/\cdots/x$) $\leq f_e(u', u)$, and (ii) $f_v(x')$ satisfies the predicate $f_v(u')$ defined on u'; similarly for desc($f_e(u, u')$, $f_v(u')$, x), for descendants of x. BGSM algorithm computes the distance matrix Matr for G in line 1. It then computes ancestor using anc() and descendent using desc() by inspecting the predicates and bounds specified in the pattern P in lines 2–3. For each node of the pattern, if $u\in V_C$, the algorithm also initializes mat(u) and demat(u) using P and Matr in lines 4–6. For each parent node u' of u, algorithm then refines mat(u') by removing those nodes in G that can not match u', namely, nodes $z \in$ demat(u) in lines 8–9. Moreover, it utilizes z to identify nodes z' that cannot match any parent u'' of u', and includes z' in demat(u') in lines 11–14. More specifically, z' is not a candidate match of u'', if z is the only descendant of z' that is within the bound $f_e(u'', u')$, satisfies the predicate $f_v(u')$, and is in mat(u'). In lines 7–15 the process iterates until no mat() can be reduced, i.e., if demat (u) is empty for all pattern node u. The nodes remaining in mat(u) are those that match u, and are collected in S, which is returned as the match in lines 16–18. If mat(u) is empty for any $u \in V_C$ in the process, u cannot find a match in G, and algorithm returns empty in line 10. Time complexity of the algorithm is depended on product of nodes and edges in pattern P and G, i.e. $O((|V| + |V_C|)(|E| + |E_C|))$, expanding $(|V| + |V_C|)(|E| + |E_C|)$, we get $(|V||E| + |V||E_C| + |E||V_C| + |V_C||E_C|)$, usually number of nodes in P is much less than that of G. $|E|$ and $|V|^2$ are approximately equal. So Time complexity of the algorithm could be simplified as $O(|V|\|E| + |E_C|\|V| + |V_C|\|V|^2)$.

2.3 Update Similar Users for Cold-Start User Based on Incremental Graph Pattern Matching

An incremental graph pattern matching is proposed to dynamically update similar users for cold-start user. Incremental graph pattern matching (IGPM) updates similar users for cold start users according to the change of path between users in G. IGPM iteratively finds change of social relationship according to length of path between users. Firstly, determine affect field AFFcs caused by change of social relationship. If length of certain path beyond a threshold value, we argue corresponding social relationship is too weak to find similar users, and could be deleted. Deleted nodes might cause length of some other paths beyond threshold value, so we have to find the deleted nodes' predecessor nodes, determine whether its path match that of P and there is only one path in maximum matching set M(P, G) satisfy P. Algorithm of incremental graph pattern matching is shown as Algorithm 2. In Algorithm 2, $G_r = (V_r, E_r)$ denotes original match set of cold-start users, Ecs stores edges in AFFcs. If $e = (v'; v)$ is a path of matching pattern then iteratively find nodes which match v', and do not match P anymore. If $e = (v'; v)$ is not a path of matching pattern, then leave G_r as it is. Time complexity of the Algorithm 2 depends on product of nodes and edges in pattern P, updated edges in E_{cs}. i.e. $O((|E_c| + |V_C|)(|E_{cs}|))$, so Time complexity of the algorithm could be $O(|E_c|\|E_{cs}| + |E_{cs}|\|V_C|)$ whose time-complexity is much less than that of updating similar users in the entire G.

```
Algorithm 2 Incremental graph pattern matching
Input: P, G_r = (V_r;E_r), E_cs .
Output: Updated similar users G'_r.
1.  if e = (v',v) ∉ E_r then delete e from G, return G_r;
2.  E_cs := ∅; E_cs:push(e);
3.  while E_cs≠∅ do
4.  e := E_cs:pop();
5.  for each ep = (u',u) match e = (v, v) do
6.  if v ' match u';
7.  if v' do not match u ' then
8.  for each e'= (v'', v') in E_r do
9.  E_r := E_r \ {e'}; E_cs:push(e');
10. V_r := V_r \ {v'}; mat(u') := mat(u') \ {v'};
11. if mat (u') = ∅ then return ∅;
12. return G'_r.
```

2.4 Give Recommendations According to Similar Users' Records

We give recommendations for cold-start user according to similar users' records of purchase and comments. Unknown ratings of items for cold-start user are predicted according to user-item matrix. Each row of user-item matrix denotes a user's ratings. Each column of user-item matrix denotes users' ratings on one item. In user-item matrix, an item with higher rating means this item has probability to be chosen. We select items with highest rating in user-item matrix as recommendations for cold-start user. The unknown ratings of cold-start user in user-item matrix are given by Eq. (1), and recommendations for cold-start user are given by Eq. (2). In Eq. (1), $r_{u,i}$ denotes rating of user u on item i. $w_{a,u}$ denotes similarity among users, the range of $w_{a,u}$ is from 0 to 1, in which 1 denotes user a and u are completely similar, while 0 means user a and u are totally dissimilar. $\bar{r}_a$ and $\bar{r}_u$ denote average ratings given by user a and u. Equation (2) gives top N items with highest rating as recommendations for cold-start user, in which N denotes the number of recommendations.

$$p_{a,i} = \bar{r}_a + \frac{\sum_{u=1}^{t} w_{a,u}(r_{u,i} - \bar{r}_u)}{\sum_{u=1}^{t} w_{a,u}} \tag{1}$$

$$Top(a, N) := \max_{N} p_{a,i} \tag{2}$$

3 Experimental Results and Analysis

In order to verify that IGPMDCR could give accurate and timely in manner recommendations, we select following state-of-art recommendation methods to compare with IGPMDCR, neural learning based recommendation method proposed by Bobadilla [1], referred to as Bobadilla's method, classification based collaborative filtering recommendation method proposed by Lika [2], referred to as Lika's method, vector cosine matrix based top-N recommendation method [4], referred to as Ling's method.

3.1 Experimental Data and Evaluation Method

The experimental dataset is chosen from Epinions.com and YouTube. Epinions.com is a consumer review site. At Epinions.com, visitors could read reviews about a variety of items to help them purchase commodity. Epinions.com contains about 131,828 nodes, 841,372 edges, which is a large scale social network. We remark users who have less than three purchase and comment records as cold-start user. Cold-start users account for 18.37 % of this dataset. We select multiple social relationships from Epinions.com, includes interest groups, comment forwarding relationship, concern blogs, trust relationship, and adoption of evaluation to describe characteristics of users.

YouTube is a video sharing site. YouTube consists of various types of interactions. We crawled 30, 522 user-profiles. Based on the crawled information, we construct 5 different relationships among 30, 522 users. These relationships include: contact network between the 30, 522 users; number of shared friends; number of shared subscriptions among users; number of shared subscribers among users; number of shared favorite videos. We remark users who have less than three shared subscriptions and shared favorite videos as cold-start user. Cold-start users account for 27.37 % of this dataset.

The entire dataset is divided into training and test sets. We choose root mean square error (RMSE) as the benchmark. RMSE measure deviation between predicted ratings and actual ratings [17], definition of RMSE is shown as Eq. (3), where $r_{u,i}$ denotes actual rating of item i given by user u. The value range of RMSE is from 0 to 5. Smaller RMSE means less deviation between predicted ratings and actual ratings.

$$\text{RMSE} = \sqrt{\frac{\sum_{i=1}^{N} (r_{u,i} - \hat{r}_{u,i})^2}{N}} \qquad (3)$$

3.2 Experimental Result

In order to verify the recommendation effectiveness of IGPMDCR during a period of time, we compare IGPMDCR, Bobadilla's, Lika's, Ling's method in a period of 640 days. We divide the entire period of 640 days into 9 stages. In dataset obtained from Epinions.com, the training data accounts for 5 %, 15 %, 25 %, 35 % respectively, and the corresponding recommendation effectiveness is shown as Fig. 1(a–d).

Users' social relationship might change as time pass by, it is reasonable to update the social relationship to get accurate recommendations. The gap between actual social relationship and predicted social relationship given by methods might grow as time pass by, so the RMSE of recommendation during early stages is less than that of the latter stages. In Fig. 1(a–d) during the first two stages, RMSE of IGPMDCR is almost

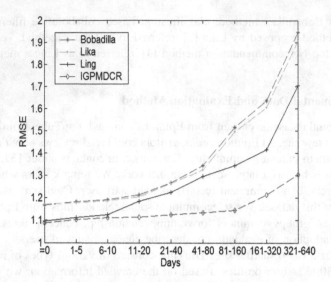

(a) Training Data accounts for 5%

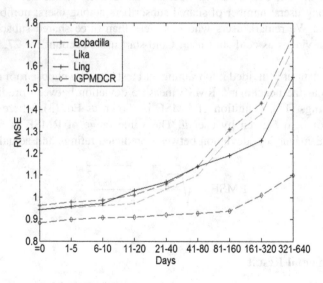

(b) Training Data accounts for 15%

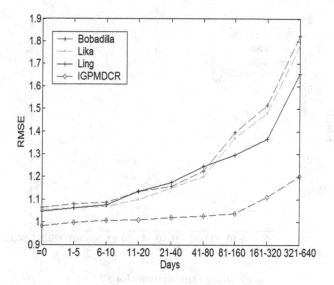

(c) Training Data accounts for 25%

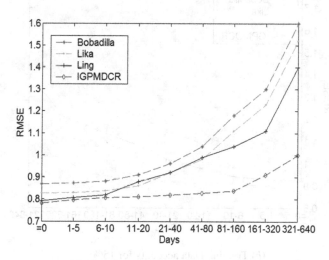

(d) Training Data accounts for 35%

Fig. 1. Comparison of recommendation effectiveness against RMSE on Epinions.com.

same with that of Bobadilla's, Lika's, Ling's method. The gap between RMSE of IGPMDCR and that of Bobadilla's, Lika's, Ling's method becomes larger from the third stage. During the fourth to the seventh stage, the gap between RMSE of IGPMDCR and that of Bobadilla's, Lika's, Ling's method continuously increases. In the last stage, the gap between RMSE of IGPMDCR and that of Bobadilla's, Lika's, Ling's method increases sharply. In the 640th day, the gap of RMSE of IGPMDCR and that of Bobadilla's, Lika's, Ling's method reaches the maximum value. In the entire

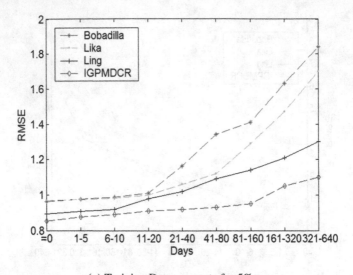

(*a*) Training Data accounts for 5%

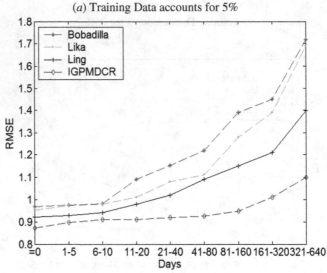

(*b*) Training Data accounts for 15%

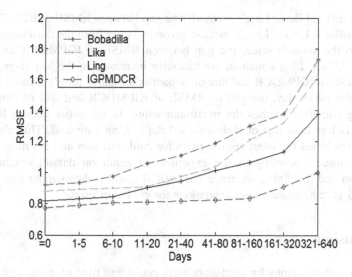

(c) Training Data accounts for 25%

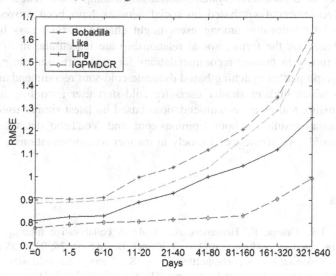

(d) Training Data accounts for 35%

Fig. 2. Comparison of recommendation effectiveness against RMSE on YouTube.

period, RMSE of IGPMDCR is less than that of Bobadilla's, Lika's, Ling's method. This means IGPMDCR could get the latest similar users for cold-start user and get more accurate recommendations.

In dataset obtained from YouTube, the training data also accounts for 5 %, 15 %, 25 %, 35 % respectively, and the corresponding recommendation effectiveness is shown as Fig. 2(a–d). In the first two stages, RMSE of IGPMDCR is almost same with that of Bobadilla's, Lika's, Ling's method. The gap between RMSE of IGPMDCR and that of Bobadilla's, Lika's, Ling's method becomes larger from the third stage. During the fourth to the seventh stage, the gap between RMSE of IGPMDCR and that of Bobadilla's, Lika's, Ling's method continuously increases. In the last stage, the gap between RMSE of IGPMDCR and that of Bobadilla's, Lika's, Ling's method increases sharply. In the 640th day, the gap of RMSE of IGPMDCR and that of Bobadilla's, Lika's, Ling's method reaches the maximum value. In the entire period, RMSE of IGPMDCR is less than that of Bobadilla's, Lika's, Ling's method. This also means IGPMDCR could get the latest similar users for cold-start user and get more accurate recommendations. According to the experimental result on dataset obtained from Epinions.com and YouTube, we argue IGPMDCR could update similar users as time pass by, and give accurate recommendations for cold-start user.

4 Conclusion

Accurate recommendations for cold-start users could win trust of users, and improve attractive force of E-commerce system. Social relationship could reflect users' characteristics. Recommendations based on social network have been proved effective. However, social relationship among users might change as time pass by. Recommendations based on the former social relationship are inaccurate. In order to give accurate and timely in manner recommendations for cold-start user, we proposed an incremental graph pattern matching based dynamic cold-start recommendation method (IGPMDCR) which updates similar users for cold-start user according to the latest social relationship, and gives recommendations based on latest similar users' records. The experimental results on both Epinions.com and YouTube dataset show that, IGPMDCR could give accurate and timely in manner recommendations.

References

1. Bobadilla, J.S., Ortega, F., Hernando, A., et al.: A collaborative filtering approach to mitigate the new user cold start problem. J. Knowl. Based Syst. **26**, 225–238 (2012)
2. Lika, B., Kolomvatsos, K., Hadjiefthymiades, S.: Facing the cold start problem in recommender systems. J. Expert Syst. Appl. **41**(4), 2065–2073 (2014)
3. Ren, Y., Li, G., Zhou, W.: Improving top-N recommendations with user consuming profiles. In: Anthony, P., Ishizuka, M., Lukose, D. (eds.) PRICAI 2012. LNCS, vol. 7458, pp. 887–890. Springer, Heidelberg (2012)
4. Ling, Y.X., Guo, D.K., Cai, F., et al.: User-based clustering with top-N recommendation on cold-start problem. In: 3rd International Conference on Intelligent System Design and Engineering Applications, pp. 1585–1589. IEEE Computer Society, New York (2013)
5. Lops, P., De Gemmis, M., Semeraro, G.: Content-based recommender systems: state of the art and trends. In: Recommender Systems Handbook, pp. 73–105. Springer, US (2011)
6. Yin, H., Cui, B., Chen, L., et al.: A temporal context-aware model for user behavior modeling in social media systems. In: 14th SIGMOD International Conference on Management of Data, pp. 1543–1554. ACM, Snowbird (2014)

7. Wang, J., De Vries, A.P., Reinders, M.J.T.: Unifying user-based and item-based collaborative filtering approaches by similarity fusion. In: 29th Annual International ACM SIGIR Conference on Research and Development in Information Retrieval, pp. 501–508. ACM, Seattle (2006)
8. Jamali, M., Ester, M.: A matrix factorization technique with trust propagation for recommendation in social networks. In: 4th ACM Conference on Recommender Systems, pp. 135–142. ACM, Barcelona (2010)
9. Ma, H., Yang, H., Lyu, M.R., et al.: Sorec: social recommendation using probabilistic matrix factorization. In: 17th ACM Conference on Information and Knowledge Management, pp. 931–940. ACM, Napa Valley (2008)
10. Wu, L., Chen, E.H., Liu, Q., et al.: Leveraging tagging for neighborhood-aware probabilistic matrix factorization. In: 21st ACM International Conference on Information and Knowledge Management, pp. 1854–1858. ACM, Maui (2012)
11. Koren, Y.: Collaborative filtering with temporal dynamics. J. Commun. ACM. **53**(4), 89–97 (2010)
12. Ren, L., Gu, J.Z., Xia, W.W.: An item-based collaborative filtering approach based on balanced rating prediction. In: 11th International Conference on Multimedia Technology, pp. 3405–3408. IEEE, Washington (2011)
13. Ma, H., King, I., Lyu, M.R.: Learning to recommend with social trust ensemble. In: 32nd International ACM SIGIR Conference on Research and Development in Information Retrieval, pp. 203–210. ACM, Boston (2009)
14. Kim, Y.A., Song, H.S.: Strategies for predicting local trust based on trust propagation in social networks. J. Knowl. Based Syst. **24**(8), 1360–1371 (2011)
15. Yuan, W.W., Guan, D.H., Lee, Y.K., et al.: Improved trust-aware recommender system using small-worldness of trust networks. J. Knowl. Based Syst. **23**(3), 232–238 (2010)
16. Jiang, W.J., Wang, G.J., Wu, J.: Generating trusted graphs for trust evaluation in online social networks. J. Future Gener. Comput. Syst. **31**, 48–58 (2014)
17. Liu, R.R., Liu, J.G., Jia, C.X., et al.: Personal recommendation via unequal resource allocation on bipartite networks. J. Phys. A Stat. Mech. Appl. **389**(16), 3282–3289 (2010)
18. Guha, R., Kumar, R., Raghavan, P., et al.: Propagation of trust and distrust. In: 13th International Conference on World Wide Web, pp. 403–412. ACM, New York (2004)
19. Sun, Y.X., Huang, S.H.: Bayesian decision-making based recommendation trust revision model in ad hoc networks. J. Softw. **20**(9), 2574–2586 (2009)
20. Kang, L., Jing, J.W., Wang, Y.W.: The trust expansion and control in social network service. J. Comput. Res. Develop. **47**(6), 1611–1621 (2010)

An Optimized Load Balancing Algorithm of Dynamic Feedback Based on Stimulated Annealing

Zhang Huyin and Wang Kan[✉]

School of Computer Science, Wuhan University, Wuhan, China
zhy2536@whu.edu.cn, 597580071@qq.com

Abstract. This article analyzed advantages and shortages of classical load balancing algorithms based on dynamic feed-back on server cluster, and combined stimulated annealing with this strategy to put forward an optimized model of dynamic load balancing. This model uses stimulated annealing algorithm to calculate accurate performance parameters of load information on every service node, then estimates the actual load of nodes by dynamic feed-back, in order to insure tasks distribution reasonable. Experimental result shows that in the case of large amount of requests, this algorithm, in comparison with classical load balancing strategy of dynamic feedback, can effectively reduce response time of tasks and ensure high throughput which could improve the whole system performance.

Keywords: Server cluster · Load balancing · Dynamic feed-back · Stimulated annealing · Optimization

1 Introduction

The rapid growth of information technology has made network resources increasingly rich. Explosive growth of number of network access arises. Network servers must offer large numbers of concurrent access service function and be equipped with strong ability of computing. Although, promoting the server hardware performance is a relatively direct and effective way, this method will cause a huge overhead. For most small and medium-sized enterprises, expensive server hardwares are a big burden. Another strategy is to use server cluster technology, by putting a group of high-performance and individual computers together, utilizing high speed network to constitute a stand-alone computing system and managing it in a simple system pattern.

In cluster technology, servers' load balancing algorithm is the key which affects the overall system performance [3, 6]. It distributes requests to service nodes according to the performance of the servers, which minimizes the time of program execution and makes task-sharing evenly [8]. Consequently, the machines achieve the goal of

Foundation item: Supported by Natural Sciences Foundation (61540059), Science and Technology Plan Project of Shenzhen (JCYJ20140603152449639).

© Springer Science+Business Media Singapore 2016
W. Che et al. (Eds.): ICYCSEE 2016, Part I, CCIS 623, pp. 196–205, 2016.
DOI: 10.1007/978-981-10-2053-7_18

improving the quality of system services and processing capacities [4, 5]. The core of cluster load balancing system is load balancing algorithms, which are mainly divided into two types: static algorithms and dynamic algorithms [13]. Static algorithms such as Round-Robin, its' principle is to distribute tasks to each server in a cyclic way. This method is simple and easy to realize for task scheduling. However, in heterogeneous server clusters, this algorithm could result in the fact that servers with weak performance will be allocated in next round [4], which may cause load unbalance. On the other side, dynamic algorithms take system's current load into consideration. Least-connection scheduling is a commonly used dynamic algorithm [14], this strategy makes system distribute tasks according to the current number of connections of each server. Which means node with minimum number of tasks will receive next request automatically. But this measure has a certain limitation that it can be very effective only in the case that servers' computing performance is close to each other. Besides, algorithm of response time conducts server cluster to send a probe request to each node(such as a ping), then sort the time of request of each server, one with the shortest time will receive next task from clients. In this strategy, the current operation state of the servers can be fully reflected, but the response time only indicates the one between load dispatching equip-ments and nodes, not between client computers and servers.

In subsequent sections of this article, we will put forward a load balancing algorithm of dynamic feedback based on stimulated annealing [1], which achieves the following goals:

(1) Adjust load balancing of servers at different time dynamically, ensure the whole system load is not apt to tilt.
(2) Raise the utilization ratio of the clusters, ensure the system throughput, shorten response delay.
(3) Reduce the complexity of this algorithm.

2   Optimized Load Balancing Algorithm

2.1   Optimized Model Based on Stimulated Annealing

Because there are many indicators of server load, when use traditional load balancing algorithms of dynamic feedback to calculate the nodes' load state, people often set weights of load information for the sake of describing the indicators' degree of impact. And then adjust the size of the weights depending on situations, to change status of each index of servers' load. However, this adjustment of weights is artificial, which is hard to be correct and reasonable in different situations. In order to reflect the influence of load information more accurately and calculate the node's actual load according to different situations automatically, in this paper, we use annealing algorithm to calculate the weights of every performance. The optimized model is as shown in Fig. 1:

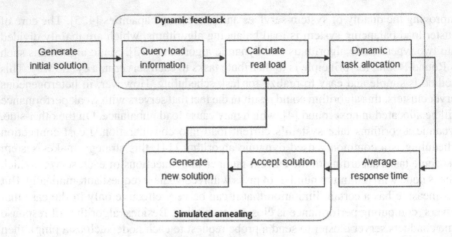

Fig. 1. Optimized model

According to the figure's description, first, randomly generates a set of load information weight values as initial state. With the operation of the system, utilization of server components change with different tasks. In order to avoid the inaccuracy of the solutions leaded by the change of resource usage during simulated annealing iterations, system has to query the servers' resource utilization periodically, and use it as a foundation of node load calculation based on simulated annealing algorithm. After each query, initializes the simulated annealing algorithm and start a new group of iteration. In this way, system with periodic load updates will ensure the weights calculated in each cycle can keep convergence. Compared with the traditional load balancing strategies, this method can seek the optimal solution according to the changes in different environments. It is self-adaptivity and maximizes the use of system resources. The model uses simulated annealing algorithm to optimize load weights with following steps:

(1) With a current solution, generates a new set of weights by the generating function. This is achieved by transforming the current solution and displacing the whole or part of the current solution. The transformation methods to produce a new solution decides the neighborhood structure of solutions, and also ensures that simulated annealing algorithm won't get into local optimal solution space.

(2) Use the new solution and server information to calculate the actual load of each node, then substitute to model for dynamic load balancing. Collect the response time of the system and get the objective function.

(3) Accept new solution: calculate the difference between new objective function value and old one, then judge whether the new solution is accepted based on the Metropolis criterion.

(4) Annealing: after each iteration, cool down so that algorithm have good convergence.

(5) Termination condition: According to the convergence speed of the model, defines the termination condition of annealing algorithm in order to make annealing algorithm can stop within a reasonable time period.

After several rounds of iteration, the system will converge to the best solution, which makes the load balancing system can accurately calculate the current load so that system will allocate resources reasonably. In following sections, simulated annealing algorithm in the application of the model and the dynamic load distribution will be introduced in detail.

2.2 Dynamic Feedback and Node Selection

Before every feedback, produce a set of weight values of performance by the generating function:

$$k = \{k_1, k_2, \ldots, k_n\}, \ \sum k = 1 \tag{1}$$

Through periodic query, get the resource utilization of server nodes:

$$L^i = \{L_1^i, L_2^i, \ldots, L_n^i\}, \ L_j^i \in (0,1), j = 1, 2, \ldots, n \tag{2}$$

Therefore, the load of node i is:

$$L_i = \sum_{j=1}^{n} L_j^i * k_j \tag{3}$$

Due to dynamic feedback queries the servers' load information at a interval T, by the formula (1), it is impossible to calculate the actual load accurately. This will directly affect the distribution of tasks scheduler. So, when evaluating the load, we should consider the change of the load in a T time interval. Here, we introduce the concept of load increment. Load incremental calculation formula:

$$INC(R) = L_i/n \tag{4}$$

n is number of node connections, when $t_1(t_0 \leq t_1 \leq t_0 + T)$, if the server receives a task, load incremental calculation formula of different services is:

$$L_i' = L_i + C/C(S_i) * INC(R) \tag{5}$$

When $t_2(t_0 \leq t_2 \leq t_0 + T)$, if the server has completed a task:

$$L_i' = L_i - C/C(S_i) * INC(R) \tag{6}$$

In the above formulas, C is the node processing power needed to complete the tasks, $C(S_i)$ is the processing capacity of node S_i.

Above all, at time t_0, use the server load information and formulas (5), (6) to calculate the modified node load and take it as current load, and use this as the basis of the selection of servers.

After getting the actual load, the next step is to choose server node to receive tasks. First set a threshold f, every time there is a new connection, calculate node S_m, and add to the collection of M, meets:

$$L_m = \min\{L_i\}, \ i = 1, 2, \ldots, n \tag{7}$$

If the difference between S_i and S_m is less than f, add S_i to M. Calculate the weight of the node:

$$W_i = C(S_i)/\sum_{j=1}^{n} C(S_j), \ i = 1, 2, \ldots, n \tag{8}$$

Select node S_k for task assignment, meets:

$$W_k = \max\{W_i\}, \ i = 1, 2, \ldots, n \cap W_i \in M \tag{9}$$

2.3 Calculation of Optimal Weight Values by Stimulated Annealing

The iteration of the model in simulated annealing algorithm has following basic steps:

(1) Initialization: initial temperature T (sufficiently large), initial solution state k (starting point of the iterative algorithm), n is iteration time.
(2) For $m = 1, 2, \ldots, n$, do step (3) to step (6).
(3) Generate new solution k'.
(4) Compute increment $\Delta t' = E(k') - E(k)$, $E(k)$ is evaluation function.
(5) If $\Delta t' \leq 0$, accept k' as a new current solution, otherwise accept k' with probability $\exp(-\Delta t'/T)$.
(6) If meets the termination conditions, output the current solution as the optimal solution and terminate the procedure. Usually, system will terminate the algorithm if several solutions are not acceptable.
(7) T decrease gradually and $T \geq 0$, go to step (2).

After selecting the task node to deal with the problem, we can get the average response time in one iteration, and use this as an objective function $E(k)$. What we should do is to compute k^i, meets:

$$E(k^i) = \min(E(k)) \tag{10}$$

i is the number of iterations. When the cluster system is in a state of k^j, create a new state k^{j+1} according to the neighborhood function. The neighborhood function results from a probability density function by random sampling.

Take the new solution k^{j+1} as the new weight values, select the node and calculate the response time $E(k^{j+1})$ of next round. Then use Metropolis criterion to determine the probability to accept the new solution as a current solution:

$$p = \begin{cases} 1 & , E(k^{j+1}) < E(k^j) \\ \exp(-\dfrac{E(k^{j+1}) - E(k^j)}{T_0}) & , E(k^{j+1}) \geq E(k^j) \end{cases} \tag{11}$$

After each iteration, cool down T_0:

$$T_{j+1} = r * T_j, r \in (0, 1) \tag{12}$$

In this article, we will set the initial temperature T as 1000, and terminate circulation when $T_j < 0.01.r = 0.9$, so after about 110 times of iteration, we will get the optimal weighting parameters.

For instance, in TSP (Traveling Salesman Problem), there are N cities, a man wants to start from one of them, iterate through all the cities only once, and back to the start point, how to find the shortest route.

The TSP belongs to the so-called NP-complete problem, the precise solution is to find all routes exhaustively, the time complexity is O(N).

By using simulated annealing algorithm, we can get an approximate optimal path relatively quickly. The thought in the process of solving Traveling Salesman Problem:

(1) Create a new traversal path P(i + 1), and calculate the length of P(i + 1), L(P(i + 1)).
(2) If L(P(i + 1)) < L(P(i)), accept P(i + 1) as the new path, otherwise accept it with the probability of formula (11).
(3) Repeat step (1) and (2) until some exit condition is satisfied.

3   Algorithm Performance Testing

In order to test the performance of this algorithm, we use a load balancing testing platform developed by ourselves. There is one device as a scheduler, and three background servers, two client computers simulate multiple requests. Configuration: CPU 2.8 GHz, memory capacity 2G, hard disk 80 GB, all the machines are connected via Ethernet. The scheduler network bandwidth is 100 Mbps, and node bandwidth is 10 Mbps. In order to obtain more accurate results, threshold value f is set to 0.3, load information query cycle is 30 s, to ensure plenty of time to do multiple iterations of simulated annealing to get the global optimal solution.

We respectively collect experimental data of three algorithms, they are simulated annealing algorithm, dynamic feedback algorithm and weighted polling algorithm. Then analyze this data. In this test, we select five parameters of load information: CPU utilization, disk I/O utilization, memory utilization, network bandwidth utilization and number of processes utilization. Before testing the dynamic feedback load balancing algorithm, we set weight values to:

$$k = \{0.3, 0.3, 0.2, 0.1, 0.1\} \tag{13}$$

After many times of query, multiple sets of annealing algorithm has been carried on, the convergence results are as shown in Table 1:

Table 1. Weight values of load information

	CPU	Disk I/O	Memory	Network bandwidth	Number of processes
Dynamic feedback	0.3	0.3	0.2	0.1	0.1
Simulated annealing 1	0.293	0.286	0.209	0.105	0.107
Simulated annealing 2	0.295	0.284	0.211	0.095	0.115
Simulated annealing 3	0.316	0.322	0.179	0.082	0.101
Simulated annealing 4	0.285	0.304	0.213	0.107	0.091

Table 1 shows that, it is inaccurate to determine the weight values of various performance artificially due to the actual situations, for example, when the type of requests have changed, the utilization of different hardware or software will also change. Therefore, the set of initial weights of dynamic feedback is lack of adaptivity. On the other side, simulated annealing algorithm's adaptive ability is stronger [7]. Let's analyze the calculation results of simulated annealing algorithm from Table 1, after the first query, the weights of CPU and I/O are 0.293 and 0.286, which are lower than 0.3 that is expected. This indicates that in the system environment during the first query, the influence of CPU and I/O is lower than before. Compared to the first one, the second query has little change of processors and disks, but the demand on network bandwidth was reduced, it was only 0.095, while the number of processes increased to 0.115, which means in the second round of simulated annealing, the network's weight got lower and the number of processes were more important. By the third round, the influence of network got further reduced, meanwhile the weight of process number decreased, CPU's weight increased by 0.021 and the weight of I/O was improved to 0.322, at this moment, the result of this algorithm converge to the utilization of CPU and disk I/O. During the fourth query, because of the change of requests' type, memory and network bandwidth are got more use. All this data shows that along with the iteration of simulated annealing, the new solution is closer to the real values, which provides more practical and effective data for load balancing strategy.

In Fig. 2, abscissa is the current number of connections of the system. In four times of measurement, the numbers of connections are 20, 80, 100, 150. Y-axis is the average response delay of the servers (millisecond). As is shown in Fig. 2, when the number of connections is 20, the load of the whole system is not big. Response time of three load balancing algorithms is short. The time of weighted polling algorithm is the shortest, this is because two other algorithms generate additional costs in the query of load information of nodes. As the connection number increases, the advantages of the dynamic feedback algorithm and annealing algorithm begin to emerge. When system has reached to 150 connections, the average response time of the simulated annealing algorithm is obviously lower than the other two strategies. This data proves that, in the case of a low number of connections, there is no clear distinction between traditional load balancing algorithm and simulated annealing algorithm. When the load becomes higher, the response time of traditional algorithm become longer. With adjusting weights adaptively, annealing algorithm can provide a basis for allocation strategy accurately so that it can give full play to the performance of the servers.

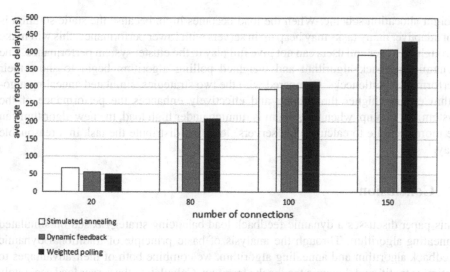

Fig. 2. Average response time

In Fig. 3, abscissa is current number of connections of the system. In four times of measurement, the numbers of connections are 20, 80, 100, 150. Y-axis is the system's throughput (MB/s). When there are a few requests, there is no are not enough to clear gap in these three strategies. With the number of tasks increases to 80, 100 and 150, server nodes' load changes. When the general load is higher, the dynamic feedback algorithm's performance to distinguish server nodes at high load is not precise enough, it can't distribute tasks to nodes at lighter load reasonably, which makes servers at high load overwhelmed and results in the decline of the whole function. The weighted polling algorithm, when under low load, has achieved a good distribution effect with effective

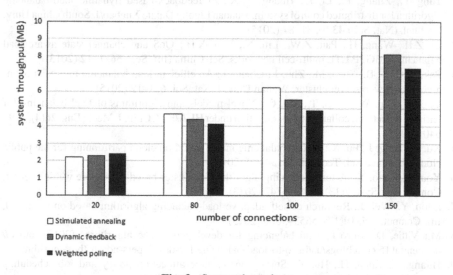

Fig. 3. System throughput

weight allocation scheme. When the load becomes high, because the static solution is not adaptive, many tasks may heap up in servers with lower performance, this will make throughout lower and users can not give full play to the cluster system performance. The dynamic feedback algorithm and weighted polling algorithm begin to show their performance bottlenecks. Compared with the two strategies, simulated annealing algorithm ensures higher throughput, and effectively enhances the performance of the system. To sum up, when the system is running under high load, the new algorithm can be more accurate to calculate the servers' load and distribute the task in a reasonable way.

4   Conclusion

This paper discusses a dynamic feedback load balancing strategy based on simulated annealing algorithm. Through the analysis of basic principle of traditional dynamic feedback algorithm and annealing algorithm, we combine both of their advantages to optimize traditional dynamic feedback algorithm. Calculating the server load accurately guarantees cluster system have higher performance under multiple connection status. The experiment shows that this algorithm in high load condition can make obvious effect.

Still, the algorithm is worth studying. Such as, how to determine the threshold and the size of the number of iteration in order to make convergence rate faster. And how to choose the best neighborhood function in different systems. These are the fields we have to explore in the future research.

References

1. Tang, F., Zhang, P., Li, F., Huang, Z.-X.: PI feedback-based dynamic load-balancing algorithm for distributed control system. Huanan Ligong Daxue Xuebao/J. South China Univ. Technol. (Nat. Sci.) **43**(9), 81–87 (2015)
2. Li, Z.H., Wang, H., Pan, Z.W., Liu, N., You, X.H.: QoS and channel state aware load balancing in 3GPP LTE multi-cell networks. Sci. China (Inf. Sci.) **56**, 1–12 (2013)
3. Pak, I., Qiao, B., Shen, M., Zhu, J., Chen, D.: An efficient load balancing approach for n-hierarchical web server cluster. Wuhan Univ. J. Nat. Sci. **6**, 1–12 (2015)
4. Yue, Y., Fan, W., Xiao, T., Ma, C.: Novel models and algorithms of load balancing for variable-structured collaborative simulation under HLA/RTI. Chin. J. Mech. Eng. **26**(4), 629–640 (2013)
5. Xu, G., Pang, J., Fu, X.: A load balancing model based on cloud partitioning for the public cloud. Tsinghua Sci. Technol. **18**, 34–39 (2013)
6. You, G., Zhao, Y.: A load-balancing algorithm for cluster-based multi-core web servers. J. Comput. Inf. Syst. **7**(13), 4740–4747 (2013)
7. Zhou, Y., Wei, J.: Research of self-adaptive load-balancing algorithm based on session. J. Inf. Comput. Sci. **10**(17), 5653–5660 (2013)
8. MacVittie, D.: Intro to load balancing for developers - The algorithms (2012). https://devcentral.f5.com/blogs/us/intro-to-load-balancing-for-developers-ndash-the-algorithms
9. Huang, L., Chen, H., Hu, T.: Survey on resource allocation policy and job scheduling algorithms of cloud computing. J. Softw. **8**(2), 480–487 (2013)

10. Li, S., Wang, F., Xiao, B., et al.: Study of load balancing technology for EAST data management. Fusion Eng. Des. **89**(5), 750–753 (2014)
11. Divanovic, S., Radonjic, M., Gardasevic, G., et al.: Dynamic weighted round robin in crosspoint queued switch. In: 21st Telecommunications Forum, TELFOR 2013, pp. 109–112. IEEE Press, San Jose (2013)
12. Moharana, S.S., Ramesh, R.D., Powar, D.: Analysis of load balancers in cloud computing. Int. J. Comput. Sci. Eng. **2**(2), 101–108 (2013)
13. Harikesh, S., Shishir, K.: Dispatcher based dynamic load balancing on web server system. Int. J. Grid Distrib. Comput. **4**(3), 89–106 (2011)
14. Wu, Y., Luo, S., Li, Q.: An adaptive weighted least-load balancing algorithm based on server cluster. In: 2013 5th International Conference on Intelligent Human-Machine Systems and Cybernetics, vol. 1, pp. 224–227. IEEE Press, San Jose (2013)
15. Adler, B.: Load balancing in the cloud: Tools, tips and techniques (2012). https://www.rightscale.com/info_center/white-papers/Load-Balancing-in-the-Cloud.pdf

App Store Analysis: Using Regression Model for App Downloads Prediction

Shanshan Wang[✉], Wenjun Wu, and Xuan Zhou

State Key Laboratory of Software Development Environment,
Department of Computer Science and Engineering,
Beihang University, Beijing, China
{wangshanshan,wwj,zhouxuan}@nlsde.buaa.edu.cn

Abstract. App store provides rich information for software vendors and customers to understand the market of mobile applications. However, app store analysis don't consider some vital factors such as version number, app description and app name currently. In this paper we propose an approach that App Store Analysis can be used to predict app downloads. We use data mining to extract app name and description and app rank information etc. from the Wandoujia App Store and AppCha App Store. We use questionnaire and sentiment analysis to quantify some app nonnumeric information. We revealed strong correlations app name score, app rank, app rating with app downloads by Spearman's rank correlation analysis respectively. Finally, we establish a multiple nonlinear regression model which app downloads defined as dependent variable and three relevant attributes defined as independent variable. On average, 59.28 % of apps in Wandoujia App Store and 66.68 % of apps in AppCha App Store can be predicted accurately within threshold which error rate is 25 %. One can observe the more detailed classification of app store, the more accurate for regression modeling to predict app downloads. Our approach can help app developers to notice and optimize the vital factors which influence app downloads.

Keywords: App store · Spearman's rank correlation analysis · Regression analysis · Regression model · App downloads prediction

1 Introduction

The App downloads increases rapidly in App Stores. As of this writing, Wandoujia App Store, AppCha App Store offer tens of thousands apps downloaded by customer daily. We can extract some crucial information from those App Stores. Then we analyze these information to find some relevant factors which influence downloads. App Developers can be noticed that those factors' optimization for high app downloads which indicates profits.

App downloads is an important indicator for usage of app and crucial factor determining the income from this app, so the App Store Downloads analysis is necessary for us.

© Springer Science+Business Media Singapore 2016
W. Che et al. (Eds.): ICYCSEE 2016, Part I, CCIS 623, pp. 206–220, 2016.
DOI: 10.1007/978-981-10-2053-7_19

The features of App Store analysis include:

(a) It provides app ranking analysis that can visualize historical ranking, daily ranking, and featured ratings for the top apps across major mobile app stores.
(b) It also gives app publishers analytics dashboard to track their app downloads, revenues, rankings and reviews across multiple app stores.

Analysis of user feedback allows application vendors to assess customer response to their mobile apps in the market and to figure out plans to enhance the products in the new release.

In our paper we take app rating and user comments into consideration.

App stores provide a rich source of information about apps concerning their name, description, version, rank, rating, price, and popularity attributes. App rating is available concerning the ratings accorded to apps by the users who downloaded them. App price is available, giving price of apps. App description information is available in the descriptions of apps, but it is in free text format, so data mining is necessary to extract the technical details. App version is app version number currently, which is available. App rank is real time app rank in app store. App Popularity is number when we search app name on Google the entries it return.

In this paper we mined the Wandoujia and AppCha App store for data to support App Store Downloads Analysis. Then we quantified data and made Spearman's rank correlation analysis to select relevant attributes with app downloads. Last, we established regression model formula to predict app downloads.

Unlike the traditional app analysis, our paper presents new attributes, such as app version number, app popularity. Some app stores although have app rating, its user comment is also important metric. We used app comment sentiment analysis algorithm to quantify user's comment as a value. To our best knowledge, very few studies were focused on make regression model for each category, even few from over one app store platform. This paper makes regression model from two app store and get some surprising conclusion according to their results comparison.

The primary contributions of this paper are:

1. We quantified App name score and descriptions via questionnaire and sentimental value of App comments (reviews) via sentiment analysis algorithm.
2. We made Spearman's rank correlation analysis to select top 3 relevant attributes with app downloads from app name score, description score, rank, version number, comment sentiment value.
3. We established a non-linear regression model while top 3 relevant attributes defined as independent variable and app downloads defined as dependent variables.

2 Related Works

Harman et al. [1] used the data mining approach to do the Spearman's rank correlation analysis of each factor, such as app prices, user ratings and downloads rank. Take the BlackBerry app store for example, 32108 non-free applications data preprocessing and Spearman's rank correlation analysis did by University College of London. All the applications classified into 19 large category according to their different types and functions explored. One can get that App Price, Rating whose spearman's rho with app downloads are positive, while the user rating's spearman's rho with app downloads is a significant positive correlation.

Pagano and Maalej [2] on the basis of the conclusions of the acquired abroad, expanding a broad study on app design. After got ratings and downloads are positively correlated and app characteristics determine app ratings, we completely set aside app price factors. The kind of app is divided into three categories, are practical (Practical), dependent (Addictive) and Efficient (Effective), then we find feature can make App got high users scores from these three categories respectively. We proposed App design philosophy for UCD (user-center design) and given design criteria for each app category.

Lim and Bentley [3], Kimbler [4] investigated the user feedback in AppStore as an empirical study. User feedback and user involvement are crucial for modern software organizations. Users increasingly rate and review apps in application distribution platforms, called app stores. While part of this feedback is superficial and at most has an impact on download numbers, others include useful comments, bug reports, user experience, and feature requests. This can help developers to understand user needs, extending the application in a 'democratic' fashion, towards crowdsourcing requirements.

User reviews become crucial information gradually, much work has focused on it. Fu [5] collected and studied over 13 million user reviews from Google Play, proposed an integrated system to analyze user reviews from three different levels. Guzman [6] presented an approach for extracting app features mentioned in user reviews and their associated sentiments.

3   App Store Analysis Framework

Figure 1 displays the four phases of our analysis on app store including data collection and parsing phase, data quantification, correlation analysis and regression analysis shown in Fig. 1.

Phase 1 (Data Collection and Parsing)
The goal of this step is to download available data from app store web sites. We implemented a web crawler to collect raw data from Wandoujia App Store [7] and AppCha [8] App Store respectively. Our crawler downloads app information in the following two steps. First, it retrieves all information of every category and sub-category from the App Stores. Then, it scans each subcategory page to collect the list of URLs of all the apps in each subcategory. Following to the URLs of the apps, the crawler can further fetch raw app data by visiting the webpage of the apps.

In the parsing phase, we extract the attributes of the apps by parsing the raw data according to customized search rules. These rules are designed based on identified HTML tags, each of which represents a unique sign for each attribute of the apps. For example, one can extract the name of the apps by searching the value of <span> HTML tag with 'class=title' in Wandoujia App Store. One can also obtain the name of the apps by searching the value of <h1> HTML tag in AppCha App Store.

This process cannot be entirely automated because some attributes field must be manually refined due to the loss or unexpected of app information. For example, the value of an app price should be numeric, but sometimes the retrieved value of the price field shows 'Free'. In such a situation, we have to manually assign a zero value to the field to meet the requirements in AppCha App Store.

Phase 2 (Data Quantification)

Questionnaire for Evaluating App Name and Description. As an important and intuitive aspect of a mobile app, the name and description of the app may have a significant influence on customers whether to download it. When a customer looks for an app for his needs, he often searches the directory of an App Store using the keywords pertaining to the app's features and locates the best app candidates. If he encounters the apps with the attractive name or satisfying app description, the good impression may stimulate him to download the app and try it out in his smart phone. Therefore, the app name and description can be regarded as potentially valuable indicators for predicting the popularity of the app.

In order to evaluate customer perception of the name and description of Apps, we took a survey approach to investigate the popularity of the name the description among potential customers. We designed and distributed a questionnaire for users on the Diaoyanbao website (http://www.diaoyanbao.com/) to rate the name of every app in our database. To get a better feedback, this questionnaire is 10-points. The scoring criteria of App's name is: are attractive (out of 3 points), is impressive (out of 3 points), if people desire to download (out of 4 points). In order to avoid interference in individual cases, this questionnaire limited a tester answering the questionnaire should not exceed three times. This survey we distributed 1050 questionnaires and tack back 927 valid questionnaires.

Quantification of App description is same as app name. The difference is scoring criteria: are articulate (out of 3 points), this description whether to allow people to know functions and features of App (out of 3 points), if people desire to download (out of 4 points).

Sentimental Analysis for User Reviews. User reviews on a specific app demonstrate their experience of using this app and significant impact the download number of this app. Some app stores implement user rating based on the summary of user reviews. It is very important for app vendors to achieve good user ratings to convince potential users to download their apps. Therefore, user ratings and reviews are closed related and can be used as primary factors to predict user downloads.

Wandoujia, one of the two app stores we investigated, doesn't provide explicit user rating to customers. To find out how user feedbacks affect app downloads in that case, we have to analyze user sentiments in user reviews about apps to quantitatively determine if users are praising an app or complaining about the apps.

We apply lexical sentiment analysis which uses dictionaries of words annotated with their semantic orientation (polarity and strength) and incorporates intensification and negation in Wandoujia App Store. The success of lexical sentiment analysis heavily depends upon the accuracy of sentiment dictionary. We conducted frequency statistics on the user reviews used Chinese sentiment dictionary called Hownet [9, 10] Dictionary.

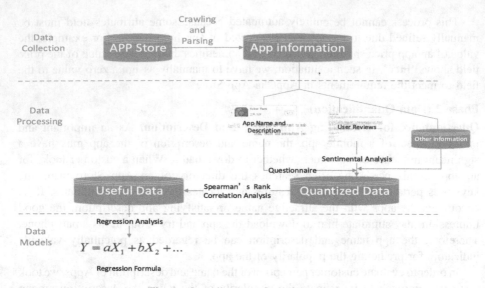

Fig. 1. Overall app analysis framework

Released by CNKI. Based on this extended dictionary, one can run the following sentiment analysis algorithm (Algorithm 1) to calculate each app review's sentiment scores [11–13].

Phase 3 (Spearman's Rank Correlation Analysis)
In order to determine the primary factors affecting app download, we adopt the Spearman's rank correlation analysis [4] for measuring statistical dependence between the downloads and other independent variables such as App Name, Description, App Ranking, App Rating and User Reviews. The Spearman's rho can indicate the significance of the correlation between two variables.

Phase 4 (Regression Analysis)
After determining the primary factors, one can develop a prediction model for estimating app downloads via regression analysis. Such a model enables vendors and customers to understand how app ranking or other factors can influence app downloads in the market. We remove irrelevant factors according to their Spearman's rho with app downloads. The remaining factors defined as independent variables, app downloads defined as dependent variable. Then, we make nonlinear regression to establish a model for app downloads prediction.

For the multiple nonlinear regression, one can use ordinary least squares for optimizing the regression formula to predict app downloads approximately. While the number of independent variables over one regression defined as multiple regression. We transform non-linear regression into a linear regression via change variable, and then use statistical device for linear processing. According to the experience acquired, we can get output and input variables expression, but its coefficient is unknown. We determine the final input-output coefficients based on several times' observations. Last, you can get a preliminary non-linear regression model based on least squares method to obtain the coefficient values.

Algorithm 1. Sentiment Analysis Algorithm

Require: Comment, Hownet Sentiment Dictionary

Output: Sentiment Value

    SentimentValue=0

    SegmentsSet= []

    SegmentsValue= []

Step 1: Perform word segmentation through the review sentences and Copy comment segmentation into SegmentsSet

    Foreach segment in SegmentsSet

        Value=0

        (1) identify the sentiment word in the current segment and compute its value based on its sentimental polarity

        Value=sentiment word value

        (2) identify the adverb for the sentiment word

        Value=Value*weight

        (3) Compute the number of privatives (such as "No")

        If number is odd:

            Value=-1.0 * Value

        End if

        (4) Compute the number of the exclamatory marks at the end of the review and multiply the sentiment value according to the number

        SegmentsValue.append (Value)

        (5) Add up all the sentimental value throughout the Segment set

    foreach value in SegmentsValue

        SentimentValue=SentimentValue+value

    End for

    Return Sentiment Value

Table 1 shows the example of user reviews' sentiment value computed by Algorithm 1.

Table 1. Example of user reviews' sentiment value

Comment	Sentiment value
It's awesome!	2.0
It's awful	−4.0
The app is chargeable? It's disappointing	−8.0
It's great	4.0
It's amazing, I will score it with five stars!	8.0
Not bad	1.6
To be honest, it's the worst app I have used	−8.0
I can't play the video	−4.0

4   Experimental Result Analysis

4.1   Spearman's Rank Correlation Analysis

Spearman's rank correlation analysis is used to identify the significant relevance between each attribute with app downloads.

For Wandoujia App Store, we investigated the relevance of the five features (i.e. App Name Score, App Description Score, Rank, Version Number, Comment Sentiment Value) with app downloads. Figure 2 shows the Spearman's rho of each attribute. Clearly, Comment Sentiment Value and App Name Score attributes have more positive impact on app downloads for every kind of app. Rank is negatively related to app downloads. The correlation values of both Version Number and Description Score have little influence on app downloads.

For AppCha App Store, we investigated the relevance of the five features (i.e. App Name Score, Price, Rating, Rank, Update Times) with app downloads. Figure 3 shows the Spearman's rho of each attribute. Clearly, Rating and App Name Score attributes have positive impact on app downloads for every kind of app. And interestingly, Rank is

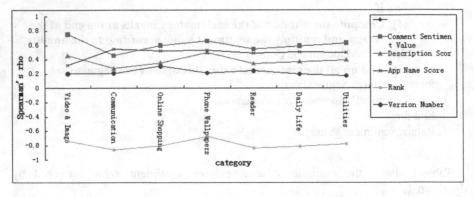

Fig. 2. Spearman's rho between each attribute and app downloads in Wandoujia App Store

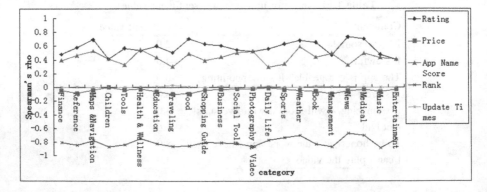

Fig. 3. Spearman's rho between each attribute and app downloads in AppCha App Store

negatively related to app downloads. The correlation values of both Price and Update Times stay close to the zero value, which indicates that both features have little inflence.

4.2   Regression Analysis

(1) Wandoujia App Store

From Fig. 2, one can observe that the three major factors including App Rank, Comment Sentiment Value, Name Score are more relevant with App Downloads than the other attributes. Therefore, we select these primary attributes to develop a multiple nonlinear regression model for predicting app downloads. Table 2 displays the variables in the regression model where App Rank, Comment Sentiment Value, Name Score are defined as independent variables and app downloads is defined as the dependent variable.

Table 2. Multiple nonlinear regression symbol table

Variable	Downloads	App comment sentiment value	App name score	App rank
Symbol	Y	X_1	X_2	X_3

We trained the regression model with 1739 apps instances in Wandoujia App Store. The multiple nonlinear regression formula is shown in:

$$Y = 11671 - 584X_1 + 4860X_2 - 1.168X_3 - 4.35X_1^2 - 479X_2^2 - 0.0196X_3^2$$
$$+ 10235X_1X_2 + 0.38X_1X_3 + 0.297X_2X_3 - 0.04X_1X_2X_3$$

The rest of 1676 app instances are utilized as testing data to verify the accuracy of regression model. The validation result shows that the prediction model has a very poor performance: only 10 % apps can achieve the prediction error rate at less 50 %. Therefore, the nonlinear regression model is not applicable for the whole Wandoujia app store, app downloads. We have to examine the effectiveness of the multiple nonlinear regression model in each category instead of the entire app store.

Based on the category defined by the Wandoujia app store, we trained non-linear regression models for every category. Table 3 displays the accuracy of these models. In average, the model is capable of giving a good estimation of download numbers for 54.42 % app instances within seven categories.

Here, we present Video & Image category as an example of the category-specific regression models as its prediction accuracy is the highest among them. The regression formula trained over 249 app instances of the Video group is shown below:

$$Y = 4475.03 + 111.7X_1 - 1915.8X_2 + 6.3X_3 + 0.42X_1^2 + 181.2X_2^2 - 0.006X_3^2$$
$$- 18.4X_1X_2 - 0.025X_1X_3 - 0.5X_2X_3 + 0.002X_1X_2X_3$$

Table 3. Prediction accuracy at rough modeling level in Wandoujia App Store

Category	Accuracy
Video & Image	0.7587
Communication	0.4465
Online Shopping	0.6772
Phone Wallpapers	0.5185
Reader	0.4309
Daily Life	0.4875
Utilities	0.4904
Average	0.5442

The rest of 249 app instances are utilized as testing data to verify the accuracy of regression model.

Figure 4 demonstrates the testing result. Although the prediction model seems to give a reasonable estimation for app downloads, the threshold for the error rate is still pretty high to 50 %, thus failing to ensure the high accuracy of the prediction.

Figure 5 shows that downloads prediction error is over large while app downloads is over 2000 thousands, this implies app prediction is not effective. Meanwhile, downloads prediction error is also large while app downloads is under 1000. So, we may remove some app instances which app downloads under 1000 and over 2000 thousands.

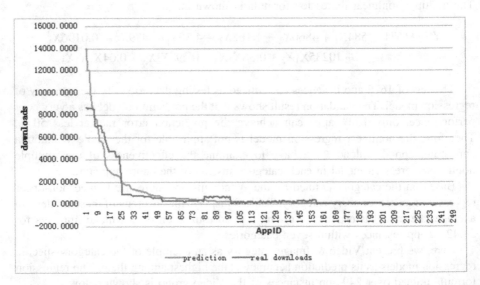

Fig. 4. Comparison between app downloads and prediction result (Apps belonging to the Video & Image category) (Color figure online)

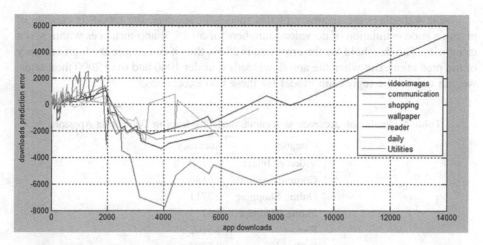

Fig. 5. Downloads prediction error of apps at category level (Color figure online)

After eliminating the data samples under 1000 and over 2000 thousands, which often yield high predication errors, we trained the regression model with 170 app instances of Video & Image category again. Then the regression formula is:

$$Y = 0.02 + 386.01X_1 - 1374.37X_2 + 22.25X_3 + 0.62X_1^2 + 162.89X_2^2 - 0.007X_3^2$$
$$- 45.28X_1X_2 - 0.767X_1X_3 - 1.99X_2X_3 + 0.0758X_1X_2X_3$$

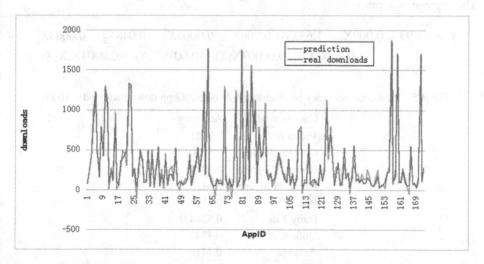

Fig. 6. Comparison between app downloads and prediction within a narrowed range of download number (Apps belong to Video & Image) (Color figure online)

The rest of 169 app instances as testing data to verify the accuracy of regression model. Figure 6 shows the improved prediction result.

Table 4 displays the accuracy of this model. In average, the model is capable of giving a good estimation of download numbers for 59.28 % app instances within seven categories. The threshold for the error rate is 25 %, thus it can ensure the high accuracy of the prediction. But when the app downloads is under 1000 and over 2000 thousands, we can trained the regression model for these two exceptions.

Table 4. Prediction accuracy at accurate modeling level in Wandoujia Appstore

Category	Accuracy
Video & Image	0.5780
Communication	0.6186
Online Shopping	0.7711
Phone Wallpapers	0.5714
Reader	0.5102
Daily Life	0.5704
Utilities	0.5301
Average	0.5928

After Spearman's rank correlation analysis, one can observe App name score, Rank, Comment sentiment value are still relevant attributes with app downloads when app downloads under 1000, Rank is not any longer a relevant attributes with app downloads while app downloads over 2000 thousands.

When app downloads is under 1000:
The regression formula is:

$$Y = -1.93 - 0.009X_1 + 1.63X_2 + 0.006X_3 + 0.0003X_1^2 - 0.048X_2^2 - 0.0003X_3^2$$
$$+ 0.27X_1X_2 - 0.00007X_1X_3 - 0.0019X_2X_3 - 0.0004X_1X_2X_3$$

Table 5. Prediction accuracy in Wandoujia Appstore (App downloads under 1000)

Category	Accuracy
Video & Image	0.8621
Communication	0.6270
Online Shopping	0.6889
Phone Wallpapers	0.6416
Reader	0.8125
Daily Life	0.8244
Utilities	0.7347
Average	0.7416

Table 5 displays the prediction accuracy when app downloads under 1000. In average, the model is able to give a good estimation of download numbers for 74.16 % app instances within seven categories. The threshold for the error rate is also 25 %, thus it can ensure the high accuracy of the prediction when app downloads under 1000.

When app downloads is under 2000 thousands:
The regression formula is:

$$Y = 35070 - 588X_1 + 3508X_2 + 3.3X_1^2 + 351.7X_2^2 - 57.9X_1X_2$$

Table 6. Prediction accuracy in Wandoujia Appstore (App downloads over 2000 thousands)

Category	Accuracy
Video & Image	0.9583
Communication	0.8186
Online Shopping	0.9711
Phone Wallpapers	0.8714
Reader	0.8102
Daily Life	0.7704
Utilities	0.6301
Average	0.8328

Table 6 shows the prediction accuracy when app downloads over 2000 thousands. In average, the model is capable of giving a perfect prediction of app downloads for 83.28 % app instances within seven categories. The threshold for the error rate is also 25 %, thus it can ensure the high accuracy of the prediction when app downloads over 2000 thousands.

(2) **AppCha App Store**

From Fig. 3, one can observe that the three major factors including App Rank, App Rating, App Name Score are more relevant with App Downloads than the other attributes. Therefore, we select these primary attributes to develop a multiple nonlinear regression model for predicting app downloads.

Table 7 displays the accuracy of regression models which app downloads between 100 and 10000 in AppCha App Store. In average, the model is capable of giving a good

Table 7. Prediction accuracy at accurate modeling level in Appcha Appstore

Category	Accuracy	Category	Accuracy
Finance	0.7833	Social Tools	0.8134
Reference	0.6528	Photography & Video	0.5442
Maps & Navigation	0.7934	Daily Life	0.6542
Children	0.5433	Sports	0.6345
Tools	0.6235	Weather	0.5412
Health & Wellness	0.7501	Book	0.6423
Education	0.8241	Management	0.6234
Traveling	0.7315	News	0.6431
Food	0.6271	Medical	0.7323
Shopping Guide	0.6526	Music	0.5633
Business	0.7631	Entertainment	0.5343
Average: 0.6668			

estimation of app downloads for 66.68 % app instances within 24 categories. The threshold for the error rate is also 25 %, thus it can ensure the high accuracy of the prediction in AppCha App Store.

One can observe that App Rank, App Rating (App Comment Sentiment Value), App Name Score is more relevant to App Downloads than other attributes. Figure 7 shows variance of each attribute's Spearman's rho with app downloads in Wandoujia and AppCha Store.

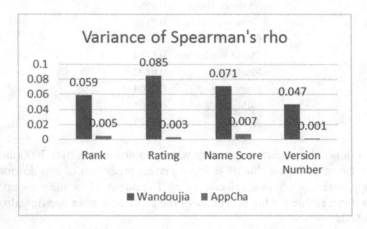

Fig. 7. Variance of each attribute's spearman's rho with downloads in Wandoujia and AppCha Store (Color figure online)

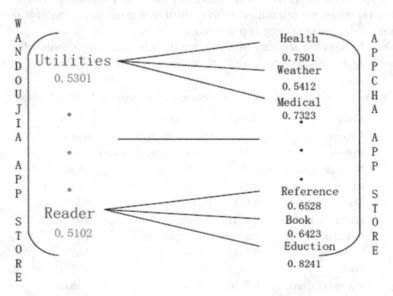

Fig. 8. Reflection of categories from Wandoujia to AppCha App Store

Apparently, the variance of Spearman's rho on Wandoujia is much higher than that on AppCha App Store. The fluctuations of each attribute on Wandoujia is much higher than AppCha App Store, implying that the more detailed classification, the more stable of various factors. Regression model established on AppCha to predict app downloads is more accurate than Wandoujia, which indicated regression model established on AppCha is closer to real downloads scenes.

Figure 8 shows the reflection of categories from Wandoujia to AppCha App Store. One can observe that prediction accuracy is higher in AppCha than Wandoujia App Store correspondingly, thus it implies the more detailed classification for App Store, the more useful for app downloads prediction using regression model.

5 Conclusion and Future Works

App stores provide a software development space and market place where we can extract some useful information and dispose it to get conclusion we want.

We have proposed an approach to extract name, comment, description, version name, rank of app in two app stores. Some information are not digital, so we quantify it via questionnaire and sentiment analysis. Then we make Spearman's rank correlation analysis about each attribute with downloads. According to spearman's rho, we select the more relevant attributes to regression modeling. In modeling phase, app downloads regarded as dependent variable, relevant attributes as independent variable.

We found a multiple nonlinear regression model formula to predict app downloads of each app category. These models are capable of giving a good estimation of download numbers for 59.28 % app instances within seven categories in Wandoujia App Store and 66.68 % app instances within 24 categories in AppCha App Store averagely. The threshold for the error rate is 25 %. But former models just suitable for appropriate app downloads. We adopt another regression models to predict app downloads when downloads too little or too large. One can observe that prediction accuracy is 0.7416 and 0.8238 when app downloads is under 1000 and over 2000 thousands respectively in Wandoujia App Store averagely.

We can conclude that the more detailed classification of app platform, the more accurate for regression modeling to predict app downloads.

In the future, we also intend to consider other attributes app image, app tag etc. because of various factor to influence app downloads. We will optimize comment sentiment analysis algorithm for basis of segment location currently. The relative sensitivity of our model regressed for App downloads predication is not high, and the model is based on the App subcategories, universality is not strong. Future research can focus on the optimization model, we will try to further improve the accuracy and universality of the model.

Acknowledgment. This work was supported in part by grant from State Key Laboratory of Software Development Environment (Funding No. SKLSDE-2015ZX-03) and NSFC (Grant No. 61532004).

References

1. Harman, M., Jia, Y., Zhang, Y.: App store mining and analysis: MSR for app stores. In: 2012 9th IEEE Working Conference on Mining Software Repositories (MSR), pp. 108–111, 2–3 June 2012
2. Pagano, D., Maalej, W.: User feedback in the appstore: an empirical study. In: 2013 21st IEEE International Requirements Engineering Conference (RE). IEEE (2013)
3. Lim, S.L., Bentley, P.J.: Investigating app store ranking algorithms using a simulation of mobile app ecosystems. In: 2013 IEEE Congress on Evolutionary Computation (CEC), pp. 2672–2679, 20–23 June 2013
4. Kimbler, K.: App store strategies for service providers. In: 2010 14th International Conference on Intelligence in Next Generation Networks (ICIN), pp. 1–5, 11–14 October 2010
5. Fu, B., Lin, J., Li, L., et al.: Why people hate your app: making sense of user feedback in a mobile app store. In: Proceedings of the 19th ACM SIGKDD International Conference on Knowledge Discovery and Data Mining, pp. 1276–1284. ACM (2013)
6. Guzman, E., Maalej, W.: How do users like this feature? A fine grained sentiment analysis of app reviews. In: 2014 IEEE 22nd International Requirements Engineering Conference (RE), pp. 153–162. IEEE (2014)
7. Wandoujia App Store [EB/OL]. http://www.wandoujia.com/apps
8. AppCha App Store [EB/OL]. http://www.appcha.net/
9. Goul, M., Marjanovic, O., Baxley, S.; Vizecky, K.: Managing the enterprise business intelligence app store: sentiment analysis supported requirements engineering. In: 2012 45th Hawaii International Conference on System Science (HICSS), pp. 4168–4177, 4–7 January 2012
10. Severance, C.: Toward developing an education app store. Computer **44**(8), 107–109 (2011)
11. Zhao, Y., Niu, K., He, Z., Lin, J., Wang, X.: Text sentiment analysis algorithm optimization and platform development in social network. In: 2013 Sixth International Symposium on Computational Intelligence and Design (ISCID), pp. 410–413, 28–29 October 2013
12. Li, S., Zhang, H., Xu, W., Chen, G., Guo, J.: Exploiting combined multi-level model for document sentiment analysis. In: 2010 20th International Conference on Pattern Recognition (ICPR), pp. 4141–4144, 23–26 August 2010
13. Sui, H., You, J., Zhang, H., Zhou, W.: Sentiment analysis of Chinese micro-blog using semantic sentiment space model. In: 2012 2nd International Conference on Computer Science and Network Technology (ICCSNT), pp. 1443–1447, 29–31 December 2012

Application Progress of Signal Clustering Algorithm

Chujie Deng, Jing Qi, Mei Li[✉], and Xuanchicheng Luo

School of Information Engineering, University of Geosciences (Beijing),
Beijing, China
maggieli@cugb.edu.cn

Abstract. Clustering algorithm, which is a statistical analysis method for research in classifications, plays an important role in data mining algorithm. Clustering algorithm based on similarity, and is easy to combine with other methods in optimization. In this review, signal clustering algorithm is introduced by discussing of the clustering parametric in different signal clustering algorithms. In order to develop traditional algorithm, we introduce a series of improvement, development and application of the methods in recent years. Finally, we make an outlook of the future direction and content of the research in this field.

Keywords: Signal · Clustering algorithm · K-means · FCM

1 Introduction

Cluster analysis was originated in taxonomy. With the development of science and technology, people have increasingly requirements towards classification. So they refer the mathematical tool in taxonomy, forming numerical taxonomy. Multivariate analysis technology was appointed into numerical taxonomy, forming the clustering algorithm.

From a practical point of view, traditional clustering algorithms are varied, it can be a stand-alone tool to obtain the distribution of data. So that we can observe each cluster data's feature, and make further analysis on a specific set of cluster. And it can also be used as a pretreatment step to other algorithms.

With the development of data, traditional clustering algorithm has been improved, which is simple and intuitive. It is widely used in data mining analysis. Combing and analyzing the literature in this field, we write this review aimed at a comprehensive understanding of the signal clustering algorithm. We have made series of ways to improve, develop and apply signal clustering algorithms. We made an outlook of the future direction and content of research in the field ultimately.

There is a large amount of sources transmit signal everywhere and every moment. In order to improve the efficiency of data applications, and discover valuable signal, we need to study and judge the signal source in advance. As a result, it is necessary to study valid signal recognition methods, such as signal clustering. To fulfill the urgent needs, many scholars [1–4] dedicated to research on large amounts of data for a fast-identification, high-accuracy, and clear-semantics clustering algorithm.

© Springer Science+Business Media Singapore 2016
W. Che et al. (Eds.): ICYCSEE 2016, Part I, CCIS 623, pp. 221–227, 2016.
DOI: 10.1007/978-981-10-2053-7_20

2   Signal Clustering Algorithm

Clustering, which is based on the similarity of the data object, polymerizes data into different clusters. In the same cluster, data objects have high similarity, but the different clusters of data objects have greater dissimilarity. The difference between clustering and classification are that clustering algorithms can process data in an unsupervised mode, and you will not to input the data label. Clustering algorithms mainly include partition-based, division-based, density-based and grid-based methods.

2.1   Clustering Algorithm Based on Partition

We assign all objects to several signal clusters, making instances of the same cluster gather around a center, and the distance between them is relatively close. At the same time, the distance between the different clusters examples is relatively far. The main representative of the algorithm are K-means, K-medoids and Clarans.

There is a difference between K-means and K-medoids. The central point of K-means is the average of all current data points, but K-medoids select the point with the closest distance to the average value. The following equations describe K-means and K-medoids.

K-means: Assign each observation to the cluster whose mean yields the least within-cluster sum of squares (WCSS). Since the sum of squares is the squared Euclidean distance, this is intuitively the "nearest" mean.

$$c^{(i)} = arg \min_j \left\| x^{(i)} - u_j \right\|^2 \tag{1}$$

K-means algorithm minimize parameter J.

$$J = \sum_{n=1}^{N} \sum_{k=1}^{K} \gamma_{nk} \left\| X_n - \mu_k \right\|^2 \tag{2}$$

Calculate the new means to be the centroids of the observations in the new clusters.

$$u_j = \frac{\sum_{i=1}^{n} 1\{c^{(i)} = j\} x^i}{\sum_{i=1}^{n} 1\{c^{(i)} = j\}} \tag{3}$$

The algorithm has converged when the assignments no longer change.

K-medoids: K-medoids changes a random dissimilarity measure Function-V to the Euclidean distance in original objective Function-J.

$$\tilde{J} = \sum_{n=1}^{N} \sum_{k=1}^{K} \gamma_{nk} V(x_n, \mu_k) \tag{4}$$

2.2 Clustering Algorithm Based on Density

We assign the field radius and the density threshold. We are looking for a target, in which field radius it contains the number of objects greater than or equal to the density threshold, and we see the target as a center point. Otherwise, we regard it as a boundary point. The main representatives of the algorithm are DBSCAN and DENCLUE.

2.3 Clustering Algorithm Based on Grid

We separate the possible values of each attribute into a number of adjacent spaces to create a collection of the grid cell. And every object falls into a grid cell, the corresponding attribute range of grid cell contains the value of the object.

2.4 Fuzzy C-means Algorithm

We square weighted the objects and its distance from the cluster center with subjection, to get a general description of the fuzzy clustering objective function [5]. FCM receives each sample point membership of the entire clustering center by optimizing the objective function. So that we can find the type of sample points, in order to achieve automatic sample classification. FCM is a clustering algorithm, which is based on membership, to determine which set the data belongs to.

3 Application Progress in Signal Clustering

3.1 Signal Clustering Algorithm in Military Field, Take Radar Signal as Example

Under normal circumstances, we cluster radar signal pulse to classify the radar signal for the following parameter estimation, emitter recognition, threat discrimination and combat situation analyzing by feature space composed by time of arrival (ToA), pulse width (PW), pulse angle of arrival (AoA), pulse amplitude (PA), radio frequency (RF) [6].

We study the different clustering algorithms for radar signal and make the following review.

3.1.1 K-means Clustering Algorithm Improved by LF Intelligent Optimization Algorithm

K-means clustering algorithm has already applied in radar signal sorting, but the clustering results are sensitive to isolated points and noise points. So the scholars usually improve the K-means algorithm, combining with other algorithm, as the literature [7] algorithm: based on ant (LF) intelligent optimization algorithms to improve the K-means clustering algorithm.

First, we use the ant colony algorithm to place a radar signal on a random plane. Then, it will generate a set of virtual code of conduct randomly. According to this criterion, we could get properties of similar signals clustering into a category eventually, and obtain the initial centroid and the number of clusters. This method can solve the difficulty of determining the number of clusters and cluster centers.

3.1.2   FCM Clustering Algorithm Based on Partition Matrix Optimization

The FCM algorithm exists problems that the number of clusters is difficult to determine, the initial value is unstable and of the local extreme instability and other issues about local extrema. So, Baraldi proposed fuzzy C-means clustering algorithm based on partition optimized matrix (FCM) [8].

To get the best partition matrix according to the principle of division of matrix and cluster center initialization at first and signal samples can be aggregated into different categories by the principle of the maximum degree of membership and use the signal characteristics of radar radiation generated by simulation as separate arrays. Then a higher similarity signal was classified as a category according to FCM clustering thought. This allows the merging of data about a large number of miscellaneous radar pulses so as to achieve a better clustering effect.

3.2   Signal Clustering Algorithm in Medical Field, Take ECG as Example

Since people raise higher requirements to identify and analyze abnormal ECG, clustering algorithm begin to be used for the sorting of the abnormal ECG process.

ECG feature extraction is mainly the detection of QRS-wave, P-wave, T-wave and R-wave. P-wave represents atrial depolarization vector. T-wave represents the process of both sides of the ventricular repolarization. Accurate detection of the R-wave represents the ability of the accuracy of QRS-wave detection.

3.2.1   Abnormal ECG Classifying Algorithm Based on FCM

The FCM clustering algorithm is widely used in ECG. As a result of lacking prior knowledge, most of fuzzy clustering algorithm needs to pre-set and depends on the number of prototype cluster-c, the weighting factor-m, and fuzzy cluster center initialization parameters. To improve this technology, Yao Cheng, in 2012, put forward a method [9] directly judged by original ECG signal, to improve the accuracy of abnormal ECG classification. First, this method uses the statistic of ECG to reflect heart disease characteristic data. Secondly, we can put forward a standard of the logical ECG by using the abnormal ECG data. Finally, we could utilize logical judgment and clustering algorithm to raise one kind of algorithm for classifying abnormal ECG, which is called LCFCM [10].

3.2.2   K-means Clustering Method Based on Simulated Annealing Algorithm

K-means clustering algorithm for ECG, the result will be easily influenced by centroid, appearing locally optimal solution and isolated point. A literature [11] raised K-means clustering method based on simulated annealing algorithm to solve above problem.

According to an initial solution, we set an initial target function. After calculating the probability, with continuous iteration, we calculate the difference of function.

By progressively determining and discarding, we will receive the optimal solution with traversing a large space.

The algorithm is of strong local search capabilities, being able to reach the global optimal solution.

3.3 Signal Clustering Algorithm in Engineering Field, Take the Rotor Rubbing, Tank Bottom Corrosion Acoustic Emission Signal as Example

There are many characteristic parameters with acoustic emission signals, for example, Waveform Amplitude (A), Rise Time (R), Energy (E), Duration (D), Pulse Index (X), Margin Index (W), Steep Wave Clicks (Z), Event Times (Y), Number of Rings (C) etc. They can be described as characteristic parameters of waveform, which can be used for clustering.

At the same time, the scholar introduces three derived AE parameters [12]: RA (Rise time/Amplitude); AF (Number of rings/duration); RD (Rise time/duration).

3.3.1 Transmission Signal of Rotor Rubbing Sound Based on K-means Clustering Algorithm

Jin et al. [13] proposed an approach to cluster rotor rub-impact acoustic emission signals by k-means. He selected the waveform amplitude, pulse index, margin indicators and kurtosis indicators as characteristic parameters to cluster and analysis acoustic emission signal. Firstly, using k-means clustering to analysis acoustic emission detection signal under the situation whether happened collision and friction and obtained the cluster center. With further analysis in conjunction with other algorithms, we finally get acoustic emission characteristics and lay the foundation for Rubbing Fault Diagnosis.

By selecting characteristic parameters, the effect of the threshold value is avoided and acoustic emission signals can be identified efficiently with high accuracy.

3.3.2 Pitting Acoustic Emission Signal Based on K-means Clustering Algorithm

Bi Haisheng apply acoustic emission technology for monitoring Tank Bottom online and fault identification [14]. For detection of Tank Bottom Corrosion acoustic emission signal, the results analyzed by a single parameter are often quite different from the activity. Therefore, title used seven characteristic parameters of A, E, D, C, RA, AF and RD to cluster. Application of K-means clustering algorithm to identify the acoustic emission signal used distance as an index from the signal target cluster and identifies three categories of typical acoustic emission signals. By clustering and identifying the fault signal, the test results will help improve the reliability and reduce the risk of running the tank, ensuring its safe operation.

This algorithm is based on the need to determine the number k of category in advance. It will make fault detection easier if this algorithm is combined with other algorithms which can change the k value adaptively.

3.4   Clustering Algorithm in FM Radio and Case in the Field of Computer Science

K-means algorithm is adopted to identify broadcasting and aviation voice signals from interfering signals in space communications through traditional methods. However, the algorithm cannot completely classify signals accurately and automatically. For this situation, it was proposed voice signals (f-kmd) algorithm based on K-medoids and FCM, which focuses on FM broadcast and aviation signal classification [15].

Characteristic parameters of the speech signal include the mean, variance, short-time average energy, average zero-crossing rate, normalized amplitude. Short voice data was clustered by K-means and K-medoids and FCM respectively at first [16]. And then we used those algorithms on entire speech data, which means three kinds of combinations of two clustering algorithms, finding the optimal clustering mode: the first FCM clustering, then K-medoids clustering.

The effect of the conjunction between FCM and K-medoids is better than individual algorithm. F-kmd clustering algorithm has good stability and the correct rate was significantly higher.

4   Development and Prospects

Clustering algorithm of signal varies, which has shown its character and charm in many fields. However, the traditional algorithm and improved algorithm theory is far from mature and practical application is far from showing its true potential. Many challenging issues waiting to be solved are listed below:

(1)   Strengthening theoretical Study for clustering algorithm [17]. Developing new mathematical analysis and modeling tools further is very important for all kinds of basic theory of algorithms in particular. We have not yet found the mathematical discourse about reasonable choice on the number of cluster centers and clusters in the absence of prior knowledge. Therefore, the mathematical theory based on a clustering algorithm applied to the signal aspect will become an important topic for future research.

(2)   The algorithm can be improved with other types of methods to develop the integrated use of hybrid-optimization method. Many authors in this respect have made a good attempt. It will become the focus of research on signal clustering to combine traditional or improved clustering algorithms with neural networks, fuzzy control, genetic algorithms, and simulated annealing algorithm and so on.

(3)   The new theory must be tested by practice in which we can find new problems, so as to promote the theory forward. Although clustering algorithms have been promoted to the application in many areas in recent years, most of them just a simple simulation algorithm in the field of application. Therefore, we should fully dig out potential of signal clustering in the practical application. Furthermore, the hardware implementation of the signal clustering algorithm will also become one of the hot research directions.

Acknowledgements. This work is financially supported by the National Natural Science Foundation of China (Grant No. 41572347).

References

1. Wang, J., Zhang, B.: A radar signal sorting algorithm based on dynamic grid density clustering. J. Mod. Electron. Tech. **36**, 1–4 (2013)
2. Li, X.Y., Yang, C.Z., Qu, W.T.: A radar signal sorting algorithm based on adaptive grid density clustering. J. Aerosp. Electron. Warfare **29**, 51–53 (2013)
3. He, X.W., Yang, C.Z., Zhang, R.: A radar signal sorting algorithm based on improved grid clustering. J. Radar ECM **31**, 43–49 (2011)
4. Zhang, C.C.: Radar Emitter Signal Deinterleaving Based on Support Vector Clustering. Xidian University, Xian (2012)
5. Xie, T.J.: Clustering Algorithm Summary (in Chinese). Beijing University of Post and Telecommunications, Beijing (2014)
6. Xiang, X.: Research of Unknown Radar Signal Sorting Algorithm. Xidian University, Xian (2011)
7. Zhao, G.X., Luo, L.Q., Chen, B.: Improved artificial fish school algorithm applied in radar signal sorting. J. Electron. Inf. Warfare Technol. **7**, 142–146 (2009)
8. Baraldi, A., Blonda, P.: A survey of fuzzy clustering algorithms for pattern precognition-part1 and part2. J. IEEE Trans. Syst. Man Cybern. Part. B **29**, 778–801 (1999)
9. Yao, C.: The Study on the Key Technology of ECG Signal Intelligent Analysis. Jilin University, Jilin (2012)
10. Lin, Z.T., Ge, Y.Z.: A study on clustering analysis of Arrhythmias. J. Biomed. Eng. **23**, 999–1002 (2006)
11. Zhang, X.R.: Study on the Improved Methodology of ECG Clustering Strategy. University of Science and Technology, Harbin (2015)
12. Giuseppe, C., Luigi, C., Edoardo, P.: Evaluation of increasing damage severity in concrete structures by cluster analysis of acoustic emission signals. In: European Conference on Acoustic Emission Testing, vol. 29, pp. 8–10 (2010)
13. Jin, Z.H., Tang, F.L., Zhao, C.M.: Extraction based on clustering analyses on rotor rubbing sound emission characteristics. J. Shenyang Univ. Chem. Technol. **29**, 342–346 (2015)
14. Bi, H.S., Li, Z.L., Hu, D.D.: Cluster analysis of acoustic emission signals during tank bottomsteel pitting corrosion process. J. China Univ. Petrol. **39**, 145–152 (2015)
15. Zhang, Z., Ma, F.L., Pei, Z.: Recognition of Aviation Interference Signal Based on K-means Clustering Algorithm. Publishing House of Electronics Industry, Beijing (2013)
16. Hu, A., Pei, Z.: K-medoids and FCM fusion clustering application research on broadcast and aviation speech signal classification. J. Univ. Jinan **30**, 1671–3559 (2016)
17. Duan, H.B., Wang, D.B., Huang, X.H.: Development on ant colony algorithm theory and its application. J. Control Decis. 19, 1322–1326, 1340 (2004)

Automated Artery-Vein Classification in Fundus Color Images

Yi Yang[1], Wei Bu[2], Kuanquan Wang[1], Yalin Zheng[3], and Xiangqian Wu[1(✉)]

[1] School of Computer Science and Technology, Harbin Institute of Technology,
Harbin 150001, China
{yangyi_cs,wangkq,xqwu}@hit.edu.cn
[2] Department of New Media Technologies and Arts, Harbin Institute of Technology,
Harbin 150001, China
buwei@hit.edu.cn
[3] Department of Eye and Vision Science, University of Liverpool, Liverpool, L7 8TX, UK
Yalin.Zheng@liverpool.ac.uk

Abstract. The estimation of Arterio-Venous ratio (AVR) is an important phase in diagnosing various vascular diseases e.g. Diabetic Retinopathy. For calculating this value, it is essential to differentiate the vessels into arteries and veins. This paper presents a novel structural and automated method for artery/vein vessels classification in retinal images. Our method is tested on DRIVE database and the classification accuracy is 88.7 % for pixels and 89.07 % for vessel lines, respectively, which demonstrate the effectives of our approach. Our method will help to achieve the fundus disease surveillance on mobile and remote medical treatment. It has a remarkable social significance.

Keywords: Diabetic Retinopathy · Vessel classification · Arteries and veins · Feature extraction · Support vector machines · Adaptive histogram equalization

1 Introduction

Retinal imaging provides a non-invasive opportunity for the diagnosis of many diseases. Diabetic Retinopathy (DR), a micro vascular complication usually happened to diabetes patients, is a major cause of visual loss in working age population of developed countries. Arterio-Venous ratio (AVR) changing is an important indicator of this disease. There are also other diseases like high blood pressure or disease of the pancreas could have a correlation with an abnormal AVR value. The classification of retinal blood vessels into arteries and veins is a tough work and has been less explored in the literature than the fundus vessel segmentation. Otherwise arteries and veins are not easily distinguished with little differences, especially two different types vessels cross over each other. But manually classification of the vessels can be time-consuming and expensive. The purpose of this research is to find an automated method for the classification of fundus vessel into arteries and veins, and easy to calculate AVR value.

The problem of classifying arteries and veins can be fallen into two categories: automated and semi-automated methods. The automated methods are based on feature

© Springer Science+Business Media Singapore 2016
W. Che et al. (Eds.): ICYCSEE 2016, Part I, CCIS 623, pp. 228–237, 2016.
DOI: 10.1007/978-981-10-2053-7_21

extraction from major vessels. First, we get the structure of the vascular network, then extract the vessel skeleton of main vessels or the centerline pixels. For each pixel or skeleton, various features are calculated, and finally each pixel or skeleton is assigned an artery or a vein label. Grisan and Ruggeri [1] performed quadrant-wise vessel classification in a concentric zone around the optic disc using fuzzy C-Mean clustering. Kondermann et al. [2] used support vector machines and neural networks combined with principal component analysis (PCA) features obtained from small vessel image paths. Relan et al. [3] automatically classify the main vessels in each optic nerve-centered quadrant based on color features using a Gaussian mixture model, while Vazquez et al. [4] employed a minimal path approach with which they connected a set of extracted vessel segments. Mirsharif et al. [5] firstly enhanced the image using different histogram equalization methods, then they estimated pixel features and finally corrected the misclassifications bifurcation points. The semi-automated methods, firstly initial pixels on the main vessels are labeled with artery or vein, then those labeled pixels treated as seed points, propagate toward smaller and thinner vessels using the structural characteristics. Rothaus et al. [6], proposed a semi-automated method for separation of artery map from veins in which they propagate some initial manual edge labels throughout the graph by solving a constraint-satisfaction problem. Joshi et al. [7] first separated their vascular graph using Dijkstra's shortest-path algorithm to find different subgraphs, then labeled each subgraph as either artery or vein using a fuzzy classifier. These methods have various problems in reality. Some methods need to choose seed points, or mark the start and end points of a vessel, or only classify a small region and have a huge limitation.

In this paper we present a novel structural and automated method for artery/vein vessels classification in retinal images. Our method consists of three main parts. First, image enhancement are employed to improve the images. Next, we extract vessel segments and a thinning algorithm applied to the vessel. Then a set of novel features is extracted to separate arteries from veins. Finally, using SVM classifier classify sample points, and a post-processing step is applied to correct the false classify points.

The rest paper is organized as follows: Sect. 2 is devoted to introduce our method. In Sect. 3 experimental results are discussed in detail and Sect. 4 concludes the research.

2   Method

Our method consists of three main steps: Image Preprocessing, Vessel Classification, and Post-processing. Image preprocessing can be divided into three steps. The first step is to enhance the image, and extract the vessel segments from background and then get vessel skeletons by thinning vessels. We cut the vessel into pieces by removing crossover and bifurcation points. Next, we calculate various features for every pixels on different color space such as RGB, HSL, LAB, etc. Last, we use the SVM classifier to classify sample points and correct false pixels.

2.1 Preprocessing

Uneven lightness and color variation through retinal images are important problems in retinal image analysis due to curved shape of retina or problems during image acquisition process [8]. Retinex image enhancement is useful for shadow removing and lightness-color constancy. In this paper, we transform RGB space to LAB space, and use adaptive histogram equalization (AHE) enhance L channel directly, and then we transform the image from LAB space to RGB. We can see that arteries and veins can be distinguished apparently in Fig. 1 (a), (b). Red channel gets good result especially when artery and vein have bright and dark red color. However, great amount of information loss in some retinal images in red channels. In order to extract the features in red channel more significantly, we apply multi-scale retinex (MSR) technique on red channel. The single-scale retinex (SSR) [10, 11] is given by

$$R_i(x, y) = \log I_i(x, y) - \log[F(x, y) * I_i(x, y)] \tag{1}$$

$$F(x, y) = Ke^{-\frac{x^2 + y^2}{\sigma^2}} \tag{2}$$

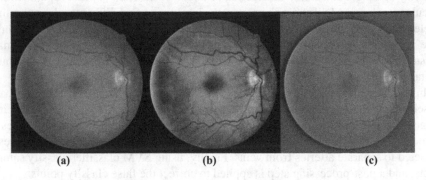

Fig. 1. (a) Comparison of original image, (b) after AHE in LAB, (c) after MSR in red channel (Color figure online)

Where $R_i(x, y)$ is the retinex output, $I_i(x, y)$ is the image distribution in the *ith* spectral band, "*" denotes the convolution operation, and $F(x, y)$ is the surround function. The MSR output is simply a weighted sum of the outputs of several different SSR outputs. Mathematically,

$$R_{MSR_i} = \sum_{n=1}^{N} w_n * R_{n_i} \tag{3}$$

Where N is the number of scales, R_{n_i} is the *ith* component of the *nth* scale, R_{MSR_i} is the *ith* spectral component of the MSR output, and w_n is the weight associated with the *nth* scale, and $w_n = 1/3$, n = 1, 2, 3 was sufficient for our application. See Fig. 1 (c).

2.2 Vessel Segmentation and Optic Disc Detection

For vessel classification, vascular structure should be extracted from original fundus image as a binary image which we indicate the location of vessel pixels. We use an automated vessel segment method proposed by Azzopardi et al. [12]. It is based on the Combination of Receptive Fields (CORF) computational model of a simple cell in visual cortex and its implementation called Commination of Shifted Filter Responses (COSFIRE). And they made forward on this method, proposed a bar-selective COSFIRE filter (B-COSFIRE). It can be effectively used to detect bar-shaped structure such as blood vessels. See Fig. 2.

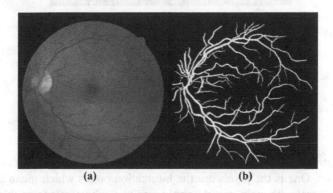

(a)   (b)

Fig. 2. (a) Original Image, (b) binary map of vascular tree by B-COSFIRE

We use B-COSFIRE to extract vessel binary images, and then thinner the vessels to get their centerlines shown in Fig. 3 (a). From the skeleton binary map, we find bifurcation and cross-over pixels and discard them for these points may mislead classifier. Cross-over and bifurcation points are the pixels in skeleton which are more than 2 pixels adjacent in the skeleton, and [20] offers a way to find cross-over and bifurcation points.

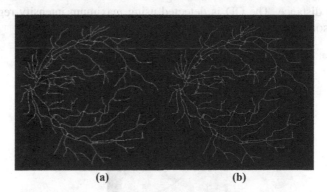

(a)   (b)

Fig. 3. (a) Skeleton of vascular tree, (b) cross points of skeleton

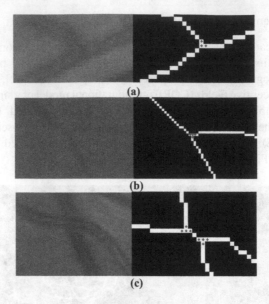

Fig. 4. Different cross and bifurcation points

We evaluate a set of cross-over and bifurcation pixels, and the pixels can be divided into three types. One is the pixels are the bifurcation points which mean main vessel split one sub vessel with main vessel or split into two sub vessels from this point. See Fig. 4 (a), (b). Another is Fig. 4 (c) shows that vein and artery cross over each other at this point. After removing those points, we segment the whole vascular structure into non-connected lines, and the pixels in the same line share the same artery/vein label. This step is meaningful when post-processing for correcting false classified points.

The detection of optic disc is a quite essential task, though what we need is just a point position to represent center location of the OD. Surround this point, we can define the ROI. We choose the ROI is a circular region away from center of OD about 40 pixels to 160 pixels distance. The OD is detected using maximum intensity region finding method with some preprocessing operator [13]. See Fig. 5.

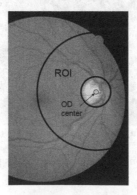

Fig. 5. ROI

2.3 Feature Extraction and Artery-Vein Classification

We treat all pixels fallen on centerlines as sample points in ROI. Then we calculate vessel radius for every pixel, and define a big zone and a small zone as feature extract zones. Figure 6 illustrates, vessel direction and search direction perpendicular to each other, so we calculate the vessel direction, and rotate the direction about 90° as search direction. We apply multi-scale gauss filters to deal with green channel of original images, and calculate Hessian matrix for every pixel in each image. For each Hessian matrix, calculate two eigenvectors' direction as the vessel direction which is corresponding to the biggest response in different scales. As we get the vessel direction, we get the search direction for vessel radius.

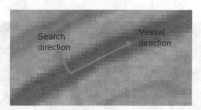

Fig. 6. Search direction and vessel direction

For radius computation, there are two ways we evaluate it. The first method is to calculate wall of vessels using some edge detection algorithm while another method is much easier. As we have got binary vessel image from background, we just search a boundary of white and black through search direction and opposite search direction. The distance of these two boundaries pixels is the diameter for this pixel.

For measurement features zone, previous work has different views. Relan et al. [3] extract four different features from the correct channels and from a circular neighborhood around each centerline pixel, with diameter 60 % of the mean vessel diameter. Mirsharif et al. [5] defined a window size of 12*12 which is located on the middle point of each sub vessel and they calculate mean and variance values for the pixels of vessels inside the window. We think that define a window as big as 12*12 may lead a confuse conclusion especially when two vessels with different labels are close to each other. An absolute size make no sense for some thin vessels while [3] define a zone which is too small to get enough information. In our paper, we calculate our features in different zones: small zone is to discriminate center reflex of artery and vein while big zone is to tell the differences of vessel and background. Big zone is a circular neighborhood around each pixel with diameter two times of the vessel diameter, and small zone is just the same diameter as the vessel diameter [15]. See Fig. 7.

At last, we choose mean, maximum, minimum, variation respectively on right channel of RGB, HSL, and LAB color spaces on both zones. Except color features in different channel, using the Gaussian derivatives of pixel intensities at different scale that are suitable for capturing the central reflex property of blood vessels. For each feature in different color space, we try to test how every feature effects the result of classify. Each feature is added to the feature vector if the accuracy for the evaluation set of images increases, that feature is added into the final feature vector.

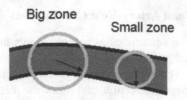

Fig. 7. Big zone and small zone for feature extraction

We introduce the widely used statistical learning method SVM in the classification stage. All vessel centerline pixels are assigned an artery label or a vein label by the classifier. Figure 8 (a) shows part results.

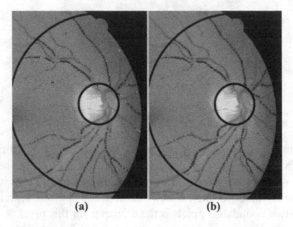

Fig. 8. (a) Results of SVM artery-vein classification, (b) post-processing results (Color figure online)

The blue pixels and red pixels are correctly classified respectively represent veins and arteries while green ones are incorrectly labeled. We can see most of the ROI points are labeled right, just some points at the start or end of the vessel segment, or the vessel is too thin and even experienced human cannot tell what the type is directly without reference to structure information.

As can be seen in Fig. 8 (a), some pixels are uncorrected classified in a single line while most of the pixels in this line are classified right. For an easy mind, we assign this single line a label by voting, and in turn to correct misclassified points. Through this simple step, we improve the classify accuracy greatly by more than 5 %. See Fig. 8 (b) shows. As we extract each centerline, we remove the bifurcation and cross-over points. Such points can lead us to post-processing further. If a bifurcation points split to three vessel segments, it means these three vessels share the same label. Another rule demonstrate that if a cross-over points connected to four vessel segments, the opposite segments should be the same type. Except for this, main arteriovenous vessels around OD appears alternately. If clockwise search labeled vessels, and successive vessels with a same label, it can be something wrong in those vessels. Although there are some experienced information, but it does not always work, the post-processing is still hard work remaining to be solved.

3   Experiment Results

For training and testing the SVM classifier, we use DRIVE database [9] which includes 20 train and 20 test images. Estrada et al. [14] have offered their labeled artery-vein data for DRIVE database publicly. We use their marked points and intend to mark whole vessel skeletons. The 20 training images are used for training classifier and the parameters, and the rest 20 images are used for testing our method.

Automated classification of blood vessels aimed at AVR measurement to indicate the diseases. In order to measure AVR previously, the ROI region is defined as major vessels in a circular region around the OD, and of a certain distance. In some proposed methods only main arteries and vein vessels parallel to each other are taken into account for artery-vein classification [20], and other methods consider major vessels as those which are distinguishable by ophthalmologist using only color features [3, 5, 19]. But actually, even the second vessel near the OD greatly decrease the recognition rate of major vessels since second vessels are hardly recognized by their colors. In this paper, we propose a full automated method, we do not select seed points or mark vessel segment start point and end point or add any manual influence into our sample selection as some semi-automated method did. We take a large region into account not just a circular region around OD with a distance from 0.5–1 DD [3, 5, 17]. So we classify much more points and vessels in a big region than those methods. Because the evaluation index and method cannot unified, or the experiment are on the different local images, most of the literatures did not compare with other's work clearly. In spite of this, we still compare with other results, and use a simple way to evaluate our results. The total accuracy of the classification method is calculated by:

$$Accuracy = \frac{n_i}{n_i + n_j} * 100 \tag{4}$$

Where n_i stands for number of vessel pixels that correct classify while n_j means the number of incorrectly labeled points. From the Table 1, we can see that our results are competitive though [3, 5] got a higher accuracy than us. Actually, [3] ignore 13.5 % vessels that are unable to be classified by their method, these vessels are not taken into account, also [3] is on their local database and cannot be compared with us fairly. For [5], they did not take the second vessel around OD thinner than 3 pixels and they calculate a smaller ROI than us, and while their methods work on a vessel thicker than 3 pixels all over, their accuracy decreases less than us.

Table 1. Performance Comparison

Proposed Systems	Database	Accuracy	Description
Ruggeri et al. [16]	Local Database of 35 images	81.6 %	
Muramatsu et al. [18]	DRIVE database	75 %	
Mirsharif et al. [5]	DRIVE database	84.05 %	Vessel thicker than 3 pixels all over
		90.16 %	Vessel thicker than 3pixels in ROI of 0.5-1 DD from O.D. margin
Relan et al. [3]	Local Database of 35 images	92 %	Total 406 vessels with 13.5 % of them unlabeled
Irshad and Akram [17]	Local Database of 25 images	81.3 %	
Proposed method	DRIVE database	88.7 %	Total 12125 pixels
		89.07 %	247 lines

4 Conclusions

In this paper, we propose an automated system for classification of retinal blood vessels into arteries and veins. The main work is on image preprocessing and analysis features for classification. This work is helpful to calculate AVR value which can indicate various diseases. Our work achieve a high accuracy in a large scale of original image in DRIVE database while we do not introduce any human factors after labeling pixels for training and testing, and it can be a reliable tool for AVR calculation.

Acknowledgment. This work was supported in part by the Natural Science Foundation of China under Grant 61472102, in part by the Fundamental Research Funds for the Central Universities under Grant HIT.NSRIF.2013091, and in part by the Humanity and Social Science Youth foundation of Ministry of Education of China under Grant 14YJC760001.

References

1. Grisan, E., Ruggeri, A.: A divide et impera strategy for automatic classification of retinal vessels into arteries and veins. In: Proceedings of the 25th Annual International Conference of the IEEE on Engineering in Medicine and Biology Society, vol. 1, pp. 890–893. IEEE (2003)
2. Kondermann, C., Kondermann, D., Yan, M.: Blood vessel classification into arteries and veins in retinal images. In: Medical Imaging. International Society for Optics and Photonics, pp. 651247–651247-9 (2007)
3. Relan, D., MacGillivray, T., Ballerini, L., et al.: Retinal vessel classification: sorting arteries and veins. In: 35th Annual International Conference of the IEEE on Engineering in Medicine and Biology Society (EMBC), pp. 7396–7399. IEEE (2013)
4. Vázquez, S.G., Cancela, B., Barreira, N., et al.: Improving retinal artery and vein classification by means of a minimal path approach. Mach. Vis. Appl. **24**(5), 919–930 (2013)

5. Mirsharif, Q., Tajeripour, F., Pourreza, H.: Automated characterization of blood vessels as arteries and veins in retinal images. Comput. Med. Imaging Graph. **37**(7), 607–617 (2013)

6. Rothaus, K., Jiang, X., Rhiem, P.: Separation of the retinal vascular graph in arteries and veins based upon structural knowledge. Image Vis. Comput. **27**(7), 864–875 (2009)

7. Joshi, V.S., Reinhardt, J.M., Garvin, M.K., et al.: Automated method for identification and artery-venous classification of vessel trees in retinal vessel networks. PLoS ONE **9**(2), e88061 (2014)

8. Mirsharif, Q., Tajeripour, F.: Investigating image enhancement methods for better classification of retinal blood vessels into arteries and veins. In: 16th CSI International Symposium on Artificial Intelligence and Signal Processing (AISP), pp. 591–597 (2012)

9. Staal, J., Abràmoff, M.D., Niemeijer, M., et al.: Ridge-based vessel segmentation in color images of the retina. IEEE Trans. Med. Imaging **23**(4), 501–509 (2004)

10. Jobson, D.J., Rahman, Z., Woodell, G.A.: Properties and performance of a center/surround retinex. IEEE Trans. Image Process. **6**(3), 451–462 (1997)

11. Jobson, D.J., Rahman, Z., Woodell, G.A.: A multiscale retinex for bridging the gap between color images and the human observation of scenes. IEEE Trans. Image Process. **6**(7), 965–976 (1997)

12. Azzopardi, G., Strisciuglio, N., Vento, M., et al.: Trainable COSFIRE filters for vessel delineation with application to retinal images. Med. Image Anal. **19**(1), 46–57 (2015)

13. Usman Akram, M., Khan, A., Iqbal, K., Butt, W.H.: Retinal images: optic disk localization and detection. In: Campilho, A., Kamel, M. (eds.) ICIAR 2010, Part II. LNCS, vol. 6112, pp. 40–49. Springer, Heidelberg (2010)

14. Estrada, R., Allingham, M.J., Mettu, P.S., et al.: Retinal artery-vein classification via topology estimation. IEEE Trans. Med. Imaging **34**(12), 2518–2534 (2015)

15. Zamperini, A., Giachetti, A., Trucco, E., et al.: Effective features for artery-vein classification in digital fundus images. In: 2012 25th International Symposium on Computer-Based Medical Systems (CBMS), pp. 1–6. IEEE (2012)

16. Ruggeri, A., Grisan, E., De Luca, M.: An automatic system for the estimation of generalized arteriolar narrowing in retinal images. In: 29th Annual International Conference of the IEEE on Engineering in Medicine and Biology Society, EMBS 2007, pp. 6463–6466 (2007)

17. Irshad, S., Akram, M.U.: Classification of retinal vessels into arteries and veins for detection of hypertensive retinopathy. In: 2014 Cairo International Biomedical Engineering Conference (CIBEC), pp. 133–136. IEEE (2014)

18. Muramatsu, C., Hatanaka, Y., Iwase, T., et al.: Automated detection and classification of major retinal vessels for determination of diameter ratio of arteries and veins. In: SPIE Medical Imaging. International Society for Optics and Photonics, pp. 76240 J–76240 J-8 (2010)

19. Niemeijer, M., van Ginneken, B., Abràmoff, M.D.: Automatic classification of retinal vessels into arteries and veins. In: SPIE medical imaging. International Society for Optics and Photonics, pp. 72601F–72601F-8 (2010)

20. Calvo, D., Ortega, M., Penedo, M.G., et al.: Automatic detection and characterisation of retinal vessel tree bifurcations and crossovers in eye fundus images. Comput. Methods Programs Biomed. **103**(1), 28–38 (2011)

Character Variable Numeralization Based on Dimension Expanding and its Application on Text Classification

Li-xun Xu[1], Xu Yu[2(✉)], Yong Wang[3], and Yun-xia Feng[2]

[1] Sino-German Faculty, Qingdao University of Science and Technology,
Qingdao 266061, China
[2] School of Information Science and Technology, Qingdao University
of Science and Technology, Qingdao 266061, China
yuxu0532@163.com
[3] College of Computer Science and Technology, Harbin Engineering University,
Harbin 150001, China

Abstract. The character variable discrete numeralization destroyed the disorder of character variables. As text classification problem contains more character variable, discrete numeralization approach affects the classification performance of classifiers. In this paper, we propose a character variable numeralization algorithm based on dimension expanding. Firstly, the algorithm computes the number of different values which the character variable takes. Then it replaces the original values with the natural bases in the m-dimensional Euclidean space. Though the algorithm causes a dimension expanding, it reserves the disorder of character variables because the natural bases are no difference in size, so this algorithm is a better character variable numerical processing algorithm. Experiments on text classification data sets show that though the proposed algorithm costs a little more running time, its classification performance is better.

Keywords: Character variable · Natural bases · Dimension expanding · Text classification

1 Introduction

Text classification [1, 2] is a very important direction of research in pattern recognition. With the development of Internet technology, text recognition is playing an increasingly crucial role. By text recognition, we can conduct public opinion analysis, which enables government to understanding aspirations of people and adjust measures in a timely manner. Text recognition can also help owners of online shopping sites to know the attitudes of consumers so that improve the service quality of their website.

Text classification typically includes the expression of texts, selection and training of classifiers, evaluation and feedback process of classification results. As in essence text classification problems belong to the scope of text classification, so a lot of typical pattern classification algorithms can be applied to text classification problems. As the effect of text classification algorithm based on statistical learning method is better, so statistical learning method is studied widely by scholars home and abroad. Statistical

W. Che et al. (Eds.): ICYCSEE 2016, Part I, CCIS 623, pp. 238–248, 2016.
DOI: 10.1007/978-981-10-2053-7_22

learning methods require a number of documents, which were accurately classified by human, as learning materials, and then computers mine rules from these documents. This process is called the training process, and the set of rules it summed up often are called classifiers. After training, the documents that computers have never seen before can be classified by the trained classifiers. Typical statistical learning methods contain Bayesian analysis method [3], KNN method [4], support vector machine method [5], artificial neural network method, the decision tree method [6] and etc. For example, Wajeed classified the textual data. In the process of classification the effects of the features distributed across the document is explored. KNN algorithm is employed and the results obtained are encouraging [7]. Sun et al. give a comparative study on text classification using SVM [8].

As mature classification methods, those methods have achieved better learning results on the text classification problems. However, text classification is a high-dimensional classification problem, and the feature vector contains a lot of character variables [9, 10]. Traditional text classification methods [11, 12] used discrete processing methods. By assigning the different values of properties to different nature numbers, the disorder character variables are undermined, and the recognition performance of text classification are affected to a certain extent.

For feature vectors in text classification problems contains a lot of character variables, this paper proposes a character variable numeralization algorithm based on dimension expanding. Firstly, the algorithm computes the number of different values which the character variable takes. Then it replaces the original values with the natural bases in the p-dimensional Euclidean space. Though the algorithm causes a dimension expanding, it reserves the disorder of character variables because the natural bases are no difference in size. Therefore, the data processing method can help classifiers to achieve a better performance. In general, the proposed data preprocessing algorithm is not limited to a particular classifier. In order to fully verify the algorithm, this paper selects KNN and support vector machine learning algorithm as the experimental classifiers. Experiments on a news text data set show that the proposed preprocessing algorithm allows the classifier to obtain a higher classification accuracy compared to a discrete numerical processing method.

This paper is organized as follows. A brief introduction to KNN algorithm and SVM algorithm is given in Sect. 2. In Sect. 3, a character variable numeralization algorithm based on dimension expanding is proposed. The KNN algorithm and SVM algorithm are used to conduct text classification experiments in Sect. 4, and the experimental results and a detailed analysis are also given in this part. Section 5 concludes the whole paper.

2   Review of the KNN Algorithm and the SVM Algorithm

2.1   The KNN Algorithm

K-nearest-neighbor is an lazy learning classification algorithm, usually denoted by KNN. For the KNN algorithm, training samples are represented by n-dimensional numerical attributes. For a label unknown sample, the KNN algorithm firstly finds the nearest k training samples from the training set. The distance between two samples can be computed by Euclidean distance. The formula is as below.

$$d(X,Y) = \sqrt{\sum_{i=1}^{n} (x_i - y_i)^2} \tag{1}$$

where $X = (x_1, x_2, ..., x_n)$ and $Y = (y_1, y_2, ..., y_n)$ denote two samples in the n-dimensional Euclidean space.

Then the label of the unknown sample is assigned with the most common class of the k neighbors. Specially, if $k = 1$, the unknown sample is assigned with the same class as its nearest neighbor.

2.2   The SVM Algorithm

Let the training sample set be $T = \{(x_1, y_1), ..., (x_l, y_l)\}$, where $x_i \in R^n$, $y_i \in \{-1, 1\}$, $i = 1, ..., l$. Assuming that the training sample set is linear separable, the SVM algorithm obtains the classification hyperplane by solving the following quadratic optimization problem.

$$
\begin{aligned}
\min_{a} \quad & \frac{1}{2} \sum_{i=1}^{l} \sum_{j=1}^{l} y_i y_j (x_i \cdot x_j) a_i a_j - \sum_{i=1}^{l} a_i \\
\text{s.t.} \quad & \sum_{i=1}^{l} a_i y_i = 0 \\
& 0 \leq a_i \leq C, \quad i = 1, ..., l
\end{aligned}
\tag{2}
$$

where a_i is Lagrange multipliers, the parameter $C > 0$ controls the trade-off between the slack variable penalty and the margin.

If the original training set is non-linear separable, the SVM algorithm converts it into a linear separable problem, and then compute the classification hyper-plane by solving the following quadratic optimization problem.

$$
\begin{aligned}
\min_{a} \quad & \frac{1}{2} \sum_{i=1}^{l} \sum_{j=1}^{l} y_i y_j K(x_i \cdot x_j) a_i a_j - \sum_{i=1}^{l} a_i \\
\text{s.t.} \quad & \sum_{i=1}^{l} a_i y_i = 0 \\
& 0 \leq a_i \leq C, \quad i = 1, ..., l
\end{aligned}
\tag{3}
$$

The decision functions corresponding to linear separable and non-linear separable are listed as below.

$$f(x) = \mathrm{sgn}(\sum_{i=1}^{l} a_i^* y_i(x_i \cdot x) + y_i - \sum_{i=1}^{l} a_i^* y_i(x_i \cdot x_j)) \tag{4}$$

$$f(x) = \mathrm{sgn}(\sum_{i=1}^{l} a_i^* y_i K(x_i \cdot x) + y_j - \sum_{i=1}^{l} y_i a_i^* K(x_i, x_j)) \tag{5}$$

where a_i^* is the optimizing solution of the corresponding optimizing problem.

3 A New Character Variable Numeralization Method

Feature vectors in text classification problems contain plenty of character variables, and most current statistical learning algorithms require the input vector must be numeric vectors. Thus the data set of text classification problems must be preprocessed. Traditional processing method for character variable is as follows. Let a character variable take m different character values, then the usual approach represents the m values of characters variables with 1,2,...,m respectively. In this paper, the above method is referred to as a character variable discrete numerical approach.

As there are no large character variable or small character variable in essence, the shortcomings of this approach lies in that it undermines the disorder of the character variables, and degrades the performance of classifiers. For this problem, this paper proposes a character variable numeralization algorithm based on dimension expanding. The proposed method replaces the p values of a character variable with the m natural bases $(0,0,...,0,1),...,(1,0,...,0,0)$. The natural base refers to a m-dimensional unit vector of which only one component is 1 and the others are 0.

Figure 1 shows the results of data processing with the proposed method when the variable has two different values.

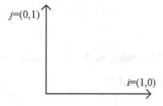

Fig. 1. Illustration of a character variable numeralization algorithm based on dimension expanding

This method replaces the variable 0 and 1 in the traditional method with two linear independent natural base i and j in the two-dimensional Euclidean space. Although it increases the dimensions of the original data, it maintains well the disorder of feature variables.

The text classification algorithm based on character variable numeralization by expanding dimensions, denoted as TCABCVNED, is given below.

Algorithm 1 The TCABCVNED algorithm

Input: text classification data set D; statistical classification algorithm A
Output: text classification rule F
Method:
  for i=1to n {

> /* n denotes the number of feature attributes in text classification data set D /*

if(Is_Character_Variable(f_i))

> /*The Is_Character_Variable function is to judge whether the attribute is a character attribute /*

{
  m=Dimension_compute(f_i)

> /*The Dimension_compute function is to compute the number of different values which a character attributes takes /*

Base_generation(f_i, m);
  }
  i++;
  }
  for i=1to n{
      for j=1to t /* t denotes the number of records in text classification data set D /*
  {
    if(Is_Character_Variable(f_i))
    Character_Variable_Numeralization ($f_i(j)$);

> /*Numerlization treatment on feature f_i by expanding dimensions /*

}
  }
  F=A(D^{new})

> /* Learning on the new training set with statistical classification
>
> algorithm A /*

Algorithm 1 shows that the proposed data preprocessing method can effectively reserve the disorder of character variables, which provides a possibility for classifiers to achieve a better performance. In Sect. 3, we will test the performance of the proposed method by several experiments.

4 Experiments

4.1 The Experimental Data Set Introduction

This paper selects a web page data set to test the performance of CABCVNED algorithm. As it is a high-dimensional text classification, containing many character attributes, it can test the performance of the proposed algorithm more precisely.

The web page data set comes from sohu news website and we extract four types news topic, including military, diplomatic, technology, and entainment, to test the classification algorithms. For each news topic, we choose randomly 600 samples for training and 300 samples for testing.

In this experiment, the data preprocessing method in reference [13] are used to obtain the training samples and the testing samples.

4.2 Classification Performances Metric

For a better performances evaluation of different classification algorithms, we choose precision and recall as the classification performances metrics. The computation formulas are as follows.

$$p = \frac{Number\ of\ correct\ predictions\ from\ one\ class}{Total\ number\ of\ samples\ predicted\ as\ one\ class} \tag{6}$$

$$r = \frac{Number\ of\ correct\ predictions\ from\ one\ class}{Total\ number\ of\ samples\ from\ one\ class} \tag{7}$$

Text classification system often needs to trade off recall for precision or vice versa. One commonly used trade-off is the F-score, which is defined as the harmonic mean of recall and precision:

$$F - score = \frac{p \times r}{(p+r)/2} \tag{8}$$

where p denotes precision, and r denotes recall.

Obviously, the algorithm can achieve a better performance, while both p and r are higher.

4.3 Detailed Experimental Method

We select KNN and SVM classification methods to conduct this experiment, and for SVM classification method, we choose the C-SVM algorithm and use Gaussian kernel functions. The formula of Gaussian kernel functions is as below,

$$K(x,y) = \exp(-g\|x - y\|^2) \tag{9}$$

where g denotes the width parameter. In this experiment, we use 10-folds cross-validation [14] to compute the best values of parameters.

As it is a multi-classification problem in this experiment, we choose the one against all (1-v-r) approach [15], which is to transform a k-class classification problem into k two-class classification problems.

4.4   The Experimental Results Analysis

We test the performance between the Character Variable Numeralization by Expanding Dimensions algorithm (CVNED) and the Character Variable Numeralization by Discreting algorithm (CVND). We first use the two algorithms to process the selected experiment data set, and then train classifiers with KNN and C-SVM algorithms. The average results about precision, recall, F-score and running time are shown in Figs. 2, 3, 4, 5, 6 and 7, and Table 1, where Training Set 1 is obtained by CVNED algorithm, and Training Set 2 is obtained by CVND algorithm.

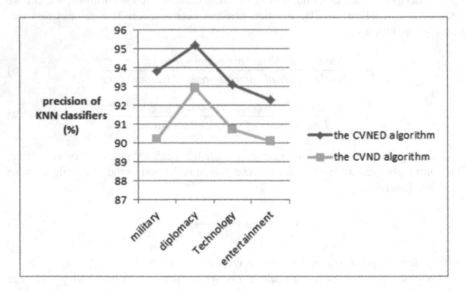

Fig. 2. Precision comparison of KNN classifiers between two data preprocessing methods

As shown in Table 1, the traditional classification algorithms cost much more time on the data sets preprocessing by the proposed CVNED. The main reason is that the CVNED algorithm increases the dimension number of samples. From Figs. 2 3, 4, 5, 6 and 7, we can see that traditional methods, like SVM and KNN, can obtain better performances in both precision and recall indexes after preprocessing by the CVNED algorithm. That is because the CVNED algorithm can reserve the disorder of character variables. The experiment results also shown that the proposed CVNED algorithm in this paper is a more reasonable character variables numeralization method than previous methods.

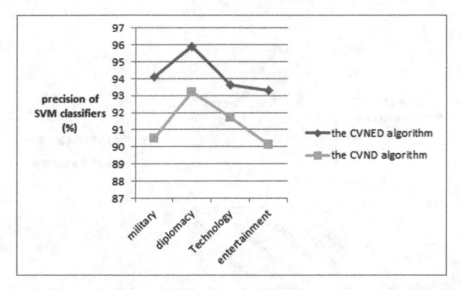

Fig. 3. Precision comparison of C-SVM classifiers between two data preprocessing methods

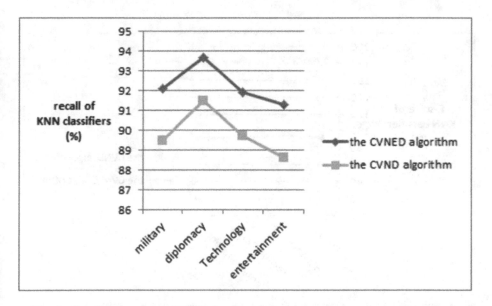

Fig. 4. Recall comparison of KNN classifiers between two data preprocessing methods

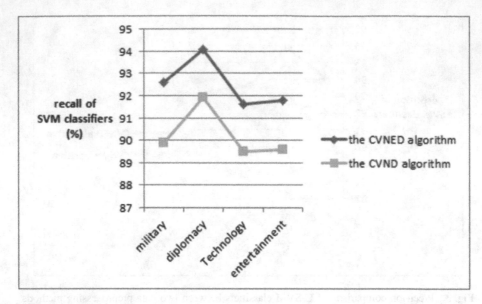

Fig. 5. Recall comparison of C-SVM classifiers between two data preprocessing methods

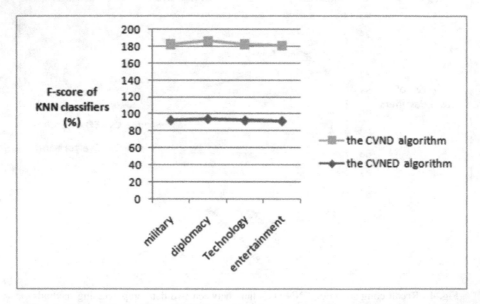

Fig. 6. F-score comparison of KNN classifiers between two data preprocessing methods

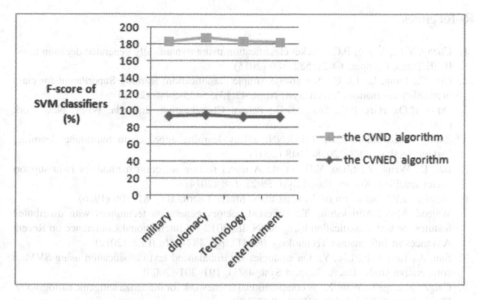

Fig. 7. F-score comparison of C-SVM classifiers between two data preprocessing methods

Table 1. Running time comparison between the two data preprocessing algorithms

Algorithms	Running time on Training Set 1 (s)	Running time on Training Set 2 (s)
The KNN algorithm	29.1	26.7
The C-SVM algorithm	30.2	27.9

5 Conclusion

For character attributes in high dimensional text classification data set, this paper proposed a character variable numeralization algorithm based on dimension expanding. The pretreated methods reserved the disorder of character variables and it is an effective data pretreated method independent of classifiers. After preprocessing, the classification performances of classifiers have been promoted largely. Experiments on text classification data sets show the effective of the proposed method.

Acknowledgement. This work is sponsored by the National Natural Science Foundation of China (Nos. 61402246, 61402126, 61370083, 61370086, 61303193, and 61572268), a Project of Shandong Province Higher Educational Science and Technology Program (No. J15LN38), Qingdao indigenous innovation program (No. 15-9-1-47-jch), the National Research Foundation for the Doctoral Program of Higher Education of China (No. 20122304110012), the Natural Science Foundation of Heilongjiang Province of China (No. F201101), the Science and Technology Research Project Foundation of Heilongjiang Province Education Department (No. 12531105), Heilongjiang Province Postdoctoral Research Start Foundation (No. LBH-Q13092), and the National Key Technology R&D Program of the Ministry of Science and Technology under Grant No. 2012BAH81F02.

References

1. Cheng, Y.C., Wang, P.C.: Packet classification using dynamically generated decision trees. IEEE Trans. Comput. **64**(2), 582–586 (2015)
2. Qiu, C., Jiang, L., Li, C.: Not always simple classification: learning SuperParent for class probability estimation. Expert Syst. Appl. **42**(13), 5433–5440 (2015)
3. Duda, R.O., Hart, P.E., Stork, D.G.: Pattern Classification, 2nd edn. Wiley, New York (2001)
4. Zhang, M.L., Zhou, Z.H.: ML-KNN: a lazy learning approach to multi-label learning. Pattern Recogn. **40**(7), 2038–2048 (2007)
5. Bai, L., Wang, Z., Shao, Y.H., et al.: A novel feature selection method for twin support vector machine. Knowl.-Based Syst. **59**(2), 1–8 (2014)
6. Quinlan, J.R.: Induction of decision trees. Mach. Learn. **1**(1), 81–106 (1986)
7. Wajeed, M.A., Adilakshmi, T.: Different vectors generation techniques with distributed features for text classification using KNN. In: 2012 1st International Conference on Recent Advances in Information Technology (RAIT), pp. 482–486. IEEE (2012)
8. Sun, A., Lim, E.P., Liu, Y.: On strategies for imbalanced text classification using SVM: a comparative study. Decis. Support Syst. **48**(1), 191–201 (2009)
9. Cai, Z., Zhang, T., Wan, X.: A computational framework for influenza antigenic cartography. PLoS Comput. Biol. **6**(10), e1000949 (2010)
10. Cai, Z., Ducatez, M.F., Yang, J., Zhang, T., Long, L.-P., Boon, A.C., Webby, R.J., Wan, X.-F.: Identifying antigenicity associated sites in highly pathogenic H5N1 influenza virus hemagglutinin by using sparse learning. J. Mol. Biol. **422**(1), 145–155 (2012)
11. Cai, Z., Goebel, R., Salavatipour, M., Lin, G.: Selecting genes with dissimilar discrimination strength for class prediction. BMC Bioinform. **8**, 206 (2007)
12. Yang, K., Cai, Z., Li, J., Lin, G.: A stable model-free gene selection in microarray data analysis. BMC Bioinform. **7**, 228 (2006)
13. Lan, J., Shi, H., Li, X., et al.: Associative web document classification based on word mixed weight. Comput. Sci. **38**(3), 187–190 (2011)
14. Kohavi, R.: A study of cross-validation and bootstrap for accuracy estimation and model selection. In: Proceedings of the 14th Joint International Conference Artificial Intelligence, pp. 1137–1145 (1995)
15. Hsu, C.W., Lin, C.J.: A comparison on methods for multi-class support vector machines. IEEE Trans. Neural Netw. **13**(2), 415–425 (2001)

Clarity Corresponding to Contrast
in Visual Cryptography

Xuehu Yan[✉], Yuliang Lu, Hui Huang, Lintao Liu, and Song Wan

Hefei Electronic Engineering Institute, Hefei 230037, China
publictiger@126.com

Abstract. The visual quality of the recovered secret image is usually evaluated by contrast in visual cryptography (VC). Precisely, the reconstructed secret image can be recognized as the original secret image when contrast is greater than zero. It is important for the user to know the clarity of the revealed secret image corresponding to different contrast values. In this paper, the clarity corresponding to contrast is investigated through conducting subjective evaluation scores and objective evaluation indexes, which can be extended to general applications.

Keywords: Visual cryptography · Visual quality · Contrast · Clarity · Social computing

1 Introduction

Visual cryptography(VC) was first introduced by Naor and Shamir [1]. VC is a kind of secret sharing scheme [1–5] that allows the decryption of the secret images without cryptographic knowledge and computational devices. In a general (k, n) threshold VC scheme, a secret image is generated into n random shares (also called shadows) which separately reveals nothing about the secret other than the secret size. The n shares are then printed onto transparencies and distributed to n associated participants. The secret image can be visually revealed based on human visual system (HVS) by stacking any k or more shares, while any $k - 1$ or less shares give no clue about the secret [1]. Figure 1 shows an application example of traditional $(2, 2)$ VC. The secret is encrypted into two random shares which have twice size of the secret image. Although some contrast loss occurs, the revealed image is clearly identified. VC can be applied in many scenes [5], such as information hiding, watermarking [6], authentication and identification, social computing security, and transmitting passwords etc.

Following Naor and Shamir's work, researchers studied the related VC problems such as contrast, different formats and the pixel expansion. An optical threshold VC with perfect black pixels reconstruction was presented by Blundo et al. [7]. Ateniese et al.[8] showed a general VC access structure. Color images schemes were studied by Liu et al.[9], Hou et al.[10], Luo et al.[11] and Krishna et al.[12]. Shyu et al.[13] proposed multiple secrets sharing. Threshold VC for different whiteness levels was proposed by Eisen [14]. Liu et al. [15] introduced step

© Springer Science+Business Media Singapore 2016
W. Che et al. (Eds.): ICYCSEE 2016, Part I, CCIS 623, pp. 249–257, 2016.
DOI: 10.1007/978-981-10-2053-7_23

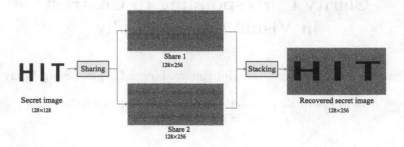

Fig. 1. An application example of traditional $(2,2)$ VC. The secret is encrypted into two random shares which have twice size of the secret image. The revealed image shows the secret image with 50% contrast loss.

construction to improve the visual quality in VC. Through choosing a column from corresponding basic matrix equally, Ito *et al.* [16] proposed probabilistic VC. Yang [2] presented probabilistic VC for different thresholds. In addition, the generalization probabilistic VCS was further extended by Cimato *et al.* [17].

Generally, the visual quality of the recovered secret image is evaluated by contrast [1, 18] in VC. Precisely, the reconstructed secret image cannot be recognized as the original secret image when contrast is equal to zero, while the recovered secret image can be recognized as the original secret image when contrast is greater than zero. Contrast is expected to be as large as possible to gain better visual quality. However, no previous research tells the user that how well human eyes could recognize the reconstructed image by contrast, i.e., the clarity of the revealed secret image corresponding to different contrast values confuses the user.

Aiming to study the clarity of the revealed secret image corresponding to different contrast values, in this paper, we investigate the clarity of contrast for the first time through conducting both subjective evaluation scores and objective evaluation indexes. The clarity with regard to contrast is given to guide the applications. The just recognition point (JRD) and security point about the contrast for the applications are discussed as well.

The rest of this paper is organized as follows. Section 2 gives the subjective evaluation and preliminary techniques as the basis for the presented work. Section 3 focuses on the subjective evaluation method. In Sect. 4, the experimental results are given and analyzed. Finally, Sect. 5 concludes this paper.

2 Preliminaries

In the section, some definitions used in the paper are presented.

Some notations used in this paper are given in Table 1. Related definitions are shown as follows.

Table 1. Notations used in this paper

Notations	Descriptions
0(resp.1)	A white(resp. black) pixel
r	Noise rate
ξ	The subjective evaluation score
S	The original binary secret image
S'	The revealed binary secret image
α	Contrast of the revealed secret image
$S(0)$ (resp. $S(1)$)	The area of all the white (resp. black) pixels in S
$SC[S(0)]$ (resp. $SC[S(1)]$)	The corresponding area of all the white (resp. black) pixels in image SC

Definition 1 (Average light transmission) [18]: For a certain pixel b in a binary image B whose size is $M \times N$, the light transmission of a transparent (resp. opaque) pixel is defined as $T(b) = 1$ (resp. $T(b) = 0$). Furthermore, the average light transmission of B is defined as

$$T(B) = \frac{\sum_{i=1}^{M} \sum_{j=1}^{N} T(B(i,j))}{M \times N} \tag{1}$$

Definition 2 (Contrast) [18]: The contrast of the recovered secret image B for original secret image A is defined as

$$\alpha = \frac{T(B[A(0)]) - T(B[A(1)])}{1 + T(B[A(1)])} \tag{2}$$

Different contrast definitions are given in the literatures. In this paper, contrast as defined in Definition 2, will be applied to evaluate the visual quality of the reconstructed secrete image, which is widely used in VC related schemes [18–20].

Remark: Contrast is more significant than evenness to evaluate visual quality. Evenness [21] can be the supplementary evaluation measurement of the visual quality when the contrast is nearly the same. Contrast in this paper is different from that contrast in image processing.

Definition 3 (Visually recognizable) [1]: The reconstructed secret image B is recognizable as original secret image A by $\alpha > 0$. Precisely, the case $T(B[A(0)]) > T(B[A(1)])$ means B could be recognized as A visually. Otherwise, it cannot be recognized by any information about A (i.e. $\alpha = 0$).

According to Definition 3, we can recognize the revealed secret image as the original secret image by $\alpha > 0$, while the reconstructed secret image cannot be recognized as the original secret image when $\alpha = 0$. Thus, we only know that contrast is expected to be as large as possible to gain better visual quality. However, we don't know how well we could recognize the reconstructed image by contrast, i.e., we don't know the clarity of the revealed secret image corresponding to different contrast values.

The subjective evaluation score is indicated by ξ. The variance of ξ, denoted as $D\xi$, is computed as $D\xi = E(\xi - E\xi)^2$, where the expectation $E\xi = \frac{1}{n}\sum_{i=1}^{n}\xi_i$. The variance can be utilized to evaluate the differences between the random variable and random variable expectation.

3   Subjective Evaluation Method

On one hand, $D\xi$ and $E\xi$ are two primary statistics of ξ, thus they are utilized to analysis the subjective evaluation scores. Generally, the subjective evaluation score follows Gaussian distribution, i.e., $\xi \sim N(E\xi, D\xi)$, hence we can apply Gaussian distribution to fit the scores for every level. On the other hand, we provide one well-known objective numerical measurement for the visual quality as a reference, the peak signal-to-noise ratio (PSNR) which is a widely used measure to evaluate the image quality.

The users don't know the clarity of the revealed secret image only corresponding to different objective evaluation values, such as contrast values. In order to study the clarity of the recovered secret image corresponding to different contrast values, herein, we conduct subjective evaluation as follows.

In the experiments, we test totally 50 different simple texture secret images with size of 512×512 and different noise rate r, where r is distributed randomly in $[0.2, 0.5]$ for each secret image. There are 22 volunteers (students aged from 18 to 35 with normal visual abilities and without experience in VC) who take part in the subjective evaluation. Every volunteer separately tests totally 50 secret images by themselves. After viewing each revealed secret image S', according to the recognition speed and recognition situation, each volunteer will mark a score following Table 2. There are four scores in subjective evaluation. The first two scores are recognizable with different recognition speed. The last two scores are unrecognizable with a little information leakage or not, where 'information leakage' means one can see little information but one can not recognize the image.

Table 2. Scoring rules in the subjective evaluation

Scores	Descriptions
3	S' is fast recognized in one second
2	S' can be recognized in more than one second
1	S' cannot be recognized while there is a little information leakage
0	S' is noise-like

4   Experimental Results and Analyses

In this section, experimental results and analyses are conducted to evaluate the clarity of the revealed secret image.

The subjective evaluation score expectations are presented in Table 3, where the corresponding r, $PSNR$, α and $D\xi$ are also given. Subjective evaluation score expectations, contrast and PSNR curves are illustrated in Fig. 2.

Remark: Since the more samples are tested, the better statistics can be obtained, the experimental result in Table 3 maybe not sufficient enough so that the related statistics are approximate results. Furthermore, although the experiment is not described in all possible aspects, we only focus on key aspects, i.e., subjective evaluation scores, contrast and PSNR.

Table 3. The subjective evaluation results

NO.	r	PSNR	α	Eξ	Dξ	NO.	r	PSNR	α	Eξ	Dξ
1	0.41	3.87	0.13	3	0	26	0.25	6.02	0.4	2.5	0.34
2	0.38	4.2	0.18	2.64	0.23	27	0.25	6.02	0.4	2.68	0.31
3	0.27	5.69	0.36	1.95	0.41	28	0.22	6.58	0.46	2.82	0.15
4	0.44	3.57	0.09	0.55	0.34	29	0.27	5.69	0.36	2.82	0.15
5	0.24	6.2	0.42	3	0	30	0.28	5.53	0.34	2.41	0.42
6	0.28	5.53	0.34	3	0	31	0.48	3.19	0.03	0	0
7	0.48	3.19	0.03	0	0	32	0.22	6.58	0.46	2.18	0.6
8	0.31	5.09	0.29	1.32	0.4	33	0.32	4.95	0.27	2.14	0.39
9	0.38	4.2	0.17	2.36	0.5	34	0.22	6.58	0.46	2.73	0.29
10	0.36	4.44	0.21	1.77	0.54	35	0.35	4.56	0.22	1.91	0.26
11	0.39	4.09	0.16	2.27	0.74	36	0.2	6.99	0.5	2.91	0.08
12	0.34	4.69	0.24	1.91	0.63	37	0.41	3.87	0.13	0.82	0.15
13	0.32	4.95	0.27	2.59	0.33	38	0.44	3.57	0.08	0.73	0.29
14	0.46	3.37	0.06	0.18	0.15	39	0.38	4.2	0.17	2.95	0.04
15	0.39	4.09	0.16	2.64	0.32	40	0.45	3.47	0.07	0.64	0.23
16	0.25	6.02	0.4	3	0	41	0.28	5.53	0.34	2.36	0.5
17	0.29	5.38	0.32	2.91	0.08	42	0.28	5.53	0.34	2.5	0.43
18	0.21	6.78	0.48	3	0	43	0.44	3.57	0.08	1	0.27
19	0.5	3.01	0	0	0	44	0.44	3.57	0.08	1.09	0.08
20	0.2	6.99	0.5	3	0	45	0.37	4.32	0.19	1.23	0.18
21	0.49	3.1	0.02	0	0	46	0.43	3.67	0.1	1.05	0.04
22	0.23	6.38	0.44	3	0	47	0.24	6.2	0.42	2.95	0.04
23	0.21	6.78	0.48	2.82	0.15	48	0.42	3.77	0.11	1	0.09
24	0.46	3.37	0.05	0	0	49	0.35	4.56	0.22	2.59	0.42
25	0.48	3.19	0.03	0	0	50	0.36	4.44	0.21	2.73	0.2

From Table 3 and Fig. 2, we can summarize the following points:

1. The contrast variable trend of the reconstructed secret image is the same as that of PSNR, which shows contrast is also an effective measure to evaluate the visual quality corresponding to objective index.
2. The contrast variable trend is nearly the same as that of $E\xi$, where few non-conformities are due to different priori knowledge of the volunteers.
3. The contrast can be utilized to evaluate the visual quality in a degree.

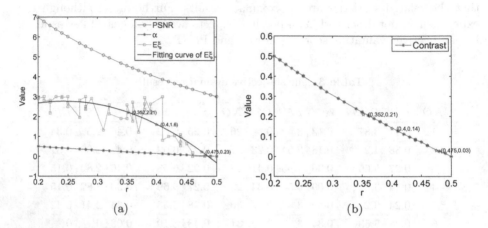

Fig. 2. Subjective evaluation score expectations, contrast and PSNR curves when possible values of r are used. (a) Subjective evaluation score expectations, contrast and PSNR curves; (b) Contrast curve.

Further analyses are exhibited in Fig. 3, where both statistical histogram of $E\xi$ and its fitting normal distribution curves are shown (as shown by the relationship of red line and blue ones in Fig. 3). According to Fig. 3, we have:

1. There are four fitting normal distribution curves corresponding to the four levels in Table 2.
2. Corresponding to four fitting curves, we can obtain three intersection points, i.e., 0.23, 1.60 and 2.21, which are also the dividing points of different score levels in Table 2.
3. According to the three intersection points, we can gain the approximate dividing contrast points, i.e., 0.03, 0.14 and 0.21, based on observing Fig. 2.

As a result, in the application, through looking up Figs. 2 and 3, the following conclusions are obtained:

1. When $\alpha \in [0, 0.03]$, there is no information leakage, i.e., the secret is secure. $\alpha = 0.03$ is the security point about contrast in VC.
2. When $\alpha \in (0.03, 0.14]$, it is hard to recognize the secret image, while there is a little information leakage. $\alpha = 0.14$ is the JRD about contrast in VC.

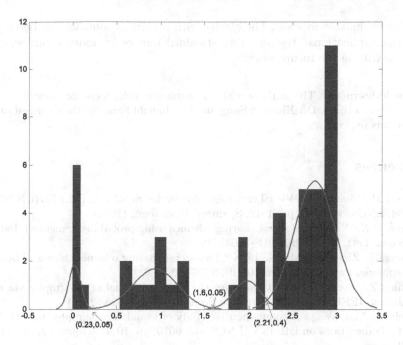

Fig. 3. Statistical histogram of the subjective evaluation score expectations. (Colour figure online)

3. When $\alpha \in (0.14, 0.21]$, the secret image can be recognized with acceptable visual quality.
4. When $\alpha \in (0.21, 1]$, the secret image is fast recognized with good visual quality.

We note that, $\alpha < 0$ when a complementary secret image is recovered, which also gives clue about the secret image. In the application, the user can consider $|\alpha|$ to apply the above conclusions.

Remark: The above values of α gained by Fig. 2 are the approximate values, which may be slightly different in different applications.

Based on the above conclusions and different application requirements, the user can select different VC methods where different contrast values are introduced. In addition, the obtained observations can use the conclusions to work for better methods with acceptable visual quality in the future. Since the paper is the first work to study the clarity about contrast in VC, there is no comparison with related work.

5   Concluding Remarks

In this paper, the clarity of the revealed secret image corresponding to different contrast values is studied, by conducting subjective evaluation scores and objective evaluation indexes. The clarity with regard to contrast can be used to guide the applications. Testing more standard images by more volunteers and the proof will be the future work.

Acknowledgement. The authors wish to thank the volunteers for their help. The authors wish to thank Dr. Xianhua Song and Dr. Jianzhi Sang for their suggestions to improve this paper.

References

1. Naor, M., Shamir, A.: Visual cryptography. In: De Santis, A. (ed.) EUROCRYPT 1994. LNCS, vol. 950, pp. 1–12. Springer, Heidelberg (1995)
2. Yang, C.N.: New visual secret sharing schemes using probabilistic method. Pattern Recogn. Lett. **25**(4), 481–494 (2004)
3. Wang, D., Zhang, L., Ma, N., Li, X.: Two secret sharing schemese based on boolean operations. Pattern Recogn. **40**(10), 2776–2785 (2007)
4. Wang, Z., Arce, G.R., Di Crescenzo, G.: Halftone visual cryptography via error diffusion. IEEE Trans. Inf. Forensics Secur. **4**(3), 383–396 (2009)
5. Weir, J., Yan, W.Q.: A comprehensive study of visual cryptography. In: Shi, Y.Q. (ed.) Transactions on DHMS V. LNCS, vol. 6010, pp. 70–105. Springer, Heidelberg (2010)
6. Lee, S., Yoo, C.D., Kalker, T.: Reversible image watermarking based on integer-to-integer wavelet transform. IEEE Trans. Inf. Forensics Secur. **2**(3), 321–330 (2007)
7. Blundo, C., De Bonis, A., De Santis, A.: Improved schemes for visual cryptography. Des. Codes Crypt. **24**(3), 255–278 (2001)
8. Ateniese, G., Blundo, C., De Santis, A., Stinson, D.R.: Visual cryptography for general access structures. Inf. Comput. **129**(2), 86–106 (1996)
9. Liu, F., Wu, C.K., Lin, X.J.: Colour visual cryptography schemes. IET Inf. Secur. **2**(4), 151–165 (2008)
10. Hou, Y.C.: Visual cryptography for color images. Pattern Recogn. **36**(7), 1619–1629 (2003)
11. Luo, H., Yu, F., Pan, J.S., Lu, Z.M.: Robust and progressive color image visual secret sharing cooperated with data hiding. In: Eighth International Conference on Intelligent Systems Design and Applications, 2008, ISDA 2008, vol. 3, pp. 431–436. IEEE (2008)
12. Prakash, N.K., Govindaraju, S.: Visual secret sharing schemes for color images using halftoning. In: International Conference on Conference on Computational Intelligence and Multimedia Applications, 2007, vol. 3, pp. 174–178. IEEE (2007)
13. Shyu, S.J., Huang, S.Y., Lee, Y.K., Wang, R.Z., Chen, K.: Sharing multiple secrets in visual cryptography. Pattern Recogn. **40**(12), 3633–3651 (2007)
14. Eisen, P.A., Stinson, D.R.: Threshold visual cryptography schemes with specified whiteness levels of reconstructed pixels. Des. Codes Crypt. **25**(1), 15–61 (2002)
15. Liu, F., Wu, C., Lin, X.: Step construction of visual cryptography schemes. IEEE Trans. Inf. Forensics Secur. **5**(1), 27–38 (2010)
16. Kuwakado, H., Tanaka, H.: Image size invariant visual cryptography. IEICE Trans. Fundam. Electron. Commun. Comput. Sci. **82**(10), 2172–2177 (1999)
17. Cimato, S., De Prisco, R., De Santis, A.: Probabilistic visual cryptography schemes. Comput. J. **49**(1), 97–107 (2006)

18. Shyu, S.J.: Image encryption by random grids. Pattern Recogn. **40**(3), 1014–1031 (2007)
19. Guo, T., Liu, F., Wu, C.: Threshold visual secret sharing by random grids with improved contrast. J. Syst. Softw. **86**(8), 2094–2109 (2013)
20. Wu, X., Sun, W.: Generalized random grid and its applications in visual cryptography. IEEE Trans. Inf. Forensics Secur. **8**(9), 1541–1553 (2013)
21. Liu, F., Wu, C., Qian, L., et al.: Improving the visual quality of size invariant visual cryptography scheme. J. Vis. Commun. Image Represent. **23**(2), 331–342 (2012)

CoGrec: A Community-Oriented Group Recommendation Framework

Yu Liu[✉], Bai Wang, Bin Wu, Xuelin Zeng, Jing Shi, and Yunlei Zhang

Beijing Key Laboratory of Intelligence Telecommunication Software and Multimedia,
Beijing University of Posts and Telecommunications, Beijing 100876, China
{liuyu,wangbai,wubin,shane}@bupt.edu.cn, {shijing_917,yunlei0518}@126.com

Abstract. Recently, group recommendation becomes substantially significant when it frequently happens that a group of users need to determine which item (e.g. movie, music, restaurant, etc.) to choose. In this paper we employ the information of friend network to propose a Community-Oriented Group Recommendation framework (CoGrec) consisting of non-negative matrix factorization based user profile generation, community detection based group identification, and overlapping community membership based group decision. Along with four inherent aggregation and allocation strategies, our proposed framework is evaluated through extensive experiments on real-world datasets. The experimental results show that the proposed framework is promising and more accurate when the given friend network is much denser, which is suitable for modern review and rating systems.

Keywords: Group recommendation · Overlapping community · Non-negative matrix factorization

1 Introduction

In the rapid development of Web 2.0, people tend to seek advices and suggestions from review and rating systems, such as Epinions[1], Ciao[2], Douban[3], etc., to help themselves to choose ideal items (e.g. movie, music, restaurants, etc.). Based on users' past reviews and rating information, individual-based recommender systems can suggest potential interesting items to individual users. However, people often form groups and recommendations to groups has its reality [13] [22]. Contrary to extensive studies on recommendation to individual users, little has been done to help grouped users find recommendations that are best selected ones in the scenario that recommended items are consumed by all members in group.

As discussed in [4], current group recommendation confronts two major problems: definition and identification of group in group recommendation, and aggregation strategy of user profile, i.e., group decision strategy.

---

[1] http://www.epinions.com/.
[2] http://www.ciao.co.uk/.
[3] http://www.douoban.com/.

© Springer Science+Business Media Singapore 2016
W. Che et al. (Eds.): ICYCSEE 2016, Part I, CCIS 623, pp. 258–271, 2016.
DOI: 10.1007/978-981-10-2053-7_24

Several grouping strategies have been used in previous studies. Occasional groups that consist of socially/temporality acquainted friends have been used in [6]. The groups are formed according to specific tasks or orientations, and group members basically retain unchanged. Random grouping has been employed in [7]. Individuals can at any time join or leave groups. Automatically identified groups that detected considering preferences of users are exploited in [3,19]. Various types of group formations result in different group decision strategies.

In individual-based recommendation, the objective is to search for items that have the highest rating scores a specific user may give [17]. Many techniques can be utilized, such as content-based recommendation [15], collaborative filtering based recommendation [20], latent-factors based recommendation [9], etc., all of which employ user's preference on different kinds of items. In group recommendation, members in a group may have different preferences on a same item [1]. Therefore, group decision strategy, i.e., aggregation strategy, is proposed to map a set of group members' preferences into group preference.

Aiming at tackling the two problems, we propose to use the social network information to identify groups based on network community so that the group formations can be well interpreted, and raise some related aggregation strategies that also exploit social network community features. In order to utilize the naturally formed community structural groups among users, in this paper we propose a Community-Oriented Group Recommendation (CoGrec) framework to perform recommendation to community(-like) groups. The contributions of this paper are as follows,

- As a group identification component of CoGrec, overlapping communities in users' friend network are discovered by using non-negative matrix factorization, where the degree of overlapping can be adjusted.
- In CoGrec, we suggest four group aggregation and allocation strategies that are elaborately designed to take advantage of user-community structural information.
- Extensive experiments are conducted to evaluate the proposed framework. Experimental results show its promising of accurate recommendation with sufficient friend network information and well interpretation.

The remaining part of this paper is organized as follows. Section 2 reviews some related work on individual-based recommendation and group-based recommendation. Section 3 gives the definition of the problem we will solve. In Sect. 4, we present the details about each component of our proposed framework, CoGrec. We conduct extensive experiments to evaluate the proposed framework in Sect. 5 and the results are analyzed to show the promise of CoGrec. Section 6 concludes our work and points out some possible future work.

2   Related Work

In this section, we will review some related work of recommendations to individual users and group users.

2.1 Individual-Based Recommendation

As aforementioned, individual-based recommendation embraces approaches like content-based recommendation, collaborative filtering based recommendation, etc. Recent recommendation based work on collaborative filtering shows its promising in mitigation the problem of data sparsity [5], some of which exploit matrix factorization to achieve better performance.

Probabilistic matrix factorization (PMF) is proposed by Salakhutdinov in [18]. For a given user-item rating matrix $\mathbf{R}$ and in order to accurately predict the missing values in it, matrix factorization is exploit. Matrix $\mathbf{R}$ thus is factorized into two matrices $\mathbf{P}$ and $\mathbf{Q}$ representing user- and item- latent-factors, respectively as follows,

$$\mathbf{R} \approx \mathbf{PQ}^{\mathrm{T}}. \tag{1}$$

It follows that a predicted user-item rating matrix is obtained by multiplying the resulting matrices, which is $\widehat{\mathbf{R}} = \mathbf{PQ}^{\mathrm{T}}$, therefore the missing values are predicted. To achieve more accurate prediction, some techniques can be elaborately added, including social regularization [12], users' implicit behavior [11], and users' trust information [21].

2.2 Group-Based Recommendation

Previous study on group-based recommendation generally focuses on user profile obtainment, group identification, and group decision strategy.

PolyLens was a recommender system proposed in [14]. In this system, groups are autonomously created by users involved in a movie rating site, MovieLens[4], and users who are interested in specific genres of movies can sign up for joining corresponding groups. The recommendation strategies used in this system lie in two aspects, one for aggregating each member's preference that represents the user's tastes into a pseudo-user preference, i.e., a group preference, and one for generating recommendation lists by merging recommendations that are for each member.

Adaptive Radio was proposed in [7], which is a music server that broadcasts to a group of users. The group is randomly formed consisting of users who access the server interface. By excluding the negative preference of each user, recommendation lists music loved by all members. Therefore, every member in the group are satisfied with the recommended music.

Baltrunas specified several rank aggregation strategies for group recommendations in [2]. Spearman footrule will find aggregations that minimize the average Spearman footrule distance for rankings. Borda count awards each individual item in the groupwise ranked list a score, and the final groupwise scores for each item is produced by adding up each individual one.

---

[4] http://movielens.umn.edu/.

3   Problem Definition

In group-based recommendation, the basis of a recommender system lies in two aspects: one for how to make the decision that the groups are meaningfully formed, and the other for how to make recommendations to group efficiently. It worth noting that in the rapid development of Web 2.0, numerous rating systems have employed social network modules that allow users to establish relationships with others. With the theory of network community, users in online social network are naturally grouped as communities. Therefore, the community can be recognized as densely grouped users, which implies interconnectivity between both users and communities.

In recommender systems, ratings by users on items (products, services, etc.) are modeled as a rating matrix $\mathbf{R}^{m \times n} = [r_{ij}]$ where r_{ij} is the rating by user p_i on item q_j. The friend network formed among users containing m users can be denoted by an adjacent matrix $\mathbf{G}^{m \times m} = [g_{ij}]$, where g_{ij} represents the relationship between user u_i and user u_j. Following formal notions and in order to simplify problems, we define $g_{ij} = 1$ if user u_i and user u_j are friends, and $g_{ij} = 0$ otherwise. For simplicity, only symmetric networks are considered in this paper.

Figure 1 shows a toy example of a social network including 6 users, the adjacent matrix of which is demonstrated in Fig. 2. And the user-item rating matrix is exemplified in Fig. 3. In this toy example, the 6 users are naturally grouped into 2 overlapping communities, one containing {User1, User2, User3, User4}, and the other consisting of {User1, User5, User6}, with User1 belonging to both communities.

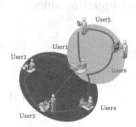

	U1 U2 U3 U4 U5 U6
U1	0 1 1 1 1 1
U2	1 0 1 0 0 0
U3	1 1 0 1 0 0
U4	1 0 1 0 0 0
U5	1 0 0 0 0 1
U6	1 0 0 0 1 0

	I1 I2 I3 I4 I5
U1	3 - 4 - 4
U2	3 - 4 - -
U3	- 3 - 2 -
U4	3 2 - 4 -
U5	- - 3 - 4
U6	1 - - - 3

Fig. 1. Friend network.   **Fig. 2.** Adjacent matrix.   **Fig. 3.** Rating matrix.

Thus, the problem we study in this paper is to generate recommendations to users in groups by using social network features. As shown in the toy example, the aim is to generate recommendations to group one containing {User1, User2, User3, User4}, and group two consisting of {User1, User5, User6}, respectively. With the friend network that naturally forms groups, this problem then is broken down into two a community detection problem and a community-oriented recommendation problem.

4   Proposed Framework

4.1   User-Item Rating Matrix Factorization

Low-rank matrix factorization (low-rank MF) is exploited in the collaborative filtering based recommendation [9]. The resulting matrix $\mathbf{P}$ in Eq. 1 representing user- latent-factors contains row vectors that are latent preferences (profiles) of users, as discussed by [19]. The regular task of matrix factorization does not constrain factorized low-rank matrices to be non-negative. However, in order to incorporate the overlapping community information into the basic MF model, the non-negativity thus should be considered since the community identifying requires non-negativity.

Following the basic objective of matrix factorization formulated in Eq. 1, the non-negative matrix factorization problem can be modeled as following objective function to be minimized,

$$\ell = \frac{1}{2} \| \mathbf{R} - \mathbf{PQ}^{\mathrm{T}} \|_{\mathrm{F}}^2$$
$$+ \frac{\lambda_{\mathrm{P}}}{2} \| \mathbf{P} \|_{\mathrm{F}}^2 + \frac{\lambda_{\mathrm{Q}}}{2} \| \mathbf{Q} \|_{\mathrm{F}}^2, \tag{2}$$
$$s.t. \ \mathbf{P} \geq 0, \mathbf{Q} \geq 0,$$

where $\mathbf{R}$ denotes the user-item rating matrix, non-negative matrices $\mathbf{P}$ of size $m \times f$ and $\mathbf{Q}$ of size $n \times f$ are user- and item- latent-factors matrix, respectively. f is the number of latent-factors.

The objective function formulated in Eq. 2 can be efficiently minimized by using an iterative multiplicative update rule [10]. Therefore, we can get the final $\mathbf{P}$ and $\mathbf{Q}$ by alternatingly updating them with the following updating rules,

$$\mathbf{P}_{ik} \leftarrow \mathbf{P}_{ik} \sqrt{\frac{[\mathbf{RQ}]_{ik}}{[\mathbf{PQ}^{\mathrm{T}} + \lambda_{\mathrm{P}} \mathbf{P}]_{ik}}}, \tag{3}$$

$$\mathbf{Q}_{jk} \leftarrow \mathbf{Q}_{jk} \sqrt{\frac{[\mathbf{R}^{\mathrm{T}} \mathbf{P}]_{jk}}{[\mathbf{QP}^{\mathrm{T}} + \lambda_{\mathrm{Q}} \mathbf{Q}]_{jk}}}, \tag{4}$$

where $[\cdot]_{ij}$ and $\cdot_{ij}$ denotes the ith row, j^{th} column element of corresponding matrix. The correctness of Eqs. 3 and 4 can be proven according to the satisfaction of the KKT condition, which is similar to those in [8], which we decide to omit here. The corresponding algorithm for rating matrix factorization is shown in Algorithm 1, where we update one matrix while fixing the other.

4.2   Community-Oriented Group Detection

Different from explicitly-defined groups that needs label information, we will define groups by leveraging the structural information of friend network. As discussed before, users in social networks are naturally grouped as communities,

Algorithm 1. ratingNMF(): Updating Procedure for Rating Matrix Factorization

---

Input:
    user-item rating matrix $\mathbf{R}$, number of latent-factors f, parameters λ_P, λ_Q
Output:
    user- latent-factors matrix $\mathbf{P}$, item- latent-factors matrix $\mathbf{Q}$
1: initialize elements of $\mathbf{P}$ and $\mathbf{Q}$ with non-negative random numbers
2: **while** not convergent **do**
3:     update $\mathbf{P}$ according to Eq. 3
4:     update $\mathbf{Q}$ according to Eq. 4
5: **end while**
6: **return** $[\mathbf{P}, \mathbf{Q}]$

---

which are well-defined groups established through user relationships. In order to incorporate detected communities into group recommendation that take advantages of matrix factorization, we adopt orthogonal non-negative matrix factorization to efficiently detect community.

As extensively studied recently, non-negative matrix factorization has been proven accurate and interpretable in overlapping community detection [16, 23]. For a given symmetric adjacent matrix $\mathbf{G}$ of size $m \times m$, minimizing the following objective function of matrix factorization will lead to a network partition,

$$\ell = \frac{1}{2}\|\mathbf{G} - \mathbf{U}\mathbf{U}^{\mathrm{T}}\|_{\mathrm{F}}^2, \quad s.t. \; \mathbf{U} \geq 0, \tag{5}$$

where $\mathbf{U}$ is the network partition indicator matrix of size $m \times d$, and d is the number of communities. Slightly different from the basic method in Eq. 5, a tuning parameter is added to control the degree of overlapping. The modified objective function can be formulated as follows,

$$\ell = \frac{1}{2}\|\mathbf{G} - \mathbf{U}\mathbf{U}^{\mathrm{T}}\|_{\mathrm{F}}^2 + \frac{\lambda_{\mathrm{UU}}}{2}\|\mathbf{U}^{\mathrm{T}}\mathbf{U} - \mathbf{I}\|_{\mathrm{F}}^2, \tag{6}$$
$$s.t. \; \mathbf{U} \geq 0,$$

where $\mathbf{I}$ is the identity matrix. The second regularization controls the degree of community overlapping. For non-overlapping communities, the inner product of indicator matrix $\mathbf{U}$ is a diagonal matrix. For overlapping communities, the result is a diagonally dominant matrix. To control the degree of community overlapping, λ_{UU} can be set from 0 to a specific positive number. We update $\mathbf{U}$ using the following multiplicative updating rule to solve the problem defined in Eq. 6,

$$\mathbf{U}_{ik} \leftarrow \mathbf{U}_{ik}\sqrt{\frac{[\mathbf{G}\mathbf{U} + \lambda_{\mathrm{UU}}\mathbf{U}]_{ik}}{[\mathbf{U}\mathbf{U}^{\mathrm{T}}\mathbf{U} + \lambda_{\mathrm{UU}}\mathbf{U}\mathbf{U}^{\mathrm{T}}\mathbf{U}]_{ik}}}, \tag{7}$$

The algorithm of community detection is presented in Algorithm 2. We also omit the proof of correctness and convergence of Eq. 7 here.

Algorithm 2. commDetNMF(): Updating Procedure for Community Detection

---

Input:
    friend network adjacent matrix $\mathbf{G}$, number of communities d, parameters λ_{UU}
Output:
    user-community membership indicator matrix $\mathbf{U}$
1: initialize elements of $\mathbf{U}$ with non-negative random numbers ranged in $[0, 1]$
2: **while** not convergent **do**
3:      update $\mathbf{U}$ according to Eq. 7
4: **end while**
5: **return** $\mathbf{U}$

4.3 Aggregation and Allocation Strategy

In group recommendation there exists several strategies to balance influences among users in group. Several aggregation strategies have been proposed in [2,19]. Inspired by prior work, here we propose several aggregation strategies along with a few allocation strategies to produce better recommendations.

The matrix $\mathbf{U}$ in Eq. 6 denotes the overlapping community membership. In $\mathbf{U}$, a specific column vector represents how much weight each user contributes to community, and a specific row vector indicates how much weight each user benefits from the overlapping communities. From the perspective of overlapping community membership, we define following aggregation and allocation strategies with respect to community-oriented group partition,

- **commMean** stands for obtaining the aggregated group profile by averaging users' profiles who are in the same community, and allocated individual profile by averaging group profiles which each user belongs to.
- **commContribution** differentially recognizes members' contributions to a community, and the weights evidence the different contributions between users are the memberships that each user has associated with the community.
- **commBenefit** distinctly regards a member's benefits from communities, and the weights demonstrate the different benefits between users are the memberships that the user has associated with overlapping communities.
- **commWeighted** considers biased both contributions to and benefits from communities.

4.4 Community-Oriented Group Recommendation

With components introduced above, we present our proposed framework in Algorithm 3. The predicted rating matrix $\hat{\mathbf{R}}$ can be evaluated by some metrics.

Algorithm 3. CoGrec: Community-Oriented Group Recommendation Framework

---

Input:
   friend network adjacent matrix $\mathbf{G}$, user-item rating matrix $\mathbf{R}$,
   number of communities d, number of latent-factors f,
   parameters λ_{UU}, λ_P, λ_Q, aggregation and allocation strategy *method*

Output:
   predicted user-item rating matrix $\widehat{\mathbf{R}}$
 1: $\mathbf{U} \leftarrow \text{commDetNMF}(\mathbf{G}, d, \lambda_{UU})$
 2: $[\mathbf{P}, \mathbf{Q}] \leftarrow \text{ratingNMF}(\mathbf{R}, f, \lambda_P, \lambda_Q)$
 3: initialize contribution matrix $\mathbf{U}_C$ and benefit matrix $\mathbf{U}_B$ of the same size as $\mathbf{U}$
 4: **switch** (*method*)
 5: **case** commMean:
 6:     $\mathbf{U}_{C_{ij}} \leftarrow 1$, if $\mathbf{U}_{ij} > 0$; 0, otherwise
 7:     $\mathbf{U}_{B_{ij}} \leftarrow 1$, if $\mathbf{U}_{ij} > 0$; 0, otherwise
 8: **case** commContribution:
 9:     $\mathbf{U}_{B_{ij}} \leftarrow 1$, if $\mathbf{U}_{ij} > 0$; 0, otherwise
10: **case** commBenefit:
11:     $\mathbf{U}_{C_{ij}} \leftarrow 1$, if $\mathbf{U}_{ij} > 0$; 0, otherwise
12: **case** commWeighted:
13:     // actually nothing to do.
14: $\mathbf{U}_C \leftarrow$ normalize $\mathbf{U}$ by column
15: $\mathbf{U}_B \leftarrow$ normalize $\mathbf{U}$ by row
16: **end switch**
17: **return** $\widehat{\mathbf{R}} \leftarrow \mathbf{U}_B \mathbf{U}_C^T \mathbf{P} \mathbf{Q}^T$

---

5 Experiments

5.1 Datasets

We use two publicly available datasets for our study, FilmTrust[5] and CiaoDVD[6]. These two datasets contain both friend network and user-item ratings.

- **FilmTrust** is a website that allows users to share movie ratings and establish trust connections with other users.
- **CiaoDVD**. Ciao is a product rating site that employs social trust features. CiaoDVD contains user ratings in "DVD" category.

In the experiments, we use the intersection of users that have both relationships and given ratings to items. Thus the datasets we conduct experiments on are subsets of original ones. The datasets specifications are shown in Table 1, and the datasets can be obtained at LibRec[7].

---

[5] http://trust.mindswap.org/FilmTrust.
[6] http://dvd.ciao.co.uk/.
[7] http://www.librec.net/datasets.html.

Table 1. Datasets specifications.

	FilmTrust	CiaoDVD
Number of users	427	733
Number of items	1827	10305
Number of ratings	11848	19621
Rating sparsity	98.48%	99.74%
Number of friend ties	807	23582
Friend network density	0.0089	0.0879

5.2 Evaluation Metric

We adopt mean absolute error (MAE) in evaluation. It is defined as follows,

$$\text{MAE} = \frac{1}{N} \sum_{i,j} |r_{ij} - \widehat{r_{g_{ij}}}|, \tag{8}$$

where r_{ij} denotes the observed rating that user i gave to item j, $\widehat{r_{g_{ij}}}$ is the rating that user i in group g gives to item j predicted by a group-based recommender system, and N denotes the number of ratings that are used as testing set. As indicated by the definitions, a smaller MAE implies a better predictive accuracy.

5.3 Comparisons

In the comparative experiments, we compare four proposed strategies in our proposed framework with a k-means based grouping method proposed in [19]. In this section, we denote strategies commMean, commContribution, commBenefit, and commWeight by CoGrec-M, -C, -B, and -W, respectively.

LGM suggests grouping by using k-means on user- latent-factors matrix $\mathbf{P}$ shown in Eq. 1. In LGM, the predicted profile of each user is the aggregated group profile by averaging user profiles in each cluster. The identified groups can be recognized as users that have similar interests.

5.4 Experiment Setup

In comparative experiments, 5-fold cross-validation is adopted to make experimental results more stable and more convincible. As extensively experimented, the number of latent-factors does not greatly affect the performance of recommender systems, and it is set as $f = 10$. For the FilmTrust dataset the number of groups d is set varied as $\{1, 5, 10, 20, 100, 200, 427\}$, and for the CiaoDVD dataset the number of groups d is set to be in $\{1, 10, 20, 100, 200, 500, 733\}$. For CoGrec, we set $\lambda_{UU} = 0$ to ensure identified communities are overlapping. λ_P and λ_Q are set as 0.01 for simplicity.

5.5   Experimental Results

Comparison Between Different Group Number in FilmTrust. The result of performance comparison for the FilmTrust dataset is shown in Table 2.

Table 2. Performance comparison between different group numbers in FilmTrust.

# of groups	CoGrec-M	CoGrec-C	CoGrec-B	CoGrec-W	LGM
1	1.52674	1.52593	1.32401	1.32267	0.95536
5	1.41679	1.42074	1.32721	1.33180	1.00121
10	1.38919	1.39289	1.32203	1.32510	1.02864
20	1.32006	1.32169	1.31516	1.31347	1.03851
100	1.16278	1.16393	1.13223	1.13100	1.06932
200	1.05752	1.07085	1.04416	1.05344	1.05765
427	1.01865	1.02086	1.01973	1.01558	1.06085

The result shows that CoGrec-B and CoGrec-W outperform the other strategies proposed in CoGrec, but perform comparatively with LGM. With the growth of the number of groups, the MAE of CoGrec-* declines: CoGrec-M improves much faster than the others, which implies that with more identified groups, averaging strategy becomes more efficiently; CoGrec-B always outperforms CoGrec-C in any experimented given number of groups, and outperforms CoGrec-W for some specific given number of groups, from which we can infer that with averaged group profiles, an individual user benefits more if he is in a community or overlapping communities; CoGrec-W treats contribution and benefit differently, the performance of which is comparative with CoGrec-B. However, LGM acts differently. As the number of groups grows, LGM becomes less efficient than CoGrec.

Comparison Between Different Group Number in CiaoDVD. Table 3 shows the result of performance comparison for the CiaoDVD dataset.

From Table 3, we can see that CoGrec-* generally outperform LGM in MAE. Similar to the situation in the FilmTrust dataset, as the group number grows, CoGrec-* improve while LGM gets worsen. It might be the reason that the friend network of CiaoDVD dataset is much denser than that of FilmTrust dataset as you can see in Table 1, though the rating sparsity is on the contrary.

For CoGrec-M and CoGrec-B, the recommendation performance improves as the number of groups increases. For CoGrec-W, the performance changes little. This also can be explained by the dense network structure. Friendship relations in the CiaoDVD dataset are enough to identify communities and thus CoGrec relatively accurately predict ratings based on group recommendation.

The results by the proposed CoGrec on the CiaoDVD dataset implies that with more network structural information, community-oriented group recommendation could achieve higher accuracy.

Table 3. Performance comparison between different group numbers in CiaoDVD.

# of groups	CoGrec-M	CoGrec-C	CoGrec-B	CoGrec-W	LGM
1	2.00683	1.99314	1.99868	**1.98487**	1.99728
10	1.99181	1.98598	1.99109	**1.98565**	2.02620
20	1.99218	**1.98489**	1.99234	1.98501	2.03498
100	1.98877	1.98574	1.98948	**1.98495**	2.06730
200	1.98781	1.98644	1.98711	**1.98490**	2.06314
500	**1.98658**	1.98671	1.98714	1.98685	2.07713
733	1.98627	1.98652	1.98730	**1.98546**	2.07981

Effect of Overlapping Communities on Group Recommendation in FilmTrust. It worth noting that for CoGrec, the number of groups can be larger than the number of users, which means that overlapping communities exist. The result of performance comparison on overlapping group for the FilmTrust dataset is demonstrated in Fig. 4.

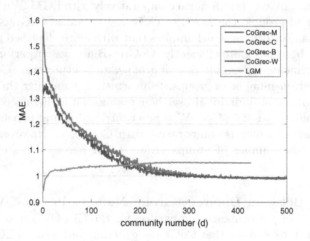

Fig. 4. Performance comparison with varied community number in FilmTrust. (Color figure online)

The first observation is that the performance of LGM in the FilmTrust dataset gets worse as soon as the number of communities arises. In the meantime, the performance of CoGrec-B and CoGrec-W become a little worse around $d = 7$ and improve quickly when d grows. Furthermore, CoGrec-M and CoGrec-C behave likewise so that the performance of them differs little, and CoGrec-B and CoGrec-W have similar effects on performance.

Last but not least, our proposed framework CoGrec can find overlapping groups, the ability of which is ensured by non-negative matrix factorization.

However, under the evaluation metric MAE, the performance among different CoGrec-* differs little, and the MAEs of them are asymptotic, as shown in Fig. 4. Thus, other evaluation metrics should be elevated for overlapping groups.

5.6   Discussions

In the experiments on the FilmTrust dataset, one can find that the proposed framework performs a little deficient while LGM shows its performance in accuracy. One possible reason could be the sparsity of friend network. As one can follow the comparative results on the CiaoDVD dataset that, with a denser friend network, CoGrec outperforms LGM. In spite of the dependency of friend network density, CoGrec will have no resolution issues that will be discussed later, when the number of groups gets larger or even exceeds the number of users.

In the experiments on both datasets, the performance of LGM decreases as the number of groups increases. It should not the case that this phenomenon happens. With more groups, the behavior of LGM group recommendation will be approaching the schema of individual recommendation. We argue that it should be a problem of implementing conventional k-means algorithm. However, LGM approach will become individual recommendation when the number of groups equals the number of users, and the groups identified by LGM are non-overlapping and not well interpretable.

6   Conclusion

In this paper, with the help of online social networks deployed in many review and rating sites, a Community-Oriented Group Recommendation framework (CoGrec) is proposed. Our proposed framework includes an overlapping community detection based grouping component to identify groups, a matrix factorization based user profile generating component, and an overlapping community membership based aggregation and allocation component to suggest predicted ratings for group users. Furthermore, four aggregation and allocation strategies exploiting overlapping community membership are proposed. Extensive comparative experiments show the effectiveness of our proposed framework. In addition to high accuracy, CoGrec has a better interpretation than most existing automatically grouping approaches that use interest-based or user- latent-factors based clustering method.

There are several interesting points for possible future work. Directed or even bidirectional social network may imply more information to identify community and assign membership. Thus the performance of proposed CoGrec should be improved by using directed friend networks. Other aggregation and allocation strategies, such as least misery, average without misery, random, or even user-defined strategy, can also be added into CoGrec and thus be evaluated. As for implementation non-negative matrix factorization, the multiplicative updating approach that requires a set of large dimension matrix multiplications, is much time-consuming. We plan to deploy non-negative matrix factorization by using projected gradient descent, which may probably be parallelized.

Acknowledgments. This research is supported by the National Natural Science Foundation of China (No. 71231002).

References

1. Amer-Yahia, S., Roy, S.B., Chawlat, A., Das, G., Yu, C.: Group recommendation: semantics and efficiency. Proc. VLDB Endow. **2**(1), 754–765 (2009)
2. Baltrunas, L., Makcinskas, T., Ricci, F.: Group recommendations with rank aggregation and collaborative filtering. In: Proceedings of the Fourth ACM Conference on Recommender Systems, RecSys 2010, pp. 119–126. ACM, New York (2010)
3. Boratto, L., Carta, S., Chessa, A., Agelli, M., Clemente, M.L.: Group recommendation with automatic identification of users communities. In: IEEE/WIC/ACM International Joint Conferences on Web Intelligence and Intelligent Agent Technologies, WI-IAT 2009, vol. 3, pp. 547–550, Sept 2009
4. Boratto, L., Carta, S.: State-of-the-art in group recommendation and new approaches for automatic identification of groups. In: Soro, A., Vargiu, E., Armano, G., Paddeu, G. (eds.) Information Retrieval and Mining in Distributed Environments. SCI, vol. 324, pp. 1–20. Springer, Heidelberg (2010)
5. Boratto, L., Carta, S.: Using collaborative filtering to overcome the curse of dimensionality when clustering users in a group recommender system. In: ICEIS, (2), pp. 564–572 (2014)
6. Campos, L.M., Fernández-Luna, J.M., Huete, J.F., Rueda-Morales, M.A.: Managing uncertainty in group recommending processes. User Model. User-Adap. Interact. **19**(3), 207–242 (2008)
7. Chao, D.L., Balthrop, J., Forrest, S.: Adaptive radio: achieving consensus using negative preferences. In: Proceedings of the 2005 International ACM SIGGROUP Conference on Supporting Group Work, GROUP 2005, pp. 120–123. ACM, New York (2005)
8. Ding, C., Li, T., Peng, W., Park, H.: Orthogonal nonnegative matrix t-factorizations for clustering. In: Proceedings of the 12th ACM SIGKDD International Conference on Knowledge Discovery and Data Mining, KDD 2006, pp. 126–135. ACM, New York (2006)
9. Koren, Y.: Factorization meets the neighborhood: a multifaceted collaborative filtering model. In: Proceedings of the 14th ACM SIGKDD International Conference on Knowledge Discovery and Data Mining, KDD 2008, pp. 426–434. ACM, New York (2008)
10. Lee, D.D., Seung, H.S.: Algorithms for non-negative matrix factorization. In: Leen, T.K., Dietterich, T.G., Tresp, V. (eds.) Advances in Neural Information Processing Systems, vol. 13, pp. 556–562. MIT Press, Cambridge (2001)
11. Ma, H.: An experimental study on implicit social recommendation. In: Proceedings of the 36th International ACM SIGIR Conference on Research and Development in Information Retrieval, SIGIR 2013, pp. 73–82. ACM, New York (2013)
12. Ma, H., Yang, H., Lyu, M.R., King, I.: Sorec: social recommendation using probabilistic matrix factorization. In: Proceedings of the 17th ACM Conference on Information and Knowledge Management, CIKM 2008, pp. 931–940. ACM, New York (2008)
13. Masthoff, J.: Group recommender systems: combining individual models. In: Ricci, F., Rokach, L., Shapira, B., Kantor, P.B. (eds.) Recommender Systems Handbook, pp. 677–702. Springer, Boston (2011)

14. O'Connor, M., Cosley, D., Konstan, J.A., Riedl, J.: PolyLens: a recommender system for groups of users. In: ECSCW 2001: Proceedings of the Seventh European Conference on Computer Supported Cooperative Work, 16–20 September 2001, Bonn, Germany, pp. 199–218. Springer, Dordrecht (2001)
15. Pazzani, M.J., Billsus, D.: Content-based recommendation systems. In: Brusilovsky, P., Kobsa, A., Nejdl, W. (eds.) Adaptive Web 2007. LNCS, vol. 4321, pp. 325–341. Springer, Heidelberg (2007)
16. Psorakis, I., Roberts, S., Ebden, M., Sheldon, B.: Overlapping community detection using bayesian non-negative matrix factorization. Phy. Rev. E **83**, 066114 (2011)
17. Ricci, F., Rokach, L., Shapira, B.: Introduction to recommender systems handbook. In: Ricci, F., Rokach, L., Shapira, B., Kantor, P.B. (eds.) Recommender Systems Handbook, pp. 1–35. Springer, Boston (2011)
18. Salakhutdinov, R., Mnih, A.: Bayesian probabilistic matrix factorization using markov chain monte carlo. In: Proceedings of the 25th International Conference on Machine Learning, ICML 2008, pp. 880–887. ACM, New York (2008)
19. Shi, J., Wu, B., Lin, X.: A latent group model for group recommendation. In: 2015 IEEE International Conference on Mobile Services (MS), pp. 233–238, June 2015
20. Shi, Y., Larson, M., Hanjalic, A.: Collaborative filtering beyond the user-item matrix: a survey of the state of the art and future challenges. ACM Comput. Surv. **47**(1), 3:1–3:45 (2014)
21. Yang, B., Lei, Y., Liu, D., Liu, J.: Social collaborative filtering by trust. In: Proceedings of the Twenty-Third International Joint Conference on Artificial Intelligence, IJCAI 2013, pp. 2747–2753. AAAI Press (2013)
22. Yu, Z., Zhou, X., Hao, Y., Gu, J.: Tv program recommendation for multiple viewers based on user profile merging. User Model. User-Adap. Inter. **16**(1), 63–82 (2006)
23. Zhang, Z.Y., Wang, Y., Ahn, Y.Y.: Overlapping community detection in complex networks using symmetric binary matrix factorization. Phy. Rev. E **87**(6), 062803 (2013)

CTS: Combine Temporal Influence and Spatial Influence for Time-Aware POI Recommendation

Hanbing Zhang[1], Yan Yang[1,2(✉)], and Zhaogong Zhang[1,2]

[1] School of Computer Science and Technology, Heilongjiang University,
Harbin 150080, China
zhbyeyu@gmail.com, Yangyan@hlju.edu.cn,
zhaogong.zhang@qq.com
[2] Key Laboratory of Database and Parallel Computing of Heilongjiang Province,
Harbin 150080, China

Abstract. As location-based social network (LBSN) services become more popular in people's lives, Point of Interest (POI) recommendation has become an important research topic.POI recommendation is to recommend places where users have not visited before. There are two problems in POI recommendation: sparsity and precision. Most users only check-in a few POIs in an LBSN. To tackle the sparse problem in a certain extent, we compute the similarity between the check-in datasets of different times. For the precision problem, we incorporate temporal information and geographical information. The temporal information will influence how the user chooses and allow the user to visit different distance point on different day. The geographical information is also used as a control for points which are too far away from the user's check-in data. Our experimental results on real life LBSN datasets show that the proposed approach outperforms the other POI recommendation methods substantially.

Keywords: Recommendation · Point-of-interest · Location-based social network · Collaborative filtering · Spatio-temporal

1 Introduction

Location-based social networks (LBSN) have been a wildly popular application of location-based services in recent years. Some of them, such as Foursquare and Facebook Places, have attracted millions of users to share their geographical locations and experiences through "check-in". Each check-in represents a user's visit to an interesting location, known as point-of-interest (POI), it can be a restaurant, a museum, a sightseeing site, etc., at a specific visiting time. We can get the user name, the visiting time and some information of the POI from the check-in data. The large volume of user check-in data in these applications makes it possible to recommend unvisited POIs to users.

To make POI recommendations, several techniques have been exploited for improve the recommendation accuracy. User-based collaborative filtering method is one of them and has performed better than others according to recent works [1, 2]. In this paper,

W. Che et al. (Eds.): ICYCSEE 2016, Part I, CCIS 623, pp. 272–286, 2016.
DOI: 10.1007/978-981-10-2053-7_25

we exploit the user-based collaborative filtering method further by investigating some important factors that have been neglected in earlier studies.

Temporal information is one of the very important factors in POI recommendation and had been exploited in many works. However, to the best of our knowledge, none of them exploit the different between workdays and weekends to make POI recommendations. The different between workdays and weekdays is a very important factor and users' activities are often influenced by it. For example, the probability of a user to go to the zoo at weekends is more than that on workdays, and he is more likely to go to the company on workdays rather than at weekends. In other words, the location that people might be interested in is different on workdays and weekends.

The other important factor in POI recommendation is spatial influence [1, 3, 4]. We find an interesting phenomenon, that is the region people visited at weekends is greater than that on workdays. We therefore divide the reach distance into two cases according to the two kinds of days, and filter some POIs that the target user can't arrive. For example, the POIs we recommend to users on weekends maybe further away than it on workdays and the candidate POIs on weekends will be more than it on workdays. Hence, the spatial influence is influenced by time in this paper.

Based upon the two important factors, we propose a new method to calculate the recommendation score of a POI, which is composed by temporal score and spatial score. The temporal score is calculated by user-based collaborative filtering method and the spatial score is calculated by a new equation. Finally, we exploit two parameters to balance the two score and combine them to make time-aware POI recommendation. The contributions of this paper are summarized as follows.

- We propose a new POI recommendation method, which considers the different of the human behavior on workdays and weekends. We also analyze the similarity between the hours and exploit it to solve the data sparse problem in a certain extent.
- We define the reach distance of users, which means the different maximum distances that a user can arrive on workdays and weekends. We can therefore filter candidate POIs to reduce the calculating time by exploit the reach distance of users.
- We conduct extensive experiments to evaluate the recommendation accuracy of CTS with real datasets. The experimental results show that CTS outperforms other time-aware POI recommendation in terms of accuracy.

The rest of this paper is organized as follows. In Sect. 2, we review related work. In Sects. 3 and 4, we describe our methods, which exploit temporal influence and spatial influence to make time-aware POI recommendation, respectively. In Sect. 5, we present a method to combine temporal influence and spatial influence. The experimental results are presented in Sect. 6 and we conclude this paper in Sect. 7.

2 Related Work

In this section, we introduce some similarity calculation methods and POI recommendation techniques as the background for our algorithm design.

2.1 Similarity Calculation

Similarity calculation is widely used in the Data Mining field. There are some classical methods, for example: Euclidean distance, Cosine similarity, Jaccard similarity coefficient and so on. Most POI recommendation has used the cosine distance to compute the similarity between the different users. Let L be the location set in an LBSN. The equation is defined below:

$$w_{u,v} = \frac{\sum_{l \in L} c_{u,l} c_{v,l}}{\sqrt{\sum_{l \in L} c_{u,l}^2} \sqrt{\sum_{l \in L} c_{v,l}^2}}. \tag{1}$$

where u and v is two different users, $c_{u,l}$ is the check-in activity for a user u and location l, when $c_{u,l} = 1$ represents that u has a check-in at l and $c_{u,l} = 0$ otherwise. The symbols used in this paper are listed in Table 1.

Table 1. Symbols

Symbol	Meaning
U, L, T	User set, POI set, time slot set of 24 h's
u, v, l, t, d	User $u, v \in U$, POI $l \in L$, time slot $t \in T$ and $d \in$ (workday, weekend)
L_u	POI set of user u $L_u \subset L$
$c_{u,l}, c_{u,t,l}$	The check-in record of u over l and u over l at t, respectively
$w_{u,v}, w_{u,v}^t$	The similarity between u and v, the similarity between u and v based time
$dis(l_i, l_j)$	Distance between l_i and l_j
$maxDis(l_i|L_u)$	The maximum distance between l_i and other POIs in the L_u
$N_l, N_{d,t,l}, N_t^u$	The set of check-ins at l, N_l at day d time t, the set of check-ins of u at time t

2.2 Collaborative Filtering

Collaborative filtering (CF) technique has been widely adopted for recommender systems and many collaborative filtering recommendation methods [5] have been proposed. The CF method exploits users' historical purchase ratings or preference to make recommendations, and it can be divided into two categories, namely memory-based CF and model-based CF [6]. Memory-based CF methods can be further divided into user-based CF and item-based CF. First, the method finds similar users based on their ratings on items using similarity calculation. Then utilizing the users' similarity and their historical ratings on the item to calculate item's recommendation score. We will detail the method in Sect. 3. In contrast, item-based CF method works by finding items that are similar to other items the user has rated or liked. Model-based CF builds models uses data mining techniques, such as matrix factorization [7] and probabilistic topic model [8], but these methods are usually computationally expensive.

2.3   POI Recommendation

Most of POI recommendation methods exploit temporal information, geographical information or social information. The work by Yuan et al. [2] is the most closely related to our work. Yuan et al. adopt a unified framework to incorporate both the temporal information and geographical information to make POI recommendation. Specifically, the work considers the temporal influence when compute the similarity between different users, because they find users' activities are often influenced by time. To exploit the geographical influence, this work assumes that the willingness that a user moves from a POI to another POI is a function of their distance, which is different with [1]. In their study, the probability that a user checks in a new POI is estimated by the popularity of the new POI and the distance between the new POI and old POIs. Although their study had been improved, there are still some problems in it. They haven't consider the difference between workday and weekend, and when they consider the geographical information, the reach distance is neglected. To exploit temporal information and geographical information better, we propose a new method which can further improve the recommendation accuracy.

There exists other work that exploits some different methods to make POI recommendation. Wang et al. [9] proposed algorithms under the framework of personalized PageRank, Yin et al. [10] proposed an LDA-based model to recommend POIs for a given user at a given city and Zhang et al. [11] proposed a POI recommendation framework based on users' opinions to the aspects of POIs in tips.

3   Method with Temporal Influence

In this section, we present the baseline user-based CF method (Sect. 3.1), and then introduce our method incorporating the temporal influence in the user-based CF (Sects. 3.2 and 3.3).

3.1   User-Based Collaborative Filtering

As we described in Sect. 2, user-based CF method is used to make POI recommendation in many studies. Specifically, given a target user u and let U be the user set. The method first finds the user that is similar with u, and then it exploits the check-in data of similar users to calculate the recommendation score. The recommendation score means the probabilistic of a user to visit a POI. The equation is as follows:

$$s_{u,l} = \frac{\sum_{v \in U} w_{u,v} c_{v,l}}{\sum_{v \in U} w_{u,v}}. \tag{2}$$

where u and v is two different users. The $w_{u,v}$ is the similarity between u and v, and that it can be calculated by Eq. 1. The $s_{u,l}$ represents the recommendation score of user u to visit POI l and the method will return the top-k POIs to user u as the recommendation results.

3.2   Incorporating Temporal Influence: Hour

The report in [2, 12] show that human have strong periodic behavior through out a day. For example, people tend to check-in hotels at night rather than during day. We can therefore exploit this property to improve the recommendation accuracy. Note that, the time and time slot are the same in this paper.

Let $\left\{ (POI_1, N_1)_{u,t}, (POI_2, N_2)_{u,t}, \ldots, (POI_l, N_t)_{u,t} \right\}$ be the check-in vector of user u at time t, and N_t represents the number of times that user u has visited POI_l at time t. We propose a new definition, namely visiting probability, which means the probabilistic that a user has visited the same POI at time t_i and t_{i+1}. For each user, we calculate the visit probabilistic between the check-in vectors of different times as follows:

$$p_{t_i,t_j}^u = \frac{\left| N_{t_i}^u \cap N_{t_j}^u \right|}{\left| N_{t_i}^u \right|}. \tag{3}$$

where $\left| N_{t_i}^u \right|$ is the number of times that user u has visited POIs at time t_i, and the $\left| N_{t_i}^u \cap N_{t_j}^u \right|$ represents the sum of the minimum number of times that user u has visited the same POIs at time t_i and t_j. For example, assume A, B, C and D are four different POI, the check-in vector of user u at time t_1 and t_2 are $\left\{ (A,3)_{u,t_1}, (B,2)_{u,t_1}, (C,1)_{u,t_1} \right\}$ and $\left\{ (A,2)_{u,t_2}, (C,2)_{u,t_2}, (D,2)_{u,t_2} \right\}$, respectively. The visiting probability between t_1 and t_2 is therefore equal to 0.5. We then calculate the average of the visiting probability values of all users between two times t_i and t_j as the similarity value between these two times t_i and t_j. Note that, this similarity is not satisfied with the reflexive law.

Figure 1(a) shows the similarity curves for three times (7:00, 12:00, and 15:00) over the Foursquare data. Each similarity curve for the three times shows the similarity between them and every other hour in a day. We find that the users' check-in behavior at two adjacent times is more similar than the users' check-in behavior at two non-adjacent times, such as the similarity between 7:00 and 8:00 is much higher than the similarity between 7:00 and 10:00. Nevertheless, the similarity between 12:00 and

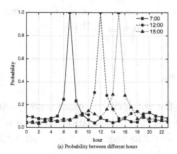
(a) Probability between different hours

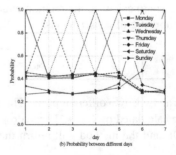
(b) Probability between different days

Fig. 1. The user behavior similarities between different times

15:00 is lower than the similarity between 12:00 and 18:00, this is because users have the same activities in 12:00 and 18:00 so that they maybe visit the same POIs.

We also tried other methods to calculate the similarity, but observed similar results. In order to avoid the occurrence of over-fitting, we just consider the similarity between two adjacent times rather than the similarity between each time in the similarity calculation. Let T be the time set. The $c_{u,t,l}$ represents a user u have visited a POI l at time t, when $c_{u,t,l} = 1$ represents that u have a check-in at l at time t and $c_{u,t,l} = 0$ otherwise. The new value for element $c_{u,t,l}$ is calculated by Eq. 4 and we propose Eq. 5 to calculate the similarity between different users. The parameter of the Eq. 4, α, is used to balance the weight of the adjacent times.

$$\bar{c}_{u,t,l} = \left(\frac{1-\alpha}{2}\right) \cdot c_{u,t-1,l} + \alpha \cdot c_{u,t,l} + \left(\frac{1-\alpha}{2}\right) \cdot c_{u,t+1,l}. \tag{4}$$

$$\bar{w}_{u,v}^{t} = \frac{\sum_{t \in T} \sum_{l \in L} \bar{c}_{u,t,l} \bar{c}_{v,t,l}}{\sqrt{\sum_{t \in T} \sum_{l \in L} \bar{c}_{u,t,l}^{2}} \sqrt{\sum_{t \in T} \sum_{l \in L} \bar{c}_{v,t,l}^{2}}} \tag{5}$$

According to Eq. 5 we can find that two users u and v are more similar if they have visited the same POIs at the adjacent time slots. This is also consistent with the objective laws of human behavior. On the other hand, the Eq. 4 can also be used to calculate the recommendation score and that solve the data sparse problem in a certain extent. The calculation equation of recommendation score is update as follows:

$$\bar{s}_{u,t,l}^{h} = \frac{\sum_{v \in U} \bar{w}_{u,v}^{t} \left(\left(\frac{1-\alpha}{2}\right) \cdot \bar{c}_{u,t-1,l} + \alpha \cdot \bar{c}_{u,t,l} + \left(\frac{1-\alpha}{2}\right) \cdot \bar{c}_{u,t+1,l}\right)}{\sum_{v \in U} \bar{w}_{u,v}^{t}}. \tag{6}$$

3.3 Incorporating Temporal Influence: Day

In this section, we will exploit the day influence to make POI recommendation, which is similar with Sect. 3.2. We split the check-in data into 7 parts based on days, and then we performed data analysis by some different similarity calculation methods. Note that, the workdays are from Monday to Friday and the weekends are Saturday and Sunday in this paper. Figure 1(b) is the similarity curves for all days, calculated by our method. The results show that the similarity between different workdays or different weekends is much higher than the similarity between workdays and weekends. We therefore split the check-in data into 2 parts based on workdays and weekends when we calculate the recommendation score. The equation is defined below:

$$\bar{s}_{u,t,l}^{h,d} = \begin{cases} \beta_{\text{workday}} \cdot \bar{s}_{u,t,l}^{h,\text{workday}} + (1 - \beta_{\text{workday}}) \cdot \bar{s}_{u,t,l}^{h,\text{weekend}} & d \in \text{workday}. \\ \beta_{\text{weekend}} \cdot \bar{s}_{u,t,l}^{h,\text{workday}} + (1 - \beta_{\text{weekend}}) \cdot \bar{s}_{u,t,l}^{h,\text{weekend}} & d \in \text{weekend}. \end{cases} \tag{7}$$

where $\bar{s}_{u,t,l}^{h,workday}$ and $\bar{s}_{u,t,l}^{h,weekend}$ is calculated by the Eq. 6 based the split-check-in data. We did not exploit the day influence and hour influence to calculate the similarity between different users to avoid the data sparse problem. The β is a control parameter, which is used to balance the weight between the recommendation score from workdays and the recommendation score from weekends.

4   Method with Spatial Influence

The geographical information is an important factor in POI recommendation and has been exploited in many studies [1, 2, 4, 13–15]. The power law distribution is used to model the probability of a user moving from one place to another. But we did not use this method, we propose a new method to utilize the geographical information of POI and show that the method can further improve the recommendation accuracy.

To utilize the geographical information better, we performed data analysis on the check-in data of users and made some new discoveries. Figure 2(a) shows the POIs that a user visited over the Gowalla data. The ID of the POIs in this figure represent the temporal order that user visited them. We find that most of the later POIs are in the area which is constructed by the previous POIs. Therefore, we propose a new equation to calculate the probability value that a user visits a new POI.

$$s_{u,l}^s = \frac{\sum_{l_c \in L_u} \dfrac{k \cdot maxDis(l_c|L_u) - dis(l_c,l)}{k \cdot maxDis(l_c|L_u)}}{|L_u|} \tag{8}$$

where $dis(l_c, l)$ is used to calculate the distance between POI l and POI l_c, and the $maxDis(l_c|L_u)$ is the maximum distance between POI l_c and other POIs in the user's check-in dataset L_u. The $|L_u|$ is the number of the POIs that user u have visited before. The k is used to control the reach distance of user u, the value of $k \cdot maxDis(l_c) - dis(l_c, l)$ will be 0 if $dis(l_c, l) > k \cdot maxDis(l_c)$. Specifically, if new POIs is near the center of the area, the probability value that the target user visits it will be high. Otherwise, if new POIs are far away from the center of the area, the probability value will be low.

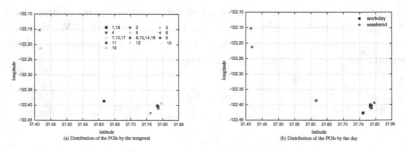

(a) Distribution of the POIs by the temporal   (b) Distribution of the POIs by the day

Fig. 2. The distribution of the POIs that a user has visited

The spatial influence will be affected by the temporal information. We performed analysis on the check-in data, which are divided into 2 parts based on the workdays and the weekends. The Fig. 2(b) is similar to Fig. 2(a), the only difference is the check-in dataset have been divided in the Fig. 2(b). The results show that the reach distance on weekends is further than that on workdays, and we have therefore different k values on different days. The reason for this is that users have more time to go to new POIs on the weekend, so that they can go somewhere that is further away from their check-in records. The Eq. 8 is then updated as follows:

$$s_{u,l}^{s,d} = \begin{cases} \sum_{l_c \in L_u} \dfrac{k_{workday} \cdot \text{maxDis}(l_c|L_u) - \text{dis}(l_c,l)}{k_{workday} \cdot \text{maxDis}(l_c|L_u)} & d \in workday. \\[4mm] \dfrac{\sum_{l_c \in L_u} \dfrac{k_{weekend} \cdot \text{maxDis}(l_c|L_u) - \text{dis}(l_c,l)}{k_{weekend} \cdot \text{maxDis}(l_c|L_u)}}{|L_u|} & d \in weekend. \end{cases} \tag{9}$$

In Eq. 9, we exploited the geographical information and some temporal information to calculate the recommendation score. Next, we will consider the popularity of a POI to improve our method. The popularity of a POI is different over time, e.g., a bar is more popular in the evening and the zoo is more popular at the weekend. We calculated the check-in probabilities of the top-5 most popular POIs and found that the popular POI will always be recommended to the target user. This situation greatly reduces the accuracy of recommendation. To balance the effect of popular POI, let $|N_{d,t,l}|$ represent the number of check-ins at POI l at time $t - 1$ to $t + 1$ on day d(workday or weekend) and $|N_l|$ represent the number of check-ins at POI l, the equation is defined as follows:

$$P_{d,t}(l) = \frac{|N_{d,t,l}|}{|N_l|} \cdot \left(\delta \cdot \frac{|N_{d,t,l}|}{\sum_{l_c \in L} |N_{d,t,l_c}|} + (1 - \delta) \cdot \frac{|N_l|}{\sum_{l_c \in L} |N_{l_c}|} \right). \tag{10}$$

the δ is a balance parameter to tunes the weight between POI l's temporal popularity and global popularity. Specifically, if the POI is a popular POI and it was always visited at the specific time, its' popularity value will be high. Otherwise, its' popularity value will be low. We then exploit the Eqs. 9 and 10 to calculate the recommendation score as follows:

$$\bar{s}_{u,t,l}^{s,d,h} = P_{d,t}(l) \cdot s_{u,l}^{s,d}. \tag{11}$$

The temporal information in this section is different with that in Sect. 3. We exploit the indirect effects of time to make POI recommendation, instead of make use of the temporal influence to calculate the similarity between different users.

5   A Combined Method

We can get two different recommendation scores based Sects. 3 and 4, respectively. But the two scores are calculated by two different methods, so we need to use min-max normalization to normalize them before we use them to make a recommendation. At last, we proposed a combined method, namely Combine Temporal Influence and Spatial Influence (CTS), which aims to exploit these two influences to make a rec-ommendation. The algorithm is shown in Algorithm 1.

The area in line 2 and line 4 is constructed by the POIs that user u has visited. First, the distribution of the POIs can be defined as a point set Q so that we can exploit the Graham's Scan method to construct the convex hull of Q. We then utilize the vertices on the convex hull to construct the area, we have a vertex as the center and we have the maximum distance between this vertex and other vertices as the radius to draw a circle. The area covered by these circles is defined as the A, which means the POIs in the A is possible to be visited by the target user. The parameter in the Eq. 9 is also used to construct the different areas on workdays and weekends. In line 4, each POI in the L' will determine whether it's in the area A. After that, we calculate and normalize the recommendation score of the POI which is in the area, and return the top ranked POIs as the recommendation results.

Algorithm1: Combine Temporal and Spatial (CTS)
Input: check-in data D, user u, time t
Output: Top-k POIs as recommendation results
1 L' the POIs that u don`t visit;
2 A   the area constructed by the POIs that u has visited;
3 foreach $l \in L'$ do
4   if l is in the A then
5   calculate and normalize the $\bar{s}_{u,t,l}^{h,d}$ and the $\bar{s}_{u,t,l}^{s,d,h}$;
6     $s \leftarrow \varepsilon \cdot \bar{s}_{u,t,l}^{h,d} + (1 - \varepsilon) \cdot \bar{s}_{u,t,l}^{s,d,h}$;
7 return top-k POIs in L' based on recommendation scores;

The time complexity of the algorithm is $O(|A| \times |U| \times |T| \times |L|)$, which related to the $\bar{s}_{u,t,l}^{h,d}$ and the $\bar{s}_{u,t,l}^{s,d,h}$. The $|A|$ is the number of POIs that in the area A, $|T|$ is the number of time slot (24 h), $|U|$ and $|L|$ is the number of users and POIs. Therefore, the time complexity is closely related to the $|U| \times |T| \times |L|$ and the algorithm is offline.

6   Experiments

In this section, we evaluate the proposed method on 2 real-world datasets and compare our method with other methods. We present experimental settings in Sect. 6.1 and analyze experimental results in Sect. 6.2.

6.1   Experimental Setup

Dataset. We use two real-world datasets in our experiments. One is the Foursquare dataset made in Singapore between Aug. 2010 and Jul. 2011 which came from [15] and the other is Gowalla dataset made in California and Nevada between Feb. 2009 and Oct. 2010 that came from [12]. Each check-in contains user ID, time information and POI information. We only consider users that had at least 10 check-ins, so we evaluated our method on 2,519 Foursquare users and 10,997 Gowalla users. For each user, we marked off 25 % of his/her recent check-ins as testing data to evaluate the performance of different methods, the other check-in data were used as training data.

Metrics. To evaluate the performance of different methods, we employ two standard metrics, namely precision@N and recall@N (denoted by Pre@N and Rec@N), where N is the number of recommended POIs.

For time-aware POI recommendation, given a user u and a time slot t, we let $tp_{u,t}$ be the number of POIs contained in both the ground truth and the top-N results produced by methods; $fp_{u,t}$ be the number of POIs in the top-N results by methods but not in the ground truth; and $tn_{u,t}$ be the number of POIs contained in ground truth but not in the top-N results by methods. Then, the Pre@N and Rec@N at time t are calculated as follows [2].

$$Pre@N(t) = \frac{\sum_{u \in U} tp_{u,t}}{\sum_{u \in U} (tp_{u,t} + fp_{u,t})} \tag{12}$$

$$Rec@N(t) = \frac{\sum_{u \in U} tp_{u,t}}{\sum_{u \in U} (tp_{u,t} + tn_{u,t})} \tag{13}$$

The average Pre@N (Rec@N) of all time slots is defined as the overall Pre@N (Rec@N) in [2, 13]. In this paper, we consider three values of N (5,10,15) in our experiments for the two metrics and set 5 as the default value.

Baseline Methods. Our method is compared with the following baseline methods.

- **UCF:** This is the baseline user-based CF method, which exploit the check-in data to calculate the similarity between different users.
- **UCF+G:** This method adopts linear interpolation to incorporate both the social and geographical influences into the user-based CF framework for POI recommendation [1].
- **UTF:** UTF is a user-based collaborative filtering method [16] and calculate the similarity between different users by weighting the check-ins with a time function.
- **LRT:** LRT incorporates temporal influence by constraining the latent factors of a user are similar in two adjacent time slots [17].

- **UTE+SE:** This method makes use of the geographical influence and temporal influence for time-aware POI recommendation [2], UTE is the method without geographical influence.
- **BPP:** This is the state-of-the-art method for time-aware POI recommendation [13], which incorporates both the geographical and temporal influences.

UCF, UTF and LRT don't exploit the geographical influence in these methods. The UTE + SE and BPP are the methods that recently proposed for time-aware POI recommendation, which we are mainly comparison and analysis.

6.2 Experimental Results

Parameter Tuning. Before we evaluate our method with others, we need to tune some parameters using the development set to optimize the effectiveness of the method. Recall that in Eqs. 4, 7 and 10, parameters α tune the weight of the adjacent times, β tune the weight of different days and δ tune the weight between a POI's temporal popularity and long-term popularity. We use the average of Pre@5 as the overall performance metric to tune them and the optimal settings of them is displayed in the Table 2. Notice that $\beta_{weekend}$ is higher than $1 - \beta_{workday}$, one reason is that the check-in datasets on the weekend is smaller than that on the workday. Another is that the POIs of user visited on workday may also be visited on the weekend, and the POIs of user visited at weekend may not be visited on workday.

Table 2. Optimal settings of α, β and δ

	α	$\beta_{workday}$	$\beta_{weekend}$	δ
Foursquare	0.5	0.8	0.4	0.6
Gowalla	0.6	0.8	0.5	0.8

The other two parameters in our method are k and ε which are used to filter out the unreachable POIs and balance the impact of temporal information and geographical information, respectively. Figure 3(a) shows that the optimal value of k for Foursquare is 2.5 during workdays and 4 on weekends; for Gowalla, k is 3 on workdays and 5 on weekends. The greater value of k means that the POIs user can visited will be more, when the k is big enough, it will be similar with the method in [2]. The decline of Pre@5 is because the popular POI, which is further away from the user's visited area, will be also recommended to the user. Figure 3(b) shows that the optimal value of ε walla are 0.6 and 0.7, respectively.

Performance of Methods. First, we compare our method with other methods which utilizing the temporal influence and the UCF as the baseline method, the precision and recall of them are reported in Fig. 4. From the figure, it is observed that the performance of LRT is not satisfactory, the reason may be the datasets in our experiments is too sparse, that the Matrix Factorization method cannot perform well. The CT is our

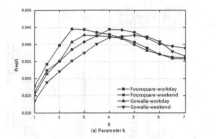
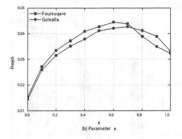

Fig. 3. Tuning parameter k

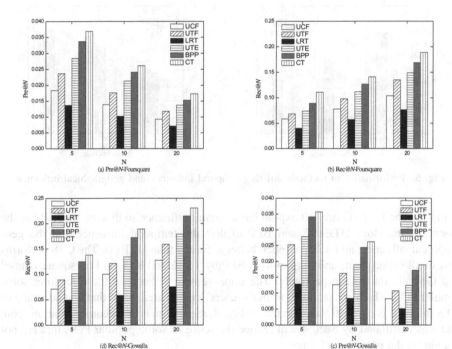

Fig. 4. Performance of methods utilizing temporal influence

method without combining the recommendation score calculated by the spatial influence. Compare to UCF, UTF and UTE, the CT performs much better than them. This is because the CT not only considers the similarity between the different times but it also exploits the difference between the workday and the weekend. Compare to BPP, the CT performs a little better, probably because the Preference Propagation that are used in the BPP represents the changes in human behavior between two different time slots in a certain extent.

We then compare our method with four methods (UCF+G, UTE+SE, BPP), which utilizes the temporal influence and the geographical influence, and show the results

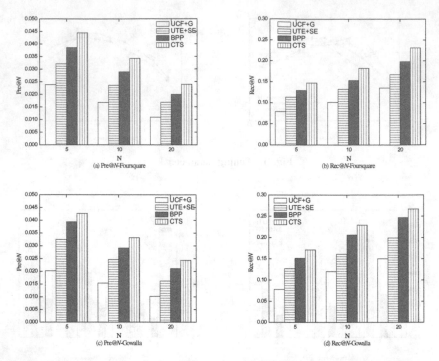

Fig. 5. Performance of methods utilizing temporal influence and geographical influence

in Fig. 5. The UCF+G doesn't exploit the temporal influence so that the method has the worst results. Both UTE+SE and BPP exploit the temporal influence and the geographical influence; and achieve much better accuracy than UCF+G. The CTS performs best in all experiment, and outperforms BPP by 15 % and 8 % on Foursquare dataset and Gowalla dataset, respectively. The improvement is because the CTS filter some popular POIs which are far away from the users' visited area. Note that, the accuracy of CTS in Fig. 5 is much higher than that in Fig. 4, the reason is the recommendation score that was calculated by Sect. 4 will reduce the score of some popular POIs that are not popular at the recommend time.

Effect of the Divide of Weekend. We compare the CTS methods that have different days as the weekend. In the CTSFS, CTSSM and CTSFM, the weekend is from Friday to Sunday, Saturday to Monday and Friday to Monday, respectively. The Fig. 6 is the comparison between these methods, and the CTS performed the best. Observe the CTSFM which have the worst results, the reason maybe the similarity between the weekend dataset is lower than others methods and the workday dataset is more sparse which is important in the recommendation score calculation. The accuracy of CTSFS and CTSSM are similar, probably because the density of workday in them is similar and both Friday and Monday have the same similarity with the weekend (Saturday and Sunday).

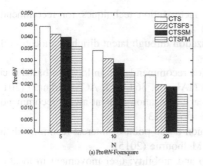 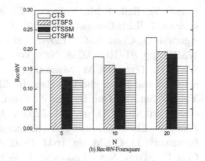

Fig. 6. Performance of the different divide of weekend

7 Conclusions and Future Work

In this paper, we introduce a new method that considers both the temporal influence and the spatial influence to make the time-aware POI recommendation, which recommend POIs to user at specific times. We first exploit the similarity between different hours and different days to calculate the temporal recommendation score. We then proposed a new method exploring the spatial influence and found the spatial influence will be influenced by the temporal influence. Finally, we combine the two methods through the impact between them. We conduct extensive experiments over two real-world LBSN datasets and compared our proposal with other time-aware methods. The experimental results show that the proposed method outperforms all other comparing methods.

For future work, we plan to consider the influence of the geographical proximity between different POIs. We are also interested in applying the proposed method to the Location Recommendation.

References

1. Ye, M., Yin, P., Lee, W.-C., Lee, D.-L.: Exploiting geographical influence for collaborative point-of-interest recommendation. In: SIGIR, pp. 325–334. ACM, Beijing (2011)
2. Yuan, Q., Cong, G., Ma, Z., Sun, A., Magnenat-Thalmann, N.: Time-aware point-of-interest recommendation. In: SIGIR, pp. 363–372. ACM, Dublin (2013)
3. Liu, B., Fu, Y., Yao, Z., Xiong, H.: Learning geographical preferences for point-of-interest recommendation. In: SIGKDD, pp. 1043–1051. ACM, Chicago (2013)
4. Li, X., Cong, G., Li, X., Pham, N., Krishnaswamy, S.: Rank-GeoFM: a ranking based geographical factorization method for point of interest recommendation. In: SIGIR, pp. 433–442. ACM, Santiago (2015)
5. Herlocker, J.L., Konstan, J.A., Borchers, A., Riedl, J.: An algorithmic framework for performing collaborative filtering. In: SIGIR, pp. 230–237. ACM, Berkeley (1999)
6. Sarwar, B.M., Karypis, G., Konstan, J.A., Riedl, J.: Item-based collaborative filtering recommendation algorithms. In: WWW, pp. 285–295. ACM, Hong Kong (2001)

7. Koren, Y., Bell, R.M., Volinsky, C.: Matrix factorization techniques for recommender systems. J. IEEE Comput. **42**(8), 30–37 (2009)
8. Agarwal, D., Chen, B.-C.: fLDA: matrix factorization through latent dirichlet allocation. In: WSDM, pp. 91–100. ACM, New York (2010)
9. Wang, H., Terrovitis, M., Mamoulis, N.: Location recommendation in location-based social networks using user check-in data. In: SIGSPATIAL, pp. 364–373. ACM Orlando (2013)
10. Yin, H., Sun, Y., Cui, B., Hu, Z., Chen, L.: Lcars: a location-content-aware recommender system. In: SIGKDD, pp. 221–229. ACM, Chicago (2013)
11. Zhang, J., Chow, C., Zheng, Y.: ORec: an opinion-based point-of-interest recommendation framework. In: CIKM, pp. 1641–1650. ACM, Melbourne (2015)
12. Cho, E., Myers, S.A., Leskovec, J.: Friendship and mobility: user movement in location-based social networks. In: SIGKDD, pp. 1082–1090. ACM, San Diego (2011)
13. Yuan, Q., Cong, G., Sun, A.: Graph-based point-of-interest recommendation with geographical and temporal influences. In: CIKM, pp. 659–668. ACM, Shanghai (2014)
14. Lian, D., Zhao, C., Xie, X., Sun, G., Chen, E., Rui, Y.: GeoMF: joint geographical modeling and matrix factorization for point-of-interest recommendation. In: SIGKDD, pp. 831–840. ACM, New York (2014)
15. Zhang, J., Chow, C.: GeoSoCa: exploiting geographical, social and categorical correlations for point-of-interest recommendations. In: SIGIR, pp. 443–452, ACM, Santiago (2015)
16. Ding, Y., Li, X.: Time weight collaborative filtering. In: CIKM, pp. 485–492. ACM, Bremen (2005)
17. Gao, H., Tang, J., Hu, X., Liu, H.: Exploring temporal effects for location recommendation on location-based social networks. In: Recsys, pp. 93–100. ACM, Hong Kong (2013)

Improvement for LEACH Algorithm in Wireless Sensor Network

Shiying Xia[✉], Minsheng Tan, Zhiguo Zhao, and Ting Xiang

School of Computer Science and Technology, University of South China,
Hengyang 421001, Hunan, China
{10826778,545832754}@qq.com, tanminsheng65@163.com

Abstract. According to the energy bottleneck of the wireless sensor network, the LEACH algorithm is improved in this paper, in the LEACH algorithm, it maybe that the cluster head nodes are far from the base station, and cluster head node has non-uniform distribution, in this case, nodes soon died after consuming energy, so the focus from three aspects of the optimal number of the cluster head and uniform distribution and many jump communication to improve, the improved algorithm through the simulation tests show that network node energy has been obviously saved, at the same time the lifecycle of the system is also effectively extension. In the application of the wireless sensor, we should always hold the principle of energy-first, therefore the LEACH which is a hierarchical routing protocol can save energy has played a crucial role, but the optimal number of the cluster head and uniform distribution can't be archived because of the change of the location of the cluster head nodes. In addition, when the cluster head far away from center sink node communicate with center sink node, the cluster head will exhaust energy quickly to die, that may affect lifecycle of whole of monitoring network and lead to worse expansibility. For some problems of the LEACH algorithm, the LEACH algorithm is improved in this paper from three aspects of the optimal number of the cluster head and uniform distribution and many jump communication, the improved algorithm through the simulation tests show that network node energy has been obviously saved.

Keywords: LEACH · Routing strategy · Effective path · Wireless sensor network

1 Introduction

LEACH (low energy adaptive clustering hierarchy) algorithm [1] is an adaptive clustering topology algorithm implemented with "round" as unit period, and a round circulation is consist of two phases, one is the implementation of initialization and another is the steady data transmission.

© Springer Science+Business Media Singapore 2016
W. Che et al. (Eds.): ICYCSEE 2016, Part I, CCIS 623, pp. 287–298, 2016.
DOI: 10.1007/978-981-10-2053-7_26

1.1 Physical Basis

The LEACH algorithm use first order radio energy dissipation model [2] which include the following several significant features as shown in Fig. 1:

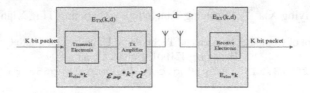

Fig. 1. First order radio energy dissipation model

(1) Base station is located where far from WSN and has enough energy to supply;
(2) The energy dissipation of radio signals in each direction of WSN is both same;
(3) In the WSN, All the nodes are exactly the same and have the same limited energy.

Thus, to transmit a kit-bit message using our radio model, the radio expend:

$$E_{send} = k * E_{elec} + k * \varepsilon_{amp} * d^{\beta} \tag{1}$$

and to receive this message, the radio expend:

$$E_{receive} = k * E_{elec} \tag{2}$$

ε_{amp} is the amplification factor and E_{elec} is the radio dissipates from transmitter or receiver circuitry. In this case, both the radio dissipates can be regarded as equivalence because the actual dissipations from transmitter and receiver are no different. β is a constant value depend on the radio channel and d is the distance which signal transmitted. From the formula $k * E_{elec} \ll k * \varepsilon_{amp} * d^{\beta}$, we can find that, the shorter distance signal transmitted, the less energy the circuitry dissipated. The variable is set as $\beta = 2$ when the distance from channel models were used to transmitter is small, and the variable is set as $\beta = 4$ when the distance is long. In this case, it is called Double path model [3].

1.2 Implementation

LEACH algorithm is implemented with "round" as unit period, the time consumed in the initialization phase must be less than the time consumed in the steady data transmission phase. The initialization phase of each round, First network to randomly selected some nodes self-organized way as cluster heads, select the cluster heads broadcast next, and ordinary nodes will be able to pass judgment signal strength to consider themselves the join of head, the principle of selection is near itself and signal stronger cluster heads. When all the nodes to join the cluster heads, each node with CSMA mechanism to inform themselves to join the cluster of cluster nodes, after the cluster formation, as the cluster head nodes will be responsible for build a TDMA time slot for cluster nodes within the table. When the cluster building work is completed, According to the rules of TDMA, ordinary nodes will be sent every needle data collected, eventually sent to the cluster head nodes, and the cluster head nodes to send data through fusion processing back-wardness to center node or base station.

(1) The stage of establishing a cluster

In the stage of establishing a cluster, the percentage of total nodes of cluster nodes in the network and the number of nodes act as cluster head nodes past determines whether a node of the cluster head nodes act as a basis. Any node n randomly generate a random number between 0, 1, joining the random number is less than the threshold T (n), the node n was elected to the current round of cluster head nodes information broadcast itself.

T (n) calculation is as follows:

$$
T_{(n)} = \begin{cases} \dfrac{p}{1 - p * (r \bmod \frac{1}{p})} & \text{if } n \in G \\ 0 & \text{other} \end{cases} \tag{3}
$$

Among them, the P said cluster head accounted for the proportion of the number of network nodes, r is said round number, $r \bmod (1/p)$ says the number of nodes, these nodes in the current cycle has been elected cluster heads, G represents a collection, the collection of all the nodes in the current loop was no cluster head.

(2) The stability of data transmission phase

In the whole network of cluster composition and also generate the TDMA time slot of cluster nodes table, begin to transmit data, which has entered the phase of stable data transmission. The nodes in the network will be constantly monitoring data, the node will be in his own time slot will be collected in each frame of monitoring data to the cluster head nodes, in order to save energy, ordinary nodes will be in its own time slot has not come close the transceiver. But the working state of the cluster head nodes must be in continuous, because to receive the quantity of nodes of cluster nodes to send over some of the data, every time after a round of data transmission, cluster head nodes to send data through fusion processing backwardness to center node. Due to Multiple clusters work together at the same time, will inevitably have an impact on each other, aiming at this problem, different cluster internal adopted CDMA (Code Division Multiple Access) system.

2   LEACH Algorithm is Improved

2.1   The Choice of the Optimal Cluster Number

Probability of cluster nodes in the LEACH, a certain effect on the lifespan of relative network. Because of too much will greatly increase the energy consumption of cluster nodes, thus shortening the life cycle of the entire network; Cluster nodes and less because of the cluster members of cluster nodes will make the node number, communication load heavy, die soon run out of energy, making the network life cycle shorten. To attain the longest network life cycle, therefore, must have a best cluster head probability value, this article first from the point of optimal number of cluster head LEACH algorithm is improved.

The optimal number of cluster head is based on the analysis of network on the basis of energy consumption [4]. First of all, we hope that each round of minimum energy consumption and energy consumption on all nodes, the goal is to will pick the optimal number of cluster heads to obtain minimum total energy consumption.

We assume that the monitoring area is the area of the M * M, Distribution in this area have N nodes, k clusters is contained in another network, This will tell you a single number of nodes in the cluster as the N/k, ordinary members of the number of nodes is N/k − 1, according to the working process of the (1) and algorithm: A single cluster of energy consumption is mainly used for ordinary nodes information after receiving, and the processing of the data fusion of data sent to the center in the process of gathering node.

So a single cluster head nodes within a frame for energy consumption:

$$E_{CH} = l\,E_{elec}(\frac{N}{K} - 1) + l\,E_{DA}\,\frac{N}{K} + l\,E_{elec} + l\,\varepsilon_{amp}\,d_{toBS}^4 \tag{4}$$

The said l single transfer data on the number of bits, d_{toBS} is the length of the cluster heads to center gathered node, E_{DA} for fusion energy consumption after processing. Because within the cluster, common member nodes and cluster heads are not far apart, we can look for a free space model.

So the ordinary node energy consumption within a frame as follows:

$$E_{non-CH} = l\,E_{elec} + l\,\varepsilon_{amp}\,d_{toCH}^2 \tag{5}$$

Among them, the cluster and d_{toCH} represents the member length between nodes.

Because the area of the monitoring area is M * M, So a single cluster is responsible for the monitoring area is about M^2/k, It is assumed that the probability of any node distribution meet $\rho(x, y)$, Cluster way can get the ordinary node of the square of the length of the mathematical expectation is as follows:

$$E\left[d_{toCH}^2\right] = \iint (x^2 + y^2)\rho(x, y)dxdy = \iint r^2\rho(r, \theta)rdrd\theta \tag{6}$$

In addition if the monitored area is a circular, and the radius of the circle is $R = (M/\sqrt{\pi k})$, can get the following formula:

$$E\left[d_{toCH}^2\right] = \int_{\theta-0}^{2\pi} \int_{r-0}^{M/\sqrt{\pi k}} \rho\,r^3\,drd\theta = \frac{\rho}{2\pi}\frac{M^4}{k^2} \tag{7}$$

In the case of uniform distribution of the nodes, the next step can be (8):

$$E\left[d_{toCH}^2\right] = \frac{1}{2\pi}\frac{M^2}{K} \tag{8}$$

We can get the following (9) by (8) and (7):

$$E_{non-CH} = l\,E_{elec} + l\,\varepsilon_{amp}\,\frac{1}{2\pi}\frac{M^2}{K} \tag{9}$$

Clusters within a single frame energy as the following Eq. (10):

$$E_{cluster} = E_{CH} + \left(\frac{N}{k} - 1\right)E_{non-CH} \approx E_{CH} + \frac{N}{K}E_{non-CH} \tag{10}$$

So the total energy consumption of k clusters together for (11) style:

$$E_{total} = k\,E_{cluster} = l\left(E_{elec}\,N + E_{DA}\,N + E_{elec}\,N + k\,\varepsilon_{amp}\,d_{toBS}^4 + \varepsilon_{amp}\,\frac{1}{2\pi}\frac{M^2}{k}N\right) \tag{11}$$

Finally, the method of derivation by seeking to obtain the optimal minimum number of clusters k as head:

$$k_{opt} = \sqrt{\frac{N}{2\pi}\frac{M}{d_{toBS}^2}} \tag{12}$$

2.2 Evenly Distributed Cluster Head Node

For LEACH, may appear far from the center of the cluster head node converge quickly energy depletion and death for this problem LEACH face, In this paper, the original algorithm was improved, by setting the type of (3) p value [5], makes from the center aggregation node distant cluster region contains more heads. Cluster establishment phase, the node type is used to calculate the threshold value (3), but for a different region, Their p value calculation method will be different, they will p value for their respective region distance monitoring center gathered node length and different, from near to far value is $(1 - x)p$ 、 p 、 $0 < x < 1$. In this way, Node cluster nodes farther away from the center will have a relatively large threshold, the probability that elected cluster head is relatively large, so as to achieve the goal of a uniform distribution of cluster heads.

2.3 Multi-hop Communication Between the Cluster Head

For LEACH protocol, all nodes in the network, including base stations can communicate directly included, but for those relatively distal from the center of the cluster head node aggregation node will be a big challenge, because the longer the transmission distance, energy consumption will increases, because the cluster head End energy consumption and death, and ultimately shorten the lifetime of the network, of course, network scalability will become poor. In this paper, the energy problem facing LEACH algorithm, a wireless multi-hop sensor networks idea that is farther from the center of the convergence of the cluster head node provides a relay station, which is the distance from the center to select some more recent collection of nodes as a distant cluster head relay station cluster head, cluster head node so far as to save some energy.

Improve ideological algorithm in a stable transfer phase, either in the choice of a cluster head node upstream forwarding node itself, it must send a message you want to own node energy consumption and the ratio of residual energy consumption into account, in addition, the cluster head node to send a large amount of information from the member nodes in the cluster, so the choice of upstream forwarding node when not only consider their distance from the issue, and to consider more efficient routing, routing efficiency level, mainly to see its forwarding information can consumption, the number is sent to the center in the convergence node.

3 Emulation

3.1 Embedded Operating System TinyOS and NesC Language Introduction

TinyOS [6] is a UC Berkeley to develop open-source wireless sensor networks designed specifically for embedded operating systems. TinyOS including distributed servers, network protocols, data identification tools and sensor-driven and some other components, it has a very good event-driven model that supports scheduling is very flexible and, in addition, there is a very good power management, currently, TinyOS gradually been used on many platforms and some sensing board, so TinyOS research has far-reaching significance and value. TinyOS was originally designed to provide a dedicated embedded operating system for wireless sensor networks, TinyOS just started using C and assembly language programming. Many workers engaged in scientific research through in-depth study found, C language can not meet the application of wireless sensor networks embedded operating system development and sensor networks. Therefore, researchers later on the basis of C's proposed NesC [7] language.

3.2 The Introduction of Simulation Software: TOSSIM

TOSSIM is a discrete event simulation software which TinyOS comes with. It can compile the application programs of TinyOS to the simulation framework in which we can use standard development tools to debug code. TOSSIM can simulate integrated application programs of TinyOS and the operating principle replacing the hardware components by simulation components. The component level which TOSSIM can replace is flexible. It could replace both the packet level communication component and radio frequency chip on bottom layer to get more exact simulation result.

3.3 The Performance Simulation of Improved Optimal Number of Cluster Heads

(1) Simulation environment and parameters

To simulate the performance of improved optimal number of cluster heads, we put 100 nodes in the monitored area of 50 m * 50 m randomly. The setting of simulation parameters is listed below in Table 1 (ε_{amp} represents amplification factor, E_{elec} represents energy consumption of transmitting and receiving, E_{DA} represents energy consumption after fusion processing)

Table 1. The simulation parameters of optimal number of cluster heads

Parameter Name	Parameter values
Area of zone	(0, 0) to (50, 50)
Number of nodes	100
Location of base station	(25, 150)
E_{elec}	50nJ/bit
ε_{amp}	0.0013pJ/bit
E_{DA}	5nJ/bit/signal
Initial energy	0.25 J

(2) The simulation results and analysis

In TOSSIM, we simulate the process with 100 nodes and get the relationship between average energy consumption per round and the number of cluster heads, as listed below (Table 2):

Table 2. The relationship between average energy consumption per round and the number of cluster heads

NCH / EC	0	1	2	3	4	5	6	7	8	9
AECPR(J)	4.5	3	2.1	1.6	1.4	1.7	2.5	3.5	4.6	6

NCH : Number of Cluster Heads
EC : Energy Consumption
AECPR : Average Energy Consumption Per Round

Converting the table of the relationship between average energy consumption per round and the number of cluster heads to the graph as below Fig. 2:

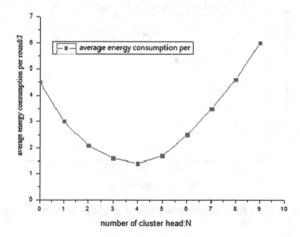

Fig. 2. The average energy consumption per round of different number of cluster heads

From Fig. 2, we know that when the number of cluster heads is 3, 4 and 5, the whole energy consumption of the network is less. Bringing the simulation parameters into the formula (12), we can get $1 < k_{opt} < 6$. So the simulation results are similar to that calculated theoretically which illustrates the optimal number of cluster heads in the area set by experiment is 3, 4 or 5, saving energy more effective. The big number of cluster heads in network will increase the expenses, while small of that will result in network paralysis.

3.4 The Performance Comparison Between Single Hop Communication and Multi-hop Communication Among Clusters

This paper simulates the network in the mode of single hop communication and multi-hop communication in energy consumption for 100, 200, 300, 400, 500, 600 and 700 rounds with the same 100 nodes distribution randomly and the same total energy 25 J. The results as below Table 3:

Table 3. Total energy consumption variation with the simulation rounds

RS / TEC	100	200	300	400	500	600	700
OLA(J)	6	12	15.5	19.5	22.5	25	
IA (J)	3.5	7.5	11.5	15.6	19	22	24

RS : Rounds of Simulation
TEC : Total Energy Consumption
OLA : Original LEACH Algorithm
IA : Improved Algorithm

Converting the table of energy consumption in the mode of single hop and multi-hop communication to graph, as below Fig. 3:

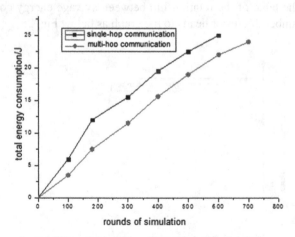

Fig. 3. The energy consumption comparison between single hop and multi-hop communication

3.5 The Life Cycle Comparison Between the Original LEACH and Improved LEACH

(1) Simulation environment and parameters

Simulate the original LEACH and improved LEACH in different energy in which the probability of taking the nodes of cluster heads is 5 % and make x equal 2. The distance between the nodes of cluster heads and base station in original LEACH algorithm uses the multipath fading model. It means $\varepsilon_{amp} = 0.0013\text{Pj/bit}/m^4$. Other parameters as below Table 4:

Table 4. Simulation parameters of network cycle

Parameter names	Parameter values
Area of zone	(0, 0) to (50, 50)
Number of nodes	100
Location of base station	(85, 90)
E_{elec}	50nJ/bit
ε_{amp}	100pJ/bitm2
E_{DA}	5nJ/bit/signal
Initial energy	0.25 J 和 0.5 J
Length of message packer	2000bit
Length of broadcast message	64bit

(2) The simulation results and corresponding analysis

For 100 nodes, when the initial energy of each node is 0.2 J, LEACH is obtained by simulation before and after improving algorithm, in which the relations of simulation time and the number of live node in the network are shown in Table 5 below.

Table 5. Correlation table of change in live node along with simulation time

ST \ LN	0	50	100	150	200	300	345	369	354	450	550
TILA(N)	100	100	100	100	99	97	50	10	0		
AAI	100	100	100	100	100	100	99	64	41	18	0

ST : Simulation Time
LN :Live Node
TILA : The Initial LEACH Algorithm
AAI : Algorithm After Improvement

LEACH algorithm in the network before and after the improvement for the node number and simulation round number relational tables into a diagram in Fig. 4 as follows:

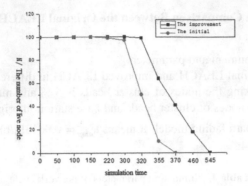

Fig. 4. Life to compare two algorithms of node

In above Fig. 4, in line with the circle said node number of survival of the original algorithm, improved algorithm with the black line of the box says the number of nodes to survive. The simulation results show that the improved algorithm makes the nodes in a network of the consumption of energy to get the uniform, thus avoiding a single node energy consumption by its larger and premature death.

3.6   The Comparison of LEACH Time Delay Before and After Improvement

In this paper, the improved algorithm respectively before and after in clusters expectations for 1 to 10 cases of packet delay are simulated, and their delays shown in the table below (Table 6):

Table 6. Packet delay change with clusters of expectations contrast figure

CE / PD	1	2	3	4	5	6	7	8	9
IA (s)	1.72	1.22	0.9	0.78	0.62	0.55	0.5	0.48	0.47
ILA(s)	1.25	0.92	0.75	0.61	0.55	0.45	0.4	0.39	0.4

CE : Clusters Expectations
PD : Packet Delay
IA : Improved   Algorithm
ILA : Initial LEACH Algorithm

Converts packet delay table to delay curve in Fig. 5 as follows:

It Can be seen from the Fig. 5, with the continuous increase of clusters expectations, before and after improvement average delay is becoming more and more small, two algorithms of cluster number is, the more the corresponding time delay is continuously decreasing, and this is also consistent with the actual wireless sensor network. But also obvious reaction from the picture has a drawback, the improved algorithm is improved algorithm delay slightly bigger than the original algorithm, because of the improved algorithm is used in many communication mode, most of the cluster head nodes not will collect the data directly to the base station, first to be sent to its upstream trunking cluster

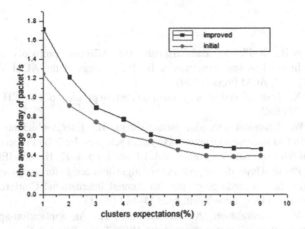

Fig. 5. Before and after the improvement of the comparison of two kinds of algorithm on the network average delay before and after improvement of two kinds of algorithm on the network average delay

head nodes, this has been more jump transmission, will produce certain delay between obviously, is a good phenomenon, increase gradually as clusters of expectations, the time delay of two algorithms are consistent.

4   Conclusions

This chapter first describes the original LEACH communication model and implementation process of the algorithm are introduced, then some problems in the face of the LEACH algorithm is analyzed, because these problems of wireless sensor network to the life cycle of caused great influence, therefore, these problems, this article from the optimal number of cluster head, cluster heads the multi-hop communication of even distribution and network of the three aspects of LEACH algorithm is improved. Simulation tests show that the improved algorithm has dramatically save the energy of each node, at the same time, the life cycle of the network has been effectively prolong, scalability of the network also has been enhanced, but throughout the network average delay, compared to the original LEACH algorithm, the improved algorithm on average delay somewhat larger than, but difference decreases continuously decreases as the clusters expectations.

Acknowledgments. This Research was supported by the National Natural Science Foundation of China under Grant 61300234, key Project of the Hunan Provincial Education Office Science Research of China under Grant 14A121, the Science Fund for Distinguished Young Scholars of Hunan Provincial Education Office under Grant 12B110, the Construct Program of the key Laboratory in University of South China (computer science and technology), the Postgraduates Scientific Research Innovation Project of University of South China under Grant 2013XCX19, the Hengyang City Science and Technology Plan under Grant 2012KG71.

References

1. Heinzelman, W.R., Kulik, J., Balakrishnan, H.: Adaptive protocols for information dissemination in wireless sensor networks. In: Proceedings of the ACM MobiCom 1999, Seattle, pp. 177–185. ACM Press (1999)
2. Tianyi, S., Di, X.: Research and improvement on routing protocol of LEACH in WSN. J. Sig. Syst. 7(1), 67–73 (2005)
3. Heinzelman, W., Chandrakasan, A., Balakrishnan, H.: Energy-efficient communication protocol for wireless microsensor networks system sciences. In: 2000 Proceeding of the 33rd Annual Hawaii International Conference, vol. 10, no. 2, pp. 4–12, Hawaii. IEEE (2000)
4. Shih, E., et al.: Physical layer driver protocol and algorithm design for energy-efficient wireless sensor networks. In: Proceedings of the 7th Annual International Conference on Mobile computing and networking, New York, pp. 134–142 (2001)
5. Heinzelman, W., Chandrakasan, A., Balakrishnan, H.: An application-specific protocol architecture for wireless microsensor networks. IEEE Trans. Wireless Commun. 7(10), 660–670 (2002)
6. TinyOSTutorial. EB/OL. http://www.tinyos.net/tinyos-1.x/doc/tutorial/,2005,5
7. Gay, D., Levis, P., Culler, D., Brewer, E.: NesC 1.1 Language Reference Manual. EB/OL (2003). http://nescc.sourceforge.net/paper/nesC-ref.pdf
8. LB05–10B03. EB/OL. http://www.mornsun.cn/UploadFiles/pdf/LB03-15%20CN.pdf
9. TinyOS web site. EB/OL. http://tinyos.net

Decentralizing Volunteer Computing Coordination

Wei Li[✉] and Eugenie Franzinelli

School of Engineering and Technology, Central Queensland University,
Rockhampton 4702, Australia
w.li@cqu.edu.au, e.a.franzinelli@gmail.com

Abstract. This paper attempted to decentralize volunteer computing (VC) coordination with the goal of reducing the reliance on a central coordination server, which had been criticized for performance bottleneck and single point of failure. On analyzing the roles and functions that the VC components played for the centralized master/worker coordination model, this paper proposed a decentralized VC coordination framework based on distributed hash table (DHT) and peer-to-peer (P2P) overlay and then successfully mapped the centralized VC coordination into distributed VC coordination. The proposed framework has been implemented on the performance-proven DHT P2P overlay Chord. The initial verification has demonstrated the effectiveness of the framework when working in distributed environments.

1 Introduction

Volunteer Computing (VC) is a type of distributed computing that uses public-donated processing and storage capacity on the Internet. Instead of relying on expensive supercomputers, VC is of significance for large-scale scientific computation due to its potential computing capacity that can be gained from millions of volunteer computers [19]. In [14], BOINC-based VC projects were about 40 and the number of volunteers was about 400,000 performing an average of over 0.5 petaFLOPS. In the 2010 report [18], there were more than 0.5 million active volunteers of BOINC-based computing, performing an average of 5.42 petaFLOPS. However, as of August 2010, the world's fastest supercomputer had a performance of about 1.76 petaFLOPS.

At the time of this writing, there were about 44 projects listed on the web page of BOINC (http://boinc.berkeley.edu/projects.php). Thus it is reasonable to treat BOINC [2] as the most widely used VC middleware. As a result, BOINC's master/worker model is the typical architecture for VC coordination, which is centralized on a project server. In a roundtrip, a volunteer (worker) downloads application executables and related data in a unit called *job* from the server (master), carries out the executables against the data and finally uploads the results back to the sever. The server plays the central role for job and result storage, job splitting and distribution and results synthesis. Belgacem et al. [4] recently introduced another coordination model named XtremWeb-CH2. Although they used different names, e.g. coordinator, warehouse node and worker, a coordinator played the central role of job management and a warehouse node was in fact the database server for managing input and output files. Thus the model was still master/worker.

© Springer Science+Business Media Singapore 2016
W. Che et al. (Eds.): ICYCSEE 2016, Part I, CCIS 623, pp. 299–313, 2016.
DOI: 10.1007/978-981-10-2053-7_27

In [2], Anderson, the leader of the BOINC development team, described that the workload of the centralized server depended on the number of volunteers and the frequency of communication. If the sever was overloaded, the whole framework failed with no response to volunteers, which in turn would become idle. He also provided an example of the bottlenecked performance and possible single point of failure of SETI@home [1], which used BOINC middleware and had quickly acquired around 400,000 volunteers since its release in May 1999. That large amount of volunteers overwhelmed its central server. As a consequence, it had to use a number of servers to share the workload. A similar problem can be found in [17] that their BOINC-based system stopped operation for 2 days on two occasions in the first 3 months and both were from central server failure. The investigation of [7] showed that given enough jobs, performance peeked at a certain number of peers for a dedicated VC system, where the VC coordination was centralized using dedicated severs. After this point, speed-up stagnated because the coordination bottlenecked the overall performance.

From another point of view, Kondo et al. [15] stated that the main costs of a centralized VC project included the purchase of expensive server hardware, high set-up costs of the hardware and installation of server software and daily maintenance of the dedicated hardware and software. They provided an example that SETI@home start-up costs were $43 K, total monthly costs were $12 K and another $10 K/month were needed to employ 2 system administrators or programmers.

All of the above suggests that centralized coordination is not an ideal or optimal model for VC although it is widely used. This paper attempted to decentralize VC coordination from master/worker structure into distributed P2P overlay, aimed to reduce the possibility of single point of failure and improve the bottlenecked performance, along with gaining the positive side effect of lowering the infrastructure and software maintenance costs. Although this paper is not the first one to use P2P for VC, its newer contribution includes: minimizing the reliance on the central severs; full consideration of the reliability of VC coordination against peer churn; mapping of master/worker structure into self-managed P2P overlay; framework implementation and verification.

This paper is structured as follows: related work is reviewed in Sect. 2. The architectural and functional properties of the typical master/worker VC coordination are defined in Sect. 3. In Sect. 4, the goal and methodology to decentralize VC coordination are proposed. In Sect. 5, a minimum set of DHT operations is justified to map VC coordination requirements. The implementation of the proposed decentralized VC coordination framework is outlined in Sect. 6. The implemented VC coordination framework is verified through the N-Queen Problem in Sect. 7. Section 8 concludes the paper by highlighting the achieved characteristics of decentralized VC coordination.

2   Related Work

As this paper aims to transform the master/worker structure into decentralized VC coordination, there are 2 types of related work. First, a variety of open source or commercial VC coordination middleware, such as BOINC [2], XtremWeb [5], Xgrid [16] and Grid MP [11] etc., shares the centralized master/worker structure as classified and reviewed in Sect. 1. The second category is shared VC coordination by using multiple dedicated or non-dedicated servers. Anderson et al. [2] optimized the original BOINC performance through sharing coordination between a dedicated job server for dispatching jobs and a dedicated database server for dispatching input files and collecting output files. Andrzejak et al. [3] exploited both dedicated and non-dedicated computing resources. They investigated the trade-off between a larger share of dedicated hosts and higher migration rates of data for non-dedicated hosts in terms of costs and service quality.

Ghafarian et al. [10] was a P2P overlay but used super nodes and knowledge (CPU speed, memory and disk space) about peers for VC coordination. Peers were classified as *reporting nodes* to keep track of computing and storage capacity of peers, *host nodes* to find suitable peers to run a job and *client nodes* to send lookup requests for running a job. However, the use of super nodes reverted the VC coordination back to centralization. The use of peers' knowledge was an estimate of a peer's capacity and thereafter to assign a job to the peer. Such strategies would become invalid when the peer committed churn.

Costa et al. [6] surpassed the master/worker structure by introducing inter-peer communication. As a consequence, the proposed model supported MapReduce style applications, in which the first step (map) was to compute jobs independently and the results were held at the *mappers* instead of uploading to the central servers. At the second step, the *reducers* obtained the jobs from the servers but downloaded inputs from the mappers. In this way, the overhead seemed to be reduced for the central servers. However, the model was unable to act on churn e.g. peer crash. In case of failure, mappers also needed to upload results back to the central servers.

Dubey and Tokekar [9] assigned roles to peers as either a job distributor, performing job distribution and result collection, or a job processor, doing the assigned jobs and returning results. A job distributor was identified because historically it was reliable and efficient. However, maintaining historical data in a P2P overlay was an issue that was not clarified in [9]. Although the job distributors were non-dedicated peers, they played the same role as the centralized severs in the master/worker model.

Ni and Harwood's work [17] was the closest one to this paper in terms of that they used peers to form a tuple space for job and result storage, where a peer was able to contribute to storage space or CPU cycles or both. The differences between their work and this paper included the following. In [17], once a new job was uploaded into the tuple space, the peer needed to periodically broadcast the job notification, which would be forwarded by the received peers to the whole peer overlay. Peer communication in such an unstructured P2P overlay had to be message flooding, which incurred significant bandwidth usage. It used redundancy for data storage. However, [17] had not proven its reliability. This paper used the structured DHT overlay Chord [20] for proven efficiency and reliability for job search and storage.

Dou et al. [8] attempted a P2P approach by forming volunteers into an unstructured cluster. A new volunteer joined the overlay by contacting an existing overlay peer, which in turn introduced the new volunteer to its neighbors. The workload of a peer was the number of jobs that was held by the peer and notified to each other. A newly uploaded job was expected to be forwarded to a neighbor that had the smallest workload. They claimed that because the clustering coefficient was very high, the job was likely to be sent to a node with lower workload. There were some issues with the model. The overhead on a peer was very high e.g. 1000 TCP connections for maintaining high clustering coefficient and flooding style messaging for workload updating. However, it was not mentioned how the model coped with job and data lost or maintained an effective neighborhood when peers left or crashed.

In this paper, we reinvestigated the master/worker model in terms of components and functionality. On this basis, we researched how to decentralize the VC coordination and data storage to the maximum extent. One important feature that distinguished this paper from other models was that our model was based on the reliability- and efficiency-proven P2P DHT overlay Chord. Its potential could not shrink in the face of churn.

3   The Typical VC Model

VC models vary. In this section, the roles/components of BOINC-based VC model are abstracted; they characterize the typical centralized VC coordination, i.e. master/worker model.

1. An advertising server: this centralized server is to advertise VC projects such as those on the BOINC project page (http://boinc.berkeley.edu/projects.php). The sever presents the description of a project and the required platforms for volunteers. The server (e.g. http://boinc.berkeley.edu/download.php) also supplies the client software for a volunteer to download.

2. A job server: each project will have a job server (e.g. http://setiathome.berkeley.edu) to create jobs, dispatch the jobs to volunteers and process the returned results. Although there could be a separate data server, at the abstract level, a job server can also supply the input file of a job for a volunteer to download.

3. Volunteers: a volunteer is a non-dedicated public computing resource on the Internet such as your desktop, laptop or even smart phone. Volunteers are fully distributed to each other or to the above servers only. A volunteer periodically communicates with the servers to get new jobs or submit completed results. Volunteers are never in contact with each other in any way e.g. by the Internet protocols or a shared state space. Volunteers' leave or crash can only be found by the central servers.

4. Jobs: a job can be treated as a self-contained computing object, in which the input data and executable can be encapsulated together and executed independently. Once a job is executed on a volunteer, the computing result can be submitted back to the server as a part of the final result in the case of embarrassingly parallel applications or as the input of another job in the case of workflow applications. In addition, a job will also need to represent its ongoing state such as *completed* or *in-progress*, or *non-splittable* or *splittable* for parallel computing.

5. Master/worker architecture: by this centralized structure, volunteers are only allowed to pull jobs from the master/server. Thus the communication is only between a volunteer and the server, which coordinates job distribution and result collection.

4   The Decentralization of VC Coordination

The centralized properties of existing VC models and the proposed decentralization goals of this paper are compared in Table 1 and justified in this section. As the decentralization is achieved through P2P overlay, every volunteer will become a peer once it joins the P2P overlay. Thus from this section the term *volunteer* and *peer* are interchangeable. In addition, because a job or sub-job can be split into multiple portions, the term job or sub-job are in fact interchangeable as well in this paper. The requirements for decentralization are as follows.

Table 1. The decentralization goals and requirements

VC coordination	Master/Worker model	Decentralization goal of P2P (this paper)	Requirement on P2P overlay
1. Project advertising	Centralized	Centralized	Maintenance of a pool of bootstrap nodes for each project
2. Project creation	Centralized	Centralized	Creation of a project overlay by the project owner; creation of a bootstrap pool of the overlay by the advertising server
3. Project participation	Centralized	Distributed	Use a random bootstrap node to join a project overlay
4. Volunteer leave	Centralized	Hybrid	Bootstrap pool updating (centralized); job check-pointing; job state updating (distributed)
5. Volunteer crash	Centralized	Hybrid	Bootstrap pool updating (centralized); searching for crashed peers (distributed)
6. Job storage	Centralized	Distributed	Distributed job storage, search and retrieval on any peer
7. Job distribution and reassignment	Centralized	Distributed	Search for jobs and peers' contact information; peer communication
8. Result return	Centralized	Distributed	Upload results onto the project overlay
9. Result storage	Centralized	Distributed	Evenly distribute the uploaded results onto the project overlay
10. Job progress monitoring	Centralized	Centralized	Search for available results and update project progress by the project owner
11. Result collection and synthesis	Centralized	Centralized	Search for and synthesize the available results by the project owner

R (Requirement) 1: The project advertising has to be centralized. Volunteers must go somewhere such as to a web site in some way such as through Google search to involve in a project. There is no other way to involve a project or it is impractical to assume that a volunteer has already known a bootstrap node through a URL or a TCP end point to involve in a project. The centralization property of project advertising cannot be decentralized but it can be minimized to only maintaining a pool of bootstrap nodes for each advertised project.

R2: The project creation has to be centralized because only the project owner knows the project. The project owner is to create a project community in terms of a P2P overlay with only one i.e. itself on the overlay initially. The creation of the project community requires the project owner to be inserted into the bootstrap pool that is maintained at the advertising server.

R3: A volunteer needs to contact the advertising server to get the bootstrap pool of a project that it wants to join. The volunteer randomly chooses a bootstrap node from the pool to join the project community.

R4: A peer is allowed to leave the project at any time. The *leave* means that the currently unfinished job within the peer will be check-pointed and updated into the overlay storage. Although in the future the same volunteer may join the same project, it will be treated as a totally new volunteer and there is not any necessity that it would be assigned the same job. The peer needs to report its leave to the advertising server because the former could be a bootstrap node and the latter needs to update the project bootstrap pool to reflect such a leave.

R5: A peer is allowed to crash at any time. A crashed peer will no longer update its job state or report anything to the advertising server. When searching for available jobs, other peers may find the crash by identifying the out-of-date job state of the peer and then report the crash to the advertising server, which needs to update the project bootstrap pool if the crashed peer was in it.

R6: P2P overlay such as Chord has proven storage capacity distributed on peers. This property can be used to decentralize the persistence requirement of VC jobs onto all peers, which form the overlay for providing not only computing but also storage capacity. Thus the jobs can be stored and later retrieved in a distributed way.

R7: To dynamically share the workload among peers, an uncompleted job needs to be found by a peer from another peer; the former can pull a portion of the job from the latter. As a consequence, the job progress state and the contact information such as a TCP end point of every peer should be searchable from the overlay.

R8: Every peer executes its own job to produce its own results independently. Every peer uploads the result of a completed job onto the overlay instead of a single sever for storage.

R9: Using the storage capacity of the underlying P2P overlay, all the uploaded results should be evenly distributed onto the overlay nodes.

R10: Job progress is the percentage that has been completed for the entire job. If a job is currently split into 100 sub-jobs for executing by peers and 80 of them have completed, the job progress is 80 % (upon the assumption that each job has the same size for not losing generality). When a sub-job's result is uploaded onto the overlay, the project owner should be able to retrieve the result and update the job progress.

R11: Although the uploaded results are distributed onto the overlay, they can be retrieved later for result synthesis or as the input for another job. The results can be retrieved by any of the peers, but result collection and synthesis is meaningful for the project owner only.

5   The Mapping of VC Coordination into DHT Operations

In this section, a minimum set of DHT operations is proposed to map the centralized VC coordination into distributed coordination. While a super set of operations could be more convenient, the proposed set is enough to fulfill the requirements as proposed in Sect. 4. The general assumption of this mapping includes:

1. Each peer has an identifier $PID = [IP\ address,\ TCP\ port\ or\ UDP\ port]$, which hash value is denoted as $hPID$.
2. Each job has a state $JOB = [job\ object,\ PID,\ jstate,\ pstate,\ timestamp]$, which hash value is denoted as $hJOB$. The *job object* is a stateful object consisting of executable, data and job scale e.g. the start point and end point of a space search problem; *PID* is the identifier of host peer (as defined in point 1) of the job; *jstate* is to describe whether the job can be spilt in terms of *splittable* or *non-splittable*; *pstate* is to describe the liveness of the host peer of this job in terms of *alive*, *left* or *crashed*. Each update of *jstate* or *pstate* is *times-stamped*.

In accordance with the requirements $R1$ to $R11$ of Sect. 4, we define the mapping of 1 to 11 as follows.

M (Mapping) 1: The maintenance of a project bootstrap pool consists of 4 operations. The insertion of a peer *put(hPID, PID)*, deletion of a peer *remove(hPID)*, retrieval of a peer *get(hPID)* and retrieval of all of the peers in the pool *peers()*. These operations enable the project advertising server to maintain the bootstrap pool or enable a volunteer to find an alive bootstrap node from the pool to join the overlay.

M2: The creation of a project overlay needs an operation *oCreate()* or *oCreate(TCP or UDP port)* if the default ports are not desirable. The project owner executes this operation and then reports itself to the project advertising server, which will create a bootstrap pool in terms of a hash table and insert the project owner as a bootstrap node. The project owner will listen to the TCP or UDP port for other peers' communication.

M3: It is assumed that there is always at least one alive bootstrap node for a project. Thus a volunteer would be able to join the project overlay through an operation *join(PID)*, where PID is the identifier of the bootstrap node.

M4: A peer needs an operation *checkpoint(hJOB, JOB)* to check point its job object's state and an operation *update(hJOB)*, which consists of an operation of *remove(hJOB)* to remove the job's old state from the overlay and an operation *put(hJOB, JOB)* to put the job's new state onto the overlay. This updates *jstate* into *non-splittable* if the current job cannot be split and/or *pstate* into *left* if the peer is leaving. A leaving peer needs to report to the advertising server, which will remove the peer from the project bootstrap pool if the former is one of the bootstrap nodes.

M5: A crashed peer will no longer do anything with its own state or its job state. If the peer restores and rejoins in the future, it will be treated as a totally new peer. Under the requirement that a peer needs to update its job state periodically, there are two conditions to ensure that a crashed peer can be found. First, every peer agrees on an updating interval e.g. 20 s. Second, every peer can search any job state stored on the overlay. If a job is found with its state out of date, it will be treated as from a crashed peer. The job can be picked up by this peer and the crashed peer can be reported to the advertising server.

M6: The most complex task is to store jobs onto or retrieve jobs from the overlay. A job is brought to the overlay by the project owner, but over time there will be the following situations to happen. A job has been split into multiple sub-jobs, which are processed by peers and stored on the overlay. Because the overlay is a DHT, each sub-job needs a key to store or retrieve it. A newly joined peer or a peer that has finished its current sub-job needs to search for another sub-job. However, the peer doesn't know the key of an available sub-job because they are all independently assigned by different peers. The proposed solution is to use containers. A container has a key and can be stored on the overlay. The number of containers and the key of each container can be made available to all peers. For example, the number is fixed at *100* and the key of a container is *jContainer_i*, where $i = 1, 2, ..., 100$. A sub-job can be stored into one of the containers, which are stored onto the plain overlay. Initially, the project owner is responsible for creating *n* job containers *jContainer_1* to *jContainer_n*. Then the standard DHT operation *put(hjContainer_i, jContainer_i)* can be used to store the containers onto the overlay, where *hjContainer_i* is the key of the container *jContainer_i*. Then each peer can retrieve any container by standard DHT operation *get(hjContainer_i)* and search for an available sub-job from the container or store a sub-job into the container.

M7: Initially, there is only one peer (project owner) and one job (the original) *JOB = [job object, PID, jstate, pstate, timestamp]*, which is stored in a container with the *jstate* of *splittable* and *pstate* of *alive* and where *PID = [IP address, TCP port or UDP port]* is the identifier of the project owner. Over time, other volunteers join the project overlay and use the following algorithm to search for a sub-job.

- Check each sub-job in each container.
- If the job is from a *crashed* peer or a *left* peer, pick up the job for execution and update *jstate*, *pstate*, *PID* and *timestamp*.
- If the job is from an *alive* peer and *splittable*, contact the peer to split the sub-job and get a portion.
- When receiving a sub-job from the requested peer, assign *PID, jstate, pstate, timestamp* to the job and update it into a job container and then execute the job.
- If all jobs are *non-splittable* from the search or there are no left jobs to pick up, leave the project.
- When a peer is contacted by another peer, it stops executing its sub-job and checks the sub-job's splittability;
- If the sub-job is *splittable*, split it and send a portion to the requesting peer; update the *job object, jstate, pstate* and *timestamp* of the job object into the job container. Otherwise notify the requesting peer that its sub-job is *non-splittable*.

M8 & 9: The result of each sub-job is stored in its job object and returned independently by peers. Each sub-job is identified by a hash key. There is no simple way to trace how many sub-jobs exist and what the hash key of each sub-job is because the sub-job splitting, reassignment and distribution are fully distributed. Only the host peer of a sub-job knows the hash key of the sub-job and can retrieve it. However, in addition to the host peer, the project owner needs to collect the result from each sub-job. The solution to this problem is still to use containers. Initially, the project owner is responsible for creating m (e.g. 100) result containers $rContainer_i$, where $i = 1, 2, ..., m$ and the standard DHT operation $put(hrContainer_i, rContainer_i)$ can be used to store the containers onto the overlay. As the key of $rContainer_i$ can be uniformly named, e.g. $hrContainer_i$, each peer can know the key and obtain the corresponding container by the standard DHT operation $get(hrContainer_i)$ and put a result into it.

M10 & 11: The project owner needs to check every result container by the $get(hrContainer_i)$ operation periodically and retrieve all available results in each container by the standard DHT operation $rContainer_i.elements()$. That enables the entire job progress to be updated and results to be synthesized when some portions or the entire job have been completed.

6   The Implementation of the Distributed VC Coordination Framework

When there are a variety of DHT protocols, Chord [20] has been chosen for this project because of the proven fault tolerance and certainty of performance. When there are a variety of implementations of the Chord protocol, Open Chord [13] has been chosen because it is a Java implementation of the full Chord protocol.

Based on the open source Open Chord API, the proposed VC coordination framework has been fully implemented with 3 types of distributed components. The *Facilitator* (i.e. advertising server) is to maintain the bootstrap pools for each project. The facilitator is executed as only one instance, with which any other components can contact. The *Job Starter* (i.e. project owner) is executed as one instance of each project. The project starter has the ability to create a Chord overlay, to request creating a bootstrap pool from the facilitator and to bring the project (original job) to the overlay. The *Peer* is executed as multiple instances. A peer has the ability to join and leave the overlay or crash at any time. A peer can get a sub-job from another peer once its current one is completed. The framework (as shown in Figs. 1, 2, 3 and 4) allows the following listed points to be verified.

- Solving a problem by only one peer i.e. the job starter. This is to verify whether the coordination framework works correctly in a non-distributed environment.

Fig. 1. The facilitator maintains the bootstrap pools for each project.

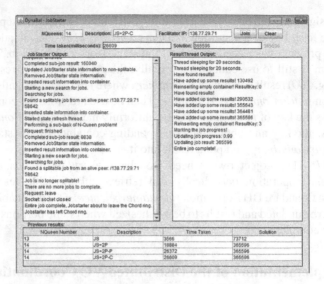

Fig. 2. The job starter interface shows it solved the 13-Queen Problem by itself (JS) only, 14-Queen Problem by itself and 2 peers (JS + 2P), 14-Queen Problem by itself and 2 peers with 1 peer left (JS + 2P–P) and 14-Queen Problem by itself and 2 peers with 1 peer crashed (JS + 2P–C).

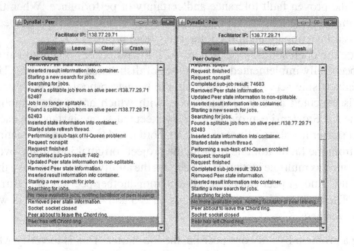

Fig. 3. 2 peers joined the 14-Queen Problem (JS + 2P), executed some jobs and left when there were no longer available jobs

- Solving a problem by multiple peers. This is to verify whether the coordination framework works correctly in a distributed environment.
- Solving a problem with joining and leaving peers. This is to verify whether the coordination framework works correctly against churn.
- Solving a problem with crashed peers. This is to very whether the VC coordination framework is fault tolerant.

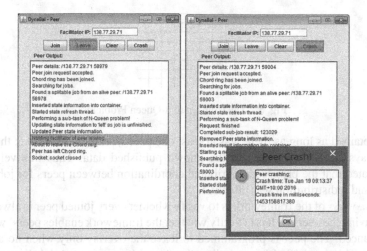

Fig. 4. Left: a peer left (JS + 2P–P) before the completion of the 14-Queen Problem. Right: a peer crashed (JS + 2P–C) before the completion of the 14-Queen Problem.

The VC coordination framework has been implemented as 2 versions. One version can be run in a single JVM for multiple peer simulation; the other can run multiple peers on different JVMs on separate physical machines in fully distributed environments. The execution of the framework is by starting the facilitator first and then the job starter. After that, a peer's join, leave and crash can be automated to happen at any time by setting proper parameters or operating manually.

7   A Case Study for Verification

To verify the implemented coordination framework, our case study is of embarrassingly parallel computing: the *N-Queen Problem*, which attempts to locate *N* chess queens on an *N*×*N* chess board so that the queens cannot attack each other [12]. A solution to the 8-queen problem is illustrated in Fig. 5. The N-Queen Problem is a space search problem; the solutions must be searched within the space of all possible positions of queens. The space search includes the initial state (e.g. the chess board has no queens on it), the goal state (e.g. the queens have been positioned without attacking each other) and the intermediate states which depend on what actions have been taken from the existing state. The N-Queen Problem is compute-intensive because the solutions increase exponentially against the number of queens. Because the N-Queen search space can be split, each peer can search a piece of the whole space and then is able to find a sub set of solutions. The result combination is to add the individual peer's results.

The verification uses 3 criteria. The *effectiveness* of the framework is to verify whether the results are correct when the original job is executed on a single peer or on multiple peers and without or with churn. We conducted such an effectiveness verification on both the single JVM simulation version and the multiple JVM distributed version. The case study was from *8* queens to *40* queens and about 80 tests with a varied combination of different numbers of peers for *join*, *leave* or *crash*. The results were

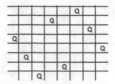

Fig. 5. The solution to 8-Queen Problem

always obtained as long as there was at least one peer working at any time; the results were always correct compared to the known published data. Thus this verification demonstrated that the proposed distributed coordination between peers for job search, splitting and redistribution was effective.

The *busyness* of the framework is to verify whether every joined peer is always busy before leaving the overlay. It is to verify whether the framework enables peers' workload to be balanced. The peer is programmed to leave the overlay only when no available jobs exist. This is verified by recording the time when every sub-job becomes non-splittable (t_1) and the time when a peer leaves the overlay (t_2). In every test, $t_1 < t_2$ was always true for any sub-jobs and peers. This meant that there was no such a case when there were still available sub-jobs but a peer left because it could not find it. Thus this verification demonstrated that the distributed coordination enabled the workload to be dynamically balanced among peers.

Although we did not have a big number of physical machines (e.g. 1000) for intensive tests, the *speed-up* of the framework was tested by using the single JVM simulation version on a laptop computer of Intel® Core™ i5-4300U@1.9 GHz (2-core CPU) and 8 GB RAM with Windows 8 of 64-bit. In Fig. 6, the test results were reported for solving 14-Queen problem with 10 peers, which joined sequentially at the time interval of 1 s, 15-Queen problem with 20 peers at the joining interval of 2 s and 16-Queen problem with 30 peers at the joining interval of 4 s and all peers' compute-capacities were the same. It was noticed that the speed-up were continuously increasing against the peer number, but it changed into flat when the task was approaching to completion.

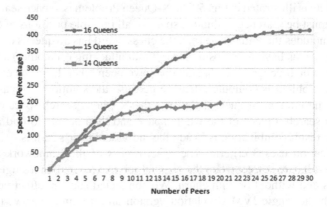

Fig. 6. The accumulative speed-up as a percentage against the number of peers in the single JVM simulation version

That meant adding peers at the early stage would speed up the whole progress significantly but did little at the late stage.

The 15-Queen Problem was solved on the following 4 physical machines to test the speed-up in a distributed environment; the compute-capacities of peers were different and they joined the project community randomly before the task was completed. The results of speed-up were an average taken from 15 tests, which were presented in Fig. 7.

- Job starter: Intel® Core™ i7-2700 K@3.50 GHz (8-core CPU) and 16384 MB RAM with Windows 7 Professional 64-bit.
- Peer 1: Intel® Core™ i7-Q740@1.73 GHz (8-core CPU) and 4096 MB RAM with Windows 7 Home Premium 32-bit.
- Peer 2: Intel® Core™ i7-870@2.93 GHz (8-core CPU) and 4096 MB RAM with Windows 7 Ultimate 64-bit.
- Peer 3: Intel® Core™ i5-3570@3.40 GHz (4-core CPU) and 8192 MB RAM with Windows 7 Home Premium 64-bit.

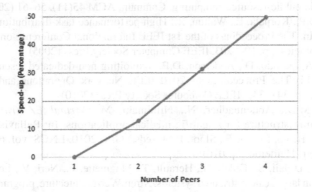

Fig. 7. The accumulative speed-up as a percentage against the number of peers in distributed environments

The results of multiple machine speed-up showed a generally linear speed-up in comparison to peer numbers. Comparing Figs. 6 and 7, the average speed-up of 49.82 % by 4 physical machines was lower than that of 4 peers on a single JVM simulation. It meant that communication costs between peers on different machines for job redistribution mattered.

8 Conclusion

While VC coordination cannot avoid maintaining a centralized job advertising point and a bootstrap pool for volunteers to start participation in a project and a project owner has to centrally monitor the project progress and collect results, all other VC coordination can be decentralized or distributed, including job, data and result storage, job search, splitting, reassignment and redistribution. Such decentralization or distribution significantly reduces the reliance on a central server and thus significantly reduces bottlenecked performance and the possibility of single point of failure.

The proposed distributed VC coordination has been practiced on the performance- and reliability-proven DHT overlay Chord, which guarantees fault-tolerance, the storage load balance among peers and search efficiency. The evaluation done in this paper demonstrated the effectiveness and the speed-up property of the implemented VC coordination framework against the number of volunteers.

Future work includes extending the framework to support workflow (data-intensive) applications beyond the currently embarrassingly parallel (compute-intensive) applications and intensive tests on a large number of physical machines.

References

1. Anderson, D.P., Cobb, J., Korpela, E., Lebofsky, M., Werthimer, D.: SETI@home: an experiment in public-resource computing. Commun. ACM **45**(11), 56–61 (2002)
2. Anderson, D.P., Korpela, E., Walton, R.: High-performance task distribution for volunteer computing. In: The Proceedings of the 1st IEEE International Conference on e-Science and Grid Technologies, pp. 196–203. IEEE Computer Society Press (2005)
3. Andrzejak, A., Kondo, D., Anderson, D.P.: Exploiting non-dedicated resources for cloud computing. In: The Proceedings of 2010 IEEE Network Operations and Management Symposium, pp. 341–348. IEEE Computer Society Press (2010)
4. Belgacem, M.B., Abdennadher, N., Niinimaki, M.: *Virtual EZ grid*: a volunteer computing infrastructure for scientific medical applications. In: Bellavista, P., Chang, R.-S., Chao, H.-C., Lin, S.-F., Sloot, P.M. (eds.) GPC 2010. LNCS, vol. 6104, pp. 385–394. Springer, Heidelberg (2010)
5. Cappello, F., Djilali, S., Fedak, G., Herault, T., Magniette, F., Neri, V., Lodygensky, O.: Computing on large scale distributed systems: XtremWeb architecture, programming models, security, tests and convergence with grid. Future Gener. Comput. Syst. **21**(3), 417–437 (2004). Elsevier
6. Costa, F., Silva, J.N., Dahlin, M.: Volunteer cloud computing: mapreduce over the internet. In: The Proceedings of 2011 IEEE International Symposium on Parallel and Distributed Processing Workshops and Ph.D. Forum, pp. 1855–1862. IEEE Computer Society Press (2011)
7. Curran, O., Shearer, A.: A workflow model for heterogeneous computing environments. Future Gener. Comput. Syst. **25**(4), 414–425 (2008). Elsevier
8. Dou, W., Jia, Y., Wang, H., Song, W., Zou, P.: A P2P approach for global computing. In: The Proceedings of 2003 IEEE International Symposium on Parallel and Distributed Processing, pp. 1–6. IEEE Computer Society Press (2003)
9. Dubey, J., Tokekar, V.: Identification of efficient peers in P2P computing system for real time applications. Int. J. Peer to Peer Netw. **3**(6), 1–12 (2012)
10. Ghafarian, T., Deldari, H., Javadi, B., Buyya, R.: A proximity-aware load balancing in peer-to-peer-based volunteer computing systems. J. Supercomputing **65**(2), 797–822 (2013). Springer
11. Grid MP (2014). http://www.univa.com/products/grid-mp.php
12. Hoffman, E.J., Loessi, J.C., Moore, R.C.: Constructions for the solution of the m queens problem. Math. Mag. **42**(2), 66–72 (1969)

13. Kaffille, S., Loesing, K.: Open Chord version 1.0.4 User's Manual. University of Bamberg, Germany (2007)

14. Kondo, D., Anderson, D.P., McLeod, J.: Performance evaluation of scheduling policies for volunteer computing. In: The Proceedings of the 3rd IEEE International Conference on e-Science and Grid Computing, pp. 415–422 (2007)

15. Kondo, D., Javadi, B., Malecot, P., Cappello, F., Anderson, D.P.: Cost-benefit analysis of cloud computing versus desktop grids. In: The Proceedings of IEEE International Symposium on Parallel and Distributed Processing, pp. 1–12. IEEE Computer Society Press (2009)

16. Kramer, D., MacInnis, M.: Utilization of a local grid of Mac OS X-based computers using Xgrid. In: The Proceedings of 13th IEEE International Symposium on High Performance Distributed Computing, pp. 264–265. IEEE Computer Society Press (2004)

17. Ni, L., Harwood, A.: P2P-Tuple: towards a robust volunteer computing platform. In: The Proceedings of 2009 International Conference on Parallel and Distributed Computing, Applications and Technologies, pp. 217–223 (2009)

18. Rodrigues, R., Druschel, P.: Peer-to-Peer systems. Commun. ACM **53**(10), 72–82 (2010)

19. Sarmenta, L.: Volunteer Computing, Ph.D. Thesis, Massachusetts Institute of Technology, USA (2001)

20. Stoica, I., Morris, R., Liben-Nowell, D., Karger, D., Kaashoek, M., Dabek, F., Balakrishnan, H.: Chord: a scalable peer-to-peer lookup protocol for internet applications. IEEE/ACM Trans. Networking **11**(1), 17–32 (2003)

Design and Implementation of Chinese Historical Text Mining System Based on Culturomics

Lin Tang[1,2(✉)] and Chonghui Guo[1]

[1] Institute of Systems Engineering, Dalian University of Technology, Dalian, China
{tanglin,dlutguo}@dlut.edu.cn
[2] Department of Software Engineer, Dalian University of Technology City Institute,
Dalian, China

Abstract. Culturomics and Chinese text mining methods are of great significance for analyzing the development and evolution of Chinese history and culture.
To help researchers analyze a large number of Chinese historical text data, a
Chinese historical text mining system based on cultruomics is designed, which
includes text data processing and analyzing subsystem, text data visualizing
subsystem, and text data clustering and retrieval subsystem. First of all, our
system preprocesses the text data, then visualizes the text data with the frequency
of words line chart and word cloud, at last selects the text data through clustering
and retrieval methods. It further supports researchers to discover knowledge from
a large number of historical text data. We demonstrate its general performance
on text data of Canton Customs into our system. The result shows that our system
is feasible and effective.

Keywords: Culturomics · Canton customs · Text mining

1 Introduction

With a long history, China remains in a variety of forms to preserve its culture, such as
textural and material researches. In the current research of Chinese culture, the traditional methods are still widely used. Most of them are used to make an empirical
summary and inferential exploration through reading and synthesizing a few documents.
But to the great extent, they are limited by researchers' subjective understanding and
thinking [1]. Because of the vast number of related materials, it is impossible to read
them all, which hinders researchers to fully study and interpret history.

Recently, with the rapid development of computer and information technology,
researchers begin to use it to investigate society and behavior through social simulation,
social network analysis and social media analysis. It gradually forms an interdisciplinary
science named computational social science(CSS) [2]. In the culture field of CSS, a new
branch cluturomics [3] has appeared. Culturomics is a compound word that consists of
culture and genomics, and it is built in mathematics method through the quantitative
analysis of digitized texts to study human behavior and cultural trends.

Erez Lieberman Aiden and Jean-Baptiste Michel, two members of the evolutionary
dynamics team, made a great contribution in the process of the development of

© Springer Science+Business Media Singapore 2016
W. Che et al. (Eds.): ICYCSEE 2016, Part I, CCIS 623, pp. 314–325, 2016.
DOI: 10.1007/978-981-10-2053-7_28

cultruomics. At first Aiden researched genomics through mathematics, then used mathematical tools in evolutionary biology to research historical culture through the quantitative analysis. Moreover, in 2005 Google Book Library, team members collected 5, 195, 769 books, accounting for about 4 % of the total number of books published between 1500 and 2008 in the world [4]. This project established a huge database and offered a visualizer named Google Ngram Viewer(abbreviated as N-Gram) [5] which is one of the most important cultruomics tools. It can query by a term which comprises one or more words. The results are the frequency of the query terms per year. Finally, the frequency is visualized by a continuous time dimension. Michel et al., applied N-Gram into the analysis of culture.

It has got a certain achievement to do the research of culture through the quantitative approach. Relatively, there is scarcity of Chinese historical articles in Google database. N-Gram's concept derives from English grammar, which differs a lot from Chinese grammar. Hence, this tool is inappropriate to direct quantitative analysis of the Chinese articles.

The integrated process of traditional Chinese text mining includes texts pre-processing, indexing and storage, intermediate representation, post-processing [6], which is mainly applied in auto-index, auto-abstract, information retrieval, information extract, document organizing, theme tracking, etc. [7]. Currently, some Chinese text data mining system products and prototype system have emerged, for example, PULSE developed by Gamon and his team in Microsoft [8], mining system of Chinese software commentary developed by Wen Tao and the team [9], etc. But the mining system products or prototype system, which is developed by the historical and culture research based on Chinese text data, has not brought much attention to the public.

We design and implement a Chinese historical text mining system that is based on Culturomics concept can efficiently help researchers to study cultural trends. Our system is designed on the basis of the entire cycle of data mining, including the data collection, text cleaning, text analysis and visualization. The corpus is Chinese historical text with time tags, which will provide multiple visual results for research. Compared with the existing Culturomics tool, our system has the following advantages:

- Our system lays the foundation on Chinese grammar, dealing with the Chinese text in a more efficient way.
- It provides abundant interaction approaches and multiple visualization results to present the results of quantitative data analysis in different aspects.
- It communicates with users to understand their intentions, and selects out what users are interested in through its partitioned clustering and search functions.

2 Design of Chinese Historical Text Mining System Based on Culturomics

According to different research requirements from various clients, including domain specialists and archivists, Chinese historical test mining system consists of three components: Text data processing and analyzing; Text data visualizing; Text data clustering and retrievalling. Our system architecture inspired by CWMS [10], a mining cloud service platform, was constructed in 2013 by Institute of Computer Science, Chinese Academy of Science, but it has some customization. The system includes three subsystems.

1. Text process and analysis subsystem. According to the word frequency statistics, it quantifies the text as the fundamentals for researchers. It has models of Chinese words segmentation, statistic of word frequency and modeling, feature extraction, stop word list maintenance, equal synonyms maintenance.
2. Text visualization subsystem uses word cloud and line chart to make the text visible, so that researchers can easily discover the pattern. It has models of word cloud, line chart of word frequency based on time series.
3. Text clustering and text retrieval system, with the function of text clustering and Full-text retrieval, helps the users find out the text they are interested in.

Figure 1 shows the structure of Chinese text mining system based on Culturomics.

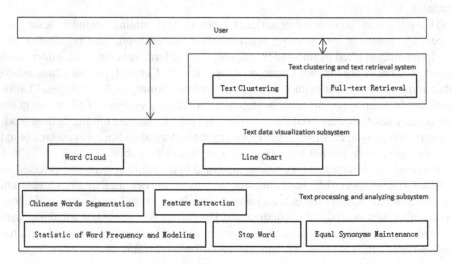

Fig. 1. Design of chinese historical text mining system based on culturomics

Text data processing section uses word segment system ICTCLAS [11]. Based on cascaded hidden markov model, ICTCLAS is developed by Dr Huaping Zhang and Dr Qun Liu from Institute of Computer technology, Chinese Academy of Science. It plays the leading part in the related domain. Data analysis uses TF-IDF(Term Frequency-Inverse Document Frequency) and normalization to process the data. The output is the quantitative analysis result with time tags. The result is presented through information visualization method such as line chart and word cloud. Users obtain their interested information as character by interacting with the system. Our system uses Lucene(a high-performance text, full-featured text search engine library) [12]. Lucene and K-means clustering algorithm is useful to filter their text.

2.1 Text Data Processing and Analyzing Subsystem

Based on text data processing and analyzing subsystem, our system is used for pre-processing Chinese text. Chinese words segmentation is executed while text is added to our system, then the results are processed in the function of word frequency statistics and modeling.

Words segmentation can break up the input text into the minimal recognizable meaningful unit through computer method. To segment English text is very simple, because of the whitespace between words. But it is difficult for Chinese text because of the lack of whitespace. As the foundation of text mining, the accuracy of Words segmentation directly determines the effect of text mining. Therefore, our system adopts the tokenizer ICTCLAS.

Word set with part-of-speech tagging is the result of the words segmentation. In the process of Words Segmentation, our system deletes the stop words list, combines the synonymous list and extracts Nouns among them.

The most commonly used algorithms [13] in text semantic analysis are Latent Dirichlet Allocation(LDA) and TF-IDF. But TF-IDF is simpler and more efficient than LDA. Thus, We chose the TF-IDF algorithm [14] to accurate the term weighting. The term weighting in each document is:

$$w_{ik} = tf_{ik} \cdot idf_{ik} = \frac{tf_{ik} \cdot \lg\left(\frac{N_i}{n_k}\right)}{\sqrt{\sum_{j=1}^{m}\left(tf_{ij} \cdot \lg\left(\frac{N_i}{n_j}\right)\right)^2}}, \quad i = 1, 2, \ldots, k = 1, 2, \ldots \tag{1}$$

Where the element tf_{ik} is the term frequency of t_k in the document d_i. The element idf_{ik} is the inverse documentation frequency of t_k. N_i is the total number of terms in the I document(the repeated terms still be count). n_k is the frequency of the computed term k.

2.2 Text Visualization Subsystem

The textual data based on a weighted metric can be visualized by different methods. The word cloud [15] can depict the representative keywords at a time. The line chart can illustrate the terms evolution over time. In the historical development, Knowledge is implied not only inside a document, but also in a document set. Therefore, our system combines word cloud and line chart through interacting with users.

The huge differences among texts lengths lead us to normalize the words frequency before the visualization.

$$tf_k^{new} = \frac{tf_k}{\sum_{j=0}^{m} tf_j}, \quad k = 1, 2, \ldots \tag{2}$$

In the equation tf_k is the term frequency of k. The total number of frequency is $\sum_{j=0}^{m} tf_j$.

We implement the word cloud with d3.js (or just D3 for Data-Driven Documents) [16], which is a JavaScript library for producing dynamic, interactive data visualizations in web browsers. Our word cloud can visualize all terms(selected the top 250 terms), all noun terms and place names. Our line chart implemented by icharjs [17], it is a graphics library based on HTML5.

Users can gain the knowledge in two ways. One way is to find interesting terms by a word cloud, then to select one or more interesting terms. The line chart will be useful to illustrate the terms evolution on time series. The other way is to find out the interesting terms from checking the hot terms' line charts, and to check them with word cloud in a particular year.

2.3 Text Cluster and Text Retrieval System

Although many visualization methods intuitively show some features of the text, they cannot be completely replaced by artificial reading. How can we select out the text needed by users and classify them? The text clustering and text retrieval system can fulfill the needs. Users can select the interesting terms as the feathers. Based on these features the text is clustered into several groups based on these features to classify the text. The full-text retrieval assists users to locate interesting words and select the text.

We use the K-means algorithm to do document clustering based on the result of similarity with angle cosine.

$$\text{Sim}(d_i, d_j) = \frac{\sum_{k=1}^{n} \omega_{ik} \cdot \omega_{jk}}{\sqrt{(\sum_{k=1}^{n} \omega_{ik}^2)(\sum_{k=1}^{n} \omega_{jk}^2)}}, \quad i = 1, 2, \ldots, j = 1, 2, \ldots \tag{3}$$

Where ω_{ik} and ω_{jk} are the term weighting of term I and term j in document k, respectively. The model of full-text retrieval uses Lucene, which is a highly scalable and super-fast open-source search engine. It offers a simple but powerful program interface. In this model, user can search the text both by the initial input term and by the splitted terms. All of the results will be highlight to help users to see clearly.

3 Case Study

China's foreign trade development is an important part of Chinese culture. Maritime transport is the most important mode of transportation. Therefore, modern coasting trade can reflect China's foreign trade in miniature. After the order of opening ports in 1685, the Qing government set up four customs. They were Canton Customs, Foochow Customs, Chekiang Customs and Shanghai Customs. Compared with the other customs, canton customs existed long and kept prosperous. We now compare our system with N-Gram in the background of Chinese culture in terms of Qing customs.

3.1 Visualization of Customs in the Qing Dynasty Based on N-Gram

N-Gram is the most common tool in cultruomics, in support of many languages such as English, Chinese, and etc. So we try to use it for mining the text about canton customs. Chinese text is still very limited in the Google's library, So we do the following experiments using English and Chinese, respectively.

Now assume we want to inspect the evolution about Canton Customs (粤海关), Foochow Customs (闽海关), Chekiang Customs (浙海关) and Shanghai Customs (江海关). First of all, we use N-Gram to query by the names of these customs. Because the tool is case sensitive, the final result is to sum the number of the result by the customs name of different cases. The key words of query are the English name of these four customs, and we query the words frequency in the period of 1800–2008. Figure 2 shows that before 1842, only canton customs have words frequency for Canton Customs was the most important customs in the early Qing dynasty. As you can see in Fig. 2, after 1842 all of words frequency of these customs became significant. We know that following the Opium War of 1840 China declined to a semi-feudal and semi-colonial country, and the four customs began to trade.

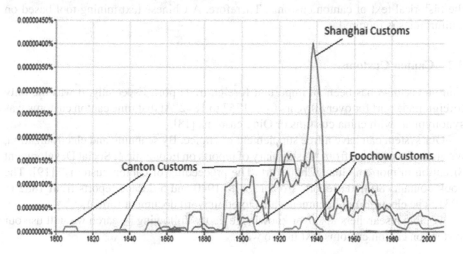

Fig. 2. English words frequency of four customs in Qing Dynasty (Color figure online)

We try to query the words frequency from 1800 to 2008. The Chinese names of these customs are the key words, which are "粤海关, 江海关, 浙海关, 闽海关". As shown in the Fig. 3, there is not any words frequency before 1921 and no any word frequency of Chekiang Customs from 1800 to 2008. It shows that the contemporary Chinese text is relatively scarce in Google Book Library. There is hardly any Chinese text about these four customs during the Qing dynasty. Because of the limitation of Google Book library, we can't study the canton customs based on Chinese text through N-Gram.

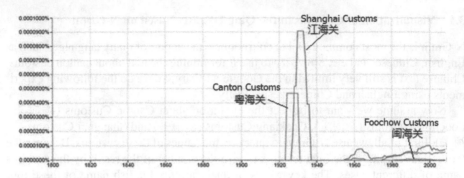

Fig. 3. Words frequency of four customs in modern times (Color figure online)

N-Gram provides only one visual way line chat. The single form limits the reflection of the trend of words frequency. Users need more diversified visual ways to observe the text. In addition, N-Gram is a tool based on the Google Book Library, which hinders users to select and supplement particular contents. We cannot use N-Gram to research the historical text of canton customs. Therefore, A Chinese text mining tool based on Culturomics is required.

3.2　Canton Customs

Canton customs has been an important foreign trade port. Especially, it was the only foreign trade port for over 80 years from 1757 to 1842. At that time canton customs was synonymous with china customs or Qing customs [18].

Our system can have a free supplement or choice. By scanning and electronization, we attain the new text from "A summary of reports on Economic & Social Development Situation in morden GuangZhou Port: The proceeding of canton customs" [19]. The book contains 80 trade reports from 1864 to 1940 and 5 decade reports from 1882 to 1931. The electronized documents are stored in word document format.

To help researchers discover knowledge, in the following research we will use our system on canton customs text in two ways.

Analysis Result of Traded Goods. Figure 4(a) and (b) respectively illustrate us the relatively active goods in 1882 and 1890 silk, including real silk, raw silk and native silk and so on, tea and cinnamon in the forms of the noun word clouds. It is easy to find them in the figures. Then, we chose these terms. Through the line chart we observe them changes from 1870 to 1892 (Fig. 5). We find the raw silk, tea are of great importance in canton customs. Silk was mainly exported as the raw material. On the contrary the occurrence of the cinnamon is lower.

In order to prove the credibility of word frequency line chart, we extracted and counted up the real port data of tea by year. The word frequency of tea is visualized by Fig. 6(a) and the data of real tea export is visualized by Fig. 6(b). We find they have the same varying trend which shows terms frequency can get a reference value results.

Fig. 4. Chinese words frequency in Qing dynasty

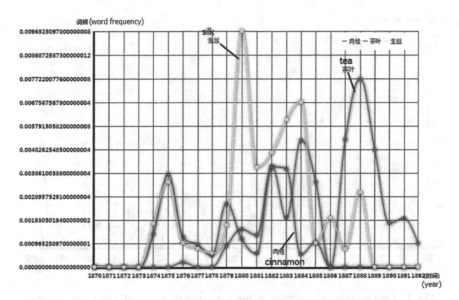

Fig. 5. "生丝" word frequency link chart from1870 to 1892 (Color figure online)

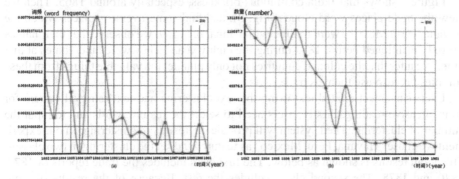

Fig. 6. Comparison chart of tea words frequency and exports data

Analysis Result of Overseas Places. Trade zone is also very important in the study of foreign trade. Our system automatically extracts the highest frequency in five overseas places, including America, Japan, India, Europe and Britain, based on the existing text and displays the results in Fig. 7. Except for Japan, other countries' words frequency fluctuates in the smooth. Before 1894 the word frequency of Japan is almost zero. While after 1894 its frequency is relatively higher. In the second half of the 19th century Japanese successful transformation on economical, increased its desire for developing broad markets. To some extends, it leaded to the outbreak of Sino-Japanese war which broke out in 1894 and China ended in failure [20]. After the war, Japan actively participated in China's foreign trade and maintained its actively for a few decades.

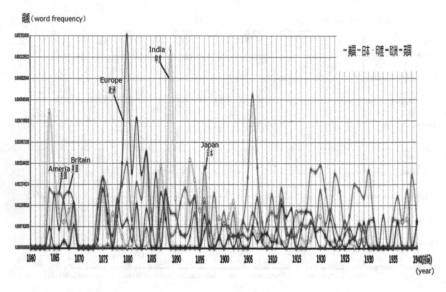

Fig. 7. Five overseas places name of highest words frequency (Color figure online)

Figure 7 shows that India continuous briskness, especially around 1865. Then we view the trade places word clouds (Fig. 8), and find some overseas places which are Havana, Peru, Mexico, were trading with China. In Fig. 9, we observe that their names are only mentioned before 1882 during the eighty years. They were all colonies at that time. We infer that the colonial countries not only plunder the valuable things from these colonies but also use them as trade ports.

Users can select related text using the text clustering and text retrieval system for their intensive reading. Firstly, users need to select the feature words and to input the number of clustering. Then system will provide the results of clustering. In our experiment we chose "Havana, Peru, Mexico" as features and set the clustering in two classes. The detail results is shown as below. The first class includes the trade reports of 1874, 1876 and 1878. The second class includes the rest. Because of the results of word frequency, we can select the first class text. So that researchers can read intensively for their further study.

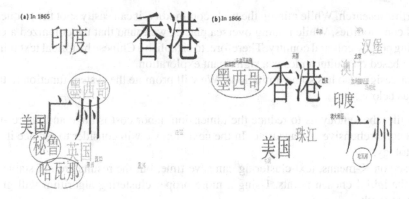

Fig. 8. Word cloud of trade places in 1865 and 1866

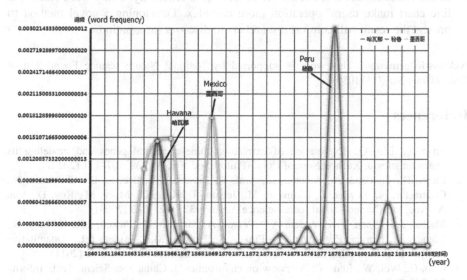

Fig. 9. Word frequency line chart of Havana, Peru, Mexico (Color figure online)

4   Conclusions

In recent years, it has become the trend to analyze and research humanities by quantitative mathematical approaches. In this paper, we established the Chinese historical mining system based on Culturomics. The system includes text data process and analysis subsystem, text data visualization subsystem and also text data collecting and research subsystem. Finally our system provides plentiful visualization results as well as text data clustering and retrievalling to help researcher discover the related information.

Through the case study, compared with N-Gram we have demonstrated the usability of our system. Firstly, we uses N-Gram tool to query the result which reflects the outstanding status of Qing Dynasty Customs. Because of the lack of Chinese resource for Qing Dynasty Customs in the Google database, it is not possible to use N-Gram too for further research. While mining the trade commodity, it can easily spot the frequently traded commodities. While mining oversea places, we found that the colonized area as a trading port by colonial country. Therefore, to establish a Chinese historical text mining system based on Culturomics is a significant exploration.

Our design also has some limitations. We will promote the system function in three ways as below.

- To filer the synonyms to reduce the dimension, labor cost is huge and the result is not comprehensive and accurate. In the next step we will consider to replace it with ontology.
- Based on K-means, text clustering can save time, but the results are not stable due to the initial chosen points. Using a more proper clustering algorithm will give a better result.
- In the visualization, it is not able to observe data efficiently. Using word cloud and line chart make users' operation more complex. Developing a novel method to present the result will comfort researchers to discover information.

Acknowledgements. This work is supported by National Nature science Foundation of China(Grant No. 71171030)

References

1. Shao, P., Lin, Q.: Exploration of extracting chinese cultural genes and modeling its characteristics. J. Xuzhou Normal Univ. (Philos. Soc. Sci. Ed.) **38**, 107–111 (2012)
2. Lazer, D., Pentland, A., Adamic, L., Aral, S., Barabási, A., Brewer, D., Christakis, N., Contractor, N., Fowler, J., Gutmann, M., Jebara, T., King, G., Macy, M., Roy, D., Van Alstyne, M.: Computational social science. Science **323**, 721–723 (2009)
3. Michel, J.B., Shen, Y.K., Aiden, A.P., Veres, A., Gray, M.K., Pickett, J.P., Hoiberg, D., Clancy, D., Norvig, P., Orwant, J., Pinker, S., Nowak, M.A., Aiden, E.L.: Quantitative analysis of culture using millions of digitized books. Science **331**, 176–182 (2011)
4. Guo, C., Wei, W., Ren, X.: A review on culturomics. J. China Soc. Scient. Tech. Inform. **33**, 765–774 (2014)
5. Aiden, E.L., Michel, J. (eds.) Uncharted Big Data as a Lens on Human Culture (2013)
6. Chen, Z., Zhang, G.: Study on the text mining and chinese text mining framework. Inform. Sci. **25**, 1046–1051 (2007)
7. Huang, X., Zhao, C.: Application of text mining technology in analysis of net-mediated public sentiment. Inform. Sci. **27**, 94–99 (2009)
8. Gamon, M., Aue, A., Corston-Oliver, S., Ringger, E.: Pulse: mining customer opinions from free text. In: Famili, A., Kok, J.N., Peña, J.M., Siebes, A., Feelders, A. (eds.) IDA 2005. LNCS, vol. 3646, pp. 121–132. Springer, Heidelberg (2005)
9. Wen, T., Yang, D., Li, J.: Design and implementation of Chinese software reviews mining system. Comput. Eng. Des. **34**, 163–167 (2013)
10. He, Q., Zhuang, F.: Big data mining and the Cloud Services. High Technol. Ind. **8**, 56–61 (2013)
11. ICTCLAS:An Open Source Chinese Lexical Analysis System. http://www.nlpir.org
12. Apache. Lucene. http://lucene.apache.org/core

13. Gansner, E.R., Hu, Y.: Stephen: visualizing streaming text data with dynamic graphs and maps. In: 20th International Symposium, Redmond, WA, USA, pp. 439–450 (2012)
14. Cheng, X., Zhu, Q.: Text Mining Principle. Science Press, Beijing (2010)
15. Lee, B., Riche, N.H., Karlson, A.K., Carpendale, S.: SparkClouds: visualizing trends in tag clouds. IEEE Trans. Vis. Comput. Graph. **16**, 1182–1189 (2010)
16. Bostock, M., Ogievetsky, V., Heer, J.: D(3): data-driven documents. IEEE Trans. Vis. Comput. Graph. **17**, 2301–2309 (2011)
17. Ichartjs Open Source Technology Group. http://www.ichartjs.com/
18. Chen, T., Liang, E.: YUEHAIGUANZHI and his researches on the history of the customs. J. Hist. **135**, 72–80 (2009)
19. Office of Local Chronicles Compilation of GuangZhou Province. A summary of reports on Economic & Social Development Situation in morden GuangZhou Port: The proceeding of canton customs. Jinan University Press, GuangZhou (1995)
20. Liu, N.: Is the outcome of the 1894 Sino -Japanese warevitable? -A review from econom ic perspective. Historical Review, pp. 105–115 (2011)

Detection of Copy-Scale-Move Forgery in Digital Images Using SFOP and MROGH

Mahmoud Emam[1,2(✉)], Qi Han[1(✉)], and Hongli Zhang[1]

[1] School of Computer Science and Technology,
Harbin Institute of Technology, Harbin 150080, China
ma7moud_emam@yahoo.com, {qi.han,zhanghongli}@hit.edu.cn
[2] Department of Mathematics, Faculty of Science,
Menoufia University, Shebin El-koom 32511, Egypt

Abstract. Social network platforms such as Twitter, Instagram and Facebook are one of the fastest and most convenient means for sharing digital images. Digital images are generally accepted as credible news but, it may undergo some manipulations before being shared without leaving any obvious traces of tampering; due to existence of the powerful image editing softwares. Copy-move forgery technique is a very simple and common type of image forgery, where a part of the image is copied and then pasted in the same image to replicate or hide some parts from the image. In this paper, we proposed a copy-scale-move forgery detection method based on Scale Invariant Feature Operator (SFOP) detector. The keypoints are then described using MROGH descriptor. Experimental results show that the proposed method is able to locate and detect the forgery even if under some geometric transformations such as scaling.

Keywords: Image forensics · Copy-move · Forgery detection · Scale invariant feature · RANSAC · MROGH descriptor

1 Introduction

Due to the existence of highly sophisticated software for editing the digital images, it became easily modify images without leaving any subtle traces. Copy-move forgery technique is the most commonly used technique where, a part of the image is copied and then pasting it into another part in the same image. Therefore, Copy-move forgery detection (CMFD) algorithms aims at detecting the same or similar regions in the forged images. Figure 1 shows an example of Copy-move forgery, where the pocket of the child's shirt is copied from his left hand side and then pasted into the other side of the shirt. Some post-processing operations can be performed on the forged images after Copy-move operation, which makes the task of forgery detection more harder. Typically, post-processing operations are applied to cover up the forgery such as geometric transformation (e.g. scaling).

Several researchers have introduced algorithms for detecting image copy-move forgery which can be found in these surveys [1,6]. Generally, these methods can be classified into two main categories: block-based methods [14] and

© Springer Science+Business Media Singapore 2016
W. Che et al. (Eds.): ICYCSEE 2016, Part I, CCIS 623, pp. 326–334, 2016.
DOI: 10.1007/978-981-10-2053-7_29

(a) Original image   (b) Forged image   (c) Detection results with
                                                          the proposed method

Fig. 1. Copy-move forgery example

keypoint-based methods [2]. Due to the limitations of block-based methods especially in the robustness against scaling manipulations and time complexity, keypoint-based methods attract many researcher's attention. Keypoint-based methods detect keypoints and then use the local features to identify duplicated regions instead of using overlapping blocks [15]. Typically, SIFT [13] and SURF [3] are used as a keypoint detectors, and their corresponding descriptors are used to find matches between these keypoints. To estimate the geometric transformations applied to the forged regions, Random sample consensus (RANSAC) algorithm [9] can be used.

Huang et al. [12] introduced a method to detect copy-move forgery based on local statistical features, known as scale invariant feature transform (SIFT). But in that method there is no estimation for the geometric transformation parameters rather than the weak performances. Another method has been proposed in [15], but that method can't manage affine transformation. Bo et al. [5] presented another CMFD method based on Speeded Up Robust Feature (SURF) descriptor to overcome geometric operations such as scaling, their experimental results were introduced visually. The algorithm appears to be promising, but it still need more improvements to automatically localize the duplicated regions. The existing CMFD methods mentioned in [1,6] can achieve an acceptable performance, but still there exist some challenges especially in the affine transformation manipulations (e.g. scaling).

In this paper, Scale Invariant Feature Operator (SFOP) detector [11] is used as a local feature detector to extract the keypoints from the forged image. After detecting the keypoints from the image, these keypoints are described using MROGH descriptor and then matched. The matched points are then clustered according to the distance between them. Then, RANSAC algorithm can be used to estimate the affine transformation parameters and remove the false matches. The rest of this paper proceeds like this: in Sect. 2, each step of the proposed method is explained; in Sect. 3, experimental results are presented and discussed; finally, Sect. 4 summarizes the paper and next research target.

2   Proposed Method

2.1   Local Features Detector and Descriptor

In our proposed method, SFOP detector is used for detecting keypoints from the forged image. The SFOP, which is a scale-space extension of the detector proposed by Förstner [10], is a local feature detector proposed by Förstner et al. in [11]. SFOP uses the general spiral feature model of [4] to unify different types of features within the same framework, and it achieves a better coverage under various geometric transformations than the other local feature detectors [7].

Given a forged image I_{forged}, Firstly we applied SFOP to detect the keypoints $P = p_1, p_2, ..., p_n$. Secondly, we used Multi-support Region Order-based Gradient Histogram (MROGH) [8] to generate a descriptive vectors $F_i, i = 1, 2, ..., n$ for each keypoint $p_i \in P, i = 1, 2, ..., n$. A two-dimensional MROGH histogram with length $\lambda_1 \times \lambda_2 \times \lambda$ can be obtained where; λ is the number of support regions, λ_1 is the number of quantifiable levels, and λ_2 is the number of order segments. We empirically choose $\lambda_1 = 8$, and $\lambda_2 = 6$ because they can achieve a good performance as presented by experiments in [8].

2.2   Feature Matching

For each feature $f_j \in F; j = 1, 2, ..., 48$, we used $kd - tree$ to obtain the k nearest neighbors $N_l, l = 1, 2, ..., k$ with corresponding distances denoted as $d_z, z = 1, 2, ..., k$ that represents the sorted Euclidean distance. The keypoints are then matched if the ratio between D_1 and D_2 is less than a threshold ($D_1/D_2 < thr$). But, this matching strategy can't deal with multiple keypoint matching. So, we used another matching procedure g2NN as presented in [2]. This method iterates the nearest neighbors test between D_r, D_{r+1} while:

$$D_r/D_{r+1} < g2NN_{thr} \tag{1}$$

Now, we obtain the set of all matched points. These matched points are then kept for further post-processing and the other mismatched keypoints are then removed.

2.3   Post-processing

In this step, the matched keypoints are clustered according to the distance between them based on a threshold D_{thr}. After that, all the clusters with members less than a minimum member number ζ in each cluster are discarded, for the others, we used RANSAC algorithm [9] to estimate the affine transformation parameters and remove the false alarms. For each estimated homography matrix, we find all inliers D less than α that fit with this transformation according to:

$$D = \left\| H \begin{pmatrix} x \\ y \\ 1 \end{pmatrix} - \begin{pmatrix} x^{'} \\ y^{'} \\ 1 \end{pmatrix} \right\|_2$$

where $(x, y, 1)^T$, $(x', y', 1)^T$ are the homogeneous coordinates of a pair of matched points and H is the estimated affine homography matrix that can be defined as follows:

$$H = \begin{bmatrix} a_{11} & a_{12} & t_x \\ a_{21} & a_{22} & t_y \\ 0 & 0 & 1 \end{bmatrix}$$

Hence, we can get some false alarms. To remove it, we again used distance-based clustering for each homography whose corresponding inlier pairs are less than γ. Then, all the clusters with members less than ζ in each cluster are removed.

Finally, we applied some morphological operations to get the final detected duplicated regions.

3 Experimental Results and Discussion

3.1 Dataset

We evaluate the performance of the proposed method by conducting a series of experiments. In the following experiments, we have used benchmark evaluation database appeared in [6]. The dataset contains 48 high resolution color images of different sizes, varying from 533×800 (*giraffe image*) to 3900×2613 (*sailing image*). The tampered images have been generated by cutting and pasting image region(s). The image region(s) selected for duplication can be geometrically transformed before being pasted. The duplicated region can vary in size (e.g., small, medium, or large). A sample of original images, forged images and its ground truth indicating the forged regions are shown in Fig. 2.

To evaluate the performance of the proposed method, precision-recall (PR) curves [16] and F_1 score are employed. Equations 2, 3, and 4 show how the precision, recall, and F_1 rates are calculated;

$$Precision = \frac{T_P}{T_P + F_P} \tag{2}$$

$$Recall = \frac{T_P}{T_P + F_N} \tag{3}$$

$$F_1 = 2.\frac{Precision.Recall}{Precision + Recall} \tag{4}$$

where;

- T_P (True Positive) represents the number of tampered pixels, which are classified as tampered.
- F_P (False Positive) represents the number of authentic pixels, which are classified as tampered.
- F_N (False Negative) represents the number of tampered pixels, which are classified as authentic.

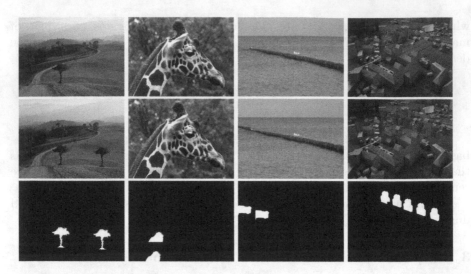

Fig. 2. Examples for copy-move forgery: the up row is the original images, middle row is the corresponding forged images, and bottom row is the ground truth map

3.2 Parameters Setup

We set up the parameters of the proposed method as in Table 1.

Table 1. Parameters setup for the proposed method

Parameters	Description	Value
$g2NN_{thr}$	ratio threshold for collecting the matched features	0.8
D_{thr}	the distance threshold	120
k	the number of nearest neighbors in $kd - tree$	8
ζ	minimum member number threshold for each cluster	3
α	distance threshold for collecting the inliers	4
λ	the number of support regions in MROGH	1
λ_1	the number of quantifiable levels	8
λ_2	the number of order segments	6

3.3 Detection Performance of the Proposed Method

Plain Copy-move Forgery. We evaluate how the proposed method can perform under plain Copy-move forgery without any post-processing operations. Figure 3 shows some visual detected examples, in which the forged regions are correctly localized by the proposed method. All images in the dataset are detected and the values of precision, recall, and F_1 score are computed.

The proposed method has a very high precision rate (exactly, 95.66 %), an acceptable recall rate value (exactly, 49.12 %), and the comprehensive assessment

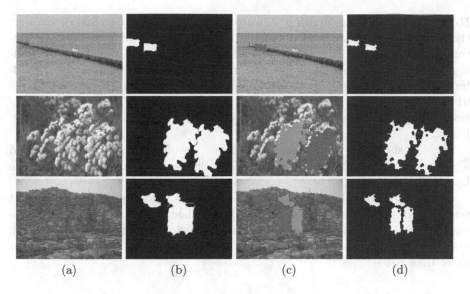

(a)   (b)   (c)   (d)

Fig. 3. Some visual results: (a)the tampered images, (b)the corresponding binary mask, (c)the correctly detected matches, (d)the detection map produced by our proposed method

F_1 score equal to 64.91 %. Therefore, the proposed method is very accurate in finding the correct matches (very slight false positives). But, it is not able to cover all the areas of the Copy-move region (more false negatives exist). Furthermore, in our experiment we found that we can observe the forgery in the images and can be easily identifiable, even when the forged regions are not detected correctly as shown in Fig. 4.(d).

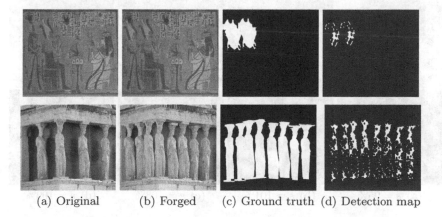

(a) Original   (b) Forged   (c) Ground truth   (d) Detection map

Fig. 4. An example of Copy-move forgery detection (proposed method)

Robustness against Scaling Manipulations. In the benchmark dataset, the cloned regions are scaled before being pasted by different scaling factors s. To evaluate the robustness of our proposed method against scaling manipulations, We test the performance of our method under 10 different scaling factors of the original size of the forged region ($s =$ 91 %, 93 %, 95 %, 97 %, 99 %, 101 %, 103 %, 105 %, 107 %, and 109 %). The detection results of some forged images with different scaling factors s are shown in Fig. 5.

Table 2 shows the performance evaluation results of the proposed method against scaling manipulations for all Copy-Scale-move forgery images in the benchmark dataset.

We noticed that the proposed method achieves a good performance against scaling manipulations. We also noticed that, when scaling factor increase, the matched points decrease due to the impact of scaling, but there are still enough matched points to be detected (see for example Fig. 5(d) and Table 2 when s is more than 101 %).

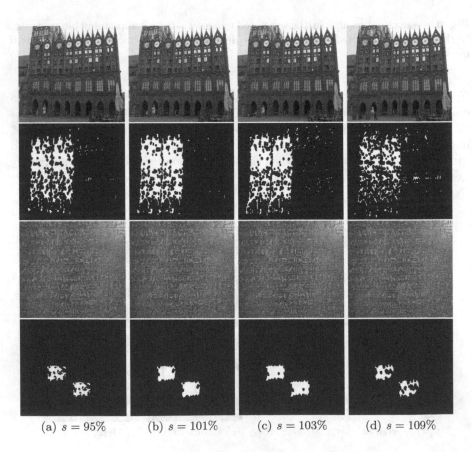

(a) $s = 95\%$ (b) $s = 101\%$ (c) $s = 103\%$ (d) $s = 109\%$

Fig. 5. The detection results with different scaling factors s

Table 2. Detection Performance of the proposed method against scaling manipulations

(%)	$s = 91$	$s = 93$	$s = 95$	$s = 97$	$s = 99$	$s = 101$	$s = 103$	$s = 105$	$s = 107$	$s = 109$
Precision	93.70	96.55	96.86	96.05	96.32	96.05	95.66	96.45	96.95	96.43
Recall	32.62	36.81	40.44	43.96	46.57	47.04	45.99	44.23	39.50	35.08
F_1 score	48.39	53.30	57.06	60.31	62.79	63.15	62.12	60.65	56.13	51.44

Hence, we can conclude that our method is of robustness to scaling manipulations, since the scale invariant feature operator (SFOP) is able to detect sufficient keypoints from the forged image with a different scaling factors. But, it still needs some improvements in the flat regions.

4 Conclusion

In this paper, a forensic method has been proposed to detect and localize copy-move regions under scaling manipulations. We used scale invariant feature operator (SFOP) as a feature points detector due to its scaling invariance. The proposed method is able to find the correct matches but, it is not able to cover all the regions of the Copy-move region due to the existence of more false negatives. Furthermore, our method can observe the forgery in the images and can be easily identifiable. The experimental results show the robustness of our method against scaling manipulations, especially in the non-flat regions. But, it still needs some improvements in the flat regions. In the future, we will try to solve this problem by using a dense interest point to find more matched keypoints. Also, we will try to use some other post-processing techniques, to recover some missing matches and hence increase the recall rate of the method.

Acknowledgment. The authors would like to thank all anonymous reviewers for their insightful comments. Additionally, This work is supported by the National Natural Science Foundation of China (Grant Number: 61471141, 61301099, 61361166006), the Fundamental Research Funds for the Central Universities (Grant Number: HIT. KISTP. 201416, HIT. KISTP. 201414).

References

1. Al-Qershi, O.M., Khoo, B.E.: Passive detection of copy-move forgery in digital images: state-of-the-art. Forensic Sci. Int. **231**(1), 284–295 (2013)
2. Amerini, I., Ballan, L., Caldelli, R., Del Bimbo, A., Serra, G.: A SIFT-based forensic method for copy-move attack detection and transformation recovery. IEEE Trans. Inf. Forensics Secur. **6**(3), 1099–1110 (2011)
3. Bay, H., Ess, A., Tuytelaars, T., Van Gool, L.: Speeded-Up Robust Features (SURF). Comput. Vis. Image Underst. **110**(3), 346–359 (2008)
4. Bigu, J., et al.: A structure feature for some image processing applications based on spiral functions. Computer Vis. Graph. Image Process. **51**(2), 166–194 (1990)

5. Bo, X., Junwen, W., Guangjie, L., Yuewei, D.: Image copy-move forgery detection based on surf. In: 2010 International Conference on Multimedia Information Networking and Security (MINES), pp. 889–892. IEEE (2010)
6. Christlein, V., Riess, C., Jordan, J., Riess, C., Angelopoulou, E.: An evaluation of popular copy-move forgery detection approaches. IEEE Trans. Inf. Forensics Secur. 7(6), 1841–1854 (2012)
7. Ehsan, S., Kanwal, N., Clark, A.F., McDonald-Maier, K.D.: Measuring the coverage of interest point detectors. In: Kamel, M., Campilho, A. (eds.) ICIAR 2011, Part I. LNCS, vol. 6753, pp. 253–261. Springer, Heidelberg (2011)
8. Fan, B., Wu, F., Hu, Z.: Aggregating gradient distributions into intensity orders: a novel local image descriptor. In: 2011 IEEE Conference on Computer Vision and Pattern Recognition (CVPR), pp. 2377–2384. IEEE (2011)
9. Fischler, M.A., Bolles, R.C.: Random sample consensus: a paradigm for model fitting with applications to image analysis and automated cartography. Commun. ACM 24(6), 381–395 (1981)
10. Förstner, W.: A framework for low level feature extraction. In: Eklundh, J.-O. (ed.) Computer Vision—ECCV 1994. LNCS, vol. 801, pp. 383–394. Springer, Heidelberg (1994)
11. Förstner, W., Dickscheid, T., Schindler, F.: Detecting interpretable and accurate scale-invariant keypoints. In: 2009 IEEE 12th International Conference on Computer Vision, pp. 2256–2263. IEEE (2009)
12. Huang, H., Guo, W., Zhang, Y.: Detection of copy-move forgery in digital images using sift algorithm. In: 2008 Pacific-Asia Workshop on Computational Intelligence and Industrial Application, PACIIA 2008, vol. 2, pp. 272–276. IEEE (2008)
13. Lowe, D.G.: Distinctive image features from scale-invariant keypoints. Int. J. Comput. Vis. 60(2), 91–110 (2004)
14. Nathalie Diane, W.N., Xingming, S., Moise, F.K.: A survey of partition-based techniques for copy-move forgery detection. Sci. World J. 2014, 13 (2014)
15. Pan, X., Lyu, S.: Detecting image region duplication using sift features. In: 2010 IEEE International Conference on Acoustics Speech and Signal Processing (ICASSP), pp. 1706–1709. IEEE (2010)
16. Powers, D.M.: Evaluation: from precision, recall and F-measure to ROC, informedness, markedness and correlation (2011)

Determining Web Data Currency Based on Markov Logic Network

Yan Zhang[1(✉)] and Rui Zhang[2]

[1] School of Computer Science and Technology, Shandong University of Finance and Economics, Jinan 250014, People's Republic of China
yanzhang_jinan@126.com
[2] Shandong Provincial Institute of Electronic Information Products Inspection, Jinan 250014, People's Republic of China
ruizhang@126.com

Abstract. This paper proposes a method based on Markov Logic Network (MLN) to determine the time order of entity attribute values. We use the characteristics of web sources' currency, web sources inter-dependency and attribute data currency in a certain web source as predicates in MLN. We define five rules (new rules can be added) to infer the currency of different values provided by different sources. On one hand, this method considers currency problem based on entity attribute instead of the entire entity, which is critical to improve the quality of data provided by Web Integration Systems; on the other hand, this method summarizes characteristics of web sources and web data based on carefully analysis. It is noteworthy that it is not complicate for the MLN model to incorporate new rules, which shows that the proposed method is extensible.

Keywords: Web data integration · Data currency · Markov logic network

1 Introduction

Web data integration systems integrate data from different sources in a certain domain to provide more comprehensive, high quality information for users. It is not rare that different sources provide conflict values for the same real world entity. In light of this, we often find that multiple values of the same entity reside in the Web Integration System (WIS). Most researches concentrate on finding a true value when this happens. However, parts of them just have become obsolete and inaccurate. That is, the value that describes the same real world entity evolves with time goes on, such as a person's address, affiliation, phone number etc. Simply finding a true one from these conflict values has following limitations:

This work is supported by the Shandong Province Natural Science Fund (No. ZR2015PF011).

© Springer Science+Business Media Singapore 2016
W. Che et al. (Eds.): ICYCSEE 2016, Part I, CCIS 623, pp. 335–349, 2016.
DOI: 10.1007/978-981-10-2053-7_30

(1) With picking one true value and discarding the others that might be out-of-data would lose the evolution information of the entity, which may be important for some analysis tasks.
(2) False fusing result would lose the real true value and lower the data quality.

In order to solve the above problems, we propose to record the currency (we use the word to represent the time order of attribute values with reference to [1]) of entity attribute values in Web data integration systems. That is, based on efficiency entity resolution, which we have achieved good results, we sort the conflict values belong to the same attribute of the same real world entity by time order. Determining the time order of different values describing the same thing has attracted much attention in the area. Most of them concentrated on relational database. The data was produced by transactions with relatively high quality compared to data from web. So the semantics and dependences in-between data were the main characteristics to infer the time order of conflict values. In Web environment there are more characteristics to indicate the currency of data, such as update frequency of web sources, update frequency of certain kinds of data in different web sources (it is possible that different sources update different kinds of data with different frequency), dependency of different web sources, and also the timestamp of collecting data in WIS. This paper provides solution to infer the time order of conflict values that describe the same attribute of the same entity. Since the conflict values came from web sources, our solution would combine the information about the sources and semantics of the data. We use Markov Logic Network (MLN) [13] to model the problem for that MLN is much better than other probability models in inferring uncertain knowledge. The main contributions of this paper are as follows:

(1) A solution based on MLN to compare the time order of Web entity attribute values was proposed. In the paper, we consider the web entity currency problem on attribute values instead of the whole entity, which is more practical in web environment.
(2) To infer the time order of conflict values from multiple sources, we combine the semantics of domain knowledge related to the entity and information about the web sources and model all the characteristics and constraints as MLN rules, which ensures the efficiency of the approach.
(3) We test the method on both real world data set and synthetic data set and the results show that our method is efficient to complete the task of sorting the conflict values by time order.

The rest of this paper is organized as follows. First, the related work is discussed in Sect. 2. In Sect. 3, we illustrate the data model used in our method and the characteristics used to measure the time order, based on which we form the problem to solve. In Sect. 4, we illustrate framework of the proposed method and discuss the process of modeling the characteristics into MLN logic rules. In Sect. 5, the experimental setup and results are shown. Finally the conclusion is presented.

2 The Problem of Determining Web Data Currency

Web Integration Systems collect web objects belong to a specific domain from different websites for further analysis and decision-making. Due to the evolving essence of entities in real world and different update frequency of web sources, it is not rare that in a Web Integration System there are conflict values describing the same attribute related to the same entity. Determining the time order of these different values help to set up the evolving evidences of the entity, which benefit further analysis. In this section, we introduce the data model we use, based on which we form the problem of determining web data currency.

2.1 Data Model

We consider all data comes from limited number of data sources $S = \{S_1, S_2, ..., S_n\}$. WIS has defined the corresponding web entity schema $ET = \{attribute_1, attribute_2, ..., attribute_m\}$. Web entities are tuples that satisfy the schema. Since we mainly concentrate on sorting conflict values describing the same aspect of an entity, we use the triple $(entityID, attributeID, value)$ to represent entity values for short. The formally definitions are as follows.

Definition 1 Data Source. Data sources provide values that describe entity attributes. They could be databases, Websites etc. We use set $S = \{S_1, S_2, ..., S_n\}$ to donate the data sources, where S_i is the ith one.

Definition 2 Entity Type. Entity type is to describe the attribute information that belong to the same type of entities. We donate it as $et = \{attr_1, attr_2, ..., attr_m\}$, where $attr_i$ is a representative description of data from the same column, like $address$ of entity $Company$. There are entity types such as Company, Book, and Paper etc.

Definition 3 Entity. Entities are objects that existing in real world and can be distinguished from each other. We use the set $E = \{e_1, e_2, ..., e_n | e_i \in e_t\}$ to represent the entities that belong to the same entity type. For example, Samsung, Apple etc. are entities of Company.

Definition 4 Entity Attribute Value. Entity attribute values are values provided by different sources describing entity attributes. In order to infer the time order of different values that describing the same entity attribute, in this paper we use a triples $t = \{e, attribute, value\}$ to represent an entity attribute value, where e is the identifier to distinguish the entity from the other, $attribute$ is one of the attributes that belongs to e and $value$ is the data provided by one of the sources regarding to the $attribute$ of entity e.

Definition 5 Observation. Observations are values provided by different sources that describe entity attribute values. We use $O = \{O_1, O_2, ..., O_n\}$ to donate observations from n sources, where $O_i = \{t | S_i \models t\}$ is set of values provided by the ith source.

Figure 1 demonstrates the relationships of the above definitions. Data currency problem is that given data sources set S and observations O on them, we discover the time order of different values belong to the same entity attribute. That is, in Fig. 1, t_1 and t_2 describe the $attr_1$ of entity e, and we need to infer which one is more up-to-date. As the same, the currency order of t_3, t_4 and t_5 needs to be inferred for that they describe the $attr_2$ of entity e.

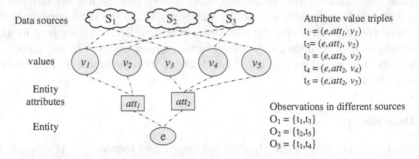

Fig. 1. Relationships between data sources, entities, entity attributes and entity attribute values

2.2 Data Currency Measuring

In this paper, we suppose the schema matching and entity resolutions have been worked out for the data in WIS, based on which, we solve the problem of attribute value currency. In the following section, we formulate the main characteristics and measurements we use to compare the currency of two values and define the problem we solve in this paper. In order to measure the currency of attribute values, we use $p(t)$ to donate the probability that the value provided in triple t is the most up-to-date.

Currency of Data Sources. Intuitively if a data source s has high precision in currency (that is, most of information it provides is fresh), it is more probable that the triples provided by it are more up-to-date than the other sources. On the other hand, if a date source s' has low recall in currency (that is, most of up-to-data information is not provided by it), it is more probable that the triples provided by it are less up-to-date than the other sources. We use the standard way to define the precise and recall: the precise of a data source s_i currency p_i is the proportion that the most up-to-date triples it provides among all observations O_i; recall r_i is the proportion that the most up-to-date triples it provides among all fresh triples. Formally, they define as follows:

$$p_i = p(t|S_i \mapsto t) \qquad (1)$$

$$r_i = p(S_i \mapsto t|t) \qquad (2)$$

It is worth to point out that the recall of data source currency is related to the scope of the data it provides. That is, if one source doesn't provide a value to attribute a of entity e, it doesn't affect the recall that the most up-to-data value doesn't come from it.

Currency of Entity Attribute in Different Data Sources. A data source doesn't update all data residing in it with the same frequency. For example, an e-commerce site updates prices more often than the other information about products. So it is reasonable that if the data source S updates attribute A of entity E more frequent, it is more probable that it provides the more up-to-data information on that attribute. Suppose observations $O_i = \{t_1, t_2, ..., t_M\}$ are values of attribute a of entity e provided by source S_i, we use p_{ia} to represent the currency of attribute a in data source S_i, which is defined as:

$$p_{ia} = \frac{\sum_{i=1}^{M} p(t_i)}{M} \tag{3}$$

Dependency of Data Sources. Except the currency of data sources themselves, the other character that is useful to infer the time order of different values is the dependency of sources. Intuitively, if we know there is copy relations between the sources S_i and S_j, then we should not increase the probability of the value to be most up-to-data when they provide the same value. For simplicity, when we compute the dependency of data sources, we mainly consider the proportion of common values to all different values. We use $A = \{attr | \pi_{attr}(O_i) \cap attr | \pi_{attr}(O_j)\}$ to donate the attributes that source S_i and S_j describe, the dependency of S_i and S_j computes as:

$$\frac{|O_i \cap O_j|}{\sum_A O_k} < \alpha \Rightarrow S_i \perp S_j \tag{4}$$

In which, $\sum_A O_k$ donates the number of different values in attribute A of a certain entity E. $S_i \perp S_j$ donates that S_i and S_j are independent. Further, in order to infer if there is dependency relationship between specified attributes, we use the following formula to compute:

$$agree_a = \frac{|\{t | \pi_{attr}(t) = a \wedge t \in O_i \cap O_j\}|}{|O_i \cap O_j|} \tag{5}$$

Semantic Constraints of Data. There is semantic constraint to some specified attributes. For example, the "marriage state" attribute, if there are values {"unmarried", "divorced", "married"} from different sources about the same person, we can conclude from the semantic constraint of "marriage state" itself that the correct time order should be {"unmarried", "married", "divorced"}. So if there are semantic constraints for some specified attributes, we can define first logic rules according to the semantic constraints as evidence predicates in MLN.

Problem Definition. Web entity attributes currency. Given data sources $S = \{S_1, S_2, ..., S_n\}$, observations from these sources $O = \{O_1, O_2, ..., O_n\}$, determine the time order of observations that describe the same attribute of an entity. That is, sort observations t_i in $O_t = \{t_1, t_2, ..., t_m | \forall i \in [1, m], t_i \in e_a\}$ to get $O'_t = \{t'_1, t'_2, ... , t'_m\}$, where for any $i < j$ $(i, j \in [1,m])$, there is $p(t'_i \geq p(t'_j))$, $p(t)$ is the probability of tuple t to be the most up-to-date value at the time.

3   Solution of Determining Data Currency

In this chapter, we will mainly discuss the method to sort the different values describing the same attribute of an entity by time order. From the definition stated in the previous section, the problem changes to calculate the probability that tuple t is the most up-to-data $p(t|O_t)$, where O_t is all observations from all sources that related to t. After obtaining these probabilities, we sort them in descendent order to get the time order of the attribution values.

3.1   Method Overview

The overview of the method proposed in this paper is illustrated in Fig. 2. First, we computer the data source context characters, including data source currency (the overall update frequency of each data source); attribute currency (the specific attribute update frequency of each data source); the dependency of data sources, which are important clues to compute the possibility of each tuple to be new. Second, we set rules for the MLN according to the characters we choose and train the MLN model to set weights to each rule. After we get the model, we use the MLN to infer the probability of each tuple in the observations and sort those describing the same attribute of an entity to get the currency order.

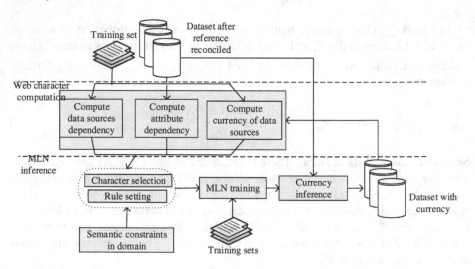

Fig. 2.   Method Overview of Inferring Web data currency using MLN

From the overview we can see that the input method is all observations O (after entity reference reconciled) from different data sources data source, the output is the data collection O' with sequential relationship. The entire algorithm can be described as follows (for convenience, we use $\{t_e\}$ to represent all tuples describing entity e and $\{t_{ea}\}$ to represent the tuples that describe the attribute a of entity e).

3.2 Choosing and Computing Predicates

With careful observation and analysis to data sources and data itself, we define features to be used in MLN rules from two aspects. (1) Web contextual feature, includes basic features, currency of data sources, currency of entity attribute in different sources and dependent among sources. (2) Semantic features of data itself. In this section we will discuss the features and measures we select to use as the MLN predicates and rules.

Basic Predicates. The basic feature mainly reflects the relationships between data sources and attribute tuples. If data source s provides attribute tuple t, the predicate can be defined as *provide(s, t);* the predicate *describe (t, e_a)* represents that attribute tuple t describes attribute a of entity e. To count the frequency of each tuple appearing in all related observation values, we use predicate *mfreq (e_a, t)* to denote that the value described in t appears most frequently among all observation values in attribute a of entity e. Predicate *current(t)* represents that tuple t contains the latest value in describing the attribute. It is the only query predicate in our MLN model.

Predicate for Data Source Currency. When it comes to decide which value is the newest for an entity attribute, it is common sense that the value provided by the data source with high updating frequency. In Sect. 1.2, we already define formula to compute the precise and recall of data source currency. We can get the corresponding probabilities by using the formula on training data. We believe that the higher the precise of a data source currency is, the higher possibility it provides the latest values; at the same time, the lower the recall is, the lower possibility it provides the latest value is. Suppose p_s represents the currency precise of data source s while r_s represents recall. Since the value of predicate in MLN is dyadic, that is $\{0, 1\}$, we use the following formula to compute the value for the predicate *CurrentSource(s)*.

$$F(s) = \frac{2 \times p_s \times r_s}{p_s + r_s} \tag{6}$$

$$CurrentSource(s) = \begin{cases} 1, \ if \cdot F(s) \geq \dfrac{\sum_{i=1}^{n} F(s_i)}{n} \\ 0, \ otherwise \end{cases} \tag{7}$$

Formula (6) represents the F value of currency for date source s, and formula (7) computes the data source currency predicate value by comparing the F-score with the average.

Predicate for Entity Attribute Currency in Data Sources. It is common that the updating frequencies for different kinds of data in the same data source may be different. For instance, for an e-commerce website, it updates prices of products more frequent than other features of products. On the contrary, for a product-evaluating website, the parameters of products are usually updated first. In Sect. 1.2.2, we define the formula to compute the currency of entity attribute currency for each data source. Like for data sources, we need to transform the value into bivariate that predicate can take. So the following formula demonstrate how we compute the value for the predicate *CurrentAttr(s,a)*.

$$CurrentAttr(s, a) = \begin{cases} 1, \ if \cdot p_{sa} \geq \dfrac{\sum_{i=1}^{N} p_{sia}}{N} \\ 0, \ otherwise \end{cases} \qquad (8)$$

Where p_{sa} is the probability defined in formula (3).

Predicate for Dependency Between Data Sources. If two data sources share many values we think that they are dependent on each other. Since different data sources focus on different sides of data, even when there is copy relationship between two data sources, they don't share all attributes of an entity type. So we mainly consider the dependency on attributes instead of the whole entity. We use predicate *depend* (s_1, s_2) to denote dependency between s_1 and s_2 and *attrDepend* $(s_1, s_2, attr)$ to donate the dependency on attribute *attr* between s_1 and s_2. We use the following formula to compute them.

$$depend(s_1, s_2) = \begin{cases} 1, \ if \ s_1 \perp s_2 \\ 0, \ otherwise \end{cases} \qquad (9)$$

$$attrDepend(s_1, s_2, attr) = \begin{cases} 1, \ if \cdot depend(s_1, s_2) \wedge agree_a(s_1, s_2)) \geq \theta \\ 0, \ otherwise \end{cases} \qquad (10)$$

Table 1. Predicates used in determining web data currency

Type of predicate	Facet	Predicate	Description
Query predicate	Data currency	*current(t)*	Tuple *t* is the latest
Evidence predicate	Basic feature	*provide(s, t)*	Data source *s* provides tuple *t*
		describe(t, ea)	Tuple *t* describes attribute *a* of entity *e*
		maxfreq (ea, t)	Tuple *t* appears most frequent among all tuples that describe attribute *a* of entity *e*
	Currency of data source	*CurrentSource(s)*	Data source *s* updates data frequent
	Currency of entity attribute in data source	*CurrentAttr(s, a)*	Data source s updates attribute *a* frequent
	Dependency between data sources	*depend(s_1, s_2)*	Data source s_1 and s_2 are dependent
	Attribute dependency between data sources	*attrDepend(s_1,s_2,a)*	Data source s_1 and s_2 are dependent in attribute *a*

Where $s_1 \perp s_2$ and $agree_a(s_1, s_2)$ are defined in formulas (4) and (5) separately. Parameter θ shows the proportion that the number of tuples s_1 and s_2 share in *attr* among all share tuples.

Predicate for Data Semantic Constraint. Data itself has some semantic constraint to show the time order in some specific domains, such as the marital status that mentioned before. If there is semantic constraint is related to time order, we can define them as predicates and set up corresponding predicate formula according to constrain regulation. In accordance with the feature of discriminant MLN, we can dynamically add the formulas easily.

Table 1 summarizes all predicates we used in our method.

3.3   Setting Rules

After selecting proper predicates, in this section we set corresponding rules for the MLN model to infer the data currency.

Rule 1: Vote. The rule of voting is the simplest and most efficient, for that most data sources try to provide fresh data. Therefore, the rule of voting can be defined as follows:

$$mfreq(ea, t) \wedge describe(t, ea) \Rightarrow current(t)$$

Rule 2: Data provided by source with higher updating frequency is the latest. When attribute values provided by several sources conflict, we believe that the value provided by the source which has higher updating frequency is more probable to be fresh. Therefore, we can give following formula based on former definition of relevant predicates:

$$CurrentSource(s) \wedge provide(s, t) \Rightarrow current(t)$$

Rule 3: Data provided by source with higher updating frequency on the attribute is the latest. According to the analysis in 1.3.3, besides the currency of data source itself, the factors affecting our judgment to data currency is related to the updating frequency of the source on the corresponding attribute. There might be data sources with low currency as a whole, but with high currency with some specific attribute. Therefore, we can define following formula of predicate:

$$CurrentAttr(s, a) \wedge provide(s, t) \Rightarrow current(t)$$

Rule 4: The rule of dependency between data sources. If there is dependency between two data sources s_1 and s, which means that they share a lot of data, then we believe that they have the same currency to other attribute values. It is noteworthy that some values appears different in different data sources, it might just have different forms but actual content is the same. This rule can be defined as follows.

$$depend(s_1, s_2) \land provide(s_1, t_1) \land provide(s_2, t_2) \land describe(t_1, ea) \land describe(t_2, ea)$$
$$\Rightarrow (current(t_1) \Leftrightarrow current(t_2))$$

Rule 5: The rule of dependency between attribute of data sources. When two data sources share values in some specific attributes, we believe that they are dependent on these attributes, and they have the same currency on other values of these attributes. The rule can be defined as follows:

$$attrDepend(s_1, s_2, a) \land provide(s_1, t_1) \land provide(s_2, t_2) \land describe(t_1, a) \land describe(t_2, a)$$
$$\Rightarrow (current(t_1) \Leftrightarrow current(t_2))$$

3.4 Inference

After setting the rules, we need to set the weights of each formula before we use the MLN to reason the currency for different observations. In the problem of determining data currency, we don't know which rules are more important over others beforehand, we need to train the model with training set to learn the weights for different rules.

Nowadays, the common method of learning weights in discriminant MLN is voting perception algorithm. This is a gradient descent algorithm. The main idea is to initialize all weights to zero, and then conducting the iterative training to weight according to whether the reasoning value and actual value in training set match. At last, for avoiding the over-fitting problem which commonly exists in machine learning, the algorithm sets weights as the average obtained in the process of learning but not the weights obtained in final round iteration. Furthermore, to use this algorithm, we also need to calculate the expectation that each formula is true. In our problem, it's hard to calculate the value, so in this paper we use MC-SAT algorithm to solve it approximately. After obtaining the weights of each formula through training, we can use MLN model to reason the test set to get the probability of different attribute values to be fresh. In the process of reasoning, pure logic algorithm usually adopts satisfiable algorithm, while traditional probability model usually uses MCMC(Markov Chain Monte Carlo) algorithm. As the reasoning of MLN has both determinacy of logic system and uncertainty of probability method, this chapter adopts MC-SAT algorithm to infer the value of query predicate.

After we get the probabilities of different attribute values, we sort the results with descendent order to get the currency order. We describe the process in Algorithm 1 in Sect. 3.1. It is noteworthy that at first the algorithm sorts the observations by entity, in the same entity, sorts by attribution, which is convenient to infer the currency order for the same attribute of the same entity. Meantime, if there are no conflict values in the same attribute, it is no need to do any reasoning.

4 Experiments

We test our method in both real world data set and synthetic data set. We use the open source tool Alchemy in our experiments. In real world data set, it's hard to obtain the ground truth of attribute value currency but it is not hard to get the fresh data from trustable websites. We analyze the experimental result from three aspects separately: (1) precision. (2) Influence of combinations of different rules. We need to point out that as it's hard to obtain real time order of attribute values in real data sets, so we just consider the precision of fresh data reasoned by the method. In synthetic data set, we simulate the attribute value currency with the process of generating data, we calculate the precision of attribute value currency.

4.1 Datasets

Real World Data Set. We collect information of 200 different restaurants in Jinan from the website dianping.com, including their names, telephones, addresses and opening time. Since dianping.com is a group purchase website, the information it provides is used for customers to contact the sellers, which means it has to be fresh at the time. We use it as ground truth for all attribute values to be up-to-date. To get more information from other data sources, we put all the restaurant names in search engine, and use our self-made extraction algorithm to collect data on the four attributes from 100 different websites, and we get 8695 records all together. After doing entity resolution we transform all records into triple observations described in Sect. 1.2.1 and get 34780 triples. We refer to this data set as restaurant data set. We choose 40 restaurants from the 100 data sources as training set, the rest as testing set.

Synthetic Data Set. Synthetic data set enables us to control the features of data and data source, for instance, dependent relationship between data sources, updating frequency of specific attributes, times of data repeating, etc. It can reflect the influence that different rule combinations to precision of the method. We generate observations from 10 different sources, with 10,000 tuples from each one and they all belong to 5,000 different entities. The average number of attributes is 6. To show the dependent relationship between data sources, when generating the observations we randomly select three data sources, where in one of the sources 70 % tuples are copied from the other two, other tuples repeat randomly once, twice, six times in different data sources respectively. We generate the observations to simulate situation that the 10 data sources provide data at a specific time. To simulate the updating of data sources, we randomly choose 1000 tuples in five times to update (data sources which have dependent relationship also reserve former dependent relationship when updating), thus, we confirm ground truth of changed attribute value based on each tuples' updating time. Table 2 illustrates data distribution of synthetic data set. We call this data set as synthetic data set.

Table 2. Data distribution in synthetic data set

#data sources |S|	#observations |Q| (including repeated)	#entities |E|	#attributes of each entity on average |A_i|	Times of simulating collecting data
10	100,000	5,000	6	5

We randomly choose 20 % tuples as training set to calculate the data source currency, dependency between data sources, currency of specific attributes in data sources. We also use the training set to train the weights of MLN rules. Regard the rest 80 % observations as testing set, reason the attribute value currencies with MLN model and compare the results with the ground truth to evaluate the s method.

4.2 Experiment Results and Analysis

To validate the effectiveness of our method, we analyze and evaluate the result of the experiments in following three aspects in this section.

Precision. For restaurant data set, we evaluate the precision of the latest value. Assuming that A is the number of attributes with multiple values in testing set, B is the number of the attributes that our method deduced the right latest value, we define the precision $p = B/A$. In synthetic data set, suppose that A is the number of all attributes in testing set, B is the number of attributes that our method deduced right attribute value sequence by time order, we define the precision $p = B/A$. Here right attribute value sequence means every value is in the correct currency position. As there is no mature method of researching data currency, the Ref. [1] only discusses theoretically how to confirm the latest value of the same entity attribute in relational database, we can't compare with it. Therefore, we compare our method with the simplest voting method and the BAYES method [14].

Influence of Different Rule Combinations on Precision: Figure 3 shows the comparison of the proposed method and other two. We can see from the figure that in restaurant data set, the proposed method is superior to Voting method, because Bayes method takes dependent between data sources into account and it has similar precision with the proposed method. In synthetic data set, as the precision relies not only on a latest value, but also the order of attribute values, so the proposed method has better results than Voting and Bayes.

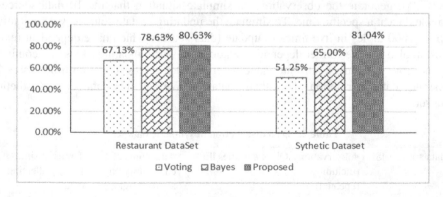

Fig. 3. Comparison of precisions

To validate the effectiveness of rules in our method, we test the influence of different rule combinations on the result precision. We mark these five rules respectively as: Voting rule-V; Data source currency-SC; Attribute currency in data sources-AC; Rule of dependency between data sources-SD; Rule of dependency between attributes in different data sources-AC. We use the voting rule as the basic, based on this we add the other four rules to constitute four rule combinations, and with the combination of all rules, that is six combinations of rules. We obtain precision by testing these six combinations with MLN reasoning.

Figure 4 shows the precisions of different combinations. From the picture we can see that the all-combination is superior to the other combinations, which means that the rules we choose are positive for determining the data currency.

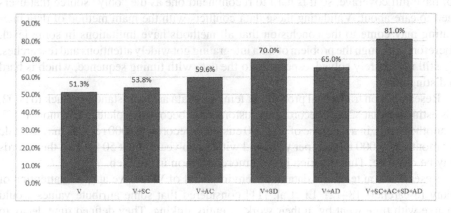

Fig. 4. Precisions of different combinations of rules

5 Related Work

Researches related to the work in this paper include data fusion and data currency. In the field of data integration, the main work is to integrate the heterogeneous data from different data sources. However, data provided by different data sources might be conflicting, which may be because of incomplete data, false data and outdated data. Providing false data in the data integration system can be misleading and even harmful one may contact a person by an out-of-date phone number, visit a clinic at a wrong address, and even make poor business decisions. So lots of researchers in field of data integration devote to fuse conflicting data from different data sources and find true value. Xin (Luna) Dong in VLDB2009 gave the definition of data fusion: data fusion means fusing records on the same real-world entity into a single record and resolving possible conflicts from different data sources.

Xin (Luna) Dong [2, 3] and other researchers [4–6] have achieved good results in data fusion, which is a part of the field of data integration. They consider copy phenomenon between Web data sources, calculate the dependent between two data sources with Bayes model, and put the result into the use of finding true value of Web data. In the meantime, they established a prototypical system called SOLOMN, which was used to represent the process of data fusing with dependent relationship between Web data sources. There are other typical researches in the field of data integration, for instance, HPI research grope in Universidad Potsdam of Germany. Scholars like Felix have already summarized existed accomplishments and provided a series of method of data fusion. Reference [2] observed that even for these domains that most people considered as highly reliable, a large amount of inconsistency: for 70 % data items more than one value was provided. Among them, nearly 50 % were caused by various kinds of ambiguity, although they had tried their best to resolve heterogeneity over attributes and instances; 20 % are caused by out-of-date data; and 30 % seem to be caused purely by mistakes. Although well-known authoritative sources, such as Google Finance for stock and Orbitz for flight, often have fairly high accuracy, they are not perfect and often do not have full coverage, so it is hard to recommend one as the "only" source that users need to care about. Validating these data conflicts with the main method of data integrating may come to the conclusion that all methods have limitations in some level. Therefore, although the problem of data integrating got widely attentions and researches, it's still not entirely solved, especially to the data with timing sequence, which is hard to distinguish.

Researches on traditional problem of temporal data are long-standing. Back to 2002, it is estimated that "2 % of records in a customer file become obsolete in one month" [7]. In another word, in a database of 500,000 customer records, 10,000 records may go stale per month, 120,000 records per year, and within two years about 50 % of all the records may be obsolete. This situation is even more common in the Web.

Researches on temporal data in the environment of Web have attracted attention of many researchers [8–12]. Li et al. [11] considered that some attribute values would change with time went by in their work of entity linking. They defined time decay to obtain the influence of time-passing on the evolution of entity values. Based on that, they provided several clustering methods to fuse different appearances of the same real-world entity into one. Reference [12] improved their approach to provide a two-step faster matching method. Paper [1] provided a data currency model in the premise of each entity having its identical mark (EID) to get all records constituted by the latest values in all attributes of entity in relationship data base. However Refs. [11, 12] were based on the assumption that the timestamps were obtainable and Ref. [1] considered their model with the condition that all records had their EIDs without considering conflicting from different sources. In the Web, either the timestamp or the EID is not always easy to get, but there are other clues about data sources and data itself help us to infer the time order of attribute values, which is the main objection of this paper.

6 Conclusion

This paper provides a method of determining the currency of entity attribute values based on MLN. We analyze the web sources and web data to use the characteristics of web sources' currency, web sources inter-dependency and attribute data currency in a certain web source as predicates in MLN. We define five rules (new rules can be added) to infer the currency of different values provided by different sources. Since MLN combines first-order logic and probabilistic graphical models in a single representation that we can use it to combine uncertain or even contradict knowledge to learn and infer, which makes our method extensible. We study the problem on the assumption that all values are true or were once true. The future direction on this study will be finding the values that are or were true with the process of determining the currency.

References

1. Fan, W., Geerts, F., Wijsen, J.: Determining the currency of data. ACM Trans. Database Syst. (TODS) **37**(4), 25 (2012)
2. Li, X., Dong, X.L., Lyons, K., et al.: Truth finding on the deep web: is the problem solved? Proc. VLDB Endow. **6**(2), 97–108 (2012)
3. Dong, X.L., Naumann, F.: Data fusion–resolving data conflicts for integration. PVLDB **2**(2), 1654–1655 (2009)
4. Galland, A., Abiteboul, S., Marian, A., Senellart, P.: Corroborating information from disagreeing views. In: WSDM, pp. 131–140 (2010)
5. Pasternack, J., Roth, D.: Knowing what to believe (when you already know something). In: COLING, pp. 877–885 (2010)
6. Pasternack, J., Roth, D.: Making better informed trust decisions with generalized fact-finding. In: IJCAI, pp. 2324–2329 (2011)
7. Eckerson, W.W.: Data quality and the bottom line: achieving business success through a commitment to high quality data. Data Warehousing Institute (2002)
8. Chiang, Y.H., Doan, A.H., Naughton, J.F.: Modeling entity evolution for temporal record matching. In: Proceedings of the ACM Conference on Management of Data (SIGMOD), pp 1175–1186 (2014)
9. Pal, A., et al.: Information integration over time in unreliable and uncertain environments. In: Proceedings of the 19th International World Wide Web Conference, pp. 789–798 (2012)
10. Christen, P., Gayler, R.W.: Adaptive temporal entity resolution on dynamic databases. In: Proceedings of the 17th Pacific-Asia Conference in Knowledge Discovery and Data Mining, pp. 558–569 (2013)
11. Li, P., Dong, X., Maurino, A., Srivastava, D.: Linking temporal records. Proc. VLDB Endow. **4**(11), 956–967 (2011)
12. Chiang, Y.H., Doan, A., Naughton, J.F.: Tracking entities in the dynamic world: A fast algorithm for matching temporal records. Proc. VLDB Endow. **7**(6), 469–480 (2014)
13. Richardson, M., Domingos, P.: Markov logic networks. Mach. Learn. (ML) **62**(1–2), 107–136 (2006)
14. Dong, X.L., Berti-Equille, L., Srivastava, D.: Truth discovery and copying detection in a dynamic world. PVLDB **2**(1), 562–573 (2009)

Efficient File Accessing Techniques on Hadoop Distributed File Systems

Wei Qu[✉], Siyao Cheng, and Hongzhi Wang

School of Computer Science and Technology, Harbin Institute of Technology,
Harbin, China
{qwei,csy,wangzh}@hit.edu.cn

Abstract. Hadoop framework emerged at the right moment when traditional tools were powerless in terms of handling big data. Hadoop Distributed File System (HDFS) which serves as a highly fault-tolerance distributed file system in Hadoop, can improve the throughput of data access effectively. It is very suitable for the application of handling large amounts of datasets. However, Hadoop has the disadvantage that the memory usage rate in NameNode is so high when processing large amounts of small files that it has become the limit of the whole system. In this paper, we propose an approach to optimize the performance of HDFS with small files. The basic idea is to merge small files into a large one whose size is suitable for a block. Furthermore, indexes are built to meet the requirements for fast access to all files in HDFS. Preliminary experiment results show that our approach achieves better performance.

Keywords: HDFS · Hadoop · Index · Small files

1 Introduction

Hadoop [1] is an open-source distributed platform under apache software foundation in which Hadoop Distributed File System [2] and Yarn [3] act as the core. It provides users with the underlying transparent distributed infrastructure. Hadoop can be deployed on cheap hardware to build a distributed system with the advantages of high fault-tolerance, high scalability, etc. With the convenience above, Hadoop can be used to deal with many complex problems—Top K problems, K-means clustering problems, Bayes classification, for example. Moreover, from the points of e-commerce, mobile-data, image-processing and so on, it has extensive applications. Then MapReduce, a distributed computing framework, allows the users without full knowledge of fundamental details to develop parallel application in distributed system, making full use of large-scale computing resources, and solve the problems in the situations where single traditional high-performance machines have low efficiency.

In the area of social computing, one of the most popular areas at present, HDFS is also widely used. The social environment is made up of social system, social network and so on. The social network is the most important one which takes people as the nodes and relationships as the edge. All the social media are based on large-scale data center and a large number of server cluster. Here the traditional relational database is

© Springer Science+Business Media Singapore 2016
W. Che et al. (Eds.): ICYCSEE 2016, Part I, CCIS 623, pp. 350–361, 2016.
DOI: 10.1007/978-981-10-2053-7_31

hard to meet the requirements for data processing and the NoSql databases perform well such as Hadoop HBase, Google Big Table and so on.

HDFS has many advantages such as: (1) supporting big data with the size of GB or even TB; (2) hardware malfunction detection and quick response; (3) paying more attentions to the throughput of data rather than the speed of data access; (4) simplified consistency model: once a file has been created or closed, it may not be modified.

However, HDFS based on the above objectives is not suitable for some scenario, it has the following two problems:

1. Large amounts of small files. Hadoop has to face such a reality that it is short of processing massive small files. Because the HDFS assigns indexes for every file which are stored in metadata on NameNode, the memory usage rate increases as the files increase, thus, making the NameNode overwhelmed. Large amounts of small files are common in many applications. Taking TaoBao as an example, 90 % flow of the whole network is caused by image access, because image is the most convincing description of goods for both buyers and sellers. There are 28.6 billion pictures stored in the backend server system in TaoBao and the average size of them is just 17.45 KB, including 61 % less than 8 KB [4]. Another example is the web crawler technology, which is web applications or scripts that automatically collect information according to certain rules. The main goal of using web crawler is to get more web pages within a few KB which are correlative with a certain topic and prepare data for users' querying [5].
2. High latency of data access. Applications interacting with users are supposed to response within several seconds or milliseconds. But Hadoop has been optimized for high data throughput, thus, the delay is relatively high. What's more, the files on HDFS are usually accessed by path which has high expectations for users' ability. If one wants to search a file under multi-level directory, the exact path of the file is needed. So the HDFS can not process the request whether a file exists or not without its path. For example, the HDFS usually throws exception when a file is uploaded because the users can not detect the existence by name. This is more obvious with the amounts of files getting larger.

In order to solve the problems mentioned above, this paper presents two approaches respectively: (1) the adaptive storage strategy and (2) name-based index.

In short, the contributions of this paper are as follows. First, an adaptive storage strategy is provided to reduce the memory usage. Second, a name-based index is proposed to accelerate file access. Third, extensive simulations and real system experiments were carried out to verify the efficiencies of all proposed methods.

The rest of this paper is organized as follows. Section 2 describes the background and preliminary knowledge of HDFS. Section 3 gives a summary of our approach. Section 4 gives the details about how an adaptive storage strategy is carried out to handle large amounts of small files. Section 5 introduces the name-based index technique. Section 6 evaluates the methods in this paper by real system experiments. Section 7 refers to some related work and Sect. 8 draws the conclusion.

2 Preliminary Knowledge

HDFS uses the master-slave structure consisting of NameNode, DataNode and Client. As the management of the whole system, NameNode manages the namespace, cluster configuration, file-based replication and so on. DataNode is the basic unit for storing files, which saves the specific data contents on HDFS and parity information in forms of block, while the client is responsible for communication with the NameNode and DataNode, data access in filesystem and document operation. In addition, there is another node named SecondaryNameNode whose function is to combine namespace image and assist daemons to edit log.

When a client has read-file operations, the DFSInputStream object will establish contact with the nearest DataNode via the stream interface. The client calls the read method repeatedly and receives the data packets. After the connection to DataNode has been closed, DFSInputStream gets information of the next block via the getBlock Locations method and receives data packets again through the interface in DataNode. Moreover, DFSInputStream may call this method several times because the client protocol will not return all the information at one turn. The client closes the stream when the read-data task is finished. The structure is shown in Fig. 1.

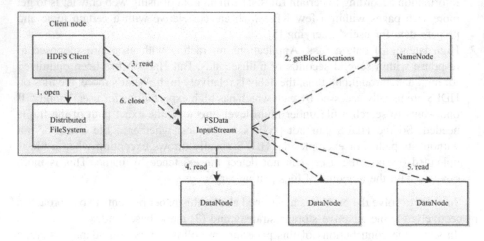

Fig. 1. Read-file process on HDFS

When the writing process begin, the client calls a method of DistributedFileSystem to create files. DistributedFileSystem creates the DFSOutputStream called from remote procedure at the same time. Then the NameNode executes the same-named method to create files in the system namespace. DistributedFileSystem packets the DFSOutputStream object into FSDataOutput instance and then sends it to client. After the empty file is created, the client applies for data blocks from NameNode. When the addBlock method is finished, it returns a LocatedBlock object which contains the identification and version number of new data blocks. Then the data in input stream is put into internal queues of DFSOutputStream object. A normal write-file process is accomplished in this way. Figure 2 shows the details [6].

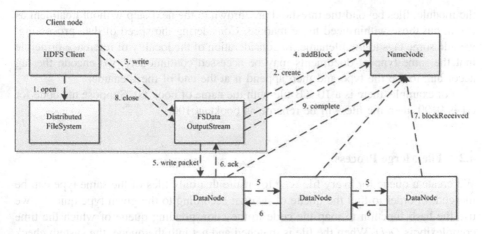

Fig. 2. Write-file process on HDFS

3 Principle of Our Approach

The basic idea of our approach is to merge small files into large ones and build indexes for them. Trie structure and hash index are well considered in the process of building index. Moreover, we follow the rule that small files are merged by type. Then the informations of the large file such as name, duplicate number and locations are send to NameNode and the data is put onto DataNode. The trie is stored in memory and the files used for map are stored on NameNode. Thus we can access a file by name rather than the path and this can be proved more effective.

The methods above can solve the problems that the memory in NameNode is not enough and can effectively manage the small files without file accessing by name.

4 Adaptive Storage Strategy

We add a preprocessing module into the traditional HDFS. The module exists in the interaction between client and NameNode. When a file is submitted, the preprocessing module first extracts the features to identify whether it is a small file or not and then renames the file for identification. Second, we merge the files into lager one by the features extracted above. The detailed processes are as follows:

4.1 Feature Extraction and Analysis

The aim of this phase is to extract features of files and make a preliminary analysis. The features including file type, file size and so forth can be stored in XML format. And thus can be transmitted between programs easily. In order to filter the big files that don't require any action, we need to determine a threshold. Here we choose the value in hdfs-site.xml as the threshold, which is 64 M by default. When the data flow through the module, files beyond the threshold are thrown to the next step without treatment as usual, but those within need to be marked. Considering the speed of data processing, we add some tags to the filename. In consideration of the locality of reference principle that the same type of documents may be accessed continuously, we encode the tag according to the file type and then append it at the end of the filename.

For example, there is a 10 KB file with the name of book.txt. Suppose the code for txt is 0100, then the file will be rename as booktag0100.

4.2 File Merge Process

We create a queue for every file type to ensure that only files of the same type can be merged. In order to find the queue of its own according to the given type quickly, we use the hash function to map the code to the corresponding queue of which the time complexity is $O(1)$. When the file is matched and get into the queue, the system check the current size of data in this queue. Once the size of data reaches the threshold of a block, all the small files in this queue are merged into a large one and then the queue is cleared. Otherwise the file is put at the end of the queue and wait for the next one. The time complexity is $O(n)$ because it scans all the files linearly. The algorithm of the file-merge process is as follows:

Algorithm 1. The file-merge algorithm

INPUT: small files; mergeThreshold
OUTPUT: merged files

```
1:    for each file in files do
2:        String name = file.getName();
3:        for (int i = 0; i < fileName.length(); i ++)        // Test whether this is a
small file according to the tag;
4:            if ("tag" == fileName.substring(i, tag.length()+i))
5:                key = fileName.substring(tag.length()+i, fileName.length());
6:                Int type = Hash(decode(key));        // Get the file type by hash the
decoded key
7:                Insert Queue[type] (file);
8:                if (Queue[type].size() > mergeThreshold)
9:                    mergeFiles(file);
10:                   Queue[type].empty();
11:               else
12:                   Queue[type].size += file.size();
13:               end if
14:           end if
15:       end for
16:   end for
```

---

5   Name-Based Index

The index is widely used in the document retrieval. It is also perfectly suitable for files on HDFS. Here we analyze the composition of a file name. A name is a string made of many characters and usually not too long in general. In order to quickly match a name, we adopt the trie technology to build indexes for the irregular string.

5.1   Trie Building Process

Trie is an ordered tree data structure used to store an associative array where the keys are usually strings. In the binomial tree, the keys are directly stored in nodes but keys in trie are determined by the location of nodes in tree. All the children of a node have the same prefix, namely the string corresponding to the node. In general, not all of the nodes have corresponding values, only the leaf node and part of the internal nodes do. The reasons why we use the trie to build index are as follows:

1. Fast and efficient. If a path is from the root to a node, then the string corresponding to this node can be achieved by connecting all the characters along the path. So the time complexity of a string with n characters is $O(n)$ at worst.
2. Less storage space. Strings with the common prefix share the ancestor nodes.
3. Keys need not to be visual when stored in nodes, so it can effectively deal with string search problems.

Based on the advantages above, we take trie as the index structure and the main idea is to trade space for time. There are three kinds of trie: standard trie, compressed trie and the suffix trie. The compressed trie is the most suitable one because it compress single child node into edges and can further reduce the query time and storage space. We can turn a trie into a compressed one by the following rules:

- The node v_i is redundant, i = 1, 2, ..., k−1
- The node v_0 and v_k are not redundant
- Then we say that the list $(v_0, v_1)(v_1, v_2) ...(v_{k-1}, v_k)$ for k $\geq$ 2 is redundant.
- Replace the list with the single edge (v_0, v_k)

Figure 3 shows the trie of words {hadoop, have, hive, mind, mine} and its compressed type.

By this way, the $O(n)$ storage space decreases to O(s), in which n is the total length of all strings in standard trie and s is the number of strings in compressed trie.

This method for index building consumes much memory and this problem is more apparent for large amounts of small files. So we put the trie on disk in DataNode and just load it into memory whenever there is a request which is the same as traditional HDFS. There is a file on NameNode acts as a directory. It stores the file name and its physical location on DataNodes. This file is stored in disk and the trie is built according to it. When users want to access a file on HDFS, the client sends request to the NameNode for the physical location. The NameNode looks up the file by name in trie and return the block information including the IP and port of the DataNode and specific

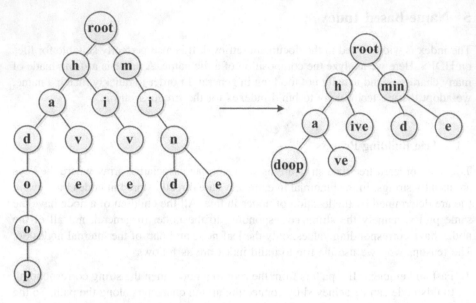

Fig. 3. Compressed trie

address. Then the client establishes connections with the clients and fetch the file according to the physical path.

5.2 Several Additional Supports

The name-based index can improve the efficiency of file access. The details are as follows. First, the client parses the extension, and makes index query in NameNode. Second, the NameNode returns a list of index blocks according to which the client access to the corresponding DataNode. Then the system searches the trie from the root node. The file is located only if the key is complete matching with the characters along one top-down path. What's more, the name-based index also supports some other operations such as store, delete and update.

The process of file storage is just the process of inserting a new string into the trie. The file is searched from the root along the edge by alphabetical order until the file name has been traversed. If the file name has not been found, the current node is extended and file is added to the corresponding block. If the size of file reaches the threshold, then the current data block is closed and another block will be created. When the NameNode is closed, the trie is written into the files on disk.

When a file is deleted from HDFS, the data block should be removed from the DataNode once the delete operations on NameNode is finished in theory. But the delete function only marks the data block to be deleted rather than remove it at once. What's more, the NameNode never contacts to the DataNodes and just listen. Only in the heartbeat response from DataNodes, the NameNode is able to delete the data block through commands. The index is marked as invalid and then the files under the index are also marked as invalid. When the files marked reach a specific threshold, the indexes and files are cleared up with the remains to be merged.

The update operation is a combination of writing and deleting files. The operations in the beginning are the same as deleting files which is finding the file locations on the basis of index. The file system returns error message. Or else it appends new data to the current file and marks the old files invalid. At the same time the index is updated and the old ones will be deleted at the right time.

6   Performance Evaluation

Our test platform is built on a cluster with three nodes. Each node has an Intel 8 CPU of 1.6 GHz, 32 GB memory and 1T SATA disk. The operating system is Ubuntu14.04. The Hadoop version is 2.6.0 with java of version 1.8.0. In the three nodes, one is master acting as the NameNode and SecondaryNameNode. The other two nodes are both slaves act as DataNode. In order to save the time of data transmission between nodes, the value of replication in hdfs-site.xml is set to 1 which means all the files need only to upload once.

6.1   Adaptive Storage Strategy Performance Evaluation

The data set is generated by the "dd" command in Ubuntu and then divided into small size. It is made up of 9536 files with the total size of 3.2 G. It has the same performance when the size is up to 6.4 G or even more. Most of the files are below 1 MB and those smaller than 100 K account for 89.25 %. We simulate the test by compressing files to nearly 64 MB manually and the performance is equal. Figure 4 shows the distribution of file size.

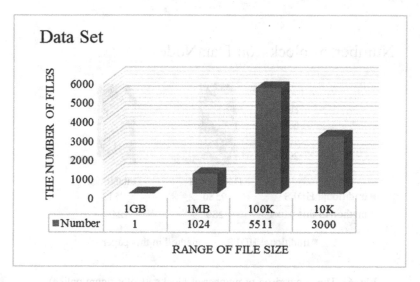

Fig. 4. Distribution of data

Figure 5 shows the time it consumes to upload the files on HDFS. It takes 1221 s to upload files on traditional HDFS and 750 s with the approach of this paper.

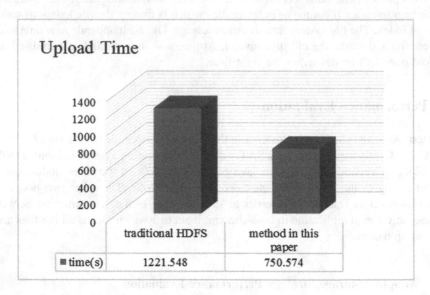

Fig. 5. The upload-time comparison

In terms of DataNode storage, the traditional HDFS applies for a data block for each small file. So the amounts of block is very large. But the number of blocks by method in this paper is much smaller. We can also infer that there is a decrease in the memory used in NameNode. The details are shown in Fig. 6.

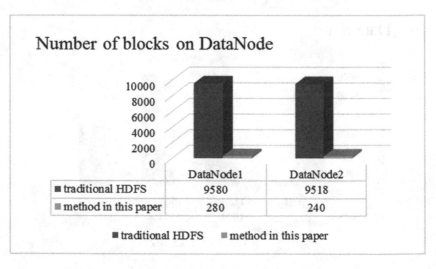

Fig. 6. The comparison of number of blocks. (Color figure online)

6.2   Name-Based Index Performance Evaluation

Figure 7 shows the time of reading file consumption by using two methods. Accessing times of the two DataNodes are 1.393 s and 1.405 s for traditional HDFS. While using the method in this paper, the times are 0.083 s and 0.171 s respectively.

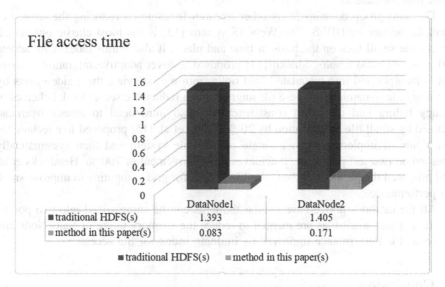

Fig. 7. The file access time comparison. (Color figure online)

7   Related Work

In the area of social computing, with the development of models and countermeasures for rumor spreading in Online Social Networks [7], and approximation algorithms for the routing problem [8–10], it is highly required that the distributed file system in cloud computing has high file-access efficiency.

HDFS can not well satisfies the current situation where large amounts of small files need to be managed. There are several methods to solve these problems at present that can be classified into three groups.

The first one is Hadoop Archives (HAR files) [11] which is file archiving tools that comes with Hadoop itself. Its emergence is to solve the memory consumption problem. HAR plays a role by building a hierarchical file system on HDFS. A HAR file can be created by the Hadoop command which runs a MapReduce task that packs small files into HAR format. But the efficiency of reading files can not be as high as reading files directly from HDFS. Moreover, the efficiency may be slightly lower to a certain extent, because each file access process contains reading operations of two layers of index and data. Although HAR files can be used as the input of MapReduce job, there is no special way for map to regard the file package as a HDFS file to handle. When users want to delete or update files, a new archive file should be created or else it will throw exceptions.

The second one is Sequence Files [12]. It is the common response to "the small file problem". This method sets filename as key and file contents as value so that many small files are organized and stored into sequence files. MapReduce can break them into chunks because the sequence files are divisible. It provides a good support for local data operation. However, the lack of maps between files makes the retrieval process of small files inefficient.

The third group contains many other research focusing on reducing the latency of small-file access on HDFS. The WebGIS system [13] is the most classic one which merges the small files on the basis of time and place. It also builds indexes for access and achieves good results. Shaikh [14] proposed a novel adaptive migration strategy that is incorporated into metadata-based optimization to alleviate these side effects by migrating file dynamically. These file migration can reduce the server load. Schemes of latency hiding and migration consistency are also introduced to reduce overhead induced by small file optimization by 20 %. Carns et al. [15] proposed five techniques which are all implemented in a single parallel file system and then systematically assessed on two test platforms. It achieves an improvement of 700 %. Hendricks et al. [16] proposed the protocol of the Chirp file system for grid computing to improve small file performance.

All the methods above have similar thoughts can be summarized into two points. The first is that small files are merged to reduce the memory usage in NameNode and the second is Performance improved by building index for file access.

8 Conclusion

Hadoop is a cluster made up of HDFS, MapReduce and Yarn which is now widely used in handling big data. But its short slab is low efficiency on dealing with large amounts of small files and high latency of data access. The NameNode is responsible for managing metadata for every block including the property and information about storage which will consume much memory. Large amounts of map tasks jam together. This paper studies the following contents to solve these problems. First we create new processing module to extract features and then merge those small files into large ones. Then we use file name to build index. The novelty of this approach is to build index by trie structure. It has a faster construction speed and files can be accessed by name easily. We achieve good result in file access operations.

For the future work, feature space are expected to build so that the compression process can be achieved according to the distance. Then the prefetch of files can be considered in which a buffer is used to further improve the access efficiency.

Acknowledgement. This paper was partially supported by National Sci-Tech Support Plan 2015BAH10F01 and NSFC grant U1509216,61472099,61133002 and the Scientific Research Foundation for the Returned Overseas Chinese Scholars of Heilongjiang Provience LC2016026.

References

1. Dong, X.: Hadoop Internals: In-depth Study of MapReduce. China Machine Press, China (2013)
2. Cai, B., Chen, X.: Hadoop Internals: In-depth Study of Common and HDFS, pp. 216–217. China Machine Press, Beijing (2013)
3. Murthy, A.C., Vavilapalli, V.K., Eadline, D., Niemiec, J., Markham, J.: Apache Hadoop YARN: Moving Beyond MapReduce and Batch Processing with Apache Hadoop 2. Pearson Education (2013)
4. Massive image storage and processing architecture in TaoBao. http://www.lxway.com/655880062.htm
5. http://baike.baidu.com/link?url=Mp9WPuBy-D0tKQeAIG-mJn9hk3UwS0fhkaQbm3xlcwx itswRHqVDjV5AfqmmJDK7NbvCzb3klb8ybFH9vkiGx_
6. Hadoop Distributed File System. http://hadoop.apache.org/docs/current/hadoop-project-dist/hadoop-hdfs/HdfsUserGuide.html
7. He, Z., Cai, Z., Wang, X.: Modeling propagation dynamics and developing optimized countermeasures for rumor spreading in online social networks. In: 2015 IEEE 35th International Conference on Distributed Computing Systems (ICDCS), pp. 205–214. IEEE (2015)
8. He, Z., Cai, Z., Cheng, S., et al.: Approximate aggregation for tracking quantiles and range countings in wireless sensor networks. Theor. Comput. Sci. **607**, 381–390 (2015)
9. Cai, Z., Lin, G., Xue, G.: Improved approximation algorithms for the capacitated multicast routing problem. In: Wang, L. (ed.) COCOON 2005. LNCS, vol. 3595, pp. 136–145. Springer, Heidelberg (2005)
10. Cai, Z., Goebel, R., Lin, G.: Size-constrained tree partitioning: a story on approximation algorithm design for the multicast k-Tree routing problem. In: Du, D.-Z., Hu, X., Pardalos, P.M. (eds.) COCOA 2009. LNCS, vol. 5573, pp. 363–374. Springer, Heidelberg (2009)
11. Hadoop archives. http://hadoop.apache.org/docs/current/hadoop-archives/HadoopArchives.html
12. Sequence File. http://hadoop.apache.org/docs/current/api/org/apache/hadoop/io/SequenceFile.html
13. Liu, X., Han, J., Zhong, Y., et al.: Implementing WebGIS on Hadoop: a case study of improving small file I/O performance on HDFS. In: IEEE International Conference on Cluster Computing and Workshops, CLUSTER 2009, pp. 1–8. IEEE (2009)
14. Shaikh, F., Chainani, M.: A case for small file packing in parallel virtual file system. http://www.andrew.emu.edu/user/mchainan/FinalPaper.pdf. Accessed 07 July 2007
15. Carns, P., Lang, S., Ross, R., et al.: Small-file access in parallel file systems. In: IEEE International Symposium on Parallel & Distributed Processing, IPDPS 2009, pp. 1–11. IEEE (2009)
16. Hendricks, J., Sambasivan, R.R., Sinnamohideen, S., et al.: Improving small file performance in object-based storage. Carnegie-Mellon Univ. Parallel Data Laboratory, Pittsburgh (2006)

Expanding Corpora for Chinese Polarity Classification via Opinion Paraphrase Generation

Da Pan, Jiaying Song, and Guohong Fu[(⊠)]

School of Computer Science and Technology, Heilongjiang University,
Harbin 150080, China
pandacs@live.cn, jy_song@outlook.com, ghfu@hotmail.com

Abstract. Although much progress has been made to date on sentiment classification, lacking annotated corpora remains a problem. In this paper we propose to expand corpora for Chinese polarity classification via opinion paraphrase generation. To this end, we first exploit three strategies for opinion paraphrase generation, namely sentences re-ordering, opinion element substitution and explicit attribution implying. To improve the quality of the generated opinion paraphrases, we define four criteria for opinion paraphrase evaluation and thus present a filtering algorithm to discard improper opinion paraphrase candidates. To assess the proposed method, we further apply the expanded corpus to a SVM classifier for polarity classification. The experimental results show that the generated opinion paraphrases are beneficial to polarity classification.

Keywords: Sentiment analysis · Polarity classification · Paraphrase generation · Supported vector machines

1 Introduction

Recently, opinion mining has been attracting lots of attention in the community of natural language processing. As a pivotal sub-problem of opinion mining, sentiment polarity classification aims to predict opinionated documents or sentences as showing positive, negative or neutral opinions. To data, much progress has been made in sentiment polarity classification. However, lacking large scale annotated corpora is still a major issue. On the one hand, statistically-based methods become the majority in sentiment polarity classification. In general, a statistically-based polarity classifier needs labeled corpora for training. On the other hand, large-scale labeled corpora are still not available for polarity classification. Furthermore, opinion mining is always domain specific. Obviously, it is time and cost consuming to manually construct a large scale labeled corpus for each domain.

Over the past years, paraphrasing has proven to be an effective tool for improve the coverage of natural language processing systems such as machine translation, information retrieval and question answering [1–4]. More recently, paraphrase generation has been applied to enhance sentiment polarity classification [5]. Unlike opinion corpus

© Springer Science+Business Media Singapore 2016
W. Che et al. (Eds.): ICYCSEE 2016, Part I, CCIS 623, pp. 362–373, 2016.
DOI: 10.1007/978-981-10-2053-7_32

annotation, paraphrases are relatively more flexible to acquire using different resources like synonym lexica, bilingual and parallel corpora, and so forth. Therefore, we believe that paraphrasing would be a feasible way to expand polarity-labeled corpora and at the same time, to alleviate the data sparse problem in statistically-base polarity classifiers.

Following the line of [5], in this paper we explore opinion paraphrase generation to expand existing opinion-labeled corpora for Chinese sentence-level polarity classification. To approach this, we first construct a large number of paraphrase candidates via three strategies, namely sentences re-ordering, opinion element substitution and explicit attribution implying. In order to improve the quality of the generated opinion paraphrases, we define four criteria for opinion paraphrase evaluation and thus present a filtering algorithm to discard the improper opinion paraphrase candidates. To assess the proposed method, we further apply the expanded polarity-labeled corpus to a supported vector machine (SVM) classifier for polarity classification. The experimental results show that the generated opinion paraphrases are beneficial to polarity classification.

The rest of the paper proceeds as follows. Section 2 provides a brief review of the literature on sentiment classification and paraphrase generation. Section 3 defines the goals of paraphrase generation. Section 4 details the proposed method to Chinese sentence polarity classification via paraphrase generation. Section 5 reports our experimental results. Finally, Sect. 6 concludes our work.

2 Related Work

Sentiment classification is a fundamental problem in opinion mining, which is usually formulated as a binary classification problem [6, 7]. Most previous studies use supervised machine learning methods, including naïve Bayes model, support vector machines (SVMs), maximum entropy models (MEMS), conditional random fields (CRFs), fuzzy set, neural networks and so forth, to perform sentiment classification on different linguistic levels such as words, phrases, sentences and documents [7–11].

Lacking large scale labeled corpora is one major problem in supervised machine learning methods. To address this problem, some recent studies exploit bootstrapping or unsupervised techniques [6, 12–16]. Unfortunately, unsupervised sentiment classifiers usually have worse performance than the supervised methods. To leverage resources in the corpora to improve the sentiment classification performance, paraphrase generation approaches have been investigated.

For paraphrase generation based sentiment classification, the pervious works are just use the opinion element substitution method to enrich the data set [5, 17]. Besides the strategy of opinion element substitution for opinion paraphrase generation in [5, 17], in the present study we also explore two other strategies, namely sentences re-ordering and explicit attribution implying so as to generate more potential candidates for opinion paraphrases. Furthermore, in order to acquire paraphrases of high quality, we propose four opinion paraphrase evaluation criteria to discard the improper opinion paraphrase candidates.

3 Goals of Paraphrase Generation

Paraphrase generation is aiming at generating a set of target sentences in the same semantic with a given source sentence. In the present study, a proper generated paraphrase should satisfy four kinds of requirements as follows.

Semantic Equivalence: We hope that the generated paraphrases would help improve performance of sentiment polarity classification. Thus, the expanded sentences must be semantically equivalent and thus have the same polarity with their source ones.

Grammaticality: The paraphrase generation task can be treated as a processes of sentence-making with a certain semantic constraints, such that the constructed sentence is required to meet the basic syntax.

Frequency of Use: From the perspective of language understanding, for ease of both humans and machines to understand the meaning of sentence, we suppose that extremely common words should be used in paraphrase generation for applicability of domains.

Diversity of Language: Since the purpose of paraphrase generation is to avoid the problem of data sparseness, we believe that the changes in style and sentence patterns can enrich original presentation of semantic and improve the quality of data.

4 The Proposed Method

Figure 1 presents the general framework for Chinese polarity classification via opinion paraphrase generation.

As shown in Fig. 1, we performance paraphrase generation over both training test datasets. In addition to opinion element substitution [5], we also introduce sentences re-ordering and explicit attribution implying to produce more potential paraphrase candidates and then employ the re-ranking model to filtering the improper ones.

4.1 Paraphrase Candidate Generation

From the viewpoint of completeness and accuracy, three strategies are exploited for opinion paraphrase candidate generation. Firstly, we re-order source sentence by semantic chunks division. Secondly, replace the evaluation phrases by paraphrase knowledge base, and then make the explicit attribution implying.

Sentence Re-ordering. Typically, product reviews are composed of several chunks of attributes and the relation between the chunks is usually parallel because of its characteristics of stronger generality and concise style. Thus we treat opinion chunks as basic unit of word order adjustment, the specific process of division as follows: (1) Cut source sentence into several clauses by punctuation marks; (2) Mark the product attributes according to the product knowledge base [5]; (3) Divide clauses without attributes into corresponding chunks; (4) Combine the chunks starting with conjunctions; (5) Full permutation of parallel chunks as paraphrase candidate.

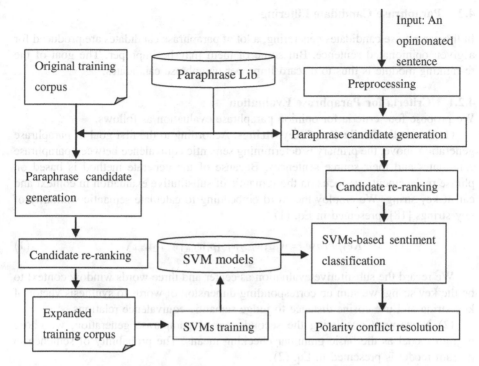

Fig. 1. The architecture of sentiment polarity classification based on paraphrase generation

Opinion Element Substitution. Since the study of domain-specific paraphrase generation is few, we construct a paraphrase knowledge base by the paraphrase recognition method by Fu *et al.* [5]. The generation approach which replacing evaluation phrase of similar attributes is simple but effective. Although the replacement method can avoid some grammatical errors and irrelevant information, but there are also some problems cannot be ignored. Due to the coarse granularity of paraphrase identification, there exists some noise in phrase knowledge base. Such as "屏幕大用着爽" (The screen is big and gives a cool feeling) and "屏幕分辨率高，看"着好清楚啊" (The resolution of screen is high and watched clearly) are both positive evaluation for the screen, but they expressed different semantics. Other noise is introduced because of the source sentence's polarity is unknown that the candidates we make may have opposite polarity.

Explicit Attribution Implying. It is observed that attributes are usually implied in real product reviews. For example, "机子挺好看的" (the machine looks nice) has an implied attribute of appearance. Usually, mining implied attributes is very difficult in information extraction in that it depends on the guess of relevance contexts. But in the task of paraphrase generation, it is very easy to construct sentences with implied attributes by just deleting original attributes during the process of evaluation phrases substitution.

4.2 Paraphrase Candidate Filtering

In the paraphrase candidates generating, a lot of paraphrase candidates are produced for a given opinionated sentence. But some of them may be improper. The goal of the re-ranking module is thus to discard improper paraphrase candidates.

4.2.1 Criteria for Paraphrase Evaluation

We propose four criteria for opinion paraphrase evaluation as follows.

(1) **Semantic Similarity of Key Strings.** According to the first goal of paraphrase generation above, the primary is determining semantic equivalence between paraphrase candidates and their source sentences. Because of the generate method is based on phrase replacement, we focus on the semantic of substitutive evaluation in context and call it key string. We employ the word embedding to calculate semantic similarity of key strings [18], presented in Eq. (1).

$$Keyphrase = w_{i-3}w_{i-2}w_{i-1}p_iw_{i+1}w_{i+2}w_{i+3} \tag{1}$$

We regard the substitutive evaluation as center and three words window context to be the key string, we sum up corresponding dimension of words to synthesis vector of key string and use cosine distance to judge semantic equivalence relations.

(2) **N-grams.** Considering the second goal of paraphrase generation, we chose n-gram model as the basic grammar checking means. The probability of sentence in n-gram model is presented in Eq. (2).

$$p(s) = \prod_{i=1}^{m} p(w_i|w_{i-n+1} \cdots w_{i-1}) \tag{2}$$

(3) **Usage Frequency of Phrases.** Follow the third goal we hope the substitute phrase is common in domain reviews and add common weight for all phrases in paraphrase knowledge base. TF-IDF is a popular representation of weight in data mining and information retrieval. At the meanwhile, the matching degree of evaluations and attributes pairs is an important factor which affect the paraphrase generation. For example, both "配置-太低" (configuration is too low) and "配置-太少" (configuration is too little) are opinion collocation in paraphrase knowledge base, but people prefer to use "太低" (too low) in reviews. We take use of TF-IDF and co-occurrence frequency of attribute-evaluation pairs to express common degree of evaluation phrases.

$$Score = TF\text{-}IDF \times count(att, val) = n_{i,j} \times \log\left(\frac{D}{d_i}\right) \times count(att, val) \tag{3}$$

Where, $n_{i,j}$ is the frequency of evaluation phrase *val* exist in corpus, D is the total sentence in corpus, d_i is count of sentences which have *val* in it and *count (att, val)* represent the co-occurrence frequency of attribute *att*, evaluation *val*.

(4) **Diversity of Key String Text.** The morphological of word is not a primary concern and is used as a secondary filtering condition. On the base of similar semantic

and correct grammar, in order to ensure the diversity of text we prefer to keep candidates whose morphological changes greatly. Thus we use Jaccard coefficient to measure the difference degree of key string (introduced in first assessment).

$$Score(morpho \log y) = 1 - jaccard(p_1, p_2) = 1 - \frac{|Set(p_1) \cap Set(p_2)|}{|Set(p_1) \cup Set(p_2)|} \quad (4)$$

Where, $Set(p_i)$ is the words set of key string p_i.

4.2.2 Strategies for Paraphrase Candidate Filtering

As we propose four evaluation criteria for opinion paraphrase evaluation above, two kinds of filtering strategies were designed to discard the candidates, namely the hierarchical filtering and the equal-Intersection filtering.

- **Hierarchical filtering**

These four evaluation criteria were sorted in line with major and minor standards. We set the most major standards called rank1, a slight secondary standards called rank2 and later followed by rank3, rank4. In filtering, we first sort all rank1 scores from highest to lowest and kept topN1 candidates. Then sort all candidates depending on rank2 scores and cut the rear ones except topN2. TopN3 and Top n-best candidates were kept later according to rank3, rank4 standards. Considering the first three criteria score are both very import, we design six kinds of order lists as follow. Using NG-score represents N-gram grammar assessment score, COS-score represents semantic similarity score, COMMON-score represents common degree score and JACC score represents difference degree score. We set N1 300, N2 150, N3 50 in experiments of this screening strategy.

    (1) order-1: NG-score → COS-score → COMMON-score → JACC-score

    (2) order-2: NG-score → COMMON-score → COS-score → JACC-score

    (3) order-3: COS-score → NG-score → COMMON-score → JACC-score

    (4) order-4: COS-score → COMMON-score → NG-score → JACC-score

    (5) order-5: COMMON-score → COS-score → NG-score → JACC-score

    (6) order-6: COMMON-score → NG-score → COS-score → JACC-score

- **Equal-Intersection screening**

We suppose that all evaluation criteria is equal and necessary in this strategy and does not distinguish major or minor standards. Thus four lists of candidates in different order received by these criteria. We cut out rear half of each list and generate the final paraphrases using the intersection of remainders in four lists.

4.3   Sentiment Polarity Classification

With the paraphrase generation, we expand annotated corpus and obtain a larger scale annotated data. Whether the expanded corpus is useful for sentiment polarity classification or introduce too much noise would be verified by the sentiment classification experiment. After paraphrase generation model, we obtain *k-best* paraphrases for a review and predict polarity of these paraphrases by the polarity classifier. We need to avoid polarity conflict when the polarities of all paraphrases are in different type. In conflict resolution, we resolve it the same as Fu *et al.* [5] by voting mechanism. We totally have $k + 1$ opinionated sentence for polarity classification. Let i $(0 \leq i \leq k)$ be the number of sentences that are classified as positive by the system and j $(0 \leq j \leq k,\ and\ i + j = k)$ be the number of sentences that are negative during polarity classification. Thus we can take the following three rules to determine the final polarity of the original sentence.

- Rule 1. if $i > j$, then the final polarity is positive.
- Rule 2. if $i < j$, then the final polarity is negative.
- Rule 3. if $i = j$, the final polarity is same as that the original polarity of the input sentence during polarity classification.

5   Experiment Results and Discussion

To assess our approach, we have exploited the proposed paraphrase generation to two corpora of product reviews in car and cellphone domains, and further developed a SVM-based sentiment polarity classifier. This section reports the relevant experimental results.

5.1   Experimental Dataset

Table 1 shows the statistic of the experiment data.

Table 1. Statistic of the experimental data

Data	Car			Cellphone		
	Total	POS	NEG	Total	POS	NEG
Training set	2667	1318	1349	2668	1318	1350
Development set	333	183	150	332	182	150
Test set	1000	600	400	1000	600	400

5.2   Effects of Different Strategies on the Number of Generated Paraphrases

We construct a large number of paraphrase candidates for train set, develop set, test set based on sentence re-ordering and phrases replacing. The statistic of expanded corpus is listed in Table 2. The statistic show that scale of paraphrase we generated is quite large and proved our paraphrase generation method can construct a more complete candidate set. Such paraphrase candidate also shows that it may contain a lot of noise, the filtering step become quite necessary.

The filtering experiment is carried out according to the two strategies above and the statistics of 10-best paraphrase corpora of two methods are show in Tables 3 and 4.

Table 2. Statistics of the corpora with paraphrase

	Datasets	Total sentence	Positive sentence	Negative sentence
Car	Training set	466015366209	318679697922	147335668287
	Development set	54022308180	46323198637	7699109543
	Test set	144248170657	104827585126	39420585531
Cellphone	Training set	447955971558	266669529917	181286441641
	Development set	138582898809	122152527307	16430371502
	Test set	99247743050	69570802365	29676940685

Table 3. Statistics of the corpora with 10-best Rank-Cut paraphrase

	Data	Total sentences	Positive sentences	Negative sentences
Car	Training set	25160	12972	12188
	Development set	3389	1851	1538
	Test set	9990	6118	3872
Cellphone	Training set	26286	13221	13065
	Development set	3352	1898	1454
	Test set	9617	5847	3770

Table 4. Statistics of the corpora with 10-best Equal-Intersection paraphrase

	Datasets	Total sentences	Positive sentences	Negative sentences
Car	Training set	20915	11083	9832
	Development set	2974	1663	1311
	Test set	7954	4931	3023
Cellphone	Training set	21131	10749	10382
	Development set	2866	1689	1177
	Test set	7504	4557	2947

5.3 Effects of Different Paraphrase Generation on Polarity Classification

Because we have six orders in hierarchical filtering and the scale of corpora is quite different when retaining different n-best strategies. We consider five kinds of n-best paraphrases are investigated, namely 20-best, 10-best, 5-best, 2-best and 1-best. The performance of these sentiment classification experiments on development data sets are shown in Tables 5, 6, 7, 8 and 9.

Table 5. Effects of 20-best paraphrase generation on sentiment classification

Paraphrase		Order-1	Order-2	Order-3	Order-4	Order-5	Order-6	Intersection
Car	F_{pos}	**0.8777**	**0.8777**	0.8647	0.8647	0.8483	0.8483	0.8679
	F_{neg}	**0.8414**	**0.8414**	0.8235	0.8235	0.7870	0.7870	0.8339
	Acc	**0.8619**	**0.8619**	0.8468	0.8468	0.8228	0.8228	0.8529
Cellphone	F_{pos}	0.8764	0.8764	**0.8883**	**0.8883**	0.8780	0.8780	0.8711
	F_{neg}	0.8571	0.8571	**0.8693**	**0.8693**	0.8475	0.8475	0.8562
	Acc	0.8675	0.8675	**0.8795**	**0.8795**	0.8645	0.8645	0.8640

Table 6. Effects of 10-best paraphrase generation on sentiment classification

Paraphrase		Order-1	Order-2	Order-3	Order-4	Order-5	Order-6	Intersection
Car	F_{pos}	0.8757	0.8757	0.8639	0.8639	0.8504	0.8504	**0.8794**
	F_{neg}	0.8446	0.8446	0.8169	0.8169	0.8000	0.8000	**0.8464**
	Acc	0.8619	0.8619	0.8438	0.8438	0.8288	0.8288	**0.8649**
Cellphone	F_{pos}	0.8858	0.8858	**0.8864**	**0.8864**	0.8798	0.8798	0.8711
	F_{neg}	**0.8656**	**0.8656**	0.8647	0.8647	0.8523	0.8523	0.8562
	Acc	**0.8765**	**0.8765**	**0.8765**	**0.8765**	0.8675	0.8675	0.8640

Table 7. Effects of 5-best paraphrase generation on sentiment classification

Paraphrase		Order-1	Order-2	Order-3	Order-4	Order-5	Order-6	Intersection
Car	F_{pos}	**0.8930**	**0.8930**	0.8602	0.8717	0.8421	0.8421	0.8825
	F_{neg}	**0.8630**	**0.8630**	0.8231	0.8356	0.7902	0.7902	0.8410
	Acc	**0.8799**	**0.8799**	0.8438	0.8559	0.8198	0.8198	0.8649
Cellphone	F_{pos}	**0.8895**	**0.8895**	0.8840	0.8840	0.8635	0.8635	0.8718
	F_{neg}	**0.8675**	**0.8675**	0.8609	0.8609	0.8393	0.8393	0.8553
	Acc	**0.8795**	**0.8795**	0.8735	0.8735	0.8524	0.8524	0.8640

Table 8. Effects of 2-best paraphrase generation on sentiment classification

Paraphrase		Order-1	Order-2	Order-3	Order-4	Order-5	Order-6	Intersection
Car	F_{pos}	0.8865	0.8865	0.8717	0.8717	0.8429	0.8429	**0.8930**
	F_{neg}	0.8581	0.8581	0.8356	0.8356	0.7887	0.7887	**0.8630**
	Acc	0.8739	0.8739	0.8559	0.8559	0.8198	0.8198	**0.8799**
Cellphone	F_{pos}	**0.8736**	**0.8736**	0.8674	0.8674	0.8571	0.8571	0.8669
	F_{neg}	0.8467	0.8467	0.8411	0.8411	0.8267	0.8267	**0.8479**
	Acc	**0.8614**	**0.8614**	0.8554	0.8554	0.8434	0.8434	0.8580

As can be seen from Tables 5, 6, 7, 8 and 9, the re-ranking model performs well while using hierarchical filtering with order-1, order-2, order-3 and order-4 strategies. This illustrates that different paraphrase evaluation criteria have different importance in filtering improper paraphrases. So we need to determine the filtering order according to

Table 9. Effects of 1-best paraphrase generation on sentiment classification

Paraphrase		Order-1	Order-2	Order-3	Order-4	Order-5	Order-6	Intersection
Car	F_{pos}	0.8721	0.8721	0.8602	0.8602	0.8610	0.8610	**0.8760**
	F_{neg}	0.8269	0.8269	0.8231	0.8231	0.8207	0.8207	**0.8362**
	Acc	0.8529	0.8529	0.8438	0.8438	0.8434	0.8434	**0.8589**
Cellphone	F_{pos}	0.8774	0.8774	0.8387	0.8387	0.8556	0.8556	**0.8785**
	F_{neg}	0.8485	0.8485	0.7945	0.7945	0.8267	0.8267	**0.8533**
	Acc	0.8645	0.8645	0.8193	0.8193	0.8424	0.8424	**0.8671**

theirs importance. Our experimental results show that the order of importance for paraphrase evaluation criteria should be grammaticality, semantic equivalence, frequency of use, and diversity of language.

The strategy of Equal-Intersection also performs very well in several experiments. Through these experiments we determined the best strategies of two domains. The following experiment intends to investigate the effects of different paraphrase generation methods in sentiment classification. The former generation of Fu [5] were used as baseline and take n-gram score as the only criterion to acquire n-best paraphrase. We applied all paraphrase generated in baseline method and found the optimal results in 1-best to 20-best of baseline.

5.4   Comparison Results of Different Methods for Polarity Classification

The results are summarized in Table 10. The result show that if generate a large number of paraphrase only without any filtering, the result is not as good as classification on original corpus. It may show that resource in paraphrase knowledge base is unbalance and result in the number of two polarity samples is too unbalanced to modeling. On the other hand, all these paraphrase generation are on the base of baseline

Table 10. Comparison of polarity classification with/without paraphrase generation

Paraphrase		No-para	Base-para	Base-n-best	Our-n-best
Car	P_{pos}	0.8582	0.8283	0.8635	0.8785
	R_{pos}	0.8067	0.8117	0.8433	0.8433
	F_{pos}	0.8316	0.8199	0.8533	**0.8605**
	P_{neg}	0.7339	0.7257	0.7729	0.7778
	R_{neg}	0.8000	0.7475	0.8000	0.8246
	F_{neg}	0.7656	0.7365	0.7862	**0.8005**
Cellphone	P_{pos}	0.9394	0.9005	0.9372	0.9579
	R_{pos}	0.8267	0.8450	0.8700	0.8717
	F_{pos}	0.8794	0.8719	0.9023	**0.9127**
	P_{neg}	0.7797	0.7872	0.8239	0.8304
	R_{neg}	0.9200	0.8600	0.9125	0.9425
	F_{neg}	0.8440	0.8219	0.8660	**0.8829**

method, but after the sentence re-ordering and adding multiple criteria, we enriched the scale of candidate to ensure completeness and the filtering help us obtained better results. The filtering of baseline only assessed the reasonableness of grammar but did not evaluate semantic similarity from the essence of paraphrase.

6 Conclusion

In this paper, we have presented a paraphrase generation based method to corpus expansion for Chinese polarity classification. In particular, we introduce three strategies, namely sentences re-ordering, opinion element substitution and explicit attribution implying to produce potential paraphrases for a given opinionated sentence, and thus exploited four criteria to opinion paraphrase evaluation and filtering. We have also evaluated the proposed method under the framework of SVMs over two corpora of product reviews. The experimental results show that using opinion paraphrase generation is of great value to polarity classification.

Acknowledgments. This study was supported by Natural Science Foundation of Heilongjiang Province under Grant No. F2016036, National Natural Science Foundation of China under Grant No. 61170148, and the Returned Scholar Foundation of Heilongjiang Province, respectively.

References

1. Bhagat, R., Hovy, E.: What is a paraphrase? Comput. Linguist. **39**(3), 463–472 (2013)
2. Heilman, M., Smith, N.A.: Tree edit models for recognizing textual entailments, paraphrases, and answers to questions. In: Proceedings of NAACL 2010, pp. 1011–1019 (2010)
3. Zhao, S., Lan, X., Liu, T., et al.: Application-driven statistical paraphrase generation. In: Proceedings of ACL-IJCNLP 2009, pp. 834–842 (2009)
4. Fader, A., Zettlemoyer, L., Etzioni, O.: Paraphrase-driven learning for open question answering. In: Proceedings of ACL 2013, pp. 1608–1618 (2013)
5. Fu, G., He, Y., Song, J., Wang, C.: Improving Chinese sentence polarity classification via opinion paraphrasing. In: Proceedings of CLP 2014, pp. 35–42 (2014)
6. Turney, P.D.: Thumbs up or thumbs down?: semantic orientation applied to unsupervised classification of reviews. In: Proceedings of ACL 2002, pp. 417–424 (2002)
7. Pang, B., Lee, L.: Opinion mining and sentiment analysis. Found. Trends Inf. Retrieval **2**(1–2), 1–135 (2008)
8. Pang, B., Lee, L., Vaithyanathan, S.: Thumbs up?: Sentiment classification using machine learning techniques. In: Proceedings of EMNLP 2002, pp. 79–86 (2002)
9. Fu, G., Wang, X.: Chinese sentence-level sentiment classification based on fuzzy sets. In: Proceedings of COLING 2010, pp. 312–319 (2010)
10. Mikolov, T., Yih, W.-T., Zweig, G.: Linguistic regularities in continuous space word representations. In: Proceedings of HLT-NAACL 2013, pp. 746–751 (2013)
11. Tang, D., Wei, F., Yang, N., et al.: Learning sentiment-specific word embedding for Twitter sentiment classification. In: Proceedings of ACL 2014, pp. 1555–1565 (2014)

12. Mihalcea, R., Banea, C., Wiebe, J.: Learning multilingual subjective language via cross-lingual projections. In: Proceedings of ACL 2007, pp. 976–983 (2007)

13. Wilson, T., Wiebe, J., Hoffmann, P.: Recognizing contextual polarity in phrase-level sentiment analysis. Comput. Linguist. **35**(3), 399–434 (2009)

14. Speriosu, M., Sudan, N., Upadhyay, S., et al.: Twitter polarity classification with label propagation over lexical links and the follower graph. In: Proceedings of the First workshop on Unsupervised Learning in NLP, pp. 53–63 (2011)

15. Mehrotra, R., Agrawal, R., Haider, S.A.: Dictionary based sparse representation for domain adaptation. In: Proceedings of CIKM 2012, pp. 2395–2398 (2012)

16. Volkova, S., Wilson, T., Yarowsky, D.: Exploring sentiment in social media: bootstrapping subjectivity clues from multilingual twitter streams. In: Proceedings of ACL 2013, pp. 505–510 (2013)

17. Song, J., He, Y., Fu, G.: Polarity classification of short product reviews via multiple cluster-based SVM classifiers. In: Proceedings of PACLIC 2015, pp. 267–274 (2015)

18. Mikolov, T., Sutskever, I., Chen, K., et al.: Distributed representations of words and phrases and their compositionality. In: Advances in Neural Information Processing Systems, pp. 3111–3119 (2013)

Feature Extraction for Effective Content-Based Cloth Image Retrieval in E-Commerce

Lingli Li$^{(\boxtimes)}$ and JinBao Li

Heilongjiang University, Harbin, China
Lilingli_grace@163.com, jbli@hlju.edu.cn

Abstract. Cloth image retrieval in E-Commerce is a challenging task. In this paper, we propose an effective approach to solve this problem. Our work chooses three features for retrieval: (1) description (2) category (3) color features. It can handle clothes with multiple colors, complex background, and model disturbances. To evaluate the proposed method, we collect a set of women cloth images from Amazon.com. Results reported here demonstrate the robustness and effectiveness of our retrieval method.

Keywords: Information retrieval · E-Commerce · Content-based image retrieval

1 Introduction

In recent years, with the rise of E-Commerce and the perfection of logistics services, online shopping has gradually replaced the traditional shopping way, and becomes a new fashion. Throughout famous shopping websites, such as http://www.like.com and http://www.amazon.com/, they all provide the users product retrieval service. However, most of them provide keyword-based or category-based interface for search. Such search interfaces are efficient for the products which could be identified with the name or the category, such as books and digital products. However, they are not suitable for clothes, since the major features of clothes are visual ones and difficult to express in words. For example, it is hard to describe some clothes designed with complex patterns and multiple colors by any keywords or category. Thus, visual information should be taken into consideration for clothes search in E-Commerce and Clothes REtrieval Based on Image(CREBI) is in demand.

CREBI brings following technical challenges.

(1) The first challenge is how to describe the visual features of a cloth image. Many works have been devoted to representing local features like sift, surf, and etc. A study from Nanjing Univ. of Aeronaut. & Astronaut. [6] analyzes the effects of different local features on product image retrieval. It was tested on 12 kinds of product, among which glasses and bikes have best performance. They only extracted the features from certain regions that best describe the product. However, unlike these products mentioned above, cloths images are more complicated since customers are willing to look at global visual features to make decision. Therefore, local features are insufficient to meet the needs in CREBI.

© Springer Science+Business Media Singapore 2016
W. Che et al. (Eds.): ICYCSEE 2016, Part I, CCIS 623, pp. 374–383, 2016.
DOI: 10.1007/978-981-10-2053-7_33

(2) Second, which global feature could have a better performance applied into cloth image retrieval? Global features include texture, color and shape, etc. Many researchers have involved studies about texture and shape and found it is easy to extract these features from cloth images [4, 7], but the extraction lacks precision since E-Commerce cloth is displayed in variant shape and its soft property greatly affects the extraction. Also, it is hard to speed up the matching process[8, 9] and has a bad user experience of real-time search which is critical for E-Commerce sites.

(3) Cloth image is complicated and its background is likely to be mistakenly considered as part of the cloth during the feature extraction. Various light conditions, complex patterns, and model's skin color also increase the difficulty of feature identifications.

According to these challenges, the crucial step for effective CREBI is to extract useful features effectively. Thus in this paper, we focus on the extraction of features that are suitable for CREBI in E-Commerce. We have following contributions.

(1) With the consideration of both effective and efficiency issues, we choose Categories, Description, and Color as the features for CREBI in E-Commerce. Experimental results demonstrate that such features are sufficient for retrieving the most relevance results.

(2) As to extract the color feature from the images in E-Commerce that may contain models and other confusing noise such as mottled gray background, we develop two methods. The former attempts to remove the color same as human skin to erase the disturbance of models. The latter removes the background of clothes by threshold-based filtering strategy. Experimental results show that these two methods could increase the quality of retrieval significantly.

(3) For efficient and effective CREBI, to reduce the size of candidates and accelerate the retrieval, we apply a hierarchical filtering method to retrieve similar cloth images by color feature, text description and category.

The paper is organized as follows. Section 2 introduces the framework of our system. Section 3 focuses on the reasons of the selected features. Section 4 highlights the process of extracting features. Experiment results are summarized in Sect. 5. Section 6 concludes the work.

2 Framework

We implement a system to search similar product image according to search inputs which could be the arbitrary combination of image, keywords and category. To balance the efficiency and effectiveness, we select three features: color, text descriptions, and categories. The details of feature selection will be discussed in Sect. 3.

Our approach first extracts an 11-dimensional color vector from each image in dataset. As illustrated by Fig. 1, we first upload a query cloth image as input. Then, the color vector V for the input image is extracted. We retrieve top-K images with the color

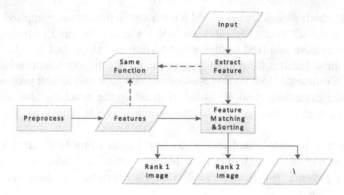

Fig. 1. Flow chart of image retrieval system

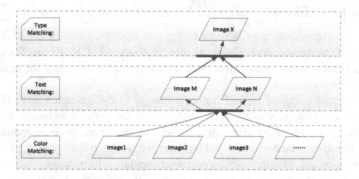

Fig. 2. Three layers of feature matching.

vectors the nearest to V in term of Euclidean distances as in Eq. 1, and could quickly output the orders of corresponding images.

$$\text{dist}(\vec{u}, \vec{v}) = \sqrt[2]{\sum\nolimits_{i=1}^{n} \left(u^{(i)} - v^{(i)}\right)^2} \tag{1}$$

Meanwhile, in order to improve the precision of searching results, we also support query by text description and category, and we have made some betterment based on that. Therefore, every retrieval result of our system is filtered in the order of the efficiency of the filtering, that is, text description, categories, and color features, as shown in Fig. 2. In practice, if we first search by words and categories, it could help us narrow the search scope which may greatly accelerate the search algorithm.

3   Feature Selection

In this section, we will discuss the motivations of selected features, that is description text, category and color feature.

First, we use global features instead of local ones. Most local features are applied into the detection of objects with some nice properties like scale invariance, rotation invariance [10–12] and etc. However, they have a bad performance in cloth images of E-Commerce for the reason that their detection could be easily affected by some unique properties of cloth image. Issues, like variant light source and cloth model, will make negative contributions to local feature detection and matching. As shown in Fig. 3, surf feature [11] matching on E-Commerce images may mislead the matching. It detection mainly concentrates on the matching of model's head and leg in cloth image, and fails to match the clothes. However, most of E-Commerce images are displayed with models wearing cloth product, so local features could not satisfy our need.

It is known that global features involve texture, shape, and color. Among them, we choose color feature as the global visual feature, due to the following two reasons.

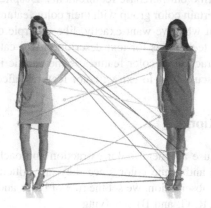

Fig. 3. SURF extraction and matching.

5588@Strapless Chiffon Short Dress Mercury.jpg   5589@Strapless Chiffon Short Dress Petal.jpg   5590@Strapless Chiffon Short Dress Clover.jpg   5591@Strapless Chiffon Short Dress Black.jpg   5592@Strapless Chiffon Short Dress Malibu.jpg   5593@Strapless Chiffon Short Dress Canary.jpg   5594@Strapless Chiffon Short Dress Apple.jpg   5595@Strapless Chiffon Short Dress Watermelon.j...

Fig. 4. Image background mixed with color

(1) Most cloth images contain both cloth and model since online shop owners ask models to try on their clothes in order to provide an overall effect for customers and motivate their purchases. In this case, the cloth image may be very diverse and have different outlooks. The model's body and pose will affect the extraction of shape feature and texture feature in a great extent.

(2) Since clothes are deformable, shadows and wrinkles may be confused as part of the texture pattern. Arbitrary shooting distance and viewing direction when taking

the cloth pictures present a variant shape of cloth. Some cloth pictures are taken as full-length portrait while some are half-length. The issues above may cause many matching errors for texture and shape. In contrast, color is invariance with the position and angle of the cloth in the image.

From above discussion, for cloth images, color feature is relative stable and the cloth shape has little influence on its color. As a result, we use color vector to represent the color feature. This vector contains 11 dimensions, each entry of which is the percent of the pixels with corresponding color. Besides 8 colors in [3], we divide yellow into orange, light orange and yellow, and blue into blue and cyan.

However, color is required but only color may lead to a low accuracy. Text descriptions and category are also in great need of specifying a cloth image. They, together, could help us improve the search precision. And the category is particularly important for us to describe our purchase requirements. Despite the fact that we could filter the clothes into a certain color group with their color feature, without category, we still could not search out what we want exactly, like a purple dress or a purple coat.

Since the extraction techniques for text description and category are well-studied [1, 2]. The effective extraction of color features is the centric issue of CREBI. In the next section, we will discuss how to extract color features effectively.

4 Feature Extraction

For a given image, we use a general color extraction approach [3]. Firstly, for RGB space, we extract black and white color since they are difficult to extract from HIS space. According to our observation, we set the rule of black and white color extraction as follows. For a color (R, G, and B) satisfying:

$$|R - B| \leq 3 \,\&\&\, |R - G| \leq 3 \,\&\&\, |G - B| \leq 3$$
$$\text{where } R, G, B \epsilon [0, 255] \tag{2}$$

It is treated as white or black. While all the values of (R, G, and B) $\in (215, 255]$ denote white, others are black.

Then we extractor other colors in HIS space. We detect colors based on saturation S, luminance I, and hue H [3]. Since the rest pixels are not white or black, we only employ the hue information for the other 9 entries in the color vector, Eq. 3 [5] shows the transferring method and θ. dotes the angle of Hue value.

$$\theta = \cos^{-1}\left(\frac{R - G + (R - B)}{\sqrt{(R - G)^2 + (R - B) + (G - B)}} \right)$$

$$H = \begin{cases} \theta & G \geq B \\ 2\pi - \theta & G < B \end{cases} \tag{3}$$

Algorithm 1. Color Percentage Computation

Input: color percentage after previous step
Output: final color percentage
For percentage (white)
IF percentage (white) > threshold (t) **THEN**

$$\text{Percentage(white)} = \frac{white - area * t}{area * (1 - t)}$$

ELSE

$$\text{Percentage(white)} = 0$$

FORother colors$c$**DO**

$$\text{Percentage}(c) = \frac{c}{area * (1 - t)}$$

return Percentage

As shown in Fig. 5, hue is displayed as a 360 °color wheel. We define the color "red" between 0–9 ° and 345 °–360 °, "orange" in the range of 9 °–42 °, "light orange" between 42 °–62 °, "yellow" between 62 °–75 °, "green" between 75 °–160 °, "cyan" between 160 °–210 °, "blue" between 210 °–280 °, "purple" between 280 °–315 °, "pink" between 315 °–345 °.

Intuitively, the extraction of color is trivial. However, the complexity in the images in E-Commerce brings following difficulties in this step.

(1) Models disturbance
(2) Confusing background noise

Fig. 5. Hue representation in HIS color (Color figure online)

Therefore, in this section, we focus on the solution of these difficulties. After our approaches are applied, the color features could be extracted directly with the methods in description, category, and color feature.

4.1 Remove Skin Color

Intuitively, clothing manufactures will not produce clothes that have the same color as people's skin. Hence, we figure out the color spectrum of people's skin and get rid of colors in this scope (Eq. 4 and 5). We define the following scopes as skin color [5].

$$R \in [215, 240] \,\&\&\, G \in [160, 190] \,\&\&\, B \in [120, 180] \,\&\&\, G - B \geq 12 \qquad (4)$$

$$R \in [100, 200] \,\&\&\, G \in [55, 160] \,\&\&\, B \in [0, 145] \,\&\&\, G - B \geq 12 \,\&\&\, R - B \geq 30$$
$$(5)$$

To sum up, a certain dimension of the color vector space satisfies equation as follows.

$$Percentage = color/(area - Backgrond\_White - skin) \qquad (6)$$

Here *Percentage* is how much a certain color accounts for in an image; *area* reflects image's pixels; *Backgroud_White* represents the pixels of white background; *skin* is the area of people's skin.

4.2 Remove Background Noise

As showed in Fig. 4, these images are bright and colorful, but their background is not pure white and mixed with some color noises. In order to solve this, we select 500 representative images for a statistical analysis. And we finally find out that this kind of "bad" image accounts for average 60 % (the threshold) pixels in an image of all the research samples. Therefore, we revise the process step above as following algorithm.

Fig. 7. Results without category.

Fig. 6. .

Fig. 8. Results with category.

Fig. 9. Sample image for model

5   Experimental Results

5.1   Databases

The robustness and effectiveness of our approach are evaluated on 21598 pieces of women cloth image from Amazon.com. Categories cover seven kinds: Wedding dresses, Wear to Work, Special Occasion dresses, Jackets & Coats, Casual, Cardigans Sweaters, and Active Shirts & Tees. We explore on colors including white, yellow, red, purple, black, and even multiple color. Some of the clothes are dressed by models, while others are not. Most of them have a pure white background, while others may be represented as mottled gray. The dimension of each image is around 385*500 pixels. Since for a large image set, the result set may be very large. Thus a user cares precision more than recall. Therefore, we only use precision to test the quality of search results.

5.2   The Effectiveness of Selected Features

We return 15 results from each retrieval tasks according to color matching. After testing for about 500 times, our method achieves 93 % accuracy rate. Some successful results of CREBI are shown in Figs. 7 and 8. Figure 6 is an example of our input image. Comparing the retrieval outputs in Figs. 7 and 8, it presents a list of images from the same retrieval with category search and without respectively. This show the selection of category could help to improve the quality of search result.

5.3   The Effectiveness of Feature Extraction

Model Disturbance. In our test, we evaluated model's skin color's influence on the proposed method. The testing results of cloth image (showed in Fig. 9) are summarized in Table 1. After our revision, the percentage of yellow in the image greatly reduces. In this case, model's disturbance is minimized.

Table 1. Comparison before and after skin color removal.

Colors	Before remove skin	After remove skin
White	0.00	0.00
Black	51.17	74.32
Yellow	43.71	18.52
Red	3.82	5.26
Blue	0.03	0.04
Green	0.14	0.21
Purple	0.12	0.17
Pink	0.69	1.00
Cyan	0.34	0.49

Fig. 11. Before removing noises.

Fig. 10. .   **Fig. 12.** After removing noises.

5.4 Background Noise

After removing the background noises, we could see an obvious improvement displayed in Fig. 11, and Fig. 12 when we upload a query image showed in Fig. 10.

5.5 Summary

In summary, we have following conclusions from the experimental results.

(1) Text description, category, and color features are suitable for CREBI in E-Commerce.
(2) Our image processing methods could improve the accuracy of CREBI in E-Commerce significantly.

6 Conclusions and Future Work

Content-based image retrieval for cloth plays an important role in E-Commerce. However, current image-retrieval techniques are not suitable for the clothes images on E-Commerce web sites. To solve this problem, we propose a Clothes REtrieval Based on Image (CREBI) method in this paper. We select suitable features for CREBI and design image processing strategies to increase the accuracy for the CREBI. Experimental results demonstrate that proposed method is suitable for CREBI. Further work include scale the methods to large image set and involve new features for more effective CREBI.

Acknowledgment. This paper was partially supported by National Sci-Tech Support Plan 2015BAH10F01 and NSFC grant U1509216,61472099,61133002. The Scientific Research Foundation for the Returned Overseas Chinese Scholars of Heilongjiang Provience LC2016026.

References

1. Tapeh, A.G., Rahgozar, M.: An ontology-based semantic extraction approach for B2C ecommerce. Int. Arab J. Inf. Technol. **8**(2), 163–170 (2011)
2. Ding, Y., Korotkiy, M., Omelayenko, B., et al.: Goldenbullet: automated classification of product data in e-commerce. In: Proceedings of the 5th International Conference on Business Information Systems (2002)
3. Yuan, S., Tian, Y.L., Arditi, A.: Clothing matching for visually impaired persons. Technol. Disabil. **23**(2), 75–85 (2011)
4. Guo, Z., Zhang, L., Zhang, D.: Rotation invariant texture classification using LBP variance (LBPV) with global matching. Pattern Recogn. **43**(3), 706–719 (2010)
5. Vezhnevets, V., Sazonov, V., Andreeva, A.: A survey on pixel-based skin color detection techniques. Proc Graphicon **3**, 85–92 (2003)
6. Zou, G., Ma, R., Ding, J., et al.: Research on product image retrieval with local features. In: 2010 Chinese Conference on Pattern Recognition (CCPR), pp. 1–5. IEEE (2010)
7. Tseng, C.H., Hung, S.S., Tsay, J.J., et al.: An efficient garment visual search based on shape context. WSEAS Trans. Comput. **8**(7), 1195–1204 (2009)
8. Wu, Q., Yu, Y.: Feature matching and deformation for texture synthesis. ACM Trans. Graph. (TOG) **23**(3), 364–367 (2004)
9. Gong, M., Yang.Y.H.,: Near real-time reliable stereo matching using programmable graphics hardware. In: IEEE Computer Society Conference on Computer Vision and Pattern Recognition, CVPR 2005, pp. 1:924–1:931. IEEE (2005)
10. Ke, Y., Sukthankar, R.: PCA-SIFT: a more distinctive representation for local image descriptors. In: Proceedings of the 2004 IEEE Computer Society Conference on Computer Vision and Pattern Recognition, CVPR 2004, vol. 2, pp. II-506–II-513. IEEE (2004)
11. Bay, H., Tuytelaars, T., Van Gool, L.: SURF: speeded up robust features. In: Leonardis, A., Bischof, H., Pinz, A. (eds.) ECCV 2006, Part I. LNCS, vol. 3951, pp. 404–417. Springer, Heidelberg (2006)
12. Juan, L., Gwun, O.: A comparison of sift, pca-sift and surf. Int. J. Image Process.(IJIP) **3**(4), 143–152 (2009)

Fundus Lesion Detection Based on Visual Attention Model

Baisheng Dai[1], Wei Bu[2], Kuanquan Wang[1], and Xiangqian Wu[1(✉)]

[1] School of Computer Science and Technology,
Harbin Institute of Technology, Harbin 150001, China
{bsdai,wangkq,xqwu}@hit.edu.cn
[2] Department of New Media Technologies and Arts,
Harbin Institute of Technology, Harbin 150001, China
buwei@hit.edu.cn

Abstract. Reliable detection of fundus lesion is important for automated screening of diabetic retinopathy. This paper presents a novel method to detect the fundus lesion in retinal fundus image based on a visual attention model. The proposed method intends to model the visual attention mechanism of ophthalmologists during observing fundus images. That is, the abnormal structures, such as the dark and bright lesions in the image, usually attract the most attention of experts, however, the normal structures, such as optic disc and vessels, have been usually selectively ignored. To measure the visual attention for abnormal and normal areas, the incremental coding length is computed in local and global manner respectively. The final saliency map of fundus lesion is a fusion of attention maps computed for the abnormal and normal areas. Experimental results conducted on the publicly DiaRetDB1 dataset show that the proposed method achieved a sensitivity of 0.71 at a specificity of 0.82 and an AUC of 0.76 for fundus lesion detection, and achieved an accuracy of 100 % for normal area (optic disc) detection. The proposed method can assist the ophthalmologists in the inspection of fundus lesion.

Keywords: Diabetic retinopathy · Fundus lesion detection · Visual attention · Incremental coding length

1 Introduction

Diabetic Retinopathy (DR) is one of the main causes of blindness, and it is usually asymptomatic until the disease is at a late stage. The early detection of DR is thus important for patients to prevent visual loss. In fundus image, the most common signs of DR are the dark lesions, such as microaneurysms and hemorrhages, and the bright lesions, such as exudates and cotton-wool spots. The existence of these lesions can reflect the severity of DR. Currently, the inspection of these lesions is usually performed with the naked eye of ophthalmologists,

© Springer Science+Business Media Singapore 2016
W. Che et al. (Eds.): ICYCSEE 2016, Part I, CCIS 623, pp. 384–394, 2016.
DOI: 10.1007/978-981-10-2053-7_34

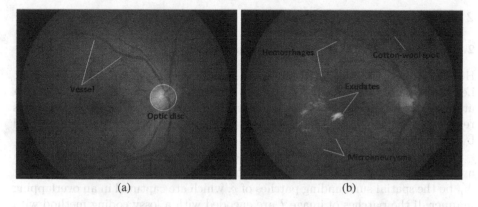

Fig. 1. Examples of retinal fundus images. (a) A normal fundus image. (b) A fundus image contains lesions.

such manual screening of DR, however, is time-consuming, subjective and error-prone. To diagnosis DR timely and reliably, the automated screening of DR is extensively investigated in past decades.

The first step of automated DR screening is the detection of fundus lesion, and because of the variability in appearance of those lesions, different algorithms have been designed to detect each type of those lesions separately. For microaneurysms detection, such as multi-scale correlation filtering [1] and diameter closing [2] have been presented. For hemorrhages detection, a splat feature classification approach has been reported in [3]. For the exudates detection, such as dynamic thresholding [4] and clustering based approach [5] have been used to detect the lesion. And with respect to the detection of cotton-wool spots, an improved Fuzzy C-Means approach has been presented in [6].

In fact, during clinically examining of the fundus image, all of those lesions will attract the most visual attention of experts, since they have irregular patterns compared with other areas. On the other hand, the normal structure such as optic disc will be selectively ignored since they are the locations that have been recently attended by the visual attention of many normal retinal images, which is so-called inhibition of return (IoR) [7]. Figure 1 shows two examples of retinal fundus images, for the task of DR screening, the lesions in Fig. 1(b) will attract the most attention of the experts.

Inspired by the biological vision mechanism, in this paper, a novel method for fundus lesion detection is presented based on a visual attention model, which intends to detect both of the dark and bright lesions in one unified framework.

The proposed method measures the "irregularity" of abnormal areas by incremental coding length (ICL) [8,9] in a local manner, and models the inhibition of return of normal areas also with ICL but in a global manner.

The rest of this paper is structured as follows. Section 2 describes our method to detect fundus lesion based on visual attention model. Experimental results are presented in Sect. 3. Section 4 concludes the paper.

2   Methodology

2.1   Abnormality Attention Computation

Here we assume that visual attention is driven by the predictive coding principle, i.e., the optimization of metabolic energy consumption in the brain [10]. The area attracted more attention of the visual system, i.e., the abnormal structure, is regarded as the location that need a more expensive neural code to be represented by its surrounding areas.

In this work, we use the incremental coding length (ICL) [8,9] to measure which area is more abnormal. Given an input image I, and p is a patch of image I. Let $\mathcal{N}_p$ be the spatial surrounding patches of p, which are captured in an overlapping manner. If the patches of image I are encoded with a lossy coding method with fixed distortion ε, $L_\varepsilon(\mathcal{N}_p)$ and $L_\varepsilon(\mathcal{N}_p \cup p)$ denotes the coding length of patches $\mathcal{N}_p$ and $\mathcal{N}_p \cup p$ respectively. The incremental coding length of patch p is then defined as [8,9]:

$$\delta L_\varepsilon(p) = L_\varepsilon(\mathcal{N}_p \cup p) - L_\varepsilon(\mathcal{N}_p) \tag{1}$$

where $\delta L_\varepsilon(p)$ is larger, the patch p is more abnormal than its surrounding areas.

To compute $\delta L_\varepsilon(p)$, we use the same scheme as suggested in [9], i.e.,

$$\delta L_\varepsilon(p) \approx L_\varepsilon(p|\mathcal{N}_p) \tag{2}$$

where $p|\mathcal{N}_p$ is the sparse representation of patch p over its surrounding patches $\mathcal{N}_p$.

Let $\mathbf{x} \in \mathbb{R}^n$ denotes the feature vector of the patch p, and $\mathbf{D} \in \mathbb{R}^{n \times m}$ is the dictionary matrix whose columns are the feature vectors of surrounding patches $\mathcal{N}_p$, as shown in Fig. 2. Finding the spare representation of p is to seek the optimal sparse code vector $\alpha \in \mathbb{R}^m$, which solves the following optimization problem:

$$\min_\alpha \frac{1}{2} \|\mathbf{x} - \mathbf{D}\alpha\|_2^2 + \lambda \|\alpha\|_1 \tag{3}$$

where the first term is the square reconstruction error, the second term is a ℓ_1 sparsity penalty on the code and λ is a coefficient controlling the sparsity penalty.

The incremental coding length $\delta L_\varepsilon(p)$ now is proportional to the number of non-zero entries of α, and in this work, it is computed as $\|\alpha\|_1$. The larger that is, the more abnormal the patch p will be. The final map of abnormality attention is then generated by accumulating $\delta L_\varepsilon(p)$ per pixel. Figure 4 illustrates examples of the map of abnormality attention, as shown in this figure, both dark and bright lesions have the higher response of the abnormality attention. However, the normal area, especially like the optic disc, also has higher response since it appears as irregular pattern in a single image. To suppress the unexpected high response in those normal areas, we model the IoR in following section.

2.2   Inhibition of Return Modelling

In the clinical task of DR screening, the normal areas, such as optic disc (OD) and vessels, will be selectively ignored by the experienced experts. At this point, we intend to model the IoR of normal areas. It is worth noting that, the normal vessel structures are networks and appearing as piece-wise linear structures, the patch of vessel generally can be sparely represented by its surroundings, the coding length $\delta L_\varepsilon(p)$ of majority of vessels is usually small. We hence mainly focus on modeling the inhibition of return for OD.

In light of the work in [10], we extend the computation of above-mentioned ICL in a global manner, as shown in Fig. 3. Instead of considering the surrounding patches of OD, we here introduce a set of reference set of OD $\mathcal{N}_{od}$, from which a dictionary $\mathbf{D}_{od} \in \mathbb{R}^{j \times k}$ is learned to yield a sparse representation of OD. Given an input patch p_{od}, its feature vector is denoted as $\mathbf{y}$, the spare code vector β is the optimal solution of the following problem:

$$\min_\beta \frac{1}{2} \|\mathbf{y} - \mathbf{D}_{od}\beta\|_2^2 + \gamma \|\beta\|_1 \tag{4}$$

where γ is a coefficient controlling the sparsity penalty.

Since ODs have a common appearance in different fundus images, the β of the patch located at the OD will be sparser than the one of the patch at other positions. Additionally, under a same level of sparsity, the square reconstruction error of the patch located at the OD will be smaller than the one of the patch at other positions. A weighted ICL of p_{od} is then computed by

$$\delta L_\varepsilon(p_{od}) \approx L_\varepsilon(p_{od}|\mathcal{N}_{od}) = \|\mathbf{y} - \mathbf{D}_{od}\beta\|_2^2 \cdot \|\beta\|_1 \tag{5}$$

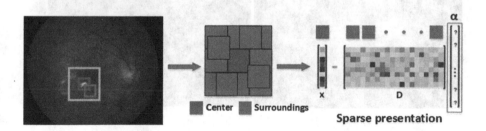

Fig. 2. Sparse representation of abnormal structures. (Color figure online)

Fig. 3. Sparse representation of optic disc.

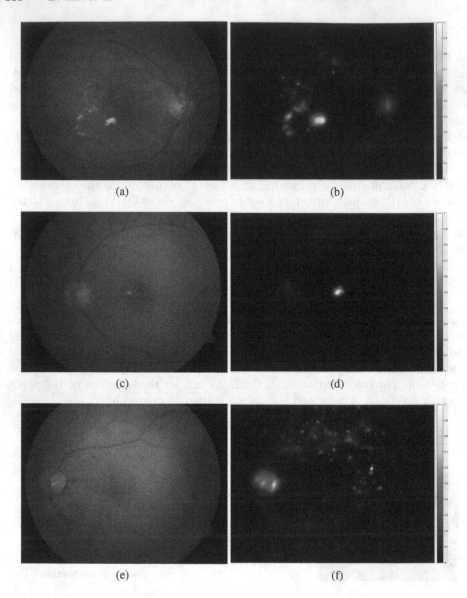

Fig. 4. Examples of abnormality attention map. (a), (c) and (e) Fundus images contains lesions. (b), (d) and (f) The map of abnormality attention.

For the patch located at the center of OD, its coding length $\delta L_\varepsilon(p_{od})$ is thus short, which can be used to reflect the level of inhibition of return. To avoid the overflow of the coding length, we introduce a sigmoid function and let

$$\delta L_\varepsilon(p_{od}) = \frac{1}{1 + \exp^{-\Lambda}} - 1, \ \Lambda = \|\mathbf{y} - \mathbf{D}_{od}\beta\|_2^2 \cdot \|\beta\|_1 \tag{6}$$

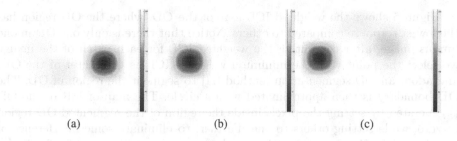

(a)   (b)   (c)

Fig. 5. The weighted ICL map of optic disc of Fig. 4(a), (c) and (e).

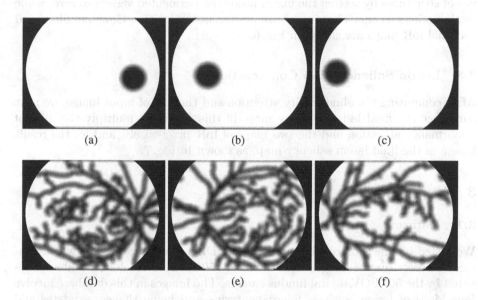

(a)   (b)   (c)

(d)   (e)   (f)

Fig. 6. IoR maps of normal areas. (a)–(b) IoR maps of the OD. (e)–(f) IoR maps of the vessels.

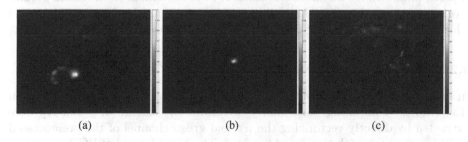

(a)   (b)   (c)

Fig. 7. Final lesion saliency map of Fig. 4(a), (c) and (e).

Figure 5 shows the weighted ICL map of the OD, where the OD region has the lowest response compared to others. Notice that there is only one OD in one fundus image, after computing the weighted ICL for each patch of the image, we select the point with the minimum weighted ICL as the center of the OD, and adopt an OD segmentation method [11] to segment the region of OD. The OD boundary is then approximated with a circle. The map of IoR of the OD is generated by setting the pixels inside the region of the segmented OD region to zero, while setting others to one. Further, to eliminate some interference of vessels, we here simply segment the vessel structures from the input fundus image with an adaptive thresholding method [1], and also construct an IoR map of the vessel structures by setting the pixels inside the segmented vessels to zero, while setting others to one. These maps are both smoothed with a Gaussian filter, and the final IoR maps are shown in Fig. 6.

2.3 Lesion Saliency Map Construction

After computing the abnormality attention and the IoR of input image, we next construct the final lesion saliency map. In this work, we multiply the map of abnormality attention and the two maps of IoR pixel-wisely, and let the result image as the final lesion saliency map, as shown in Fig. 7.

3 Experiments

3.1 Dataset

We tested our method on the public DiaRetDB1 V2.1 dataset [12], which contains 89 color fundus images with the fixed 1500×1152 resolution and are captured by the $50°$ FOV digital fundus camera. The images in this database involve four kinds of lesion, such as microaneurysms and hemorrhages, exudates and cotton-wool spots. For each image, four ground truth annotated by four different medical experts are provided. Since there are disagreements among four experts' annotations, we take a consensus of 75 % agreement as the fusion ground truth, as suggested in [12]. According to this ground truth, there are 51 images which contain lesions and 38 images without any lesions.

3.2 Implementation Details

In our experiments, the input image is contrast enhanced with the method in [13]. And for computing abnormality attention, the feature of p and $\mathcal{N}_p$ were extracted by directly vectorizing the red and green channel of the preprocessed RGB image patch with the size of $8 \times 8 \times 2$. The blue channel of RGB patch was abandoned since it is rather dark and short of useful information. In addition, because both dark and bright lesions in the fundus image have a variety of size, the computation of coding length was performed on image pyramid with 10 levels.

For computing the IoR of the OD region, the feature vector included three local features, i.e., the Dense SIFT [14], LIOP [15] and HoG [16], which were extracted from the green channel of RGB image patch, and the dictionary $\mathbf{D}_{od}$ is learned with the K-SVD algorithm [17].

3.3   Results

The performance of the proposed lesion detection was evaluated at image level as most of previous works in bright lesion detection. In this work, a detection of the image is considered as a true positive (TP) if finding out at least one kind of lesions in the image with lesions; a false negative (FN) if there is no founding of any lesions or only finding out the normal objects in the image with lesions; a false positive (FP) if finding out some objects in the image without any lesions; a true negative if there is no founding of any lesions in the image without any lesions. The sensitivity (SE) of the proposed method was then computed as TP/(TP + FN), the specificity (SP) was computed as TN/(TN + FP).

The proposed method achieves a SE of 0.71 at a SP of 0.82. To the best of our knowledge, no corresponding quantitative results have been reported for

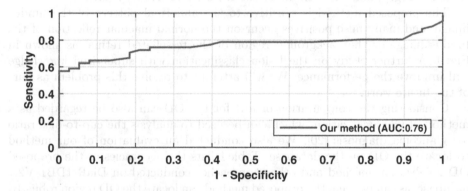

Fig. 8. ROC curve of fundus lesion detection.

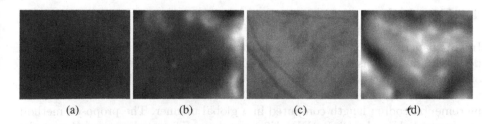

(a)   (b)   (c)   (d)

Fig. 9. The interference of lesion detection on normal images. (a) Image patch with normal macular reflection. (b) The corresponding saliency map patch of (a). (c) Image patch with the irregular background of tessellated retina. (b) The corresponding saliency map patch of (d).

Table 1. The performance of OD detection.

Methods	#Images	Success	ACC
Soares *et al.* [21]	89	88	98.88%
Mahfouz and Fahmy [22]	89	87	97.75%
Ramakanth and Babu [23]	89	88	98.88%
The proposed method	**89**	**89**	**100%**

detecting the dark and bright lesion at the same time. For comparison, here we simply exhibit some results of the methods only detect the dark or bright lesions. For bright lesion (exudates) detection, [18] reported a SE of 0.70 at a SP of 0.85 conducted on the same dataset, while for dark lesion (hemorrhages) detection, [19] reported a SE of 0.85 at a SP of 0.21 conducted on a private dataset. Noted that our method detects the dark and bright lesions at the same time, and can achieve a competitive performance with these methods, which proved that the proposed method is competent for the fundus lesion detection. The receiver operating characteristic (ROC) curve of the proposed lesion detection is also shown in Fig. 8, and the area under the ROC curve (AUC) is 0.76.

The proposed method is sensitive to the abnormal pattern of the fundus image, and many false positives occur on the normal macular reflection of the fundus image or the background region of the tessellated retina, as shown in Fig. 9. A further study on the lesion classification with supervised knowledge will improve the performance. We will attempt to resolve this problem as part of the future work.

Considering the computation of IoR for the OD can also be regarded as a method of OD localization, which can be used to analysis the cup-to-disc ratio for glaucoma diagnosis [20]. We also conducted an evaluation of our method to locate the OD on the database. Table 1 lists the accuracies of the proposed OD localization method and other approaches conducted on DiaRetDB1 V2.1 database, as can be seen, the proposed method can locate the OD region robustly, and outperforms other OD localization approaches.

4   Conclusions

In this paper, we proposed a novel method based on visual attention model to detect fundus lesions from the color retinal image. The abnormality attention of fundus lesion was measured with the incremental coding length computed in a local manner, while the IoR of the OD area was measured with the weighted incremental coding length computed in a global manner. The proposed method was evaluated on the DiaRetDB1 V2.1 database. The results revealed that the proposed method was an efficient scheme for automated lesion detection, and it was also an alternate method to locate the OD for the purpose of automated glaucoma diagnosis.

Acknowledgments. The authors would like to thank those who provided materials that were used in this study. This work was supported in part by the Natural Science Foundation of China under Grant 61472102, in part by the Fundamental Research Funds for the Central Universities under Grant HIT.NSRIF.2013091, and in part by the Humanity and Social Science Youth foundation of Ministry of Education of China under Grant 14YJC760001.

References

1. Zhang, B., Wu, X., You, J., Li, Q., Karray, F.: Detection of microaneurysms using multi-scale correlation coefficients. Pattern Recognit. **43**(6), 2237–2248 (2010)
2. Walter, T., Massin, P., Erginay, A., Ordonez, R., Jeulin, C., Klein, J.C.: Automatic detection of microaneurysms in color fundus images. Med. Image Anal. **11**(6), 555–566 (2007)
3. Tang, L., Niemeijer, M., Reinhardt, J.M., Garvin, M.K., Abràmoff, M.D.: Splat feature classification with application to retinal hemorrhage detection in fundus images. IEEE Trans. Med. Imaging **32**(2), 364–375 (2013)
4. Sánchez, C.I., García, M., Mayo, A., López, M.I., Hornero, R.: Retinal image analysis based on mixture models to detect hard exudates. Med. Image Anal. **13**(4), 650–658 (2009)
5. Hsu, W., Pallawala, P., Lee, M.L., Eong, K.G.A.: The role of domain knowledge in the detection of retinal hard exudates. In: Proceedings of the 2001 IEEE Computer Society Conference on Computer Vision and Pattern Recognition, 2001, CVPR 2001, vol. 2, pp. II-246–II-251. IEEE (2001)
6. Xiaohui, Z., Chutatape, O.: Detection and classification of bright lesions in color fundus images. In: 2004 International Conference on Image Processing, 2004, ICIP 2004, vol. 1, pp. 139–142. IEEE (2004)
7. Pratt, J., Abrams, R.A.: Inhibition of return to successively cued spatial locations. J. Exp. Psychol. Hum. Percept. Perform. **21**(6), 1343 (1995)
8. Wright, J., Ma, Y., Tao, Y., Lin, Z., Shum, H.Y.: Classification via minimum incremental coding length. SIAM J. Imaging Sci. **2**(2), 367–395 (2009)
9. Li, Y., Zhou, Y., Xu, L., Yang, X., Yang, J.: Incremental sparse saliency detection. In: 2009 16th IEEE International Conference on Image Processing (ICIP), pp. 3093–3096. IEEE (2009)
10. Hou, X., Zhang, L.: Dynamic visual attention: searching for coding length increments. In: Advances in Neural Information Processing Systems, pp. 681–688 (2009)
11. Wong, D., Liu, J., Lim, J., Jia, X., Yin, F., Li, H., Wong, T.: Level-set based automatic cup-to-disc ratio determination using retinal fundus images in argali. In: 30th Annual International Conference of the IEEE Engineering in Medicine and Biology Society, 2008, EMBS 2008, pp. 2266–2269 (2008)
12. Kauppi, T., et al.: The DiaRetDB1 diabetic retinopathy database and evaluation protocol. In: Proceedings of BMVC, pp. 1–10 (2007)
13. Farbman, Z., Fattal, R., Lischinski, D., Szeliski, R.: Edge-preserving decompositions for multi-scale tone and detail manipulation. ACM Trans. Graph. **27**(3), 67:1–67:10 (2008)
14. Yang, J., Yu, K., Gong, Y., Huang, T.: Linear spatial pyramid matching using sparse coding for image classification. In: IEEE Conference on Computer Vision and Pattern Recognition, 2009, CVPR 2009, pp. 1794–1801. IEEE (2009)
15. Wang, Z., Fan, B., Wu, F.: Local intensity order pattern for feature description. In: Proceedings of ICCV, pp. 603–610. IEEE (2011)

16. Dalal, N., Triggs, B.: Histograms of oriented gradients for human detection. In: Proceedings of CVPR, vol. 1, pp. 886–893. IEEE (2005)
17. Aharon, M., Elad, M., Bruckstein, A.: K-SVD: an algorithm for designing overcomplete dictionaries for sparse representation. IEEE Trans. Signal Process. **54**(11), 4311–4322 (2006)
18. Rocha, A., Carvalho, T., Jelinek, H.F., Goldenstein, S., Wainer, J.: Points of interest and visual dictionaries for automatic retinal lesion detection. IEEE Trans. Biomed. Eng. **59**(8), 2244–2253 (2012)
19. Hatanaka, Y., Nakagawa, T., Hayashi, Y., Mizukusa, Y., Fujita, A., Kakogawa, M., Kawase, K., Hara, T., Fujita, H.: CAD scheme for detection of hemorrhages and exudates in ocular fundus images. In: Medical Imaging, pp. 65142M–65142M. International Society for Optics and Photonics (2007)
20. Lowell, J., Hunter, A., Steel, D., Basu, A., Ryder, R., Fletcher, E., Kennedy, L.: Optic nerve head segmentation. IEEE Trans. Med. Imaging **23**(2), 256–264 (2004)
21. Soares, I., Castelo-Branco, M., Pinheiro, A.: Optic disk localization in retinal images based on cumulative sum fields (2015)
22. Mahfouz, A.E., Fahmy, A.S.: Fast localization of the optic disc using projection of image features. IEEE Trans. Image Process. **19**(12), 3285–3289 (2010)
23. Ramakanth, S.A., Babu, R.V.: Approximate nearest neighbour field based optic disk detection. Comput. Med. Imaging Graph. **38**(1), 49–56 (2014)

Identifying Transportation Modes from Raw GPS Data

Qiuhui Zhu[1], Min Zhu[1(✉)], Mingzhao Li[2], Min Fu[1], Zhibiao Huang[3],
Qihong Gan[4], and Zhenghao Zhou[5]

[1] College of Computer Science, Sichuan University, Chengdu, China
zhumin@scu.edu.cn
[2] RMIT University, Melbourne, Australia
[3] Chengdu Institute of Computer Application, Chinese Academy of Sciences,
Chengdu, China
[4] Modern Education Technology Center, Sichuan University, Chengdu, China
[5] High School No. 7, Chengdu, China

Abstract. Raw Global Positioning System (GPS) data can provide rich context information for behaviour understanding and transport planning. However, they are not yet fully understood, and fine-grained identification of transportation mode is required. In this paper, we present a robust framework without geographic information, which can effectively and automatically identify transportation modes including car, bus, bike and walk. Firstly, a trajectory segmentation algorithm is designed to divide raw GPS trajectory into single mode segments. Secondly, several modern features are proposed which are more discriminating than traditional features. At last, an additional postprocessing procedure is adopted with considering the wholeness of trajectory. Based on Random Forest classifier, our framework can achieve a promising accuracy by distance of 82.85 % for identifying transportation modes and especially 91.44 % for car mode.

Keywords: GPS · Transportation mode · Random forest classifier

1 Introduction

Due to ever-growing traffic congestion, human activities have become more complex and associated life trajectories more intensive. User behaviour extraction, trajectory analysis and traffic pattern recognition are particularly significant for service provider and decision maker [3]. Normally, urban transportation modes are classified as road ones (car, bus, bike and walk) and rail ones (subway and train). It's obviously easy for researchers to distinguish rail ones from road ones by using such relatively simple methods as velocity modelling [2]. But our work focuses on identifying different means of road transportation modes which are more complicated depending on the raw GPS data.

In the past several years, researchers collected the data information of transportation modes through questionnaires and telephone interviews recorded by

© Springer Science+Business Media Singapore 2016
W. Che et al. (Eds.): ICYCSEE 2016, Part I, CCIS 623, pp. 395–409, 2016.
DOI: 10.1007/978-981-10-2053-7_35

participants, which often resulted in inaccurate and incomplete data under easy overlooked or short trips [10]. Nowadays, urban sensing technology enables us to collect scientific data in a new and innovative way. Being two of the lowest-power sensors available on the phone, accelerometer and GPS are the predominantly used sensors in transportation mode identification. Accelerometer can detect acceleration in the phone's 3 axial directions. However, its readings are phone orientation and position-dependent, as well as vehicle-dependent. To identify transportation modes accurately, sampling rate is typically 10HZ and above. The high sampling rate, 3 axial directions, and position dependence make the classification complicated and increases power consumption [12]. Compared with accelerometer, GPS devices are becoming more popular for urban transportation mode identification in that its advantage of mobility and low sample rate.

In this paper, we present a robust framework to identify urban road transport including car, bus, bike and walk from raw GPS data. The contributions of the paper lies in three aspects: (1) A trajectory segmentation algorithm is designed based on logical assumptions, which can find almost 90 % single mode segments. (2) Several modern features are defined such as acceleration change rate, timeslice type, 85 % percentile velocity and acceleration, which were more discriminating than traditional features. (3) Relying on the wholeness of trajectory, a postprocessing procedure is developed to further improve the precision of mode identification without geographic information.

The remainder of the article is organized as follows. Related work is discussed in Sect. 2. Section 3 describes the dataset for our study. In Sect. 4 the classification model is introduced, followed by a presentation of postprocessing procedure in Sect. 5. In Sect. 6, the result of experiment is reported. Finally, we draw a conclusion, closed with corresponding discussion in Sect. 7.

2   Related Work

Identifying hybrid transportation modes from context information is still a relatively popular study. Biljecki et al. [2], Lin et al. [9], Shin et al. [14], Witayangkurn et al. [17] and Zheng et al. [19], present different approaches for identifying transportation modes. Table 1 shows a summary of the reviewed methods for transportation mode identification using GPS data. As shown in Table 1, a common processing step is applied to divide GPS logs into single mode segments based on criteria, such as transition points which denote a transition of transportation modes from one segment to another.

Precise identification of transportation mode is attributed to the high quality recognition of transition points. Many existing approaches of finding transition points require fine-grained acceleration data or geographic data. Usually, fine-grained acceleration data is generated by accelerometer embedded in mobile phone. Shin et al. [14] detected walking activity through acceleration data as a separator to partition the data stream into other activity segments. With the increase in sampling rate and time complexity, the accuracy of transportation mode identification can not be significantly improved. In addition, many researchers explored transition points relying on geographic data instead of acceleration data. For instance, Liao et al. [8] segmented multi-modal trajectories by analysing the proximity to potential transition locations such as bus stops. Biljecki et al. [2] used OpenStreetMap data to help the segmentation process in a two-step process, partition of trajectories to single-journey segments based on two meaningful locations, and segmentation of journeys into single-mode segments. Geographic data such as road networks, bus stops and parking lots are not widely used by current approaches, because it can add to the cost and complexity of the system and increase calculation consumption. It is beneficial to develop approaches that do not rely on such data. Mountain and Raper [11] indicated that transition points mainly appeared in a rapid and sustained change in direction or speed when one user ceased one activity and began another. Zheng et al. [19] found transition points by a logical assumption that the start point and end point of walk segment can be a transition point in very high probability. Compared with their researches, we design a novel processing method which is robust for noise and perform better in finding transition points.

For each segment generated by transition points, most of the previous work was accomplished by building classification models with extracting significant features. Many of these models regard velocity as the significant feature for mode identification. Bolbol et al. [4] concluded the velocity variable could contribute positively to the classification. Due to the measurements of noise, researchers noted that approximately maximum values should be used [13,16]. Zheng et al. [18] proposed the method which is still robust for noise using top two maximum values of velocity. Besides, Stenneth ct al. [15] derived features related to transportation network to improve classification effectiveness. In spite of high accuracy, their work needs great calculation consumptions. Zheng et al. [18] considered features that characterize changes in movement direction, velocity and acceleration. However, modern cities show more characteristics along with the development of times, such as traffic congestion, changes of people's behavior. Our work extract more powerful features to achieve high quality transportation mode identification, which can also fix and enhance traditional methods.

Table 1. A summary of the reviewed methods for transportation mode identification using GPS data

Study	Sensor	Geographic information	Modes	Accuracy
Zheng et al. [18]	GPS	No	4	76.2%
Witayangkurn et al. [17]	GPS	No	5	77.4%
Lin et al. [9]	GPS	Yes	4	76.3%

3 Data Preprocessing

In this section, we first introduce the GPS trajectory dataset in Geolife project from Microsoft Research Asia and define several terms used in this paper. Then we describe the procedure of trajectory segmentation in detail.

3.1 Data Survey

The GPS trajectory dataset used in this paper was collected in Geolife project [18,20,21] from Microsoft Research Asia by 182 users in a period of over five years (from April 2007 to August 2012). The majority of the data was created in Beijing, the capital city of China, which has an integrated urban land use and a composite transportation network including the complex road network. This dataset recorded a broad range of users' outdoor movements, including not only life routines like go home and go to work but also some entertainments and sports activities. In each day of data collection, every user can label their data in the following way, 2011/03/19, 06:43:50-06:52:14, bus. Each trajectory in this dataset is represented by a sequence of time-stamped point. Every point contains the specific information of latitude, longitude and time. The uneven time sampling rate is set to be 2 s or 5 s. 73 users of total 182 users have labelled their trajectories with transportation mode. In this experiment, we use the dataset of version 1.3 which contains abundant information about transportation modes.

3.2 Travel Survey Definitions

In order to better understand GPS trajectories, some terms have been defined to describe different fragments of the trajectory. The total trajectory about a specific user in one day is called a trip. A trip is consist of a number of segments (such as car segment, walk segment, etc.). A new segment is generated when a user changes transportation mode or the time between two consecutive points exceed the specified threshold. A transition point is the point whose previous and posterior points belong to different segment. For instance, in Fig. 1, a transition point is generated when a user changes transportation mode from bus to walk. The time interval of two consecutive point is 2 s or 5 s, in Fig. 1 Δt is 2 s.

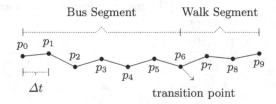

Fig. 1. Trip, segment and transition point

3.3   Data Preprocessing Procedure

Just as Mountain and Raper [11] stated that, the situation of one user may have ceased one activity and begun another mainly appears in a rapid and sustained change in direction or speed. We adopt the concept and presented a new trajectory segmentation method. Our approach is comprised of two portions, label specification and segmentation procedure. Algorithm 1 gives a detailed description of our trajectory segmentation procedure.

Label Specification. In this procedure, we first specify the points as *walk-point* if its velocity and acceleration is under the appointed threshold values, or *non-walk-point*. Owing to the error specification caused by noise points, it would lead to the following observations. First, the abnormal points may appear in the trajectory, which are usually far away from the nearest adjacent points. Second, the sharp-pointed points may occur with sharp turnings which would deviate from the trajectory trend and generate a zigzag segment. The second phenomenon usually happens when GPS points accumulated together, such as walk segments. The aforementioned two phenomenons would result in some points with high speed which belong to walk segment actually, thus may classify the walk segment as the bus segment or other modes mistakenly. So we put forward label revision method to eliminate the abnormal points specification. The loop in line 9 to 19 of Algorithm 1 describes the procedure in detail. The state of point (walk point or non-work point) depends on the states of its previous and posterior points. Our procedure for label specification can not only keep high points specification, but also maintain the trajectories original.

Segmentation. In segmentation procedure, we adopt the concept that the transition points usually appear in the situation that velocity encounter with sudden change. So we first get candidate transition points collection and many segments. Line 20 to 27 in Algorithm 1 describes the processing procedure in detail. Then, the segments with a length under the specified value d_{thd} should be merged with its nearby segments. Firstly, the merged segment usually continues almost 90 s through statistical analysis. This statistic time is consistent with the maximum tolerable time of pedestrians in red light. Most of pedestrians achieve 50 m with normal walk speed in 90 s. An example of this practice is the situation that, a car stops for a few seconds, then goes on the trip with high velocity, and the interval usually does not exceed 90 s. Secondly, the common users would not frequently change their transportation modes within such a short distance. For instance, within a short distance, it is impossible for a person to take the following transition, Bus→Walk→Bus→Walk→Bus. Hence this type of segments generated by interval should be merged with its nearby segments.

Algorithm 1. Trajectory Segmentation algorithm

Input: GPS logs T, velocity threshold v_{thd}, acceleration threshold a_{thd}, distance threshold d_{thd}, positive integer N represent point number, *scale* represent coefficient of proportionality, $0 < scale \leq 1$, candidate transition points collection CTP, segment collection $CSEG$ divided by CTP

Output: a set of segment

```
 1: function SEGALG(T, v_thd, a_thd, d_thd, N, scale)
 2:     for each p_i ∈ T do
 3:         if p_i.v < v_thd and p_i.a < a_thd then
 4:             label p_i as walk-point
 5:         else
 6:             label p_i as non-walk-point
 7:         end if
 8:     end for
 9:     Initialize positive integer M ← ⌈(N * scale)⌉
10:     repeat
11:         for each p_i ∈ T do
12:             if at least M in N of both adjacent previous points and posterior points
                    of p_i is labelled as walk-point then
13:                 label p_i as walk-point
14:             end if
15:             if at least M in N of both adjacent previous points and posterior points
                    of p_i is labelled as non-walk-point then
16:                 label p_i as non-walk-point
17:             end if
18:         end for
19:     until all points' label keep unchanged
20:     Initialize candidate transition points collection CTP
21:     CTP ← φ
22:     for each p_i ∈ T do
23:         if at least M in N of adjacent points in front of p_i is labelled as non-walk-
                point and at least M in N of adjacent posterior points of p_i is labelled
                as walk-point or opposite then
24:             CTP ← CTP ∪ p_i
25:         end if
26:     end for
27:     generate a segment collection CSEG based on the transition points collection
            CTP
28:     for each segment seg_j ∈ CSEG do
29:         if distance of segment seg_j < d_thd then
30:             merge seg_j with its previous and posterior segments into one segment
31:         end if
32:     end for
33: end function
```

4   Methodologies

This section is organized as follows. Firstly, features used to identify transportation modes are extracted from raw GPS logs. Secondly, the remainder of analysis in this section will focus on the inference model.

4.1   Feature Selection

Timeslice type (TS). The dataset was collected in Beijing which has a complex road network. Individual activity makes their moving trajectories interweave together due to daily routine. According to the time-statistical analysis, the rush hours mainly distributed in the time slot 7:00–10:00 and 16:00–21:00. During these two timeslices, people are more likely to encounter traffic congestions. When the average velocity of car is as slow as bike or in other uncommon situations, transportation modes may be labelled as other improper modes, then this mistaken information will result in inaccurate mode identification. Therefore, we divide the whole daily time into two timeslice types, as T_busy and T_idle. Specifically, we denote the timeslice type value of segment as T_busy if its timeslice falls into the specified time slots described above, otherwise, the type will be set to T_idle.

Acceleration change rate (ACR). Nowadays, a majority of modern cities have built the private passageways for buses in order to economize the time wasted on the roadway, especially in the heavy traffic city of Beijing. In rush hours, transportation modes always line up together under the red light. Even though, bus drivers can drive straight in special bus lane regardless of the states of cars or pedestrians in the arterial road. Also, taxi drivers always shift down or speed up frequently according to the drivers' personal behaviors, skills and preferences. For example, under the tempt of profit, a taxi driver would continually change velocity in a very small time slot to keep high speed, slow down or speed up suddenly. Therefore, there are many swings in the acceleration distribution of single car mode. However, the bus drivers or pedestrians are prone to keeping a small acceleration change. This phenomenon implies the potential mode difference among bus, car and walk. ACR modeling this principle is defined as Eqs. (1) and (2). First we can calculate the $ARate$ of each GPS point based on Eq. (1), in which A_i is the acceleration of point i. Then we can get the statistics of the number of GPS points whose $ARate$ are greater than a certain threshold A_r, and calculate ACR based on Eq. (2), in which $Distance$ is the total distance of single segment.

$$p_1 : ARate = |A_2 - A_1|/A_1; \tag{1}$$

$$ACR = |P_v|/Distance; \tag{2}$$

Where $P_v = \{p_i | p_i \in P, p_i.ARate > A_r\}$. Generally speaking, ACR makes it clear that the change frequency of acceleration in different transportation modes, which can be identified from each other.

85th percentile of velocity and acceleration (85thV, 85thA). As the primary variables for mode identification, velocity and acceleration play important roles in transportation mode identification. Due to sensor noise and data drift, the points always deviate from their original trend in the raw GPS trajectory. In Fig. 2, we use box plot to present the distribution of velocity and acceleration. The box plot uses the median, the approximate quartiles, and the lowest and highest data points to convey the level, spread and symmetry of a distribution of data values. Figure 2(a) and (b) make a comparison between 85 % percentile velocity and the maximum velocity. The comparison indicates the robustness of 85th percentile velocity, which is different from the maximum velocity that is prone to being disturbed by positioning errors. Also, as shown in Fig. 2(c) and (d), Fig. 2(d) describes the actual distribution of acceleration compared to Fig. 2(c). Noticeable mode shift of car reflects the potential characteristic when used in distinguishing car from other modes. These two features will get supports from later experiment results.

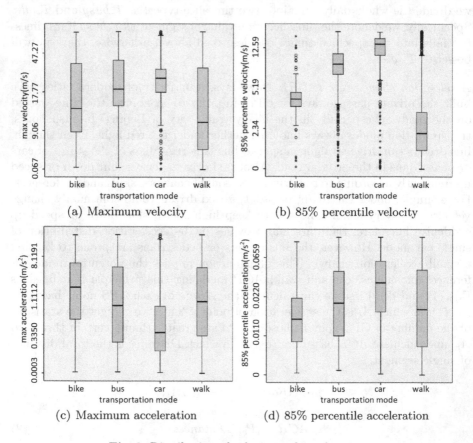

(a) Maximum velocity (b) 85% percentile velocity

(c) Maximum acceleration (d) 85% percentile acceleration

Fig. 2. Distribution of velocity and acceleration

Besides the features described above, we also extract other prominent features about velocity and acceleration including features between second maximum velocity and acceleration ($MaxV_2$, $MaxA_2$), mean velocity and acceleration (MeanV, MeanA), median velocity and acceleration (MedianV, MedianA), minimum velocity and acceleration (MinV, MinA), expectation of velocity and acceleration (Ev, Ea), covariance of velocity and acceleration (Dv, Da). Considering the details of GPS trajectory, features introduced by Zheng et al. [18] such as Heading Change Rate (HCR), Stop Rate (SR) and Velocity Change Rate (VCR), are extracted from the raw GPS trajectory.

4.2 Classification Model

The GPS trajectory dataset used in this paper was collected in Geolife project from Microsoft Research Asia. From this type of dataset, experimental results have demonstrated that the segmentation method based on transition points followed by a Decision Tree algorithm showed the highest identification accuracy of the transportation modes [18, 19]. In our work, we decided to employ Random Forest as a model because the works by Stenneth et al. [15] showed that performance of Random Forest is better than Decision Tree in transportation mode identification. Random Forest Classifier, developed by Breiman et al. [6], is an ensemble classification and regression method that constructs a number of decision trees at the training level, predicts the class using each tree and outputs the final class as the mode of the individually predicted classes. One of the major advantages of RF model is that it can handle high dimension data, obtain important features automatically. It is obvious to improve the training speed by extracting a few attributes from original features set every time. It is more stable and less prone to prediction errors as a result of data perturbations [7]. Therefore, RF model is viewed as one of the most accurate general-purpose learning techniques available [1].

5 Postprocessing Procedure

After applying the former inference model, we can obtain the predicted transportation modes of segment divided by transition points. Considering the wholeness of trajectory and walk mode logical assumption [19], and the analysis that car, bike and bus mode have similar velocity in the heavy traffic. Therefore, we view the whole trajectory as a chain of predicted modes, and modify the predicted modes as the high probability modes which follow the general trend. This practice can solve the above issues to some extent. Figure 3(a) gives the predicted mode sequence after the first classification. According to experience, the segment classified as bus mode in the line is likely the segment with improper classification. So we change the predicted bus mode to car mode, reasoning that the person is impossible to switch car mode to bus mode, or bike mode directly. Moreover, this segment is surrounded with predicted car mode and has similar characteristics with non-walk mode. After processing, we 'repair' the original predicted mode sequence as the available final-result mode sequence. Figure 3(b) shows the modified transportation mode line.

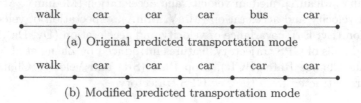

(a) Original predicted transportation mode

(b) Modified predicted transportation mode

Fig. 3. General case of segment postprocessing

6   Results and Discussion

In this section, we firstly describe how we select the parameters for each procedure. Secondly, we verify the efficiency of presented features and get the corresponding results about our overall inference model.

6.1   Parameter Setting

In the preprocessing step, two consecutive GPS points are divided into two different segments if the time gap is more than $20\,min$. When labelling the points in Algorithm 1, the value of velocity and acceleration threshold v_{thd}, a_{thd} is $1.8\,m/s$ and $0.6\,m/s^2$ [19]. From Fig. 4, when variable N and $scale$ is set to be 10 and 0.8, we can get highest recall of transition points. Referring to the situation that most of pedestrians achieve $50\,m$ with normal walk speed in $90\,s$, then interval distance d_{thd} is set to be $50\,m$. About the features, we set the threshold value for HCR, SR and VCR of 15, 3.2 and 0.36 respectively [18]. Figure 5 shows the inference accuracy changing over the threshold value A_r when ACR is used alone to identify transportation modes. Obviously, when A_r equals to 0.25, ACR shows its greatest advantages in identifying transportation modes.

Besides, our Random Forest classifier is the combination of 100 randomized decision trees. At each node in decision tree, a subset of features is randomly selected. Typically the size of every subset is $\sqrt{n}$, where n is the total number of features. In our experiment, the number of feature set used in the inferring is 19, i.e., n equals to 19. Thus, we set the number of subset features k is 4. With regard to the toolkit we used in the experiments, Weka (Waikato Environment for Knowledge Analysis) 3.7 toolkit [5] is selected to implement Decision Tree and Random Forest. About 70 % of all the segments are trained, and the remaining are used for testing.

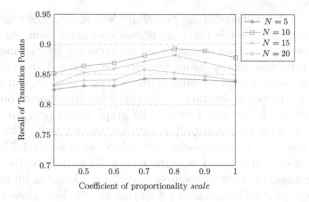

Fig. 4. Selecting value for variable N and *scale* through recall of transition points

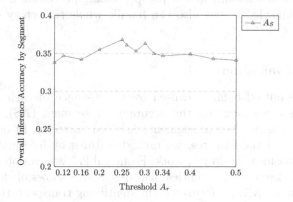

Fig. 5. Selecting threshold (A_r) for ACR. ACR is the only feature used in the inference model

6.2 Effectiveness of Preprocessing and Postprocessing Step

The preprocessing step divides the GPS trajectory into single mode segments based on transition points collection. In the first part, we evaluate the effectiveness of our label specification procedure by the precision of points specification. According to the statistics of the number of walk points and non-walk points, the precision of points specification rises to 78.79 % from 76.42 % after the procedure of label specification, demonstrating 2.37 % improvment in labelling the state of points. Then the second part of preprocessing step is measured in terms of the recall of transition points. While the recall of transition points has higher priority over their precision mainly because we hope to obtain all the transition points. Therefore, if the distance between an inferred transition points and its ground truth is within 150 m, we regard the transition point as a correct inference. As a result, we retrieved 89.3 percent of the actual transition points from the corresponding GPS data.

Meanwhile, in the postprocessing procedure, we consider the wholeness of trajectory to further improve the precision of mode identification. As it turns out, the postprocessing procedure has achieved an accuracy by distance of 82.85 % based on the inference model using features we explored in the experiment, while 81.13 % without postprocessing. This indicates that we can make almost 2 percent improvement in accuracy by distance for transportation mode identification. Zheng et al. [18] put forward the graph-based postprocessing which bring 3.4 % promotion over the preliminary inference result. This comparison makes clearly that our postprocessing procedure accuracy is 1.7 % lower than the performance of graph-based postprocessing. Nevertheless, the graph-based postprocessing mentioned above requires spatial knowledge as input, we have to know most geographic information of urban region. For the regions which are not covered by trajectory data, it can not perform better. What is more, it needs a lot of statistical calculation. But our postprocessing procedure not only doesn't rely on spatial knowledge, but also can handle whole trajectory of individuals anywhere.

6.3   Feature Evaluation

Considering the out-off-balance caused by the distance of each segment with its characteristics, we focus on the accuracy by segment (AS), which means the accuracy of the number of segment classified correctly. In order to evaluate the efficiency of the features, we ranked features by information gain and single feature classification in our work. From Table 2, we can observe that two ranking methods keep top 11 identical features, 85thV shows obvious advantage over other features, ACR performs well in identifying transportation modes and 85thA outperforms other features related to acceleration.

Table 2. Classification features ranking

(a)   Information gain ranking

Rank	Features
1	85thV
2	MedianV
3	MeanV
4	Ev
5	Dv
6	SR
7	HCR
8	ACR
9	VCR
10	MaxV$_2$
11	85thA

(b) Single feature for segment accuracy ranking

Rank	Feature	AS
1	85thV	45.70%
2	MeanV	43.23%
3	MedianV	41.57%
4	Ev	41.52%
5	HCR	40.87%
6	SR	39.74%
7	VCR	38.83%
8	Dv	37.53%
9	ACR	36.80%
10	MaxV$_2$	36.66%
11	85thA	29.78%

Table 3. Feature comparison

Combine features	Traditional features	New features
52 %	58.4 %	60 %

Another, we evaluate entire features in three different combination ways. ACR, 85thV, 85thA and TS make up the *combine features*. The *traditional features* include top two maximum velocity (MaxV$_1$, MaxV$_2$), top two maximum acceleration (MaxA$_1$, MaxA$_2$), MedianV, MedianA, MinV, MinA, MeanV, MeanA, Ev, Dv, Ea, Da, SR, HCR and VCR, while the *new features* is the feature set which we explored in this experiment. The results are shown in Table 3, from which the overall accuracy by segment of *combine features* indicates that the new features are enough to identify transportation mode in the proposed work. It means that our approach will not lose too much performance when applying our feature combination only. Meanwhile, considering the whole features, the overall accuracy by segment of *new features* rises from 58.4 % to 60 % compared the *traditional features*.

Finally, when Decision Tree is selected to perform the inference, the overall accuracy by segment can achieve 58.9 %, about 1.1 % lower than Random Forest which can perform 60 %. To summarise, Random Forest outperforms other classification model while considering the *new features*.

6.4 Mode Identification

For transportation mode identification, we evaluated the classification by using two well-known performance measures: Precision and Recall. As described in the previous section, we used Random Forest classifier as the mode identification classifier. We could achieve an overall accuracy by distance of 82.85 %, demonstrating a better discrimination about transportation modes than previous researches [9,14,17,18]. As shown in Table 4, walk mode identification achieves about 66.11 % accuracy while 91.23 % of recall. Very few actual walk segments are classified as other modes because of label revision in data preprocessing step. In addition, both accuracy and recall of car mode identification in matrix remain high scores. Among many reasons, the most important one is that car segments hold very long distance with the characteristic of high velocity. Also, the overall accuracy by distance of bike mode can reach about 81.82 %.

Table 4. Matching matrix of the detecting results in terms of distance and percentage

Detected results (KM and percentage)					
Mode	Walk	Car	Bus	Bike	Recall
Walk	1665.0 (91.23 %)	50.7 (2.78 %)	72.2 (3.96 %)	37.2 (2.03 %)	91.23 %
Car	400.5 (3.22 %)	11547.3 (92.91 %)	469.6 (3.78 %)	11.7 (0.9 %)	92.91 %
Bus	192.0 (6.25 %)	626.8 (20.40 %)	1931.7 (62.89 %)	321.3 (10.46 %)	62.89 %
Bike	261.0 (10.00 %)	403.8 (15.48 %)	276.8 (10.63 %)	1666.3 (63.89 %)	63.89 %
Precision	66.11 %	91.44 %	70.24 %	81.82 %	

From the misclassification between car and bus, 3.78 % length of car mode are misclassified as bus mode and 20.4 % length of bus mode are misclassified as car mode. For the dataset we explored in this paper, it was created in Beijing, China. Being the capital city of China, it has the complex road network. During the daytime, bus and car are more likely to encounter traffic congestions and perform the similar behavior. In spite of this phenomenon, the overall accuracy by distance of bus mode can perform 70.24 percent, which is better than similar researches [9,18].

7 Conclusions

In this paper, we presented a new robust framework for identifying road transportation modes focusing on raw GPS data. Firstly, without geographic information, we design a trajectory segmentation algorithm which can find almost all the transition points. Secondly, we propose some features which are more discriminating in transportation mode identification than the features which existing works [4,13,18] used. Additionally, the natural flow of whole GPS trajectory is considered when processing segments after classification. As a result, our work maintains relatively high precision when comparing with previous work [9,14,17,18], especially for car mode detection. The overall accuracy by distance of our framework can perform 82.85 % in transportation mode identification.

The application of our framework may be useful for user behavior analysis. However, many issues remain open and they are worthy of further study. First, the wholeness of trajectory maybe play a more significant role in identifying transportation modes instead of each segment individually. We can not ensure the natural flow of the travel pattern of every participant. Secondly, segmentation method generates many little segments segmented by transition points and influences the effectiveness of new introduced features. So combing different segmentation method is also potential work to do.

Acknowledgments. The research was supported by the project of Key Technology R&D program of Sichuan province (2013GZ0015). Special thanks to Zhaoyang Xie, Binbin Lu, Ruoyu Jia and Lei Gong for their inspiring discussions on the design of framework. Furthermore, we would also like to thank all of the reviewers for their valuable and constructive comments, which greatly improved the quality of this paper.

References

1. Biau, G.: Analysis of a random forests model. J. Mach. Learn. Res. **13**(1), 1063–1095 (2012)
2. Biljecki, F., Ledoux, H., Van Oosterom, P.: Transportation mode-based segmentation and classification of movement trajectories. Int. J. Geogr. Inf. Sci. **27**(2), 385–407 (2013)
3. Bohte, W., Maat, K.: Deriving and validating trip purposes and travel modes for multi-day GPS-based travel surveys: a large-scale application in the netherlands. Transp. Res. Part C Emerg. Technol. **17**(3), 285–297 (2009)

4. Bolbol, A., Cheng, T., Tsapakis, I., Haworth, J.: Inferring hybrid transportation modes from sparse GPS data using a moving window SVM classification. Comput. Environ. Urban Syst. **36**(6), 526–537 (2012)
5. Bouckaert, R.R., Frank, E., Hall, M.A., Holmes, G., Pfahringer, B., Reutemann, P., Witten, I.H.: Weka–experiences with a Java open-source project. J. Mach. Learn. Res. **11**, 2533–2541 (2010)
6. Breiman, L.: Random forests. Mach. Learn. **45**(1), 5–32 (2001)
7. Gislason, P.O., Benediktsson, J.A., Sveinsson, J.R.: Random forests for land cover classification. Pattern Recogn. Lett. **27**(4), 294–300 (2006)
8. Liao, L., Patterson, D.J., Fox, D., Kautz, H.: Building personal maps from GPS data. Ann. N. Y. Acad. Sci. **1093**(1), 249–265 (2006)
9. Lin, M., Hsu, W.J., Lee, Z.Q.: Detecting modes of transport from unlabelled positioning sensor data. J. Location Based Serv. **7**(4), 272–290 (2013)
10. McGowen, P., McNally, M.: Evaluating the potential to predict activity types from GPS and GIS data. In: Transportation Research Board 86th Annual Meeting, Washington (2007)
11. Mountain, D., Raper, J.: Modelling human spatio-temporal behaviour: a challenge for location-based services. In: GeoComputation, Brisbane (2001)
12. Sankaran, K., Zhu, M., Guo, X.F., Ananda, A.L., Chan, M.C., Peh, L.S.: Using mobile phone barometer for low-power transportation context detection. In: Proceedings of the 12th ACM Conference on Embedded Network Sensor Systems, pp. 191–205. ACM (2014)
13. Schuessler, N., Axhausen, K.: Processing raw data from global positioning systems without additional information. Transp. Res. Rec. J. Transp. Res. Board **2105**, 28–36 (2009)
14. Shin, D., Aliaga, D., Tunçer, B., Arisona, S.M., Kim, S., Zünd, D., Schmitt, G.: Urban sensing: using smartphones for transportation mode classification. Comput. Environ. Urban Syst. **53**, 76–86 (2015)
15. Stenneth, L., Wolfson, O., Yu, P.S., Xu, B.: Transportation mode detection using mobile phones and GIS information. In: Proceedings of the 19th ACM SIGSPATIAL International Conference on Advances in Geographic Information Systems, pp. 54–63. ACM (2011)
16. Stopher, P., FitzGerald, C., Zhang, J.: Search for a global positioning system device to measure person travel. Transp. Res. Part C Emerg. Technol. **16**(3), 350–369 (2008)
17. Witayangkurn, A., Horanont, T., Ono, N., Sekimoto, Y., Shibasaki, R.: Trip reconstruction and transportation mode extraction on low data rate GPS data from mobile phone. In: Proceedings of the International Conference on Computers in Urban Planning and Urban Management (CUPUM 2013), pp. 1–19 (2013)
18. Zheng, Y., Li, Q., Chen, Y., Xie, X., Ma, W.Y.: Understanding mobility based on GPS data. In: Proceedings of the 10th International Conference on Ubiquitous Computing, pp. 312–321. ACM (2008)
19. Zheng, Y., Liu, L., Wang, L., Xie, X.: Learning transportation mode from raw GPS data for geographic applications on the web. In: Proceedings of the 17th International Conference on World Wide Web, pp. 247–256. ACM (2008)
20. Zheng, Y., Xie, X., Ma, W.Y.: Geolife: a collaborative social networking service among user, location and trajectory. IEEE Data Eng. Bull. **33**(2), 32–39 (2010)
21. Zheng, Y., Zhang, L., Xie, X., Ma, W.Y.: Mining interesting locations and travel sequences from GPS trajectories. In: Proceedings of the 18th International Conference on World Wide Web, pp. 791–800. ACM (2009)

Image Segmentation: A Novel Cluster Ensemble Algorithm

Lei Wang[✉], Guoyin Zhang, Chen Liu, and Wei Gao

College of Computer Science and Technology, Harbin Engineering University,
Harbin 150001, China
343771666@qq.com

Abstract. Cluster ensemble has testified to be a good choice for addressing cluster analysis issues, which is composed of two processes: creating a group of clustering results from a same data set and then combining these results into a final clustering results. How to integrate these results to produce a final one is a significant issue for cluster ensemble. This combination process aims to improve the quality of individual data clustering results. A novel image segmentation algorithm using the Binary k-means and the Adaptive Affinity Propagation clustering (CEBAAP) is designed in this paper. It uses a Binary k-means method to generate a set of clustering results and develops an Adaptive Affinity Propagation clustering to combine these results. The experiments results show that CEBAAP has good image partition effect.

Keywords: Cluster ensemble · Binary k-means · Adaptive affinity propagation clustering · Image segmentation

1 Introduction

Image segmentation can be defined as a process that partitions an image into several non-intersected areas and each area is homogeneous and related. The relevant section is the essential topic in computer vision. [1, 2]. Since color image contains more image information than gray one. Academics have been paying more attention to color image segmentation. They have developed many image segmentation algorithms based on clustering [3]. In essence, clustering analysis belongs to the problem of an unsupervised pattern recognition. It can be viewed as a process of clustering and unmarked data points are devided into k groups with a few clustering criteria so that the intercluster dissimilarity is maximized while the intracluster dissimilarity is minimized [4]. The goal of cluster analysis is to capture the underlying structure of specific properties in the data and several clustering criteria. Many kinds of clustering approaches have been presented over the past few decades, such as Expectation Maximization (EM), k-means, graph-based approaches and hierarchical clustering approaches like Single-Link, Fuzzy c-Means [5, 6].

However, a clustering approach that can accurately capture the underlying structure of all data sets has not been proposed yet. It imposes an organization to the data following the data, the characteristics of an utilized dissimilarity function and an internal standard after we apply the clustering approach to a data set. The idea of

W. Che et al. (Eds.): ICYCSEE 2016, Part I, CCIS 623, pp. 410–417, 2016.
DOI: 10.1007/978-981-10-2053-7_36

integrating different clustering results becomes an alternative pattern. The clustering methods of improving the quality are the results which are challenging. The successful combination of supervised classifiers is the foundation of cluster ensemble. When it is given a group of objects, a cluster ensemble approach consist of two significant stages: Generation and combination (Ensemble or Fusion). The first stage is about the producing of a set of partitions of the objects. The second stage, a new partition which is computed, is the combination of all partitions created in the generation stage. For cluster ensemble, it is a challenge to integrate all the clustering results. In recent years, scholars have put forward many cluster ensemble methods. For example, Su et al., proposed fuzzy-link-based clustering ensemble approach, where the clusters generated by fuzzy c-means are fuzzy sets; next a fuzzy graph was employed to obtain the final ensemble clustering results [7, 8]. Banerjee et al., developed an unsupervised cluster-ensemble-based land-cover classification method that introduced cluster ensembles to generate a consistent labeling scheme for each clustering of the consensus [9, 10]. Li et al., proposed a double partition around medoids based cluster ensemble method. It repeatedly applied PAM to cluster attributes to produce multiple partitions with which an N-cut method was used to integrate these partitions to gain a final result [11]. However, some of these proposed cluster ensemble methods can not guarantee the optimal ensemble effect in the step of combination. To solve this problem, a novel Cluster Ensemble method using the Binary k-means and the Adaptive Affinity Propagation clustering (CEBAAP) is proposed to solve the image segmentation problem in this paper. CEBAAP performs a k-means method using the binary thought to generate more appropriate partitions and develop an adaptive affinity propagation clustering to combine these partitions. Based on the original affinity propagation clustering, the adaptive affinity propagation clustering can adaptively pick out the optimal ensemble result. Experimental results show the feasibility of the proposed algorithm on image segmentation. The remaining of the paper is organized as the following: Sect. 2 gives the contributions on which this paper builds. Section 3 is devoted to presenting the proposed CEBAAP to solve the color image segmentation problem. Section 4 evaluates CEBAAP in several real color images, and we conclude this paper in Sect. 5.

2 Related Works

2.1 Checking the PDF File

The data set with the number of clusters that are greater than two is suitable for being clustered by the binary thought. Because k-means method is fast and stable when it is applied to data with partitioning class $k = 2$.

The binary thought can be depicted as the following: Data points will be firstly divided into two clusters and then each cluster will be segmented into two, and repeat. Algorithm 1 showed the k-means approach using the binary thought.

Algorithm 1. The k-means method using the binary thought (KMBT) [12].
   Input: Data set   , number of desired classes k.
   Step 1. Calculate the iteration times T, T = int(lg k / lg2)+1.
   Step 2. for t = 1 to T do
                Calculate and renew clusters number K, K = 2t-1.
                Calculate and renew clusters size   ns, ns = n/2t-1.
           for j = 1 to K do
                Call kmeans (ns,2) to segment these clusters.
           end for
       end for
   Step 3. Calculate the leaves number K, K = 2T.
   Step 4. Merge K leaves into k.
   Output: Cluster membership for each data point.

2.2   Affinity Propagation Clustering

Instead of requiring that the number of clusters is specified, the number of clusters gained by affinity propagation is effected by the values of the input w(k, k) and the message passed on the iteration process; thus affinity propagation clustering gets over the drawback of many cluster approaches that is very sensitive to the initial choice of exemplars. The following is the main procedures of the affinity propagation clustering.

Algorithm 2. Affinity Propagation Clustering [13]
   Input: Data set $X = \{x_1, x_2, ..., x_n\}$, $\lambda = 0.5$, w(i, k): the similarity between point i and point k ($i \neq k$).
   Step 1. Initialize availabilities and availabilities, g(i, k), f(i, k), $g(i, k) = f(i, k) = 0$.
   Step 2. Update responsibilities,

$$f^{old}(i,k) \leftarrow f(i,k)$$
$$f^{new}(i,k) \leftarrow w(i,k) - \max_{k' \neq k}\{g(i,k') + w(i,k')\}$$
$$f(i,k) \leftarrow (1-\lambda)f^{new}(i,k) + \lambda f^{old}(i,k)$$

   Step 3. Update availabilities,

$$g^{old}(i,k) \leftarrow g(i,k)$$
$$if \ i = k, g^{new}(k,k) \leftarrow \sum_{i' \neq k}\max\{0, f(i',k)\}\}$$
$$if \ i \neq k, g^{new}(k,k) \leftarrow \max_{i' \neq k}\{0, f(i',k)\}$$
$$g(i,k) \leftarrow (1-\lambda)g^{new}(i,k) + \lambda g^{old}(i,k)$$

   Step 4. In order to make decisions of the exemplar, we should combine availabilities and responsibilities, and then stop message-passing procedure till a fixed iteration number is reached or the message change falls below a threshold; otherwise go to step 2.
   Output: Number of the clusters, apk, and cluster membership for each data point.

3   The Description of the Proposed Cebaap

According to the above affinity propagation clustering, this paper presents the process of the adaptive affinity propagation clustering.

Equation (1) indicates the preference step with $\alpha\hat{I}[0.8, 1]$ that is significant to scan the space flexibility.

$$p_s = \alpha \frac{\sum_{i,j=1}^{n} w(i,j)}{n^2 \sqrt{k + 55}} \tag{1}$$

For the purpose of sampling the whole space, the base of scanning step is set as $\sum_{i,j=1}^{n} w(i,j)/n^2$. The reason is that less-cluster case is less sensitive than that of a more-cluster case, so the fixed increasing step cannot match the different demand of different cases such as less and more clusters. Equation (1) shows that we can dynamically set the value of p_s with the count of clusters (k). If k is small, p_s will be large, and vice versa.

Furthermore, we regard global silhouettes index as the validity indices in this paper. Silhouettes provides a measure of how to classify the data point well when it is assigned to a cluster in accordance with the separation between the clusters and the tightness, which is introduced as a general graphical method for interpretation and validation of cluster analysis.

The following equation represents the global silhouette index [14],

$$GS = \frac{1}{n} \sum_{j=1}^{n} S_j \tag{2}$$

where local silhouette index can take the form,

$$S_j = \frac{1}{r_j} \sum_{i=1}^{r_j} \frac{y(i) - x(i)}{\max\{x(i), y(i)\}} \tag{3}$$

Here r_j is the count of the objects in class j, y(i) is the minimum average distance between object i and objects in or close to class j, x(i) is the average distance between object i and the objects in the same class j.

Algorithm 3 presents the process of adaptive affinity propagation clustering method. In accordance with Ref. [15], we should calculate a series of GS values, and the largest GS value corresponds to both the optimal number of clusters and the best clustering quality.

Algorithm 3. Adaptive Affinity Propagation Clustering

Input: Data set $\{x_1, x_2, ..., x_n\}$, $\lambda = 0.5$, $p = 0.5 p_m$, p_m: the median of the similarity, w(i, k): the similarity between points i and k ($i \neq k$)

Step 1: Call the Algorithm 2 to generate k clusters.

Step 2: Check whether these k clusters converge

If converge, use global silhouette index, GS, to evaluate the clustering result, and let $p = p - p_s$.

Otherwise, let $\lambda = \lambda + 0.05$. Moreover, if it does not converge until $\lambda \geq 0.85$, let.

Step 3: stop til a fixed iteration number is reached or the cluster number equals two, otherwise go to step 1.

Output: Number of the clusters, apk, and cluster membership for each pixel.

In addition, to accurately describe a pixel, we give a introduction of the three color spaces, RGB, LUV and HSV. A description of the gray value and spatial locations are also contained in the description of a pixel. Note: we briefly use horizontal and vertical coordinates to describe the spatial locations. To sum up, a pixel can be represented by a 12 dimensional row vector.

Given an image matrix $X \subseteq R^{n \times d}$, let $P = \{p_1, p_2, ... p_r\}$ denote a group of partition results of X. We use the idea of obtaining a hypergraph proposed in Ref. [16] to create a hypergraph (denoted by $H = \{h_1, h_2, ... h_r\}$) of P with n vertices and $t = rk(t \ll n)$ hyperedges.

Therefore, the detailed steps of the CEBAAP algorithm can be described as the following.

Algorithm 4. Cluster Ensemble method using the Binary k-means algorithm and the Adaptive Affinity Propagation clustering, CEBAAP

Input: an image matrix $X \subseteq R^{n \times d}$, m, k: number of sampling points and desired clusters, respectively, m > k.

Step 1. Call the Algorithm 1, KMBT, to partition the data set X into k groups, thus obtaining one partition result pi.

Step 2. Run KMBT r times on the data set X to obtain P,　and construct its hypergraph H,　.

Step 3. Call the Algorithm 3 to partition the hypergraph H into apk groups.

Step 4: Use the edge detection algorithm to gain the outline of the image according to these apk obtained groups.

Output: The final segmentation results.

4　Experiment and Analysis

The experimental environment: PC equipped with a 2.4-GHz Intel Core 2 Quad Q6600 CPU and 500-G Hitachi HDP 725050GLA360 Hard Disk, Matlab2007b. The testing images are chosen from the Berkeley image database. The size of all the images is 381×421 or 421×381. Furthermore, we will compare the following three image segmentation algorithms to verify the validity of the CEBAAP.

k-means: All the experimental algorithms use it to cluster the membership for each data finally. We randomly select k points as the initial cluster centers, set the number of iterations $n_r = 10$ and set the maximum number of iterations $n_i = 100$.

Mean Shift and N-cut algorithm (MSNcut) [17]. It incorporates the MS segmentation and the normalized cut methods. The normalized cut method is applied to solve the global optimal partition problem. Additionally, we directly call the MS segmentation method where the selection of the bandwidth parameter $h = (h_r, h_s) = (6, 8)$ and the minimum number of pixels in the regions M = 2000.

CEBAAP: It incorporates the binary k-means method and the adaptive affinity propagation clustering method to perform the image partition.

The comparison results are shown in Fig. 1. In this figure, from top to bottom, the number of desired classes k of each row is 5, 6, 6 and 4 respectively. The first column depicts the original color images. The second to forth columns show the segmentation results of the k-means algorithm, the MSNcut, and the CEBAAP, respectively.

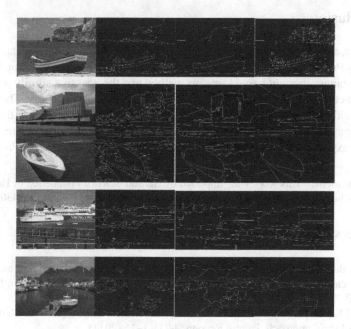

Fig. 1. Segmentation results of color images of different comparison algorithms.

We can have different observations from Fig. 1: (1) The segmentation results of the k-means algorithm are not good. For example, in the first row, the body of the boat is not completely partitioned, in the remaining three rows, we can hardly recognize the boats from the segmentation results. For k-means algorithm, it is easy to obtain local optimal solutions and has clear deficiencies in grasping the outlines of images. (2) The segmentation effect of the MSNcut is better than that of the k-means algorithm. For example, in the former two rows, the boats are completely segmented. For MSNcut, it uses MS to preserve the salient characteristics of the whole image and applies the Ncut method to perform the global optimal segmentation. In addition, as is shown in column three, MSNcut may generate balanced partitions. (3) CEBAAP can gain the best image segmentation results. For example, in the first and second rows, CEBAAP may completely segment the contours of the boats. In the third row, the theme color of the two ships is white, what's more, the stern of front cruise ship block the bow of the behind cruise ship, thus it is difficult to distinguish these two ships according to the color features. Although CEBAAP can not tell these two ships, it obtains the contour of main body of the ships. In the forth row, CEBAAP may accurately divide the image into four categorizes, the sky, mountain, water and architecture. However, CEBAAP clusters the dark red part of the boat and the reflection of the boat in the water into one group because the colors of these two parts are close. Similarly, in vision, the boundary between the boat and the shore is relatively vague. As a result, CEBAAP falls to accurately segment the boat. Although this failure is a fact, CEBAAP may tell us that there is a target in that area. To sum up, CEBAAP may obtain good segmentation effect.

5   Conclusion

In this paper, an image segmentation algorithm using the Binary k-means and the Adaptive Affinity Propagation clustering (CEBAAP) is designed. By incorporating the advantages of the binary thought based k-means method and the adaptive affinity propagation clustering method, CEBAAP can have good image segmentation quality. The Binary k-means algorithm is applied to generate a series of partition results, and the adaptive affinity propagation clustering is adopted to obtain the optimal partition results adaptively. The experimental results indicate the effectiveness of the CEBAAP.

Acknowledgement.   This work was supported by Natural Science Foundation of Heilongjiang province of China (F201406) and Liaoning Science and Technology Project (2014302006).

References

1. Gui, Y., Bai, X., Li, Z., Yuan, Y.: Color image segmentation using mean shift and improved spectral clustering. In: 12th International Conference on Control Automation Robotics & Vision (ICARCV), pp. 1386–1391. IEEE (2012)
2. Siang, T.K., MatIsa, N.A.: Color image segmentation using histogram thresholding-fuzzy c-means hybrid approach. Pattern Recogn. **44**(1), 1–15 (2011)
3. Gillet, A., Macaire, L., Botte-Lecocq, C., Postaire, J.G.: Color Image Segmentation by Analysis of 3D Histogram with Fuzzy Morphological Filters, pp. 153–177. Springer, Berlin Heidelberg, Fuzzy Filters for Image Processing (2003)
4. Lu, Z.M., Yang, P., Wang, R., Li, C.: A novel spectral clustering algorithm using low-rank approximation. ICIC Express Lett. **6**(12), 3125–3130 (2012)
5. Jain, A.K., Murty, M., Flynn, P.: Data clustering: a review. ACM Comput. Surv. (CSUR) **31**(3), 264–323 (1999)
6. Xu, R., Wunsch II, D.: Survey of clustering algorithms. IEEE Trans. Neural Netw. **16**(3), 645–678 (2005)

7. Su, P., Shang, C., Shen, Q.: Link-based pairwise similarity matrix approach for fuzzy c-means clustering ensemble. In: IEEE International Conference on Fuzzy Systems (FUZZ-IEEE), 2014, pp. 1538–1544. IEEE (2014)
8. Cai, Z.P., Ducatez, M.F., Yang, J.L., Zhang, T., Long, L.P., Boon, A.C., Webby, R.J., Wan, X.F.: Identifying antigenicity associated sites in highly pathogenic H5N1 influenza virus hemagglutinin by using sparse learning. J. Mol. Biol. **422**(1), 145–155 (2012)
9. Cai, Z.P., Goebel, R., Salavatipour, M., Lin, G.H.: Selecting genes with dissimilar discrimination strength for class prediction. BMC Bioinform. **8**, 206 (2007)
10. Banerjee, B., Bovolo, F., Bhattacharya, A., et al.: A new self-training-based unsupervised satellite image classification technique using cluster ensemble strategy. IEEE Geosci. Remote Sensing Lett. **12**(4), 741–745 (2015)
11. Li, L, You, J, Han, G, et al.: Double partition around medoids based cluster ensemble. In: International Conference on Machine Learning and Cybernetics (ICMLC), 2012, pp. 1390–1394. IEEE (2012)
12. Tian, Y., Yang, P.: Cluster ensemble algorithm using the binary k-means and spectral clustering. J. Comput. Inf. Syst. **10**(12), 5147–5154 (2014)
13. Xiao, Y., Yu, J.: Semi-Supervised clustering based on affinity propagation algorithm. J. Softw. **19**(11), 2803–2813 (2008)
14. Dudoit, S., Fridlyand, J.: A prediction-based resampling method for estimating the number of clusters in a dataset. Genome Biol. **3**(7), research0036 (2002)
15. Velamuru, P.K., Renaut, R.A., Guo, H., et al.: Robust clustering of positron emission tomography data. In: Joint Conference of the Classification Society of North America and Interface Foundation of North America (2005)
16. Strehl, A., Ghosh, J.: Cluster ensembles - a knowledge reuse framework for combining partitionings. In: Proceedings of Conference on Artificial Intelligence (AAAI 2002), Edmonton, AAAI/MIT Press, pp. 93–99 (2002)
17. Tao, W., Jin, H., Zhang, Y.: Color image segmentation based on mean shift and normalized cut. IEEE Trans. Syst. Man Cybern. Part B Cybern. **37**(5), 1382–1389 (2007)

Influence Maximization for Cascade Model with Diffusion Decay in Social Networks

Zhijian Zhang[1,2], Hong Wu[1,3], Kun Yue[1(✉)], Jin Li[4], and Weiyi Liu[1]

[1] School of Information Science and Engineering, Yunnan University, Kunming, China
kyue@ynu.edu.cn
[2] College of Science, Kunming University of Science and Technology, Kunming, China
[3] College of Computer Science and Engineering, Qujing Normal University, Qujing, China
[4] School of Software, Yunnan University, Kunming, China

Abstract. Maximizing the spread of influence is to select a set of seeds with specified size to maximize the spread of influence under a certain diffusion model in a social network. In the actual spread process, the activated probability of node increases with its newly increasing activated neighbors, which also decreases with time. In this paper, we focus on the problem that selects k seeds based on the cascade model with diffusion decay to maximize the spread of influence in social networks. First, we extend the independent cascade model to incorporate the diffusion decay factor, called as the cascade model with diffusion decay and abbreviated as CMDD. Then, we discuss the objective function of maximizing the spread of influence under the CMDD, which is NP-hard. We further prove the monotonicity and submodularity of this objective function. Finally, we use the greedy algorithm to approximate the optimal result with the ration of $1 - 1/e$.

Keywords: Social networks · Influence maximization · Cascade model · Diffusion decay · Submodularity · Greedy algorithm

1 Introduction

With the popularity of online social networks, such as Facebook, Twitter, WeChat, etc., the online social networks play an increasingly important role in daily communication among people. Many researchers have studied the diffusion phenomenon in social networks, such as the diffusion of news and opinions [1, 2], the adoption of products [3], the spread of infectious diseases [4–6], etc. Influence maximization is a fundamental problem of the diffusion in social networks. An application of influence maximization is viral marketing [3, 7, 8]. There have been extensive commercial instances of viral marketing succeed in real life, such as Nike Inc., used orkut.com, and facebook.com to market products successfully [9] and the Hotmail phenomenon [10].

Focusing on how to model the diffusion process, some researchers have proposed various diffusion models for the diffusion of innovations, ideas, etc. [12–15]. Randomness, the cumulative effect and the decay characteristic are the main characteristics of propagation. Most of the existing models describe the first two characteristics. But few researchers focus on the decay characteristics of influence diffusion. In short, the

© Springer Science+Business Media Singapore 2016
W. Che et al. (Eds.): ICYCSEE 2016, Part I, CCIS 623, pp. 418–427, 2016.
DOI: 10.1007/978-981-10-2053-7_37

diffusion decay refers to the decay of influence during the diffusion. For example, Tom read an interesting piece of news, and then he may forward it to his friends with probability p on the first day. But if not, he still may forward it with probability p' on the second day ($p' < p$), and p' decreases with time. That is, in the real diffusion process, the influence will decrease with time going and is reflected by the decreasing of the activate probability of node. Thus, a model that does not consider diffusion decay cannot simulate the actual spread process well. Furthermore, it is critical to model the spread process with diffusion decay for analyzing the influence maximization problem of social networks, which exactly we will solve in this paper.

In our study, we focus on the problem of selecting the seeds to maximize the influence spread considering diffusion decay in a social network. For this purpose, we consider the following problems:

(1) How to model the spread process with diffusion decay?
(2) How to select k seeds to maximize the influence spread?

For the problem (1), it is natural to consider extending the classic independent cascade model (IC model) [8] to incorporate the influence probability decaying with time, which is called as cascade model with diffusion decay, abbreviated as CMDD. In the CMDD, the activate probability is influenced by the following three factors: the previous cumulative effects, the influence power of new activated neighbor nodes and the decay factor. The probability of node v is a function of these three factors, which can well reflects realistic characteristics of influence spread in a social network.

For the problem (2), selecting k seeds to maximize the influence spread under the CMDD is NP-hard. Whether the CMDD model defined upon the IC model still keeps the monotonicity and submodularity is the key and difficult part in our work. We prove the monotonicity and submodularity of the objective function, and thus the greedy algorithm can be used to approximate the optimal result with the ration of $(1 - 1/e)$ based on the theoretic conclusion given by Nemhauser et al. [11].

In order to test the feasibility of the method proposed in this paper, we implement our algorithms and make corresponding experiments.

The reminder of this paper is organized as follows. In Sect. 2, we introduce related work. In Sect. 3, we give CMDD to model the influence spread of node. In Sect. 4, we obtain the objective function of influence maximization under the CMDD, and prove the monotonicity and submodularity of this objective function. In Sect. 4.2, we exploit the approximation algorithm to maximize the influence spread. In Sect. 5, we show the experimental results and performance studies. Finally in Sect. 6, we conclude and discuss the further work.

2 Related Work

Domingos et al. [7] discussed the influence maximization as an algorithm problem for the first time, and they modelled custom network as a graph and used a Markov random filed to calculate the influence probabilistic among them. In the aspect of modeling the diffusion process of influence, many researchers proposed various methods of influence maximization from various perspectives [8, 12–15].

Kempe et al. [8] formulated the problem of selecting a set of influence individuals to maximize the influence spread as a discrete optimization problem and proposed independent cascade model (IC model) and linear threshold model (LT model) based on earlier works [16–19]. The key feature of the model is that diffusion events along every arc in the social graph are mutually independent [20]. The LT model reflects the influence cumulative effect during the process of propagation, and the IC model can reflect the randomness of node activation. In this paper, our CMDD is based on the IC model, and it not only retains the cumulative effect in LT model, but also describes the influence decay during the diffusion. Our CMDD retais the monotonicity and submodularity in both LT and IC models.

Saito et al. [12] presented a method for predicting diffusion probabilities by using the Expectation Maximization algorithm based on the IC model. Yang and Leskovec [14] presented the linear influence model (LIM) to model the global influence of nodes. Goyal et al. [13] proposed three models: static model, continuous time model, discrete time model, in which the influence probabilities are relative to the action log instead of the discrete time step. In their works, dynamic activate probability is not discussed. In our work, we employ *Inf* to describe node influence power that can reflect not only the graphic characteristics but also some actual factors. The node's activate probability is changing with recently activated neighbours and the decay factor during the diffusion.

In the time-critical influence maximization problem, Chen et al. [23] extended the IC model and the LT model to incorporate the time delay aspect of influence diffusion, but the diffusion decay is not considered. Liu et al. [24] defined time constrained activate probability which is an assumed value at different times. In CMDD, we mainly consider that the influence diffusion probabilities of nodes decay with varying time step. Actually, the probability at time t is a function of the probability at time $t-1$, the decay factor and the influence by neighbours which are activated at $t-1$, which is dynamic.

In the aspect of how to select seeds, many researches proposed heuristics and tried to solve the influence problem more efficiently [8, 21, 22]. In terms of algorithm design, our work follows the idea given in [8]. To select the optimal seeds is NP-hard under the CMDD, and then the greedy algorithm is used to approximate the optimal result based on the mathematical theory given in [11].

3 Cascade Model with Diffusion Decay

A social network is denoted as an undirected graph $G = (V, E)$, where V is the set of nodes representing individuals and E is the set of edges representing the relationships among individuals. There are two classic diffusion models. One model describing how influence spreads in social network is LT model [8], which considers the influence accumulation of diffusion with time steps. Another model is IC model, in which an activated node u tries to activate its neighbor v with initialized p_{uv} only once [8].

In this paper, we propose the CMDD based on the IC model. CMDD combines the time step characteristics of influence diffusion and the influence accumulation. In this model, each node is either active or inactive. At step t, the node v is activated with probability p_v^t, which can be described as follows:

$$p_v^t = \alpha \times p_v^{t-1} + \frac{\sum_{w \in A_{t-1} \cap N(v)} Inf_w}{Inf_v + \sum_{u \in N(v)} Inf_u} \qquad (1)$$

where A_{t-1}, $N(v)$ and α denote the activated nodes at step $t-1$, the neighbors of node v and the decay parameter of influence respectively, where $0 \le \alpha \le 1$. This decay parameter can be denoted as a constant or an exponential function with parameters depending on the time. In order to facilitate the discussion, we employ a constant to denote the decay parameter. For $\alpha = 0$, this model is similar to the IC model. For $\alpha > 0$, this model can reflect the random property and the influence accumulation of the LT model. The greater the value of α, the slower the process of influence decay. Inf_w denotes the influence power of node w, such as the node's importance degree. $N(v)$ denotes neighbors of v and A_{t-1} denotes the nodes set which are activated at time $t-1$.

Example 1. Figure 1 shows an example of the diffusion process of CMDD. We assume $\alpha = 0.8$ and Inf_v is the degree of node v. Initially at $t = 0$, one seed v_6 is activated. At step $t = 1$, v_6 tries to activate its inactive neighbors with probabilities $p_{v_1}^1 = 0.308$, $p_{v_3}^1 = 0.267$, $p_{v_7}^1 = 0.4$ and $P_{v_8}^1 = 0.4$ respectively. At step $t = 2$, v_1 and v_7 are randomly activated, but v_3 and v_8 are inactive, and then the activated probabilities of v_2, v_3 and v_8 are $p_{v_2}^2 = 0.375$, $p_{v_3}^2 = 0.547$ and $p_{v_8}^2 = 0.32$. Similarly, we can obtain the activated probabilities of nodes at step $t = 3$.

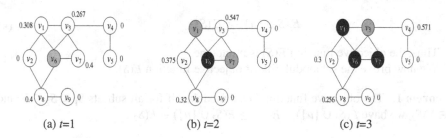

(a) $t=1$ (b) $t=2$ (c) $t=3$

Fig. 1. The diffusion process of the CMDD ($\alpha = 0.8$). In (a), (b) and (c), grey nodes denote the newly activated nodes, and the black nodes denote the activated nodes that lose the activated ability, other nodes are inactive.

4 Maximizing Influence Spread Under CMDD

In this section, we define the objective function of influence maximization problem under the CMDD, which is NP-hard. Then, we show that the objective function is monotone and submodular, which leads to a greedy approximation based on the theory given by Nemhauser et al. [11].

4.1   Objective Function of Influence Maximization Problem

The influence maximization problem is an optimal problem, in which given a graph $G = (V, E)$, the number of the seed k, we want to find a seed set S of the size k such that the expected number of nodes is maximized. Now, we first consider the objective function of influence maximization problem.

At step $t = 0$, $A_0(S) = S$, the expected activated value of influence maximization under the CMDD is $E_{t=0}(S) = A_0(S)$. We can obtain the expected activated value at step t as follows:

$$E_t(S) = \alpha \times E_{t-1}(S) + \sum_{i \in V \backslash A_{t-1}(S)} \frac{\sum\limits_{k \in A_{t-1}(S) \cap N(i)} Inf_k}{Inf_i + \sum\limits_{j \in N(i)} inf_j} \tag{2}$$

The overall expected activated values in t steps is equal to the sum of the expected activated value with t steps, that is,

$$E(S) = \sum_{t=0}^{t} E_t(S) = \sum_{t=0}^{t} (\alpha \times E_{t-1}(S) + \sum_{i \in V \backslash A_{t-1}(S)} \frac{\sum\limits_{k \in A_{t-1}(S) \cap N(i)} Inf_k}{Inf_i + \sum\limits_{j \in N(i)} inf_j}) \tag{3}$$

To select the optimal seed set to maximize the influence spread with the objective function and under the CMDD is NP-hard. We can prove the monotonicity and submodularity of the objective function.

Obviously, we have

$$E_t(S \cup \{u\}) \geq E_t(S) \tag{4}$$

Thus, the objective function $E(S)$ is monotone.

We now prove the submodularity of objective function $E(S)$.

Theorem 1.   The objective function is submodular, if for all subsets $S_1 \subseteq S_2 \subseteq V$ and $u \in V \backslash S_2$, we have $E(S_1 \cup \{u\}) - E(S_1) \geq E(S_2 \cup \{u\}) - E(S_2)$.

Proof.   We employ the Mathematical Induction to prove Theorem 1.

At step $t = 1$, the objective function $E(S)$ is obviously submodular.

At step $t - 1$, if the objective function is submodular, then we have

$$E_{t-1}(S_1 \cup \{u\}) - E_{t-1}(S_1) \geq E_{t-1}(S_2 \cup \{u\}) - E_{t-1}(S_2) \tag{5}$$

At step t, we have

$$E_t(S_1 \cup \{u\}) - E_t(S_1)$$

$$= \alpha(E_{t-1}(S_1 \cup \{u\}) - E_{t-1}(S_1)) + \sum_{i \in V \backslash A_{t-1}(u)} \frac{\sum\limits_{k \in A_{t-1}(u) \cap N(i)} Inf_k}{Inf_i + \sum\limits_{j \in N(i)} inf_j} - \left(\sum_{i \in V \backslash A_{t-1}(u \cap S_1)} \frac{\sum\limits_{k \in A_{t-1}(u \cap S_1) \cap N(i)} Inf_k}{Inf_i + \sum\limits_{j \in N(i)} inf_j} \right) \quad (6)$$

We have the similar expression of $E_t(S_2 \cup \{u\}) - E_t(S_2)$. We can see the activated process as flip the coin. Based on Equality (5) and (6), we have

$$E_t(S_1 \cup \{u\}) - E_t(S_1) \geq E_t(S_2 \cup \{u\}) - E_t(S_2) \quad (7)$$

The linear combination of a submodular function is also submodular, so we have $E(S_1 \cup \{u\}) - E(S_1) \geq E(S_2 \cup \{u\}) - E(S_2)$.

4.2 Greedy Algorithm for Influence Maximization Problem

We have proven that the objective function of influence maximization problem under the CMDD is monotone and submodular. According to the result proposed in [11], the greedy algorithm given in Algorithm 1 can be used to approximate the optimal result with the relation of $1 - 1/e$. The algorithm selects the node that provides the largest marginal gain to the seed set, and each time one node will be selected as a seed.

Algorithm 1: Greedy (k, S)

Input: A social network: $G=(V, E)$, number of seeds: k,
Output: Seed Set S,
Steps:
  1: initialize $S=\Phi$
  2: for i=1 to k do
  3:   select $u = \text{argmax}_{v \in V \backslash S}(E(S \cup \{v\}) - E(S))$
  4:   $S=S \cup \{u\}$
  5: end for
  6: return S

The running time of Algorithm 1 is determined by the greedy part at step 3. The time complexity of Algorithm 1 is $O(knd_1d_2)$, where k is the number of seeds, n is the number of nodes in network, d_1 is the average degree of nodes and d_2 is the max distance from other inactive nodes to node v, when we calculate the contribution of v.

5 Experimental Results

To test the feasibility and effectiveness of selecting seeds under the time cascade decay model, we implemented our method and made corresponding performance studies.

5.1 Experimental Setup

The ca-HepTh and ca-GrQc are HEP-TH (High Energy Physics-Theory) collaboration network extracted from the e-print (http://arXiv.org/). The former is extracted from the "High Energy Physics" and the latter is extracted from the "General Relativity". The nodes in these two networks are *authors* and an edge between two nodes means the *two coauthored at least one paper*. The p2p-Gnutella08 record the Gnutella peer to peer network from August 8 2002, where nodes represent hosts in the Gnutella network topology and edges represent connections between the Gnutella hosts (Table 1).

Table 1. Statistics of the two real-world networks in resulting graph

Dataset	ca-HepTh	ca-GrQc	p2p-Gnutella08
Number of nodes	9878	5242	6301
Number of edges	51971	28980	20777

5.2 Performance Studies

First, we tested the convergence rate of influence spread in ca-HepTh. In this experiment, we tested the influence spread with $\alpha = 0.4$ and $\alpha = 0.8$ under the CMDD respectively where spread time steps $t = 5$ and node $user_{ID} = 1441$ with high degree for obvious experiment result. Figure 2 shows that the convergence rate with $\alpha = 0.4$ was faster that than $\alpha = 0.8$, and the number of the convergence of influence spread with $\alpha = 0.4$ and $\alpha = 0.8$ were 1000 and 3000 respectively. This was because that the influence accumulation of node decreased slowly when the value of α is greater.

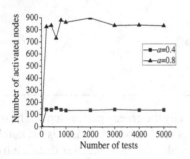

Fig. 2. The convergence rate of influence spread with different α value

Fig. 3. The expectation value of active nodes with different α values

Then, we tested the relationship of the expectation value of influence spread with different α in ca-HepTh. We compared the expectation value of influence spread with $\alpha = 0.2$, $\alpha = 0.4$, $\alpha = 0.6$, $\alpha = 0.8$ and $\alpha = 0$, where we assigned the spread time step from 1 to 10 and node $user_{ID} = 63113$. The comparison is shown in Fig. 3. It can be seen that greater α, the greater expectation value under the CMDD, since the value of α is greater, the value of influence probabilities of nodes decreases slower.

It is known that the max-degree algorithm [25] is well regarded as the effective algorithm for the networks with power law distributions, and it sorts the nodes by the degree, and it selects k max degree nodes as seeds. Random algorithm selects k seeds randomly. Finally, we tested the effectiveness of Algorithm 1. In this experiment, we selected 20 seeds with Algorithm 1 with Depth = 1 and Depth = 2, the max-degree algorithm (denoted as Max-degree) and random algorithm (denoted as Random) to maximize the influence spread in ca-GrQc and p2p-Gnutella08 and set $\alpha = 0.4$, $t = 3$. The depth in greedy algorithm means the max nodes distance we consider. If Depth = 1, we only consider the neighbors of active nodes. If Depth = 2, we consider not only the neighbors of active nodes but also the neighbors of their neighbors. Figure 4(a) shows that the greedy algorithm (denoted as Greedy) is better than Max-degree and outperforms Random. But in Fig. 4(b) and (c), the greedy algorithm is close to Max-degree, since the Inf_v of node v is calculated by degree in our experiments, which verifies that our proposed CMDD model and the corresponding algorithm are feasible.

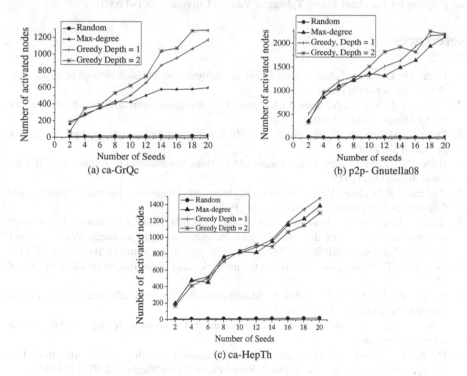

Fig. 4. Expectation value of active nodes with different algorithms inca-GrQc, p2p-Gnutella08 and ca-HepTh,

6   Conclusions and Future Works

In this paper, we redefined the node activate probability and proposed the CMDD, which is close to the real diffusion process. The CMDD reflects the change of probability with time step and new activated nodes, meanwhile it retains the cumulative effect and randomness. Then we proved the monotone and submodularity of this objective function and the greedy algorithm is used to approximate the optimal result.

However, our algorithm is not far superior to max-degree algorithm on some datasets. It is because the Inf_v of node v is calculated by degree in our experiments. We will extend our experiments to some real networks in which the Inf_v is determined by some actual factor. Furthermore, employing a constant to describe the diffusion decay parameter has its limitations. The decay factor function that can better describe the real spread process in a social network is still worth discussing. These are our next research directions.

Acknowledgement. This paper was supported by the National Natural Science Foundation of China (61562091), Natural Science Foundation of Yunnan Province (2014FA023, 201501CF00022), Program for Innovative Research Team in Yunnan University (XT412011), and Program for Excellent Young Talents of Yunnan University (XT412003).

References

1. Gruhl, D., Guha, R., Liben-Nowell, D., et al.: Information diffusion through blogspace. In: WWW, pp. 491–501 (2004)
2. Leskovec, J., Krause, A., Guestrin, C., et al.: Cost-effective outbreak detection in networks. In: KDD, pp. 420–429 (2007)
3. Datta, S., Majumder, A., Shrivastava, N.: Viral marketing for multiple products. In: ICDM, pp. 118–127 (2010)
4. Bailey, N.T.J.: The Mathematical Theory of Infectious Diseases and Its Applications. Haffner Press, Royal Oak (1975)
5. Anderson, R.M., May, R.M., Anderson, B.: Infectious Diseases of Humans: Dynamics and Control. Oxford University Press, Oxford (1992)
6. Kim, L., Abramson, M., Drakopoulos, K., Kolitz, S., Ozdaglar, A.: Estimating social network structure and propagation dynamics for an infectious disease. In: Kennedy, W.G., Agarwal, N., Yang, S.J. (eds.) SBP 2014. LNCS, vol. 8393, pp. 85–93. Springer, Heidelberg (2014)
7. Domingos, P., Richardson, M.: Mining the network value of customers. In: KDD, pp. 57–66 (2001)
8. Kempe, D., Kleinberg, J., Tardos, É.: Maximizing the spread of influence through a social network. In: KDD, pp. 137–146 (2003)
9. Johnson, A.: Nike-tops-list-of-most-viral-brands-on-facebook-twitter (2010). http://www.kikabinkcom/news/
10. Hugo, O., Garnsey, E.: The emergence of electronic messaging and the growth of four entrepreneurial entrants. New Technol. Based Firms New Millenium **2**, 97–123 (2002)
11. Nemhauser, G., Wolsey, L., Fisher, M.: An analysis of approximations for maximizing submodular set functions—I. Math. Program. **14**(1), 265–294 (1978)
12. Saito, K., Nakano, R., Kimura, M.: Prediction of information diffusion probabilities for independent cascade model. In: Lovrek, I., Howlett, R.J., Jain, L.C. (eds.) KES 2008, Part III. LNCS (LNAI), vol. 5179, pp. 67–75. Springer, Heidelberg (2008)
13. Goyal, A., Bonchi, F., Lakshmanan, L.V.S.: Learning influence probabilities in social networks. In: WSDM, pp. 241–250 (2010)

14. Yang, J., Leskovec, J.: Modeling information diffusion in implicit networks. In: ICDM, pp. 599–608 (2010)
15. Gomez, R.M., Leskovec, J., Krause, A.: Inferring networks of diffusion and influence. In: KDD, pp. 1019–1028 (2010)
16. Durrett, R.: Lecture Notes on Particle Systems and Percolation. Wadsworth Publishing, Boston (1988)
17. Liggett, T.M.: Interacting Particle Systems. Springer, Heidelberg (1985)
18. Granovetter, M.: Threshold models of collective behavior. Am. J. Sociol. **83**(6), 1420–1443 (1978)
19. Schelling, T.: Micromotives and Macrobehavior. Norton, New York (1978)
20. Chen, W., Lakshmanan, L., Castillo, C.: Information and Influence Propagation in Social Networks. Morgan & Claypool, California (2013)
21. Horel, T., Singer, Y.: Scalable methods for adaptively seeding a social network. In: WWW, pp. 441–451 (2015)
22. Wang, C., Chen, W., Wang, Y.: Scalable influence maximization for independent cascade model in large-scale social networks. Data Min. Knowl. Disc. **25**(3), 545–576 (2012)
23. Chen, W., Lu, W., Zhang, N.: Time-critical influence maximization in social networks with time-delayed diffusion process. In: AAAI, pp. 1–5 (2012)
24. Liu, B., Cong, G., Zeng, Y., et al.: Influence spreading path and its application to the time constrained social influence maximization problem and beyond. IEEE Trans. Knowl. Data Eng. **26**(8), 1904–1917 (2014)
25. Wasserman, S., Faust, K.: Social Network Analysis: Methods and Applications. Cambridge University Press, Cambridge (1994)

Link Mining in Online Social Networks with Directed Negative Relationships

Baofang Hu[1,2(✉)] and Hong Wang[2]

[1] School of Information Technology, Shandong Women's University,
Jinan 250014, China
hbf0509@126.com
[2] School of Information Science and Engineering, Shandong Normal University,
Jinan 250014, China
30900607@qq.com

Abstract. One of the most important work to analyse online social networks is link mining. A new type of social networks with positive and negative relationships are burgeoning. We present a link mining method based on random walk theory to mine the unknown relationships in directed social networks which have negative relationships. Firstly, we define an extended Laplacian matrix based on this type of social networks. Then, we prove the matrix can be used to compute the similarities of the node pairs. Finally, we propose a link mining method based on collaboration recommendation method. We apply our method in two real social networks. Experimental results show that our method do better in terms of sign accuracy and AUC for mining unknown links in the two real datasets.

Keywords: Social networks · Link mining · Collaborative filtering · Random walk

1 Introduction

In recent years, a new type of networks called signed social networks is burgeoning, such as Slashdot news review site, Epinions consumer review site and Wikipedia vote site. The relationships in these networks can be positive (friendly, like) or negative (hostile, dislike) and are more complicated than the relationships in traditional social networks whose links are all positive. So the research in social networks needs more comprehensive analysis of the two types of relationships and the research results in traditional unsigned networks are not applicable. Link mining in signed social networks is more complicated than that in general networks and can offer us more information. We can mine not only the possibilities of future links between unrelated nodes but also the future relationships (friendly or hostile).

A popular type of methods analysing the structure of general social networks is to calculate the commute distance based on random walk theory. These methods show good performance both in terms of accuracy and time complexity [1,2].

© Springer Science+Business Media Singapore 2016
W. Che et al. (Eds.): ICYCSEE 2016, Part I, CCIS 623, pp. 428–440, 2016.
DOI: 10.1007/978-981-10-2053-7_38

We aim at mining the sign and direction of future links in directed social networks using the relationships between Laplacian matrix and commute distance. However, the commute distance should be symmetric and traditional Laplacian matrix in directed graph is asymmetric. We define an extended Laplacian matrix and prove that it can be a legal similarity distance in directed signed networks. We also mine the sign and direction of links based on the idea of collaborative filtering which is usually used in recommendation systems.

The rest of the paper is organized as follows. We review related works about link mining in signed social networks in Sect. 2 and introduce some definitions and basic theories in Sect. 3. In Sect. 4, we present our exact definition of commute distance in signed social networks. And link mining process is proposed in Sect. 5. In Sect. 6, we design different experiments and show the experimental results. Finally, we provide the conclusions in Sect. 7.

2 Related Works

Link mining in signed social networks became popular through the work of Guha and Kumar [3]. They proposed a framework of trust propagation schemes to mine the sign of links in undirected signed networks. Kunigis et al. [4,5] studied the resistance distance in signed networks and mined the friend/foe relationship, however, they did not mine the directions of the links. Leskovec et al. [6] proposed status theory in signed networks and used it to mine positive and negative links. Chiang et al. [7] gave a definition of social imbalance (MOIs) based on l-cycles in signed social networks and proposed a link mining method.

Although there are large bodies of works involving negative relationships in on-line domains, they pursue directions different from our work focus here. In this paper, we focus on commute distance property in directed signed social networks to mine not only the sign but also the direction of the unknown relationships. It is well known that the commute distance is related to the spectrum of the graph Laplacian in general undirected social networks [8]. Our work focuses on the relationship between the commute distance and graph Laplacian in directed signed networks and using the relationship to mining the unknown links.

3 Preliminaries

3.1 Mathematical Model

We begin our work by describing the method for directed unweighted signed social networks, and then extend it to weighted networks. Given a directed graph $G = (V, E)$ with a sign (positive or negative) on each edge, we let adjacent matrix $A := (a_{ij})_{i:j=1,2,\cdots,n}$ denote the adjacent matrix of graph G. The element a_{ij} indicates the sign of the edge from node i to j. That is, $a_{ij} = 1$ when i marks j as a friend, -1 when i marks j as a foe, 0 when i doesn't mark j. Because G is a directed graph, A is asymmetric and $a_{ij} \neq a_{ji}$. And because there are too many users in social networks, the matrix A is a sparse matrix.

3.2 Commute Distance in General Social Networks

Fouss et al. [8] present a method to compute the similarities of node pairs based on a Markov-chain model of random walk through the undirected general graph. They prove that the square root of average commute distance is an Euclidean distance and provide similarities between any pair of nodes, having the nice property of increasing when the number of paths connecting those elements increases and when the 'length' of paths decreases.

- **The average first hitting time** $h(i, j)$ is defined as the expect time that a random walker, starting in state i hits the state j for the first time. It can be computed as shown in formula (1).

$$h(i, j) = \sum_{k \in nbs(i)}^{n} p_{ik} + \sum_{k \in nbs(i)}^{n} p_{ik} h(k, j) \tag{1}$$

In general networks, p_{ik} means the transition probability from state i to state k. It can be expressed as shown in formula (2).

$$p_{ik} = \frac{a_{ik}}{\sum_{k \in nbs(i)}^{n} a_{ik}} = \frac{a_{ik}}{\sum_{k=1}^{n} a_{ik}} \tag{2}$$

Where, a_{ik} is the element of the adjacency matrix A of the graph which is defined as usual as: $a_{ik} = w_{ik}$ if node i is connected to node k and $a_{ik} = 0$ otherwise.

- **The average commute distance** $n(i, j)$ is defined as the expect time that a random walker, starting in state i, enters state j for the first time and goes back to i.

Fouss et al. [8] prove the average commute distance can be computed by the Moore-Penrose pseudoinverse of the Laplacian matrix in undirected general networks as shown in formula (3).

$$n(i, j) = V_G(l_{ii}^+ + l_{jj}^+ - 2l_{ij}^+) \tag{3}$$

l_{ij}^+ is the element of the Moore-Penrose pseudoinverse (L^+) of the Laplacian matrix (L) of the graph. $L = D - A$, D is the degree matrix of the graph and A is the adjacent matrix of the graph. V_G means the volume of the graph. $V_G = \sum_i d_{ii}$, d_{ii} is the diagonal element of the degree matrix D.

4 Commute Distance in Directed Signed Social Networks

In this section we define the commute distance in directed signed networks and prove the distance is a legal kernel to compute the similarities of node pairs.

In signed networks, the weight a_{ik} may be negative and the transition probability p_{ik} may be negative by the previous definition in formula (2). It conflicts with the traditional non-negative probability. However, if we simply see the negative weight a_{ik} as zero, the network would degenerate into an general network and lose a lot of important information.

Feynman in [9] proposed the definition of negative probability. In his work, probabilities may be negative under certain assumed conditions. In recent years, negative probabilities theory has been widely used in quantum field. Here, if we consider probabilities as an intermediary probability from one state to another state and showing some posture (hostile or friend), the transition probability between states can be negative.

In signed networks, transition probabilities between states in Markov chain can be involved in three situations as shown in Fig. 1. By definition, the transition probability from state S_0 to state S_n is equal to the product of every transition probability. There is no difference between the first situation and traditional general graph. The transition probability from state S_0 to state S_n in the second situation and the third may be negative or positive. It depends on the number of negative edges in these situations. The final probability is negative when the number of negative edges is $(-1)^{2k+1} (k \geq 0)$, but positive when the number is $(-1)^{2k} (k \geq 0)$. The product coincides with the structure balance theory that the foe of my foe is more likely to be my friend.

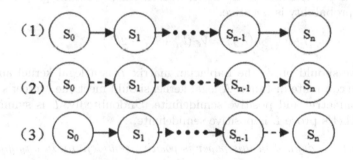

Fig. 1. Transition probabilities in signed graph(dotted line means -1, solid line means $+1$, (1)all edges are positive (2)all edges are negative (3)some edges are positive and some edges are negative)

In some special cases, the denominator in formula (2) ($\sum\limits_{j \in nbs(i)}^{n} a_{ij}$) may be zero when the number of positive edges is equal to that of negative edges. To avoid this meaningless situation, we use the absolute value of the out-degree to calculate the sum of the weights. We denote the extended transition probability $\tilde{P}^{(1)}$ in signed networks as shown in formula (4).

$$\tilde{p}_{ik} = \frac{a_{ik}}{\sum\limits_{j \in nbs(i)}^{n} |a_{ij}|} \quad (-1 \leq \tilde{p}_{ik} \leq 1) \tag{4}$$

We also extend the definition of the diagonal degree matrix D denoted as $\tilde{D}$, and $\tilde{d}_{ii} = \sum\limits_{(i,j)\in E} |a_{ij}|$. According to the definition of transition probability matrix, $\tilde{P}$ is equal to $\tilde{D}^{-1}A$. In most cases, $\sum\limits_{i} \tilde{p}_{ik}$ in formula (4) is not equal to 1 and clashes with the property of transition probability matrix. Hence, we should normalize the matrix $\tilde{P}$.

The corresponding Laplacian matrix is asymmetric because matrix A is asymmetric. In this paper, we use the normalize Laplacian matrix which is proposed by Chung [11], $\tilde{L} = (L + L^T)/2$, to define Laplacian matrix in directed signed network as shown in formula (5).

$$\tilde{L} = \frac{L + L^T}{2} = \frac{(\tilde{D} - A) + (\tilde{D}^T - A^T)}{2} = \frac{\tilde{D} + \tilde{D}^T}{2} - \frac{A + A^T}{2} = \tilde{D}' - B \quad (5)$$

Note: $\tilde{D}^T$ is not the transposed matrix of $\tilde{D}$ and $\tilde{d}_{ii}^T = \sum\limits_{(j,i)\in E} |a_{ji}|$.

In the paper of Fouss [8], the derivation process of the relationship between commute distance and Laplacian matrix doesn't make any requirements for the value of transition possibility matrix P. The relationship between commute distance and Laplacian matrix $\tilde{L}$ shown in formula (6) is still proper when the transition probability is negative.

$$n(i,j) = V_G(\tilde{l}_{ii}^+ + \tilde{l}_{jj}^+ - 2\tilde{l}_{ij}^+) \quad (6)$$

Now, we should prove the Laplacian matrix $\tilde{L}$ is a legal kernel and it can express the commute distance. A legal kernel should meet the Mercer's theorem and be symmetric and positive semidefinite. Undoubtedly, $\tilde{L}$ is symmetric by definition. Let's prove $\tilde{L}$ is positive semidefinite.

Theorem 1. $\tilde{L}$ defined in this paper is positive semidefinite in any graph G.

Proof. Let $\tilde{L}$ be the sum over the edges of graph G.
$\tilde{L} = \sum\limits_{i} \sum\limits_{j} \tilde{L}^{(i,j)}$. Where, $\tilde{L}^{(i,j)} \in R^{V*V}$ has four non-zero elements:
$\tilde{l}_{ii}^{(i,j)} = \tilde{l}_{jj}^{(i,j)} = \frac{|a_{ij}| + |a_{ji}|}{2} \geq \frac{|a_{ij} + a_{ji}|}{2}$, $\tilde{l}_{ij}^{(i,j)} = \tilde{l}_{ji}^{(i,j)} = -b_{ij} = -\frac{a_{ij}+a_{ji}}{2}$

Let $x \in R^V$ be a vertex column vector. Considering the bilinear of $\tilde{L}^{(i,j)}$, we find $\tilde{L}^{(i,j)}$ is positive semidefinite.

$x^T \tilde{L}^{(i,j)} x$
$= x_i^2 * \frac{|a_{ij}| + |a_{ji}|}{2} - 2x_ix_j * \frac{a_{ij}+a_{ji}}{2} + x_j^2 * \frac{|a_{ij}| + |a_{ji}|}{2}$
$\geq x_i^2 * \left|\frac{a_{ij}+a_{ji}}{2}\right| - 2x_ix_j * \frac{a_{ij}+a_{ji}}{2} + x_j^2 * \left|\frac{a_{ij}+\tilde{a}_{ji}}{2}\right|$
$= \left|\frac{a_{ij}+a_{ji}}{2}\right| (x_i - \text{sgn}\left(\frac{a_{ij}+a_{ji}}{2}\right) x_j)^2$
≥ 0
$x^T \tilde{L} x = \sum\limits_{i} \sum\limits_{j} x^T \tilde{L}^{(i,j)} x \geq 0$

Hence $\tilde{L}$ is positive semidefinite.

The extended Laplacian matrix $\tilde{L}$ can be used to calculate the similarity of node pairs. The time complexity in calculating the inverse of Laplacian matrix is $O(n^2)$. The time cost of commute distance method is huge for social networks which have millions of users. Matrix factorization method such as singular value decomposition (SVD) [12] is one of the effective ways to reduce the computational cost.

5   Link Mining Based on Collaborative Recommendation

Our proposed link mining method is based on the idea of collaborative filtering. We regard the target node as an item and the edges from the top-k nodes to the target node as the ratings given by the users. The top-k similarities of node i are used to predict the edge form i to j as shown in formula (7).

$$r(i,j) = \frac{\sum\limits_{m \in i' stop-knodes} sim(i,m) * a_{mj}}{\sum\limits_{m \in i' stop-knodes} sim(i,m)} \tag{7}$$

Where, $r(i,j)$ is denoted as average attitude (friendly/hostile) from the top-k nodes of i to the node j. $sim(i,m)$ is nodes similarity which can be calculated with the commute time proposed in formula (6). $sim(i,j) = n(i,j)$. a_{mj} is the element of the adjacent matrix A of the directed signed network.

6   Experiments

6.1   Experiment Process

The link mining algorithm proposed in this paper consists of three steps:

(1) Compute the extended Laplacian matrix $\tilde{L}$ as shown in formula (5) and it's Moore-Penrose pseudo-inverse.
(2) Compute the commute distance as shown in formula (6).
(3) Mine the direction and sign of the node pair in test set.

We compare experimentally our algorithm with four existing link mining algorithms, the low rank modelling with matrix factorization algorithm [7], transitive node similarity algorithm [13] and resistance distance method [4] denoted as LR-ALS, FriendTNS+ and A-sym, respectively. Henceforth, our proposed link mining approach based on commute distance similarity is denoted as CSLP.

6.2   Datasets and Metrics

To evaluate the performance of the algorithms, we adopt two real social signed networks: the Slashdot Zoo dataset and the Epinions dataset (downloaded from snap.stanford.edu). The two real datasets show high local clustering coefficients and low average shortest path lengths. These features can be mainly discovered in small-world networks.

In the experiment process, we adopt rand walk methodology to select 4000 nodes from each dataset and divide each dataset into two sets: (i) the training set E^T is regarded as known information and, (ii) the test set E^P is used for testing and no information in the test set is allowed to be used for link mining. We evaluate and compare these algorithms using a 10-fold cross-validation methodology.

In this paper, we use sign accuracy, AUC, precision and recall as evaluation metrics.

- **Sign accuracy.** Sign accuracy is the ratio of the number of right edges in the predicting results, which have the same signs as the corresponding edges in the test set, to the number of edges in the test set E^P.
- **AUC.** AUC is equivalent to the area under the receiver-operating characteristic (ROC) curve. It is the probability that a randomly chosen missing edge (an edge in E^P) is given a higher similarity value than a randomly chosen non-existent edge (an edge in $E - E^T$, where E denotes the universal set). In the implement process, among n times of independent experiments, if there are m times the missing edge having higher similarity value and p times the missing edge and non-existent edge having the same similarity value, we define AUC, as follows: $AUC = (m + 0.5 * p)/n$.
- **Precision.** Precision is the ratio of the number of right edges in the predicting results, having the right signs and direction, to the number of edges in the predicting results. As there is no existing method for predicting links' direction, the precision measure is only for the method proposed in this paper.
- **Recall.** Recall is the ratio of the number of right edges (both sign and direction) in the predicting results to the number of edges in the test set E^P. This measure is also only for the method proposed in this paper.

In this section we compare CLSP with other methods in terms of sign accuracy and AUC. Sign prediction accuracies of various methods are calculated with different values of k and the accuracies are averaged by 10-fold cross validation. The detailed results are shown in Fig. 2. In our algorithm, k is the top-k similarities. In FriendTNS+, k means the top-k transitive similarities. k is equal to the reduced dimension in LR-ALS and A-sym algorithm. The line charts in Fig. 2 show the accuracy for signs with different values of k. We can see that CLSP consistently achieves the highest accuracy for most of thresholds T in two real datasets, and the CLSP algorithm gets obviously higher sign accuracy in the Slashdot dataset.

We also compare CLSP with other algorithms in terms of AUC as shown in Fig. 3. We use a pure chance predictor as baseline algorithm which simply randomly selects pairs of nodes to be friends. The AUC value of pure chance predictor is 0.5. We use the AUC metric, which pays attention to an algorithm's overall ability to rank all the missing links over non-existent ones. We plot a

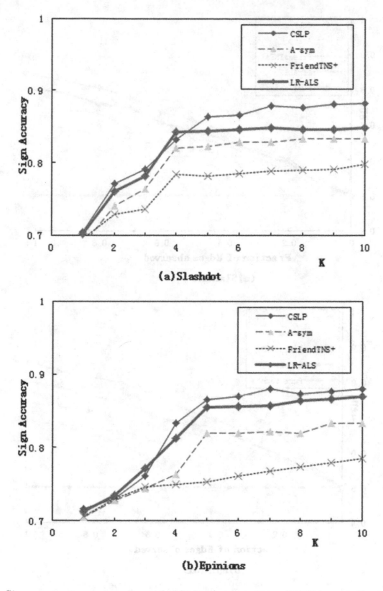

Fig. 2. Sign accuracy comparison of CSLP, A-sym, FriendTNS+ and the LR-ALS algorithm for Slashdot (b) Epinions datasets.

curve for AUC vs. the fraction of observed edges used in the training set. As shown, CLSP does better than pure chance and other algorithms, indicating that it is a strong predictor of missing structure. The main reason is the method seizes the edges' sign and direction and the link predicting process is based on the idea of collaborative filtering.

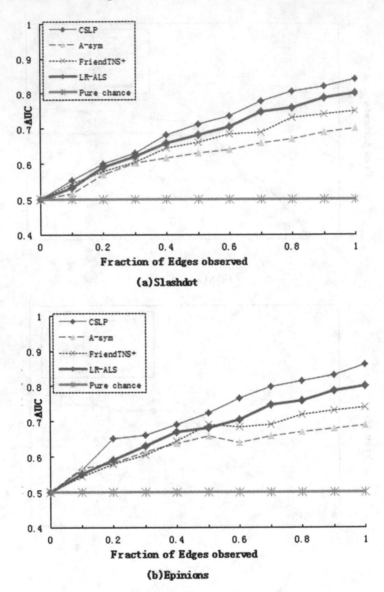

Fig. 3. AUC comparison of CSLP, A-sym, FriendTNS+, LR-ALS algorithm and pure-chance for (a)Slashdot (b) Epinions datasets.

We present the precision performance of CLSP when we take into account both right direction and sign in predicting results. As shown in Fig. 4, the precision is lower than sign accuracy shown in Fig. 2 at corresponding different levels of k. However, the precision is close to 80 percent in the two real datasets. In the Slashdot dataset, the precision is high to 0.7813 while it is 0.8010 in Epinions dataset.

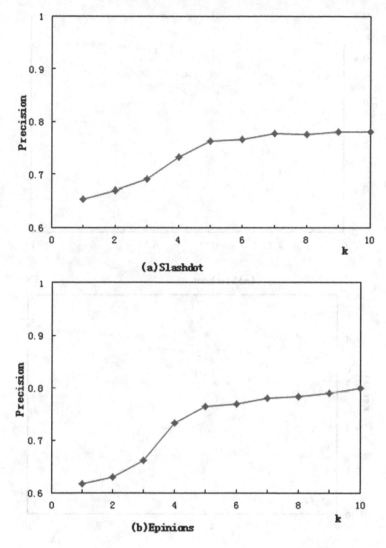

Fig. 4. Precision (right sign and direction) in CLSP proposed in this paper for (a)Slashdot (b) Epinions datasets.

Next, we proceed to examine the performance of our algorithm in terms of recall. The experiment result is shown in Fig. 5. The maximum value of recall is 0.4332 in the Slashdot dataset while it is 0.430 in the Epinions dataset.

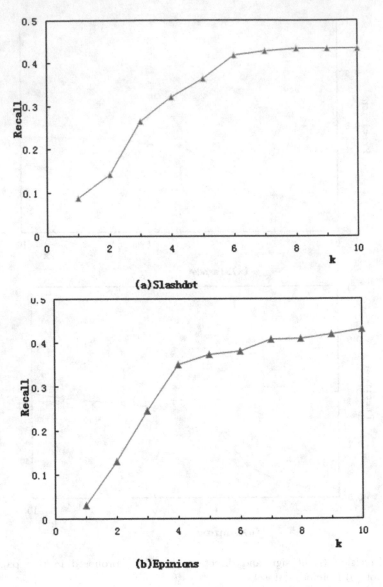

Fig. 5. Recall (right sign and direction) in CLSP proposed in this paper for (a)Slashdot (b) Epinions datasets.

7   Conclusion

We introduce a definition of the commute distance similarity in directed signed networks and use the similarity to mine the direction and sign of links. Our core mining process is based on the idea of collaborative filtering between friendships and links. The study extends the research approach of link mining and can be the necessary supplement of link mining in directed networks. The experiment results show the method gaining better performance in terms of sign accuracy and AUC measures than several existing algorithms.

In the future, we will continue our research on link mining in directed signed networks. With further study, we will explore the refinement of corresponding parameters and evaluation measure [15]. In addition, we will extend the predicted information by considering the bi-direction link prediction [14].

Acknowledgement. This work is supported by National Natural Science Foundation under Grant (No. 61373149, 61472233, 61572300), Technology Program of Shandong Province under Grant (No.2014GGB01617, ZR2014FM001), Taishan Scholar Program of Shandong Procince(No.TSHW201502038), Exquisite course project of Shandong Province (No. 2012BK294, 2013BK399, and 2013BK402), and Education scientific planning project of Shandong province (No. ZK1437B010).

References

1. Leicht, E.A., Holme, P., Newman, M.E.J.: Vertex similarity in networks. Phys. Rev. E **73**(3), 026120–026130 (2006)
2. Sarukkai, R.R.: Link prediction and path analysis using markov chains. Comput. Netw. **33**(1–6), 377–386 (2000)
3. Guha, R., Kumar, R., Raghavan, P., Tomkins, A.: Propagation of trust and distrust. In: Proceedings of the 13th International Conference on World Wide Web, pp. 403–412. ACM Press (2004) (doi:10.1145/988672.988727)
4. Kunegis, J., Lommatzsch, A., Bauckhage, C.: The Slashdot Zoo: mining a social network with negative edes. In: Proceedings of the 18th International Conference on World Wide Web, ESP, pp. 741–750. ACM Press (2009) (doi:10.1145/1526709.1526809)
5. Kunegis, J., Preusse, J., Schwagereit, F.: What is the added value of negative links in online social networks? In: Proceedings of the 22nd International Conference on World Wide Web, pp. 727–736 (2013)
6. Leskovec, J., Huttenlocher, D., Kleinberg, J.: Predicting positive and negative links in online social networks. In: Proceedings of the 19th International Conference on World Wide Web, pp. 641–650 (2010)
7. Chiang, K.-Y., Hsieh, C.-J., Natarajan, N., Dhillon, I.S., Tewari, I.S.: Prediction and clustering in signed networks: a local to global Perspective. J. Mach. Learn. Res. **15**(1), 1177–1213 (2014)
8. Fouss, F., Pirotie, A., Renders, J.M., et al.: Random-walk computation of similarities between nodes of a graph with application to collaborative recommendation. IEEE Trans. Knowl. Data Eng. **19**(3), 355–369 (2007)
9. Cartwright, D., Harary, F.: Structure balance: a generalization of Heiders theory. Psych. Rev. **63**(5), 277–293 (1956)
10. Hiley, B.J., Peat, F.D.: Quantum Implications: Essays in Honour of David Bohm. Psychology Press, London (1991)
11. Chung, F.: Laplacians and the Cheeger inequality for directed graphs. Ann. Comb. **9**(1), 1–19 (2005)
12. Koren, Y., Bell, R., Volinsky, C.: Matrix factorization techniques for recommender systems. IEEE Comput. **42**(1), 30–37 (2009)

13. Symeonidis, P., Tiakas, E., Manolopoulos, Y.: Transitive node similarity for link prediction in social networks with positive and negative links. In: Proceedings of the 2010 ACM Conference on Recommender Systems, pp. 183–190 (2010)
14. Wang, H., Yuan, W., Yu, X.: Bi-direction link prediction in dynamic multi-dimension networks. J. Comput. Inform. Syst. **10**(3), 1333–1340 (2014)
15. Zheng, Q., Skillicorn, D.B.: Spectral embedding of signed networks. In: SDM (2015)

Lossless and High Robust Watermarking of Electronic Chart for Copyright Protection

Jianguo Sun[1,2], Junyu Lin[1], Liguo Zhang[1], Shouzheng Liu[1(✉)],
Qian Zhao[1], Chao Liu[1], and Kou Liang[1]

[1] Institute of Information Engineering,
Chinese Academy of Sciences, Beijing 100859, China
5843825@qq.com
[2] Harbin Engineering University, Harbin 15001, China

Abstract. With the progression of sea exploration and offshore engineering, electronic charts have come to see widespread use in many intelligent applications. Like other digital products, electronic charts are easy to duplicate and distribute. Some watermarking solutions have proven defective to prevent copying of electronic charts because it's as easy to forge as it is to redistribute. If the problems of copyright infringement cannot be solved, the creation of these electronic charts will be limited. The most important characteristic of electronic charts is the topological relationships among vertices, but few algorithms can control this feature. A new watermarking algorithm is here proposed as a means of copyright protection, in which the watermarks will be hosted in the electronic chart by taking into account the preservation of the topology. Sometimes, additional vertices are inserted into the middle of two adjacent vertices, sometimes not, which are governed by the value of the watermark. Experiments show that the improved algorithm is better than similar algorithms; it was found to resist geometric attacks and format exchange attacks.

Keywords: Digital watermarking · Electronic charts · Copyright protection · Redundancy embedding · Topology

1 Introduction

In general, electronic charts are one kind of vector map, which is the core data in the electronic chart display and information system [1]. They are considered necessary to safe navigation. Raster charts are also used in these areas. Electronic charts have become a fundamental means of representing data structures in several marine navigational information systems. Especially, electronic charts, which have both land and marine coordinate systems (Fig. 1(c)), are more flexible than raster maps (Fig. 1(a)) and regular vector map (Fig. 1(b)). During the past twenty years, electronic charts have been widely used and distributed. This has raised concerns regarding the copyright protection of electronic charts. However, most existing algorithms can distort original maps after the watermarks are extracted [2]. This is highly undesirable because any disruption of the accuracy of the data will make the map unusable. It is difficult to recover the land and marine coordinate systems at the same time.

© Springer Science+Business Media Singapore 2016
W. Che et al. (Eds.): ICYCSEE 2016, Part I, CCIS 623, pp. 441–452, 2016.
DOI: 10.1007/978-981-10-2053-7_39

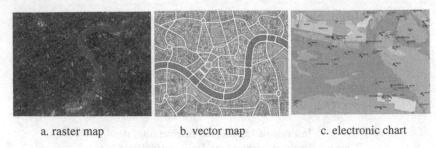

a. raster map   b. vector map   c. electronic chart

Fig. 1. Three types of digital maps

2 Features of the Watermarking Algorithm

Generally, the watermark can be embedded into the frequency domain or spatial domain. In spatial domain scenario, the watermark is directly embedded into the coordinate data, which visibly distorts the map. The map should be converted into the frequency domain, at which point the watermark is embedded into the frequency coefficients, which only slightly distorts the map.

There are few digital watermarking algorithms for electronic charts, but the digital watermarking methods for vector maps have been widely investigated in the last decade. Because of dual coordinate systems, not all algorithms for vector maps are capable of protecting the copyright of electronic charts. For example, by changing the vector data to embed the watermark in the spatial domain, the precision loss of land coordinate system is different from marine coordinate systems. To improve the watermarking of electronic charts, the topology of the maps was taken into account.

Few works have been proposed on topology: Lee et al. proposed the first water-marking algorithm for vector maps. It worked by modifying the vector coefficients in the DCT domain [3]. Considering the characteristic of vector geo-spatial data provided in two previous works, Zhang et al. presented a new watermarking scheme based on spectral coefficients in each factitious divided mesh [4]. With the goal of creating lossless watermarking for vector maps, Sun et al. proposed an algorithm based on recursive embedding [5]. Researchers also proposed several improved algorithms based on topological structure in frequency domains [6–8]. The topological characteristics of the map graphics are extracted, and then the relative coefficients of these vertices are calculated. Finally, the watermark is embedded into the coordinates. Lai and Zhang were the first to propose a solution to security issues regarding electronic charts [7]. In one study, they describe the methods they used to evaluate the performance of watermarking algorithms, and in another, they provided a security cryptography scenario to protect the content of electronic charts.

In general, the performance of digital watermarking for electronic charts means robustness, capacity, and invisibility. Information must be encoded into the watermark and embedded it into the electronic chart to indicate the copyright owner and allow detection of tampering.

A robust watermarking technique should be able to withstand different types of attacks and be easily read from the attacked map.

Producers generally prefer robust watermarking algorithms with high total information.

Invisibility is a key factor that determines the capability of the watermarking scenario to embed the watermarks. Sometimes, capacity and invisibility are mutually exclusive (Fig. 2).

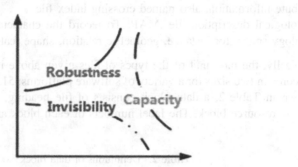

Fig. 2. Main performance factors

3 Characteristics of the Vector Map

Generally, the broadcasted vector map consists of single points, lines, and regions, which make up three types of letters, sometimes the maps only contain two or three types of objects. This makes these maps more complex than other digital products, such as images, audio, and video.

The object in the map is described with three types of information, which are attribute, topology, and property. These instructions describe the location, name, color, and spatial relationships with other unities in the map.

Because vector maps are widely used across many areas, such as geographic information systems (GIS), military management systems, intelligent transportation systems, and similar systems. There are several storage formats, and .SHP (Table 1) is one of main standard structures for vector map. The software environment was an ArcGIS programming platform.

Table 1. Structure of SHP format for vector map

No.	File name	Definition
1	.TAB FILE	Definition of variable
2	.DAT FILE	Description of resource index
3	.ID FILE	Description of object index
4	.MAP FILE	Description of data index for object

The details introductions of format files are as follows:

1. Structural definition file .TAB: To define the la, basic layers, color rendering, and other patterns.
2. Attribute description file .DAT: To describe the attribute information for every object in the map, including single vertexes, lines, and regions.
3. Object index file .ID: To create the one-to-one match between the object and its attribute information, also named crossing index file.
4. Topological description file .MAP: To record the characteristics of object in the topology space, for instance, geometric relation, shape features, and graphic data.

Typically, the base unit of the types of files given above is the data block. These blocks come in two sizes for a variety of software platforms: 512 bytes and 1024 bytes. As shown in Table 2, a data block consists of file heading, index block, definition block, and resource block. The label numbers of each block are marked as follows:

Table 2. Definitions of data block

Marked ID	Name
0	HEADER BLOCK
1	INDEX BLOCK
2	OBJECT DEFINITION BLOCK
3	COORDINATE DEFINITION BLOCK
4	DELETED BLOCK
5	RESOURCE BLOCK

4 Proposed Watermarking Scheme

The watermarking scheme consists of three steps: pre-processing, embedding, and extraction.

4.1 Pre-processing

To prevent various types of attacks, before embedding the operation, the vertices to be used to embed watermark must first be chosen. This procedure is called pre-processing. Vertices will be selected using the topological characteristics of the host map.

As mentioned in Sect. 3, the vertex is described by data blocks for object definition and object attributes. By tracking the ID, some features including the name, location and attributes of the specified vertex were retrieved from the map file.

The algorithm used for pre-processing is as follows:

Input: vector map V, vertex v, the coordinate value of which is (v_x, v_y), connected relation R in the line, connected relation Q in the region;

Output: feature vertices set FS.

In the map V, all of connected vertices can be denoted linked list L_i. The elements belonging to L_i, are sorted connected vertex set $\{v_1, v_2, \ldots, v_j\}$.

1. If vertices v_i, v_j that make up the $v_i R v_j$ are established and $v_j R v_i \equiv v_i R v_j$, then definite $L_i = L_i \cup \{v_i, v_j\}$;
2. Similarly, if some vertices connect to a closed area, such as a region in the map, denoting $v_i Q v_{i+1} Q \ldots v_{i+n-1} Q v_{i+n} Q v_i$, then $L_i = L_i \cup \{v_i, v_{i+1}, \ldots, v_{i+n-1}, v_{n+1}\}$;
3. As shown in Fig. 1(a), for some cross lines in the map, if $L_i \cap L_j = v_m$, $L_l \cap L_k = v_n$, $v_m R v_n$, indicating that some lines have common vertices. This defines $L' = L_i \cap \{v_m, v_n\}$.
4. As shown in Fig. 1(b) for some neighborhood regions, if $L_i \cap L_j \neq \Phi$, then define $L' = L_i \cap L_j$. Let L' be L' the critical set shared by multiple adjacent regions.
5. The linked list $L' L'$ is the result used to embed the watermark, and the elements of $L' L'$ also occupy the most important topological locations in the vector map.
6. $FS = L'$.

Before embedding, the map is pre-processed using the scenario given above, and one dataset FS is given to embed the watermark binary (Fig. 3).

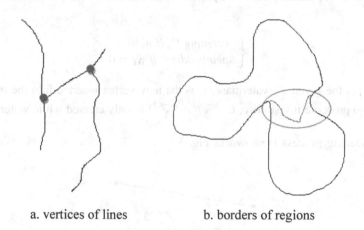

a. vertices of lines   b. borders of regions

Fig. 3. Critical objects in the vector map

4.2 Embedding Procedures

Restructure linked list $L' L'$: load the vertex elements from header of $L' L'$, and dived $L' L'$ into some sections by two vertices one pair. As shown in Fig. 4, every line curve consists of many micro-straight lines, so is the region in the map. And the pair of vertices is two terminal points of the straight line.

Define the coordinate values of two terminal vertices as $(x_a, y_a), (x_b, y_b)$ (x_a, y_a) and (x_b, y_b) and the coordinate value of the medium position C denotes $x_c = \frac{|x_b - x_a|}{2}$, $y_c = \frac{|y_b - y_a|}{2}$.

The embedding process takes place as follows:

Step 1: Group the pair vertices from linked list.

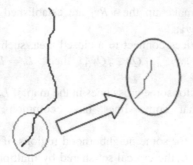

Fig. 4. Group theory for linked list

Step 2: Select the specified pair of vertices, and compute the coordinate value of medium point C.

Step 3: Using the watermarking bit value, embed watermark by creating the new vertex C or not creating one. The process is carried out on the pre-processed vector map:

$$\begin{cases} creating\ C,\ if\ w_i = 1 \\ unembedding,\ if\ w_i = 0 \end{cases} \tag{1}$$

Where w_i is the given i^{th} watermark, C is the new vertex inserted into the middle of the line, to preserve invisibility, $C(\frac{|x_b - x_a|}{2}, \frac{|y_b - y_a|}{2})$ is only created while watermark bit equals 1.

The embedding process is shown in Fig. 5.

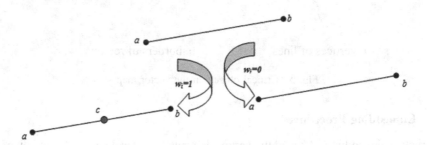

Fig. 5. Diagram of the watermark embedding process

4.3 Extraction Procedures

For a watermarked electronic chart, the procedure of watermark extraction and recovery can be demonstrated as follows:

Step 1: According to the topological distribution of the marked map, recover the feature dataset of vertices $L_{ex}\ L_{ex}$, the elements are all critical to the topology, if vector map is available, then $L_{ex} \approx L'\ L_{ex} \approx L'$ is established.

Step 2: After keys grouped by the pair of vertices for linked list L_{ex} L_{ex}, by secret key, the watermark bit could be extracted from every pair vertices.

Step 3: If there is one medium vertex C' between one pair of vertices in the same line, then remove C', and $W_{ti} = 1$, else, $W_{ti} = 0$.

5   Experimental Validation

To evaluate the performance of the proposed watermarking scheme, we use three .shp format maps, here called 'Harbin Services map', 'Road map' and 'Urban map'. As shown in Fig. 6, the original maps used in the current experiments represent three types of topological maps: single vertex sets (Fig. 6(a)), linear vertex sets (Fig. 6(b)), and regional vertex sets (Fig. 6(c)).

Abbas et al. proposed the robust algorithm for vector map, in which additional vertex coordinates are inserted between adjacent vertices whose locations are governed by the value of the watermark data [9]. From the point of realization, it can be compared to the algorithm used in this paper. The watermark image is still 70×60 pixels in size, like the one used by Abbas et al., and also black and white. The vector maps are in .shp format files. The parameter described by Abbas et al. is here specified as 0.03.

The watermark image used is shown in Fig. 7.

Fig. 6. Vector maps used in the experiments

Fig. 7. Watermark used in the experiments

The performance of the proposed algorithm was here measured using three criteria: robustness, watermarking capacity, and invisibility. The experiment exploits Visual Studio 6.0 and ArcGIS AO plug platform.

5.1   Robustness Evaluation

To assess the ability of the proposed algorithm to withstand attacks, the watermarked maps were here subjected to several rules. Then attempts were made to extract the watermark from the attacked host map. To quantify the extracted watermarks integrity, normalization correlation (NC) was calculated as described in Eq. 2 the algorithm.

$$Sim(W, W_t) = \frac{\sum\limits_{i=1}^{n} <W_i, W_{t_i}> /N}{\sqrt{\sum\limits_{i=1}^{n} W_i^2} \cdot \sqrt{\sum\limits_{i=1}^{n} W_{t_i}^2}}, \quad <W_i, W_{t_i}> = \begin{cases} 1, & W_i = W_{t_i} \\ 0, & W_i \neq W_{t_i} \end{cases} \quad (2)$$

Here n denotes the length of the watermark and W_{t_i} and W_i are the extracted and original watermark binary values, respectively.

Table 3 shows the results of geometric and format exchange attacks.

Table 3. Results of geometric and format exchange attacks

Attacks		NC	
		Our algorithm	Abbas et al.
Geometric attacks	Shearing ¼	0.95	0.87
	Rotation 30 degrees to the right	0.931	0.927
	Translation	1	1
Format exchange attacks	From SHP format to DWG format	1	1
	From SHP format to DWG format and back to SHP format	1	1

The results show that the proposed algorithm can preserve watermark information perfectly. It performed better when some of the topography was removed or changed.

Noise attacks made on the marked vector map, and the noise was applied to the vertices in the map. Two algorithms were used under the same experimental conditions (Table 4). The thresholds of noise attacks were specified as 0.015 and 0.03.

Because the algorithm proposed here has less information to be embedded into the vector map, the watermark does not distort the image as much as other watermarks do.

Sometimes, the map is simplified for convenient use. Here, a watermarked vector map was simplified using a compression algorithm [10]. By converting coordinates to different types, mapping the vector data, and then filtering the vertex on the curve from the map, the compression rate could exceed 40 %.

Table 5 shows the extracted watermarks after compressed attack on the vector map. The proposed algorithm performed visibly better than the one described by Abbas et al. [9]. Because the newly created vertices in the embedding algorithms were all

Table 4. Results for noise attacks

Noise strength	Threshold	Extracted watermark		NC	
		Current algorithm	Abbas et al.	Current algorithm	Abbas et al.
$5 \cdot 10^{-11}$	0.015	UCB	UCB	1	1
$5 \cdot 10^{-10}$	0.015	UCB	UCB	1	1
$5 \cdot 10^{-5}$	0.015	UCB	UCB	0.89	0.87
$5 \cdot 10^{-11}$	0.03	UCB	UCB	0.98	0.94
$5 \cdot 10^{-10}$	0.03	UCB	UCB	0.93	0.92
$5 \cdot 10^{-5}$	0.03	UCB	UCB	0.90	0.89

redundant, any compression algorithm was prone to remove these non-terminal points in some straight lines. Because the Abbas algorithm requires embedding almost twice as many watermarks as the current algorithm doing, it showed extracted watermark images of poorer quality [9].

5.2 Watermarking Capacity

To evaluate the capacity of the algorithm to carry the watermarks, the proposed algorithm was compared to the existing algorithm under the same conditions in [9]. We chose the same watermark to embed them into maps for 2, 5, and 10 times.

As shown in Table 6, the proposed algorithm had a larger capacity and better performance than the earlier algorithm. The medium vertex in the map was created when the watermark bit was equal to 1, showing the current algorithm to have twice the theoretical capacity as the Abbas algorithm [9].

Table 5. Results for compression attacks

| Compression Rate | Extracted watermark | | NC | |
	Current algo-rithm	Abbas et al.	Current algo-rithm	Abbas et al.
5%			0.92	0.83
10%			0.86	0.74
15%			0.79	0.69
20%			0.72	0.61

Table 6. Results for capacity testing

| Names of maps | NC | | Capacity (bits) | Recursions |
	Current algorithm	Abbas et al.		
Harbin services map	0.99	0.99	24039	2 times
Harbin services map	0.98	0.94	49021	5 times
Harbin services map	0.95	0.89	57284	10 times
Road map	0.99	0.99	18092	2 times
Road map	0.95	0.91	30927	5 times
Road map	0.91	0.87	42413	10 times
Urban map	0.98	0.97	1502	2 times
Urban map	0.94	0.87	2693	5 times
Urban map	0.89	0.78	4098	10 times

5.3   Invisibility

Figure 8 shows the original map (Fig. 8(a)) and watermarked map (Fig. 8(b)). As shown, the partial watermarks do not distort the map (Fig. 8(d)). Visually, the imperceptibility is very good. Figure 8(c) shows results produced using their algorithm in [9]. The fidelity is also perfect. However, as discussed, if the capacity is increased, the mark becomes more visible.

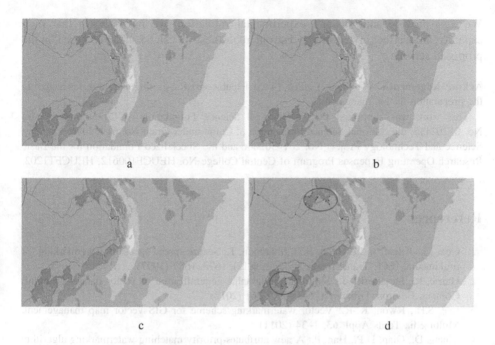

Fig. 8. Results of invisibility testing

The coordinates of the regular vertices in the map showed no changes after the watermarks were embedded. None of the original vertices were moved. Only limited medium vertexes were created. For this reason, all changes were within the allowable range of distortion. The watermarked map is also shown, and has the same function as original map.

For vector maps, the precision of geographical coordinates is very important. That is, if the coordinates of some key vertices are wrong, the map cannot be used. For this reason, if the critical vertices are used to anchor the watermarks, the watermarked map can reach a sweet spot, striking a balance between performance of the algorithm and usability of the map. That is, if some watermarks are destroyed, the critical vertices that maintain the topology of the map are removed or changed, rendering the map useless.

6   Conclusion

In this paper, an improved robust embedding scheme is proposed for the electronic chart safety. Our method could keep the less distortion to the map and enhance the performance of the watermarking algorithm. To handle the first issue, we decrease the embedding amounts to preserve the topology space on the map, which results in the high invisibility and preserved topology for the watermarked map. To handle the second issue, we proposed one improved means, in which fewer vertices are created and embedded between two terminal points in some important straight lines or regions.

We also provide analysis and experimental validation, we have shown that high capacity, and perfect invisibility and strong robustness all can be achieved based on the proposed scheme.

Acknowledgements. We thank LetPub (www.letpub.com) for its linguistic assistance during the preparation of this manuscript.

This work was supported by the National Science Foundation of China under Grant No. 61202455, the National Science Foundation of China under Grant No. 61472096, Liaoning Science and Technology Project No. 2014302006 and the Specialized Foundation for the Basic Research Operating Expenses Program of Central College No. HEUCF100612, HEUCFT1202.

References

1. Cox, I.J., Kilian, J., Leighton, F.T., Shamoon, T.: Secure spread spectrum watermarking for multimedia. IEEE Trans. Image Process. **6**(12), 1673–1687 (1997)
2. Harrie, L., Sarjakoski, J.: Simultaneous graphic generalization of vector datasets. GeoInf. Comput. Environ. Urbansyst. **6**(3), 233–261 (2012)
3. Lee, S.H., Kwon, K.-R.: Vector watermarking scheme for GIS vector map management. Multimedia Tools Appl. **63**, 1–34 (2011)
4. Zhang, D., Qian, D.P., Han, P.: A new attributes-priority matching watermarking algorithm satisfying topological conformance for vector map. In: 3rd International Conference on Intelligent Information Hiding and Multimedia Signal Processing, Kaohsiung, Taiwan, pp. 469–472 (2007)
5. Sun, J., Men, C.: Wavelet neural network based watermarking technology of 2D vector maps. High Technol. Lett. **17**(3), 259–262 (2011)
6. Clementine, E., Difelice, P.: A comparison of methods for representing topological relationships. Inf. Sci. **80**(1), 1–34 (2004)
7. Lai, M., Zhang, G.: A novel vector map watermarking evaluation based on electronic chart. Appl. Mech. Mater. **12**(11), 622–627 (2014)
8. Zhao, Y., Li, G., Li, L.: Electronic chart encryption method based on chaotic stream cipher. J. Harbin Eng. Univ. **28**(1), 60–64 (2007)
9. Abbas, T.A., Jawad, M.J., Sudirman, S.: Robust watermarking of digital vector maps for copyright protection, Liverpool, United Kingdom, pp. 1–6 (2013)
10. Xu, G., Tan, J., Zhong, J.: Adaptive efficient non-local image filtering. J. Image Graph. **17**(4), 471–479 (2012)

MapReduce for Big Data Analysis: Benefits, Limitations and Extensions

Yang Song, Hongzhi Wang[✉], Jianzhong Li, and Hong Gao

Harbin Institute of Technology, P.O. Box 750, Harbin 150001, China
wangzh@hit.edu.cn

Abstract. Big data becomes a hot topic. MapReduce is a popular programming paradigm for big data analysis with many benefits. Even though it has widely applications in industry, MapReduce still has limitations in some applications. For these limitations, some extensions have been proposed. In these brief communications, we discuss the benefits and limitations of MapReduce programming paradigm and also its extensions to make MapReduce go beyond the limitations.

Keywords: Big data · MapReduce · Parallel computation · Analysis

1 Introduction

We are at the new frontier of information explosion, in which industry, academia, business and governments are accumulating data in an unimaginable, unpredictable and unprecedentedly high speed. Thus big data processing techniques are in demand, whose goal is to accomplish computation tasks over huge datasets in a reasonable time.

To handle big data, it is a natural way to adopt parallel techniques. In the past few year, many parallel computing platforms have been built, which involve hundreds or even thousands of machines to process big data. Each platform may have a special programming paradigm. One of the most notable one is MapReduce [9], which is proposed by Google. Google also put up a computational diagram Pregel for graph. Spark made some improvements. Others like Hyacks and Graphlab is based on distributed system.

MapReduce provides an easy way to accomplish computational tasks on big data and can scale to large computer clusters. Meanwhile, it also supports the hardware fault tolerance during the process of calculation. To use this paradigm, the user only needs to write two functions, Map and Reduce. The system can manage the task parallel execution and coordination of mappers and reducers, and handle the failures. A MapReduce algorithm instructs these functions to perform a computational task collaboratively.

The MapReduce process a computational task as follows. It involves more than one map job. The input of each map job is one or more DFS files. A Map job converts file blocks to a key-value sequence. The main controller collects a series of key-value pairs from every Map task and sorts them by key sizes. For each time, a reducer process key-value pairs sharing the same key value, and combines the value according to requirement.

© Springer Science+Business Media Singapore 2016
W. Che et al. (Eds.): ICYCSEE 2016, Part I, CCIS 623, pp. 453–457, 2016.
DOI: 10.1007/978-981-10-2053-7_40

In MapReduce, grouping and aggregation are according to the same criterion. The Map function converts all input elements into key-value pairs. A map job can generate multiple key-value pairs from the same key. The Reduce function combines the values of a series of key-value table on a demand basis. The output of each reducer is also a key-value sequence, since the reduce task receives the key K of each key pairs as input keys. The outputs of all Reducers are merged into a single file as the final results.

As the earliest big data computing tools, MapReduce unavoidably has some foibles. However, they do not influence its important role in the research and development of big data. It has many advantages beyond other methods. The parallel big data processing improves the performance in big data analysis, which follows by other big data paradigms. This model is easy to use, even for programmers with no experience of distributed development. Also, it hides the details of parallel computing, disaster error, the optimization and load balance. MapReduce can easily handle large-scale parallel computing. For example, Google uses MapReduce to provide web search services, sorting, data mining, machine learning, and other systems. Through MapReduce, applications can run in more than 1000 large cluster nodes which provide optimized disaster error. What's more, the scalability of MapReduce is very awesome. For every server, MapReduce puts computing ability access to the cluster. For most of the past distributed processing frameworks [11, 12], their scalability inferiors to MapReduce a lot. Of course, there are still some limitations about MapReduce. However, with some effective proper algorithm, these problems can be solved completely.

To prove that MapReduce still alive, we will discuss a few representative examples of limitation in MapReduce and give some extensions to solve these problems. MapReduce is unique and can't be replaced. We will discuss in this paper and give evidence to demonstrate this thought.

2   Limitations

MapReduce is becoming a basic framework for processing massive data, due to its excellent scalability, reliability and elasticity. After released, MapReduce went through years of improvement into a mature paradigm. Although this is not a perfect approach, many ways have been proposed to advance it and contribute it to a more efficient, practical and convenient tool for big data analysis. First, we list some limitations to illustrate present problems. Here are some limitations in MapReduce and we take them as examples to share our view about MapReduce.

Limitation 1: incremental iterative MapReduce

Cloud intelligence applications often perform iterative computations on constantly changing datasets [4]. Many data-centric algorithms require efficient iterative computations, such as the well-known PageRank [5] algorithm in web search engines, gradient descent [6] algorithm for optimization, and many other iterative algorithms for applications including k-means algorithm, recommender systems [7] and link prediction [8]. Previous studies are too expensive to perform an entirely new large-scale MapReduce iterative job. It's difficult to timely accommodate new changes to the underlying datasets. To handle sophisticated iteration algorithm, the problem is eager to be solved.

Limitation 2: multiple inputs MapReduce

MapReduce is not designed to directly support operations with multiple inputs such as joins [2]. Many studies on join algorithms including Bloom join [10] in MapReduce have been conducted but they still have non-joining data generated and transmitted over the network. The join operation is considered as a paradigm of such operations. Although there have been many studies on join algorithms in parallel DBMSs, a MapReduce environment is not straightforward to implement joins. To improve MapReduce competence with multiple inputs, the solutions of this problem are in demand.

Limitation 3: minimal algorithms MapReduce

Ideally, a MapReduce system should achieve a high degree of load balancing among the participating machines, and minimize the space usage, CPU and I/O time, and network transfer at each machine. Although these principles have guided the development of MapReduce algorithms, limited emphasis has been placed on enforcing serious constraints on the aforementioned metrics simultaneously [3]. To guarantee the best parallelization and up to small constant factors, many studies about minimal algorithms of MapReduce have been performed.

3   Extensions

With many attentions about MapReduce, researchers have carried on many research work and come up with many extensions to go beyond the limitations. These extensions improve the performance of MapReduce, making it more powerful. Here we introduce some extensions that can effectively make up for the inadequacy of MapReduce.

Extensions 1: incremental iterative MapReduce

I2 MapReduce puts up to deal with incremental iterative MapReduce.

I2 MapReduce introduces a Map-Reduce Bipartite Graph model to represent iterative and incremental computations, which contains a loop between mappers and reducers [4]. A converged iterative computation means that a Map-Reduce Bipartite Graph state is stable. In many cases, only a very small fraction of the underlying dataset has changed, and the newly converged state is quite close to the previously converged state. Based on this observation, the design and implementation of I2 MapReduce to efficiently utilize the converged a Map-Reduce Bipartite Graph state to perform incremental updates is proposed.

An existing MapReduce application needs only slight code modification to take advantage of I2 MapReduce.

Extensions 2: multiple inputs MapReduce

As to multiple inputs MapReduce, a new type of bloom filter [10] called Intersection Bloom filter which represents the intersection of the datasets to be joined is proposed.

Unlike previous solutions that filter only on one dataset, the method filters out disjoint elements or non-joining tuples from both datasets. Each tuple from the input datasets is queried into the intersection filter and will be removed if it is a disjoint element. There are three approaches for building the intersection filter and show their feasibility in two-way joins and join cascades.

It is proved that the join operation using the intersection filter is more efficient than other solutions since it significantly reduces redundant data, and thus produces much less intermediate data [2]. Moreover, the intersection filter provides an extremely important characteristic for a join cascade in which intermediate join results generated from component joins only contain actual joining tuples without filtering. Although the intersection filter has false positives and an extra cost for the pre-processing step, its efficiency in space-saving and filtering often outweighs these drawbacks.

Extensions 3: minimal algorithms MapReduce

In recent years, optimum algorithms of MapReduce are studied. MapReduce has grown into an extremely popular architecture for large-scaled parallel computation. Even though there have been a great variety of algorithms developed for MapReduce, few are able to achieve the ideal goal of parallelization: balanced workload across the participating machines, and a speedup over a sequential algorithm linear to the number of machines [3].

"Minimal MapReduce algorithm" puts together for the first time four strong criteria towards the highest parallel degree. At first glance, the conditions of minimality appear to be fairly stringent. Nonetheless, the existences of simple yet elegant algorithms that minimally settle an array of important database problems.

The proposed algorithms demonstrate that the immediate benefit brought forward by minimality and significantly improve the existing state of the art for all the problems tackled.

4 Discussions

In this section, we discuss important drawbacks and solutions about MapReduce. In recent days, there is a debate whether MapReduce should be abandon. Some new methods have been proposed to take the place of MapReduce and claimed it is not wise to follow MapReduce which has many foibles and weakness. However, as discussed in Sect. 2, MapReduce can be improved through the above algorithm that we found. With these improvements, it is still lively in big data analysis. Through the extensions, MapReduce is becoming more and more mature and one of its implementation, Hadoop, has won more and more users. Besides, many proposed techniques on MapReduce give us more confidence on MapReduce. We have many reasons to believe that MapReduce will not vanish and it will have broad prospects, especially in offline social network. Hadoop has advantages in offline calculation with great accuracy. We are looking forward to breakthrough about MapReduce.

Acknowledgments. This paper was partially supported by National Sci-Tech Support Plan 2015BAH10F01 and NSFC grant U1509216, 61472099, 61133002 and the Scientific Research Foundation for the Returned Overseas Chinese Scholars of Heilongjiang Provience LC2016026.

References

1. Rajaraman, A., Ullman, J.D.: Mining of Massive Datasets. Posts & Telecom Press, Beijing (2012)
2. Phan, T.-C., d'Orazio, L., Rigaux, P.: Toward intersection filter-based optimization for joins in MapReduce. In: Cloud-I, p. 2 (2013)
3. Tao, Y., Lin, W., Xiao, X.: Minimal MapReduce algorithms. In: SIGMOD Conference, pp. 529–540 (2013)
4. Zhang, Y., Chen, S.: i2MapReduce: incremental iterative MapReduce. In: Cloud-I, p. 3 (2013)
5. Brin, S., Page, L.: The anatomy of a large-scale hypertextual web search engine. Comput. Netw. ISDN Syst. **30**, 107–117 (1998)
6. Avriel, M.: Nonlinear Programming: Analysis and Methods. Courier Dover Publications, Mineola (2003)
7. Baluja, S., Seth, R., Sivakumar, D., Jing, Y., Yagnik, J., Kumar, S., Ravichandran, D., Aly, M.: Video suggestion and discovery for youtube: taking random walks through the view graph. In: Proceedings of the WWW 2008, pp. 895–904 (2008)
8. Liben-Nowell, D., Kleinberg, J.M.: The link-prediction problem for social networks. JASIST **58**(7), 1019–1031 (2007)
9. Dean, J., Ghemawat, S.: Mapreduce: simplified data processing on large clusters. In: Proceedings of the OSDI 2004 (2004)
10. Bloom, B.H.: Space/time trade-offs in hash coding with allowable errors. Commun. ACM **13**(7), 422–426 (1970)
11. Borkar, V.R., Carey, M.J., Grover, R., Onose, N., Vernica, R.: Hyracks: a flexible and extensible foundation for data-intensive computing. In: ICDE, pp. 1151–1162 (2011)
12. Jiang, D., Chen, G., Ooi, B.C., Tan, K.-L., Wu, S.: epiC: an extensible and scalable system for processing big data. PVLDB **7**(7), 541–552 (2014)

Measurement of Nodes Importance
for Complex Networks
Structural-Holes-Oriented

Hui Xu[1,2(✉)], Jianpei Zhang[1], Jing Yang[1], and Lijun Lun[3]

[1] College of Computer Science and Technology,
Harbin Engineering University, Harbin, China
hzytsg2009@163.com
[2] Heilongjiang University of Chinese Medicine Library, Harbin, China
[3] College of Computer Science and Information Engineering,
Harbin Normal University, Harbin, China

Abstract. Mining important nodes in the complex network should not only consider the core nodes, but also consider the locations of the nodes in the network. Despite many researches on discovering important nodes, the importance of nodes in the structural holes is still ignored easily. Therefore, this paper proposes a method of local centrality measurement based on structural holes, which evaluates the nodes importance both by direct and indirect constraints caused by the lack of structural holes around the nodes. In this method, the attributes and locations of the nodes and their first-order and second-order neighbors are taken into account simultaneously. Deliberate attack simulation is carried out through selective deletion in a certain proportion of network nodes. Calculating the decreased ratio of network efficiency is to quantitatively describe the importance of nodes in before-and-after attacks. Experiments indicate that this method has more advantages to mine important nodes compared to clustering coefficient and k-shell decomposition method. And it is suitable for the quantitative analysis of the nodes importance in large scale networks.

Keywords: Complex networks · Structural holes · Nodes importance · Constraints

1 Introduction

Complex network is the abstract model established in reality in a variety of complex systems, with small-world characteristics [1] and scale-free property [2]. Complex network theory is widely used in the fields of Internet, power network, social network, scientific collaboration network and so on. Evaluating and measuring the nodes importance are of great significance to improve system robustness and design efficient system structure, such as effectively controlling the spread of diseases, rumors and computer viruses.

At present, the discovery of the important nodes in complex networks has become a hot research topic. Quantitative measurement of nodes importance in the network will usually use some centrality metrics, such as degree centrality [3], closeness centrality [4],

© Springer Science+Business Media Singapore 2016
W. Che et al. (Eds.): ICYCSEE 2016, Part I, CCIS 623, pp. 458–469, 2016.
DOI: 10.1007/978-981-10-2053-7_41

flow betweenness centrality [5], Katz centrality [6] and so on. But these indicators are strongly dependent on the network topology. From a new perspective, Kitsak et al. [7] uses the k-shell decomposition method to study the key nodes in the network, which is related to the locations of these nodes. K-shell decomposition method has low computational complexity, but the ranking results are too coarse-grained, which makes nodes have little difference. In addition, in the field of search engines, there are the famous Google PageRank algorithms [8] and LeaderRank algorithms [9]. And there are many other methods, such as nodes deletion, nodes contraction, mutual information and so on. These studies are respectively from different angles to sort the key nodes in the network. It is easy to see, mining important nodes should not only consider the core nodes, but also consider the locations of the nodes in the network. But as above algorithms, the capability of identifying the nodes in structural holes is limited.

In this paper, a new local centrality method based on structural holes is proposed, which is used to quantitatively evaluate the nodes importance by direct network constraints and indirect network constraints. And we propose a measuring method of nodes importance, which meditates the attributes and locations of the nodes and their first-order and second-order neighbors. This method uses total constraints as indicators to assess nodes importance. The greater the node's total constraint is, the less important it is. Namely, structural holes are difficult to form around it. Deliberate attack simulation is carried out through selective deletion in a certain proportion of network nodes. Calculating the decreased ratio of network efficiency is to quantitatively describe the importance of nodes in before-and-after attacks. This method is more advantageous than clustering coefficient and K-shell decomposition method for ranking accuracy. And it is also suitable for the quantitative analysis of the nodes importance in large scale network.

The remainder of this paper is organized as follows. Section 2 discusses some related works on measurement of nodes importance. We introduce structural holes theory and define the value of indirect constraints in Sect. 3. We propose a measuring method of nodes importance. And deliberate attack simulation is carried out through selective deletion in a certain proportion of network nodes. Calculating the decreased ratio of network efficiency is to quantitatively describe the importance of nodes in before-and-after attacks in Sect. 4. A summary of the contributions and future work is given in Sect. 5.

2 Related Works

Identifying important nodes in the complex network is of great significance to network optimization and robustness enhancements. So far, researchers have put forward a variety of mining important nodes based on the specific problems. Next, some related works on measurement of nodes importance will be summarized in the undirected graph networks [19].

Measurement of nodes importance is based on nodes adjacent. This method is the most simple and intuitive method. Degree centrality [3] only considers the node's local information without contemplating the environment around the node, such as its network location and more high order neighbors. So degree centrality is not precise enough in many cases. Chen et al. [20] present semi-local centrality, and its computational complexity is growing linear with network scale. It consumes much less computation time and its ranking results are far better than the degree centrality and betweeness centrality. Kitsak et al. [7] proposed k-shell decomposition to determine the positions of the nodes in the network. This method has low computational complexity and has many applications on analysis of hierarchy structure in large scale networks. However, this method has some limitations. K-shell decomposition can't play a role in some scenes, such as a tree diagram, regular network and the BA scale-free network. The ranking results are too coarse-grained, which makes the distinctions of nodes are small. K-shell decomposition method only deliberates the influence of residual degrees, and it is obviously unreasonable to think that the same layer nodes' outer layer has the same number of neighbors.

Measurement of nodes importance is based on paths. This method discusses the control of information flow of nodes, which are closely related to the path of the network. Closeness centrality [4] calculates the average value of the distance between one node and all other nodes in the network in order to eliminate the interference of special values. But its time complexity is relatively high. Katz centrality [6] not only pays attention to the shortest paths, but also considers the other non-shortest paths. But its time complexity is also relatively high. Information indices consider all paths, and can simplify the complicated calculation process by using a resistance network. The proposed method can be easily extended to the power network, and non-connected networks. Betweenness centrality [22, 23] describes the control of the network flow along the shortest path in the network. Its application is limited because of high time complexity. As flow betweenness centrality [5] is defined, much more the ratio of passing non-duplicate paths of a node is, the more important it is. Betweenness centrality and flow centrality consider two extremes. The former only discusses the shortest paths, and the latter considers all paths which are considered the same effect. Communicability centrality [24] takes into account all the paths, and gives each longer path the smaller weights. Based on this, Newman [25] proposed a random walk betweenness centrality. From the global perspective, Subgraph centrality [26] examines the enhancement effect of all the reachable neighbors in the network to the node centrality, and thinks that the enhancement effect will decrease with the increase of the distance.

Measurement of nodes importance is based on eigenvector. This method studies the number of nodes' neighbors and the influence of the nodes importance. Eigenvector centrality [27] and cumulative nomination [28] are generally used in undirected networks, which converge faster. PageRank algorithm [8] and LeaderRank algorithm [9] are used to identify web pages importance by simulating user surfing and increasing the nodes' score along access paths. Experimental results show that LeaderRank algorithm performs better than PageRank algorithm. HITs algorithm [29], automatic resource compilation (ARC) [30] and stochastic approach for link structure analysis (SALSA) [31] take into account the dual role of the node: authority and hub, which influence each other. This kind of method has received great attention in both theory and business and it is very helpful for reference.

Measurement of nodes importance is based on nodes removal and shrinkage. This method's most notable feature in the ranking process is that the structure of the network will be in the midst of the dynamic changes and the nodes importance often reflects the network destruction after nodes removed.

Currently, there are not many researches on measurement of nodes importance for complex networks based on structural holes. On the one hand, the nodes in the location of structural holes have unique characteristics and these characteristics will separate the structural holes nodes and the core nodes. So even if the nodes in the structural hole play an important role, they are still neglected. On the other hand, complex network researches on structural holes are not thorough, and how to scientifically measure structural holes nodes in the complex network are also worth exploring [32].

3   Foundations

3.1   Structural Holes

Inspired by "weak ties theory" [10] and "network closure theory" [11], Burt proposed structural holes theory in "Structural Holes: The Social Structure of Competition" in 1992 [12]. It studies tertius relations, namely, one party of structural holes gains social capital and benefits through controlling the other two parties. According to Burt's definition, a structural hole is a non-redundant contact between two players [21]. In other words, if there is no direct connection between the two players who both connect one player, this player occupies the network location called structural holes from the perspective of relationship missing. As showed in Fig. 1 [13], there are no redundant contracts among nodes 2, 3, and 4, which are all associated with node 1. Node 1 occupies the three structural holes 23, 34 and 24, and gains social capital and benefits, which take the cumulative yields but not overlapping network benefits. Therefore, node 1 is more important than its connected nodes.

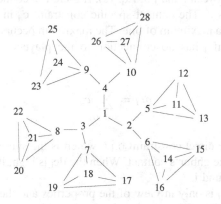

Fig. 1. Structural holes

3.2   Secondary Holes

Secondary only refers to the removal of a hole from the central player. Generally, structural holes are limited to focus on an egocentric network of the central player. And in the egocentric network of their direct contacts around the central player, there will be structural interval between the direct contacts and their direct contacts. Formed structural holes are called secondary holes for the centric player. Primary holes are between a player's direct contacts, and secondary holes are between indirect contacts. Of the two kinds of holes, the latter is the more intense. By analogy, we can define n grade structural holes (n $\geq$ 2). Burt [13] discussed the role of secondary holes in the focus of central players. As showed in Fig. 1, there are primary structural holes between node 1 and its direct contacts 2–4, and there are secondary holes between node 1 and its non-direct contacts 5–10.

3.3   Direct Constraint and Indirect Constraint

Burt uses the network constraint index to measure the constraint of the network node to form a structural hole. Equation (1) defines j's constraint on i:

$$c_{ij} = \left(p_{ij} + \sum_q p_{iq}p_{qj}\right)^2, \quad i \neq j \neq q \tag{1}$$

$$p_{ij} = z_{ij} \Big/ \sum_{k \in \Gamma(i)} z_{ik}, \quad j \in \Gamma(i), \tag{2}$$

where $\Gamma(i)$ is a finite set of i's neighbors, p_{ij} is the proportion of i's network time and energy invested in the relationship with j (interaction with j divided by the sum of i's relations), p_{iq} and p_{qj} are the proportion of i's and j's network time and energy invested in the relationship with their common neighbor q respectively, z_{ij} indicates whether there is a connection between i and j in Eq. (2). If there is a connection between i and j, z_{ij} is 1. Otherwise z_{ij} is 0. The contact-specific constraint c_{ij} in Eq. (1) varies from a minimum of zero up to a maximum of one. The maximum occurs if j is i's only contact. The minimum occurs if j has no connection to the players with whom you could replace him.

$$dc_i = \sum_{j \in \Gamma(i)} c_{ij} \tag{3}$$

In Eq. (3), dc_i is the direct constraint on i, which is the sum of constraints from i's relationship with each neighbor's contact. When the dc_i is high, it is difficult to develop the structural holes around i.

Direct constraint dc_i is only in view of the properties and the network location of i and its first-order neighbors, but does not take into account the properties and the network location of i's second-order neighbors. Therefore, we define the indirect constraint of i caused by forming a secondary hole in Eq. (4). sc_i is the arithmetic mean

of the aggregate constraints of the i's direct contacts, which is used to measure the lack
of the secondary holes around i.

$$sc_i = \sum_{k \in \Gamma(j)} dc_k \Big/ N_j, \quad j \in \Gamma(i) \tag{4}$$

Where sc_i is the indirect network constraint, k is the finite set of i's second-order
neighbors. N_j is the number of i's first-order neighbors.

As showed in Fig. 1, we will calculate 1's and 2's indirect network constraints.
Node 2's, 3's and 4's first-order neighbors sets are $\Gamma(1) = \{2, 3, 4\}$, $\Gamma(3) = \{1, 7, 8\}$
and $\Gamma(4) = \{1, 9, 10\}$ respectively and 1's direct constraint is $dc_1 = c_{12} + c_{13} +$
$c_{14} = (1/9 + 1/9 + 1/9) = 1/3$. Similarly, node 2's, 3's and 4's direct constraints are
$dc_2 = dc_3 = dc_4 = 1/3$. Then node 1's indirect constraint is $sc_1 = (c_2 + c_3 + c_4)/$
$3 = 1/3$ and 2's indirect constraint is $sc_2 = (c_1 + c_5 + c_6)/3 = 1/2$. Here the direct
constraint of 1 and 2 is equal, but in the actual network as showed in Fig. 1, node 1 is
more important than node 2. Namely, only considering the direct network constraint is
not enough. We should also pay close attention to the indirect constraint, in order to
better evaluate the nodes importance. In the following section, we will introduce a
measuring methods of nodes importance based on structural holes and simulation
experiments.

4 Methods and Experiments

4.1 Measurement of Nodes Importance

From the complex network perspective, the network nodes with more structural holes
are more conducive to the dissemination of information, so these nodes are more
important. Through the analysis of the previous section, node 1's and 2's direct constraints are equal, and we cannot distinguish which node is more important. Therefore
we propose a measuring method of nodes importance TCM (total constraints method),
which evaluates the nodes importance by the total constraints of direct and indirect
constraints caused by the lack of structural holes around the nodes. This method does
not only focus on the direct constraints of the central player to form primary structural
holes, but also takes into account the indirect constraints of the central player to form
secondary holes, which better distinguish the node importance.

$$TCM_i = \alpha \, dc_i + \beta \, sc_i \tag{5}$$

In Eq. (5), TCM_i is the total constraints on i, dc_i is the direct network constraint, sc_i
is the indirect network constraint, α and β are adjustable parameters respectively, which
are used to adjust nodes importance by the proportion of direct constraints and indirect
constraints. From the perspective of spatial autocorrelation, the closer two objects are,
the more they are dependent on each other. Using spatial autocorrelation theory, it can
be considered that the direct network constraint is more important than the indirect

network constraint. Studies have shown that some characteristics of complex systems go down exponentially with the increase of distance. Here α equals 1 and β equals 1/2.

As showed in Fig. 1, the total constraints of node 1 and 2 are respectively: $TCM_1 = dc_1 + 1/2) * sc_1 = (1/3) + (1/2) * (1/3) = 0.5$, $TCM_2 = dc_2 + (1/2) * sc_2 = (1/3) + (1/2) * (1/2) = 0.53$. Node 1's total constraints are less than node 2's total constraints, which mean that the node 1 is easier to form structural holes and is more favorable to communicate information, and vice versa. The measuring method of nodes importance can overcome the limitations of the direct constraints, which considers both the properties and location of the node and its neighbors to better measure the nodes importance.

4.2   Network Efficiency

Network efficiency [14, 15] is the indicator to express the network connectivity being good or bad. The better network connectivity is, the higher the network efficiency is. In a network, one node is removed when it is subjected to network attacks. This means that at the same time, all edges connected to the node may be removed, which may cause certain paths in the network to be interrupted. If there are multiple paths between i and j, interrupting some paths may cause the length of the shortest path d_{ij} increased. And the average path length of the entire network is also increased, which makes the network connectivity become worse.

The efficiency between node i and the other nodes in the network is $1/d_{ij}$, and network efficiency ε is expressed as:

$$\varepsilon = \frac{1}{N(N-1)} \sum_{i \neq j \in G} (1/d_{ij}) \tag{6}$$

where the network efficiency ε's values range is between zero and one. If ε is one, it indicates the network connectivity is best. If ε is zero, it indicates the network is consisted of isolated nodes.

Deliberate attack simulation is carried out through selective deletion in a certain proportion $p \in [0,1]$ of network nodes using different methods such as TCM, clustering coefficient and k-shell decomposition method. Calculating decreased proportion (before-and-after attacks) of network efficiency is to quantitatively describe the importance of nodes. Assume that before suffering network attacks, network efficiency is ε_0, and network efficiency is ε after attacks by selectively deleting some important nodes in a certain proportion p. Decreased ratio of network efficiency e is expressed as:

$$e = 1 - \frac{\varepsilon}{\varepsilon_0} \tag{7}$$

where e varies from a minimum of zero up to a maximum of one. If e equals one, it indicates network efficiency is down to zero after attacks. Namely the network is composed of isolated nodes. If e equals zero, it indicates network efficiency is not changed after attacks. In Eq. (7), it can be seen that the greater e is, the worse the network efficiency gets after attacks by selectively deleting some important nodes. So this method is more accurate to measure nodes importance.

4.3 Simulation Experiments

The experimental data used in this paper are open for public data sets,such as dolphin social network, American college football, books about US politics and neural network. The dolphin social network [16] is an undirected social network of frequent associations between 62 dolphins in a community living off Doubtful Sound, New Zealand. American college football [17] is the network of American football games between Division IA colleges during regular season Fall 2000. Books about US politics are the network of books about US politics published around the time of the 2004 presidential election and sold by the online bookseller Amazon. The network was compiled by V. Krebs and is unpublished, but can be found on Krebs' website. Neural network [1, 18] is the network representing the neural network of C. Elegans.

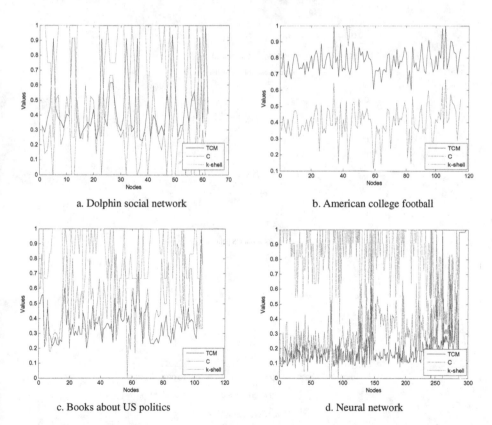

a. Dolphin social network   b. American college football

c. Books about US politics   d. Neural network

Fig. 2. a–d Values distributions of TCM, clustering coefficient and k-shell of dolphin social network, American college football, books about US politics and neural network (Color figure online)

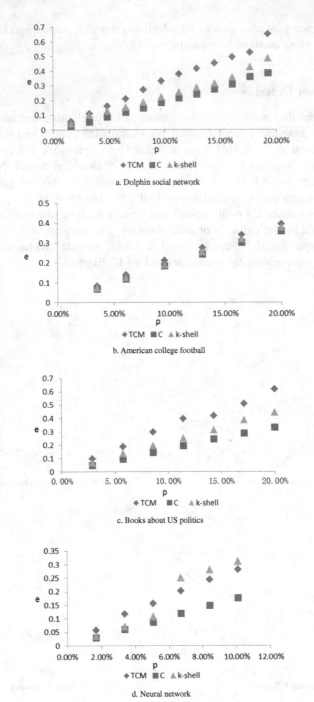

a. Dolphin social network

b. American college football

c. Books about US politics

d. Neural network

Fig. 3. (a–d) Values of decreased network efficiency ratio e with different proportions p of deleting nodes

The Physical significance and measuring unit of clustering coefficient, k-shell decomposition method and TCM are different, which leads to the difference of data dimensions and orders of magnitude. Therefore, the three indicators are normalized. Figure 2a–d display the values distributions of TCM, clustering coefficient and k-shell of the four networks. It is easy to see that the importance of each node is different with different indicators.

In order to further verify the effect of identifying the important nodes of TCM presented in this paper, we choose the clustering coefficient, k-shell decomposition method and TCM to execute a deliberate attack simulation by deleting nodes with different proportions. Calculating the decreased ratio of network efficiency is to quantitatively describe the importance of nodes in before-and-after attacks. We select the twenty percent important nodes of the four different networks and the experimental results are showed in Fig. 3a–d, where e is the decreased ratio of network efficiency and p is the certain proportion of deleting nodes.

Obviously, TCM's experimental results are better and it has more stability than clustering coefficient and k-shell. And TCM can accurately measure the nodes importance in complex networks. But in Fig. 3d, there is a jump in k-shell's decreased network efficiency ratio, which means that k-shell decomposition method doesn't accurately locate the core of the network in all real networks.

5   Conclusions

This paper proposes a new method TCM for local centrality, which is used to evaluate the nodes importance in the network. This method uses direct network constraints and indirect network constraints to quantitatively evaluate nodes importance based on structural holes. Through the simulation, this TCM method is proved to be a good balance between the complexity and the ranking accuracy compared to clustering coefficient and k-shell decomposition method. TCM is more accurate than direct constraints, and its calculation is only based on the local information of the network.

This paper studies the effect of nodes importance both by direct and indirect constraints caused by the lack of structural holes around the nodes. The next step we will study the interaction between the local structural holes and the global structural holes.

Acknowledgments. The work was supported by The National Natural Science Foundation of China (Nos. 61402126, 61073043, 61370083).

References

1. Watts, D., Strogatz, S.: Collective dynamics of small world networks. Nature **393**, 440–442 (1998)
2. Barabási, A., Bonabeau, E.: Scale free networks. Sci. Am. **288**, 60–69 (2003)
3. Bonacich, P.: Factoring and weighting approaches to status scores and clique identification. J. Math. Sociol. **2**(1), 113–120 (1972)

4. Freeman, L.C.: Centrality in social networks conceptual clarification. Soc. Netw. **1**(3), 215–239 (1979)
5. Yan, G., Zhou, T., Hu, B., et al.: Efficient routing on complex networks. Phys. Rev. E **73**, 046108 (2006)
6. Katz, L.: A new status index derived from sociometric analysis. Psychometrika **18**(1), 39–43 (1953)
7. Kitsak, M., Gallos, L.K., Havlin, S., et al.: Identification of influential spreaders in complex networks. Nat. Phys. **6**, 888–893 (2010)
8. Brin, S., Page, L.: The anatomy of a large-scale hypertextual web search engine. In: Seventh International World-Wide Web Conference, vol. 30, pp. 107–117 (1998)
9. Lü, L., Zhang, Y.C., Yeung, C.H., et al.: Leaders in social networks, the delicious case. PLoS ONE **6**, e21202 (2011)
10. Granovetter, M.S.: The strength of weak ties. Am. J. Sociol. **78**(6), 1360–1380 (1973)
11. Coleman, J.S.: Social capital in the creation of human capital. Am. J. Sociol. **94**, 95–120 (1988)
12. Burt, R.S.: Structural Holes: The Social Structure of Competition. Harvard University Press, Boston (1992)
13. Burt, R.S.: Secondhand brokerage: Evidence on the importance of local structure for managers, bankers and analysts. Acad. Manag. J. **50**(1), 119–148 (2007)
14. Vragovic, I., Louis, E., Díaz, A.: Efficiency of informational transfer in regular and complex networks. Phys. Rev. E **71**, 036122 (2005)
15. Ren, Z.M., Shao, F.L., Jian, G., et al.: Node importance measurement based on the degree and clustering coefficient information. Acta Phys. Sin. **62**(12), 128901 (2013)
16. Lusseau, D., Schneider, K., Boisseau, O.J., et al.: Behav. Ecol. Sociobiol. **54**, 396–405 (2003)
17. Girvan, M., Newman, M.E.J.: Community structure in social and biological networks. Proc. Natl. Acad. Sci. USA **99**(12), 7821–7826 (2002)
18. White, J.G., Southgate, E., Thompson, J.N., Brenner, S.: Phil. Trans. R. Soc. Lond. **314**, 1–340 (1986)
19. Ren, X.L., Lü, L.Y.: Review of ranking nodes in complex networks. Chin. Sci. Bull. (Chin. Ver.) **59**, 1175–1197 (2014). (in Chinese)
20. Chen, D.B., Lü, L., Shang, M.S., et al.: Identifying influential nodes in complex networks. Physica A **391**(4), 1777–1787 (2012)
21. Stephenson, K., Zelen, M.: Rethinking centrality: methods and examples. Soc. Netw. **11**(1), 1–37 (1989)
22. Brandes, U.: A faster algorithm for betweenness centrality. J. Math. Sociol. **25**, 163–177 (2001)
23. Zhou, T., Liu, J.G., Wang, B.H.: Notes on the algorithm for calculating betweenness. Chin. Phys. Lett. **23**(8), 2327 (2006)
24. Estrada, E., Hatano, N.: Communicability in complex networks. Phys. Rev. E **77**(3), 036111 (2008)
25. Newman, M.E.J.: A measure of betweenness centrality based on random walks. Soc. Netw. **27**, 39–54 (2005)
26. Estrada, E., Rodriguez, J.A.V.: Subgraph centrality in complex networks. Phys. Rev. E **71**, 056103 (2005)
27. Bonacich, P.: Factoring and weighting approaches to status scores and clique identification. J. Math. Sociol. **2**(1) (1972)
28. Poulin, R., Boily, M.C., Mâsse, B.: Dynamical systems to define centrality in social networks. Soc. Netw. **22**(3), 187–220 (2000)

29. Kleinberg, J.M.: Authoritative sources in a hyperlinked environment. J. ACM (JACM) **46**(5), 604–632 (1999)
30. Chakrabarti, S., Dom, B., Raghavan, P., et al.: Automatic resource compilation by analyzing hyperlink structure and associated text. Comput. Netw. ISDN Syst. **30**(1–7), 65–74 (1998)
31. Lempel, R., Moran, S.: The stochastic approach for link-structure analysis (SALSA) and the TKC effect. Comput. Netw. **33**(1–6), 387–401 (2000)
32. Han, Z.M., Wu, Y., Tan, S.X., et al.: Ranking key nodes in complex networks by considering structural holes. Acta Phys. Sin. **64**(5), 058902, 1–9 (2015)

Method of Consistency Judgment for App Software's User Comments

Meng Ran[1,2], Ying Jiang[1,2(✉)], Qixin Xiang[1,2], Jiaman Ding[1,2], and Haitao Wang[1,2]

[1] Yunnan Key Lab of Computer Technology Application, Kunming 650500, China
{374085783,346760502}@qq.com, {jy_910,tjom2008,kmwht}@163.com
[2] Faculty of Information Engineering and Automation,
Kunming University of Science and Technology, Kunming 650500, China

Abstract. With the popularity of mobile intelligent terminal, user comments of App software is viewed as one of the research interests of social computing. Faced with the massive App software, most users usually view the other users' comments and marks to selecting the desired App software. Due to the freedom and randomness of the network comments, the inconsistence between the user's comment and mark makes it difficult to choose App software. This paper presents a method by analyzing the relationships among user's comment information, the user's mark and App software information. Firstly, the consistency between user's comment information and App software information is judged. Then, through analyzing the grammar relationships among the feature-words, adverbs and the feature-sentiment-words in App software's feature-sentiment-word-pairs, the user's emotional tendency about App software is quantified combining with the dictionary and the network sentiment words. After calculating the user's comprehensive score of App software, the consistency of App software's user comment is judged by comparing this score and the user's mark. Finally, the experimental results show that the method is effective.

Keywords: Social computing · App software · The consistency of user comment · User's comment information · User's mark · Feature-sentiment-word-pairs · Network sentiment word

1 Introduction

With the development of information technology, social computing is known as one of the data-intensive science which catches the scholars' attentions, especially social networks analysis, generative use and user comments of product [1]. In recent years, In recent years, App software is regarded as a novel experience-based product [2], especially Android and iOS system increased rapidly. According to the data, the number of active mobile devices has reached 899 million up to December 2015 in China.

© Springer Science+Business Media Singapore 2016
W. Che et al. (Eds.): ICYCSEE 2016, Part I, CCIS 623, pp. 470–483, 2016.
DOI: 10.1007/978-981-10-2053-7_42

Different from other applications, App software has the following characters: (1) because App's development cycle is usually short, the developer can make the software development strategy after requiring user's requirements through their comments [3]. (2) Before selecting App software, users cannot get the quality information about the software by an advertisement or a brand, which will makes it difficult to choose App software [4]. (3) In order to make the rational choice, users would like to know the quality of App software through user's comment information [4]. User comments of App software include rating marks (usually five marks) scored by user and comment descriptions after using App software. User comments often imply potential information such as the preference and attention to some specific properties of App software. However, as the freedom and random of network comments, there are some inconsistencies among App software information, user's comment information and user's mark, which will bring great difficulties to evaluation of App software's quality for users. So it is an important problem to judge the consistency of user comments of App software.

2 Related Work

Currently, the study of users' comments in social computing has been mature, especially aiming at user comments of online communities, hedonic use, etc. [5]. But user comments of App software are still on study. Gao et al. [6] establish a theme dynamics update model through extracting theme comments and sorting based on different App software user comments. AlQuwayfili et al. [7] divide the user comments into credible comments and incredible comments based on the analysis. The above research shows that the theme and credibility of user comments have influences on user's choice of App software. However, there is lack of comprehensive analysis for user's comment information, user's mark and App software information. Other researches focus on the influences of the sentiment tendency [8] of users to App software on App software's quality in user comments. Islam [9] divides the user comments into different marks by analyzing sentiment and optimizing probability of user comments. Guzman et al. [10] identify the 'coarse feature' of App software by establishing a software mining warehouse and defining feature patterns, then the positive and negative attitudes towards App software in user comments are explored based on the greedy clustering algorithm. In the above researches, the tendencies of user comments are divided into two levels. Usually, users also give five marks when they comment App software. However, the current researches ignore whether the user's comment information and user's mark are consistent when user is selecting App software.

In this paper, we aim at how to judge the consistency of user comments on App software through analyzing the relationship among user's comment information, user's mark and App software information comprehensively. The consistency between the user's comment information and the App software information is judged. When the user's comment information is consistent with the App software information, the consistency between the user's comment information and the user's mark is judged by the following steps: (1) the grammatical relations among feature words, adverbs and sentiment words in feature-sentiment-word-pairs of App software are analyzed. (2) The sentiment tendency of users to App software is quantified with dictionaries and network sentiment words. (3) The comprehensive score of App software in each user comment is calculated.

The main contributions in this paper are as follows:

(1) The internal relations among user's comment information, user's mark and App software information are mined through comprehensively analysis. In order to identify whether the user's comment information is aim at certain App software, the App software information is used as the basis to judge the consistency between the user's comment information and the App software information.

(2) In order to improve the coarse-grained qualitative evaluation method which divided the sentiment into satisfaction and dissatisfaction, the comprehensive score in the user's comment information of App software is calculated. The comprehensive score is divided into five marks, which are compared with five marks to judge whether the user's comment information conform to the user's marks.

The method in this paper will identify whether user comments are consistent with the App software information and whether the degree of sentiment tendency of user comments is consistent with user's marks. It is helpful for users to select App software.

3 The Consistency Judgment Between the User's Comment Information and the App Software Information

The App software information is published by the developers. Because there are the most features of App software, the App software information can be used as the basis to judge whether the user comments are consistent with the App software information. Most users would like to make comments containing one or more features because the user comments for App software is relatively free on the network. However, some user's comment information is not concerned with the App software's features, which leads to the inconsistency between user's comment information and the App software information and the inconsistency between user's comment information and user's mark. So whether the user's comment information aims at App software must be judged at first.

Currently, App software information and user's comment information usually are described by natural language. We cannot identify the consistency of their information directly. Hence the software features of the App software information and the feature-sentiment-word-pairs in the user comments are extracted to judge the consistency.

3.1   Feature-Sentiment-Word-Pairs Extraction of App Software

Currently, most researches acquire the description or comment data about product characteristics through extracting feature and sentiment words from user comments. In this paper, user name, user ID, comment software type, user's mark and user's comment information are extracted from the massive user comments data. After analyzing the above information, users usually use 'sentiment words' or 'adverbs + sentiment words' to modify the software features. For example, a user gives comment on App software '微信' as follows: 下载很麻烦'. In this comment, the sentiment word '麻烦' modifies the feature word '下载' and the degree adverb '很' modifies the sentiment word '麻烦'. Therefore, there are corresponding relations among feature word, sentiment word and adverb. Features words and sentiment words usually occur in pairs. We name them as feature-sentiment-word-pairs of App software, which is defined as follows.

Definition 1. Feature-sentiment-word-pairs of App software $f = (Wh, Wd, Wa)$. Wh is App software feature focused by user, such as '下载', '画面' and so on. Wa is sentiment word modifying the feature Wh, such as '麻烦', etc. Wa includes network sentiment words which express user's objective impression on App software feature, such as '神器', etc. As adverb modifying the sentiment word Wa, Wd expresses degree of user's sentiment tendency to feature words, such as '很', '非常' and so on.

The App software information includes software name, software type, software ID, software introduction, and software source. It is released by the developer, and can describe most features of App software. In order to judge the consistency of user comments on App software, we extract nouns, verbs and noun phrases as feature set of App software information. Based on feature words and sentiment words extraction as in [11], we extract feature-sentiment-word-pairs f at the same time.

3.2   The Consistency Judgment Between the User's Comment Information and the App Software Information

In order to judge the consistency between user's comment information and App software information, we compare Wh in f with software feature in App software information and calculate their similarity. Based on the current user comments on App software, it is can be shown that most users give comprehensive comments. Wh is usually omitted in the user's comment information, such as '很好，很满意，很不错'. We regard this type of comments as for the whole App software, which means it is consistent between the user's comment information and the App software information.

Additionally, a user comment may be involved in many feature-sentiment-word-pairs of App software, which every feature is different. The degrees of user's sentiment tendency are varied as well. This sentiment tendency can influence the user's mark on App software. Therefore, the set of App software feature-sentiment-word-pairs is defined as follows.

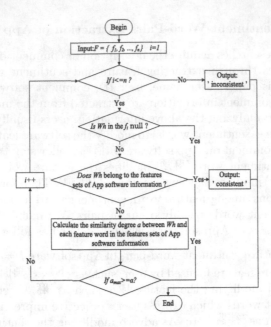

Fig. 1. Flow diagram of the consistency Judgment between the user's comment information and the App software information.

Definition 2. The set of App software feature-sentiment-word-pairs $F = \{f_1, f_2, ..., f_n\}$. $f_i = (Wh, Wd, Wa)$ $(i = 1, 2, ..., n)$ indicates feature-sentiment-word-pairs of App software.

Based on the set of App software feature-sentiment-word-pairs, the consistency between user's comment information and App software information is judged. The flow diagram is as shown in Fig. 1.

The similarity calculation method in Fig. 1 uses the method described in [12]. And a_{max} indicates the maximum value in similarity a, threshold value $\alpha = 0.1$.

4   The Consistency Judgment Between the User's Comment Information and the User's Mark

When the user's comment information is consistent with the App software information, the users would still be confused about selecting App software if the user's comment information is inconsistent with user's mark. The users cannot judge whether the user's comment information or user's mark are the focus. So the consistency between the user's comment information and user's mark should be judged further. Because the set of App software feature-sentiment-word-pairs F is natural language, its consistency with user's mark cannot be judged directly. Therefore, the emotional tendency of each f in App software feature-sentiment-word-pairs set F is quantified at first. Then the composite score of App software feature-sentiment-word-pairs set F is calculated. Finally, the score will be divided into five level marks corresponding with the user's mark. Accordingly, the consistency between the emotional tendency in the user's comment information and the uses' mark of this App software can be judged.

4.1 Sentiment Tendency Degree Quantification

Currently, the user's emotional tendency analysis includes dictionaries-based method and corpus-based method. The dictionaries-based method identify the vocabulary similarity degree between the undetermined polarity words and the fiducially polarity words based on dictionaries. For example, after giving a seed polarity word, the expanded semantic lexicon of synonyms and synonym are found using *Word Net* [13]. The corpus-based method makes the identification through the co-occurrence model or modified model of the undetermined words and the fiducially words, such as word frequency mutual information, syntactic dependencies and association rules [14].

In user's comment information of App software, the intensity difference between network sentiment words and adverbs will influence user's sentiment tendency degree. And the semantic difference result from the co-occurrence words order of adverbs and negatives also will influence user's sentiment tendency degree. So, we further analyze the polarity words of sentiment words. Combining the grammatical relations among Wh, Wa and Wd in f, the sentiment corpus in *How Net* and network sentiment words, the degree of the sentiment tendency is quantified gradually. Finally, the comprehensive score of each user comment is calculated. Wh include nouns, adverbs and noun phrases. Wa include adjectives, verbs and network sentiment words. Wd include adverbs and negatives. The flow diagram of sentiment tendency degree quantification for user's comment information is as shown in Fig. 2.

(1) Processing of Network Sentiment Words. As user's comment information on App software is a typical network comment, most users would like to use network sentiment words (including adjectives, nouns and verbs) to express sentiment tendency to App software. For example, '这就是个SB软件',, where 'SB' is a noun expressing a strong negative sentiment to the current App software. Another example, '这款软件超赞',, where the adverbs '超' and the network sentiment word '赞' express a strong positive sentiment to App software. Currently, these network sentiment words cannot be found in *How Net*. Due to they can express user's sentiment tendency to the features of App software, the consistency judgment of user comments on App software will be influenced. Therefore, we build a network sentiment lexicon including 137 high-frequency network sentiment words, such as 'TMD', 'SB', '神器' and so on. We define the weight and polarity of these network sentiment words, which express user's strong attitude to features of App software. We define the quantitative formula of sentiment tendency degree as follows.

$$F(nr) = F(nd) * F(na) \tag{1}$$

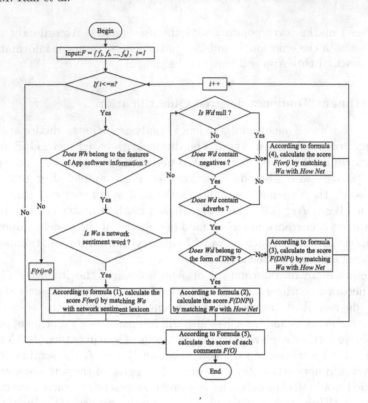

Fig. 2. The flow diagram of sentiment tendency degree quantification for user's comment information.

$F(nr)$ represents the score of emotional tendency degree that made by the current network word on certain feature comment. $F(nd)$ is the polar parameter of adverbs, and $F(na)$ indicates the original polarity of emotional words in network sentiment lexicon.

(2) Processing of Degree Adverb and Co-occurrence of Degree Adverb and Negative Adverb. As the adverbs and negatives in user comments on App software have important impacts on the quantitative result of the sentiment tendency degree and on the accuracy of consistency judgment for user comments, the adverbs and negatives must be processed. The existing researches show that adverbs have different intensity grades. For example, Xu et al. [13] divide adverbs into two categories and set four intensity grades. Lin [14] divide adverbs into six categories based on *How Net* dictionary, where the polarity parameters of '最' and '超' are '1.6'. So we combine the '最' and '超' categories into the '超' category. Finally we divide adverbs in user comments on App software into five different polarity parameter categories, which as shown in Table 1.

When processing negatives, Yao et al. [15] adopt a sentiment polarity processing method for negatives prefix, which negated and then divided by 2. According

Table 1. Degree adverbs categories and polarity parameters

Category	Polarity Parameter	Word(some examples)	Amount
1.最	1.6	非常、特别	99
2.很	1.4	很、太	39
3.较	1.2	蛮、挺	36
4.稍	0.8	有点、略	18
5.欠	0.6	相对、大概	8

to this algorithm, the sentiment tendency degrees of '不很满意' or '很不满意' are the same. In fact, they are different because this algorithm does not consider the semantic difference resulted from the order when negatives and adverbs co-occur. Aimed at the user comments on App software, we set the polarity parameter of negatives as -0.5 based on [16]. The co-occurrence order of negatives and adverbs can be divided into two categories.

(1) When adverbs are prior to negatives, it affirms negatives degree of adverb, and the negative degree is increasing. For example, '画面不好' and '画面很不好'. So we calculate the degree of the sentiment tendency described in [14], which formula is as follows.

$$F(DNP) = -0.5 * F(d) * F(a). \tag{2}$$

$F(DNP)$ indicates the score of emotional tendency degree that made by adverbs prior to negatives. -0.5 is the polar parameter of negatives. $F(d)$ is the polar parameter of adverbs. $F(a)$ is the original polarity of emotional words.

(2) When negatives are prior to adverbs, it negates the degree of adverbs, and the degree of adverbs is weakening. For example, '画面不很好' and '画面相对好'. The sentiment degrees of the two sentences are identical in principle. In the case, they are capable of inferring with each other semantically [16]. So we set the polarity parameter of this type of negative adverb as '欠' category in Table 1, which formula is as follows.

$$F(NDP) = 0.6 * F(d) * F(a). \tag{3}$$

$F(NDP)$ shows that the score of emotional tendency degree made by negatives prior to adverbs. 0.6 is polar parameter of negatives.

(3) Sentiment Tendency Degree Quantification. For App software, the degree of user's sentiment tendency on features depends on adverbs, negatives and sentiment words. We calculate the degree of user's sentiment tendency in user's comment information of the App software using the following formula.

$$F(or) = F(oa) * F(od) * F(on) \tag{4}$$

$F(oa)$ is the original polarity of emotional word Wh. $F(od)$ is the polar parameter of adverbs. $F(on)$ is the polar parameter of negatives. When adverbs or negatives are null, the polar parameter is set to 1. When the adverbs and negatives are co-occurrence, then the score should be calculated according formulas (2) and (3).

(4) Comprehensive Score Calculation of Sentiment Tendency Degree. In F, there are differences in user's sentiment tendency degree of each f. In order to judge the consistency between user's comment information and user's mark more exactly, we calculate the sentiment tendency degrees of features in F comprehensively. The formula is as follows.

$$F(O) = \sum_{i=1}^{m} F(ri)/m \qquad (5)$$

Wherein, $F(ri)$ means the score of emotional tendency degree for the ith features comment in the users' comments. $F(ri)$ includes $F(nr)$, $F(or)$ and other conditions. m represents the number of f which is consistent with the feature in the App software information, that is the number of $f(ri)! = 0$.

4.2   Range Division of Comprehensive Score

Based on Sect. 4.1, the user's mark of App software depends on the comprehensive score $F(O)$. Accordingly, we divide $F(O)$ into five marks, which respectively corresponding to user's mark of App software. User's mark include five-mark (very good), four-mark (good), three-mark (general), two-mark (bad) and one-mark (very bad). Then we judge the consistency of user's comments on App software by comparing the grades of $F(O)$ and user's mark.

The range of $F(O)$ can be calculated in $[-1.6, +1.6]$ based on Sect. 4.1. In Table 1, the maximum value of polarity parameters is 1.6, the polarities of positive and negative sentiment words are respectively $+1$ and -1, and the polarity of negative words is -0.5. Therefore, the maximum value of five-mark is $1.6 * 1 = 1.6$. The maximum value of four-mark is 1 and the minimum is $-1 * (-0.5) = 0.5$. So the value range of five-mark is $(1, 1.6]$, the range of four-mark is $[0.5, 1]$. The range is as shown in Table 2.

Table 2. Range division of comprehensive score

User's mark	5	4	3	2	1
$F(O)$ Range	$(1, 1.6]$	$[0.5, 1]$	$(-0.5, 0.5)$	$[-1, -0.5]$	$[-1.6, -1)$

5 Experimental Result and Analysis

5.1 Experimental Data Source and Processing

In order to verify the effectiveness of the method in this paper, 20296 pieces of user's comment information of 7 categories and 27 types of App software are picked up randomly from Android electronic market. The comment time is from March 2014 to November 2015. We establish App software information database and user's comment database. Some information is shown in Tables 3 and 4.

Table 3. App software information (some examples)

App Types	App Software Name	Introduction	Version No.	Comment Amount
影音	酷狗音乐	【核心功能】1.更人性化的界面...	7.3.9	980
	百度视频	【活动主题】：登录就送好...	7.12.0	960
社交	QQ	-主要功能-聊天消息：随时随地聊天...	5.6.1	970
	同城蜜恋	同城蜜恋是一个专业的婚恋相亲...	2.8	1000

Table 4. User's comment information of App software (some examples)

User Name	User ID	App Software Name	User's Mark	User's Comment Information
周隆	601014127921598465	QQ	4	SB软件
圆音	599514748127678479	QQ	4	您透婆婆十块钱

As shown in Table 4, some users give high user's mark on App software, the sentiment tendency of user's comment information expresses negative attitude, or the comment is not aimed at features of App software. For example, user '周隆' gives user's mark '4' on App software 'QQ', but the sentiment tendency of its user's comment information 软件' expresses negative attitude. And user '圆音' gives user's mark '4' on software 'QQ', but the sentiment tendency of its user's comment information '您透婆婆十块钱' is not aimed at features of 'QQ'. Therefore, there is inconsistency in the user comments on App software.

Table 5. Data processing results (some examples)

Before word segmentation	After word segmentation	The set of Feature-sentiment-word-pairs
不太好	不/d 太/d 好/a	{ (null, (不/d 太/d), 好/a) }
软件很赞	软件/n 很/d 赞/v	{ (软件/n, 很/d, 赞/v) }

ICTCLAS 2015 is used as data preprocessing tool in this paper, which completes comment word segmentation and part-of-speech marking. Then feature-sentiment-word-pairs of App software are extracted. Some data processing results are shown in Table 5.

5.2   Experimental Result and Analysis

Aiming at the above experimental data, the consistency of user comments is judged. In order to verify the effectiveness of our methods, some App software users are invited to manual mark the consistency of comment information. Some judgment results of consistency are shown in Table 6 ('Consistency between UCI and UM(manual marking)' means the result that artificial marking is used to judge whether the 'User's Comment Information' is consistent with the 'User's Mark'. 'Consistency between UCI and ASI' means the result judged by the method set out in Sect. 3. 'UM after Quantification' means the corresponding mark converted from F (O) in Sect. 4. 'Consistency between UCI and UM' means the comparison result between 'User's Mark after quantization' and 'User's Mark'.).

Table 6. Conformity Judgment Results (some examples)

User ID	App Software Name	User's Mark	User's Comment Information	Consistency between UCI and UM(manual marking)	Consistency between UCI and ASI	UM after Quantification	Consistency between UCI and UM
589651758 389792771	美图秀秀	5	很好，很满意,很不错	Yes	Yes	5	Yes
601724559 133118478	QQ	4	劳资日穿墙	No	No	-	No
543218047 436066818	畅读	4	TMD, SB软件	No	Yes	1	No

From the above experiment, the average consistency between user's comment information and user's mark only cover 40.06%. The result of user comments consistency on various App software is shown in Fig. 3, in which '社交类' only cover 32.98%. It is obvious that most of user's comment information in the current App software comments is inconsistent with user's mark.

In comparison with the manual marking, the accuracy of consistency judgment using our methods is averagely up to 78.78%, in which the accuracy of '浏览器类' is the highest, up to 81.67%, while that of '资讯类' is the lowest, only 76.84%.

When the method in Sect. 3 is used to judge whether user's comment information is consistent with App software information, Wh of some user's comment information is mistaken as the App software feature. The reason is that Wh is not

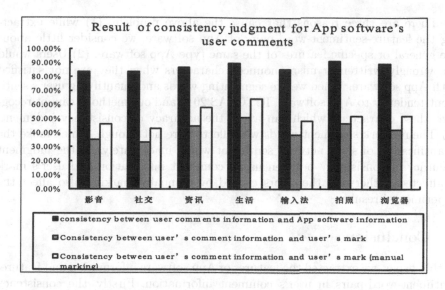

Fig. 3. Result of consistency judgment for App software's user comments.

precise enough to calculate the similarity between Wh and feature words in App software information, which results in a higher consistency between user's comment information and App software information. Then the judgment accuracy of consistency between user's comment information and user marks is influenced.

The consistency of user marks for '浏览器类' and '资讯类' are shown in Fig. 4. The accuracy of two-mark and one-mark are a little low, respectively 66.67 % and 57.89 %, 59.26 % and 60.00 %. This is the common in 7 types of App software.

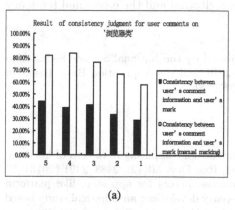

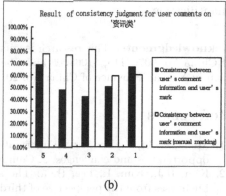

(a)   (b)

Fig. 4. (a) Result of consistency judgment for user comments on '浏览器类'; (b) Result of consistency judgment for user comments on '资讯类'.

There are three reasons that cause the above problem: (1) while extracting the feature-sentiment-word-pairs of App software, we consider little about the general or specific features of the same type App software. (2) Users would use wrongly written or mispronounced characters when they are not satisfied with App software. When we are segmenting words and quantifying user's sentiment tendency to App software, ICTCLAS 2015 and our methods cannot recognize these characters, which can affect the accuracy of consistency judgment. (3) The network sentiment words we collected are not enough, which make the quantification of some network sentiment words inaccurately. So the judgment accuracy of consistency between user's comment information and user's mark is affected. Otherwise, the errors caused by manually marking also affect the experimental results.

6   Conclusion

In this paper, we extracted the features of App software information and feature-sentiment-word-pairs in user's comment information. Firstly, the consistency between user's comment information and App software information is judged. Secondly, user's sentiment tendency to App software is quantified by analyzing the grammatical relations in feature-sentiment-word-pairs of App software combining with dictionaries and network sentiment words. Thirdly, the comprehensive score of App software in each user's comment information is calculated. In order to judge the consistency between user's comment information and user's mark, the comprehensive score is compared with user's mark.

The experimental result shows that our method is practicable. To improve the judgment accuracy of the consistency, the follow issues will be researched further: the feature-sentiment-word-pairs extraction on the general and specific characters of the same type App software, feature similarities of the App software calculation, the network sentiment words collection and the emotional tendency quantification.

Acknowledgments. This research is sponsored by the National Science Foundation of China No. 60703116, 61063006 and 61462049, and the Application Basic Research Plan in Yunnan Province of China No. 2013FZ020.

References

1. Meng, X., Li, Y., Zhu, J.: SO information. Social computing in the era of big data: opportunities and challenges. J. Comput. Res. Dev. **50**(12), 2483–2491 (2013)
2. Kim, H.J., Kim, I., Lee, H.G.: The success factors for app store-like platform businesses from the perspective of third-party developers: an empirical study based on a dual model framework. In: Pacific Asia Conference on Information Systems, PACIS 2010, Taipei, Taiwan (2010)
3. Hao, L., Li, X., Tan, Y.: The economic role of rating behavior in third-party application market. In: Proceedings of the International Conference on Information Systems, ICIS, Shanghai, China (2011)

4. Liu, Y., Liao, X., Liu, Y.: The impact of online comment on software and platforms pricing strategies. J. Syst. Eng. **29**(4), 560–570 (2014)
5. Zhou, J., Sun, J., Kumaripaba, A., Dinesh, W., Mika, Y.: Pervasive social computing: augmenting five facets of human intelligence. J. Ambient Intell. Hum. Comput. **3**(2), 153–166 (2012)
6. Gao, C., Xu, H., Hu, J., Zhou, Y.F.: AR-tracker: track the dynamics of mobile apps via user comment mining. In: 2015 IEEE Symposium on Service-Oriented System Engineering (SOSE), pp. 284–290. IEEE (2015)
7. AlQuwayfili, N., AlRomi, N., AlZakari, N., Al-Khalifa, H.S.: Towards classifying applications in mobile phone markets: the case of religious apps. In: 2013 International Conference on Current Trends in Information Technology (CTIT), pp. 177–180. IEEE (2013)
8. Khalid, H.: On identifying user complaints of iOS apps. In: 35th International Conference on Software Engineering (ICSE), pp. 1474–1476. IEEE (2013)
9. Islam, M.R.: Numeric rating of apps on google play store by sentiment analysis on user reviews. In: 1st International Conference on Electrical Engineering and Information Communication Technology, pp. 1–4 (2014)
10. Guzman, E., Maalej, W.: How do users like this feature? a fine grained sentiment analysis of app reviews. In: 22nd International Requirements Engineering Conference (RE), pp. 153–162. IEEE (2014)
11. Hu, Z., Zheng, X.: Product recommendation algorithm based on users' comments mining. J. Zhejiang Univ. (Eng. Sci.) **47**(8), 1475–1485 (2013)
12. Zhu, Z., Sun, J.: Improved vocabulary semantic similarity calculation based on *HowNet*. J. Comput. Appl. **33**(8), 2276–2279, 2288 (2013)
13. Xu, L., Lin, H., Yang, Z.: Text orientation identification based on semantic comprehension. J. Chin. Inform. Procession **21**(1), 96–101 (2007)
14. Lin, Q.: Design and Implementation of the Product Comments Analysis System Based on Affective Computing. Fudan University (2013)
15. Yao, T., Lou, D.: Research on semantic orientation analysis for topics in Chinese sentences. J. Chin. Inform. Procession **21**(5), 73–79 (2007)
16. Yin, H.: Research on Syntax and Semantics for Co-occurring of Negatives and Adverbs. Graduate School of Chinese Academy of Social Sciences, Beijing (2008)

Multi-GPU Based Recurrent Neural Network Language Model Training

Xiaoci Zhang(✉), Naijie Gu, and Hong Ye

School of Computer Science and Technology,
University of Science and Technology of China, Hefei, China
{zxiaoci,yehong91}@mail.ustc.edu.cn, gunj@ustc.edu.cn

Abstract. Recurrent neural network language models (RNNLMs) have been applied in a wide range of research fields, including nature language processing and speech recognition. One challenge in training RNNLMs is the heavy computational cost of the crucial back-propagation (BP) algorithm. This paper presents an effective approach to train recurrent neural network on multiple GPUs, where parallelized stochastic gradient descent (SGD) is applied. Results on text-based experiments show that the proposed approach achieves 3.4× speedup on 4 GPUs than the single one, without any performance loss in language model perplexity.

1 Introduction

Statistical language models (LMs) play a vital role in modern nature language processing (NLP) systems. To address the data sparsity problem in conventional n-gram LMs, neural network language models (NNLMs) [1–6] have been proposed, which use shallow neural network to extract features from input words, and back-propagation (BP) [14] algorithm to update the network parameters. Depending on the network architecture, NNLMs can be classified into two groups: feed-forward NNLM [1,2] where a finite length of context is considered, and recurrent NNLMs (RNNLMs) [3–6] where a recurrent vector is used to preserve longer and variable length context. In recent years, RNNLMs have been reported to obtain significant performance improvements over n-gram LMs and feed-forward NNLMs on a wide range of tasks [3,4].

To build language models based on recurrent neural networks (RNNs), a practical issue is the heavy computational complexity in training the neural network. In typical tasks with huge quantities of input data, the training process exceeds days or even weeks. Such long training periods are uncomfortable for practical use. One direction to accelerate RNN training speed is to use parallel algorithm to train the network on many-core machines or clusters. However, while BP algorithm plays a vital role in RNN training, it is difficult to develop an effective parallel training algorithm, for BP involves a full model update after each iteration.

This paper presents a effective data parallelism approach to train RNNs on multiple GPUs, using a ring-oriented graph-structure. Experimental results on

© Springer Science+Business Media Singapore 2016
W. Che et al. (Eds.): ICYCSEE 2016, Part I, CCIS 623, pp. 484–493, 2016.
DOI: 10.1007/978-981-10-2053-7_43

the famous Penn Treebank Wall Street Journal corpus show a 3.4× speedup on a cluster with 4 GPUs compared with single one, without degrading performance of the trained language model.

The rest of this paper is organized as follows. Section 2 discusses related works. Section 3 describes the architecture of typical recurrent neural networks and critical factors in training. Section 4 introduces the parallel RNN training approach and shows how to apply it on multiple GPUs. In Sect. 5 experimental results on text-based dataset are presented. Conclusions and future works are discussed in Sect. 6.

2   Related Works

Inspired by the huge success of parallel training in deep neural networks, many works to accelerate RNN training algorithm focus on distributing the training process to multi-core machines or clusters. The parallel algorithms of training neural networks can be classified into two categories, i.e. model parallelism and data parallelism. The major difference between the two algorithms is whether the network model or training data is split across computing clients.

Data parallelism based works like [4] apply asynchronous stochastic gradient descent (ASGD) on multi-core machines, which share the same structure in [10]. In this scheme, the working clients run asynchronously and accumulate the overall gradient to the model using a centralized parameter server.

Model parallelism approaches split the network model across multiple clients, like [8] used pipelined model parallelism on two GPUs and led to a 1.6× speed-up over one GPU. However, pipeline algorithm requires relatively equal computational loads on each device, thus it is difficult to extend to large clusters.

GPUs are recently attracting emphasis in speeding up the training of large-scale neural networks [10–13], therefore it is advisable to also exploit the parallelization power of GPU or even GPUs in RNN training. For example, [7] implemented a single-GPU based RNN training approach using spliced sentence bunch and obtained decent acceleration ratio over CPU.

3   Model Architectures

3.1   Recurrent Neural Network in Language Modeling

A recurrent neural network is similar to a multi-layer perception except for the recurrent vector to restore history information, as shown in Fig. 1.

To predict LM probabilities $P(w_t|w_{t-1}, s_{t-2})$ for word w_t, the input to the network at time t consists of two parts: w_{t-1} is the previously seen word, and s_{t-2} is the reserved history information. w_{t-1} is encoded using 1-of-k coding (index of current word in the vocabulary is 1 and all rests are 0), which is essentially a word vector of vocabulary size (e.g. 100 k). The hidden layer s_{t-1} compresses the two inputs and applies sigmoid function to avoid linearity. The term *recurrent* in RNN means that the hidden layer is recurrently fed back into

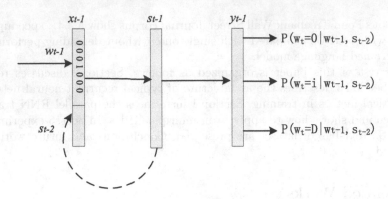

Fig. 1. Recurrent neural network

the input layer, as the 'future' history in computing $P(w_{t+1}|w_t, s_{t-1})$. The hidden layer usually consists of a few thousands of units (e.g. 1024). The output layer y_t is also of vocabulary size, y_t shows the predicted LM probability distribution of the current word (i.e. $P(w_t|w_{t-1}, s_{t-2})$). The neural network is trained using standard BP algorithm.

3.2 Stochastic Gradient Descent

In BP training, one simple way to find a global minimum of training error is to take steps in the steepest direction of the error curved surface, called gradient descent (GD), as

$$W_{t+1} = W_t - \alpha \cdot \nabla E(W_t) \tag{1}$$

where α is called learning rate which determines the learning step size, E is the error function, W is the network parameter. The gradient of $E(W_t)$ is calculated based on all input samples, called batch update. However, batched SD is quite slow for it needs to accumulate gradients over the whole dataset. Moreover, for a none-convex problem, such a learning process may not be able to find the global minimum of the error function when there are multiple local minima on the curved surface.

A variant of gradient descent called incremental gradient descent or stochastic gradient descent (SGD) has been proven useful to alleviate those problems. Instead of updating upon the whole training dataset, the error gradient $\nabla E_n(W_t)$ is calculated on a small portion of samples called mini-batch, so that

$$E(W) = \Sigma_{n=1}^N E_n(W) \tag{2}$$

$$W_{t+1} = W_t - \alpha \cdot \nabla E_n(W_t) \tag{3}$$

where N is the total number of mini-batches in the training set.

Compared with batched GD, SGD is more likely to escape from local minimum, and costs less time to reach convergence, therefore it is widely used in training neural networks including DNNs and RNNs [3,4,10–13].

On the other hand, the introduction of incremental update in SGD also makes it difficult to effectively parallelize the BP algorithm. Consider an RNN model with d units for input layer and h units for hidden layer, the total parameter number in the network is $((2 \cdot d + h) \cdot h)$. Let $d = 100\,\mathrm{K}$, $h = 1024$, the number of parameters will be in the order of 10^8. Since SGD involves a full model update after processing each mini-batch, it becomes a tricky problem to deal with the heavy communication loads between training clients.

4   Parallelization of RNNLM Training

In this paper we focus on investigating a effective and entensive parallel SGD algorithm, and applying it in training RNNLMs on multi-GPUs. In the rest of this section we will show that this algorithm has low bandwidth request independent of client number, therefore it is easy to extend it to other machine learning tasks which evolve BP training.

4.1   Parallel Stochastic Gradient Descent

The skeleton of a simple parallel SGD algorithm is shown in Algorithm 1. Multiple computing devices serve as clients in the system and are serially numbered. In the beginning of the training process, weight parameters (or model, denoted as W in Algorithm 1) on all clients are randomly initialized. At every time step, each client fetches one mini-batch from the input data, and computes gradient G_{train}^{id} of local model W^{id}. Then each client broadcasts G_{train}^{id} to all other clients in the cluster, and accumulates gradients from others (G_{recv}^{id}) to update local model. In this way the gradient calculated from each mini-batch of input data is shared within the whole cluster, therefore synchronism is achieved over the training dataset.

Despite the availability, this algorithm involves a exceedingly high complexity in communication ($O(N^2)$ where N is the number of clients), for each client

---

Algorithm 1. Parallel Stochastic Gradient Descent

Require : $0 \leq id < client\_num$
randomly initiate W on all clients
loop
  $X \leftarrow$ *get next mini-batch*
  if X is empty **then**
    exit
  end if
  $G_{train}^{id} \leftarrow$ *calculate gradient of W^{id} based on X*
  send G_{train}^{id} to other clients
  $G_{recv}^{id} \leftarrow$ *accumulate gradients from other clients*
  $W^{id} \leftarrow W^{id} + G_{train}^{id} + G_{recv}^{id}$
end loop

Algorithm 2. Modified Parallel Stochastic Gradient Descent

Require: $0 \leq id < client\_num$
   $pre \leftarrow (id + client\_num - 1)\%client\_num$
   $suc \leftarrow (id + 1)\%client\_num$
   set $signal[0]$ **true**, $signal[1..client\_num - 1]$ **false**
   randomly initiate W on all clients
   loop
     $X \leftarrow$ *get next mini-batch*
     if X is empty **then**
       exit
     end if
     $G^{id} \leftarrow$ *compute gradient of W^{id} based on X*
     wait until signal[id] is **true**
     $W^{id} \leftarrow G^{id} + W^{pre}$
     set $signal[id]$ **false**
     set $signal[suc]$ **true**
   end loop

---

needs to send gradient to other $N - 1$ clients. To address this problem, a modified parallel SGD algorithm is investigated, as shown in Algorithm 2. In this algorithm, clients pass **models** instead of **gradients**.

In Algorithm 2, clients are connected in a ring-oriented graph-structure, each with a predecessor and a successor node. At first, only one client is activated. This first client reads a mini-batch from the train set and computes gradient of W^{id}, denoted as G^{id}. Meanwhile it fetches a copy of model from its predecessor, denoted as W^{pre}. Then it uses G^{id} and W^{pre} to update W^{id}. Subsequently, the first client sends a signal to activate its successor client, this client computes its gradient, updates its local model using the model from the first client, and activates the next client. After all clients in the ring are activated, the data

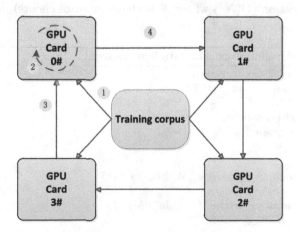

Fig. 2. Parallel stochastic gradient descent in ring-oriented structure

flow in the cluster is established, and each client needs only to periodically copy remote data from its predecessor client.

While this training approach looks like a variant of pipeline technique, the basic parallel strategy of this approach is data parallelism instead of model parallelism (notice this difference from [8,12]), which indicates it can be applied without considering the network structure. Under this scheme the gradients are still accumulated over all clients while the communication complexity is reduced to $O(N)$.

4.2 GPU Implementation and Communication Optimization

We implemented the parallel SGD algorithm on multiple GPUs to further accelerate RNN training, letting each GPU work as a client. In order to fully exploit the computing power of GPUs, we extend the training algorithm to operate on many input words in parallel, using the same spliced sentence bunch technique as [7]. In this way the core computation in the algorithm turns into matrix operations, making it feasible to accelerate using CUBLAS, the basic linear algebra subprograms (BLAS) library developed for Nvidia GPUs.

The structure of the multi-GPU training approach is shown in Fig. 2. In the beginning, the network model is randomly initialized on CPU and distributed to all GPUs in the cluster, then the GPUs are activated in sequence as described in Algorithm 2. For each GPU, step 1 in Fig. 2 means that it fetches a minibatch from the training corpus, consistency of which is guaranteed by POSIX mutex. The gradient of the local model, G^{id}, is then calculated according to local model W^{id} (Step 2). Each GPU waits until its predecessor's local model W^{pre} is updated, then it downloads W^{pre} and uses W^{pre} and G^{id} to update W^{id} (Step 3), and signals its successor to fetch the new model (Step 4).

In case the four steps in this parallel algorithm run serially, it is easy to infer that the speed-up ratio of this system is

$$K = \frac{N \cdot T_{calc}}{T_{calc} + T_{tran}^{recv} + T_{add}} \tag{4}$$

where T_{calc} is the time per mini-batch on calculating local gradient G^{id}, T_{tran}^{recv} is the combination of waiting time before W^{pre} is ready and the data transfer time of W^{pre}, T_{add} is the time on adding G^{id} to W^{pre} and updating W^{id}.

Since modern GPUs facilitate concurrent kernel execution and memory copies between devices, it is allowed to overlap data transfers with kernel executions using asynchronous CUDA kernels [15]. We utilize this feature to eliminate communication costs of the parallel SGD algorithm on GPUs. The asynchronous CudaMemcpyAsync function is used for data transfers among GPUs, so gradient computing and model fetching process can run simultaneously on each GPU. By carefully arranging the execution flow among all clients, the time of fetching W_{pre} on one client can be completely overlapped by the computing process of G^{id} provided $T_{tran}^{recv} \leq T_{calc}$. In this case the speed-up ratio is then

$$K = \frac{N \cdot T_{calc}}{T_{add} + \max\left(T_{tran}^{recv}, T_{calc}\right)} \tag{5}$$

5   Experiments

5.1   WSJ Experiment

To evaluate performance of parallel SGD based RNNLMs, we choose the Wall Street Journal (WSJ) data in the Penn Treebank (PTB) corpus[1], which was the same dataset used in [5,9]. In common with previous researches, section 0–20 were used as training data, section 21–22 were used as validation data, and section 23–24 used as test data. On a computer with dual Intel Xeon E5-2620 2.00 GHz processor and 24 physical cores, we trained Kneser-Ney smoothed 3-gram model (KN3) with default cut-offs using SRILM [16], as the performance baseline. The parallel SGD based RNN version was implemented based on Mikolov's toolkit [9] and trained using 4 NVIDIA Tesla K40m GPUs. For comparison purpose, a single GPU version of RNNLM was also tested. This experiment used a 10 K word vocabulary with out-of-vocabulary words being mapped to a special token $\langle unk \rangle$. The number of tokens in the train, validation and test sets were 930 K, 74 K, and 82 K words, respectively. For all RNNLMs, the number of classes and hidden layer nodes were fixed as 100, the number of BPTT steps was set 5. The test results of all language models are shown in Table 1.

Table 1. Perplexity and Speed of various LMs trained on WSJ data

Model	Valid PPL	+KN3	Test PPL	+KN3	Speed(w/s)
Baseline	162.4	-	152.9	-	-
Single GPU	157.8	132.2	150.3	126.4	11.2k
Parallel SGD	158.6	132.9	150.5	126.5	37.4k

The first row in Table 1 shows results on the baseline n-gram LM, where perplexities on both validation and test sets are evaluated. The second row shows results on single GPU based RNNLM, which gives 2.9 % and 1.7 % relative improvements in perplexity (PPL), on validation and test sets respectively. There is a further 10–15% relative reduction in perplexity when interpolated with the KN3 baseline LM. The third row shows results on the proposed parallel SGD RNNLM trained on 4 GPUs. The learning rate was auto-tuned during the whole training process in order to get better convergence, with large size in early iterations and small size in later ones. Test results showed no performance degrading comparing the multi-GPU based implementation with the CPU-based RNNLM, also an approximate 3.4× speed up was obtained using 4 GPUs.

---

[1] http://www.cis.upenn.edu/treebank/.

5.2 Leipzig Corpora Experiment

In this experiment the proposed algorithm was evaluated on a large text corpora from Leipzig Corpora Collection[2]. The Leipzig Corpora Collection presents corpora in different languages. We chose one corpora of a million English sentences collected from newspaper texts, named *eng_news_2008_1M-text*. We used 900 K sentences in the corpora as the train set, which consists about 17M words. The left 100 K sentences were evenly divided into the validation set and test set. The text was lowercased and words and punctuation were tokenized before further processing. The 71 K vocabulary was automatically generated from the corpora, where low-frequency words were discarded to restrict vocabulary size. 200 hidden layer nodes and 200 classes in output layer were used. The size of mini-batch used in training was proved to have a major impact in both training speed and performance. Figure 3 shows the training data entropy of the multi-GPU RNNLM against wall clock time, with different mini-batch sizes (represented by 'batch k'). The single GPU version with a batch size of 16 is used for comparison purpose. The parallel algorithm was implemented on 4 Nvidia K40m GPUs.

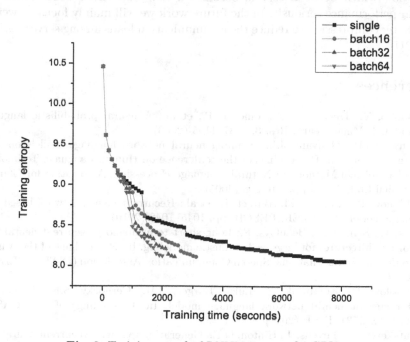

Fig. 3. Training speed of RNNLMs on multi-GPUs

As shown in Fig. 3, the convergence speed became fairly slow when using undersized mini-batch, due to the poor utilization of GPU computing power. By increasing the batch size, a growth of convergence speed was observed. However, very large mini-batch may also make the convergence less stable and lead to

---

[2] available at http://corpora2.informatik.uni-leipzig.de/download.html.

performance degradation. We found a warm-start beneficial in training multi-GPU based RNNLMs, i.e. the initial network model is firstly trained on a single client for a certain number of mini-batches, then the trained model is copied to other clients and all clients start parallel training using the pre-trained model. Results of the parallel ones in Fig. 3 were obtained using this warm-start strategy.

6   Conclusions and Future Work

In this paper a effecitve RNNLM training approach is introduced. We apply data parallelism on multiple GPUs and utilize asynchronous transmission feature to overlap the data transmission between GPU clients. Experimental results show that the proposed approach yields a 3.4× speedup on 4 GPUs compared with the single GPU version without degrading LM performance. We believe that accelerating RNN training further will result in more breakthroughs in research areas such as nature language processing and speech recognition.

When extending our work to multi-server multi-GPU architecture, the network bandwidth between servers becomes a new bottleneck which hurts the system performance seriously. In the future work we will mainly focus on weight parameter compression to reduce the communication loads among servers as well as GPUs.

References

1. Bengio, Y., Ducharme, R., Vincent, P., et al.: A neural probabilistic language model. J. Mach. Learn. Res. **3**, 1137–1155 (2003)
2. Schwenk, H., Gauvain, J.L.: Training neural network language models on very large corpora. In: Proceedings of the conference on Human Language Technology and Empirical Methods in Natural Language Processing. Association for Computational Linguistics, pp. 201–208 (2005)
3. Mikolov, T., Karafiàt, M., Burget, L., et al.: Recurrent neural network based language model. In: INTERSPEECH, pp. 1045–1048 (2010)
4. Sak, H., Senior, A., Beaufays, F.: Long short-term memory recurrent neural network architectures for large scale acoustic modeling. In: Proceedings of the Annual Conference of International Speech Communication Association (INTERSPEECH) (2014)
5. Mikolov, T., Kombrink, S., Lukas Burget, J.H., Cernocky, S.K.: Extensions of recurrent neural network language model. In: Proceedings of the ICASSP, pp. 5528–5531. IEEE (2011)
6. Sutskever, I., Martens, J., Hinton, G.E.: Generating text with recurrent neural networks. In: Proceedings of the 28th International Conference on Machine Learning (ICML-11), pp. 1017–1024 (2011)
7. Chen, X., Wang, Y., Liu, X., et al.: Efficient GPU-based training of recurrent neural network language models using spliced sentence bunch. Submitted to Proceedings of the ISCA Interspeech (2014)
8. Chen, X., Liu, X., Gales, M.J.F., et al.: Improving the training and evaluation efficiency of recurrent neural network language models. In: Proceedings of the IEEE ICASSP, Brisbane, Australia (2015)

9. Mikolov, T., Kombrink, S., Deoras, A., et al.: RNNLM-Recurrent neural network language modeling toolkit. In: Proceedings of the 2011 ASRU Workshop, pp. 196–201 (2011)
10. Hinton, G., Deng, L., Yu, D., et al.: Deep neural networks for acoustic modeling in speech recognition: the shared views of four research groups. IEEE Sig. Process. Mag. **29**(6), 82–97 (2012)
11. Jaitly, N., Nguyen, P., Senior, A.W., et al.: Application of pretrained deep neural networks to large vocabulary speech recognition. In: INTERSPEECH (2012)
12. Dahl, G.E., Yu, D., Deng, L., et al.: Context-dependent pre-trained deep neural networks for large-vocabulary speech recognition. IEEE Trans. Audio Speech Lang. Process. **20**(1), 30–42 (2012)
13. Dean, J., Corrado, G., Monga, R., et al.: Large scale distributed deep networks. In: Advances in Neural Information Processing Systems, pp. 1223–1231 (2012)
14. Hinton, G., Rumelhart, D., Williams, R.: Learning representations by back-propagating errors. Nature **323**, 533–535 (1986)
15. Nvidia, C.: Programming guide (2008)
16. Stolcke, A.: SRILM-an extensible language modeling toolkit. In: INTERSPEECH 2002 (2002)

Negation Scope Detection with Recurrent Neural Networks Models in Review Texts

Lydia Lazib, Yanyan Zhao, Bing Qin[✉], and Ting Liu

Research Center for Social Computing and Information Retrieval, Harbin Institute of Technology,
Harbin 150001, China
lydialazib@gmail.com, bing.qin@gmail.com,
{yyzhao,tliu}@ir.hit.edu.cn

Abstract. Identifying negation scopes in a text is an important subtask of information extraction, that can benefit other natural language processing tasks, like relation extraction, question answering and sentiment analysis. And serves the task of social media text understanding. The task of negation scope detection can be regarded as a token-level sequence labeling problem. In this paper, we propose different models based on recurrent neural networks (RNNs) and word embedding that can be successfully applied to such tasks without any task-specific feature engineering efforts. Our experimental results show that RNNs, without using any hand-crafted features, outperform feature-rich CRF-based model.

Keywords: Negation scope detection · Natural language processing · Recurrent neural networks

1 Introduction

Negation, as simple as it can be in concept, is a complex and an essential phenomenon in any language. It has the ability to inverse the meaning of an affirmative statement into its opposite meaning. In a sentence, the presence of negation is indicated by the presence of a negation cue. The negation cue is a lexical element that carries negation meaning that can occur in different forms: as an explicit negation, using negation words (e.g., "no", "not", "neither...nor", etc.), or as an implicit negation, where syntactic patterns imply negative semantics (e.g., "This movie was below my expectations."). The scope of negation is the sequence of words in the sentence that is affected by the negation cue [20].

For many Natural Language Processing applications, distinguishing between affirmative and negative information is an important task. A system that does not deal with negation would treat the facts in these cases incorrectly as positive. For example, in sentiment analysis detecting the negation is a critical process, as it may change the polarity of a text and results with a wrong prediction. And in query answering systems,

Specialized Research Fund for the Doctoral Program of Higher Education (No. 20122302110039)

W. Che et al. (Eds.): ICYCSEE 2016, Part I, CCIS 623, pp. 494–508, 2016.
DOI: 10.1007/978-981-10-2053-7_44

failing to account for negation can results in giving wrong answers. These systems are widely used in the social media domain, where a wrong interpretation of a user's comments or questions may results in giving wrong recommendations or answers.

The negation scope detection task can be tackled as a sequence labelling problem, where the task is to label each word in a sentence using the IO scheme (Inside or Outside the scope).

The scope detection problem started to be solved using machine learning approaches since the creation of the Bioscope Corpus [26]. The common approaches used memory-based engines [21], K-nearest neighbor [20], SVM classifier [20], or even tree kernel methods [28] to detect negation cues and their scope in a sentence. Later approaches used CRF-based models to solve these tasks [1, 2, 7]. These CRF-based approaches gave quite successful results, and proved to be more effective in solving some sequence labelling problems, by outperforming all the other machine learning methods. However, this success depends heavily on the use of an appropriate feature set, which often requires a lot of engineering efforts for each task in hand.

Deep learning methods are alternative approaches that can automatically learn latent features, and have recently been shown to outperform CRFs-based models on different sequence labelling tasks. [27] proposed to use a BLSTM-RNN (Bidirectional Long Short Term Memory) for a unified tagging solution that can be applied for various tagging tasks including part-of-speech tagging, chunking and named entity recognition. Here, instead of exploiting specific features carefully optimized for each task, their solution only uses one set of task-independent features and internal representations learnt from unlabeled text for all tasks. [16] also proposed a general class of discriminative models based on recurrent neural network (RNN) architecture and word embedding to solve fine-grained opinion mining tasks. The experimental results of these methods all show that RNNs, without using any hand-crafted features, outperform the feature-rich CRFs-based models.

RNNs have been also successfully applied to more specific sequence prediction tasks, such as language modeling [18, 24], spoken language understanding [17], and speech recognition [9, 22].

Motivated by all these recent success of recurrent neural networks, we propose in this paper various RNNs-based models (LSTM, BLSTM and GRU), to solve the problem of negation scope detection, without using any hand-crafted feature, and prove that these methods outperform feature-rich models. We also use word embedding as an input feature in some models, to show the performance of word embedding on the RNNs models.

Our results on negation scope detection task show that RNNs models outperform the CRF-based model, and including word embedding as an input feature improves the RNNs models performance. Our best result remains the BLSTM-based model using word embedding as input feature, with an F1 score of 89.38 % on the test data.

This paper is organized as follows, we first present in Sect. 2 the different RNNs models that we use in our experiments, then describe our experiments settings in Sect. 3 and analyze the results obtained in Sect. 4. Finally, we summarize our contribution with future directions in Sect. 5.

2   Models

In this section we describe the different models we use in this paper to identify the scope of negation in a sentence, namely: LSTM, BLSTM, GRU and CRF.

2.1   Long Short Term Memory (LSTM)

Recurrent neural networks are a kind of artificial neural networks that perform the same task for every element of a sequence, by maintaining a memory based on history information, which enables the network to predict the current output depending on the previous computations. Figure 1 shows the RNN structure [8] which has an input layer x, a hidden layer h and an output layer y. In negation scope detection context, x represents input features and y represents tags. The RNN computes the values in the hidden layer h and the output layer y as follows:

$$h(t) = f(Ux(t) + Wh(t - 1)) \tag{1}$$

$$y(t) = g(Vh(t)) \tag{2}$$

where U, W and V are the connections weights to be computed in training time, and $f(z)$ and $g(z)$ are the sigmoid and the softmax functions.

Figure 1 illustrates a negation scope detection system, in which each word is tagged with Inside (I) or Outside (O) the scope. The sentence *Do not talk about it.* is tagged as O $O I I I O$.

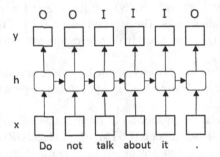

Fig. 1. A simple RNN architecture

The recurrent layer in a RNN architecture is designed to store history information, thus, RNNs introduce the connection between the previous hidden state and the current hidden state. However, in practice, the range of input history that can be reached is limited, since the influence of a given input word decay or blow up exponentially as it circulates around the hidden states, which is known as vanishing gradient problem [12]. The most efficient alternative for this problem so far is the LSTM (Long Short Term Memory) network [13]. LSTM networks are formed like RNNs, except that the hidden units are replaced by special designed units called *memory blocks*. These units allow the network to remember information over long periods of time and avoid the vanishing gradient problem. Each memory block contains, one or more recurrently connected memory cells c (i.e., a neuron), an input

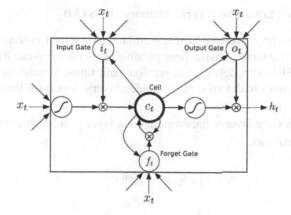

Fig. 2. A Long Short-Term Memory block

gate I to control the flow of input signal into the neuron, an output gate o to control the effect of neuron activation on other neuron, and a forget gate f to allow the neuron to adaptively reset its current state through the recurrent connection. Figure 2 illustrates all these components [10].

The output of a LSTM hidden layer h given an input x is computed as follows:

$$i_t = \sigma\left(W_{xi}x_t + W_{hi}h_{t-1} + W_{ci}c_{t-1} + b_i\right) \tag{3}$$

$$f_t = \sigma\left(W_{xf}x_t + W_{hf}h_{t-1} + W_{cf}c_{t-1} + b_f\right) \tag{4}$$

$$c_t = f_t c_{t-1} + i_t \tanh\left(W_{xc}x_t + W_{hc}h_{t-1} + b_c\right) \tag{5}$$

$$o_t = \sigma\left(W_{xo}x_t + W_{ho}h_{t-1} + W_{co}c_{t-1} + b_o\right) \tag{6}$$

$$h_t = o_t \tanh\left(c_t\right) \tag{7}$$

where σ is the logistic sigmoid function, and i, f, o and c are respectively the input gate, forget gate, output gate and cell activation vectors, all of which are the same size as the hidden vector h. Figure 3 shows a LSTM negation scope detection model.

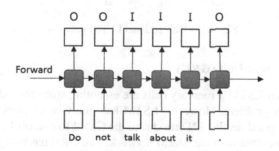

Fig. 3. A LSTM network

2.2   Bidirectional Long Short Term Memory (BLSTM)

For many sequence processing tasks, it is useful to analyze the future as well as the past of a given point in the series, this has been possible using Bidirectional Recurrent Neural Networks [23]. BRNN models exploit previous and future context by processing the data in both directions with two separate hidden layers, which are then fed forward to the same output layer.

The forward hidden layer $\vec{h}$, backward hidden layer $\overleftarrow{h}$ and their combined output y_t are computed as follows:

$$\vec{h}_t = f\left(\vec{U}h_{t-1} + \vec{V}x_t + \vec{b}\right) \tag{8}$$

$$\overleftarrow{h}_t = f\left(\overleftarrow{U}h_{t-1} + \overleftarrow{V}x_t + \overleftarrow{b}\right) \tag{9}$$

$$y_t = W\vec{h}_t + W\overleftarrow{h}_t + b_y \tag{10}$$

where $\vec{U}, \vec{V}$ and $\vec{b}$ are the forward weight matrices, and $\overleftarrow{U}, \overleftarrow{V}$ and $\overleftarrow{b}$ are their backward counterparts. The concatenated vector $h_t = \left[\vec{h}_t, \overleftarrow{h}_t\right]$ is passed to the output layer.

Similarly, the unidirectional LSTM can be extended to a Bidirectional LSTM by replacing the hidden states in BRNN with LSTM memory blocks [11], which allows bidirectional connections in the hidden layers. This amounts to having a backward counterpart for each of the equations from 3 to 7. Figure 4 below illustrates the BLSTM model.

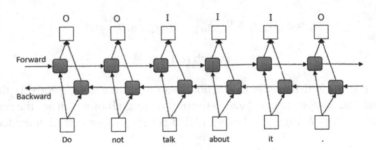

Fig. 4. A Bidirectional LSTM network

2.3   Gated Recurrent Unit (GRU)

It is easy to see that LSTM's memory cells are essential, since they allow to avoid the vanishing gradient problem. However, LSTM has many other units that may not be needed to achieve good results. [3] introduced the Gated Recurrent Unit (GRU), which has a similar architecture with the LSTM network. The GRU has two gates, a reset gate r and an update gate z. Intuitively, the reset gate determines how to combine the new input with the previous hidden state, and the update gate defines how much information from the previous hidden state will carry over to the current hidden state. This acts similarly to the memory cell in the LSTM network and helps the GRU to remember long-term information and avoid the vanishing gradient problem. The GRU is defined by the following equations [5]:

$$r_t = \sigma \left(W_{xr} x_t + W_{hr} h_{t-1} \right) \tag{11}$$

$$z_t = \sigma \left(W_{xz} x_t + W_{hz} h_{t-1} \right) \tag{12}$$

$$\tilde{h}_t = \tanh \left(W_{xh} x_t + W_{hh} \left(r_t \odot h_{t-1} \right) \right) \tag{13}$$

$$h_t = z_t \odot h_{t-1} + \left(1 - z_t \right) + \tilde{h}_t \tag{14}$$

2.4 Conditional Random Fields (CRFs)

Conditional random fields (CRFs) [15] are probabilistic models for labelling and segmenting structured data, such as sequences. The underlying idea is that of defining a conditional probability over label sequences Y given a particular observation X, rather than a join distribution over both label and observation sequences. The primary advantage of CRFs over hidden Markov models is their conditional nature, resulting in the relaxation of the independence assumptions required by HMMs in order to ensure tractable inference. Additionally, CRFs avoid the label bias problem, a weakness exhibited by maximum entropy Markov models (MEMMs) and other conditional Markov models based on directed graphical models. Note that the inputs and outputs are directly connected, as opposed to LSTM and BLSTM networks where memory blocks are employed in between. Figure 5 below illustrates the CRF model.

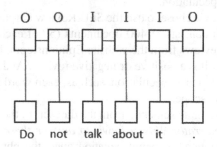

Fig. 5. A CRF network

3   Experiments Settings

In this section, we first describe our task, then, introduce the dataset used to compute our models, and finally, present our experimental settings of our different models.

3.1 Task Description

In this paper, negation scope detection is regarded as a sequence labelling problem, where each word in a sentence is marked as being INSIDE or OUTSIDE the scope. We conduct several experiments using different sequence labelling models: LSTM, BLSTM, GRU and CRF, to show the model that solves the best the negation scope detection problem, and whether hand-crafted features engineering are essential to achieve good results.

To train our recurrent neural network models we have used a recently implemented library called KERAS [4].

KERAS is a minimalist, highly modular neural networks library, written in Python and capable of running on top of either TensorFlow or Theano. It was developed with a focus on enabling fast experimentation. KERAS is a deep learning library that:

- Allows for easy and fast prototyping (through total modularity, minimalism, and extensibility).
- Supports both convolutional networks and recurrent networks, as well as combinations of the two.
- Supports arbitrary connectivity schemes (including multi-input and multi-output training).
- Runs seamlessly on CPU and GPU.

3.2 Data Set

The negation detection studies has mainly focused on the biomedical domain and has been neglected in general domains, leading to a lack of freely available corpus. There are two freely available corpus in general domains for negation detection. One was distributed for the Shared Task 2012 [19], containing sentences from Conan Doyle stories, and the other is the Simon Fraser University (SFU) Review Corpus [25] annotated for negation and speculation.

For our experiments, we chose to use the SFU Review corpus, because of the larger dataset provided in it. It consists of 400 documents (50 of each type) of movie, book and consumer product reviews from the website Epinions.com. Each text was assigned a label based on whether it is a positive or negative review. And each sentence in a text was annotated for negation and speculation, such as, each word is annotated as being a cue, inside or outside the scope.

Figure 6 shows an example sentence from the dataset. In the sentence *"Don't count on Nicolas Sparks, or narrator Wilson Lewis to answer these questions."* the words *"Do"* and *"n't"* are annotated as being negation cues, the phrase *"count on Nicolas Sparks"* as being inside scope, and the phrase, *"or narrator Wilson Lewis to answer these questions."* as being outside the scope.

The SFU Review corpus was annotated for negation and speculation following the Bioscope corpus annotation guidelines [25], bringing some modifications. The main changes are summarized below:

- The cue words are not included in their scope;
- A different scheme for annotating coordination was used;

- Embedded scopes were quite a frequent case;
- There is a case of "no scope" both in the case of negation and speculation.

For our experiments, we preprocessed the original dataset to keep only the annotations related to the negation, and remove all those related to the speculation. When an embedded scope is met, where a sentence contains more than one negation cue, we treat each cue of the sentence separately, and create a new instance of the same sentence for each negation cue.

```
- <SENTENCE>
  - <cue type="negation" ID="34">
      <W>Do</W>
      <W>n't</W>
    </cue>
  - <xcope ID="38">
      <ref ID="44" SRC="34"/>
      <W>count</W>
      <W>on</W>
      <W>Nicholas</W>
      <W>Sparks</W>
    </xcope>
      <W>,</W>
      <W>or</W>
      <W>narrator</W>
      <W>Wilson</W>
      <W>Lewis</W>
      <W>to</W>
      <W>answer</W>
      <W>these</W>
      <W>questions</W>
      <W>.</W>
  </SENTENCE>
```

Fig. 6. Example sentence from the SFU Review corpus

Table 1. Statistics of the SFU Review Corpus

Domain	#Sentences	#Negated sentences	%Negated sentences
Books	1,640	405	24.69
Cars	3,087	569	18.43
Computers	3,104	589	18.97
Cookware	1,559	375	24.05
Hotels	2,169	384	17.70
Movies	1,865	480	25.75
Music	3,162	465	14.70
Phone	1,085	231	21.29
Total	**17,671**	**3,498**	**19.79**

Statistical analysis of the annotated corpus, after preprocessing, revealed that out of the total amount of 17,671 sentences 20 % contained negation cues, as shown in Table 1.

More information about the corpus and the annotation guidelines can be found in [14].

We split the original dataset into three subsets, a Training set, a Development set and a Test set. We employed the ratio 3:1:1 to split the dataset (60 % for Training, 20 % for Development and 20 % for Test). The original dataset is split into 8 folders (Books, Cars, Movies, etc.), and each folder is also split into two subfolders (positive and negative reviews). We proceed by partitioning each subfolder of each folder with the ratio 3:1:1, to have a balanced data from each domain (Books, cars, etc.), and avoid a dissociation between the different sub-datasets. The first 60 % of each folder were assigned to the Training data, the first 20 % of the rest to the Development data, and the last 20 % to the Test data. A k-fold cross-validation experiment couldn't be done in our experiments, because of the time-consuming of each RNN model.

3.3   CRF Baseline

We use a linear-chain CRF of order 2 as our baseline. The features used in the baseline model includes:

- The current token.
- POS of the current token.
- The relative position of the current token to the cue: before, after or none (no cue).
- The distance between the current token and the cue: the number of tokens between them.
- Is a cue: set to 1 if the current token is a cue, 0 otherwise.

The CRF considers the features of the current token, the two previous and two forward tokens. It also uses features conjunction by combining features of neighboring tokens, and finally makes use of bigram features. The process of feature selection started by combining different feature sets found in some previous work similar to our task [1, 7]. This feature set has been polished from the useless features, and added some new features, that we found to be helpful for the scope detection task. Our final feature set has the advantage in capturing the potential relationship between cues and their scopes in a sentence.

To train and test our model we use the CRF++ toolkit, which is an open source software for CRFs. This tool can be used for many NLP tasks, like POS tagging, Chunking, Named Entity Recognition and Information Extraction.

3.4   Word Embedding

Word embeddings are distributed representations of words, represented as real-valued, dense, and low-dimensional vectors. Each dimension potentially describes syntactic or semantic properties of the word [16].

It has been shown in [6] that word embeddings play a vital role to improve sequence tagging performance, and to be very helpful for many other NLP tasks. For our experiment, we use word embedding to improve the performance of our different RNN models in solving the negation scope detection problem. We use a task-specific word embedding model, because several misspellings words, abbreviations, and compositions of words occur in the dataset. These words are identified as UNKNOWN words by a pre-trained word embedding model. And we noticed during training that these UNKNOWN words are always labelled as *Outside* the scope.

We build our task-specific word embedding model using the *"Embedding"* layer of KERAS library. This layer takes as input a 2 dimensional matrix of integers representing each word in the corpus, and outputs a 3 dimensional matrix, which represents the word embedding model. This layer is on the top of our RNN architecture.

3.5 RNN Settings

For our different experiments, we do not perform any pre-processing on the dataset, except for the case of embedded scopes. We first start to build a vocabulary directly on the raw data of the entire corpus, and then represent each word in a sentence using its index in the vocabulary. We note that there is no UNKNOWN word in the vocabulary, for the same reasons mentioned earlier.

We use the index representation of each sentence of the dataset as the input data of the *"Embedding"* layer of our model. The resulting word embedding model will be used then by the RNN layer. In order to show the impact of word embedding on the RNN models performance, we proceed in two different ways. The first way is to use the word embedding as the input data of the RNN layer, and the other is to use the index representation as the input data of the RNN layer, and the word embedding as an input feature.

We train each RNN model with a batch size equal to 1, because of the variability in sentences' length. We run each model with one epoch and calculate the F1 score on the Development data to tune the parameters. The size of the hidden layers and the dimension of the word embedding model are empirically set based on the performance on the Development data. The hidden layer sizes we experimented with are 50, 100, 150 and 200, and the word embedding dimensions are 40, 80, 120, 160, 200, 240, 280, 320, 360 and 400 (the dimension gives more optimal results when it is a multiple of 4). We report all the optimal results of each model in Table 2.

4   Results

In Table 2 we present the results of our experiments using the CRF model and the different RNNs models mentioned before. We trained the models on our Training data, fine-tuned the parameters on the Development data, and tested the overall on the Test data.

The evaluation metric measures the standard precision, recall and F1 score based on the exact match of the token. This means that a classification of a token is considered to be correct only if it exactly matches with the annotation by the human. We calculate all the metrics using the evaluation script conlleval, distributed within the CoNLL 2000 Shared task.

Table 2. Results of negation scope detection using different RNNs models and a CRF model.

System	Development data			Test data		
	P	R	F	P	R	F
CRF	86.65	91.07	88.81	86.23	89.03	87.61
LSTM +word_emb as input +size hidd_layer = 200 +dim word_emb = 280	88.66	90.71	89.68	90.14	88.03	89.07
LSTM +words index as input +word_emb as feature +size hidd_layer = 200 +dim word_emb = 200	87.4	92.21	89.74	88.69	90.04	89.36
BLSTM +word_emb as input +size hidd_layer = 100 +dim word_emb = 360	88.92	90.01	89.46	90.8	87.35	89.04
BLSTM +words index as input +word_emb as feature +size hidd_layer = 100 +dim word_emb = 200	89.38	90.31	**89.84**	91.21	87.56	**89.38**
GRU +word_emb as input +size hidd_layer = 200 +dim word_emb = 200	88.17	90.98	89.54	88.82	88.07	88.95
GRU +words index as input +word_emb as feature +size hidd_layer = 150 +dim word_emb = 40	88.33	90.9	89.6	90.61	87.95	89.26

In the following, we highlight our main findings.

4.1   CRF vs RNNs

We can see from Table 2 that all of our RNNs models outperform the CRF model on the negation scope detection task, with a maximum gain of 1.03 % and 1.77 % by the BLSTM model, using word embedding as feature, on the Development data and the Test data, respectively. This is remarkable, knowing that these RNNs models do not use any hand-crafted feature, unlike the feature-rich CRF model which needs a minimum of features to achieve acceptable results.

These results demonstrate that RNNs as sequence labelers are more effective than the CRFs for negation scope detection task. This can be attributed to the RNN's proficiency to learn features automatically and to capture long-range sequential dependencies between the output labels.

4.2 Comparison Among RNN Models

A comparison among the RNNs models in Table 2 shows that the BLSTM model, using word embedding as input feature, capture better the relation between the cue and the tokens in a sentence with an F1 score of 89.38 % on the Test data. But the LSTM model stay behind with just 0.02 % loss in performance, which tells that the unidirectional links of the LSTM may be sufficient to capture the relationship between the cue and the other tokens of the sentence, without going through the time consuming of the bidirectional links of the BLSTM model.

We can also notice that, the GRU, also using word embedding as input feature, has just 0.10 % loss in performance comparing to the LSTM model. This loss can be compensated by the time saving in the execution of the GRU model, which is comparing to the LSTM model much faster; about 67 min for the GRU model against 127 min for the LSTM model. The choice of the model may depend then on personal expectations; faster model or more accurate one.

4.3 Contribution of Word Embedding in RNNs

As we can see in Table 2, adding word embedding as input feature instead of direct input give a small but a non-negligible improvement. It improves all the RNN models with almost 0.30 % on the Test data, this proves that the model learn more information using word embedding as feature combined with word index as input, than just using it as a direct input.

The results using pre-trained word embeddings have not been reported in this paper, because of the unsuccessful and poor results of the models.

4.4 Error Analysis

After a deep analysis of the predicted labels of our different models, we have noticed a problem that has actually occurred in the original dataset. We found an ambiguity in the annotation of the punctuation marks. In some sentences they are annotated as inside the scope, and in others as outside the scope. In other corpus, like the Shared task 2012's corpus [19], all punctuation marks are assumed to be outside the scope.

The two sentences below are both extracted from the dataset, and both contain a negation cue, and end with a punctuation mark. In the first sentence the punctuation is marked as inside the scope, but in the second sentence it is marked as outside. But there isn't any evident reason for such a divergent annotation. The phrases between brackets are the negation scopes.

1. I understand that on Monday and Tuesday, the hotel was without [air conditioning at all].
2. Unfortunately the winds were too high each night that we were there and we never [got to see it live].

Because of this ambiguity, during training our different models learn to always label the punctuation marks as being outside the scope, which decrease drastically the performance of our results. The punctuation marks represent about 10 % of the Test data. We assume that fixing the role of the punctuation marks as always inside the scope, or even always outside the scope, may improve remarkably the performance of our models.

Also, as mentioned in the Data Set section, there are cases of "no scope" in the data, this means that there are sentences that contain annotated cues, but have no scope corresponding to them. In this case, the model incorrectly try to label words in the sentence as being inside the scope.

5   Conclusion

In this paper we proposed an approach to identify the scope of negation by tackling it as a sequence labelling problem, and use a variety of recurrent neural networks (RNNs) models and word embedding to solve it.

Our results on negation scope detection task prove that RNN models outperform the CRF model, without using any hand-crafted specific task feature. They also demonstrate that word embeddings improve the performance of RNNs models when they are used as input feature. Our best result on the Test data remains the BLSTM model using word embedding as feature. The other RNNs models (LSTM and GRU), also gave promising results, with a negligible loss in their performances. The GRU may be a great choice for an accurate and a faster model, noticing the time-consuming in the execution of the LSTM and the BLSTM models.

For our future work, we intend to use the current system to solve some problems related to the negation. We believe that this kind of system can improve the accuracy of several works that are sensitive to polarity in the information extraction and natural language processing domain, like sentiment analysis or question answering systems.

References

1. Abu-Jbara, A., Radev, D.: UMichigan: a conditional random field model for resolving the scope of negation. In: Proceedings of the First Joint Conference on Lexical and Computational Semantics-Volume 1: Proceedings of the Main Conference and the shared task, and Volume 2: Proceedings of the Sixth International Workshop on Semantic Evaluation, pp. 328–334. Association for Computational Linguistics (2012)
2. Agarwal, S., Yu, H.: Biomedical negation scope detection with conditional random fields. J. Am. Med. Inform. Assoc. **17**(6), 696–701 (2010)
3. Cho, K., van Merrienboer, B., Bahdanau, D., Bengio, Y.: On the properties of neural machine translation: encoder-decoder approaches, arXiv preprint arXiv:1409.1259 (2014)
4. Chollet, F.: Keras. http://keras.io/

5. Chung, J., Gulcehre, C., Cho, K.H., et al.: Empirical evaluation of gated recurrent neural networks on sequence modeling, arXiv preprint arXiv:1412.3555 (2014)
6. Collobert, R., Weston, J., Bottou, L., et al.: Natural language processing (almost) from scratch. J. Mach. Learn. Res. **12**, 2493–2537 (2011)
7. Councill, I.G., McDonald, R., Velikovich, L.: What's great and what's not: learning to classify the scope of negation for improved sentiment analysis. In: Proceedings of the Workshop on Negation and Speculation in Natural Language Processing, pp. 51–59. Association for Computational Linguistics (2010)
8. Elman, J.L.: Finding structure in time. Cogn. Sci. **14**(2), 179–211 (1990)
9. Graves, A., Jaitly, N.: Towards end-to-end speech recognition with recurrent neural networks. In: Proceedings of the 31st International Conference on Machine Learning, pp. 1764–1772 (2014)
10. Graves, A., Mohamed, A., Hinton, G.: Speech recognition with deep recurrent neural networks. In: IEEE International Conference on Acoustics, Speech and Signal Processing (ICASSP), pp. 6645–6649 (2013)
11. Graves, A., Schmidhuber, J.: Framewise phoneme classification with bidirectional LSTM and other neural network architectures. Neural Netw. **18**(5), 602–610 (2005)
12. Hochreiter, S., Bengio, Y., Frasconi, P., et al.: Gradient flow in recurrent nets: the difficulty of learning long-term dependencies. In: A Field Guide to Dynamical Recurrent Neural Networks, pp. 237–243. IEEE press (2001)
13. Hochreiter, S., Schmidhuber, J.: Long short-term memory. Neural Comput. **9**(8), 1735–1780 (1997)
14. Konstantinova, N., De Sousa, S.C.M.: Annotating negation and speculation: the case of the review domain. In: RANLP Student Research Workshop, pp. 139–144 (2011)
15. Lafferty, J., McCallum, A., Pereira, F.C.N.: Conditional random fields: probabilistic models for segmenting and labeling sequence data. In: Proceedings of ICML, pp. 282–289 (2001)
16. Liu, P., Joty, S., Meng, H.: Fine-grained opinion mining with recurrent neural networks and word embeddings. In: Conference on Empirical Methods in Natural Language Processing, EMNLP 2015 (2015)
17. Mesnil, G., He, X., Deng, L., et al.: Investigation of recurrent-neural-network architectures and learning methods for spoken language understanding. In: Proceedings of INTERSPEECH, pp. 3771–3775 (2013)
18. Mikolov, T., Karafiát, M., Burget, L., et al.: Recurrent neural network based language model. In: Proceedings of INTERSPEECH, pp. 1045–1048 (2010)
19. Morante, R., Blanco, E.: *SEM 2012 shared task: Resolving the scope and focus of negation. In: Proceedings of the First Joint Conference on Lexical and Computational Semantics-Volume 1: Proceedings of the main conference and the shared task, and Volume 2: Proceedings of the Sixth International Workshop on Semantic Evaluation, pp. 265–274. Association for Computational Linguistics (2012)
20. Morante, R., Daelemans, W.: A metalearning approach to processing the scope of negation. In: Proceedings of the Thirteenth Conference on Computational Natural Language Learning, pp. 21–29. Association for Computational Linguistics (2009)
21. Morante, R., Liekens, A., Daelemans, W.: Learning the scope of negation in biomedical texts. In: Proceedings of the Conference on Empirical Methods in Natural Language Processing, pp. 715–724. Association for Computational Linguistics (2008)
22. Sak, H., Senior, A.W., Beaufays, F.: Long short-term memory recurrent neural network architectures for large scale acoustic modeling. In: Proceedings of INTERSPEECH, pp. 338–342 (2014)

23. Schuster, M., Paliwal, K.K.: Bidirectional recurrent neural networks. IEEE Trans. Sig. Process. **45**(11), 2673–2681 (1997)
24. Sundermeyer, M., Schlüter, R., Ney, H.: LSTM neural networks for language modeling. In: Proceedings of INTERSPEECH, pp. 194–197 (2012)
25. Taboada, M., Anthony, C., Voll, K.: Methods for creating semantic orientation dictionaries. In: Proceedings of the 5th Conference on Language Resources and Evaluation, pp. 427–432 (2006)
26. Vincze, V., Szarvas, G., Farkas, R., et al.: The BioScope corpus: biomedical texts annotated for uncertainty, negation and their scopes. BMC Bioinform. **9**(11), 1 (2008)
27. Wang, P., Qian, Y., Soong, F.K., et al.: A unified tagging solution: bidirectional LSTM recurrent neural network with word embedding. arXiv preprint arXiv:1511.00215 (2015)
28. Zou, B., Zhou, G., Zhu, Q.: Tree kernel-based negation and speculation scope detection with structured syntactic parse features. In: Empirical Methods in Natural Language Processing, pp. 968–976 (2013)

Numerical Stability of the Runge-Kutta Methods for Equations $u'(t) = au(t) + bu([\frac{K}{N}t])$ in Science Computation

Yingchun Song$^{(\boxtimes)}$ and Xianhua Song

School of Applied Science, Harbin University of Science and Technology,
Harbin 150080, China
songyingchun_syc@sohu.com

Abstract. Differential equation has widely applied in science and engineering calculation. Runge Kutta method is a main method for solving differential equations. In this paper, the numerical properties of Runge-Kutta methods for the equation $u'(t) = au(t) + bu([\frac{K}{N}t])$ is dealed with, where K and N is relatively prime and $K < N, K, N \in \mathbb{Z}^+$. The conditions are obtained under which the numerical solutions preserve the analytical stability properties of the analytic ones and some numerical experiments are given.

Keywords: The unbounded retarded differential equations · Piecewise continuous arguments · Runge-Kutta methods · Asymptotic stability

1 Introduction

We consider the following differential equations with piecewise continuous arguments (EPCA):

$$\begin{cases} u'(t) = f(t, u(t), u(\alpha(t))), t \geqslant 0, \\ u(0) = u_0, \end{cases} \tag{1.1}$$

where u_0 is a given initial value, the argument $\alpha(t)$ has the intervals of constancy. For example, $\alpha(t) = [\frac{K}{N}t]$, $[\cdot]$ denotes the greatest integer function.

EPCA describe hybrid dynamics systems, combine properties of both differential and difference equations, and have applications in certain biomedical models in the work of Busenberg and Cooke [16]. The general theory and basic results for EPCA have by now been thoroughly investigated in [2,3] and the book of Wiener [12]. In recent years, much research has been focused on numerical solutions of EPCA [1,4–9].

In [4], the authors consider the stability properties of analytical and numerical solutions of the following EPCA with unbounded retarded delay

$$\begin{cases} u'(t) = au(t) + bu([\frac{1}{N}t]), t \geqslant 0, \\ u(0) = u_0, \end{cases} \tag{1.2}$$

where $N \in \mathbb{Z}^+, a + |b| < 0, a, b \in \mathbb{R}, u_0 \in \mathbb{R}$ is a given initial value.

© Springer Science+Business Media Singapore 2016
W. Che et al. (Eds.): ICYCSEE 2016, Part I, CCIS 623, pp. 509–519, 2016.
DOI: 10.1007/978-981-10-2053-7_45

In this paper, we will generalize (1.2) to the following form

$$\begin{cases} u'(t) = au(t) + bu([\frac{K}{N}t]), t \geqslant 0, \\ u(0) = u_0, \end{cases} \tag{1.3}$$

where K and N is relatively prime and $K < N, K, N \in \mathbb{Z}^+, a + |b| < 0, a, b \in \mathbb{R}$, $u_0 \in \mathbb{R}$ is a given initial value.

Definition 1. *A solution of Eq. (1.3) on $[0, \infty)$ is a function $u(t)$ that satisfies the conditions:*

(1) $u(t)$ *is continuous on $[0, \infty)$.*

(2) *The derivative $u'(t)$ exist at each points $t \in [0, \infty)$, with the possible exception of the point $t = \frac{N}{K}n, n = 0, 1, 2, \cdots$, where one-side derivatives exist.*

(3) *Eq. (1.3) is satisfied on each interval $[\frac{N}{K}n, \frac{N}{K}(n+1)) \subset [0, \infty)$.*

In the following theorems, we will give the existence of solutions and then provide sufficient conditions for the asymptotic stability of all solutions of (1.3).

Theorem 1 [12]. *Assume that $K \leqslant N, K, N \in \mathbb{Z}^+$ and that $a + |b| < 0$. Then the analytical solutions of*

$$\begin{cases} u'(t) = au(t) + bu([\frac{K}{N}t]), t \geqslant 0, \\ u(0) = u_0, \end{cases} \tag{1.4}$$

satisfies

$$\lim_{t \to \infty} u(t) = 0$$

for all $u_0 \in \mathbb{R}, a, b \in \mathbb{R}$.

Remark 1. In [4], the special case of (1.3) with $K = 1$ has been discussed. The sufficient conditions under which the numerical solutions preserve the stability of the analytic solutions are obtained. Hence, we only consider (1.3) with $K \neq 1$, $K < N, K, N \in \mathbb{Z}^+$ and K, N is relatively prime in this paper.

We will investigate the stability of Runge-Kutta methods for (1.3). In Sect. 2 we consider the adaptation of the Runge-Kutta methods. In Sect. 3 we give the conditions of the stability of the Runge-Kutta methods. In Sect. 4 some numerical experiments are illustrated.

2   Runge-Kutta Methods

In this section we consider the adaptation of the Runge-Kutta methods (A, B, C), where $A = (a_{ij})_{v \times v}$ and $B = (B_1, B_2, \cdots, B_v)^T$ and $C = (C_1, C_2, \cdots, C_v)^T$. For the Runge-Kutta methods we always assume that $B_1 + B_2 + \cdots + B_v = 1$ and $0 \leqslant C_1 \leqslant C_2 \leqslant, \cdots, \leqslant C_v \leqslant 1$.

Let $h = \frac{1}{Km}$ be a given stepsize with integer $m \geqslant 1$ and $t_n = nh(n = 0, 1, 2, \cdots)$. The adaptation of the Runge-Kutta methods to (1.1) leads to a numerical process generating approximations u_n to the analytical solution $u(t)$ of (1.1) at the gridpoints $t_n (n = 0, 1, 2, \cdots)$, i.e.,

$$u_{n+1} = u_n + h\Sigma_{i=1}^{v} B_i f(t_n + C_i h, y_i^n, z_i^n), \tag{2.5}$$

where $y_1^n, y_2^n, \cdots, y_v^n$ satisfy

$$y_i^n = u_n + h\Sigma_{j=1}^{v} a_{ij} f(t_n + C_j h, y_j^n, z_j^n), i = 1, 2, \cdots, v, \tag{2.6}$$

and the argument z_i^n denotes the given approximation to $u(\alpha(t_n + C_i h)), i = 1, 2, \cdots, v, n = 0, 1, \cdots$.

We are interested in the application of (2.5) and (2.6) to (1.3). The application of the process (2.5) and (2.6), in the case of (1.3), yields

$$\begin{aligned} u_{n+1} &= u_n + h\Sigma_{j=1}^{v} B_i(ay_i^n + bz_i^n) \\ y_i^n &= u_n + h\Sigma_{i=1}^{v} a_{ij}(ay_j^n + bz_j^n), i = 1, 2, \cdots, \end{aligned} \tag{2.7}$$

where the z_i^n is the approximation to

$$u([\frac{K}{N}(t_n + C_i h)]), i = 1, 2, \cdots, v, n = 0, 1, 2, \cdots.$$

If we denote $n = mNj + l, (l = 0, 1, 2, \cdots, Nm - 1, j = 0, 1, 2, \cdots$, then $[\frac{K}{N}(t_{mNj+l} + C_i h)] = t_{mKj}$. It is suitable to define

$$z_i^{mNj+l} = u_{mKj},$$

where $l = 0, 1, \cdots, Nm - 1. i = 1, 2, \cdots, v, j = 0, 1, 2, \cdots$.

Let $Y^n = (y_1^n, y_2^n, ..., y_v^n)^T$. Then (2.7) reduced to

$$\begin{aligned} u_{mNj+l+1} &= u_{mNj+l} + haB^T Y^{mNj+l} + hbu_{mKj}, \\ Y^{mNj+l} &= u_{mNj+l}e + haAY^{mNj+l} + hbAeu_{mKj}, \end{aligned} \tag{2.8}$$

where $e = (1, 1, \cdots, 1)^T$, which is equivalent to

$$u_{mNj+l+1} = R(x)u_{mNj+l} + \frac{b}{a}(R(x) - 1)u_{mKj}, \tag{2.9}$$

where $x = ah, R(x) = 1 + xB^T(I - xA)^{-1}e$ is the stability function of the Runge-Kutta methods, provided that $I - xA$ is invertible.

Theorem 2. *Assume that the Runge-Kutta method is of order p. Then numerical solutions of (1.3) converge to the analytical ones with accuracy of order p.*

Proof. Let the R-K methods is of order p. There exist constants C such that for sufficiently small h

$$|e^x - R(x)| \leqslant Ch^{p+1}. \tag{2.10}$$

It is easy to see from (2.9) that if $u(t_{mNj+l}) = u_{mNj+l}$ and $u(t_{mKj}) = u_{mKj}$, then

$$\begin{aligned} &|u(t_{mNj+l+1}) - u_{mNj+l+1}| \\ &= |e^x u(t_{mNj+l}) + \frac{b}{a}(e^x - 1)u(t_{mKj}) - [R(x)u_{mNj+l} + \frac{b}{a}(R(x) - 1)u_{mKj}]| \\ &\leqslant Ch^{p+1}(1 + |\frac{b}{a}|)\max_{j \leqslant t \leqslant \frac{N}{K}(j+1)} |u(t)|, \end{aligned}$$

which implies that for (1.3) the Runge-Kutta method is also convergent of order p.

3   Numerical Stability

In this section we will discuss the stability of the Runge-Kutta methods. In view of $R(x) = 1 + xB^T(I - xA)^{-1}e$, there exists a $\delta < 0$ such that $R(x)$ is continuous function, for $\delta < x < 0$. Define $M = \frac{a}{\delta}$, where $a < 0$.

Definition 2. *Process* (2.5) *for Eq.*(1.3) *is called asymptotically stable at* (a, b) *if for all* $m \geqslant \frac{M}{K}$ *and* $h = \frac{1}{Km}$

1. $(I - xA)$ *is invertible.*
2. *For any given* u_0, *relation* (2.8) *defines* u_n *that satisfies* $u_n \to 0$ *as* $n \to \infty$.

Let $G(l) = (R^l(x) + |\frac{b}{a}|(1 - R^l(x)))$. When $0 < R(x) < 1$, by [4], we know that $G(l)$ is monotonically decreasing and $0 < G(l) < 1$.

Firstly, we will give the following lemma.

Lemma 1. *Let* $M_0 = 0, M_1 = 1, A_0 = [0, \frac{N}{K})$ *and* $A_k = [\frac{NM_k}{K}, \frac{NM_{k+1}}{K})$, $k = 1, 2, \cdots$. *Define* M_{k+1} *be the unique integer in* $[\frac{NM_k}{K}, \frac{NM_k}{K} + 1)$, $k = 1, 2, \cdots$. *Then*

$$[\frac{K}{N}t] \in A_{k-1}, t \in A_k, (k = 1, 2, \cdots). \tag{3.11}$$

Proof. If $t \in A_1$, then

$$M_1 \leqslant \frac{Kt}{N} < M_2.$$

By the definition of M_k and $M_1 = 1$, we have

$$1 \leqslant [\frac{Kt}{N}] \leqslant M_2 - 1 < \frac{N}{K},$$

namely, $[\frac{Kt}{N}] \in A_0$.

We assume that $[\frac{Kt}{N}] \in A_{k-2}$ hold for $t \in A_{k-1}$.

For $t \in A_k$, we obtains that

$$M_k \leqslant \frac{K}{N}t < M_{k+1},$$

which implies by the definition of M_k that

$$M_k \leqslant [\frac{K}{N}t] \leqslant M_{k+1} - 1, \tag{3.12}$$

moreover,

$$\frac{NM_{k-1}}{K} \leqslant M_k, \quad M_{k+1} - 1 < \frac{NM_k}{K}. \tag{3.13}$$

From (3.12) and (3.13), we can easily obtain

$$\frac{NM_{k-1}}{K} \leqslant M_k \leqslant [\frac{K}{N}t] \leqslant M_{k+1} - 1 < \frac{NM_k}{K}.$$

Then

$$[\frac{K}{N}t] \in A_{k-1}, \text{for } t \in A_k.(k = 1, 2, \cdots).$$

By induction, the proof is completed.

Let $\overline{A_k} = (\frac{NM_k}{K}, \frac{NM_{k+1}}{K}], k = 0, 1, 2, \cdots$, we have the following theorem.

Theorem 3. *If $K \neq 1$ and $0 < R(x) < 1$, then*

$$(1)\ |u_{mNj+l}| \leqslant G^{k+1-r(k-1)}(1)|u_0|,$$
$$for\ j = M_k, M_k + 1, \cdots, M_{k+1} - 1, l = 1, 2, \cdots, Nm. \qquad (3.14)$$
$$k = 0, 1, 2, 3, \cdots,$$

where $r(k - 1)$ is the total number of $\frac{NM_i}{K} = M_{i+1}$ and $r(-1) = 0$, $i \leqslant k - 1$.
$$(2)\ \lim_{j\to\infty} u_{mNj+l} = 0.$$

Proof. (1). In view of (2.9), we can obtain

$$u_{mNj+l} = R^l(x)u_{mNj} + \frac{b}{a}(R^l(x) - 1)u_{mKj}, \qquad (3.15)$$

where $j = 0, 1, 2, \cdots, l = 1, 2, \cdots, Nm$.
If $t_n \in \overline{A_0}$, by virtue of (3.15), we have

$$|u_l| \leqslant (R^l(x) + |\frac{b}{a}|(1 - R^l(x)))|u_0| \leqslant G(1)|u_0|,\ j = M_0 = M_1 - 1 = 0, l = 1, 2, \cdots, Nm.$$

We assume that (3.14) holds, for $t_n \in \overline{A_k}$. When $t_n \in \overline{A_{k+1}}$, by Lemma 3.2, we have

$$t_{mKj} \in A_k, j = M_{k+1}, M_{k+1} + 1, \cdots, M_{k+2} - 1.$$

When $j = M_{k+1}$, we have $t_{mNM_{k+1}} \in \overline{A_k}$. There are two cases.

Case (1): If $\frac{NM_k}{K} = M_{k+1}$, we have $r(k) = r(k - 1) + 1$.
Furthermore, we obtain $t_{mKM_{k+1}} \in \overline{A_{k-1}}$, by (3.15),

$$|u_{mNM_{k+1}+l}| = |R^l(x)u_{mNM_{k+1}} + (R^l(x) - 1)u_{mKM_{k+1}}|$$
$$\leqslant G(l)G^{k-r(k-2)}(1)|u_0|$$
$$\leqslant G^{k+1-r(k-1)}(1)|u_0| = G^{k+2-r(k)}(1)|u_0|.$$

Case (2): If $\frac{NM_k}{K} \neq M_{k+1}$, we have $r(k) = r(k - 1)$ and $t_{mKM_{k+1}} \in \overline{A_k}$. By (3.15), we have

$$|u_{mNM_{k+1}+l}| \leq G(l)G^{k+1-r(k-1)}(1)|u_0| \leq G^{k+2-r(k)}(1)|u_0|.$$

By the same method, we can obtain

$$|u_{mNj+l}| \leq G^{k+2-r(k)}(1)|u_0|, j = M_{k+1} + 1, \cdots, M_{k+2} - 1. \qquad (3.16)$$

By induction, (3.14) holds.

(2) We only need to proof that $\lim_{k\to\infty}(k - r(k)) = \infty$ because $G(1) < 1$. It is obvious that $k - r(k)$ denotes the total number of $\frac{NM_i}{K} \neq M_{i+1}, i < k$ and $k - r(k)$ is an increasing sequence of number.

Assume that $\lim_{k\to\infty}(k - r(k)) = C(C \neq \infty)$, then there exists $\widetilde{K}$ such that $\frac{NM_k}{K} = M_{k+1}$, for $k > \widetilde{K}$. If $k_0 > \widetilde{K}$, then $M_{k_0+1} = \frac{NM_{k_0}}{K}$. There exist constant L and s such that $M_{k_0+1} = \frac{NM_{k_0}}{K} = LK^s$, where L and K is coprime. Furthermore,

$$M_{k_0+2} = \frac{NM_{k_0+1}}{K} = NLK^{s-1},$$

by recursion,

$$M_{k_0+s} = \frac{NM_{k_0+s-1}}{K} = N^s L,$$

hence,

$$\frac{NM_{k_0+s+1}}{K} = \frac{N^s L}{K},$$

where N, L and K are coprime, so $\frac{NM_{k_0+s+1}}{K}$ is not integer. It is a contradiction.

3.1   The Padé Approximation to the Exponential Function

In this subsection, we will consider the stability of the Runge-Kutta methods with the stability function which is given by Padé approximation to e^z. The following lemmas will be useful to prove our theorems in the paper.

Lemma 2 ([4,10]). *The* $(r, s) - $ *Padé approximation to* e^x *is given by*

$$R(x) = \frac{P_r(x)}{Q_s(x)}, \tag{3.17}$$

where $x \in \mathbb{R}$,

$$P_r(x) = 1 + \frac{r}{r + s}x + \frac{r(r - 1)}{(r + s)(r + s - 1)}\frac{x^2}{2!} + \cdots + \frac{r!s!}{(r + s)}\frac{x^r}{r!}$$

$$= \frac{r!}{(r + s)!}\sum_{j=0}^{r}\frac{(r + s - j)!}{j!(r - j)!}x^j,$$

$$Q_s(x) = 1 - \frac{s}{r + s}x + \frac{s(s - 1)}{(r + s)(r + s - 1)}\frac{x^2}{2!} + \ldots + (-1)^s\frac{r!s!}{(s + r)}\frac{x^s}{s!}$$

$$= \frac{s!}{(s + r)!}\sum_{j=0}^{s}\frac{(s + r - j)!}{j!(s - j)!}(-x)^j,$$

where error

$$e^x - R(x) = (-1)^s \frac{r! s!}{(r+s)!(r+s+1)!} x^{r+s+1} + O(x^{r+s+2}). \qquad (3.18)$$

It is the rational approximation to e^x of order $r + s$, such that the degrees of numerator and denominator are r and s, respectively.

Lemma 3 ([10,13,15]). Define $D = \{x \in \mathbb{C} : |R(x)| > |e^x|\}$. If the Runge-Kutta method is of order p, then for $x \to 0$, D behaves like a star with $p + 1$ sectors of equal width $\frac{\pi}{p+1}$, separated by $p + 1$ similar white sectors of the complement of D, each of the same width.

Lemma 4. *If $R(x)$ is the (r, s)-Padé approximation to e^x, then*

(1) there are s star sectors in the right-half plane, each containing a pole of $R(x)$;
(2) there are r white sectors in the left-half plane, each containing a zero of $R(x)$;
(3) all sectors are symmetric with respect to the real axis.

Lemma 5 [4]. *If $|x| < 1$ and $x \in R$, then $R(x) > 0$.*

From Lemmas 3.4, 3.5, 3.6 and 3.7, we have the following results.

Theorem 4. *Assume that the stability function of the Runge-Kutta methods is given by the (r, s)-Padé approximation to the exponential. If r is odd and $-1 < x < 0$, then $0 < R(x) < e^x < 1$, i.e., the Runge-Kutta methods is asymptotically stable.*

Theorem 5. *Assume that the stability function of the Runge-Kutta methods is given by the (r, s)-Padé approximation to the exponential. If $r \leq s$ and $x < 0$, then the Runge-Kutta methods is asymptotically stable.*

Theorem 6 *(Theorem and Conjecture of Ehle)* [15]. *Any Padé approximation $\frac{P_r(x)}{Q_s(x)}$ to the exponential function is A-stable if and only if*

$$s - 2 \leq r \leq s.$$

For the higher order Runge-Kutta methods, we have the following results from Theorem 3.9 or Theorem 3.10.

Corollary 1. *For the A-stable higher order Runge-Kutta method, we can get*

(1) The ν-stage Gauss-Legendre is asymptotically stable.
(2) The ν-stage Radau IA and IIA methods are asymptotically stable.
(3) The ν-stage Lobatto IIIA, IIIB and IIIC methods are asymptotically stable.

3.2   The θ-methods

Applying the one-leg θ-method and the linear θ-method to (1.3), we obtain the same recurrence relation. Hence the stability function of the two θ-methods are the same.

Theorem 7. *The θ-methods is asymptotically stable if*

$$0 \le \theta \le 1 \text{ and } \frac{1}{\theta - 1} < x < 0.$$

Proof. Clearly, if $0 \le \theta \le 1$ and $\frac{1}{\theta-1} < x < 0$, then

$$0 < R(x) = \frac{1 + (1 - \theta)x}{1 - \theta x} < 1.$$

Theorem 8. *The θ-method is asymptotically stable if $\frac{1}{2} \le \theta \le 1$.*

Proof. If $\frac{1}{2} \le \theta \le 1$, then

$$0 < R(x) = \frac{1 + (1 - \theta)x}{1 - \theta x} < 1.$$

So, we can know that the θ-method is asymptotically stable.

4   Numerical Experiments

In this section, we give two numerical examples. We consider the following equations:

Example 1:

$$\begin{cases} u'(t) = -3u(t) + u([\frac{3}{10}t]), t \ge 0, \\ u(0) = 1. \end{cases} \tag{4.19}$$

Example 2:

$$\begin{cases} u'(t) = -200u(t) + 80u([\frac{3}{10}t]), t \ge 0, \\ u(0) = 1. \end{cases} \tag{4.20}$$

where $t_n = nh, n = mNj + l, j = 0, 1, 2, \cdots, l = 0, 1, 2, \cdots Nm - 1, h = \frac{1}{3m}$.

From Fig. 1, we can see that the numerical solutions are asymptotic stability, which is consistent with our theoretical analysis.

We shall use several methods listed in Table 1 with the $h = \frac{1}{3m}$ to get the numerical solution at $t = 2$ for (4.19), where the exact solution is $u(2) \approx 3.34985834784444E - 1$. In Table 1, we list the absolution errors at $t = 2$ and the ratio of the errors of the case $m = 30$ over that of $m = 60$. We can see from the table that the methods preserve their order of convergence.

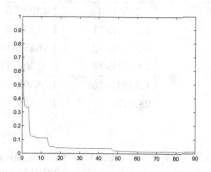

a) The numerical solution of (4.19) with
m = 1.

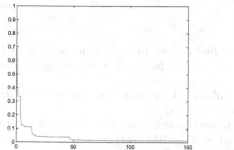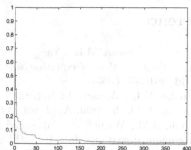

b) The numerical solution of (4.19) with c) The numerical solution of (4.20) with
m = 2.   m = 2.

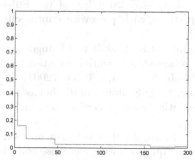

d) The numerical solution of (4.20) with
m = 60.

Fig. 1. The numerical solution of (4.19) and (4.20).

Table 1. Absolute errors of problem (4.19)

	2-Gauss-Legendre AE	3-Radau IA AE	3-Lobatto IIIC AE	$\theta = \frac{1}{4}AE$
$m = 1$	1.4651E-5	1.2156E-6	1.4300E-5	1.6E-3
$m = 10$	1.3779E-9	1.3543E-11	1.9823E-9	2.4328E-4
$m = 20$	8.6080E-11	4.2699E-13	1.2645E-10	1.2285E-4
$m = 30$	1.7002E-11	5.6011E-14	2.5150E-11	8.2149E-5
$m = 60$	1.0630E-12	1.9984E-15	1.5830E-12	4.1196E-5
Ratio	15.9946	28.0278	15.8874	1.9941

Acknowledgments. This work is supported by the Research Fund of the Natural Science Foundation of Heilongjiang Province (No. A201214) and the National Natural Science Foundation of China(61501148).

References

1. Liu, M.Z., Song, M.H., Yang, Z.W.: Stability of Runge-Kutta methods in the numerical solution of equation $u'(t) = au(t) + bu([t])$. J. Comput. Appl. Math. **166**, 361–370 (2004)
2. Cooke, K.L., Wiener, J.: Retarded differential equation with piecewise constant delays. J. Math. Anal. Appl. **99**, 265–297 (1984)
3. Shah, S.M., Wiener, J.: Advanced differetial equations with piecewise constant argument deviations. Int. J. Math. Sci. **6**, 671–703 (1983)
4. Liu, M.Z., Ma, S.F., Yang, Z.W.: Stability analysis of Runge-Kutta methods for unbounded retarded differetial equations with piecewise continuous arguments. Comput. Math. Appl. **191**, 57–66 (2007)
5. Lv, W.J., Yang, Z.W., Liu, M.Z.: Stability of the Euler-Maclaurin methods for neutral differential equations with piecewise continuous. Comput. Math. Appl. **186**, 1480–1487 (2007)
6. Lv, W.J., Yang, Z.W., Liu, M.Z.: Stability of Runge-Kutta methods for the alternately advanced and retarded differential equations with piecewise continuous arguments. Comput. Math. Appl. **54**, 326–335 (2007)
7. Liang, H., Liu, M.Z., Lv, W.J.: Stability of θ-schemes in the numerical solution of a partial differential equation with piecewise continuous arguments. Appl. Math. Lett. **23**, 198–206 (2010)
8. Yang, Z.W., Liu, M.Z., Nieto, J.: Runge-Kutta methods for first-order periodic boundary value differetial equations with piecewise constant argument. J. Comput. Appl. Math. **233**, 990–1004 (2009)
9. Song, M.H., Yang, Z.W., Liu, M.Z.: Stability of θ-methods advanced differetial equations with piecewise continuous argument. Comput. Math. Appl. **49**, 1205–1301 (2005)
10. Wanner, G., Hairer, E., Nørsett, S.P.: Order stars and stability therems. BIT **18**, 475–489 (1978)
11. Wiener, J.: Differetial equations with piecewise continuous delays. In: Lakshmikantham, V. (ed.) Trends in the Theory and Practice of Nonlinear Differetial Equations, pp. 547–552. Marcel Dekker, New York (1983)

12. Wiener, J.: Generalised Solutions of Differetial Equations. World Scientific, Singapore (1993)
13. Iserles, A., Nørsett, S.P.: Order stars and rational approximations to EXP(Z). Appl. Numer. Math. **5**, 63–70 (1989)
14. Bucher, J.C.: The Numerical Analysis of Ordinary Differential Equations: Runge-Kutta and General Linear Methods. Wiley, New York (1987)
15. Dekker, K., Verwer, J.G.: Stability of Runge-Kutta Methods for Stiff Nonlinear Differential Equations. North-Holland, Amsterdam (1984)
16. Busenberg, S., Cooke, K.L.: Models of vertically transmitted diseases with sequential-continuous dynamics. In: Lakshmikantham, V. (ed.) Nonlinear phenomena in Mathematical Sciences. Academic Press, New York (1982)

Optimization Analysis of Hadoop

Jinglun Li[✉], Shengfei Shi, and Hongzhi Wang

School of Computer Science and Technology, Harbin Institute of Technology, Harbin, China
belanhd@outlook.com, {shengfei,wangzh}@hit.edu.cn

Abstract. Hadoop is a distributed data processing platform supporting MapReduce parallel computing framework. In order to deal with general problems, there is always a need of accelerating Hadoop under certain circumstance such as Hive jobs. By outputting current time to logs at specially selected points, we traced the workflow of a typical MapReduce job generated by Hive and making time statistics for every phase of the job. Using different data quantities, we compared the proportion of each phase and located the bottleneck points of Hadoop. We make two major optimization advices: (1) focus on using combine and optimizing Net Work and Disk IO when dealing with big jobs having a large number of intermediate results; (2) optimizing map function and Disk IO when dealing with short jobs.

Keywords: Hadoop · Hive · Optimization

1 Introduction

MapReduce [1] parallel computing framework was proposed by Google in 2004, then Apache Software Foundation developed open source distributed computing platform Hadoop based on Google's paper and released Hadoop2.X version in 2013 [2]. Hadoop2.X introduced YARN resource manager to Hadoop framework and made YARN along with distributed file system-HDFS, parallel computing framework-MapReduce the three core components of Hadoop.

Because of Hadoop's good performance in reliability, scalability, efficiency and fault tolerance, it has been widely used in processing large scale datasets. Yahoo! uses Hadoop to support it's advertising system and research of web searching; Facebook uses Hadoop to support it's data analysing and machine learning; Baidu uses Hadoop on analysing searching logs and data mining of web pages. Hadoop has been a widely-used solutions in processing big data.

Hive is an open source volunteer project under the Apache Software Foundation. Hive allows using HiveQL, a SQL-like language, to operate data and that makes Hive very easy to use. It allows developers without any knowledge about distributed system to use Hadoop. Hive is more than an interface of Hadoop, it has now graduated to become a top-level project of its own [3].

Hive costs much more time than traditional database on normal querying and that makes how to optimize Hadoop and Hive a very important question. Because Hadoop is a generic distributed data processing platform, it always can't achieve the highest

W. Che et al. (Eds.): ICYCSEE 2016, Part I, CCIS 623, pp. 520–532, 2016.
DOI: 10.1007/978-981-10-2053-7_46

efficiency under certain circumstance such as Hive jobs. There is a need of accelerating Hadoop under Hive jobs. We try to find out the bottleneck points of Hadoop to make an assessment about how much can a Hive job be accelerated. By outputting current time to logs at specially selected points, we traced the workflow of a typical MapReduce job generated by Hive and making time statistics for every phase of the job. Using different data quantities, we compared the proportion of each phase and located the bottleneck points of Hadoop.

Social computing is an interdiscipline of computer science and social science. As it always works on large scale of social data, traditional database and serial computing can't meet its needs. Hadoop and Hive becomes a good solution for social computing on big data. Optimization of Hadoop would also be helpful to social computing.

Our experiment not only applies to jobs generated by Hive, but also applies to normal MapReduce jobs running on MRv2 because we care about the internal efficiency of Hadoop, not the concrete realization of function map and reduce. We make two major optimization advices: (1) focus on using combine and optimizing Net Work and Disk IO when dealing with big jobs having a large number of intermediate results; (2) optimizing map function and Disk IO when dealing with short jobs.

2   Background

Hadoop2.X consists of MapReduce parallel computing framework, YARN resource manager and HDFS distributed file system. MapReduce programming model is easy to use, developers only need to implement map function and reduce function to write a MapReduce App. Complicated tasks in parallel programming such as: job submission, run time environment configuration, tasks schedule, load balancing and fault tolerance are automatically done by Hadoop. User can also check running progresses of MapReduce jobs on a web browser. YARN is introduced to Hadoop2.X to make it a future oriented computing platform. YARN allows other computing frameworks run on hadoop like Spark-On-YARN [4], Storm-On-YARN [5] and Tez-on-YARN [6]. That makes it a dynamic parallel computing platform. HDFS is a distributed file system. It is well suited for distributed storage and distributed processing using commodity hardware. It is fault tolerant, scalable, and extremely simple to expand [7].

Hive is a data warehouse based on Hadoop platform. It provides many functions on managing data warehouse such as: tools of data ETL, data storage management and abilities of searching and analyzing on large scale data sets. Hive allows using HiveQL, a SQL-like language, to operate with Hive. HiveQL commands will then be converted into MapReduce job running on Hadoop cluster.

2.1   MapReduce

MapReduce computing model is a classic and effective parallel computing model. It's totally different from many previous parallel programming model like message passing interface [8] and parallel virtual machine [9]. It divides computing progress into two phases, map and reduce. A MapReduce job usually splits the input data-set into independent chunks which are processed by the map tasks in a completely parallel manner. The framework sorts the outputs of the maps, which are then input to the reduce tasks. Typically both the input and the output of the job are stored in a file-system. The framework takes care of scheduling tasks, monitoring them and re-executes the failed tasks [10].

In a typical MapReduce application, map phase takes key-value as input, after computing it output intermediate results in a key-value form too. The intermediate results will then be gathered by key, after gathering, key-value pairs with the same key will be handled by the same reduce function. Finally, the collector of reduce phase will write the final key-value results into HDFS. There are some differences between Mrv1 and Mrv2. After involving YARN, there is no JobTracker or TaskTracker in Hadoop2.X, YARN and MRAppMaster takes their job, YARN takes the resource schedule and MRAppMaster takes the task schedule.

2.2   YARN

YARN is the resource manager of Hadoop, involved by Hadoop2.x, solving problems on scalability, reliability and resource utilization ratio of MRv1. The fundamental idea of MRv2 is to split up the two major functionalities of the JobTracker, resource management and job scheduling/monitoring, into separate daemons. The idea is to have a global ResourceManager (RM) and per-application ApplicationMaster (AM) [11]. YARN takes master-slave architecture, letting node manager managing specific node while resource manager managing all node manager. By abstracting computing resource as containers, YARN deliver computing resources to each Application Master and collect them when container is run out.

The ResourceManager has two main components: Scheduler and Application-Master. The Scheduler is responsible for allocating resources to the various running applications subject to familiar constraints of capacities, queues etc. The Scheduler is pure scheduler in the sense that it performs no monitoring or tracking of status for the application. Also, it offers no guarantees about restarting failed tasks either due to application failure or hardware failures. The Scheduler performs its scheduling function based the resource requirements of the applications; it does so based on the abstract notion of a resource Container which incorporates elements such as memory, CPU, disk, network etc. In the first version, only memory is supported. The ApplicationsManager is responsible for accepting jobsubmissions, negotiating the first container for executing the application specific ApplicationMaster and provides the service for restarting the ApplicationMaster container on failure [11].

2.3   HDFS

HDFS is short for Hadoop Distributed File System, and it's Hadoop's default file system. HDFS uses master-slave architecture, letting Namenode, the master, manage the metadata while DataNode, the slave, manage the storage of actual data. Clients contact

NameNode for file metadata or file modifications and perform actual file I/O directly with the DataNodes [7].

HDFS is well suited for distributed storage and distributed processing using commodity hardware. It is fault tolerant, scalable, and extremely simple to expand. HDFS is highly configurable with a default configuration well suited for many installations.

HDFS is good for files that are wrote once and read a lot. It is not designed for many little files because they will quickly enlarge the meta-data in Namenode. HDFS is not suitable for low latency data access and it is not a real time storage system. Hadoop file system is pluggable, an optional replacement schema is Openstack Swift [12].

2.4 Hive

Hive is a data warehouse based on Hadoop platform supplying SQL like interactive mode. After inputting a HiveQL command, Hive would convert this command into a MapReduce job and run it. When job is over, the results will be passed to user. Hadoop and Hive together becomes a popular data warehouse solution.

Hive is very different from traditional database, the operation usually costs more time and it doesn't support data sorting or data cache. It also not support online transaction processing and real time querying. Hive is not designed for real time querying, it's designed for batch jobs on a large scale invariant data. Hive's advantages are scalability, fault tolerance and low constraint data entry format.

3 Test Plan

Our goal was finding out the bottleneck points of Hadoop so we just ignored the time that Hive converting a HiveQL command into a MapReduce job. Because of the time from the submission of a MapReduce job to the first map task started is very short, we focused on the time after the first map task is started. By tracing invocation relationship of map tasks and reduce tasks, we did statistics of every function.

3.1 Workflow

On Hadoop2.X, MapReduce computing model is built on YARN resource manager. Through ApplicationClientProtocol, a job is sent to ResourceManager from MapReduce client. Then ResourceManager will assign a container to MRAppMaster whose job is scheduling all tasks. After assignment, ResourceManager orders NodeManager to start MRAppMaster in the corresponding container and waits for MRAppMaster's register.

MRAppMaster invokes InputFormat to do logic splits on input data and generates map tasks and reduce tasks. Map tasks number equals to input splits number while reduce tasks number is configured in job's configuration info. After that, MRAppMaster starts to apply computing resource for its tasks, once it get a container, the computing resource, from ResourceManager, it notices NodeManager to start its TaskAttempt. The corresponding NodeManager will set task's running environment and start the task. YARNChild corresponds to the Child in MRv1 and its method main() is the entrance of a task. It gets startup parameters like host, port, TaskAttempID and jvmId. After some set up work, it creates a task object and invoke task's method run().

For MapTask, its method run() first determines is this job a new API job or an old API job. As we are using old API, we only concern invocations of old API. Then runOldMapper() is called to get input splits, create RecordReader and Collector objects and call method MapRunner.run(). Finally, MapRunner.run() get key-value pairs from InputSplit and pass them to function map() programmed by user recurrently until every key-value pair is done.

For ReduceTask, its method run() first start Shuffle and Merge phase to fetch input. Then it determines whether it's a new API job nor an old API one. Again we only concern about old API so it calls runOldReducer(). Finally runOldReducer() creates Collector object and passes key-iterator of values to function reduce() programmed by user.

3.2 Test Plan

We do our experiment by inserting points that output current time to logs to do statistics. For MapTask, it is divided into Read phase, Map phase, Collect phase, Spill phase and Combine phase. The methods invocation relationship is YARNChild.main()-MapTask.run()-MapTask.runOldMapper()-MapRunner.run(). We insert begin point and end point in every method, besides we do statistics of the total amount of map, count and next in MapRunner.run() as summap, sum-count and sum-next respectively.

For ReduceTask, it is divided into Shuffle phase, Merge phase, Sort phase, Reduce phase and Write phase. The invocation relationship is YARNChild.main()-ReduceTask.run()-ReduceTask.runOldReducer(). Again we insert begin and end points into every function to do statistics.

3.3 Time Statistics

Interval between first Begin and last End(IBE): the IBE of MapTask(ReduceTask) means the time interval between the first begin time and the last end time of all map(reduce) tasks. For example, MapTask IBE = max{t|t ∈ map tasks end time} − min{t|t ∈ map tasks begin time}.

Average Time: the average time of MapTask(ReduceTask) means the total time of MapTasks(ReduceTasks) divided by the number of all MapTasks(ReduceTasks).

Time Without Embedded Functions(TWEF): if function A() calls function B() which is also a function we want to test, then the TWEF of function A() means the total run time of A() without the run time of function B(). For example, YARNChild.main() TWEF = total time of YARNChild.main() − total time of ReduceTask.run().

4   Test Result

4.1   Environment

Hive edition: 0.14.0.

Hadoop edition: 2.6.0.

Cluster: One Master, Two Slaves.

Hardware Information: Processor: intel i7-3770 8 cores; Memory: 32G; Disk: 1T;

Test Data Set: 400 M, 7.78G, 23G.

Data source: Generated by TPC-H.

Test HiveQL: select l_parkty from table group by l_parkty.

4.2   Map and Reduce Proportion

We did the same HiveQL on three data sets which are only different in sizes. The average time shows the proportion of map task and reduce task on how long does one task need to finish. With the increase of data size, map tasks creates more intermediate results and that makes reduce task's time increase faster than map task's time which results in a bigger proportion of reduce task. At 400 M dataset reduce tasks only take up about 32 % of the whole proportion while it fast increase to 37 % at 7.78 G dataset and 50 % at 23 G dataset.

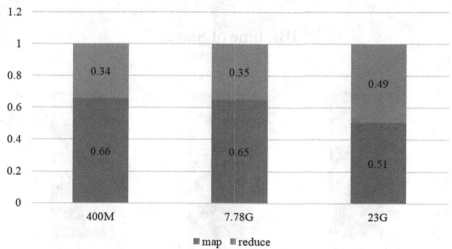

Fig. 1. Average time of Slave1 (Color figure online)

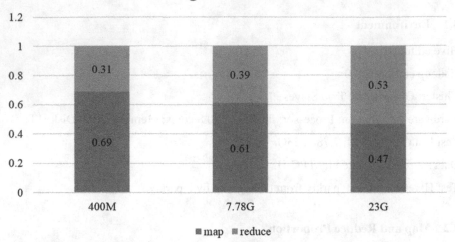

Fig. 2. Average time of Slave2 (Color figure online)

The IBE shows the total proportion of map tasks and reduce tasks. When data sets gets bigger, reduce tasks proportion gets bigger too. It increases from 37 % in 400 M to 45 % in 7.78 G and ends 55 % in 23 G data sets. Though it costs more time for both map tasks and reduce tasks while data set gets bigger, reduce tasks' time cost rises faster in both average time and IBE. When dealing with small jobs we should focus on optimizing the map task. If the data set is bigger than 23 G, it will be strongly commend to optimize

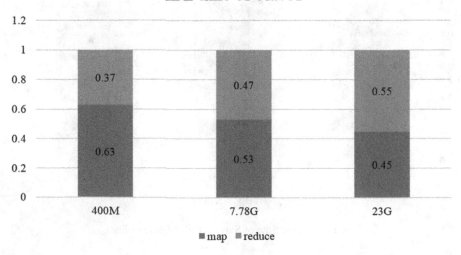

Fig. 3. IBE time of Slave1 (Color figure online)

the reduce task. Only having the proportion of map task and reduce task is not enough, after further testing, we will have a more detailed proposal (Figs. 1, 2, 3 and 4).

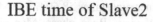

IBE time of Slave2

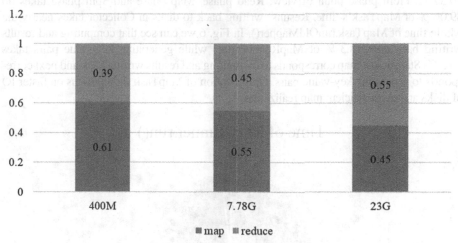

Fig. 4. IBE time of Slave2 (Color figure online)

4.3 MapTask Proportion

To locate the bottleneck points of map task and reduce task, we insert output points into functions of MapTask and ReduceTask. We do our experiments on 400 M, 7.78 G and 23 G

MapTask time

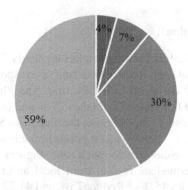

Fig. 5. MapTask time (Color figure online)

data sets. The proportion of each function in MapTask is stable on the three different data sets, so we take proportion of 400 M data set as an example. As Fig. 5 shows that MapRunner.run() and MapTask.runOldMapper() take up nearly 90 % of the whole MapTask time while MapRunner.run() takes 58.69 % and MapTask.runOldMapper() takes 30.37 %. From phase point of view, Read phase, Map phase and Spill phase takes up 89.06 % of MapTask's time. Results' writing back to disks in Collector takes nearly the whole time of MapTask.runOldMapper(). In Fig. 6 we can see that computing and results writing back take 78.3 % of MapRunner.run() while generating key-value pairs takes 20.3 %. Statistics of map corresponds to computing and results writing back and next corresponds to generating key-value pairs. Optimization of MapTask should focus on faster IO of disks and more efficient map realization.

Time of MapRunner.run()

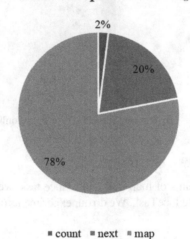

■ count ■ next ■ map

Fig. 6. Time of MapRunner.run() (Color figure online)

4.4 ReduceTask Proportion

Figures 7 and 8 show that ReduceTask.run() takes up nearly 70 % of the whole ReduceTask progress in 400 M data set. ReduceTask.run() corresponds to Shuffle phase and Merge phase which means it only costs little CPU time. The other function taking a large proportion is MapTask.runOldReducer(). It takes 27 % and corresponds to Sort phase, Reduce phase and Write phase. Figure 9 shows the change of Shuffle&Merge's proportion when data set gets bigger. Then we insert points into thread Fetcher to find out how does fetching intermediate results cost so much time. Figures 10 and 11 show the time proportion of fetching intermediate results from local and remote. We can see that fetching from remote takes 83.45 % of ReduceTask and 41.72 % of the whole job time. Because of fetching intermediate results from remote costs about 8 times more than fetching from local, a local combine will be very useful in reducing Shuffle and Merge. Network IO is another important way to optimize ReduceTask. Fetching from remote costs more than local because of Network IO's blocking. Optimizing disk IO and function reduce() will also improve ReduceTask's efficiency but cannot be as good as optimizing Network IO.

ReduceTask time of Slave1 400M

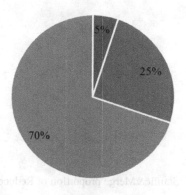

■ YarnChild.main()   ■ ReduceTask.runOldReducer()   ■ ReduceTask.run()

Fig. 7. ReduceTask time of Slave1 400 M (Color figure online)

ReduceTask time of Slave2 400M

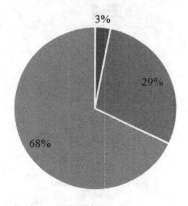

■ YarnChild.main()   ■ ReduceTask.runOldReducer()   ■ ReduceTask.run()

Fig. 8. ReduceTask time of Slave2 400 M (Color figure online)

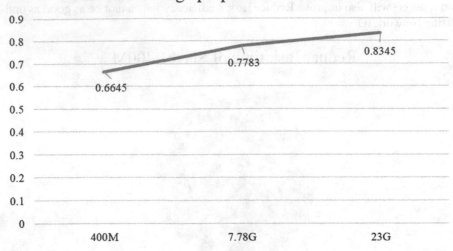

Fig. 9. Shuffle&Merge proportion of ReduceTask

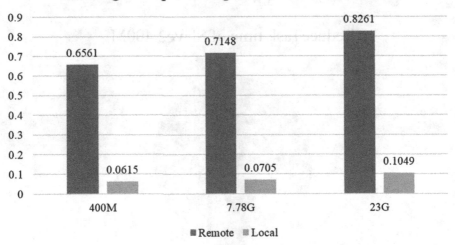

Fig. 10. Fetching time percentage of Task-Slave1 (Color figure online)

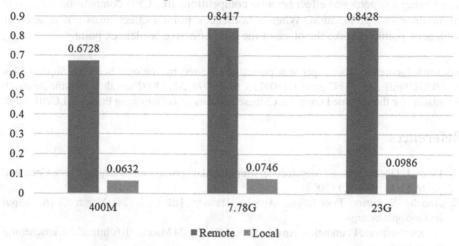

Fig. 11. Fetching time percentage of ReduceTask-Slave2 (Color figure online)

5 Conclusion

As we care about the internal efficiency of Hadoop, not the concrete realization of function map and reduce, our conclusions not only applies to jobs generated by Hive, but also applies to normal MapReduce jobs running on MRv2. We did our test on three different data sets with size 400 M, 7.78 G and 23 G respectively. We found the proportion of ReduceTask only takes up less than 40 % in small jobs and it gets bigger quickly with data set's growth. So when dealing with small jobs, we will recommend a MapTask optimization. Maptask optimization should be focused on function map and disk IO according to our research within the MapTask. By reducing half of the function map's time, we will get a 27.57 % optimization of the whole time in a 400 M job while reducing half of the IO will give us an optimization of more than 20 %. As data set increase, the MapTask optimization perform will get worse and worse. When data set come to 23 G, reducing half of function map may only make an 18.9 % optimization.

When data set gets bigger, we should change our focus to optimizing ReduceTask. According to our research, Shuffle and Merge cost 83.45 % of ReduceTask and 41.72 % of the whole job in a 23 G job. Optimizing ReduceTask should be focused on reducing the intermediate results of map to reduce Shuffle and Merge time. Because of fetching intermediate results from remote costs about 8 times more than fetching from local, a local combine will be very useful in reducing Shuffle and Merge. The other important way to optimize ReduceTask is accelerate Network IO. Fetching from remote costs more than local because of Network IO's blocking. If fetching time can be reduced by a half, there would be at most a 41.72 % optimization of ReduceTask and 21.27 % optimization of the whole job.

Accelerating Disk IO will optimize both MapTask and ReduceTask. Taking 23 G results as an example, if we can make the disk two times faster, we would get a 28.4 %

optimization of the whole job at most. The simultaneous optimization of multiple items won't bring a superposed effect because competitions like CPU competition will erode the revenue of optimization. When a bottleneck point's effect turns down, another bottleneck point will take the place of the most effective bottleneck point.

Acknowledgement. This paper was partially supported by National Sci-Tech Support Plan 2015BAH10F01 and NSFC grant U1509216, 61472099, 61133002 and the Scientific Research Foundation for the Returned Overseas Chinese Scholars of Heilongjiang Province LC2016026.

References

1. Dean, J., Ghemawats, S.: MapReduce: simplified data processing on large clusters. Commun. ACM **51**(1), 107–113 (2008)
2. Apache Software Foundation. Apache Hadoop [EB/OL], 24 March 2016. http://hadoop.apache.org/
3. Apache Software Foundation. Apache Hive [EB/OL], 24 March 2016. http://hive.apache.org/index.html
4. Apache Software Foundation. Spark-on-Hadoop [EB/OL], 24 March 2016. http://spark.apache.org/docs/0.6.0/running-on-yarn.html
5. Apache Software Foundation. Storm-on-Hadoop [EB/OL], 24 March 2016. http://storm.apache.org/index.html
6. Apache Software Foundation. Tez-on-Hadoop [EB/OL], 24 March 2016. http://tez.apache.org/
7. Apache Software Foundation. Hadoop HDFS [EB/OL], 24 March 2016. http://hadoop.apache.org/docs/r2.6.4/hadoop-project-dist/hadoop-hdfs/HdfsUserGuide.html#Overview
8. Argonne National Laboratory. Message passing interface standard [EB/OL], 24 March 2016. http://www.mcs.anl.gov/research/projects/mpi
9. Computer Science and Mathematics Division of Oak Ridge National Laboratory. Parallel virtual machine [EB/OL], 24 March 2016. http://www.csm.ornl.gov/pvm/
10. Apache Software Foundation. Hadoop MapReduce [EB/OL], 24 March 2016. http://hadoop.apache.org/docs/r2.6.4/hadoop-mapreduce-client/hadoop-mapreduce-client-core/MapReduceTutorial.html
11. Apache Software Foundation. Hadoop YARN [EB/OL], 24 March 2016. http://hadoop.apache.org/docs/r2.6.4/hadoop-yarn/hadoop-yarn-site/YARN.html
12. OpenStack. OpenStack Swift [EB/OL], 24 March 2016. http://docs.openstack.org/developer/swift/#overview-and-concepts

Outsourcing the Unsigncryption of Compact Attribute-Based Signcryption for General Circuits

Fei Chen[1], Yiliang Han[1,2(✉)], Di Jiang[1], Xiaoce Li[1],
and Xiaoyuan Yang[1]

[1] Department of Electronic Technology, Engineering University of CAPF,
Xi'an 710086, China
[2] College of Information Science and Technology, Northwest University,
Xi'an 710127, China
yilianghan@hotmail.com

Abstract. In the driven of big data, social computing and information security is undergoing rapid development and beginning to cross. This paper describes a key-policy attribute-based signcryption scheme which has less computation costs than existing similar schemes by utilizing secure outsourcing of scientific computation in cloud computing and eliminates overhead for users, the ciphertext is short, compact, the correctness of transformation algorithm is verifiable. The decrease of ciphertext is 17 %. Additionally, new scheme remits the key escrow problem and is proven selective security in the standard model, it could be verified publicly, applied in mobile devices.

Keywords: Social computing · Public key cryptosystem · Attribute-based signcryption · Outsourced computation · Multilinear maps · Selective security

1 Introduction

Social computing [1] is an interdisciplinary research area, it could improve the effectiveness of using dynamic, intensive and open data. Cloud computing [2] is an important technical support for social computing and it is applied widely in human life. With the advent of the era of big data, social computing and information security is beginning to cross, crowdsourcing is a typical embodiment, and outsourced computation is viewed as an application of crowdsourcing in cloud computing, it could eliminate the user's computation and storage from terminal devices to cloud. So outsourced computation is widely used in the encryption and decryption of cryptography.

Attribute-based encryption(ABE) [3] is a promising public key encryption technique to protect the end-to-end security and privacy of big data in the cloud. As a new cryptographic primitive, attribute-based signcryption (ABSC) [4] is a combination of attribute-based encryption and attribute-based signature (ABS) [5]. ABSC could provide not only the combined security of confidentiality and unforgeability, but also a granular and non-interactive access control mechanism of encrypted and signed data which is fit for cloud environment. Recently, there is a wide range of research on

W. Che et al. (Eds.): ICYCSEE 2016, Part I, CCIS 623, pp. 533–545, 2016.
DOI: 10.1007/978-981-10-2053-7_47

security and practicality. Rao et al. [6] proposed a ABSC scheme with the constant size signcryption using the technique of key-policy ABS [7]. Then, Han et al. [8] proposed a new attribute-based generalized signcryption, the scheme was flexibility to achieve encryption, signature or signcryption.

Boneh et al. [9] first exploited multilinear maps to construct cryptography in 2003. Then, the development of this mathematic tools was at a nascent stage [10, 11], Garg et al. [12] provided the first construction of ABE for general circuits in 2013. They investigated the techniques "move forward and shift" to prevent the backtracking attacks and did not rely on any sophisticated linear secret sharing scheme. Because of these excellent properties, ABE based on multilinear maps raised comprehensive concern in efficiency and security.

In 2014, Attrapadung [13] and Garg [14] contributed in security study and the proposed schemes were adaptively security, ciphertext in [13] associated with the number of boolean variables while [14] associated with size of the decryption policy circuits. In 2015, Chandran [15] set the decryption policy circuits into two parts, l_{OR} and l_{AND} respectively, they paid attention to the reduction of k in constructions of access control design, and the notions were based on k-linear maps which was postulated by Boneh, the result was that they achieved an efficient improvement and constructed ABE for arbitrary circuits based on $(l_{OR} + 1)/(l_{AND} + 1)$-linear maps not k-linear maps, Datta [16] used the skill of [17] in developing a full domain hash from multilinear maps. They generated a sequential product of string input which was called "hash value" of the encryption input string, the "value" was masked in the ciphertext, and the feature of ciphertext was compact and short. In 2016, Brakerski [18] adopted Learing with Errors (LWE) problems to construct a unbounded circuit-ABE, the security was semi-adaptive. In contrast to previous schemes where the length of the public parameters (PP) increased linearly with the maximal attribute length, the size of the PP in [18] was a fixed polynomial.

However, the computational cost in encryption (signature) and decryption (verification) commonly grows with the complexity of access structure construction in known ABE/ABS schemes, which is a bottleneck limiting its application especially in mobile equipment that the ability of computation is limited. In the above literature, decryptor has to perform heavy decryption computation, it is very inconvenient especially to mobile devices.

Cloud computation [19, 20] has a great deal of computational resources, which enable clients with limited computational resources to outsource their mass computing to the cloud. So, in this paper, we focus on decreasing the computational cost to receiver side.

Identity-based cryptography (IBC) was presented by Shamir [21] in 1984. Then, Lee [22] proposed an Identity-based Signcryption (IBSC) in 2002. It was the former of ABSC. The same as IBC, user's secret key is generated by Key Generation Center (KGC) in ABSC. So it emerges a serious security risk, the evil KGC can create secret key with arbitrary access structure, it can sign and decrypt the message for any clients with its master key. The challenge which comes from the security of KGC is called as key escrow problem. Here, an efficient outsourced ABSC (OABSC) is proposed, the construction is inspired by [16], we utilize cloud service providers as third party provider, and perform heavy decryption through "borrowing" the computation resources from the provider. And in our scheme, the users need to set and keep secret value which is known by herself, the secret value would be used to decrypt ciphertext, so the scheme remits the key escrow problem which is an inherent issue in IBC. The ciphertext is stored in cloud, when users want to gain message, they provide their attribute set S to verify, then they send transformation key to cloud service providers to compute ciphertext, cloud providers return a transformation ciphertext, after receiver verifies the outsourced computation correctness, retrieves plaintext by private key and secret value.

2 Preliminary

2.1 Leveled Multilinear Maps

Here, we introduce notion of the leveled multilinear map [11]. Suppose exist a group generator G, then it inputs a security parameter 1^λ and a positive integer k. Then, the algorithm outputs a sequence of groups $\vec{G} = (G_1, \ldots, G_k)$, here, we require $p > 2^\lambda$ (p is the prime order). g_i is the generator of G_i respectively.

Suppose exist a set $\{e_{i,j} : G_i \times G_j \to G_{i+j} | i, j \geq 1; i+j \leq k\}$ that satisfies the relation $e_{i,j}(g_i^a, g_j^b) = g_{i+j}^{ab}$ $(\forall a, b \in Z_p)$, the consequence of this relation is $e_{i,j}(g_i, g_j) = g_{i+j}$. Sometimes, we could omit the subscripts i, j to pursue a formal briefness, for instance, may write $e(g_i^a, g_j^b) = g_{i+j}^{ab}$.

2.2 The Notion of OABSC

In this section, we give our definition for OABSC. To be honest, our definition mirrors the literature [16, 23]. A key-policy OABSC scheme consists of six algorithms $\Omega = ($**Setup**, **SignKeyGen**, **DecryKeyGen**$_{out}$, **Signcrypt**, **Transform**$_{out}$, **Unsigncrypt**).

Setup: KGC takes security parameter λ, for the general circuits, sets n as the length of boolean inputs to decryption policy circuits, n' as the length of boolean inputs to signing policy circuits, l as the allowed depth of both types, then outputs the public parameters PP, keeps the master secret key (MK).

SignKeyGen: For each user's private key request on signature string y, KGC sets MK, PP, the description of a signing policy circuit g (g is defined by signcryptor). Then, it sends a signing key SK$^{(SIG)}$ to the signcryptor.

DecryKeyGen$_{out}$**:** The algorithm inputs PP and MK, the description of a decryption policy circuit f (f is acquired by receiver). KGC computes the decryption private key SK and transformation key TK, then hands to the decryptor.

Signcrypt: The user runs signcryption algorithm and inputs PP, SK$^{(SIG)}$, an encryption input string $x \in \{0,1\}^n$ which generated by the legitimate decryptor, a signature input string $y \in \{0,1\}^{n'}$ (y is generated by signcryptor and needs to satisfy $g(y) = 1$), a message M. Then, computes and generates a ciphertext CT.

Transform$_{out}$: Cloud service providers run the ciphertext transformation algorithm, inputs a TK and CT. It exports the partially decrypted ciphertext CT', or the symbol $\perp$ otherwise.

Unsigncrypt: The decryptor runs unsigncryption algorithm and inputs PP, CT' and its decryption private key SK. It checks CT' to judge the correctness of outsourced computation. If it is successful, verifies and outputs M; otherwise, exports $\perp$. If outsourced computation is wrong, it means the transformation ciphertext is invalid and user rejects to unsigncrypt the transformation ciphertext.

2.3 Security Requirement for OABSC

Here, we briefly analysis the confidentiality and unforgeability about scheme. We mirror and reform the notions in [16] and assume that all the outsourced computation is semi-honest, it means cloud service providers would implement the algorithm rigidly, but attempts to acquire extra information. Compare with traditional ABSC, In our OABSC scheme, adversary could corrupt the cloud seriver and query any transformation result.

3 Construction

3.1 Setup $(1^\lambda, k = l + n + n' + 1)$

KGC runs the setup algorithm and takes λ, n, n', l as input. λ is used to generate a security parameter 1^λ, n is the length of inputs to decryption policy, n' is the length of inputs to signing policy, l is the allowed maximum depth, gains a sequence of groups $\vec{G} = (G_1, \cdots, G_{k+2})$, the generators are $g_1, \cdots, g_{k+2}$, chooses a cryptographic hash functions $H : G_{k+1} \rightarrow \{0,1\}^\Delta$, here, the symbol Δ represents the length of message M.

Next, it chooses $\alpha_1, \alpha_2 \in_R Z_p$ along with $(a_{1,0}, a_{1,1}), \cdots, (a_{n,0}, a_{n,1}) \in_R Z_p^2$ and $(b_{1,0}, b_{1,1}), \cdots, (b_{n',0}, b_{n',1}) \in_R Z_p^2$, then, the KGC sets $\alpha = \alpha_1 + \alpha_2$, computes $\varepsilon = g_{l+2}^\alpha$, $\varepsilon_1 = g_{l+2}^{\alpha_1}$, $A_{i,\beta} = g_1^{a_{i,\beta}}$, $B_{j,\beta} = g_1^{b_{j,\beta}}$ for $i = 1, \cdots, n; j = 1, \cdots, n'; \beta \in \{0,1\}$.

Then, chooses $\theta \in_R Z_p$, generates $\varpi = g_n^\theta$, $X = g_{n+l+1}^{\theta \alpha_2}$. The PP consists of the $\vec{G}$ together with H, ϖ, X, the master secret key MK $= \left(g_l^{\alpha_1}, g_l^{\alpha_2} \right)$ is kept.

3.2 DecryKeyGen(PP, MK, f)

First, we briefly describe our notion about circuit. The circuit has $n + q$ wires, $\{1, \ldots, n\}$ are the set of input wires, $\{n + 1, \ldots, n + q\}$ are the set of gates, wire $n + q$ is the presentation of output wire. KGC inputs the PP, MK and a description $f = (d, l, q, A, B, Gat - eType)$ of a decryption policy.

1. The user chooses secret value $t \in_R Z_p$, keeps secret value t, computes $T = g_1^t$ and conveies T to KGC.

2. After KGC received T, KGC creates secret key for users, it chooses $r_1, \cdots, r_{n+q} \in_R Z_p$, the random value r_w is associated with wire $w \in \{1, \ldots, n+q\}$, computes user's secret key $SK = e\left(T, g_1^\alpha g_l^{-r_{n+q}}\right) = g_{l+1}^{t(\alpha - r_{n+q})}$.

3. The KGC runs transformation algorithm, it creates transformation key TK by the category of w.

①INPUT wire: If $w \in \{1, \ldots, n\}$, it computes the transformation key $TK_w = e\left(A_{w,1}, g_1, T\right) = g_3^{r_w a_{w,1} t}$.

②OR gate: If wire $w \in \{n+1, \ldots, n+q\}$, $GateType(w) = OR$, the depth of w is $j = depth(w)$, it selects $\mu_w, v_w \in_R Z_p$, gains the transformation key TK, TK is divided into four parts. $TK_{w,1} = e\left(g_1^{\mu_w}, T\right) = g_2^{\mu_w t}$, $TK_{w,2} = e\left(g_1^{v_w}, T\right) = g_2^{v_w t}$, $TK_{w,3} = e\left(g_j^{r_w - \mu_w r_{A(w)}}, T\right) = g_{j+1}^{t\left(r_w - \mu_w r_{A(w)}\right)}$, $TK_{w,4} = e\left(g_j^{r_w - v_w r_{B(w)}}, T\right) = g_{j+1}^{t\left(r_w - v_w r_{B(w)}\right)}$.

③AND gate: $w \in Gates$, $GateType(w) = AND$, $j = depth(w)$, KGC selects $\mu_w, v_w \in_R Z_p$, creates the TK, TK is divided into three parts $TK_{w,1} = e\left(g_1^{\mu_w}, T\right) = g_2^{\mu_w t}$, $TK_{w,2} = e\left(g_1^{v_w}, T\right) = g_2^{v_w t}$, $TK_{w,3} = e\left(g_j^{r_w - \mu_w r_{A(w)} - v_w r_{B(w)}}, T\right) = g_{j+1}^{t\left(r_w - \mu_w r_{A(w)} - v_w r_{B(w)}\right)}$.

Then KGC sends the TK $= \left(f, \{TK_w\}_{w \in \{1, \ldots, d+q\}}\right)$ and SK to user.

3.3 SignKeyGen(PP, MK, g)

Here, the describe is similar to [16], KGC inputs PP, MK and the description $g = (n', l, q', A, B, GateType)$ of a signing policy.

1. It picks $r'_1, \ldots, r'_{n'+q'-1} \in_R Z_p$, then defines $r'_{n'+q'} = \alpha_2$.
2. It computes the key components:

①INPUT wire. If $w \in \{1, \ldots, n'\}$, computes the key component $K'_w = e\left(B_{w,1}, g_1\right) = g_2^{r'_w b_{w,1}}$.

②OR gate. It selects $\mu'_w, v'_w \in_R Z_p$, computes K'_w is divided into four parts as follows. $K'_{w,1} = g_1^{\mu'_w}$, $K'_{w,2} = g_1^{v'_w}$, $K'_{w,3} = g_j^{r'_w - \mu'_w r'_{A(w)}}$, $K'_{w,4} = g_j^{r'_w - v'_w r'_{B(w)}}$.

③AND gate. It picks $\mu'_w, v'_w \in_R Z_p$, the key component is $K'_w = \left(K'_{w,1} = g_1^{\mu'_w}, K'_{w,2} = g_1^{v'_w}, K'_{w,3} = g_j^{r'_w - \mu'_w r'_{A(w)} - v'_w r'_{B(w)}}\right)$.

3. KGC sends the signing key $SK^{(SIG)} = \left(D, g, \{K'_w\}_{w \in \{1, \ldots, n'+q'\}}\right)$ to the signcryptor.

3.4 Signcrypt(PP, SK$^{(SIG)}$, x, y, M)

A signcryptor runs the signcryption algorithm and inputs PP, SK$^{(SIG)}$, strings $x = x_1 \cdots x_n \in \{0,1\}^n$ and $y = y_1 \ldots y_{n'} \in \{0,1\}^{n'}$, they indicate encryption input and signature input, we use this two strings to "replace attribute", here, the string y is required to satisfie $g(y) = 1$, the plaintext $M \in G_k$, it generates the ciphertext as follows.

1. In order to read conveniently, here, we define $\delta(x) = \prod_{i=1}^{n} a_{i,x_i}$ and $\delta'(y) = \prod_{i=1}^{n'} b_{i,y_i}$, computes $D = e(A_{1,x_1}, \cdots, A_{n,x_n}) = g_n^{\delta(x)}$ and $D' = e(B_{1,y_1}, \cdots, B_{n',y_{n'}}) = g_{n'}^{\delta'(y)}$. Here, the procedure of the ciphertext is still depends upon the category of the wire w.

   ①INPUT wire. If $w \in \{1, \ldots, n'\}$, moreover, it corresponds to the w-th input. According to the rule of general circuit, we required $y_w = 1$, the signcryptor computes $E'_w = e(K'_w, B_{1,y_1}, \cdots, B_{w-1,y_{w-1}}, B_{w+1,y_{w+1}}, \cdots, B_{n',y_{n'}}) = g_{n'+1}^{r'_w \delta'(y)}$.
   ②OR gate. Suppose that $g_w(y) = 1$, then, we talk about the value of $g_{A(w)}(y)$ and $g_{B(w)}(y)$, if $g_{A(w)}(y) = 1$, computes $E'_w = e(E'_{A(w)}, K'_{w,1})e(D', K'_{w,3}) = g_{n'+j}^{r'_w \delta'(y)}$. Alternatively, if $(g_{A(w)}(y) = 0) \wedge (g_{B(w)}(y) = 1)$, computes $E'_w = e(E'_{A(w)}, K'_{w,2}) \cdot e(D', K'_{w,4}) = g_{n'+j}^{r'_w \delta'(y)}$. Else, the signcryptor considers the AND gate.
   ③AND gate. The same as the former. The value of $g_{A(w)}(y)$ and $g_{B(w)}(y)$ is 1. Then computes $E'_w = e(E'_{A(w)}, K'_{w,1})e(E'_{B(w)}, K'_{w,2})e(K'_{w,3}, D') = g_{n'+j}^{r'_w \delta'(y)}$.

2. Then signcryptor chooses $s \in_R Z_p$, generates $C = H\big[(e(\varepsilon_1, A_{1,x_1}, \cdots, A_{n,x_n}, D') \cdot e(E'_{n'+q}, A_{1,x_1}, \cdots, A_{n,x_n}, g_2))^s\big] \oplus M = H\big(g_{k+1}^{\alpha s \delta(x)\delta'(y)}\big) \oplus M, C' = g_1^s, C'' = e(\varpi, E'_{n'+q}) = g_{k-1}^{\theta \alpha_2 \delta'(y)}$. The ciphertext is CT $= (x, y, C, C', C'')$.

3.5 Outsourcing Computation (PP, TK, x, y, CT)

The user sends herself attribute input string x and transformation key TK to the server, and server judges user's attribute input string x whether match the condition $f(x) = 1$. If x does not, it outputs $\perp$. Suppose x satisfies, server uses TK to transform the ciphertext CT as follows.

1. Extracting $\{A_{i,x_i}\}$ and $\{B_{i,y_i}\}$ from PP, computes $D = e(A_{1,x_1}, \cdots, A_{n,x_n}) = g_n^{\delta(x)}$ and $D' = e(B_{1,y_1}, \cdots, B_{n',y_{n'}}) = g_{n'}^{\delta'(y)}$.
2. The servers deals with each case on the category of w.

   ①INPUT wire. If $w \in \{1, \ldots, n'\}$, Suppose $x_w = f_w(x) = 1$, It computes $E_w = e(\text{TK}_w, A_{1,x_1}, \cdots, A_{w-1,x_{w-1}}, A_{w+1,x_{w+1}}, \cdots, A_{n,x_n}, C) = e(g_3^{r_w a_{w,x_w} t}, g_1^{a_{1,x_1}}, \cdots, g_1^{a_{w-1,x_{w-1}}}, g_1^{a_{w+1,x_{w+1}}}, \cdots, g_1^{a_{n,x_n}}, g_1^s) = g_{n+4}^{r_w s \delta(x) t}$.

②OR gate. If $(f_w(x) = 1) \wedge (f_{A(w)}(x) = 1)$, computes $E_w = e(E_{A(w)}, \mathrm{TK}_{w,1})$.
$e(\mathrm{TK}_{w,3}, C, D) = e\left(g_{n+j}^{sr_{A(w)}\delta(x)}, g_2^{\mu_w t}\right) \cdot e\left(g_{j+1}^{r_w - \mu_w r_{A(w)}}, g_1^s, g_n^{\delta(x)}\right) = g_{n+j+2}^{r_w s\delta(x)t}.$

If $f_{A(w)}(x) = 0$, then the $f_{B(w)}(x)$ must satisfy $(f_{B(w)}(x) = 1) \wedge (f_w(x) = 1)$, computes $E_w = e(E_{A(w)}, \mathrm{TK}_{w,2})e(\mathrm{TK}_{w,4}, C, D) = g_{n+j+2}^{r_w s\delta(x)t}.$

③AND gate. If $f_w(x) = 1$ and $f_{A(w)}(x) = f_{B(w)}(x) = 1$, computes $E_w = e(E_{A(w)},$
$\mathrm{TK}_{w,1})e(E_{B(w)}, \mathrm{TK}_{w,2})e(\mathrm{TK}_{w,3}, C', D) = g_{n+j+2}^{sr_w\delta(x)t}, \bar{C} = e(D, C') = g_{n+1}^{s\delta(x)}.$

3. Then server sends the transformation ciphertext $CT' = (C, E_{n+q}, C'', \bar{C}, D')$ to user.

3.6   Unsigncrypt(PP, CT', t, x, SK)

A receiver gains the transformation ciphertext CT', in order to verify and decrypt the ciphertext. It still inputs PP and operates as following.

1. Checking the value of function $f(x)$, if the value is 0, outputs $\perp$, otherwise, runs the verification algorithm and computes $\hat{E} = e(\mathrm{SK}, \bar{C})$, $E = \hat{E}E_{n+q}$, checks the equation $e(\bar{C}, \varepsilon)^t = e(E, g_1)$, if the equation is valid, it means the outsourced computation is correct and receiver could use the transformation ciphertext CT' to gain plaintext, otherwise, outsourced computation is wrong, outputs $\perp$.
2. The receiver computes $e(C'', g_1) = e(X, D')$ to judge whether or not that the ciphertext is signcrypted by signcryptor. If the equation is valid, proceeds to implement next step, otherwise, outputs $\perp$.
3. Receiver computes $\bar{E} = e(E, D')$, $C \oplus H(\bar{E}^{1/t}) = M$ to gain message.

4   Analysis of the Proposal

4.1   Correctness

Verification of the outsourced computation

$$\hat{E} = e(\mathrm{SK}, \bar{C}) = g_{n+l+2}^{t(\alpha - r_{n+q})s\delta(x)}; \quad E = \hat{E}E_{n+q} = g_{n+l+2}^{t(\alpha - r_{n+q})s\delta(x)}g_{n+l+2}^{r_{n+q}s\delta(x)t} = g_{n+l+2}^{t\alpha s\delta(x)};$$

$$e(\bar{C}, \varepsilon)^t = e\left(g_{n+1}^{s\delta(x)}, g_{l+2}^{\alpha}\right)^t = g_{k+2}^{s\delta(x)\alpha t} = e\left(g_{k+1}^{s\delta(x)\alpha t}, g_1\right) = e(E, g_1).$$

Unsigncryption

$$e(C'', g_1) = e\left(g_{n+n'+l}^{\theta\alpha_2\delta'(y)}, g_1\right) = e\left(g_{n+l+1}^{\theta\alpha_2}, g_{n'}^{\delta'(y)}\right) = e(X, D'); \quad \bar{E} = e(E, D') = g_{k+1}^{t\alpha s\delta(x)\delta'(y)};$$

$$C \oplus H(\bar{E}^{1/t}) = H\left(g_{k+1}^{\alpha s\delta(x)}\right) \oplus M \oplus H\left(\left(g_{k+1}^{\alpha s\delta(x)t}\right)^{1/t}\right) = M.$$

4.2 Confidentiality

Assumption 1 (*k*-Multilinear Decisional Diffie-Hellman, *k*-MDDH). We describe the assumption briefly. First, generates $\vec{G} = (G_1, \ldots, G_k)$, sets and computes a number of parameters, such as $C_1 = g_1^{c_1}, \cdots, C_k = g_1^{c_k}, \Re, \bar{S} = g_1^{\bar{s}}(\bar{s}, c_1, \cdots, c_k \in_R Z_p)$, set $\Re_0 = g_k^{s\prod_{j\in[1,k]} c_j}$, $\Re_1$ is set of random element, the problem that judging the value of $\Re$ is $\Re_0$ or $\Re_1$ is difficult.

Theorem 1 (Message Confidentiality of OABSC). To reasonable circuits setting, the OABSC scheme supports arbitrary input length n, n' and depth l, achieves selective message confidentiality against Chosen Plaintext Attack (CPA) as per the model under the *k*-MDDH assumption.

Proof. B is the challenger and interacts with a Probabilistic Polynomial Time (PPT) adversary A as follows:

Init: A publishs $\sigma_{\bar{b}} = (\vec{G}, g_1, C_1, \ldots, C_k, \Re_{\bar{b}}, \bar{S})$ such that $\bar{S} = g_1^{\bar{s}}$, $C_1 = g_1^{c_1}, \cdots$, $C_k = g_1^{c_k}$, challenge string $x^* = x_1^* \cdots x_n^* \in \{0,1\}^n$.

Setup: According to the value of β, we define two functions $A_{i,\beta} = \begin{cases} g_1^{c_i}, & \text{if } \beta = x_i^*, \text{set } a_{i,\beta} = c_i; \\ g_1^{z_i}, & \text{if } \beta \neq x_i^*, \text{set } a_{i,\beta} = z_i. \end{cases}$ and $b_{t,\beta} = \begin{cases} c_{n+t} + z_t'\beta, & \text{if } \beta = 1; \\ c_{n+t}, & \text{if } \beta = 0. \end{cases}$. We utilize the function $\gamma(u,v)$ to stand for $\prod_{h=u}^{v} c_h$. B picks $\theta, \alpha_2, \xi_1 \in_R Z_p$, lets $\alpha_1 = \xi_1 + \gamma(n + n' + .1, n + n' + l + 1)$ and computes $\varepsilon = e(C_{n+n'+1}, \ldots, C_{n+n'+l+2})g_{l+2}^{\xi_1} = g_{l+2}^{\alpha_1}$, $\varpi = g_n^{\theta}, X = g_{n+l+1}^{\alpha_1}$.

Phase 1: B acquires α_2 and provides the signing keys with signing policy $g = (n', l, q', A, B, GateType)$ $(g(y) = 1)$ then A queries a decryption key $(f(x^*) = 0)$, the challenger B computes transformation key TK. The key components Kw corresponding to the input wires, OR, and AND gates are simulated in an identical manner. Thus, B will ultimately view $r_{n+q} = \gamma(n + n' + 1, n + n' + l + 1) + \eta_{n+q}(\eta_{n+q} \in_R Z_p)$ and computes $K = g_l^{\xi_1 + \alpha_2 - \eta_{n+q}}$. B returns the queried decryption key SK to A. A queries a signcryption of M. B knows ξ_1, α_2, extracts ϖ, $\{A_{i,x_i}\}$, $\{B_{j,y_j}\}$ from PP, and uses the elements $\{C_{n+n'+h}\}_{h=1,\ldots,l+1}$ from the given *k*-MDDH instance to simulate the query as follows: B picks $s \in_R Z_p$ and computes $C = H[e(C_{n+n'+l}, \cdots, C_{n+n'+l+1}, A_{1,x_1}, \cdots, A_{n,x_n}, B_{1,y_1}, \cdots, B_{n',y_{n'}})^s \cdot e(g_{l+1}^{\xi_1+\alpha_2}, A_{1,x_1}, \cdots, A_{n,x_n}, B_{1,y_1}, \cdots, B_{n',y_{n'}})^s] \oplus M = H(g_k^{\alpha s\delta(x)\delta'(y)}) \oplus M$, $C' = g_1^s$, $C'' = e(g_l^{\alpha_2}, \varpi, B_{1,y_1}, \cdots, B_{n',y_{n'}}) = g_{k-1}^{\theta\alpha_2\delta'(y)}$, then, B sends ciphertext CT $= (x, y, C, C', C'')$ to A. Phase 2 is the same as phase 1.

Challenge: According to the f, inputs two challenge messages $M_0^*, M_1^* \in G_k$. B chooses $b \in_R \{0,1\}$, treats s as the randomness in creation of the challenge ciphertext and computes components of the challenge ciphertext as $C^* = H(T_{\bar{b}}Z) \oplus M_b^*$, $C'^* = \bar{S}$, $C''^* = e(g_l^{\alpha_2}, \varpi, B_{1,y_1}, \cdots, B_{n',y_{n'}})$, here, we define $Z = g_k^v$, $v = \alpha\bar{s}\delta(x^*)\delta'(y^*) - \bar{s}\gamma$

$(1,k) = (\gamma(n+n'+1, n+n'+l+1) + \xi_1 + \alpha_2)\bar{s} \cdot \gamma(1,n)\delta'(y^*) - \bar{s}\gamma(1,k) = \bar{s}\sigma$. Suppose $w \in \{1, \ldots, d+q\}$, $TK'_w = \left\{ \left(e(C_{n+n'+1}, C_{n+n'+2}, T)g_3^{\eta''_w} \right)^{z_w^*}, \text{ if } y_w^* = 1, \right.$

$\eta''_w \in_R Z_p$; $e(C_{n+w}, g_1, T)^{r'_w}$, if $y_w^* = 0$, $r'_w \in_R Z_{p \cdot}$, when GateType$(w) = $ OR, if $g_w(y^*) = 1$, $g_{A(w)}(y^*) = 1$ or $g_{B(w)}(y^*) = 1$, B picks $\mu'_w, v'_w, r'_w \in_R Z_p$, computes the key component K'_w which is divided into four parts as follows. $K'_{w,1} = g_1^{\mu'_w}$, $K'_{w,2} = g_1^{v'_w}$, $K'_{w,3} = g_j^{r'_w - \mu'_w r_{A(w)}}$, $K'_{w,4} = g_j^{r'_w - v'_w r'_{B(w)}}$. If $g_w(y^*) = 0$, $g_{A(w)}(y^*) = 0$ and $g_{B(w)}(y^*) = 0$, B chooses $\psi'_w, \phi'_w, \eta'_w \in_R Z_p$, sets $\mu'_w = c_{n+n'+j+1} + \psi'_w$, $v'_w = c_{n+n'+j+1} + \phi'_w$, $r'_w = \gamma(n+n'+1, n+n'+j+1) + \eta'_w$, $TK'_{w,1} = e(C_{n+j+n'+1}, T)g_2^{\psi'_w} = g_2^{\mu'_w t}$, $TK'_{w,2} = e(C_{n+n'+j+1}, T) = g_2^{v'_w t}$, $TK'_{w,3} = e(C_{n+j+n'+1}, g_{j-1}, T)^{-\eta'_{A(w)}} e(C_{n+n'+1}, \cdots,$

$C_{n+n'+j}, T)^{-\psi'_w} g_{j+1}^{\eta'_w - \psi'_w \eta'_{A(w)}} = g_{j+1}^{t\left(r'_w - \mu'_w r'_{A(w)}\right)}$, $TK'_{w,4} = e(C_{n+j+n'+1}, g_{j-1}, T)^{-\eta'_{B(w)}}$

$e(C_{n+n'+1}, \cdots, C_{n+n'+j}, T)^{-\phi'_w} g_{j+1}^{\eta'_w - \phi'_w r'_{B(w)}} = g_{j+1}^{t\left(r'_w - v'_w r'_{B(w)}\right)}$, $E_w = g_{n+j+2}^{r_w s \delta(x) t}$. When GateType$(w) = $ AND, if $g_w(y^*) = 1$, $g_{A(w)}(y^*) = 1$ and $g_{B(w)}(y^*) = 1$, B picks $\mu'_w, v'_w, r'_w \in_R Z_p$, the key component is $K'_w = \left(K'_{w,1} = g_1^{\mu'_w}, K'_{w,2} = g_1^{v'_w}, K'_{w,3} = g_j^{r'_w - \mu'_w r'_{A(w)} - v'_w r'_{B(w)}} \right)$. If $g_w(y^*) = 0$, $g_{A(w)}(y^*) = 0$ or $g_{B(w)}(y^*) = 0$, then the analysis is divided into two parts as follows. If $g_{A(w)}(y^*) = 0$, B chooses $\mu'_w, v'_w, r'_w \in_R Z_p$, sets $\mu'_w = c_{n+n'+j+1} + \psi'_w$, $v'_w = c_{n+n'+j+1} + \phi'_w$, $r'_w = \gamma(n+n'+1, n+n'+j+1) + \eta'_w$, the key component K'_w is $K'_{w,1} = C_{n+n'+l+1} \cdot g_1^{\mu'_w} = g_1^{b'_w}$, $K'_{w,2} = g_1^{v'_w}, K'_{w,3} = g_j^{r'_w - \mu'_w r'_{A(w)} - v'_w r'_{B(w)}}$. $\bar{C}^* = e(D^*, C'^*)$. B sends the challenge ciphertext $CT' = (C^*, E_{n+q}^*, C''^*, \bar{C}^*, D'^*)$ to A.

Guess: At last, A outputs the guess about b'. If $b' = b$, B outputs 1 to guess $\nabla = g_k^{s \prod_{j \in [1,k]} C_j}$; otherwise, outputs 0 to guess that ∇ is a random group element. So, if there exists adversary could break the scheme with probability $\Pr\left[\left(\{g^{C_i}\}_{i=1}^k, \nabla = \Re_0 \right) = 1 \right] = 1/2 + \partial$, then, it could solve the k-linear maps problems with same probability.

4.3   Unforgeability

Proof. Here, consider the length of this paper, we prove the unforgeability briefly. Forger A' promulgates the challenge strings $y^* = y_1^* \cdots y_{n'}^* \in \{0,1\}^{n'}$ and access structure g^*, the challenger B' chooses $a_{i,\beta} \in_R Z_p$.

If $w \in \{1, \ldots, n'\}$, define: $K'_w = \begin{cases} e(C_{n+w}, g_1)^{r'_w} = g_2^{r'_w b_{w,1}}, \text{if} y^*_w = 1; \\ \left(e(C_{n+n'+1}, C_{n+n'+2})g_2^{\eta'_w}\right)^{z'_w} = g_2^{r'_w b_{w,1}}, \text{if} y^*_w = 0 \end{cases}$.

Then, if $\text{GateType}(w) = \text{OR}$, $K'_{w,1} = g_2^{\mu'_w t}$, $K'_{w,2} = g_2^{v'_w t}$, $K'_{w,3} = g_{j+1}^{t\left(r'_w - \mu'_w r'_{A(w)}\right)}$,

$K'_{w,4} = g_{j+1}^{t\left(r'_w - v'_w r'_{B(w)}\right)}$. Suppose that $\text{GateType}(w) = \text{AND}$, $K'_{w,1} = g_1^{b'_w}$, $K'_{w,2} = g_1^{v'_w}$,

$K'_{w,3} = g_j^{r'_w - \mu'_w r'_{A(w)} - v'_w r'_{B(w)}}$, B' provides $\text{SK}^{(\text{SIG})}$ to A'. A' eventually produces a valid forgery CT^* for M^* with string x^* and y^*. Then B' solves the k-Multilinear Computation Diffie-Hellman assumptions (MCDH) problem by outputting C''. Since CT^* is a valid forgery, we have $C'' = g_{k-1}^{\gamma(1,k)}$ which is the answer of the k-MCDH problem instance given to B'. We verify the equation $e(C''', g_1) = e(X, D^*)$.

Suppose there exists a forger could forgery the valid signcryption CT, the probability is $\Pr\left[\left(\{g^{C_i}\}_{i=1}^k, \Re = g_k^{\beta \prod_{j=1}^k C_j}\right) = 1\right] = \partial$, so, if it could forgery signcryption with probability ∂, then it could solve the k-linear maps problem with probability ∂.

4.4 Public Verifiability

When user verifies the equation $e(C'', g_1) = e(X, D')$, parameter X, D', g_1 and C'' are all public, so, it does not reveal the information of decryptor's private key and plaintext, in other word, decryptor does not need involving in this process, the scheme satisfies public verifiability.

4.5 Efficiency

We analyze the efficiency of scheme. The improvement of our scheme is mainly in unsigncryption, so, we prevailingly pay attention to decryptor's communication overheads and computation cost. The comparison of communication overheads is displayed in Table 1, the comparison of computation is displayed in Table 2. There are two computational costs, include multilinear pairing computation p and exponent arithmetic e. In the comparison table, $|G|$ represents the length of G_i elements, $|Z|$ is the length of finite field Z_p^* elements, N represents the number of attribute, d represents the value of threshold, N_m represents the length of message.

Observation from the Table 1, we can see that the length of ciphertext and key are both fixed and substantially decreasing, we have compared OABSC scheme with scheme [16], the decrease of ciphertext length and key length respectively are $|G| - |Z|$ and $(n + 4q)|G| - |Z|$. In Table 2, it is easy to find that the scheme has less computation cost than homogeneous similar schemes, so, the construction achieves our original intention.

Table 1. Comparison of communication overheads

Scheme	Key length	Ciphertext length
Gagne [4]	$3\,N|G|$	$(2\,N + 3)|G| + N_m$
Emura [24]	$(4\,N + 1)|G|$	$(4\,N + 3)|G| + (N + 1)|Z|$
Han [25]	$(N + 2)|G|$	$6|G| + |Z|$
Wang [26]	$3\,N|G|$	$(4\,N + 3)|G| + |Z|$
Datta [16]	$(n + 4q + 1)|G|$	$3|G|$
Ours	$|G| + |Z|$	$2|G| + |Z|$

Table 2. Comparison of computation

Scheme	Unsigncrypt
Gagne [4]	$(3d + 2)p + (N_m/2 + 2d + 1)e$
Emura [24]	$(5\,N + 2)p + Ne$
Han [25]	$8p + (N + 4)e$
Wang [26]	$(4\,N + 1)p + (1 + 2\,N)e$
Datta [16]	$(n + 3q + 7)p$
Ours	$6p + 2e$

In order to intuitively show the advantage of scheme, we set $|Z| = 192$ bits and $|G| = 384$ bits[1], have compared ours with Datta [16], the decrease of ciphertext is 17 %. Apparent, comparing with other schemes, the advantage is more obvious. We proceed to set the number of attribute[2], suppose it is 2, we have computed the key length and compared with Han [18], the decrease of key length is 63 %.

Because the scheme satisfies public verifiability. So, we continue to study that whether equation $e(C'', g_1) = e(X, D')$ could be outsourced in the future, if succeed, the efficiency to reciver's verification algorithm would be further improved.

5   Conclusions

In this paper, we further study the attribute-based signcryption scheme proposed by Datta [16], present an efficient outsourced ABSC scheme in cloud computing, the decrease of ciphertext and key length respectively are 17 % and 63 % after setting logical parameter. And the scheme has less computation costs. There is an effective remedy given to reduce the trust level of KGC and remit the key escrow problem. Our proposed scheme is proven secure in the standard model under k-MDDH assumptions and k-MCDH assumptions. Additionally, scheme could be verified publicly, it means we also could outsource verification to cloud service provider to save receiver's computing cost, but the checking of correctness need to be considered in next step.

---

[1] The setting of parameter references the Elliptic Curve Public Cryptography.

[2] We set $N = 2$, in real life applications, N is far greater than 2, here, we choose the minimum.

Acknowledgments. This work is supported by National Natural Science Foundation of China (61572521, 61272492), Natural Science Basic Research Plan in Shaanxi Province of China (2015JM6353) and Foundation Funding Research Project of Engineering University of Chinese Armed Police Force (WJY201523).

References

1. Nepal, S., Bouguettaya, A., Paris, C.: Guest editorial: special issue on clouds for social computing. IEEE Trans. Serv. Comput. Serv. Comput. **7**(3), 329–332 (2014)
2. Pasupuleti, S.K., Ramalingam, S., Buyya, R.: An efficient and secure privacy-preserving approach for outsourced data of resource constrained mobile devices in cloud computing. J. Netw. Comput. Appl. **64**, 12–22 (2016)
3. Goyal, V., Pandey, O., Sahai, A., Waters, B.: Attribute-based encryption for fine grained access control of encrypted data. In: Proceedings of the 13th ACM Conference on Computer and Communications Security, pp. 89–98 (2006)
4. Gagné, M., Narayan, S., Safavi-Naini, R.: Threshold attribute-based signcryption. In: Garay, J.A., De Prisco, R. (eds.) SCN 2010. LNCS, vol. 6280, pp. 154–171. Springer, Heidelberg (2010)
5. Maji, H., Prabhakaran, M., Rosulek, M.: Attribute-based signatures: achieving attribute-privacy and collusion-resistance. Technical report, IACR Cryptology ePrint Archive (2008)
6. Rao, Y., Dutta, R.: Expressive attribute based signcryption with constant-size ciphertext. In: Pointcheval, D., Vergnaud, D. (eds.) AFRICACRYPT. LNCS, vol. 8469, pp. 398–419. Springer, Heidelberg (2014)
7. Rao, Y., Dutta, R.: *Expressive* bandwidth-efficient attribute based signature and signcryption in standard model. In: Susilo, W., Mu, Y. (eds.) ACISP 2014. LNCS, vol. 8544, pp. 209–225. Springer, Heidelberg (2014)
8. Han, Y.L., Bai, Y.C., Fang, D.Y., Yang, X.Y.: The new attribute generalized signcryption scheme. In: Wang, H., et al. (eds.) ICYCSEE 2015, CCIS 503, pp. 353–360. Springer, Heidelberg (2015)
9. Boneh, D., Silverberg, A.: Applications of multilinear forms to cryptography. Contemporary Mathermatics **324**(1), 71–90 (2003)
10. Lee, H.T., Seo, J.H.: Security analysis of multilinear maps over the integers. In: Garay, J.A., Gennaro, R. (eds.) CRYPTO 2014, Part I. LNCS, vol. 8616, pp. 224–240. Springer, Heidelberg (2014)
11. Coron, J.S., Lepoint, T., Tibouchi, M.: New multilinear maps over the integers. In: Gennaro, R., Robshaw, M. (eds.) CRYPTO 2015. Part I, LNCS, vol. 9215, pp. 267–286. Springer, Heidelberg (2015)
12. Garg, S., Gentry, C., Halevi, S., Sahai, A., Waters, B.: Attribute-based encryption for circuits from multilinear maps. In: Canetti, R., Garay, J.A. (eds.) CRYPTO 2013, Part II. LNCS, vol. 8043, pp. 479–499. Springer, Heidelberg (2013)
13. Attrapadung, N.: Fully secure and succinct attribute based encryption for circuits from multilinear maps. Technical report, IACR Cryptology ePrint Archive (2014)
14. Garg, S., Gentry, C., Halevi, S., Zhandry, M.: Fully secure attribute-based encryption from multilinear maps. Technical report, IACR Cryptology ePrint Archive (2014)
15. Chandran, N., Raghuraman, S., Vinayagamurthy, D.: Reducing depth in constrained PRFs: from bit-fixing to NC^1. In: Cheng, C.-M., et al. (eds.) PKC 2016. LNCS, vol. 9615, pp. 359–385. Springer, Heidelberg (2016). doi:10.1007/978-3-662-49387-8_14

16. Datta, P., Dutta, R., Mukhopadhyay, S.: Compact attribute-based encryption and signcryption for general circuits from multilinear maps. In: Biryukov, A., Goyal, V. (eds.) INDOCRYPT 2015, LNCS, vol. 9264, pp. 3–24. Springer, Heidelberg (2015)
17. Hohenberger, S., Sahai, A., Waters, B.: Full domain hash from (leveled) multilinear maps and identity-based aggregate signatures. In: Canetti, R., Garay, J.A. (eds.) CRYPTO 2013, Part I. LNCS, vol. 8042, pp. 494–512. Springer, Heidelberg (2013)
18. Brakerski, Z., Vaikuntanathan, V.: Circuit-ABE from LWE: unbounded attributes and semi-adaptive security. http://eprint.iacr.org/2016/118
19. Kawai, Y.: Outsourcing the Re-encryption Key Generation: Flexible Ciphertext-Policy Attribute-Based Proxy Re-encryption. In: Lopez, J., Wu, Y. (eds.) ISPEC 2015. LNCS, vol. 9065, pp. 301–315. Springer, Heidelberg (2015)
20. Tang, Q., Pejo, B., Wang, H.S.: Protect both integrity and confidentiality in outsourcing collaborative filtering computations. http://eprint.iacr.org/2016/079
21. Shamir, A.: Identity-based cryptosystems and signature schemes. In: Blakely, G.R., Chaum, D. (eds.) CRYPTO 1984. LNCS, vol. 196, pp. 47–53. Springer, Heidelberg (1985)
22. Malone-Lee, J.: Identity-based signcryption. Technical report, IACR Cryptology ePrint Archive (2002)
23. Green, M., Hohenberger, S., Waters, B.: Outsourcing the decryption of ABE ciphertexts. In: Proceedings of the 20th USENIX Conference on Security, SEC 2011, pp. 34–49. USENIX Association, Berkeley (2011)
24. Emura, K., Miyaji, A., Rahman, M.S.: Dynamic attribute-based signcryption without random oracles. Int. J. Appl. Crypt. 2(3), 199–211 (2012)
25. Han, Y.L., Lu, W.Y., Yang, X.Y.: Attribute-based signcryption for circuits from multi-linear maps. J. Sichuan Univ. (Eng. Sci. Edn.) 45(6), 27–32 (2013)
26. Wang, C.J., Huang, J.S.: Attribute-based signcryption with ciphertext-policy and claimpredicate mechanism. In: 7th IEEE International Conference on Computational Intelligence and Security, pp. 905–909. IEEE Press, New York (2011)

Projecting Distortion Calibration and Evaluation of Coding Fringes in Structured Light System

Haibin Wu[✉], Qing Xu, Xi Wu, Guanglu Sun, Xiaoyang Yu, and Xiaoming Sun

The Higher Educational Key Laboratory for Measuring and Control Technology and Instrumentations of Heilongjiang Province, Harbin University of Science and Technology, Harbin 150080, China
woo@hrbust.edu.cn

Abstract. Coded structured light is an accurate, fast 3D measurement approach with high sampling density, of which the encoded fringes are distorted when projected to curved surface. Focused on the demand of encoding, decoding, multi-view registration and system calibration, we expect to obtain undistorted fringes from camera image. Therefore, in this paper, we analyze the accuracy and sampling density of projecting distortion calibration approach based on control point and fitting surface. Moreover, combining the characteristic of coded structured light system, we design encoded fringe projecting distortion calibration scheme based on simplified encoded structured light model. Primarily, we neglect the minor parameters that affect the calibration in structured light model to reduce complexity. Then, we build the correspondence between camera image points and projector image points and achieve the calibration. Finally, we design evaluation scheme of projecting distortion calibration with parallelism and equal interval, and verify the effectiveness and accuracy of the approach through visual effect and experimental data.

Keywords: Coded structured light · Projecting distortion · Geometric calibration · Encoded fringe

1 Introduction

A typical coded structured light system consists of a projector and a camera, of which the encoded fringes are distorted on the curved surface. Focused on the demand of encoding, decoding, multi-view registration and system calibration, we expect to obtain the undistorted fringes from the image captured by the camera.

Raskar is the first to study on the projecting distortion calibration on quadric surfaces like spherical surface, cylindrical surface and paraboloid. When studying on multi-projector imaging, he used interpolation to make up for the differences among each projecting images. Besides, within in a small scale, Raskar's research is also representative [1]. Yuichiro tried to calibrate the projecting distortion on irregular surface [2], which has a positive result as a whole but not good in detail, and then Michael proposed a distortion calibration approach which implements block mapping to the 2D projecting images. This approach does not need to calibrate the camera but it has a disadvantage

© Springer Science+Business Media Singapore 2016
W. Che et al. (Eds.): ICYCSEE 2016, Part I, CCIS 623, pp. 546–555, 2016.
DOI: 10.1007/978-981-10-2053-7_48

that the calibration precision relies on the intensity of the image blocks. With the improvement of the calibration and iteration algorithm, the accuracy and robustness of calibration approaches are improved [3, 4]. Zhang divided the projected surface in equal angle under the polar coordinates and achieved the distortion calibration of the image projected on the curved surface with the texture mapping [5]. This approach is fast but the precision need to be improved. Liu used the corresponding image coordinates under the two cameras of a certain point on the projected surface to get the conversion matrix [6]. The precision is improved but the complexity of the system is increased. Yang uses cubic Beizer to solve the conversion matrix [7], which is suitable for complex curved surfaces. However, it reduces the calculation accuracy and speed.

Recently, the projecting distortion calibration approaches mainly focus on normal curved surfaces and emphasizes the visual effect. In this paper, the research focuses on the structured light 3D measurement and simultaneously considers the encoded fringe distortion calibration approach, which is able to improve the accuracy of the system calibration and measurement.

2 Coded Structured Light and Projecting Distortion

In this paper, we use Infocus 82 projector and HV-F22F camera to construct the coded structure light sytem, which is shown in Fig. 1. The projector projects the patterns orthographically and the camera captures images from the side. According to the projecting range of the projector, the measuring range in z direction is 120 mm and the measuring ranges in x and y directions are respectively 350 mm and 250 mm.

Fig. 1. Coded structured light measuring device

In order to fit the measurement of the surfaces with strong reflection and various colors, we use the binary encoding/decoding principle, which is shown in Fig. 2. Combing the line-shifting fringe center and the Gray code fringe edge to encode and decode is an effective way to improve the sampling density when guaranteeing the decoding accuracy and resolution.

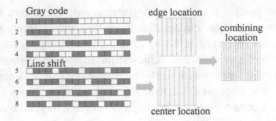

Fig. 2. Encoding and decoding principle by combining Gray code and line shift fringe (Color figure online)

In the process of encoding/decoding, multi-view registration and system calibration, the Gray code or line-shifting fringe will distort when projected to the measured surface, which is shown in Fig. 3. If we want to obtain the undistorted image, we need to project pre-calibrated patterns. The following will analyze the projecting distortion calibration approach that is suitable for coded structured light system.

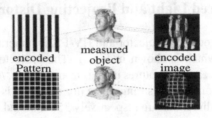

Fig. 3. Encoded pattern distortion

3   Practicability Analysis of Projecting Distortion Calibration Approach in Coded Structured Light

Typical approaches mainly include control point calibration approach and fitting surface calibration approach.

3.1   Control Point Calibration Approach

The control point calibration approach sets grid to the projector image, determines the corresponding grid intersection of the projector and camera images and fit the conversion matrix of them. Finally, according to the undistorted camera image, calculate the pre-calibrated projector image through the conversion matrix. When projecting the pre-calibrated image, the camera is able to capture the undistorted image [8, 9].

The conversion matrix can be presented with a bivariate polynomial, which is shown in (1). To quickly get the conversion matrix, we can let $n = 1$, namely, linear conversion, which is shown in (2).

$$x' = \sum_{i=0}^{n} \sum_{j=0}^{n-i} a_{ij} x^i y^j \tag{1}$$

$$y' = \sum_{i=0}^{n} \sum_{j=0}^{n-i} b_{ij} x^i y^j$$

$$x' = a_{00} + a_{10}x + a_{01}y \quad y' = b_{00} + b_{10}x + b_{01}y \tag{2}$$

In the equations, a and b represent the conversion coefficients of the projector image and the camera image respectively. x, y, x' and y' are the coordinates of the projector and camera image before and after the conversion.

To accurately obtain the conversion matrix, we can take advantage of higer-degree polynomial such as Bezier surface, which is shown in (3).

$$P(u,v) = \sum_{i=0}^{N} \sum_{j=0}^{M} P_{ij} B_i^N(u) B_j^M(v); u, v \in [0, 1] \tag{3}$$

In (3), (u, v) is the image coordinates, P is the conversion coefficient of the image and B is the Bezier function. When solving, we determine the Bezier surface with $(N+1) \times (M+1)$ control points. Each point on the surface has a corresponding u and v value. When the projected surface is a complex surface, to describe the surface more accurately, we need to increase the degree of the Bezier surface, which consequently increases the calculation work.

3.2 Fitting Surface Calibration Approach

In the fitting surface calibration approach, The converting relation T_{pc} of the projector and camera coordinate systems to the projected surface Q can be presented with function Ψ, which is a 4×4 symmetric matrix. Q can be obtained according to cloud point X measured with the structured light, namely, $X^TQX = 0$. Then, for a point x on Q, the relation between its projector coordinate x_p and its camera coordinate x_c can be denoted by (4).

$$x_p \cong (B - eq^T)x_c \pm \sqrt{x_c^T(qq^T - Q_{33})x_c} \cdot e \tag{4}$$

In (4), B and e are the rotation matrix and transformation matrix of T_{pc}, which are obtained through calibration. Q_{33} and q can be obtained according to $Q = \begin{bmatrix} Q_{33} & q \\ q^T & g \end{bmatrix}$. The image coordinate system is a special case of the world coordinate system. Therefore, function Ψ can also be treated as the relation between the projector image and observation image, and consequently, we can calculate the projector image.

The control point calibration approach has a simple process and the control point calibration is accurate. However, those which are not control points are obtained through fitting, so they are not accurate enough. Despite the fact that we can use higher-degree

Beizer surface to improve the fitting accuracy, for a complex surface, it is still hard to achieve accurate calibration with one Beizer expression. We need multiple Beizer surfaces to achieve that, which lower the calibration speed.

The fitting surface calibration approach comprehensively considers the projected surface. Points in and out of the point cloud are all accurately calibrated. But for complex non-quadric surfaces, the fitting error is large, especially in complex part and edges.

Combining the features of the two approaches, we propose a calibration scheme based on simplified encoded structured light.

4 Calibration Scheme Based on Simplified Coded Structured Light Model

According to the coded structured light model which is shown in Fig. 4, the world coordinate (X^w, Y^w, Z^w) and the camera image coordinate (m^c, n^c) of the spatial point P has a converting relation which is shown in (5).

$$Z^c \begin{bmatrix} n^c \\ m^c \\ 1 \end{bmatrix} = \begin{bmatrix} (N^c/2) \cdot c \tan \beta_1^c & (\tan \alpha \cdot f_1^c)/d_1^c & n_0^c \\ 0 & (M^c/2) \cdot c \tan \beta_2^c & m_0^c \\ 0 & 0 & 1 \end{bmatrix} \cdot [R^c, T^c] \cdot \begin{bmatrix} X^w \\ Y^w \\ Z^w \\ 1 \end{bmatrix} \quad (5)$$

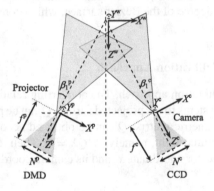

Fig. 4. Spatial converting relation

In the equation, $\tan a \cdot f_1^c$ indicates the skew angle α caused by the incomplete orthogonality of the principle optic axis and CCD image surface. Furthermore, n^c and m^c are the corresponding column order and row order of the spatial point P in the camera image. d_1^c is the width of a single pixel of the CCD. N^c and M^c are the column and row numbers of the CCD. Besides, β_1^c and β_2^c are half of the horizontal and vertical field angles. (n_0^c, m_0^c) is the pixel position of the principle point on CCD image surface. R^p and T^p are the rotation matrix and translation matrix, respectively. If we ignore the two position error between the principle optic axis and the CCD image surface, (5) can be simplified as (6).

$$Z^c \begin{bmatrix} n^c \\ m^c \\ 1 \end{bmatrix} = \begin{bmatrix} (N^c/2) \cdot c \tan \beta_1^c & 0 & 0 \\ 0 & (M^c/2) \cdot c \tan \beta_2^c & 0 \\ 0 & 0 & 1 \end{bmatrix} \cdot [R^c, T^c] \cdot \begin{bmatrix} X^w \\ Y^w \\ Z^w \\ 1 \end{bmatrix} \tag{6}$$

Likewise, the world coordinate (X^w, Y^w, Z^w) and the projector image coordinate (m^p, n^p) has a converting relation which is shown in (7).

$$Z^p \begin{bmatrix} n^p \\ m^p \\ 1 \end{bmatrix} = \begin{bmatrix} (N^p/2) \cdot c \tan \beta_1^p & 0 & n_0^p \\ 0 & (M^p/2) \cdot c \tan \beta_2^p & m_0^p \\ 0 & 0 & 1 \end{bmatrix} \cdot [R^p, T^p] \cdot \begin{bmatrix} X^w \\ Y^w \\ Z^w \\ 1 \end{bmatrix} \tag{7}$$

In (7), m_0^p of the principle point coordinate (n_0^p, m_0^p) is not zero and even goes beyond the DMD chip. Through connecting (6), (7) and the encoding/decoding principle (8), we can establish the correspondence among the projector image point, the camera image point and the projected surface point and then build the calibration conversion matrix.

$$n^p = g[\Phi v(m^c, n^c), N^p] \tag{8}$$

In (8), $\Phi_V(n^c, m^c)$ is the encoding value of the camera image point.

Compared with the control point calibration approach, the fitting of the proposed approach is more accurate because the number of the point cloud obtained by coded structured light measurement is much more than the number of the control point. Moreover, we simplified the minor parameters of the system and therefore the calculation work is reduced. Then, compared with the fitting surface calibration approach, the proposed approach has no fitting error in complex part and edges of the surface and consequently increase the accuracy of the calibration.

5 Projecting Distortion Calibration Experiments on Encoded Fringes

5.1 Visual Effect Evaluation

To verify the effectiveness of the proposed approach, we use 3dsmax to build the coded structured light system and implement the projecting distortion calibration experiment.

Figure 5(a) shows a standard projecting image. The distortion situation of projecting the image to the sphere is shown in Fig. 5(b). What's more, the pre-calibrated image obtained with the proposed approach is shown in Fig. 5(c) and the image captured by the camera is shown in Fig. 5(d). From the figure, we can see that the chessboard is effectively calibrated and that is equivalent to the fact that there is no visual effect distortion in Fig. 5(a).

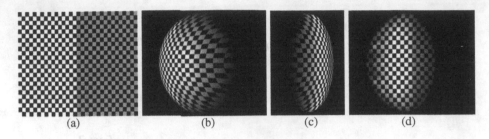

(a)   (b)   (c)   (d)

Fig. 5. Visual effect on sphere

Furthermore, the same calibration effect is also obtain on complex surface. Figure 6(a) and (b) are the camera images before and after the calibration.

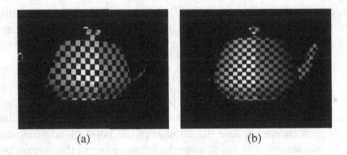

(a)   (b)

Fig. 6. Visual effect on complex surface

5.2 Quantitative Evaluation

In order to verify the accuracy of the proposed approach, we implement projecting distortion calibration experiment on the sphere with both the proposed approach and the fitting surface calibration approach.

Theoretically, the edges of the encoded fringes in the camera image are parallel and equal-width straight lines. Therefore, we evaluate the calibration accuracy by judging the straightness and equal-width degree of the fringe edges after the distortion calibration. First of all, we use the gray curve intersection method to locate the edges [10]. Then, we fit the discrete edge points into edge lines and use the discrete degree to evaluate the straightness of the edge. What's more, we calculate the average interval according to each interval of the adjacent edges and use the maximum difference between the average interval and each edge interval to evaluate the equal-width degree.

The encoded fringe of which the width is 64 pixels is shown in Fig. 7. The captured camera image after the calibration is shown in Fig. 7(a) and the fitted edge is shown in Fig. 7(b). Taking the largest point straightness error of each edge as the straightness error of that edge, which is shown in Fig. 7(c). Also, take the largest difference between the average interval and each edge interval as the equal-width error, which is shown in Fig. 7(d).

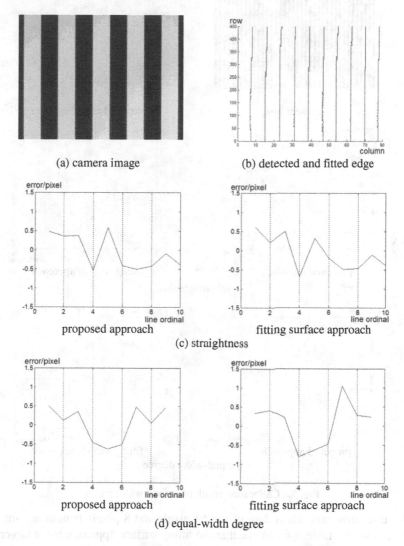

(a) camera image

(b) detected and fitted edge

proposed approach

fitting surface approach

(c) straightness

proposed approach

fitting surface approach

(d) equal-width degree

Fig. 7. Calibration result of 64 pixel-width fringe

The encoded fringe of which the width is 8 pixels is shown in Fig. 8.

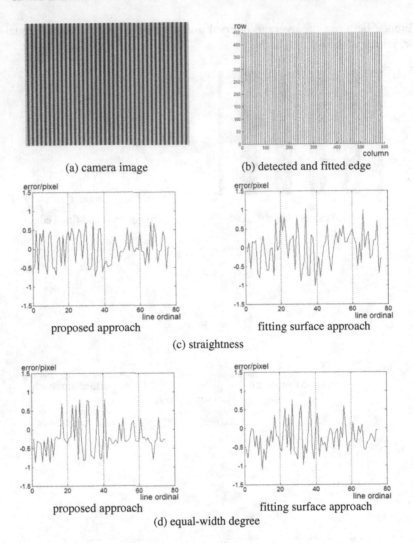

(a) camera image

(b) detected and fitted edge

proposed approach   fitting surface approach

(c) straightness

proposed approach   fitting surface approach

(d) equal-width degree

Fig. 8. Calibration result of 8 pixel-width fringe

The distortion calibration data of the 64 pixels and 8 pixels fringes are shown in Table 1. From the table, we can see that the fitting surface approach has a larger error than that of the proposed approach.

Table 1. Error of projecting distortion calibration

Fringe width	Approach	Straightness error/pixel	Equal-Width error/pixel
64	Proposed approach	0.62	0.65
	Fitting surface approach	0.86	1.24
8	Proposed approach	0.60	1.01
	Fitting surface approach	0.82	1.19

6   Conclusion

Focused on the demand of the structured light system distortion calibration, we propose a calibration scheme based on simplified coded structured light system and according to the control point approach and the fitting surface approach. Moreover, we analyze the accuracy and sampling density advantage of the proposed approach to the complex surface. Besides, we verify the effectiveness of the proposed approach through visual effect. We also design the encoded fringe distortion calibration evaluating scheme based on the straightness and equal-width degree. In the experiments of 64-pixel width fringe and 8-pixel width fringe, the straightness error and equal-width error of the proposed approach are all smaller than those of the fitting surface approach, which proves that the proposed approach has a smaller comprehensive error. In the future, we will focus on the fringe distortion calibration of highly complex surface and improve the measuring accuracy of the system.

Acknowledgments. The support of National Science Foundation of China (61571168, 61401126), Leading Talent Team Backup Leader Foundation of Heilongjiang Province are gratefully acknowledged.

References

1. Raskar, R., Baar, J., Willwacher, T., et al.: Quadric transfer for immersive curved screen displays. Eurographics **23**(3), 451–460 (2004)
2. Yuichiro, F., Ross, S., Takafumi, T., et al.: Geometrically-correct projection-based texture mapping onto a deformable object. IEEE Trans. Vis. Comput. Graph. **20**(3), 540–549 (2014)
3. Li, X., Tang, D., Yang, T., et al.: Multi-projector geometric calibration error analysis. J. Comput.-Aided Des. Comput. Graph. **27**(1), 106–113 (2015)
4. Zhu, B., Xie, L., Yang, T., et al.: An adaptive calibration algorithm for projected images in dailyenvironment. J. Comput.-Aided Des. Comput. Graph. **24**(7), 941–948 (2012)
5. Zhang, H., Jia, Q., Sun, H.: New solution of geometrical calibration based on immersive multiprojector displays. J. Syst. Simul. **18**(2), 493–496 (2006)
6. Liu, G., Yao, Y.: Correction of projector imagery on curved surface. Comput. Aided Eng. **15**(2), 11–14 (2006)
7. Yang, J., Wang, S.: Auto-nonlinear geometry calibration and edge blending of multi-projector display system. J. Shanghai Jiaotong Univ. **42**(4), 574–583 (2008)
8. Madi, A., Ziou, D.: Color correction problem of image displayed on non-white projection screen. In: Kamel, M., Campilho, A. (eds.) ICIAR 2013. LNCS, vol. 7950, pp. 255–263. Springer, Heidelberg (2013)
9. Robert, U., Rahul, S.: Correcting luminance for obliquely-projected displays. In: IEEE International Conference on Consumer Electronics, pp. 367–368 (2005)
10. Haibin, W., Xiaoyang, Yu., Guan, C.: Structured light encoding stripe edge detection based on grey curve intersecting point. Acta Optica Sinica **28**(6), 1085–1090 (2008)

Recognizing Visual Attention Allocation Patterns Using Gaze Behaviors

Cheng Wu, Feng Xie[✉], and Changsheng Yan

School of Urban Rail Transportation, Soochow University, Suzhou 215011, China
{cwu,fengxie}@suda.edu.cn, chang-sheng.yan@hp.com

Abstract. Human Attention Allocation Strategy (HAAS) is related closely to operating performance when he/she is interacting a machine through a human-machine interface. Gaze behaviors, which is acquired by eye tracking technology, can be used to observe attention allocation. But the performance-sensitive attention allocation strategy is still hard to measure using gaze cue. In this paper, we attempt to understand visual attention allocation behavior and reveal the relationship between attention allocation strategy and interactive performance in a quantitative manner. By using a novel Multiple-Level Clustering approach, we give some results on probabilistic analysis about interactive performance of HAAS patterns in a simulation platform of thermal-hydraulic process plant. It can be observed that these patterns are sensitive to interactive performance. We conclude that our Multiple-Level Clustering approach can extract efficiently human attention allocation patterns and evaluate interactive performance using gaze movements.

Keywords: Bioinformatic · Human Attention Allocation Strategy (HAAS) · Multiple-Level Clustering Algorithm · Human-Machine Interface

1 Introduction

One of the important issues in Human-Machine Interaction (HMI) is to evaluate whether an interface design is optimal or not. An excellent HMI interface should enhance human operators' ability, that is, extracting useful information from an amount of chaotic information and further improving interactive performance for accomplishing a complex task. It concerns with human attention, which is interpreted conventionally as the ability to select a topic of interest (a goal) for extracting information useful for a given task [1]. However, human attention is a scarce resource [2,3]. At any moment, human operators pay attention to only one or a few objects. Paying attention to something makes the operator to see these objects more clearly, more accurately, and remember better. Thus, whether allocation of human attention is reasonable is related closely to the performance of

Project supported by the National Nature Science Foundation of China (No. 61471252) and the Natural Science Foundation of Jiangsu Province (No. BK20130303).

© Springer Science+Business Media Singapore 2016
W. Che et al. (Eds.): ICYCSEE 2016, Part I, CCIS 623, pp. 556–567, 2016.
DOI: 10.1007/978-981-10-2053-7_49

Human-Machine Interaction. Just based on this viewpoint, allocation of attention has become one of hot topics in the research of Human-Machine Interface in recent years.

In fact, there are still some difficulties when we conduct the study about human attention allocation on human machine interaction. The root causes for this difficulties come from the inherent characteristics of human attention such as [4–6]:

(1) *Human Attention is Unobservable by Itself.* In the real world, Human attention is intangible. There is no reliable way to monitor the status of human attention [1,7].
(2) *Human Attention can Hardly be Measured.* There is neither a physical unit to describe human attention nor a method to infer the value of the specific attention on an object [8,9].

Recently, human behavior symptoms triggered by human attention are used as indicators of human attention. While interacting with the interface through viewing the interface, eye movements play a more salient role than other symptoms. Eye movements can implicitly indicates the focus area of the user's attention. There are several excellent studies on using the eye movement parameters to understand the operator's visual attention focus (or area of interest) [10]. Thus, there seems to be reasonable consensus for eye movements as indicators of visual attention focus. Theoretically, the consensus comes from the scientific rationale and advantages for the use of eye movements in observing human attention allocation focus.

(1) eye movements may yield important clues to human attention focus at a fine temporal grain size, typically on the order of 10 ms;
(2) the data about eye movement is collected non-intrusively, such that data collection in no way affects task performance;
(3) eye movement parameters can serve as the sole source of data or as a supplement to other sources like Verbal Protocols (VP), Electromyography (EMG), Electroencephalogram (EEG) etc.

The measurement of human attention is based on the above consensus that eye movements can act as indicators of visual attention focus. It is well known that information from eye movements includes eye fixations, eye saccades, pupil size, blink rate, and eye vergence etc. Eye fixations are pauses in the eye scanning process over informative regions of interest [11]. Measures of fixation include two attributes: the fixation duration (i.e. the time spent investigating a local area of the visual field; also known as dwell time) and the number of fixations (i.e. the number of times the eye stops on a certain area of the visual field). Longer fixation duration implies more time spent on interpreting, processing or associating a target with its internalized representation. Fixation duration is negatively correlated to the efficiency of task execution [12]. A larger number of fixations imply that more information is required to process a given task [13].

In cognitive sciences, researchers have studied eye movements to elicit and analyze the cognitive processes in a variety of task domains. This research has achieved notable and significant benefits in domains such as reading [14–16], arithmetic [17], and word problems [18]. In neurobiology, researchers have focused on the relation between the underlying mechanisms of eye movements and human visual processing [11,19]. There are a number of important reasons why eye movements have become popular indicators of human attention in so many research fields. Firstly, eye movements yield important clues to human behavior at a fine temporal grain size, typically on the order of 10 ms. Secondly, eye-movement data is collected non-intrusively, such that data collection in no way affects task performance. Thirdly, eye movements can serve as the sole source of data or as a supplement to other sources like verbal protocols.

In recent years, some researchers show the allocation of attention has close correlation with the performance of interaction between human operators and computers. Michael and Kim [20] presented the correlation between visual attention allocation and various measures of performance using regression analysis, and Wu et al. [8] proposed a notion of effective attention allocation as well as its measure and demonstrated that it is sensitive to operator's performance variations on different interfaces and different operators in a qualitative manner. Although these researches have wonderfully revealed that there exists strong links between attention allocation and performance, an important fact is ignored, that is, different interaction processes with similar performance from different operators often possess similar patterns of visual attention allocation. Therefore, it is imperative to extract the patterns of visual attention in relation to the performance of interaction.

Intuitively speaking, different patterns of visual attention is correlated with different performance characteristics obtained from different interaction processes on a specific interface in Human-Computer Interaction. Some patterns might lead to superior performance while others patterns might be deemed to result in bad performance. Given an interface design, evaluation of the design requires that human subjects perform task on the interface. It is possible that we can find the patterns of operators visual attention allocation which are correlated to task performances. Such a relationship can help the human-computer interface design and human-Machine interaction management in two ways. The first way is that we can select operators based on whether their attention allocation strategy patterns match with the pattern which corresponds to the best task performance. The second way is that we can optimize the interface design in terms of the best task performance achieved by the operator with a certain attention allocation strategy pattern.

Despite its importance, there is still few published work to show the relationship between the patterns of visual attention and performance characteristics, our study just starts to show the relationship in a quantitative manner. It is noted that the relationship between attention patterns and performance should be shown in the form of probabilities because of the uncertainty of human attention itself and external interference. In addition, the extraction of visual attention patterns depends to a great extend on specific task and specific interface.

The remainder of this paper is organized as follows. Section 2 proposes a novel approach - Multiple-Level Clustering Algorithm (MLCA) for extracting the patterns of HAAS. In Sect. 3, we evaluate our approach in a simulation platform of thermal-hydraulic process plant and give probabilistic analysis about interactive performance of HAAS patterns. At last, we concludes the paper and discusses the future work.

2 Multiple-Level Clustering Algorithm

2.1 Two Levels of Clustering for Solving Two Key Questions

As mentioned above, the goal of the study is to extract the patterns of HAAS from dynamic interactive process in HMI. Clearly, there are two key questions need to be solved.

(1) How to represent a HAAS?
(2) How to extract common features of HAAS?

With regard to the first question, it has already been well known that the information of eye movement can provide a possible way to represent a HAAS. The sampling data of eye movement at a moment, specifically, Eye Fixation Point Coordinate (EF-PC), Eye Fixation Dwell Time (EF-DT) reflect current attention focus in a real-time manner. A series of EF-PCs and EF-DTs reveal the allocation of human attention during a dynamic interactive process. Then further combining the detail interface display, all EF-PCs are grouped into main components on the interface display using K-means clustering algorithm. At last, an encoding procedure is issued in order to produce a feasible HAAS. K-means Clustering is chosen because cluster number (the number of main components on an interface display) can be predetermined easily.

With regard to the second question, it is difficult to extract common features from mass HAASes. These features either are independent on these HAASes or have relation with them. For the reason, unsupervised clustering procedure is applied based on its merit to probe the potential rules behind chaotic data. Considering the difficulty of foreknowing the number of clusters, hierarchical clustering becomes a reasonable choice for grouping these HAASes in our experiment. Then, the patterns of HAAS are extracted from different groups.

2.2 Hybrid Clustering Algorithm for Extracting HAAS Patterns

In the section, a novel hybrid clustering algorithm is proposed for extracting HASS patterns. The algorithm includes four phases:

(1) Grouping EF-PC using K-means Clustering;
(2) Encoding a HAAS;
(3) Grouping HAASes using Hierarchical Clustering;
(4) Extracting common features.

Phase I: Grouping EF-PCs: in this phase, K-means Clustering Procedure is used as the first-level clustering for grouping Eye Fixation Points. In order to apply K-means Clustering, the following parameters need to be determined in advance:

- Determining the number of cluster: it depends on the number of key components on interface display.
- Determining cluster distance: it depends on data characteristic. Because the EF-PCs represent some points on a two-dimension space, Squared Euclidean distance is a feasible choice during the course of clustering the EF-PCs. Let P_1 and P_2 be two points on a two-dimension space, then the Squared Euclidean distance D_{se} between P_1 and P_2 is:

$$D_{se} = (x_{p_1} - x_{p_2})^2 + (y_{p_1} - y_{p_2})^2$$

- Determining initial cluster centroids: in order to make clustering procedure more orientable, initial cluster centroids are set to the centers of key components.
- Avoiding local Minima: an iterative algorithm is employed to minimize the sum of distances from each point to its cluster centroid.

Phase II: Encoding HAAS: it is well known that an interface display consists of certain key components and human attention always shuttles across these components. If these components are labeled numerically and are arranged in sequence, the codeword for describing interface components distribution can be constructed. Further, if the total EF-DT of every component is determined and assigned into the corresponding component field of this codeword, the codeword will describe a HAAS. The construction of a HAAS codeword is shown in Fig. 1(b).

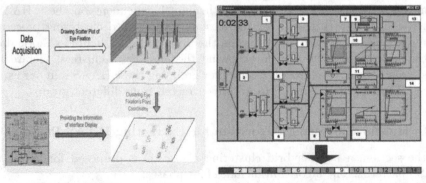

(a) HAAS acquisition procedure   (b) HAAS encoding procedure

Fig. 1. The representation of a HAAS using EF-PCs. (Color figure online)

Now the problem is how to determine the total EF-DT of every component. The total EF-DT of every cluster can be achieved easily by accumulating all EF-DTs in the cluster. Then the distances between all cluster centroids and all component centers are calculated respectively. The total EF-DT of a cluster is accumulated to the nearest component and then the total EF-DTs of all components can be determined. The codeword is built using absolute EF-DTs, which is called *Codeword With Absolute EF-DTs* (CWAE). It should be seen that CWAE has its limitation. Because the duration of each trial on the experiment is different, what is most meaningful to HAAS is the percentage of the EF-DT on each component over the total duration. Hence, the *Codeword with relative EF-DTs* (CWRE) is designed by replacing absolute EF-DTs with relative EF-DTs, where

$$relative\ EFDT = \frac{absolute\ EFDT\ on\ the\ component}{total\ duration\ of\ the\ trial} \tag{1}$$

Phase III: Grouping HAASes: in this phase, Hierarchical Clustering Procedure is used as the second-level clustering for grouping HAASes. The detail steps are listed as the following.

– Determining the Cluster distance and calculating the similarity between every pair of HAASes: it is standardized Euclidean distance that is used as cluster distance. The reason is to eliminate measure difference on different scales, which may result from different participants, different operation time, and different equipment calibration etc. Assume $P_1, P_2, ..., P_m$ are the points on a N-dimension space, the coordinates of the points construct a mXn matrix. To any two points P_i and P_j, the standardized Euclidean distance D_{st} is:

$$D_{st} = (p_i - p_j)D^{-1}(p_i - p_j)'$$

where D is the diagonal matrix with diagonal elements given by v^2, which denotes the variance of the variable p over m point vectors.
– Drawing cluster tree of HAASes.
– Determining division points of clusters and grouping all HAASes: according to the above cluster tree, an inconsistency coefficient threshold is specified for dividing the tree into several parts. Each part corresponds to a cluster respectively.

Phase IV: Extracting common features: In the phase, common features of each cluster are extracted, such as: which component are the operators paying more attention for? Which components are the operators concerned with during the whole interactive process? How about are the frequencies of these components being visited? Note that these common features would be deemed to have strong relations with the performance of interactive process. The Multiple-Level Clustering Algorithm (MLCA) is illustrated and outlined as shown in Fig. 2.

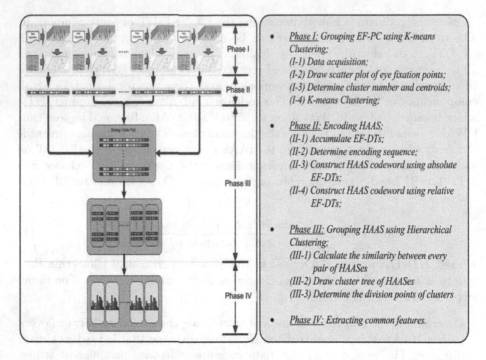

- *Phase I: Grouping EF-PC using K-means Clustering;*
  - *(I-1) Data acquisition;*
  - *(I-2) Draw scatter plot of eye fixation points;*
  - *(I-3) Determine cluster number and centroids;*
  - *(I-4) K-means Clustering;*

- *Phase II: Encoding HAAS;*
  - *(II-1) Accumulate EF-DTs;*
  - *(II-2) Determine encoding sequence;*
  - *(II-3) Construct HAAS codeword using absolute EF-DTs;*
  - *(II-4) Construct HAAS codeword using relative EF-DTs;*

- *Phase III: Grouping HAAS using Hierarchical Clustering;*
  - *(III-1) Calculate the similarity between every pair of HAASes*
  - *(III-2) Draw cluster tree of HAASes*
  - *(III-3) Determine the division points of clusters*

- *Phase IV: Extracting common features.*

Fig. 2. Multiple-Level Clustering Algorithm for extracting HAAS

3   Experimental Evaluation

The section will give an example for demonstrating how to conduce the above algorithm, which is divided into two parts. First, HAAS Patterns extracted using MLCA are presented and some important phases for extracting these patterns are illustrated. Second, combining the performance of interactive process, an evaluation is given to prove the validity of these extracted patterns.

In order to track and record real observers eye movements, experiments have been conducted using an X50 eye tracker from Tobii Technology. This apparatus is mounted on a rigid headrest for greater measurement accuracy (less than $0.5°$ on the fixation point). Experiments were conducted in a simulation platform of a thermal-hydraulic process plant, which was called the dual reservoir simulation system (DURESS). Its interface: EID-DURESS was designed based on Ecological Interface Design (EID) interface design framework. There were 20 subjects (6 female and 14 males) involved in the experiment. All of them had an engineering background. They were asked to control the system (including adjust the openings of the valves and the heaters) until reaching the dynamic equilibrium (the demanded temperature and flow rate of the water out of the reservoirs) as quickly as possible. The participants were required to perform three replications for each trial. As a result, in total there were 60 runs for the entire experiment. The collected data corresponds to 60 different pieces of human visual attention allocation strategies.

3.1   The Extraction of HAAS Patterns

The experimental data of the scenario L01 on the EID-DURESS is analyzed to demonstrate how to extract HAAS patterns using MLCA. There are 60 trials from 20 participants, which correspond to 60 different HAASes.

In the first phase of MLCA, the number of clusters and initial cluster centroids are preset. For EID-DURESS, the number of cluster is 14 and initial cluster centroids are the centers of 14 key components shown as in Fig. 1(b). Then the scatter plot of eye fixation points is draw. Further, a K-Means clustering procedure whose cluster distance is Squared Euclidean distance is issued. After 500 iterations, the cluster result with the minimal total sum of point-to-centroid distances is chosen. As an example, the clustering result of the second trial of Participant 4 is illustrated in Fig. 3(b). The diagram in Fig. 3(c) and (d) give 3D effects of the above cluster result, whose X-Y plant represents interface surface and Z axis represents total EF-DT of a cluster.

In the second phase, an encoding sequence is determined in term of the partition of EID-DURESS. Then, HAAS codewords with absolute EF-DT and relative EF-DT are constructed.

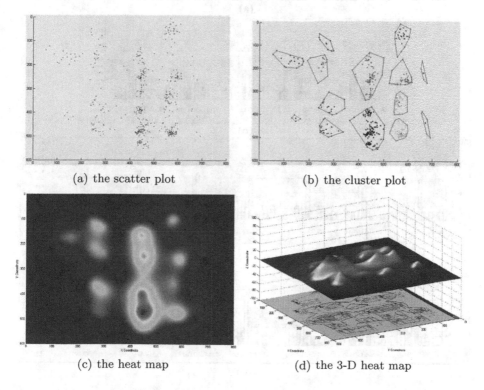

(a) the scatter plot   (b) the cluster plot

(c) the heat map   (d) the 3-D heat map

Fig. 3. The clustering result of the second trial of Participant 4

In the third phase of MLCA, hierarchical clustering procedure is performed for grouping the HAASes produced in Phase 2 into different clusters. Once the inconsistency coefficient threshold is set to 0.225, the HAASes are divided into four clusters: Cluster 1, Cluster 2, Cluster 3 and Cluster 4. In Cluster 4, there are three sub-clusters: Cluster 41, Cluster 42 and Cluster 43. In the last phase, a statistic analysis is done and some patterns are extracted from each cluster. For example: for Cluster 1, the HAASes adopted by the operators embody the following patterns: (1) the most attention focus is allocated on H2, whose percentage is near or beyond 25 %; (2) M-VOL1, M-VOL2, H1 and Principle2 pull some attention focus on themselves. For Cluster 2, the patterns of the HAAS are: (1) the

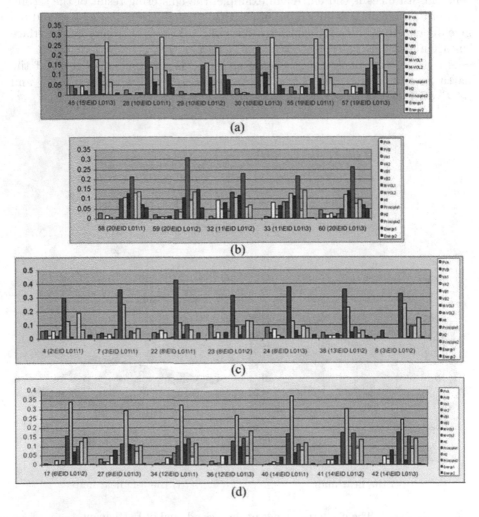

(a)

(b)

(c)

(d)

Fig. 4. This is the caption of the figure displaying a white eagle and a white horse on a snow field

most attention focus is allocated on Principle1, whose percentage is beyond 20 %; (2) part of attention is attracted by H1 and Principle2. For Cluster 4, a common pattern of HAAS is that attention focus has a large distribution on M-VOL2, whose percentages are near 20 % for Cluster 41 and beyond 25 % for Cluster 42 and Cluster 43 differently. But Cluster 41, Cluster 42, and Cluster43 have still their own particular patterns. For Cluster 41, M-VOL1 has the same or even more attraction than M-VOL2 to the operators, and at the same time, H1, H2, Principle1 and Principle2 are also pulling some attention focus. For Cluster 42, although most of the HAASes' attention focus percentages on M-VOL2 are clearly higher than those on the other components, M-VOL1, H1, H2, Principle1 and Principle2 play the role as the second attractants. The above patterns about the HAASes for scenario L01 on EID-DURESS can be achieved easily by observing the diagrams shown in Fig. 4.

3.2   Evaluation of HAAS Patterns

By analyzing mass HAASes using MLCA, the patterns of the HAASes have been extracted successfully. However, whether these extracted patterns are valid, that is, have a strong relation with the performance of interaction between human and machine system is still a pending question.

As mentioned in Sect. 2.2, different patterns of HAAS correspond to different interactive performances and the kind of corresponding relationship exists in the form of probabilities. Based on the view, a probabilistic analysis about all interactive performances from the HAASes in each cluster is issued to solve the above question, that is, to prove these patterns' validity. Table 1 describes the statistical result of the HAASes for scenario L01 on EID-DURESS.

Table 1. My caption

Pattern Source		Weight	Interactive Performance		
			Reach Dynamic Balance	Below Average Duration	Above 80% of Maximum Duration
Cluster 1		10.00%	83.33%	80.00%	20.00%
Cluster 2		8.33%	60.00%	100%	0%
Cluster 3		11.67%	28.57%	50.00%	50.00%
Cluster 4	Cluster 5	20.00%	88.67%	50.00%	37.50%
	Cluster 6	35.00%	90.48%	63.16%	15.79%
	Cluster 7	11.67%	85.71%	33.33%	33.33%

From Table 1, it is clear that the patterns with large weight (e.g. over 10 %) have their distinctive performance characteristics. First, the percentages of the HAASes of reaching dynamic balance over the total HAASes concentrate on two contrasting positions, that is, the one from Cluster 3 is very small (only 28.57 %) and the ones from Cluster 1 and Cluster 4 are all very large (beyond 80 %). Specially, the pattern extracted from Cluster 42 with the largest weight has over 90 % probability to make its HAASes reach dynamic balance. Second, among the clusters that can reach dynamic balance, the pattern from Cluster1 corresponds to the highest performance because most of its HAASes (80 %) is below average duration and only 20 % of the HAASes have the large duration. The pattern from Cluster 42 is secondary and its HAASes? percentages below average duration and above large duration are about 63.16 % and 15.79 % of the totals respectively. The pattern from Cluster 43 is the worst and its performance is deemed to be unsatisfactory because of the lowest percentage of "below average duration" and the higher percentage of "above large duration".

4   Conclusion and Discussion

In this paper, an experiment-oriented study is issued to extract the patterns of HAAS from dynamic interactive process of HMI. First, a novel approach - Multiple-Level Clustering Algorithm is proposed and illustrated. Further, a probabilistic analysis about interactive performance of HAAS patterns is performed, which provides evidence for the validity of the extracted patterns.

With an eye to the necessity of extracting the patterns of HAAS on HMI, the paper has purposed a preliminary study on its algorithm and application. Indeed, still a lot of works need to be further explored. First, the algorithm itself about extracting HAAS patterns may be modified using more complex clustering processes. Second, although eye movement can provide main clues about human attention allocation, more clues from hand movement, head movement and fMRI etc. perhaps should be fused for more accurate and more comprehensive inference. Third, the application about this study would be extended to more complicated cases, such as attention analysis of drivers and pilots, human behavior prediction on intelligent adaptive interface, etc.

References

1. Gonalves, M.G., Luiz, A.A.F., Oliveira, R.A., Grupen, D.S., Wheeler, A.H.F.: Tracing patterns and attention: Humanoid robot cognition. IEEE Intell. Syst. Mag. **15**, 70–75 (2000)
2. Gabaix, X., Laibson, D.I., Moloche, G., Weinberg, S.E.: The allocation of attention: theory and evidence. MIT Department of Economics Working Paper No. 03–31 (2003)
3. Admoni, H., Dragan, A., Srinivasa, S.S., Scassellati, B.: Deliberate delays during robot-to-human handovers improve compliance with gaze communication, pp. 49–56 (2014)
4. Lin, Y., Zhang, W.J., Wu, C., Yang, G., Dy, J.: A fuzzy logics clustering approach to computing human attention allocation using eyegaze movementcue. Int. J. Hum. Comput. Stud. **67**(5), 455–463 (2009)
5. Brown, R., Pham, B., Maeder, A.: Visual importance-biased image synthesis animation. In: Proceedings of the 1st International Conference on Computer Graphics and Interactive Techniques, New York, pp. 63–70 (2003)

6. Wu, C.: Extracting performance-related attention pattern for human-computer interaction. J. Comput. Inf. Syst. 10(17) (2014)
7. Harr, R.: Cognitive Science. Sage Publications, New Jersey (2002)
8. Wu, C., Lin, Y., Zhang, W.J.: Human attention modeling in a human-machine interface based on the incorporation of contextual features in a Bayesian network. In: Proceedings - IEEE International Conference on Systems, Man and Cybernetics, vol. 1, pp. 760–766 (2005)
9. Horvitz, E., Kadie, C., Paek, T., Hovel, D.: Models of attention in computing and communication: from principles to applications. Commun. ACM 46(3), 52–59 (2003)
10. Hoque, M.M., Onuki, T., Kobayashi, Y., Kuno, Y.: Effect of robots gaze behaviors for attracting and controlling human attention. Adv. Robot. 27(11), 813–829 (2013)
11. Erkelens, C.J., Vogels, I.: The initial direction and landing position of saccades. In: Findlay, J.M., Walker, R., Kentridge, R.W. (eds.) Eye Movement Research: Mechanisms, Processes, and Applications, pp. 133–144. Elsevier Science Publishing, New York (1995)
12. Asteriadis, S., Karpouzis, K., Kollias, S.: Visual focus of attention in non-calibrated environments using gaze estimation. Int. J. Comput. Vis. 107(3), 293–316 (2014)
13. Sorostinean, M., Ferland, F., Dang, T.-H.-H., Tapus, A.: Motion-oriented attention for a social gaze robot behavior. In: Beetz, M., Johnston, B., Williams, M.-A. (eds.) ICSR 2014. LNCS, vol. 8755, pp. 310–319. Springer, Heidelberg (2014)
14. Schilling, H.E.H., Rayner, K., Chumbley, J.I.: Comparing naming, lexical decision, and eye fixation times: word frequency effects and individual differences. Mem. Cognit. 26, 1270–1281 (1998)
15. Pomarjanschi, L., Dorr, M., Bex, P.J., Barth, E.: Simple gaze-contingent cues guide eye movements in a realistic driving simulator, vol. 8651 (2013)
16. Pejsa, T., Andrist, S., Gleicher, M., Mutlu, B.: Gaze and attention management for embodied conversational agents. ACM Trans. Interact. Intell. Syst. 5, 3 (2015)
17. Suppes, P.: Eye-movement models for arithmetic and reading performance. In: Kowler, E. (ed.) Eye Movements and their Role in Visual and Cognitive Processes, pp. 455–477. Elsevier Science Publishing, New York (1990)
18. Hegarty, M., Mayer, R.E., Green, C.E.: Comprehension of arithmetic word problems: evidence from students eye fixations. J. Educ. Psychol. 84, 76–84 (1992)
19. Bampatzia, S., Vouloutsi, V., Grechuta, K., Lallée, S., Verschure, P.F.M.J.: Effects of gaze synchronization in human-robot interaction. In: Duff, A., Lepora, N.F., Mura, A., Prescott, T.J., Verschure, P.F.M.J. (eds.) Living Machines 2014. LNCS, vol. 8608, pp. 370–373. Springer, Heidelberg (2014)
20. Michael, E.J., Kim, J.V.: Attention allocation within the abstraction hierarchy. Int. J. Hum. Comput. Stud. 48, 521–545 (1992)

Research of Constructing 3D Digital River Based on ArcEngine and SketchUp

Xueyan Tang[✉], Yuansheng Lou, Feng Ye, and Xiaorong Yan

College of Computer and Information, Hohai University, Nanjing, China
1540013226@qq.com

Abstract. With the development of the 3D GIS and the digital basin, it is important for us to deal with the rivers from the 3D space. Compared to traditional forms of 2D GIS, the 3D GIS system brings better user experience for us. After introducing the current development status of the 3D digital river, the paper proposes a novel method for constructing 3D digital river system. It explores the methods in 3D digital river modeling by SketchUp, driving 3D scene by ArcScene and secondary developing by ArcGIS Engine. It discusses the specific issues of 3D river modeling with a river in Nanjing, and sums up a complete 3D modeling of operational process. Finally, that proves the feasibility of the method.

Keywords: 3D digital river · Arcscene · Sketchup · Arcengine · 3D modeling

1 Introduction

3S [1] (GIS, RS, GPS) technology with the development of visualization techniques has also been a great development. Due to the continuous development of 3S technology and computer technology, which have been more mature, people have proposed the idea of a digital implementation of national, digital city. Digital concept has become firmly rooted in the hearts of people. In recent years, various industry areas start research on digital technologies, and the water conservation industry is no exception, which has already obtained a lot of achievements for digital river basin. In numerous studies in the field of digital river basin, 3D visualization technology is a key research direction and also a foundation of constructing digital river basin visualized information platform. Thus, it is necessary for us to combine 3D GIS [2] technology and river management. However, there are still many constraints during the process in the construction of 3D river information systems: (1) the data used to build a model of a 3D scene need high production costs but get low productivity; (2) the limited capacity of the 3D spatial analysis functions, the lack of high level development of functional analysis, especially for 3D space applications are the biggest limitations. Thus, it is a beneficial experiment for us to combine SketchUp and ArcGIS to realize the coupling of 3D rivers scene visualization and 3D object management.

This paper took Chu River as an example, making a research on using SketchUp to realize modeling rapidly, and using Arc Scene [3] to drive the whole 3D scene, developing a 3D digital visualization system to achieve the 3D display and roaming, property

© Springer Science+Business Media Singapore 2016
W. Che et al. (Eds.): ICYCSEE 2016, Part I, CCIS 623, pp. 568–576, 2016.
DOI: 10.1007/978-981-10-2053-7_50

query and other functions by the Arc GIS Engine [4], and also providing a better river management basic information platform for the government.

The organization of this paper is as follows. In the Sect. 2, related work is introduced, including the advantages and disadvantages the existing works in the 3D visualization field. In the Sect. 3, we propose the overall design which described the technology roadmap and key technologies. The Sect. 4, secondary development is introduced. This part mainly introduces ArcEngine—advanced components, and the main function of this system. The last section is use case introduction, took Chu River as an example and described the result of the development.

2 Related Work and Key Technologies

2.1 Related Work

Recently, the research of 3D GIS has made great progress. In terms of 3D GIS a lot of foreign scholars have made a larger number of researches. Kofler use R-tree to realize the 3D GIS database management and visualization capabilities [5]. Goktas HH and so on to achieve a 3D virtual city automatic simulation system based on the Maps [6]. Shi, W.Z., C.K. Fung, B.S. Q.Q.Li and others made intensive studies on the 3D data model and structure optimization from the perspective of the data structure of view [7–9]. Guo Shan, Yu Kai, Yan Penglian [10], who utilized Arc GIS collaborated with Google SketchUp to convert the two-dimensional GIS data into Multipatch conveniently, and the Multipatch data can be read and write directly by Arc GIS through the way of the combining the ESRI plug with Arc GIS, modeling the Shape data, however, this conversion method can be used only for ArcGIS9 series, does not apply to 10 series. Sun He, Feng ZhongKe, Wang Haiping, Liying [11], who utilize SketchUp modeling software, to achieve the single wooden 3D visualization in Arc-GIS platforms, combined with remote sensing image and DEM to realize the 3D visualization expression of all the trees. On the base of 2D data drawing by Auto CAD accurately, Wuzhao Yan, Tang Mengping [12] use SketchUp ESRI plug-in between SketchUp and ArcGIS to convert the shape file into.skp format which can be identified by SketchUp, in the ArcScene, superimposed the forest scene, small lines and 3D feature model to realize the function of roaming, query, data management and spatial analysis so that the forests landscape can be visualized.

Above-mentioned methods realized the 3D description for features and buildings, the texture of the building can also be meticulous expressed and the 3D scene can also be driven. However, due to the complexity during the process of 3D modeling, difficulty for operating, and also requires a lot of manpower, material, the method of combining ArcGIS and SketchUp is necessary, which can quickly and easily create, modify and render the 3D creativity so that can achieve the establishment of 3D model and the use of visual space ArcEngine platform to develop a GIS platform-independent 3D digital applications.

2.2 Key Technologies

(1) SketchUp [13]. SketchUp is a humane, intelligent 3D design software, which possesses the characteristics of a simple interface and easy to learn. Using SketchUp to conduct 3D modeling is just like painting with a pencil on the drawing, modeling process is simple, drawing lines into the surface, sliding into the body, and we can quickly build and edit 3D building models. In addition to the creation of 3D model, SketchUp also has its own powerful mapping capabilities, which can be used for texture mapping function of rendering 3D model built, landscaping, create a 3D model with a strong sense of reality.

(2) ArcScene. ArcScene is using the terrain data to display 3D feature data, analysis and operate the related data, in ArcScene environment, a variety of data need to be 3D and realize the interaction between the vector and raster data. ArcScene is a platform which based on OpenGL to realize the display of a variety of topographic and non-topographic data by analysis and operating. In ArcScene environment of terrain on 3D display, it can be loaded directly to the platform and make it to be 3D, and the image data is the same, where image data can be reduced to enhance the efficiency of the computer through the image resolution.

(3) ArcEngine. ESRI's ArcGIS Engine is based on the original version ArcGIS 9 launched, is a simple and independent senior GIS component, the main component of a software development kit ArcGIS Engine Developer kit [14] and a run-time Arc Engine Runtime [15], applying the both components and tools can provide not only the function of two-dimensional graphics, and can achieve a 3D display, information and advanced features such as 3D analysis.

3 The Proposed Architecture and Overall Design

3.1 The Architecture Proposed

The system is divided into four layers, user layer, business logic layer, data services layer and the support layer (Fig. 1): user layer distribute the operating authority by system administrators and ordinary users; business logic using Visual Studio 2010 platform and ArcGIS Engine Component Development Technology complete development system functional modules; data services layer data indicates the organization and management model, using File Geodatabase and MySQL database distributed storage; support layer is the system set up hardware and software environment.

3.2 Overall Design of 3D River

The basic process of constructing a 3D digital river is as follows.

(1) Pre-processing and blocking the existing CAD data;
(2) Importing the CAD data which have already been dealt to SketchUp and using its convenience and efficiency to achieve a 3D scene modeling and texturing the surface of the model image;
(3) Transferring the 3D model into Multipatch (* mdb) format, and adding the 3D model in mdb format into ArcScene to realize a 3D digital visualization river;

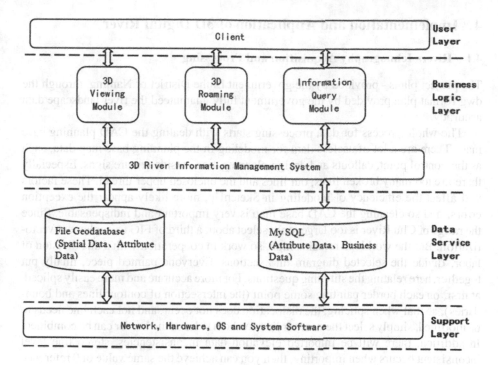

Fig. 1. The architecture proposed

(4) Utilizing the ArcGIS Engine Development Kit, combined visual programming language C# to conduct a secondary development by obtaining the required GIS functionality, and a customized GIS application can be constructed. Figure 2 showed the progress.

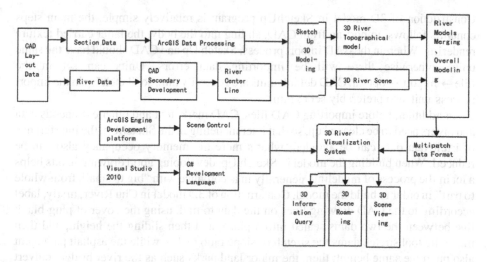

Fig. 2. Technology roadmap

4   Implementation and Application of 3D Digital River

4.1   Rivers Ichnography Acquisition and Processing

The project plans - provided by the government Liuhe District of Nanjing, through the dwg format plan provided by the government fully guaranteed the river landscape data accuracy.

The whole process for data processing starts with dealing the CAD planning base map. There are a lot of useless data for modeling in the planning base map data, such as the control point, callouts and design planning area outside feature signs. Especially there are too many broken lines, out lines and the unclosed upper thread. These factors will affect the efficiency of modeling in SketchUp, more likely appear the exception errors, and so cleaning the CAD base map is very important and indispensable. Since the range of Chu River is too large, only select about a third of FIG to implement vectorization. But the workload remains high, so work in cooperation with a due division of labor, divide the selected diagram into sections. Everyone painted piece, finally put together, here relating the stitching questions. For more accurate and more easily spliced, at first, on each border painting some point (the intersection of contour lines and boundaries), so that when splicing, the dislocation does not occur, and not each line needs to be modified, simply select the contour lines on both sides of the border can be combined. In addition, there will be imported elevation data inconsistencies, stereo line level inconsistent occurs when importing, then you can achieve the same value of 0 reference plane by a unified set of z.

Secondly, the texture data processing, usually by high-resolution digital photo camera taking photos, then the feature texture information obtained after the adoption of Photoshop and other image processing software.

4.2   3D Modeling

Construction of 3D model in SketchUp program is relatively simple, the main steps contains follows, importing from CAD, sliding into the body, the latter carried texture rendering. Wherein the CAD import process, Sketch Up and CAD should have the same units, otherwise there will be importing data errors. Units can set in the File → import → option. CAD default units are generally mm, so Sketch Up in the import process unit also preferably set to mm.

In addition, before importing CAD files, CAD mark, text and other extraneous data and layers need to be cleaned up, so that later modeling in the SketchUp, the interference of irrelevant data can be avoided; what's more the memory occupancy also can be reduced. When building the model in Sketch Up, developing good drawing habits helps a lot in the process of modeling, generally in accordance with "big to small, from whole to part" in order to build the model. Construction of 3D model in Chu River, firstly, label according to the CAD drawing lines on the dam to find, using the cover of plug-blank line between the two dams sealed into a plane, and then sliding the height, and then move the tool stretch dams the stretch of slope ratio of 1:3, while the asphalt pavement also push the same height; then, the major landmarks such as the river bridge, culvert and drainage station so through SketchUp produced; Finally, add the appropriate texture. Drainage station rendering and sluice are shown in Fig. 3.

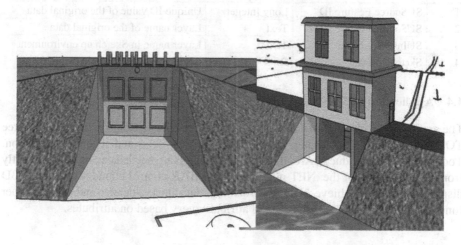

Fig. 3. Chu river drainage station rendering and sluice model

4.3 3D Scene Building

After completing the river's modeling, convert the model format to Arc GIS supported polyhedral model Multipatch format and to add some attribute information for the models. By using ArcGIS tools, import SketchUp model Multipatch data. First a Personal Geodatabase database need to be created in ArcCatalog module, while establishing a Multipatch of Feature Class in its database and set the layer essential fields (Table 1), and then by Arc GIS software tools Import3Dfile import SketchUp to implement Multipatch conversion. In this process which needs to be noted is that after the model importing into ArcScene, the original texture mapping may be lost. In ArcScene, opening the 3D Editor /start editing /selected to replace an object /3D Editor /replace with model /selecting model exported in the SketchUp, replace the original model, stop editing, refresh, textures will appear. Landscape on the river like the bridge, culverts, drainage sluice station as well as some sections mark places will be as style symbol library importing into ArcScene. The specific ideas of the SketchUp 3D model to be symbolic are as follows: First, the 3D model skp format stored in the file. Then, in the corresponding coordinate point of Arc GIS model add point elements, and add the name of the field and the bottom field as well as the high point feature. In the file list box, click the point feature need to be modified, click the Properties button in the Type drop-down list box, select the 3D Marker Symbol option and the desired 3D model skp format symbolic. Click Set Actual Size button, makes the model in line with the actual size of the space coordinates. Dimensions and Rotate axis adjustment parameters to offset model, rotation, direction and GIS consistent 3D scene. This method of importing the model can be more realistic, but must be independent of the imported 3D model. After all models are arranged in place, it will eventually save the modified data model file sxd format.

Table 1. Feature class necessary layer fields

NO.	Field name	Field type	Remark
1	SUSource Feature ID	Long Integer	Unique ID value of the original data
2	SUSource Feature Class	Text	Layer name of the original data
3	SUInstance Name	Text	Layer name in Sketch up environment
4	SketchUp Data	Blob	model field saved into Sketch Up

4.4 Application of 3D Scene Control

The application's main interface uses Arc GIS control Scene Control, the directory tree TOCControl control to display all the layers in the scene and legend information. Toolbar directly introduced ArcGIS tools, tools which are needed can be added directly from Arc GIS tools of the .NET platform. TOOLBAR control bound scenecontrol 3D display window to achieve 3D scene roaming, data query, delete, and modify other functions, zoom, browse attributes, and attribute query based on attributes.

4.5 Main Functions

The main function of the system is divided into two parts: a 3D scene browsing module, attributes query module, capable of two-dimensional plan of Chu River and 3D scene display, having a 3D landscape and dynamic observation flight, support for marquee zoom in and out, in order to from global and local two different angles were better browsing scene, and scenes can be exported in the form of pictures; providing information query, the system has a certain practicality.

(1) Realization of 3D scene browsing. 3D viewing capabilities give users an immersive feeling of the river, sitting in the boat in the river swimming sensation, so by a 3D model of the unmanned aircraft, make it roam in the river. In this paper, through the ArcGIS Engine development components, programming related procedures to make the 3D model in 3D scene roam along a specified path, in order to achieve the function of the river cruise.

(2) Attributes Query. The display of 3D scene can provide people with real intuitive feel, through features such as zoom and pan functions can view detailed scenes, but for the river section attribute information, such as the section name, location, river width, design and so on, this information cannot be directly displayed in the form of scene. Therefore, in order to combine 3D scene display, a better understanding of Chu River, we need to provide information query. Depending on the river landscape objects, primarily provides information section includes some attribute name, location, river width, rainfall and water level of the current query.

4.6 Application in Chu River

Study area is Liuhe Chu River, as an important tributary of the Yangtze River section of "Golden Waterway", flows through the ancient capital of Nanjing North Portal - Liuhe District. The water in Chu River is very rich, flourishing industry and agriculture along the river, a huge potential for development. However, since the Liuhe locate in the Yangtze River, Chu River downstream, when in the flood season, Chu River bears over 6600 km² water from upstream, midstream runoff, large flood came, and by Koko top barrier, flood attack, known as "flood corridor". In this applications, use Liuhe WCB as the background, by constructing "smart Chu River" as a typical demonstration applications, to achieve the goal of "perception Chu River, reproduce Chu River, decision Chu River, ecological Chu River", and then the constructing green trees, blue surrounded by flowers, and cultural heritage of lovable homes to reflect the advantages of wisdom city. Therefore, based on SketchUp quickly realize Chu River landscape visualization effect. SketchUp make good use of 3D renderings and import into the Arc Scene and ArcGIS Engine secondary developed the 3D digital river management system based on ArcGIS Engine platform. Staff may be performed by a network or a touch screen and rotate 3D perspective Chu River roaming, and information inquiry, the river of government staff to facilitate maintenance work (Fig. 4).

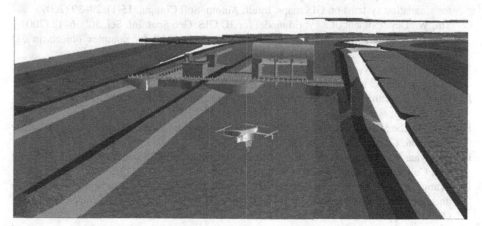

Fig. 4. Interface of 3D Chu river

5 Conclusion

This paper utilize professional SketchUp to model and use Ruby Script Editor mass model, using existing arctoolbox tool to convert model data, and import the mass model into arcscene platform, then build an application which is independent of ArcGIS platform based the development platform. This application achieves the display of scene and preliminary 3D analysis, but how to analysis the huge amount of data and fine 3D display are for further study.

Acknowledgement. This work is supported by (1) National Natural Science Foundation of China (61300122); (2) A Project Funded by the Priority Academic Program Development of Jiangsu Higher Education Institutions; (3) Water Science and Technology Project of Jiangsu Province (2013025).

References

1. Mu, Y., Gao, H.: Research on GIS 3D visualization based on ArcGIS and SketchUp. In: Zhong, Z. (ed.) Proceedings of the International Conference on Information Engineering and Applications (IEA) 2012. Lecture Notes in Electrical Engineering, vol. 216, pp. 493–500. Springer, Heidelberg (2013)
2. Wei, B.S., Liang, H., Wang, T.L.: 3D modeling and visualization of city drain-pipes network based on ArcGIS and sketchup. Adv. Mater. Res. **726–731**, 3030–3034 (2013)
3. Wang, L., Luo, L.: Research on 3D digital city based on 3DSMax and ArcEngine. Geomatics Spat. Inf. Technol. **4**, 138–140 (2015)
4. Yue, H.: Research and implementation of 3D visualization of drilling and geological body based on ArcGIS. Henan Sci. **3**, 435–439 (2015)
5. Michael, K., Michael, G., Michael, G.: R-trees for organizing and visualizing 3D GIS databases. J. Visual. Comput. Animation **11**(3), 129–143 (2000)
6. Göktaş, H.H., Çavuşoğlu, A., Şen, B.: Auto city: A system for generating 3D virtual cities for simulation systems on GIS maps. Intell. Autom. Soft Comput. **15**(1), 29–39 (2009)
7. Shi, W.: Development of a hybrid model for 3D GIS. Geo Spat. Inf. Sci. **3**(2), 6–12 (2004)
8. Shi, W., Yang, B., Li, Q.: An object-oriented data model for complex objects in 3D geographical information systems. Int. J. Geogr. Inf. Sci. **17**(5), 411–430 (2003)
9. Yang, B., Shi, W., Li, Q.: An integrated TIN and Grid method for constructing multi-resolution digital terrain models. Int. J. Geogr. Inf. Sci. **19**(10), 1019–1038 (2005)
10. Guo, Q.: Research of 3D GIS visualization by using google SketchUp. CitySurvey **6**, 51–53 (2010)
11. Sun, H.: The 3D visualization of campus trees based on ArcGIS and SketchUp. For. Inventory Plann. **06**, 17–20 (2011)
12. Zhaoyan, W.: The forest landscape visualization based on GIS and SketchUp. J. Zhejiang For. Coll. **29**(03), 352–358 (2012)
13. Wang, Q., Zhu, X.: The implementation of campus 3D electronic map based on SketchUp and ArcGIS. In: 2010 International Conference on Audio Language and Image Processing (ICALIP), pp. 1031–1034. IEEE (2010)
14. Wang, Q., Zheng, W.M., Zhao, X.M., et al.: Construction of virtual campus of gansu normal university for nationalities based on ArcGIS. Appl. Mech. Mater. **411–414**, 514–517 (2013)
15. Ma, J.M.: The design and realization of 3D virtual campus. Appl. Mech. Mater. **667**, 171–176 (2014)

Research of the DBN Algorithm
Based on Multi-innovation Theory
and Application of Social Computing

Pinle Qin[1(✉)], Meng Li[1], Qiguang Miao[2], and Chuanpeng Li[1]

[1] School of Computer and Control Engineering,
North University of China, Taiyuan 030051, China
QPL@nuc.edu.cn
[2] School of Computer Science and Technology,
Xidian University, Xian 710075, China

Abstract. Aimed at the problems of small gradient, low learning rate, slow convergence error when the DBN using back-propagation process to fix the network connection weight and bias, proposing a new algorithm that combines with multi-innovation theory to improve standard DBN algorithm, that is the multi-innovation DBN(MI-DBN). It sets up a new model of back-propagation process in DBN algorithm, making the use of single innovation in previous algorithm extend to the use of innovation of the preceding multiple period, thus increasing convergence rate of error largely. To study the application of the algorithm in the social computing, and recognize the meaningful information about the handwritten numbers in social networking images. This paper compares MI-DBN algorithm with other representative classifiers through experiments. The result shows that MI-DBN algorithm, comparing with other representative classifiers, has a faster convergence rate and a smaller error for MNIST dataset recognition. And handwritten numbers on the image also have a precise degree of recognition.

Keywords: DBN algorithm · Convergence error · Multi-innovation theory · MI-DBN algorithm · Social computing

1 Introduction

With the accelerating progress of information digitization and information networking, humans' behavior are recorded more frequently. As a result, it's possible to observe and research the society by computer technology. Computational Social Science is growing up and people will automatically collect and utilize information, the depth and scope of which has never seen before, to serve the social science research. Social computing [1] refers to such a research field combined with computing systems and social behavior, which is investigated how to use computing systems to help people communicate and collaborate, and how to study the rules and trends of social functioning by utilizing computing technology. Social networking services, collective wisdom, social network analysis, content computing and artificial society are included in this kind of research. The issues of content computing consists of public opinion

© Springer Science+Business Media Singapore 2016
W. Che et al. (Eds.): ICYCSEE 2016, Part I, CCIS 623, pp. 577–590, 2016.
DOI: 10.1007/978-981-10-2053-7_51

analysis, interpersonal relationships excavation and weibo applications. Microblog simultaneously has the properties of social network and media platform, which triggers the revolution of information production and propagation mode. So images with numbers such as phone numbers and geographic coordinates may include important and useful information. We accurately and quickly recognize the numbers by means of deep learning [2].

Deep learning has achieved great breakthroughs and success in the recent 10 years in many fields such as artificial intelligence, speech recognition, natural language processing, computer vision, Image and visual analysis, multimedia and so on. There are many kinds of models in Deep Learning. One of them, called Deep Belief Networks (DBN) [3, 4], which is an unsupervised algorithm, can create layers of feature detection without requiring labelled data. And the network layers can be used for the reconfig-uration and the modeling for the movement of feature detection. During the process of the pre-training, the weights of a deep network can be initialized to meaningful values. After that, a final layer of output units will be added to the top of the network and the whole deep system could be using standard back propagation algorithm. All this operations will be prominently effective in handwritten digit recognition.

Hinton presented DBNs and used it in the task of digit recognition on MNIST dataset [3]. He put forward the DBN network which takes "784-500-500-2000-10" as its structure. Based on the images with 28*28 pixel resolution in MNIST dataset, we can get 784 features in the first layer of the network. The last layer is related to 10 digit labels and other three layers are hidden layers with stochastic binary neurons. Finally this paper achieved 1.25 % classification error rate on MNIST test dataset. Keyvanrad [5] improved the sampling method of Restricted Boltzmann Machines (RBM) based on the DBN proposed by Hinton and get a 1.11 % classification error rate by changing the original Contrastive Divergence (CD) algorithm to Free Energy in Persistent Con-trastive Divergence (FEPCD) algorithm. Liu [6] put forward a new categorizer based on DBN, which is the Discriminative Deep Belief Network (DDBN), which integrates the abstract ability of DBN with the resolution capability of back propagation strategy.

Hinton used DBN as a nonlinear model for feature extraction and dimension reduction [7]. Indeed the DBN may be considered as a model that can generate features in its last layer with the ability to reconstruct visible data from generated features. When a general Neural Network is used with many layers, the Neural Network becomes trapped in local minima and the performance will decrease. As a result, it is vital to ensure the initial weights of the nerve net.

According to the multi-innovation identification theory proposed by Ding [8, 9], this paper puts forward a new algorithm–DBN algorithm based on multi-innovation theory. Compared with the previous one, this algorithm accelerates the convergence of errors and increase the accuracy of numerical recognition. The rest of this paper is organized as follows: Sect. 2, Introduce the procedure of original DBN algorithm; Sect. 3, Introduce the multi-innovation identification theory and the improvement of original DBN algorithm; Sect. 4, Introduce the MNIST dataset and the image of the weibo, and test the error date; Sect. 5, Conclude this paper.

2   DBN

DBN is a probability generation model, which is compared with the traditional model of neural network discrimination, and generation model is to establish a joint distribution between the observed data and labels. DBN network structure is composed of several layers of RBM [10] which each layer of output as the input vector to the next layer. The DBN of connection is determined by top-down generation weights, RBM is more easy to connect the weight of learning [11]. DBN stratified training, a tag set is attached to the top. Through a bottom-up, the learned identify the right values obtained a network classifier. Using back propagation technique and using the entire classification equipment adjust the weights optimization classification. Structure is as shown in Fig. 1:

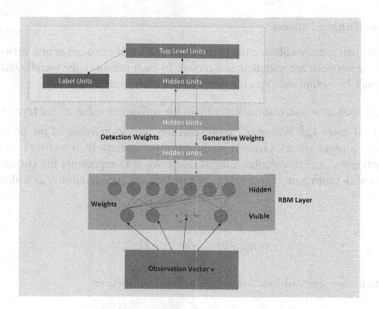

Fig. 1. DBN network structure

The feature extraction of DBN network model requires three process. They are pre-training process, fine-tuning process and the testing process.

2.1   Pre-training Process

Pre-training process is that after extracting input data characteristics through each layer of the network obtain excitation response and add it to the output layer. The specific process is as follows:

(1) First pre-propagation and single unsupervised training of each layer of RBM network. Make sure the feature vector mapped to different characteristics space. Feature information are preserved as much as possible.

(2) The last layer of DBN set a classifier, receiving an output feature vector of RBM as its input feature vector. It supervised to train the entity relationship classifiers. And each layer of RBM network can confirm the weight of its own layer, on the layer of the feature vector to achieve the best, not to the entire DBN feature vector mapping to achieve optimal. Back propagation will transmits the wrong message to every layer of RBM from the top-down, so fine-tune the entire DBN network. The process of RBM network training model can be seen as initialization of weights parameter for logistic regression network. So DBN overcame the short-comings of logistic regression network that the random initialization parameter weights are easy to fall into the disadvantaged of local optimum and long training time.

2.2 Fine-Tuning Process

Fine-tuning can greatly enhance the performance of a self-encoded neural network. All layers of the network are regarded as a model. In each iteration, the weight and bias of network will be optimized. Specific process is as follows:

(1) Pre-propagation calculations can obtain the activation value $a_i^{(l)}$ of layer L_2, layer L_3 until layer L_{n_l}. Where $a_i^{(l)}$ represents the activation value of the unit i of the layer l (output value). Given a given set of parameters W, b, network can follow function $h_{W,b}(x)$ to calculate output result. $h_{W,b}(x)$ represents the corresponding desired output result. Calculating step of forward propagation is as follows:

$$z^{(l+1)} = W^{(l)} a^{(l)} + b^{(l)} \tag{1}$$

$$h_{W,b}(x) = a^{(l+1)} = f\left(z^{(l+1)}\right) \tag{2}$$

Which uses sigmoid function as the activation function:

$$f(z) = sigmoid(z) = \frac{1}{1 + e^{-z}} \tag{3}$$

(2) The error function of each output unit can be calculated
(3) The error function of each layer can be calculated
(4) According to the references [12, 13], we know that the update of parameters W and b can be written as:

$$W_{ij}^{(l)} = W_{ij}^{(l)} - \alpha a_j^{(l)} \delta_i^{(l+1)} \tag{4}$$

$$b_i^{(l)} = b_i^{(l)} - \alpha \delta_i^{(l+1)} \tag{5}$$

2.3   Testing Process

After the pre-training and fine-tuning process, DBN network model has extracted feature layer by layer. At the same time parameters W and b have been optimized. By testing sample of dataset we can test DBN network.

3   DBN Algorithm Based on Multi-innovation Theory

This paper introduces the multi-innovation theory aiming at the situation that standard DBN algorithm mentioned in the paper only uses the current data and loses all the previous epochs data when updating the weight and bias of the parameters. Multi-innovation identification has developed a new field of identification. This method has been widely applied in all kinds of estimations of model parameters because it can improve the fall rate of the gradient.

3.1   Multi-innovation Theory

Common identification algorithm is usually single innovation identification methods which only use single innovation to modify the parameters. [14] As for scalar system: [15]

$$y(t) = \varphi^T(t)\theta + v(t) \tag{6}$$

Where $y(t)$ is the output data of the system, $\varphi^T(t)$ is a vector combined by the output data and input data, θ is the final recognized parameter which is a vector, $v(t)$ is the random noise in system with the zero average, which shows the existence of noises in system.

As for parameter vector θ, using the stochastic gradient identification algorithm has the following formula:

$$\hat{\theta}(t) = \hat{\theta}(t-1) + L(t)e(t) \tag{7}$$

Where $L(t)$ is a gain vector, $e(t)$ is used to describe the output of the error at time t, $e(t)$ can be showed as:

$$e(t) := y(t) - \varphi^T(t)\hat{\theta}(t-1) \tag{8}$$

Where $\hat{\theta}(t)$, modifying the identity parameter at time $t-1$ by multiplying gain vector $L(t)$ by scalar innovation $e(t)$, is the estimation at time t based on the recurrence of $\hat{\theta}(t-1)$.

Developing it, according to the formulas (21) and (22), extending the scalar innovation $e(t)$ to a form of vector. That is multi-innovation. Simultaneously, the other matrix and vector will change in system. To extend the form of gain vector $L(t)$ to matrix. So we get the multi-innovation identification. It can write as:

$$\widehat{\theta}(t) = \widehat{\theta}(t-1) + \Gamma(p,t)E(p,t) \tag{9}$$

$\Gamma(p,t)$ is gain matrix, $E(p,t)$ is innovation vector, p $(p \geq 1)$ is the length of the innovation. When the p is equaled with 1, it degenerated as a standard innovation.

Multi-innovation make full use of some of the useful information before, which can have better identification effect. This method increases the computation of the algorithm itself, but it can improve the convergence rate, the precision of the experiment in a small time complexity. The specific recurrence and demonstration of Multi-innovation theory can be found in the reference [16].

3.2 The DBN Algorithm Combined with Multi-innovation—MI-DBN

DBN makes the network reach a great initial point with unsupervised and trained layer by layer [17, 18] before using the back propagation algorithm to optimize the network globally. So it can reach a better local minimum point after training. There probably are thousands of the samples and weights in study system. The top layer is added with the labelled sample to train machine. In order to adjust the weight vector correctly, the gradient vector of each weight is calculated by the back-propagation algorithm, which indicates that the error will increase or decrease if the weight is increased by a very small value. Back propagation algorithm can be used to transmit the gradient through every layer of multilayered network repeatedly, from the output of the top layer to the lowest layer, the gradient vector of weight in every layer can be solved after working out the input derivative in every layer. The weight vector can be adjusted in the opposite direction of the gradient vector. The calculation of the stochastic gradient algorithm is small and the convergence rate is slow. In order to improve the convergence rate of the stochastic gradient identification method, the length of innovation is introduced. The weights updating every time are all based on the weight in last epoch in reverse fine-tuning of DBN. So it improves the convergence rate of the error.

If the weight of the network dynamic changes according to some certain law, the mapping relations between input and output can be changed with time when we identify the time-varying system, so it is possible to identify the time-varying system. Because of the difficulty to get the rules of time variation, the length of innovation p is the whole epoch of the network fine-tuning. Time t is the number of epochs. Define the sequence of positive integer $\{t_0, t_1, t_2, \ldots, t_s\}$, satisfy $0 < t_0 < t_1 < t_2 < \ldots < t_s$, and $1 \leq t_s^* = t_s - t_{s-1}$. In the distance of iteration t_s^* seconds, we probably get the useful data in layer l and node i during every $1/N$ s. The innovation vector $E(p,t)$ is produced to correct the parameter θ using the group of data p. At this time, from $t = t_s - p + 1$ to $t = t_s$, there are p groups of sample data, so the output of node j at time t in layer l which is a visual layer :

$$A_j^{(l)}(p, t_s) = \left[a_j^{(l)}(t_s), a_j^{(l)}(t_s - 1), a_j^{(l)}(t_s - 2), \ldots, a_j^{(l)}(t_s - p + 1) \right]^T \tag{10}$$

The expected output:

$$H_j^{(l)}(p,t_s) = \left[h_j^{(l)}(t_s), h_j^{(l)}(t_s-1), h_j^{(l)}(t_s-2), \ldots, h_j^{(l)}(t_s-p+1) \right]^T \quad (11)$$

The nolinear cost function:

$$J(W,b) = \frac{1}{2} \sum_{j=1}^{l} \left\| H_j^{(l)}(p,t_s) - Y_j^{(l)}(p,t_s) \right\|^2 \quad (12)$$

$Y_j^{(l)}(p,t_s)$ is the P output values of J node in the output layer. $H_j^{(l)}(p,t_s)$ is the corresponding expected output. When calculating output nodes, J will decline, following the rules of network traning, in accordance with gradient in every training cycle. The update of W and b in every iteration are as follows:

$$W_{ij}^{(l)} = W_{ij}^{(l)} - \alpha \frac{\partial J(W,b)}{\partial W_{ij}^{(l)}} \quad (13)$$

Where α is the learning rate. Parameter $W_{ij}^{(l)}$ represents the weights between the unit j of the layer l and the unit i of the layer $l+1$. For each node i of the layer l, we calculate the "residual" $\delta_i^{(l)}$, which shows that the amount of import that the node produces on the error of the final output value. The calculation of $\delta_i^{(l)}$ is based on the error of the weighted average of the layer $l+1$ node. These nodes use $a_i^{(l)}$ as the input value. $\delta_i^{(l)}$ has the following formula:

$$\frac{\partial J(W,b)}{\partial W_{ij}^{(l)}} = \frac{\partial J(W,b)}{\partial z_i^{(l+1)}} \frac{\partial z_i^{(l+1)}}{\partial W_{ij}^{(l)}} \quad (14)$$

It is seen by the formula (1),

$$z_i^{(l+1)} = \sum_{k=1}^{s_l} W_{ik}^{(l)} a_k^{(l)} + b_i^{(l)} \quad (15)$$

So there is that:

$$\frac{\partial z_i^{(l+1)}}{\partial W_{ij}^{(l)}} = a_j^{(l)} \quad (16)$$

Then, the formula (14) is extended to a vector as follows:

$$\frac{\partial J(W,b)}{\partial W_{ij}^{(l)}} = \frac{\partial J(W,b)}{\partial z_i^{(l+1)}(p,t_s)} A_j^{(l)}(p,t_s) \quad (17)$$

There are:

$$Z_j^{(l)}(p, t_s) = \left[z_j^{(l)}(t_s), z_j^{(l)}(t_s - 1), z_j^{(l)}(t_s - 2), \ldots, z_j^{(l)}(t_s - p + 1) \right] \tag{18}$$

The back propagation of error in the output layer nodes:

$$\delta_i^{(l)} := \frac{\partial J(W, b)}{\partial z_i^{(l)}} \tag{19}$$

The formula (19) is extended to a vector

$$\delta_i^{(l)}(p, t_s) = \frac{\partial J(W, b)}{\partial Z_i^{(l+1)}(p, t_s)} \tag{20}$$

(1) Based on the inference [19], in every the unit of output i error formula can be derived as:

$$\delta_i^{(n_l)} = \frac{\partial J(W, b; x, y)}{\partial z_i^{(n_l)}} = -(y_i - a_i^{(n_l)}) \cdot f'(z_i^{(n_l)}) \tag{21}$$

Formula (3) on both sides at the same time derivative as:

$$f'(z_i^{(n_l)}) = f(z_i^{(n_l)})[1 - f(z_i^{(n_l)})] = y_i \cdot (1 - y_i) \tag{22}$$

Combination (21) and (22)

$$\delta_i^{(n_l)} = \frac{\partial J(W, b)}{\partial z_i^{(n_l)}} = -(y_i - a_i^{(n_l)}) \cdot y_i \cdot (1 - y_i) \tag{23}$$

The formula (23) is extended to a vector

$$\overline{\delta}_i^{(n_l)}(p, t_s) = \frac{\partial J(W, b)}{\partial Z_i^{(n_l+1)}(p, t_s)} = -(Y_j^{(n_l)}(p, t_s) - A_j^{(n_l)}(p, t_s)) \cdot Y_j^{(n_l)} \cdot (1 - Y_j^{(n_l)}) \tag{24}$$

There are

$$\overline{\delta}_i^{(n_l)}(p, t_s) = \left[\delta_i^{(n_l)}(t_s), \delta_i^{(n_l)}(t_s - 1), \delta_i^{(n_l)}(t_s - 2), \ldots, \delta_i^{(n_l)}(t_s - p + 1) \right] \tag{25}$$

Weight of the output layer

$$W_{ij}^{(n_l)}(t_s) = W_{ij}^{(n_l)}(t_s - 1) - \alpha A_j^{(n_l)}(p, t_s)\overline{\delta}_i^{(n_l)}(p, t_s) \tag{26}$$

(2) According to the inference of reference [19], layer l $(l = n_l - 1, n_l - 2, n_l - 3, \ldots, 2)$ can be known. The deviation formulation of the i node in layer l can be derived as:

$$\delta_i^{(l)} = \left(\sum_{j=1}^{s_{l+1}} W_{ji}^{(l)} \delta_j^{(l+1)}\right) \cdot f'(z_i^{(l)}) \tag{27}$$

The same as formula (22):

$$f'(z_i^{(l)}) = a_i^{(l)} \cdot (1 - a_i^{(l)}) \tag{28}$$

Referring to the formulas (26) and (27), the deviation function can be extended to the vectors:

$$\overline{\delta}_i^{(l)}(p, t_s) = \left(\sum_{j=1}^{s_{l+1}} W_{ji}^{(l)} \delta_j^{(l+1)}\right) \cdot A_i^{(l)}(p, t_s)[1 - A_i^{(l)}(p, t_s)] \tag{29}$$

So the weight of each hidden layer is corrected:

$$W_{ij}^{(l)}(t_s) = W_{ij}^{(l)}(t_s - 1) - \alpha A_j^{(l)}(p, t_s)\overline{\delta}_i^{(l+1)}(p, t_s) \tag{30}$$

(3) Just like the process of correcting and derivating the weight, we can deduce the bias correction function in output layer

$$b_i^{(n_l)}(t_s) = b_i^{(n_l)}(t_s - 1) - \alpha\overline{\delta}_i^{(n_l)}(p, t_s) \tag{31}$$

The bias correction function of each hidden layer:

$$b_i^{(l)}(t_s) = b_i^{(l)}(t_s - 1) - \alpha\overline{\delta}_i^{(l+1)}(p, t_s) \tag{32}$$

Compared with the standard DBN, advantages of the MI-DBN mentioned in this paper are as follows:

(a) When updating the weight and bias, standard DBN just uses the innovation of current epoch while the MI-DBN mentioned in this paper also uses the innovation of several past epochs, which increases the convergence rate of deviation.

(b) Adding multi-innovation to standard DBN can accelerate the convergence of deviation and increase the accuracy of the algorithm. However, adding too much innovation will increase the calculation quantity and decrease the real-timing instead of making the algorithm better. As a result, the amount of innovation should be considered.

4 Experiment and Analysis

In this paper, the improved algorithm is applied to the MNIST handwritten dataset [20]. And the result of the experiment can be achieve by using the toolbox-DeeBNet toolbox [21] in deep learning.

4.1 Experimental Environment

Experiment of software environment: Windows 7 system, VisualStudio2013, Matable2013.Hardware platform: 2.80 GHz Intel (R) CPU E5-2680 V2, 32 GB of memory.

4.2 MNIST Dataset

In our experiments, the used data is based on digital MNIST handwritten dataset. The image pixels have discrete values between 0 and 255 that most of them have the values at the edge of this interval [22]. The image pixel values were normalized between 0 and 1. The dataset was divided to train and test parts including 60,000 and 10,000 images respectively. As shown in Fig. 2:

Fig. 2. MNIST dataset sample

The judging standard involved in this paper are mainly in the following aspects:

(1) Error rate

Error rate of index is mainly used to measure different classification model for MNIST dataset the degree of error classification. The index is an important indicator used to measure the degree of classification error and the direct relationship between the reliability of the algorithm.

Table 1 shows the different algorithms for MNIST dataset classification error rate [4], a bold data is MI-DBN algorithm of MNIST dataset after the classification error rate. It can be seen that the standard DBN algorithm and two DBN improved algorithm in 60000 sample training, 10000 sample testing, 50 epochs training, 150 epochs testing, compared with a much smaller error rate.

Table 1. Classification error rate of MNIST dataset with different classification algorithms

Classification algorithm	Labels	Error rate (%)
SVM	10	1.4
KNN	10	1.6
DBN	10	1.24
FEPCD-DBN	10	1.11
MI-DBN	**10**	**0.54**

(2) Error convergence curve

The simulation data of the experiment error, in the process of experimental training epochs of 50, fine-tuning the epochs of 150. The following chart is the error convergence analysis of different algorithms. As shown in Fig. 3:

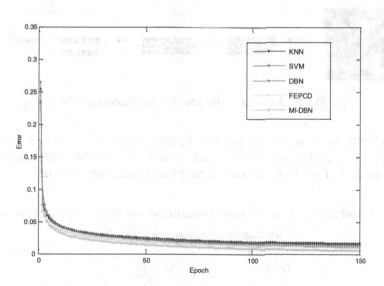

Fig. 3. Error convergence analysis of different algorithms

4.3   Image in Weibo Classification

In recent years, weibo has been developing rapidly. There is a quantity of information in weibo, and it is very hard to efficiently extract useful information we need from massive images in weibo. We created a dataset called WeiBoImage by ourselves. This dataset contains 1000 images downloaded from weibo. Some images of this dataset are as follows Fig. 4. We respectively use trained MI-DBN network and DBN network to extract and recognize handwritten numbers in images.

Fig. 4. Some images of WeiBoImage dataset

First we locate the image and extract information we need. Then we do pixel filling of the image and make the image a 28*28 pixels one after completing the binarization and segmentation of handwritten numbers. Afterwards, the images will be recognized in trained model of MI-DBN or DBN. Specific process is shown as the following Fig. 5:

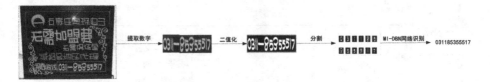

Fig. 5. An image of the specific identification process

The 28*28 image will be the input of the trained network model of MI-DBN and DBN, it is recognized by our trained model. Table 2 shows the MI-DBN algorithm and DBN algorithm of WeiBoImage dataset after the classification error rate.

Table 2. Classification error rate of WeiBoImage dataset with different classification algorithms

Classification algorithm	Error rate (%)
MI-DBN	3.85
DBN	4.5917

It can be concluded from the experiment that when WeiBoImage dataset is tested by network model trained by MI-DBN, the precision of the result is high and the loss of weibo image information can be solved quickly. This experiment, utilized as a plug-in in social competing software, applies computers to social network service. As a result, the research of this experiment is of great significance.

5 Conclusions

In this paper, based on the deficiency of DBN algorithm, we have presented a new Deep Belief Networks learning algorithm which combines with multi-innovation theory of stochastic gradient identification. Improved algorithm MI-DBN, utilizing historical epochs data, makes the best of useful information implied in the past data, and in this way, the connection weight and offset of the network can reach the predictive value faster. Simulation results show that the error rate of convergence is improved. In the process of MNIST handwritten dataset recognition, the error rate of classification is reduced after applying MI-DBN algorithm. We upload images on weiboa, and identified useful information from the images, we had obtained better effect. It is useful for us, because we will study the images of weibo where reflect the person's emotional state.

Acknowledgment. This work is partially supported by Shanxi Nature Foundation (No.2015011045). The authors also gratefully acknowledge the helpful comments and suggestions of the reviewers, which have improved the presentation.

References

1. Choi, B.: Multiagent social computing. Int. J. Web Portals **3**(4), 56–68 (2011)
2. Lecun, Y., Bengio, Y., Hinton, G.E., et al.: Deep learning. Nature **521**, 436–444 (2015)
3. Hinton, G.E., Osindero, S., Teh, Y.W.: A fast learning algorithm for deep belief nets. Neural Comput. **18**(7), 1527–1554 (2006)
4. Larochelle, H., Erhan, D., Courville, A., et al.: An empirical evaluation of deep architectures on problems with many factors of variation. In: ICML 2007: 2007 International Conference on Machine Learning, pp. 473–480 (2007)
5. Keyvanrad, M.A., Homayounpour, M.M.: Deep belief network training improvement using elite samples minimizing free energy. Int. J. Pattern Recogn. Artif. Intell. **29**(5), 1411–4046 (2014)
6. Liu, Y., Zhou, S., Chen, Q.: Discriminative deep belief networks for visual data classification. Pattern Recogn. **44**(10–11), 2287–2296 (2011)
7. Hinton, G.E., Salakhutdinov, R.: Reducing the dimensionality of data with neural networks. Science **313**(5786), 504–507 (2016)
8. Ding, F., Tao, X., Tao, D.: Multi-innovation stochastic gradient identification methods. Control Theory Appl. **20**(6), 870–874 (2003)
9. Ding, F.: Several multi-innovation identification methods. Digital Sig. Process. **20**(4), 1027–1039 (2010)
10. Hinton, G.E.: A practical guide to training restricted boltzmann machines. In: Montavon, G., Orr, G.B., Müller, K.-R. (eds.) Neural Networks: Tricks of the Trade, 2nd edn. LNCS, vol. 7700, pp. 599–619. Springer, Heidelberg (2012)

11. Swersky, K., Chen, B., Marlin, B., et al.: A tutorial on stochastic approximation algorithms for training Restricted Boltzmann Machines and Deep Belief Nets. In: Information Theory and Applications Workshop (ITA), pp. 1–10. IEEE (2010)

12. Schölkopf, B., Platt, J., Hofmann, T.: Greedy layer-wise training of deep networks. Adv. Neural Inform. Process. Syst. **19**, 153–160 (2007)

13. Sarikaya, R., Hinton, G.E., Deoras, A.: Application of deep belief networks for natural language understanding. IEEE/ACM Trans. Audio Speech Lang. Process. **22**(4), 778–784 (2014)

14. Ding, J., Xie, L., Ding, F.: Performance analysis of multi-innovation stochastic gradient identification for non- uniformly sampled systems. Control Decis. **26**(9), 1338–1342 (2011)

15. Ding, F.: Multi-innovation identification theory and methods. J. Nanjing Univ. Inform. Sci. Technol. **4**(1), 1–28 (2012)

16. Ding, F.: Convergence of multi innovation identification under attenuating excitation conditions for deterministic systems. J. Tsinghua Univ. **9**, 111–115 (1998)

17. Lee, H., Ekanadham, C., Ng, A.: Sparse deep belief net model for visual area V2. Adv. Neural Inform. Process. Syst. **20**, 873–880 (2008)

18. Krizhevsky, A.: Learning multiple layers of features from tiny images. University of Toronto, Canada, 17 (2009)

19. Salakhutdinov, R., Hinton, G.E.: Using deep belief nets to learn covariance kernels for gaussian processes. Adv. Neural Inform. Process. Syst. **20**, 1249–1256 (2007)

20. Lecun, Y., Cortes, C.: The MNIST database of handwritten digits [DB/OL] (2011). http://yann.lecun.com/exdb/mnist/index.html

21. Keyvanrad, M.A., Homayounpour, M.M.: A brief survey on deep belief networks and introducing a new object oriented MATLAB toolbox (DeeBNetV2.2). Comput. Vis. Pattern Recogn. **12**, 1408–3264 (2014)

22. Tieleman, T., Hinton, G.: Using fast weights to improve persistent contrastive divergence. In: ICML 2009: Proceedings of the 26th Annual International Conference on Machine Learning, Montreal, Quebec, Canada, pp. 1033–1040 (2009)

Research on Feature Fusion Technology of Fruit and Vegetable Image Recognition Based on SVM

Yanqing Wang[1(✉)], Yipu Wang[1,2], Chaoxia Shi[3], and Hui Shi[4]

[1] School of Computer Science and Technology,
Harbin University of Science and Technology, Harbin 150080, China
wyq0325@126.com
[2] ZhongJing FuDian (Shanghai) Electronic Technology Co., Ltd.,
Shanghai 201203, China
[3] School of Computer Science and Engineering,
Nanjing University of Science and Technology, Nanjing 210094, China
[4] Henan Electric Power Transmission and Transformation Engineering
Company, Zhengzhou 450051, China

Abstract. In order to improve the accuracy and stability of fruit and vegetable image recognition by single feature, this project proposed multi-feature fusion algorithms and SVM classification algorithms. This project not only introduces the Reproducing Kernel Hilbert space to improve the multi-feature compatibility and improve multi-feature fusion algorithm, but also introduces TPS transformation model in SVM classifier to improve the classification accuracy, real-time and robustness of integration feature. By using multi-feature fusion algorithms and SVM classification algorithms, experimental results show that we can recognize the common fruit and vegetable images efficiently and accurately.

Keywords: Feature extraction · Multi-feature fusion · Support vector machine · Fruit and vegetable image recognition

1 Introduction

With the development of technology in the field of pattern recognition, computer vision and image processing has become the hot point of research. Because the technology can identify various types of images automatically and get result precisely and efficiently. So domestic and foreign researchers conduct to identify fruit and vegetable mostly based on contour feature, color feature, texture feature and so on. When researchers identify fruit and vegetable images in uncontrolled lighting conditions and the context of complex environments, it will be lead to increase the likelihood of misclassification. Although the multi-feature algorithms achieved some results, but most of multi-feature algorithms just combine some features simply without effective integration. So, the number of feature vectors' dimension is too large and the accuracy of image recognition is poor. Currently, there are some kinds of shortcomings in fruit and vegetable recognition filed. Some algorithms can find the position [1, 2] or grade [3] of fruit or vegetable, but these algorithms can't identify the type of fruit and vegetable. Some algorithms can identify

© Springer Science+Business Media Singapore 2016
W. Che et al. (Eds.): ICYCSEE 2016, Part I, CCIS 623, pp. 591–599, 2016.
DOI: 10.1007/978-981-10-2053-7_52

the type of fruit and vegetable by only a single feature vector [4–6], so the recognition accuracy of these algorithms is low. This paper proposes an improved heterogeneous features compatibility feature fusion algorithm and an improved SVM algorithm to achieve multi-classification of fruits and vegetables.

2 Feature Extraction and Feature Fusion

This paper used mature feature extraction algorithms to extract features. And then, we combined multiple features with one feature. The feature fusion algorithm can improve the performance of the system. An important issue is the integration of features compatibility between heterogeneous features. Therefore, we need use normalization techniques to solve the problems [7] before connecting vector. This project introduces the Reproducing Kernel Hilbert space to improve the multi-feature compatibility and improve the accuracy of multi-feature fusion algorithm.

2.1 Feature Extraction

This paper mainly extracted contour feature, color feature and texture feature. OTSU algorithm [8, 9] can segment original image into foreground image and background image by a threshold. The OTSU threshold can be used as the input threshold of Canny algorithm [10, 11], and then we can get contour extraction. This paper used K-means algorithm [12, 13] to extract color range of fruit and vegetable images. We can obtain the color features by the cluster of similar color. This paper used roughness component, contrast components, and orientation components of Tamura texture feature [14] as texture features of images.

2.2 Feature Fusion

Change feature compatibility. This paper introduces the Reproducing Kernel Hilbert space to improve the multi-feature compatibility and improve multi-feature fusion algorithm, and then we can fuse features. We can get two sets of heterogeneous features $X = \{x_i \in \mathbb{R}^{D_x}, i = 1, \ldots, N\}$, $Y = \{y_i \in \mathbb{R}^{D_y}, i = 1, \ldots, N\}$ and constraints set S. If $x_i \in X$ and $y_j \in Y$ have the same kind of labels, then $(i, j) \in S$. We defined feature spaces $\mathcal{F}_1$ and $\mathcal{F}_2$ with nonlinear mapping $\phi, \mathbb{R}^{D_x} \to \mathcal{F}_1, \mathbb{R}^{D_y} \to \mathcal{F}_2$. We introduces the Reproducing Kernel Hilbert space $K(\cdot, \cdot)$. $K(\cdot, \cdot)$ is a semi-definite kernel function, and then $\langle \phi(x_i), \phi(x_j) \rangle = K(x_i, x_j)$ and $\langle \phi(x_j), \phi(x_i) \rangle = K(y_i, y_j)$ are established. The purpose of introduce Reproducing Kernel Hilbert space is to find a pair of mapping P_x and P_y, then heterogeneous data sets are the closest set S after mapping to get the optimal solution of formula (1).j

$$\underset{P_x, P_y}{\arg\min} J(P_x, P_y) = \sum_i \sum_j \left\| P_x^T \phi(x_i) - P_y^T \phi(y_j) \right\|^2 S_{ij} \tag{1}$$

Formula (1) can be obtained the formula (2) after rewriting and simplify.

$$J(P_x, P_y) = Tr\left\{ \begin{bmatrix} P_x \\ P_y \end{bmatrix}^T \begin{bmatrix} \phi(X) \\ & \phi(Y) \end{bmatrix} \left(D - \begin{bmatrix} 0 & S \\ S^T & 0 \end{bmatrix} \right) \begin{bmatrix} \phi(X) \\ & \phi(Y) \end{bmatrix}^T \begin{bmatrix} P_x \\ P_y \end{bmatrix} \right\} \quad (2)$$

In formula (2), $\Phi(X) = [\phi(x_1), \phi(x_2), \ldots, \phi(x_N)]$, $\Phi(Y) = [\phi(y_1), \phi(y_2), \ldots, \phi(y_N)]$. We defined that $W = \begin{bmatrix} 0 & S \\ S^T & 0 \end{bmatrix}$, then D is a diagonal matrix which values in each row equals the sum value of the row of W. We defined $P_x = [p_x^1, p_x^2, \ldots, p_x^m]$ and $P_y = [p_y^1, p_y^2, \ldots, p_y^m]$, p_x^i and p_y^j can be extended as $p_x^i = \sum_{j=1}^{N} \alpha_j^i \phi(x_i)$ and $p_y^i = \sum_{j=1}^{N} \beta_j^i \phi(y_i)$ linearly according to theorem. We defined that $\alpha^i = [\alpha_1^i, \alpha_2^i, \ldots, \alpha_N^i]^T \in \mathbb{R}^{N \times 1}$, $\beta^i = [\beta_1^i, \beta_2^i, \ldots, \beta_N^i]^T \in \mathbb{R}^{N \times 1}$, $A = [\alpha^1, \alpha^2, \ldots, \alpha^m] \in \mathbb{R}^{N \times m}$ and $B = [\beta^1, \beta^2, \ldots, \beta^m] \in \mathbb{R}^{N \times m}$, then we can get formulas (3) and (4)

$$P_x = \Phi(X)A \quad (3)$$

$$P_y = \Phi(Y)B \quad (4)$$

We can get formula (5) by formulas (2), (3) and (4).

$$J(A, B) = Tr\left\{ \begin{bmatrix} A \\ B \end{bmatrix}^T \begin{bmatrix} K_x \\ & K_y \end{bmatrix} \left(D - \begin{bmatrix} 0 & S \\ S^T & 0 \end{bmatrix} \right) \begin{bmatrix} K_x \\ & K_y \end{bmatrix}^T \begin{bmatrix} A \\ B \end{bmatrix} \right\} \quad (5)$$

K_x and K_y are nuclear matrixes, $K_x(i,j) = \mathcal{K}(x_i, x_j)$, $K_y(i,j) = \mathcal{K}(y_i, y_j)$. We defined that $P = \begin{bmatrix} A \\ B \end{bmatrix}$ and $K = \begin{bmatrix} K_x \\ & K_y \end{bmatrix}$, and then we can get formula (6).

$$J(P) = Tr\{P^T K (D - W) K^T P\} \quad (6)$$

Finally, in order to remove any scaling factor we add a constraint $P^T K D K^T P = I$. Then we can get formula (7).

$$P^* = \underset{P^T KDK^T P = I}{\arg\min} \ Tr\{P^T K (D - W) K^T P\} = \underset{P^T KDK^T P = I}{\arg\max} \ Tr\{P^T KWK^T P\} \quad (7)$$

By solving the eigenvalue of maximum problem of formula (8), we can get the optimal solution of P.

$$(KWK^T)P = (KDK^T)P\Lambda \quad (8)$$

Feature Fusion. We defined two sets of heterogeneous features are $X \in \mathbb{R}^{D_x \times N}$ and $X \in \mathbb{R}^{D_x \times N}$, then we can get features $X^p \in \mathbb{R}^{m \times N}$ and $Y^p \in \mathbb{R}^{m \times N}$ through mapping by formulas (9) and (10).

$$X^p = P_x^T \Phi(X) = A^T \Phi(X)^T \Phi(X) = A^T K_x \tag{9}$$

$$Y^p = P_y^T \Phi(Y) = B^T \Phi(Y)^T \Phi(Y) = B^T K_y \tag{10}$$

We can simply get the fusion feature $Z \in \mathbb{R}^{m \times N}$ of train set through calculating the average feature points of the training set as formula (11).

$$Z = \frac{1}{2}(X^p + Y^p) \tag{11}$$

We defined heterogeneous feature set of test $X_t = \{x_i^t \in \mathbb{R}^{D_x}, i = 1, \ldots, N_t\}$ and $Y_t = \{y_i^t \in \mathbb{R}^{D_y}, i = 1, \ldots, N_t\}$. The test feature also can mapping into the same space according to formulas (12) and (13).

$$X_t^P = P_x^T \Phi(X_t) = A^T \Phi(X)^T \Phi(X_t) = A^T K_x^t \tag{12}$$

$$Y_t^P = P_y^T \Phi(Y_t) = B^T \Phi(Y)^T \Phi(Y_t) = B^T K_y^t \tag{13}$$

K_x^t and K_y^t are nuclear matrixes, $K_x^t(i,j) = \mathcal{K}\left(x_i, x_j^t\right)$, $K_y^t(i,j) = \mathcal{K}\left(y_i, y_j^t\right)$. We can calculate fusion features by the formula (14).

$$Z_t = \frac{1}{2}\left(X_t^p + Y_t^p\right) \tag{14}$$

Finally, this paper used SVM classification $\hat{f}$ to train fusion train set Z and label. For test features X_t and Y_t, we can calculate the fusion feature Z_t and predict the classification label $\hat{f}(Z_t)$.

3   Improved SVM Algorithm

This paper used SVM classifier to classify. Since we used multi-feature fusion, we should improve SVM algorithm and improve the accuracy of results. This paper introduces the thin-plate splines (TPS) conversion model in SVM classifier, which has a strong versatility and power performance in terms of higher order distortion.

We defined the mapping function of TPS conversion model is $f(x) : \mathbb{R}^d \to \mathbb{R}^d$. So the minimizing the penalty function for the smooth TPS is formula (15).

$$J_m^d(f) = \int \|D^m f\|^2 dX = \sum_{\alpha_1 + \ldots + \alpha_d = m} \frac{m!}{\alpha_1! \ldots \alpha_d!} \int \cdots \int \left(\frac{\partial^m f}{\partial x_1^{\alpha_1} \ldots \partial x_d^{\alpha_d}}\right)^2 \prod_{j=1}^d dx_j \tag{15}$$

$D^m f$ is the m order partial derivatives of matrix f. α_k is a positive number. $dX = \prod_{j=1}^d dx_j$, x_j is a part of vector x. The classic solution of formula (15) has radial basis function form (TPS interpolation function) as shown in formula (16).

$$f_k(x) = \sum_{i=1}^{n} \psi_i G(\|\mathbf{x} - \mathbf{x}_i\|) + \ell^T \mathbf{x} + c \tag{16}$$

$\|\cdot\|$ is the Euclidean norm. $\{\psi_i\}$ is the weight of nonlinear part. ℓ and c are the weight of linear part. We can simplify formula (15) by the radial distance of TPS kernel, as shown in formula (17).

$$G(\mathbf{x}, \mathbf{x}_i) = G(\|\mathbf{x} - \mathbf{x}_i\|) \propto \begin{cases} \|\mathbf{x} - \mathbf{x}_i\|^{2m-d}\ln\|\mathbf{x} - \mathbf{x}_i\| & 2m - d \, iseven \\ \|\mathbf{x} - \mathbf{x}_i\|^{2m-d} & others \end{cases} \tag{17}$$

The input space of interpolation point TPS transformation model (as shown in formula (16)) is the deformation of nonlinear learning distortion. This shift should be able to ensure the desired smoothness. Accordingly, as shown in formula (15) we reduced the bending energy $J_m^d(f)$ on maximum degree. In the settings of learning measure, x is one of train fusion feature. x is converted to $f(x)$. By calculation, the matrix can be obtained in the form of formula (18).

$$f(\mathbf{x}) = L\mathbf{x} + \psi \begin{pmatrix} G(\mathbf{x}, \mathbf{x}_1) \\ \ldots \\ G(\mathbf{x}, \mathbf{x}_p) \end{pmatrix} = L\mathbf{x} + \psi \mathbf{G}(\mathbf{x}) \tag{18}$$

L is the d*d linear transformation matrix, ψ is the d*p nonlinear weight matrix, p is the number of anchors. By nonlinear TPS transformation model, SVM paradigm can be defined by the Margin-Radius-Ratio (MRR).

We defined train set $\chi = \{\mathbf{x}_i | \mathbf{x}_i \in \mathbb{R}^d, i = 1, \ldots, n\}$ and classification label y_i. We can get formula (19) by formula (18).

$$\min_{L, \psi, w, b} J = \frac{1}{2}\|\mathbf{w}\|^2 + C_1 \sum_{i=1}^{n} \xi_i + C_2 \|\psi\|_F^2$$

$$s.t. \quad \begin{aligned} & y_i(\mathbf{w}^T f(\mathbf{x}_i) + b) \geq 1 - \xi_i, \xi_i \geq 0, \forall i = 1 \ldots n(I \,\&\, II) \\ & \|f(\mathbf{x}_i) - \mathbf{x}_c\|^2 \leq 1, \ \forall i = 1 \ldots n(III) \\ & \sum_{i=1}^{P} \psi_i^k = 0, \sum_{i=1}^{P} \psi_i^k \mathbf{x}_i^k = 0, \ \forall k = 1 \ldots d(IV) \end{aligned} \tag{19}$$

ψ_i^k is k-th column of $\psi.\mathbf{x}^k$ is k-th part of $\mathbf{x}$. In addition to traditional soft margin of SVM components, another component is $\|\psi\|_F^2$. Joining squared Frobenius norm ψ in the objective function can be regularized objective function to prevent overfitting. C_1 and C_2 are two weigh hyperparameters. The first two non-constraint (I and II) are the same with traditional SVM same. The third constraint is a non-closed unit ball constrained, so its transformation space in a unit radius sphere can avoid the trivial solution, and it is the center of all samples. The last two equivalent constraint (IV) can be used to maintain TPS transform properties at infinity. This paper introduces TPS model and improve the classification accuracy of the results.

4   Experimental Results

In classification step, we chose 11 kinds of fruits and vegetables image (apple, banana, carrot, cucumber, kiwi fruit, orange, yellowish orange, pear, eggplant, tomato, date). The number of training samples and test samples image is shown in Table 1.

Table 1. The number of training samples and test samples image

Recognition type	Training sample	Test sample
Apple	96	48
Banana	98	48
Carrot	78	39
Cucumber	67	33
Kiwi fruit	53	26
Orange	118	58
Yellowish orange	104	52
Pear	110	55
Eggplant	87	43
Tomato	51	25
Date	46	23
Total	908	450

We can get contour feature, color feature and texture feature. We used traditional SVM algorithm to recognition by single features and fusion features respectively. We used traditional SVM algorithm and improved SVM algorithm respectively to recognition by fusion features. During the experiment, we selected polynomial kernel as the SVM kernel function, and used PSO algorithm to determine punishment variables c and gamma function g: $c = 54.3$, $g = 64$.

The result of classification algorithms is shown in Table 2.

We can know some information through Table 2.

1. The recognition rate of the same test sample are very different between different single feature. Figure 1 is the contour feature comparison figure of test sample. Figure 1 (a) and (c) are the carrot and the cucumber test image respectively. Figure 1(b) and (d) are the carrot and the cucumber contour image respectively. We can find the contour feature of the two kinds vegetables are very similar from the figure, so it's so easy to confuse by contour feature. Figure 2 is the color feature comparison figure of test sample. Both of Fig. 2(a) and (c) are apple test images. Figure 2(b) is the pear test image. Figure 2(d) is the tomato test image. We can find that different kinds of apples have great color differences, and then the difference increase the difficulty of classification.

2. As 1. above, some features have high similarity and easily confused. There are many kinds of similar images, so the classification is very difficult. An improved SVM classification algorithm in this paper can distinguish similar images, and the classification accuracy rates are over 95 %.

Table 2. The result of classification algorithms

Recognition type	Recognition rate of single feature			Recognition rate of fusion feature	
	Contour	Color	Texture	Traditional SVM	Improved SVM
Apple	93.75 %	91.66 %	89.58 %	93.75 %	95.83 %
Banana	100 %	95.83 %	91.66 %	100 %	100 %
Carrot	94.87 %	97.43 %	92.30 %	100 %	100 %
Cucumber	90.90 %	93.93 %	90.90 %	93.93 %	96.96 %
Kiwi fruit	88.46 %	92.30 %	84.61 %	92.30 %	96.15 %
Orange	91.37 %	91.37 %	89.65 %	91.37 %	94.82 %
Yellowish orange	92.30 %	94.23 %	90.38 %	94.23 %	96.15 %
Pear	96.36 %	98.18 %	92.72 %	98.18 %	100 %
Eggplant	95.34 %	97.67 %	93.02 %	97.67 %	100 %
Tomato	92 %	92 %	88 %	92 %	96 %
Date	86.95 %	91.30 %	86.95 %	91.30 %	95.65 %
Average	92.93 %	94.17 %	89.97 %	94.97 %	97.41 %

Fig. 1. The contour feature comparison of test sample

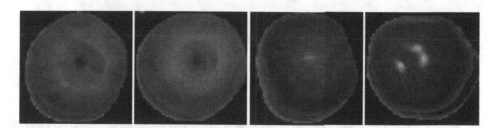

Fig. 2. The color feature comparison of test sample

3. In case of using the same kind of SVM algorithm, the classification accuracy rates with fusion feature are higher than the classification accuracy rates with single feature. The result indicates that the fusion feature utilize more information and enhance the ability to identify.

4. Improved SVM algorithm added TPS transformation model, since this model has a stronger versatility and power performance in higher-order deformation terms, as compared with the traditional SVM algorithm, classification accuracy increased by nearly 4 %.

Thus, the following conclusions:

1. The accuracy rate of single feature recognition is lower, and the reliability and stability of classification is poor. The accuracy rate of fusion feature recognition is higher than simple feature recognition and it has good robustness.
2. Multi-feature fusion algorithm can obtain more information. The fusion feature recognition rate is greatly improved compared with the single feature recognition rate.
3. The classification accuracy of improved SVM algorithm with fusion features is improved greatly, and it can compare very similar images. The average recognition rate with fusion features is 97.41 %.

5   Conclusion

This project not only introduces the Reproducing Kernel Hilbert space to improve the multi-feature compatibility and improve multi-feature fusion algorithm, but also introduces TPS transformation model in SVM classifier to improve the classification accuracy, real-time and robustness of integration feature. The average recognition rate with fusion features is 97.41 %.

Acknowledgments. This paper has been supported by the National Natural Science Foundation of China (Grant No. 61371040).

References

1. Gatica, C., Best, S., Ceroni, J., Lefranc, G.: A new method for olive fruits recognition. In: San Martin, C., Kim, S.-W. (eds.) CIARP 2011. LNCS, vol. 7042, pp. 646–653. Springer, Heidelberg (2011)
2. Amosh, L.S., Sheikh Abdullah, S.N.H., Che Mohd, C.R., Jameson, J.: Adaptive region growing for automated oil palm fruit quality recognition. In: Zaman, H.B., Robinson, P., Olivier, P., Shih, T.K., Velastin, S. (eds.) IVIC 2013. LNCS, vol. 8237, pp. 184–192. Springer, Heidelberg (2013)
3. Fang, C., Hua, C.: Advances in the application of image processing fruit grading. In: Li, D., Chen, Y. (eds.) CCTA 2013. IFIP AICT, vol. 419, pp. 168–174. Springer, Heidelberg (2013)
4. Zawbaa, H.M., Abbass, M., Hazman, M., Hassenian, A.E.: Automatic fruit image recognition system based on shape and color features. In: Hassanien, A.E., Tolba, M.F., Taher Azar, A. (eds.) AMLTA 2014. CCIS, vol. 488, pp. 278–290. Springer, Heidelberg (2014)

5. Potter, M.C., Hagmann, C.E.: Banana or fruit? Detection and recognition across categorical levels in RSVP. J. Psychon. Bull. Rev. **22**, 578–585 (2015)
6. García-Lamont, F., Cervantes, J., Ruiz, S., López-Chau, A.: Color characterization comparison for machine vision-based fruit recognition. In: Huang, D.-S., Bevilacqua, V., Premaratne, P. (eds.) ICIC 2015. LNCS, vol. 9225, pp. 258–270. Springer, Heidelberg (2015)
7. Jain, A., Nandakuma, K., Ross, A.: A. Score normalization in multimodal biometric systems. J Pattern Recogn. **38**, 2270–2285 (2005)
8. Cheng, C.-C., Tsai, C.-M.: Using red-otsu thresholding to detect the bus routes number for helping blinds to take bus. In: Ali, M., Pan, J.-S., Chen, S.-M., Horng, M.-F. (eds.) IEA/AIE 2014, Part I. LNCS, vol. 8481, pp. 321–330. Springer, Heidelberg (2014)
9. Martín-Rodríguez, F.: New tools for gray level histogram analysis, applications in segmentation. In: Kamel, M., Campilho, A. (eds.) ICIAR 2013. LNCS, vol. 7950, pp. 326–335. Springer, Heidelberg (2013)
10. Ramachandran, D., Kam, M., Chiu, J.: Social dynamics of early stage co-design in developing regions. J. Proc. Chi 1087–1096 (2015)
11. Shah, Z.H., Kaushik, V.: Performance analysis of canny edge detection for illumination invariant facial expression recognition. In: IEEE International Conference on Industrial Instrumentation and Control (ICIC), pp. 101–106 (2015)
12. Kaur, R., Heena, H.: An approach defining gait recognition system using k-means and MDA. Int. J. Comput. Appl. **117**, 1–4 (2015)
13. Zhao, C., Li, X., Yan, C.: Bisecting k-means clustering based face recognition using block-based bag of words model. J Optik – Int. J. Light Electron Optics **126**, 1761–1766 (2015)
14. Tiwari, D., Tyagi, V.: Dynamic texture recognition based on completed volume local binary pattern. J. Multidimension. Syst. Signal Process. **27**, 1–13 (2015)

Selecting Seeds for Competitive Influence Spread Maximization in Social Networks

Hong Wu[1,2], Weiyi Liu[1], Kun Yue[1(✉)], Jin Li[3], and Weipeng Huang[1]

[1] School of Information Science and Engineering,
Yunnan University, Kunming, China
kyue@ynu.edu.cn
[2] School of Computer Science and Engineering,
Qujing Normal University, Qujing, China
[3] School of Software, Yunnan University, Kunming, China

Abstract. There exist two or more competing products in viral marketing, and the companies can exploit the social interactions of users to propagate the awareness of products. In this paper, we focus on selecting seeds for maximizing the competitive influence spread in social networks. First, we establish the possible graphs based on the propagation probability of edges, and then we use the competitive influence spread model (CISM) to model the competitive spread under the possible graph. Further, we consider the objective function of selecting k seeds of one product under the CISM when the seeds of another product have been known, which is monotone and submodular, and thus we use the CELF (cost-effective lazy forward) algorithm to accelerate the greedy algorithm that can approximate the optimal with $1 - 1/e$. Experimental results verify the feasibility and effectiveness of our method.

Keywords: Social networks · Competitive influence spread · Possible graph · Submodularity · CELF algorithm

1 Introduction

Social networks play an important role as a media for the spread of various information. For example, the diffusion of disease [1], viruses and even malicious rumors propagation [2–4], the information of product diffusion through the viral marketing [5, 6], etc. Understanding the dynamics of these networks may help us to control the disease (or computer viruses), minimize the spread of rumors and promote products. In this paper, we take the product promotion of social networks (i.e. viral marketing) as the background. Viral marketing takes the advantage of "word-of-mouth" among the relationships of individuals to spread the influence of products. The spread of viral marketing in a social network can be described as follows. First, we select some initial nodes (i.e., seeds) with free samples or provide the information of products. Then these initial nodes will tell the information of products to their friends, who then tell it to their friends and so on, which is called as cascade spread. Finally, a large portion of nodes will be influenced by these seeds [5].

© Springer Science+Business Media Singapore 2016
W. Che et al. (Eds.): ICYCSEE 2016, Part I, CCIS 623, pp. 600–611, 2016.
DOI: 10.1007/978-981-10-2053-7_53

It is known that how to select these seeds to maximize the influence spread is the problem of influence maximization. Domingos and Richardson [7] considered the influence maximization as an algorithmic problem, where the customer network was modeled as a graph and a Markov random field was used to calculate influence propagation among them. Kempe et al. [8] formulated the influence maximization as a discrete optimization problem and proposed two diffusion models based on the early work [9–12]: Independent cascade model (ICM) and linear threshold model (LTM), under which the focus is to select k seeds to maximize the influence spread. The problem is clearly NP hard, but the greedy algorithm can be used to approximate the optimal result based on the submodularity.

Focusing on how to design a new heuristic algorithm that is easily scalable to large-scale social networks, some researchers have improved the scalability of the Kempe et al.'s greedy algorithm for influence maximization [14–19]. From the perspective of challenges in the studies of influence maximization, there frequently exists competition among the influences of two or more ideas or product information in a social network [20–23], such as the same product of competing companies Apple and Samsung or two political candidates of the opposing parties Bush and Hillary and so on. They all want to attract people's attention and spread their influence as much as possible in a social network. Thus, it is necessary to select the initial nodes to spread the influence via the relationships among individuals, exactly the problem that we will solve in this paper.

In our study, we focus on the problem of selecting the seeds for maximizing the competitive influence spread in a social network, that is, how to select k seeds to maximize the competitive influence spread under certain diffusion model given the seed set of competing product I_B and the budget k of one product? For this purpose, we consider the following problems:

(1) How to construct the spread model of competitive influence spread?
(2) How to select the k seeds?

We first denote the social network as a directed graph $G = (V, E)$, where V is the set of nodes representing individuals and E is the set of directed edges representing relationships among the individuals. Each edge $e(u, v)$ in G is associated with a propagation probability $p(u, v)$, where $0 < p(u, v) \leq 1$.

For the problem (1), it is natural to consider the classical IC model, a popular influence diffusion model that describes how influence is propagated throughout the network starting from the initial seed nodes. Chen et al. [25] has proved that computing the influence spread given a seed set under the IC model is #P-hard, where the hardness of calculating the influence is due to the probability $P(u, v)$ of edge $e(u, v)$. If the probability of each edge is deterministic (i.e., the probability of each edge is exactly 1), then the breadth-first-search (BFS) can be used to obtain the influenced nodes incurred by a seed set. Therefore, the linear-time algorithm for computing the influence spread can be obtained in a deterministic graph [26]. In this paper, we can take advantage of possible graphs to effectively obtain the active nodes of the competitive influence spread. We first select top possible graphs from all possible graphs to effectively approximate the optimal result. We further give the competitive influence spread model (CISM) to describe the competitive diffusion process in a possible graph, where the competitive information diffusion process can be well reflected. The construction of CISM can be described as follows. Initially, two sets of nodes in the social network are selected as the seeds of A and B respectively, which are then activated, denoted as A-activated and B-activated respectively. At each step, the nodes of A-activated and B-activated try to activate their out-neighbors with probability 1 by the "live-edge" in possible graph, and the influence that A dominates.

For the problem (2), the optimization problem for selecting the most effective k seeds given the seed set of B is NP hard under the CISM. This objective function is monotone and submodular, and propose the CELF algorithm to approximately solve the problem of maximization competitive influence with $1 - 1/e$. The CELF algorithm is an accelerated algorithm, which can avoid evaluations when they are not necessary. The CISM with the CELF algorithm selects a currently best seed iteratively from $V - I_B$ starting from an empty set, which can maximize the competitive influence spread until k seeds are selected.

To test the efficiency and effectiveness of the proposed CELF algorithm for the CISM under the possible graphs, we implement our algorithms and make corresponding experiments to show the feasibility.

The remainder of this paper is organized as follows. In Sect. 2, we introduce the idea to obtain the possible graphs. In Sect. 3, we give the competitive influence spread model of the possible graph. In Sect. 4, we exploit the approximate algorithm to maximize the competitive influence spread. In Sect. 5, we show experimental results and performance studies. Finally in Sect. 6, we conclude and discuss further work.

2   Generating Possible Graphs

In a social network, the process of calculating the influence spread under the IC model and LT model is #P-hard when the seed set has been given. Similarly, in this paper, the process of calculating the influence of A under the IC model and LT model is also #P-hard given the seed set I_B and I_A. The hardness derives from the calculation of P_{uv}^A and P_{uv}^B of edge $e(u, v)$. Therefore, we exploit the approximate algorithm to calculate the influence of A and that of B simultaneously.

Hu et al. [17] proposed possible graphs, similar to the subgraphs proposed by Chen et al. [24], where the possible graphs are generated by the following idea.

For a given directed graph $G = (V, E, P)$, the number of nodes (or resp. edges) is n (or resp. m), and there are 2^m "live-edge" and "block-edge" possible graphs in G. Let $G' = (V', E')$ denote a possible graph of G, where $V' = V$, $E' \subseteq E$, $P'(e) = 1$ for all $e \in E'$. The existence probability of possible graph G' is as follows [24].

$$P(G') = \prod_{e \in G'} P(e) \times \prod_{e' \in G \setminus G'} (1 - P(e')) \tag{1}$$

Based on the work in [17, 24], we propose the method for calculating the probability of each possible graph of competitive influence spread, in which each edge $e(u, v)$ has two diffusion probabilities, P_{uv}^A and P_{uv}^B. In order to generate the possible graphs of competitive influence spread, we consider the following two situations: $P_{uv}^A \neq P_{uv}^B$ and $P_{uv}^A = P_{uv}^B$.

If $P_{uv}^A \neq P_{uv}^B$, then the existence probability of possible graph G_A' (or resp. G_B') equals to the product of probabilities of all the edges G_A' (or resp. G_B'), formally described as

$$P(G_A') = \prod_{e \in G_A'} P(e) \times \prod_{e' \in G \setminus G_A'} (1 - P(e')) \qquad (2)$$

$$P(G_B') = \prod_{e \in G_B'} P(e) \times \prod_{e' \in G \setminus G_B'} (1 - P(e')) \qquad (3)$$

If $P_{uv}^A = P_{uv}^B$, then the probability of each possible graph can be described as

$$P(G_A') = Pr(G_B') = \prod_{e \in G_A'} P(e) \times \prod_{e' \in G \setminus G_A'} (1 - P(e')) = \prod_{e \in G_B'} P(e) \times \prod_{e' \in G \setminus G_B'} (1 - P(e')) \qquad (4)$$

In this paper, we assume $P_{uv}^A = P_{uv}^B$, which can be easily extended to $P_{uv}^A \neq P_{uv}^B$.

Example 1. In Fig. 1(a), v_1 is selected as the seed at step $t = 0$. At step $t = 1$, v_1 tries to active v_2 and v_3, and v_1 successfully active v_3, but v_1 fails to active v_2. In Fig. 1(b), the edge $e(v_1, v_3)$ is called as "live-edge", and the edge $e(v_1, v_2)$ is called as "block-edge".

Based on Eq. (1), we can obtain the probability of possible graph G' as follows:

$$P(G') = P(v_1, v_3) \times P(v_3, v_2) \times (1 - P(v_1, v_2)) \times (1 - P(v_2, v_3)) \times (1 - P(v_2, v_4)) \times (1 - P(v_3, v_4))$$
$$= 0.06048$$

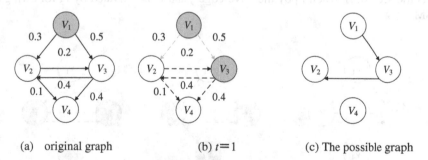

(a)   original graph   (b) $t=1$ (c) The possible graph

Fig. 1. Example of "live-edge" or "block-edge" graph and possible graph. In (a) and (b), green nodes denote the active nodes, and white nodes denote inactive nodes. A solid green arc from node v_1 to v_3 means that v_1 successfully activates v_3 through this arc. A dotted green arc from node v_1 to v_2 means that v_1 fails activates v_2. (c) is the possible graph of G obtained by (a), where $V' = V$, and $E' = \{e(v_1, v_3), e(v_2, v_3)\} \subseteq E$. (Color figure online)

3 The Competitive Influence Spread Model

We now give the diffusion model of the possible graph for competitive influence spread.

In a possible graph $G' = (V', E')$, the diffusion probability of each product by "live-edge" is $P^A_{(u',v')}$ and $P^B_{(u',v')}$, where $P^A_{(u',v')} = P^A_{(u',v')} = 1$. We can describe the competitive influence spread model (CISM) of possible graph as follows.

In the CISM, each node has three states, A-activated (i.e., individual buys product A), B-activated (i.e., individual buys product B), and inactive. In every step, each activated node tries to active its out-neighbors by the live-edge of A and B based on the following rules. The discrete time step $t = 0, 1, 2, ..., n$ is used to describe the diffusion process.

At step $t = 0$, the seed sets $I_A, I_B \subseteq V$ are activated and $I_A \cap I_B = \phi$.

Let $I^t_A \subseteq V$ and $I^t_B \subseteq V$ be the sets of nodes activated by I_A and I_B respectively at step t.

At step $t + 1$, for any node $v \in N^{out}(I^t_A \cup I^t_B)$, where $v \in N^{out}(I^t_A \cup I^t_B)$ denotes out-neighbors of $I^t_A \cup I^t_B$. We consider the following four situations:

(a) If the node v is only be reached by the "live-edges" from I^t_A, then v is added into I^{t+1}_A.
(b) If the node v is only be reached I^t_B by the "live-edges" from I^t_B, then v is added into I^{t+1}_B.
(c) If the node v can be reached by "live-edges" from I^t_A and I^t_B, then the influence of A dominates and thus node v is added into I^{t+1}_A.
(d) If the node v cannot be reached by "live-edges" from I^t_A and I^t_B, then the influence of A and B to node v is "block", and thus v is inactive.

The activation process stops when there are no new active nodes in a time step.

Example 2. Now we give an example of CISM in Fig. 2. Figure 2(a) is the possible graph G', where $e(v_1, v_3)$, $e(v_2, v_3)$, $e(v_2, v_4)$, and $e(v_3, v_4)$ are the "live-edges", and $e(v_1, v_2)$ and $e(v_3, v_2)$ are the "block-edges". In Fig. 2(b), v_2, as the seed of B, can reach nodes v_3 and v_4 by the "live-edge" following the CISM. In Fig. 2(c), v_2, as the seed of B, and v_3, as the seed of A, reach v_4 by the "live-edge", and v_4 is activated by v_3 following the CISM.

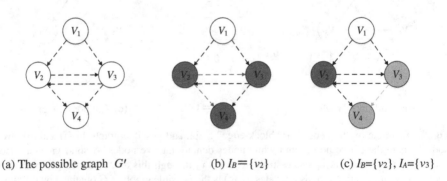

(a) The possible graph G' (b) $I_B = \{v_2\}$ (c) $I_B = \{v_2\}$, $I_A = \{v_3\}$

Fig. 2. Example of CISM

4 Maximizing the Competitive Influence Spread

In this Section, we discuss the objective function that selects k seeds of A under the CISM, when the seeds of B have been known.

4.1 Objective Function for Competitive Influence Spread

We use Formula (4) to compute the probability $P(G_i)$ of each G_i obtained in Sect. 2. In a possible graph G_i, we compute the expectation value $\sigma_{G_i}(I_A, I_B)$ of each G_i, that is, we compute the number of nodes activated by I_A under the CISM when the seed set I_A and I_B are spread simultaneously. The objective function of graph G can be formally described as follows.

$$\sigma_G(I_A, I_B) = \sum_{i=1}^{m} Pr(G_i) \times \sigma_{G_i}(I_A, I_B) \tag{5}$$

Selecting k optimal nodes to maximize their influence when the initial nodes I_B have been known under the CISM is NP hard. The objective function $\sigma_G(S_A, I_B)$ is monotone and submodular under the CISM and $\sigma_G(\phi, I_B) = 0$. Based on the theorem given by Nemhauser et al. [13], we can use a greedy algorithm to approximate the optimal result with $1 - 1/e$ (where e is the base of natural logarithm).

4.2 Approximate Algorithm for Maximizing the Competitive Influence Spread

According to the conclusion in Sect. 4.1, we adopt the CELF algorithm proposed by Leskovec et al. [14] to select the seeds of A. In Algorithm 1, we select the most effective seed of A from $V - I_B$ in each iteration, until k seeds are selected.

First, we describe the basic idea of the algorithm as follows.

Let $\sigma_G(u|I_A, I_B)$ denote the marginal gain of node u added into the seed set I_A when the seed set I_B spreads simultaneously, that is, $\sigma_G(u|I_A, I_B) = \sigma_G(I_A \cup \{u\}, I_B) - \sigma_G(I_A, I_B)$.

In the first iteration, we select the first element of Q into the seed set I_A.

In the i-th iteration ($1 < i \leq k$), if $\sigma_G(v_j^i|I_A, I_B)$ is not smaller than the margin gain of all the other nodes v_l ($v_l \in V \backslash v_j \cup I_{A'} \cup I_B$) added into the set $I_{A'}$ in the earlier iteration, i.e., $\sigma_G(v_j^i|I_A, I_B) \geq \sigma_G(v_l^{i=1:i-1}|I_{A'}, I_B), I_{A'} \subset I_A$, which means that v_l does not need to be computed in the i-th iteration, and thus v_j is added into I_A, where $\sigma_G(v_l^{i=1:i-1}|I_{A'}, I_B)$ represents the margin gain of v_l added into the subset of $I_{A'}$ in the earlier iteration.

In the i-th iteration ($1 < i \leq k$), if $\sigma_G(v_j^i|I_A, I_B)$ is not smaller than the margin gain of some nodes v_l ($v_l \in V \backslash v_j \cup I_A \cup I_B$) added into $I_{A'}$ in the earlier iteration, but it is not larger than the margin gain of other nodes v_m ($v_m \in V \backslash v_j \cup I_A \cup I_B$) added into the set $I_{A'}$ in the earlier iteration, i.e., $\sigma_G(v_l^{i=1:i-1}|I_{A'}, I_B) < \sigma_G(v_j^i|I_A, I_B) < \sigma_G(v_m^{i=1:i-1}|I_{A'}, I_B)$. Then we need compute the margin gain of these nodes v_m added into I_A in the i-th iteration, and select the node with the maximal value of $\sigma_G(v_j^i|I_A, I_B)$ as the seed of A.

Algorithm 1:The CELF algorithm for competitive influence spread: (k, σ)

Input: A social network, I_B, $K=|I_A|$

Output: the seed set of $A : I_A$

Steps:

1. Initialization:

    $I_A \leftarrow \phi$

    priority queue $Q \leftarrow \phi$

    Variables:

    R: the number of possible graph $R=10$

    k: the number of A seeds $k=30$

2. **for1** $i=1$ to k **do**

3. $u \leftarrow \arg\max_{w \in V \backslash I_A \cup I_B} (\sigma_G(u | I_A, I_B))$

4. $I_A \leftarrow I_A \cup \{u\}$

5. **end for1**

6. function $\sigma_G(u|I_A, I_B)$

7. **for2** $r=1$ to R **do**

8.   **for3** each G^r, $j=1$ to n **do**

9. $\sigma_G(v_j | I_A, I_B) \leftarrow \sum P(G^r) \times \sigma_{G^r}(v_j | I_A, I_B)$

10. $v_j.mg \leftarrow \sigma_G(v_j| I_A, I_B)$

11.   insert element v_j into Q with $v_j.mg$ as the key

12.   **if** $i=1$ **then**

13.   select the first element of this Q into I_A, and delete this element in Q

14.   **for4** $i=2$ to k **do**

15.   compute the margin revenue $\sigma_G(v_j^i | I_A, I_B)$ of first element v_j of this Q

16.   **if** $\sigma_G(v_j^i | I_A, I_B) > \sigma_G(v_l^{i=1:i-1} | I_{A'}, I_B), I_{A'} \subset I_A$ **then**

17. $I_A \leftarrow I_A \cup \{v_j\}$

18.   **else if** $\sigma_G(v_l^{i=1:i-1}|I_A, I_B) < \sigma_G(v_j^i | I_A, I_B) < \sigma_G(v_m^{i=1:i-1}|I_A, I_B)$, $v_m, v_j, v_l \in V \backslash I_A \cup I_B$

19.   **then**

                compute the margin revenue $\sigma_G(v_m^i | I_A, I_B)$;

                select $\arg\max \sigma_G(v_m^i | I_A, I_B)$ as the i seed

20.   **end for4**

21.   **end for3**

22. **end for2**

Leskovec et al. [14] empirically showed that the CELF algorithm can provide 700 times of speed-up for greedy algorithm. Therefore, Algorithm 1 is 700 times faster when compared to the greedy algorithm. The time complexity of greedy algorithm is $O(RKnm)$, where R, k, n and m is the number of possible graphs, seeds, nodes and edges respectively.

5 Experimental Results

To test the feasibility and effectiveness of maximizing the competitive influence spread under the CISM in the possible graphs, we conduct experiments on four real-world datasets.

5.1 Experiment Setup

The NetHEPT and Ca-GrQc are Collaboration networks extracted from the ePrint arXiv (http://www.arXiv.org), which is the same source used in the experimental study in [8]. The former is extracted from the "High Energy Physics-Theory" and the latter is extracted from the General Relativity. The nodes in these two networks are authors and an edge between two nodes means the two coauthored at least one paper. The p2p-Gnutella08 record the Gnutella peer to peer network from August 8 2002 where nodes represent hosts in the Gnutella network topology and edges represent connections between the Gnutella hosts. The Wiki-Vote is directed graph that Wikipedia users vote the administrators, where the nodes represent Wikipedia users and a directed edge from node u to node v represents that user u voted on user v.

We use the trivalency cascade model [16] to generate the influence weight of edges. On each edge, we uniformly at random select a probability from $\{0.33, 0.66, 0.99\}$, corresponding to high, medium, low influences.

In order to measure the spread effectiveness of influence for different target sets, we compared the CELF algorithm for competitive influence spread with the max-degree heuristic and random heuristic on the above four datasets. The CELF algorithm for competitive influence spread chooses the seeds by Algorithm 1. The max-degree heuristic chooses nodes with the largest degree as the product seeds of A. The random heuristic randomly chooses nodes as the product seeds of A.

In order to exploit the relationship of ration of the number of B (i.e., product B) seeds to the number of A (i.e., product A) seeds (i.e., $|I_B|/|I_A|$) with the A-activated nodes, we considered the A-activated nodes under the value of $|I_B|/|I_A|$ from 0.1 to 1 when the seed set I_B is fixed.

First, we tested the effectiveness of Algorithm 1. In this experiment, we select 10 seeds with random algorithm as the initial seeds of B and select 30 seeds of A with the Algorithm 1, max-degree and random algorithm to maximize the spread of A in NetHEPT, ca-GrQc, p2p-Gnutella08 and Wiki-Vote networks. Figure 3 shows that the CELF algorithm outperforms the max-degree algorithm and the random algorithm.

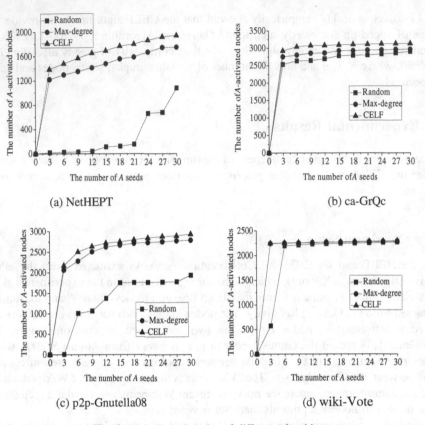

Fig. 3. A-activated nodes of different algorithms

This is because some of max-degree seed nodes may be clustered, and selecting all of them as the seeds of A cannot effectively spread the influence of A. By the random heuristic, as a baseline heuristic algorithm, some of selected seeds cannot spread the influence effectively.

Then, we tested the number of I_B to the number of I_A (i.e., $|I_B|/|I_A|$) with the A-activated nodes. In this experiment, we chose 3000, 2598, 3000, 1369 nodes and 7494, 9958, 9014, 16373 edges from NetHEPT, ca-GrQc, p2p-Gnutella08 and wiki-Vote networks respectively and generate four synthetic networks, which are called as NetHEPT-new, ca-GrQc-new, p2p-Gnutella08-new, and Wiki-Vote-new respectively. We select 5, 10 and 15 seeds by the random heuristic as the initial seeds of B respectively and set the value of $|I_B|/|I_A|$ from 0.1 to 1 to maximize the influence spread of A with the Algorithm 1. Figure 4 shows that the number of A-activated nodes is decreased with the increase of the value of $|I_B|/|I_A|$ when the seeds of B are fixed. This is because the value of $|I_B|/|I_A|$ is decreased when the seed set of I_A is increased, and thus the number of A-activated nodes is increase.

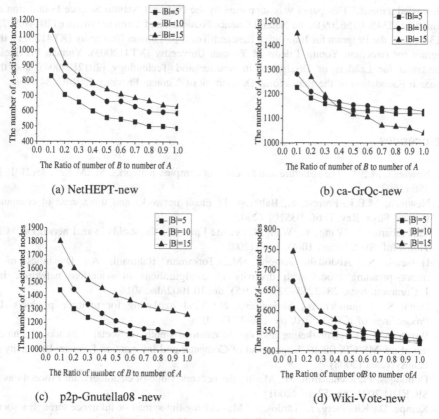

Fig. 4. The *A*-activated nodes with the value of $|I_B|/|I_A|$ from 0.1 to 1

6 Conclusions and Future Work

Aiming at the effective of selecting seeds for the competitive influence maximization, we proposed the CISM under the possible graph, under which we can obtain the active nodes by the BFS. The possible graph can overcome the hardness of calculating the influence probability of a social network, and the CISM can well reflect the process of competitive influence. Further, we gave the submodular function and CELF algorithm for solving the problem of competitive influence maximization, which exploit the submodularity to accelerate the Greedy algorithm.

The CELF algorithm for a possible graph proposed in this paper can select seeds for competitive influence maximization. However, the CELF algorithm is not effective for large scale social networks. For our future work, we plan to explore more effective algorithms for the competitive influence maximization under the possible graphs. Other than the effectiveness, one interesting direction is to consider the influence quality of competitive products, and another interesting direction is to consider asynchronous product spread in a social network.

Acknowledgement. This paper was supported by the National Natural Science Foundation of China (61472345, 61562091), the Natural Science Foundation of Yunnan Province (2014FA023, 2013FB010), the Program for Innovative Research Team in Yunnan University (XT412011), the Program for Excellent Young Talents of Yunnan University (XT412003), Yunnan Provincial Foundation for Leaders of Disciplines in Science and Technology (2012HB004), and the Research Foundation of the Educational Department of Yunnan Province (2014C134Y).

References

1. Newman, M.E.J.: The structure and function of complex networks. SIAM Rev. **45**(2), 167–256 (2003)
2. Newman, M.E.J., Forrest, S., Balthrop, J.: Email networks and the spread of computer viruses. Phys. Rev. E **66**, 035101 (2002)
3. Chakrabarti, D., Wang, Y., Wang, C., et al.: Epidemic thresholds in real networks. ACM Trans. Inf. Syst. Secur. **10**(4), 1–26 (2008)
4. Hosseini, S., Abdollahi Azgomi, M., Torkaman Rahmani, A.: Dynamics of a rumor-spreading model with diversity of configurations in scale-free networks. Int. J. Commun. Syst. **28**, 2255–2274 (2015). doi:10.1002/dac.3016
5. Datta, S., Majumder, A., Shrivastava, N.: Viral marketing for multiple products. In: Proceedings of ICDM 2010, pp. 118–127 (2010)
6. Wortman, J.: Viral marketing and the diffusion of trends on social networks. Technical report no. MS-CIS-08-19, Department of Computer and Information Science, University of Pennsylvania (2008)
7. Domingos, P., Richardson, M.: Mining the network value of customers. In: Proceedings of SIGKDD 2001, pp. 57–66 (2001)
8. Kempe, D., Kleinberg, J., Tardos, É.: Maximizing the spread of influence through a social net work. In: Proceedings of SIGKDD 2003, pp. 137–146 (2003)
9. Durrett, R.: Lecture Notes on Particle Systems and Percolation. Wadsworth Publishing, Boston (1988)
10. Liggett, T.M.: Interacting Particle Systems. Springer, New York (1985)
11. Granovetter, M.: Threshold models of collective behavior. Am. J. Soc. **83**(6), 1420–1443 (1978)
12. Schelling, T.: Micromotives and Macrobehavior. Norton, New York (1978)
13. Nemhauser, G., Wolsey, L., Fisher, M.: An analysis of approximations for maximizing submodular set functions—I. Math. Program. **14**(1), 265–294 (1978)
14. Leskovec, J., Krause, A., Guestrin, C., et al.: Cost-effective outbreak detection in networks. In: Proceedings of SIGKDD 2007, pp. 420–429 (2007)
15. Chen, W., Wang, Y., Yang, S.: Efficient influence maximization in social networks. In: Proceedings of SIGKDD 2009, pp. 199–208 (2009)
16. Wang, C., Chen, W., Wang, Y.: Scalable influence maximization for independent cascade model in large-scale social networks. Data Min. Knowl. Disc. **25**(3), 545–576 (2012)
17. Hu, J., Meng, K., Chen, X., et al.: Analysis of influence maximization in large-scale social networks. SIGMETRICS Perform. Eval. Rev. **41**(4), 78–81 (2014)
18. Galhotra, S., Arora, A., Virinchi, S., et al.: ASIM: a scalable algorithm for Influence maximization under the independent cascade model. In: Proceedings of WWW 2015, pp. 35–36 (2015)
19. Shi, Q., Wang, H., Li, D., et al.: Maximal influence spread for social network based on MapReduce. In: Proceedings of ICYCSEE 2015, pp. 128–136 (2015)

20. He, X., Song, G., Chen, W., Jiang, Q.: Influence blocking maximization in social networks under the competitive linear threshold model. In: Proceedings of SIAM ICDM 2012, pp. 463–474 (2012)
21. Budak, C., Agrawal, D., El Abbadi, A.: Limiting the spread of misinformation in social networks. In: Proceedings of WWW 2011, pp. 665–674 (2011)
22. Carnes, T., Nagarajan, C., Wild, S.M., et al.: Maximizing influence in a competitive social network: a follower's perspective. In: Proceedings of ICEC 2007, pp. 351–360 (2007)
23. Wu, H., Liu, W., Yue, K., Huang, W., Yang, K.: Maximizing the spread of competitive influence in a social network oriented to viral marketing. In: Li, J., Sun, Y., Dong, X.L., Yu, X., Sun, Y., Dong, X.L. (eds.) WAIM 2015. LNCS, vol. 9098, pp. 516–519. Springer, Heidelberg (2015). doi:10.1007/978-3-319-21042-1_53
24. Chen, W., Collins, A., Cummings, R., et al.: Influence maximization in social networks when negative opinions may emerge and propagate. In: Proceedings of SDM 2011, vol. 11, pp. 379–390 (2011)
25. Chen, W., Wang, C., Wang, Y.: Scalable influence maximization for prevalent viral marketing in large-scale social networks. In: Proceedings of SIGKDD 2010, pp. 1029–1038 (2010)
26. Long, C., Wong, R.C.W.: Minimizing seed set for viral marketing. In: Proceedings of ICDM 2011, pp. 427–436 (2011)

Sentence-Level Paraphrasing for Machine Translation System Combination

Junguo Zhu, Muyun Yang$^{(\boxtimes)}$, Sheng Li, and Tiejun Zhao

Computer Science and Technology, Harbin Institute of Technology,
92 West Dazhi Street, Harbin 150001, China
{jgzhu,ymy}@mtlab.hit.edu.cn, {lisheng,tjzhao}@hit.edu.cn
http://www.hit.edu.cn

Abstract. In this paper, we propose to enhance machine translation system combination (MTSC) with a sentence-level paraphrasing model trained by a neural network. This work extends the number of candidates in MTSC by paraphrasing the whole original MT translation sentences. First we train a neural paraphrasing model of Encoder-Decoder, and leverage the model to paraphrase the MT system outputs to generate synonymous candidates in the semantic space. Then we merge all of them into a single improved translation by a state-of-the-art system combination approach (MEMT) adding some new paraphrasing features. Our experimental results show a significant improvement of 0.28 BLEU points on the WMT2011 test data and 0.41 BLEU points without considering the out-of-vocabulary (OOV) words for the sentence-level paraphrasing model.

Keywords: Machine translation · System combination · Paraphrasing · Neural network

1 Introduction

Machine translation (MT) has made great progress in the past decades of years. Various kinds of MT systems, as represented by Google Translate and Bing Translator, provide diverse translation candidates, which can help people master a rough idea. Each candidate translation is still far from perfect though bearing its own comparative advantage in some condition. Due to the complementary between each other, fusing the translation candidates will be a reasonable solution promoting the quality of MT outputs.

Focusing on integrating multiple MT outputs, many approaches of machine translation system combination (MTSC) have been developed. One of the state-of-the-art solution is a confusion network based method [6,8,9,11,17,20], which splits the candidate sentences into words and reconstruct them as a directed graph. Generating a final translation output is finding an optimized path on the

J. Zhu—This paper is supported by the project of Natural Science Foundation of China (Grant No. 61272384&61370170).

© Springer Science+Business Media Singapore 2016
W. Che et al. (Eds.): ICYCSEE 2016, Part I, CCIS 623, pp. 612–620, 2016.
DOI: 10.1007/978-981-10-2053-7_54

graph. Since it implement the combination in word level, the coherence and consistency between words in the original translation is missing. To address the problem, some phrase-level approaches are proposed to add context constrains in combination. Rosti et al. extract source-to-target phrases pairs from the alignments between source sentences and MT system outputs to construct a re-decoding model [19]. Huang and Papineni introduce a hierarchical phrase model is used to handle the source syntax information [10]. A target-to-target phrase-level system combination is construct on word lattice, in which each edge is associated with a phrase (a word or a sequence of words) [4,5,15]. The phrases pairs are extracted form word alignments between a selected best MT candidate and other candidates. One problem of the approach is that the context constrains only prune the branch on existing paths, can not produce the better paths. Ma and McKeown [16] use a hierarchical paraphrasing model to rewrite MT outputs, but they select the most probable one in the rewritten sentences without further combination. In addition, the paraphrases are extracted the alignment between MT outputs, so the approach cannot yet produce new path out of the existing space.

In this paper, a pipeline framework is developed to joins the paraphrasing and translation combination. When paraphrasing, we introduce an encoder-decoder architecture of bidirectional recurrent neural network (RNN) to retain the information of whole sentence. The MT outputs and references are leveraging to train the paraphrasing model, which can extend the space of existing space. As an added benefit, learning the changes from the raw MT outputs to perfect references enable the paraphrasing model to have the ability to correct the translation errors, although it can make new errors. And in the combination process, we introduce a confusion network based combining method with paraphrasing features, which can Our experimental results shows that the effectiveness of our approach is demonstrated on the WMT2011 data of system combination task.

The rest of this paper is organized as follows: Sect. 2 introduces previous studies related to this work. Section 3 presents our method in detail. Section 4 describes and analyzes the experimental results. Finally, Sect. 5 concludes the work with possible future work.

2   Related Works

Focusing on enhancing machine translation system combination, there are two kinds of work related with our method: confusion network based combination and paraphrasing for system combination.

2.1   Confusion Network Based Combination

Bangalore et al. [3] first introduced confusion network into MT system combination using a multiple string alignment algorithm. Confusion network is a weighted directed graph. Each edge is labeled with a word. A translation candidate is represented as a path from the start node to the end node goes through all the other nodes. There are two major directions in current confusion network research. One is developing monolingual alignment algorithm, such as GIZA++ [17], TER [21], incremental TER [20], ITG [11], IHMM [8], and METEOR [9]. Another one is selecting backbone which is a base of alignment. Sim et al. [21] select the most similar sentence with others according to minimum bayes risk (MBR). Karakos et al. [11] take each sentence as a backbone in one time, and then combine them. Heafield et al. [9] develop a combination system MEMT, which allows the backbone changing in decoder, so the search space is on the top of all the candidates, which is the largest in all previous works.

All above works are based on the existing candidates, extending the number of candidate translations is still an open issue. We aim at further extending the search space to provide more translation candidates by introducing extra paraphrase knowledge. So we use a combination system MEMT [9] as our baseline.

2.2   Paraphrasing for System Combination

Ma and McKeiwn [15] introduce paraphrasing model into combination framework. Mapping the bilingual phrase extraction [14], the paraphrases are extracted according to the word alignments between multiple MT translations. TERp [22] is adopted to obtain word alignments. The phrasal decoder used in the phrase-level combination is based on standard beam search, guided by a log-linear model score. And in their another work, they use a hierarchical paraphrases model to rewrite the MT outputs via CKY algorithm with a Viterbi approximation. Then they combine the combination outputs of the two methods in a sentence-level selection.

Unlike the their works, we rewrite the MT outputs in sentence level. So paraphrasing model can learn the whole context information. And another difference from their works, we extract the paraphrase between MT outputs and references. Since this rewriting is from raw translation to perfect one, it has the ability of correcting errors.

3   Sentence-Level Paraphrasing for System Combination

In this work, we integrate the paraphrasing and system combination in a pipeline style (Fig. 1). Each raw machine translation (MT) candidate is rewritten into a new one. Then we combine MT outputs and rewritten outputs into a final translation.

3.1   Sentence-Level Paraphrasing

Aiming at leveraging the whole sentence information, in our neural paraphrasing work, each translation candidate is represented by a vector $x_1, x_2, ..., x_{T_x}$, also the target translation represented by $y_1, y_2, ..., y_{T_y}$. Similar to a neural machine translation work [1][1], a RNN encoder-decoder architecture is implemented to translate the candidate into target translation.

---

[1] An open source Toolkit is available at https://github.com/lisa-groundhog/GroundHog.

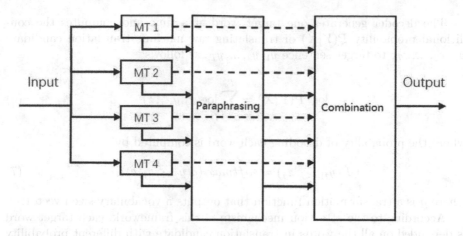

Fig. 1. A Pipeline Framework of Paraphrasing for Translation Fusion

Encoder. In the encoder-decoder framework, a translation candidate $X = (x_1, x_2, ..., x_{T_x})$ is encoded into a fixed-length vector c. $h = (h_1, h_2, ..., h_{T_x})$ by a bi-directional recurrent neural network such that

$$c = q(\{h_1, h_2, ..., h_{T_x}\}) \tag{1}$$

where

$$h_t = \left[\overleftarrow{h_t}; \overrightarrow{h_t} \right] \tag{2}$$

and

$$\overleftarrow{h_t} = f(x_t, \overleftarrow{h}_{t+1}), \overrightarrow{h_t} = f(x_t, \overrightarrow{h}_{t-1}) \tag{3}$$

where h_t is a hidden state at time t, $\overleftarrow{h_t}$ is a forward hidden state and $\overrightarrow{h_t}$ is a backward hidden state. where f and q is some nonlinear functions.

Decoder. In decoder, the context vector c is compute as a weighted sum of these hidden states $h_1, h_2, ..., h_{T_x}$:

$$c_i = \sum_{j=1}^{T_x} \alpha_{ij} h_j \tag{4}$$

The weight α_{ij} of each hidden state is computed as

$$\alpha_{ij} = \frac{\exp(a(h_j, z_{i-1},))}{\sum_{k=1}^{T_x} \exp(a(h_k, z_{i-1}))} \tag{5}$$

$a()$ is a score function of an position alignment model between the inputs and the outputs. z_{i-1} is a RNN hidden state, which emits y_i.

The decoder generates one target word at a time and computes the conditional probability $P(Y|X)$ of translating raw machine translation candidate $x_1, x_2, ..., x_{T_x}$ to target sentence $y_1, y_2, ..., y_{T_y}$ as follows:

$$\log P(Y|X) = \sum_{j=1}^{T_y} \log(y_j|y_{<j}, z) \tag{6}$$

where the probability of decoding each word is computed by

$$P(y_j|y_{<j}, z_j) = softmax\ (g(y_{j-1}, z_j, c)) \tag{7}$$

where g is a transformation function that outputs a vocabulary-sized vector.

According to the attention mechanism of this framework, each target word is depended on all the words in translation candidate with different probability α_{ij}. We relieve that all information in translation candidate are encoded into a fixed-length vector.

3.2   System Combination with Paraphrases

To refine the neural paraphrasing outputs into one translation, we connect a translation fusion module in a pipeline style. Since we are not emphasized on the mechanism of system combination itself, here we just chose an existing CN-based combination system MEMT [9] to combine all the MT candidates and paraphrasing candidates. In MEMT framework, the word alignments between difference candidates are implemented by METEOR [2]. The search space is defined on top of all the aligned sentences. A hypothesis starts with the first word of some sentence, and continue to follow the sentence or switch any other sentence. Thus, this pattern can extending the search space as large as possible. A group of features in MEMT are used for scoring partial and complete hypotheses.

- **Length:** the count of words in the hypothesis;
- **LM:** a log probability of language model;
- **Backoff:** average n-gram length found in the language model.
- **Match:** For each n and each system, the number of n-gram matches between the hypothesis and system.

In addition to these original features in MEMT, we also introduce some new features as:

- **Paraphrasing Indicator:** a binary feature in our model to represent if a word is from paraphrasing candidate or not.
- **Paraphrasing Word Counts:** the number of words which from paraphrasing in one sentence.

4 Experiments

4.1 Experimental Settings

In our experiment, we use 2.56 million parallel sentences (machine translation texts and its gold references) to training our neural paraphrasing model. The machine translation texts is provided by a phrase-based machine translation system Moses [13], which is trained on the LDC Chinese-English news corpus[2] In training the RNN encoder-decoder model, the encoder consists of forward and backward recurrent each having 1,000 hidden units, and the decoder also has 1000 hidden units. the word of inputs and outputs is represented by a vector of 620 dimensions. The size of vocabulary in inputs and outputs is set 30,000 words. A minibatch stochastic gradient descent (SGD) algorithm is used to train the neural paraphrasing model. Each SGD update direction is computed using a minibatch of 80 sentences. 350 thousands mini-batches cost 4 days on Taitan X GPU. A beam search is used to find a translation that approximately maximizes the conditional probability [7].

We test our method on the Czech-to-English test data of WMT2011 system combination task. The development set contains 1,003 sentence and test set contains 2,000 sentences. Each sentence has a group of four MT candidates. The translation quality is evaluated by BLEU-4 score [18]. And a statistical significance test is computed by a bootstrap re-sampling method [12] on a 0.1 significant level.

4.2 The Results of Neural Paraphrasing

The BLEU scores of the four MT system on the development and test set are listed in Table 1. And the BLEU score of their rewritten outputs are also listed.

From Table 1, we can find that the paraphrasing model can not improve the raw machine translation for each single MT systems. The greatest cause of these results is that in our methods the whole candidates are completely rewritten by neural paraphrasing model leading to produce some new errors.

But we leverage MEMT to combine 4 way MT translation and 4 way paraphrasing translations, the result is showed in Table 2. For comparison, we also show the highest BLEU score of single system and the BELU score of MEMT.

From Table 2, we can find that our method has an improvement of 1.10 BLEU point than Single best, and an improvement of 0.28 BELU point than MEMT. Clearly, although the neural paraphrasing model can not have an advantage on single MT translations, but in combination process, it can generate helpful candidates with translation knowledge.

However, the size of vocabulary in neural paraphrasing model is a fixed value since a large vocabulary can increase training complexity as well as decoding

---

[2] LDC2002E18, LDC2002L27, DC2002T01, LDC2003E07, LDC2003E14, LDC2004T07, LDC2005E83, LDC2005T06, LDC2005T10, LDC2005T34, LDC2006E24, LDC2006E26, LDC2006E34, DC2006E86, LDC2006E92, LDC2006E93, LDC2004T08 (HK News, HK Hansards).

Table 1. The Results of Neural Paraphrasing

Systems	Dev	Test
System 1	19.45	18.14
Paraphrasing 1	19.62	18.03
System 2	21.13	20.80
Paraphrasing 2	20.69	20.39
System 3	18.03	17.62
Paraphrasing 3	17.75	17.66
System 4	21.61	21.71
Paraphrasing 4	21.31	21.30

Table 2. The Results of Translation Fusion

	Dev	Test
Single Best	19.45	18.14
MEMT	23.28	22.74
MEMT+Paraphrasing	23.60	23.02

Table 3. The Results of Translation Fusion without OOV

	Test
Single Best	21.18
MEMT	21.87
MEMT+Paraphrasing	22.28

complexity. Hence, the OOVs will affect the final translation performance. We filter those sentence with OOV words from the test set. We list the results Single best, MEMT, and MEMT+Paraphrasing, on test set without OOVs in Table 3.

In Table 3, we present that our method has a further improvement of 0.41 BLEU points. Obviously, neural paraphrasing is able to expanding the decoding space for translation fusion by generating new candidates, while our combination model is able to grasp the useful information by consistency decoding.

5   Conclusion

In this work, we propose a new pipeline framework to integrate the paraphrasing and MT system combination. We train a RNN encoder-decoder paraphrasing model to provide more translation candidates for system combination. Our neural paraphrasing model is trained on the whole sentence, which can retain sentence-level information. And we also use a state-of-the-art combination system with some paraphrasing features to combine the original translation and their rewritten translations. The experimental results show an effective performance at an improvement of 0.28 BLEU point on the WMT11 test and 0.41 on the test set without OOVs compared with MEMT.

Our contributions are as following: (1) We extending the number of candidates for system combination via paraphrasing to extend the search space of the final translations. (2) We introduce a neural network to exploit the sentence-level information. And (3) we use the MT outputs and references to train the paraphrasing model, which can correct the errors in MT outputs.

In this work, we find that the RNN-based paraphrase model can not improve the translation directly, so further work will try more neural network frameworks to address the task of MT system combination.

References

1. Bahdanau, D., Cho, K., Bengio, Y.: Neural machine translation by jointly learning to align and translate (2014). arXiv:1409.0473
2. Banerjee, S., Lavie, A.: Meteor: An automatic metric for MT evaluation with improved correlation with human judgments. In: Proceedings of the ACL Workshop on Intrinsic and Extrinsic Evaluation Measures for Machine Translation and/or Summarization, pp. 65–72 (2005)
3. Bangalore, S., Bordel, G., Riccardi, G.: Computing consensus translation from multiple machine translation systems. In: IEEE Workshop on Automatic Speech Recognition and Understanding, ASRU 2001, pp. 351–354. IEEE (2001)
4. Du, J., Way, A.: Using terp to augment the system combination for SMT. In: Proceedings of the Ninth Conference of the Association for Machine Translation. Association for Machine Translation in the Americas (2010)
5. Feng, Y., Liu, Y., Mi, H., Liu, Q., Lü, Y.: Lattice-based system combination for statistical machine translation. In: Proceedings of the 2009 Conference on Empirical Methods in Natural Language Processing, vol. 3, pp. 1105–1113. Association for Computational Linguistics (2009)
6. Freitag, M., Peter, J.T., Peitz, S., Feng, M., Ney, H.: Local system voting feature for machine translation system combination. In: Proceedings of the Tenth Workshop on Statistical Machine Translation, pp. 467–476 (2015)
7. Graves, A.: Sequence transduction with recurrent neural networks (2012). arXiv:1211.3711
8. He, X., Yang, M., Gao, J., Nguyen, P., Moore, R.: Indirect-HMM-based hypothesis alignment for combining outputs from machine translation systems. In: Proceedings of the Conference on Empirical Methods in Natural Language Processing, pp. 98–107. Association for Computational Linguistics (2008)
9. Heafield, K., Lavie, A.: Combining machine translation output with open source: the Carnegie Mellon multi-engine machine translation scheme. Prague Bull. Math. Linguist. **93**, 27–36 (2010)
10. Huang, F., Papineni, K.: Hierarchical system combination for machine translation. In: Proceedings of the 2007 Joint Conference on Empirical Methods in Natural Language Processing and Computational Natural Language Learning (EMNLP-CoNLL), pp. 277–286. Association for Computational Linguistics, Prague, Czech Republic, June 2007

11. Karakos, D., Eisner, J., Khudanpur, S., Dreyer, M.: Machine translation system combination using ITG-based alignments. In: Proceedings of the 46th Annual Meeting of the Association for Computational Linguistics on Human Language Technologies: Short Papers, pp. 81–84. Association for Computational Linguistics (2008)

12. Koehn, P.: Statistical significance tests for machine translation evaluation. In: EMNLP, pp. 388–395. Citeseer (2004)

13. Koehn, P., Hoang, H., Birch, A., Callison-Burch, C., Federico, M., Bertoldi, N., Cowan, B., Shen, W., Moran, C., Zens, R., et al.: Moses: open source toolkit for statistical machine translation. In: Proceedings of the 45th Annual Meeting of the ACL on Interactive Poster and Demonstration Sessions, pp. 177–180. Association for Computational Linguistics (2007)

14. Koehn, P., Och, F.J., Marcu, D.: Statistical phrase-based translation. In: Proceedings of the 2003 Conference of the North American Chapter of the Association for Computational Linguistics on Human Language Technology, vol. 1, pp. 48–54. Association for Computational Linguistics (2003)

15. Ma, W.Y., McKeown, K.: Phrase-level system combination for machine translation based on target-to-target decoding. In: Proceedings of the 10th Biennial Conference of the Association for Machine Translation in the Americas (2012)

16. Ma, W.Y., McKeown, K.: System combination for machine translation through paraphrasing. In: Proceedings of the 2015 Conference on Empirical Methods in Natural Language Processing, pp. 1053–1058 (2015)

17. Matusov, E., Ueffing, N., Ney, H.: Computing consensus translation for multiple machine translation systems using enhanced hypothesis alignment. In: EACL, pp. 33–40 (2006)

18. Papineni, K., Roukos, S., Ward, T., Zhu, W.J.: Bleu: a method for automatic evaluation of machine translation. In: Proceedings of the 40th Annual Meeting on Association for Computational Linguistics, pp. 311–318. Association for Computational Linguistics (2002)

19. Rosti, A.v.I., Ayan, N.F., Xiang, B., Matsoukas, S., Schwartz, R., Dorr, B.J.: Combining outputs from multiple machine translation systems. In: Proceeding NAACL-HLT 2007, pp. 228–235 (2007)

20. Rosti, A.V.I., Matsoukas, S., Schwartz, R.: Improved word-level system combination for machine translation. In: Annual Meeting-Association for Computational Linguistics, pp. 312–319 (2007)

21. Sim, K.C., Byrne, W.J., Gales, M.J.F., Sahbi, H., Woodland, P.C.: Consensus network decoding for statistical machine translation system combination. In: IEEE International Conference on Acoustics Speech Signal Processing, ICASSP 2007, vol. 4, pp. 2–5 (2007)

22. Snover, M.G., Madnani, N., Dorr, B., Schwartz, R.: TER-Plus: paraphrase, semantic, and alignment enhancements to translation edit rate. Mach. Transl. **23**(2–3), 117–127 (2009)

Social Computing in Open Source Community: A Study of Software Reuse

Mengwen Chen[✉], Tao Wang, Cheng Yang, Qiang Fan,
Gang Yin, and Huaimin Wang

National Laboratory for Parallel and Distributed Processing,
National University of Defense Technology, Changsha, China
{chenmengwen1991,delpiero701}@126.com, taowang.2005@outlook.com,
{fan1864150,jack_nudt,whm_w}@163.com

Abstract. Software projects are not developed in isolation but often build upon other open source resources. These projects form a kind of reference ecosystem regarded as a software world. Most of social computing works focus on social networks such as Facebook and weibo to mine information. However, few previous works analyze Open Source Community which could help developers conduct collaborative development. In this paper, we model the Java reference ecosystem as a network based on the reuse relationships of GitHub-hosted Java projects and analyze the characteristics and the patterns of this reference ecosystem by using community detection and pattern discovery algorithms. Our study indicates that (1) Developers prefer to reuse software limited in only a small part of projects with cross cutting functionality or advanced applications. (2) Developers usually select software reused with similar function widely depending on different requirements, resulting to different patterns. Based on these collective intelligence, our study opens up several possible future directions of reuse recommendation, which are considered as guidance of collaborative development.

Keywords: Social computing · Open source software · Reuse relationship · Characteristic and pattern

1 Introduction

Open Source Software (OSS for brief) has become one of the most important resources in software development. More and more well-known organizations and companies have made their efforts in open source community, such as opening source of Google's deep learning system [1], which is suitable for software reuse. In this manner, the emergence of OSS extends the scope of software reuse and increases the level of reuse-object.

The OSS and their reuse relationships form a kind of reference ecosystem regarded as a software world. Like the real world, there are two major problems about this virtual world. One is how it looks like, and the other is how to exchange information in it. From the perspective of software reuse, we further state these two problems: (i) What is the status quo of software reuse in the open source community from a macro perspective, and (ii) How dose each project reuse or be reused in software development from a micro

W. Che et al. (Eds.): ICYCSEE 2016, Part I, CCIS 623, pp. 621–631, 2016.
DOI: 10.1007/978-981-10-2053-7_55

perspective. These two questions can collect collective intelligence in order to give developers guidance of collaborative software development.

Most of the existing works have focused on social networks to mine information in a particular area. [2] applied social network analysis to examine the communication characteristics of travel-related electronic world-of-mouth, which may have influence on decision-making. [3] presented a social network analysis system component in order to use data from bank statements during an investigation into money laundering cases. In addition, some other works often analyze reference ecosystem formed by a typical project and its surroundings. The reference ecosystem of each typical Java project and its dependency packages was studied in [4], where the authors found less significant community structure. However, unfortunately, few previous works consider open source community as social community to gain collective intelligence and study a language-specific software word at a large scale (130,078 projects).

In this paper, we analyze the characteristics and the patterns of the Java reference ecosystem modeled as a network based on the reuse relationships of GitHub-hosted Java projects by using community detection and pattern discovery algorithms. Our study indicates that, for Java projects: (1) Developers prefer to reuse software limited in only a small part of projects with cross cutting functionality or advanced applications. (2) Developers usually select software reused with similar function widely depending on different requirements, resulting to different patterns. Our study opens up several possible future directions of guidance of collaborative software development. Compared with previous works, our study focuses on a specific software area instead of common social networks. Moreover, our study on this software area is at the scale of more than 100,000 Java projects, which makes it valuable for collecting collective intelligence and guiding collaborative software development.

This paper is organized as follows. Section 2 describes the definition, modeling and analysis methods of Java reference ecosystem. Experiments designs and settings are proposed in Sect. 3. Analysis of the results is given in Sect. 4. Sections 5 and 6 discuss the threats that could affect the validity and the existing works. Last but equally important, Sect. 6 concludes this paper and opens up several possible future directions of guidance of collaborative software development.

2   Java Reference Ecosystem in GitHub

In this section, we discuss the definition, modeling and analysis methods of Java reference ecosystem in GitHub.

2.1   Definition of Reference Ecosystem

Using more software reuse is a method of promoting software development rapidly. The development of one project often reuses some other projects while some other projects are widely reused by some projects, which reflects two kinds of relationships: *reusing and being reused.* Thus, we propose the concept of reference ecosystem in this paper that is "**a collection of software projects which have the relationships of reusing or** being reused". And we further define two types of projects: *library-project and application-project.* Library-project means that it is a reusable software project reused by some other projects. Application-project means that it is an application software project reusing some other projects. But a library-project reusing other library-projects by itself can also be seen as an application-project.

2.2   Modeling of Java Reference Ecosystem in GitHub

In this part, we overview the process of modeling Java reference ecosystem in GitHub, including data extraction, data pre-processing and ecosystem modeling, as is shown in Fig. 1. The details of these three major steps are presented.

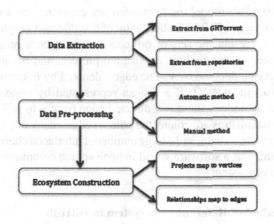

Fig. 1. The process of constructing reference ecosystem

Data Extraction: We winnow out Java software projects from GHTorrent [5] as initial project dataset such as "Apache/Spark". Meanwhile, we do not consider the projects forked from other projects, since the reuse relationships of these projects are the same as that of the original projects. Moreover, software repositories containing a large amount of available data about software projects and developers' performances, can be considered as collective intelligence to be collected and analyzed to speed up software collaborative development and improve development quality. Thus, we develop a tool to scan the files of each repository of the projects in the initial project dataset to identify a list of library-projects from the artifacts of build managers and dependency managers such as Maven's pom.xml. For example, "<groupId> org.eclipse.jetty </groupId> and <artifactId> jettyservlet </artifactId>" are written under the label of "<dependency>" in "Apache/Spark" repository's pom.xml, which reflects that developers develop "Apache/Spark" by reusing "elipse.jetty". More reuse relationships like this consist of collective intelligence. In order to describe easily, the projects using manager tools or extracted from manager tools are regarded as application-projects and library-projects respectively. They and their reuse relationships are comprised of initial experiment dataset.

Data Pre-processing: Even though the reuse relationship identification can be performed automatically, project duplication exists within the initial experiment dataset. In order to identify and remove duplication, we need both automatic and manual efforts. We identify all groups of the projects with the same name in initial experiment dataset and sort them by the number of fork. For each group, if the number of a project's fork is far greater than others', we only keep this project. Otherwise, 20 students in my research group identify and remove duplication through browsing description information. In particular, to ensure the accuracy of the results, we also delete some projects with common names such as "android" from the initial experiment dataset, because these projects may implement different functions, but their names cannot reflect their characteristics better. After these efforts, we obtain the final experiment dataset.

Ecosystem Modeling: we model the Java reference ecosystem as a network, denoted as a directed graph $G = \langle V, E \rangle$ The instantiation of this directed graph is that the sets of vertices and edges represent the sets of projects and their reuse relationships respectively. In details, the vertices include application-projects and library-projects in our final experiment dataset, denoted by V. The edges, denoted by E, connect many pairs of vertices $E(V) = \{(u,v)|u, v \in V\}$. If a project represented by node u reuses another project represented by node v, there is a directed edge from u to v. Thus, two vertices having the reuse relationships are connected with an edge, many edges having different public vertices constitute a chain, and a large number of interlaced chains form a complex network, as is regarded as a software world in open source community taking the form of a Java reference ecosystem.

2.3   Analysis Method of Reference Ecosystem in GitHub

In this part, we introduce the methods used for analyzing the Java reference ecosystem in GitHub, including the characteristics and the patterns, which are used to analyze social networks.

Analysis of Characteristics. Usually, connectivity and cohesiveness are two common properties of social network. Connectivity means that the large network is divided into some connected components. Cohesiveness means community detection, which divides the network nodes into groups within which the connections are dense and between which they are sparse [6].

Analogy to the social network, the structure of a reference ecosystem formed by a collection of projects and their reuse relationships in an open source community also presents developers' experience. Basedon this fact, we use Depth First Search and Louvain community detection algorithms [7] to analyze the structure. The target is to identify developers' intelligence reflected by characteristics of reference ecosystem, which implies the status quo of software reuse in the open source community. In order to analyze better, we also use the Gephi [8], which is a social network graphing tool, to visualize the results.

Analysis of Patterns. Recent researches indicate that there are frequent patterns and special patterns in a certain network. Frequent patterns have the top-N highest occurrence frequencies in this actual network, and special patterns called network motif [9] have a higher occurrence frequency in this actual network than that in some randomized networks with the same characteristics as this actual network. These characteristics include that the number of nodes is the same and the in-degree and out-degree of each node are also the same.

Similarly, analogy to the social network, Unidirectional or bidirectional edges between two nodes also reflect collaborative development by developers among projects. Based on this fact, we use sub-graph discovery and matching algorithms, and corresponding open source tool [10] to mine frequent patterns and special patterns. The target is to identify collaborative development patterns, which implies how to reuse or be reused for each project in software development.

3 Experiments Design

In this section, we further refine the research questions as well as corresponding experimental designs and settings.

3.1 Research Question

There are a large amount of OSS in open source community in which analyzing reuse relationships is a method of collecting collective intelligence to give guidance of collaborative software development. In this setting, one of the most important problems is how they are reused. This problem can be further divided into two sub-problems with different granularity.

- RQ1: What is the status quo of software reuse in the open source community from a macro perspective?
- RQ2: How does each project reuse or be reused in software development from a micro perspective?

We identify the characteristics representing the status quo of software reuse in open source world and the patterns reflecting how to reuse or be reused for each project in software development. Since these two sub-problems are in macro and micro perspectives respectively. We implement a progressive analysis from global to local, from the whole network to each node in it.

3.2 Datasets and Experimental Settings

Overall, we use the GHTorrent data 2015-09-01. After the data extraction and preprocessing stages described in Sects. 2.1 and 2.2, the final number of application-projects considered in our work is 92,400 and 44,273 for the library-projects found in the pom.xml or ivy.xml. The number of their intersection part is 6595.

For RQ1, since we study from a macro perspective, we consider the reference ecosystem covering the entire data of 130,078 projects and 802,413 reuse relationships. They form the global reference ecosystem of Java, denoted as a reuse network.

For RQ2, since we study from a micro perspective, we take a local and typical reference ecosystem, Apache, as an example. It contains 757 nodes and 3,343 edges, which is also because of the limitation on storage space and running time.

4   Research Results

In this section, we discuss two sub-problems mentioned above and present several phenomena about networks, components, communities and patterns which reflect collective intelligence.

4.1   RQ1: Characteristics of Reference Ecosystem

As discussed in Sect. 3.2, we use the entire data including 130,078 projects and 802,413 reuse relationships among them, to model the global reference ecosystem of Java as a network. The analysis is as follows.

First of all, the in-degree and out-degree distribution of this network are depicted in Fig. 2. A few nodes have a relatively high in-degree or out-degree. It indicates that developers often focus on a few projects to conduct reuse software development. For example, the project with the highest in-degree is "junit". It is a simple framework project to write repeatable test reused by 55,983 projects in our work. The word "wildfly", the name of "WildFly Application Server" renaming "JBOSS", is the project with highest out-degree. It reuses 454 projects according to our study.

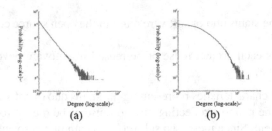

(a)   (b)

Fig. 2.   Distribution trend: in-degree (a) and out-degree (b).

Secondly, this network is divided into 398 connected components. There are 99 % of these projects that belong to the largest connected component where each node is connected to some nodes by certain paths. All of the other components except the largest are primarily comprised of the projects from the same organization or company respectively. An example about "Qcadoo" framework from a certain isolated component is shown in Fig. 3.

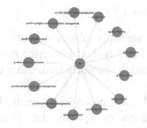

Fig. 3. An example of reuse within the same organization or company

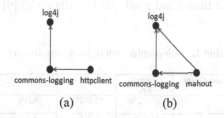

Fig. 4. Two examples of log: cayenne (a) and continuum (b).

The central project "mes" reuses 13 library-projects based on "Qcadoo" framework with a prefix "qcadoo". Thus, developers often use the projects from the same organization or company combined into a complete function together. In fact, 98.18 % of the components except the largest have the connection relationships like Fig. 3, which have only one application-project and some library-projects, but there are no links among these library-projects. The functions of these projects in each isolated simple component are not very common. Thus, the possibility of being reused by developers is relatively smaller, but when requiring these functions, finding corresponding collection of OSS by them is a very easy thing.

Thirdly, 96 communities are identified from the largest component. Meanwhile, we also obtain a modularity score of 0.4 representing less significant community structures, because values above 0.30 are commonly regarded as an indication of community structure [8]. The largest component is dense and communities are thus only loosely identified. Furthermore, after deleting the top-50 projects ranking by in-degree, the modularity score increases up to about 0.6. Since all of them have a higher in-degree than out-degree, they can be seen as library-projects. The results indicate that library-projects are really a bridge between communities. Developers usually reuse a large number of library-projects with cross cutting functionality in many different domains.

For example, "junit" is on the top-1 of in-degree rank. The log-managing project named log4j is on the top-5. The more the library-projects with higher in-degree are, the less modular structure they are expected to convey. Thus, some mechanisms of the discovery and the recommendation of other resources between communities except these library-projects with higher in-degree are needed to be provided to developers based on intelligence mentioned above.

4.2 RQ2: Patterns of Reference Ecosystem

As also discussed in Sect. 3.2, we use a subset, including 757 projects and 3,343 reuse relationships, to model a local and typical reference ecosystem as a network, which contains the Apache projects and the projects from other organizations they reuse. The analysis is as follows.

Taking the sub-graphs with three nodes as an example, we obtain five sub-graphs with the top-5 occurrence frequencies. Specific data is shown in Table 1. The results contain two types of sub-graphs. One is the frequent pattern such as No. 1 and No. 2, whose frequency in this actual network is relatively high. The other is the special pattern, whose Z score is greater than 2 and p value is less than 0.05 [9], such as No. 1, No. 4 and No. 5.

Table 1. Sub-graphs with three nodes discovery

No.	Sub-Graph	Original-Fre	Mean-Fre	Z-Score	P-Value
1		55.122%	55.072%	5.518	0
2		37.312%	37.441%	-5.669	1
3		5.0493%	5.501%	-5.708	1
4		2.479%	1.943%	5.772	0
5		0.008%	0.004%	2.769	0.001

According to our statistical analysis, No. 1 and No. 2 have the highest occurrence frequency, which indicates that developers' favorite reuse pattern is that a project is reused by other projects or a project reuses other projects, but other projects have no reuse relationships. Especially, No. 2 conform to Lego hypothesis [11], which states that a software project is constructed out of other smaller projects that are relatively independent of each other. The analysis of No. 3 and No. 4 are discussed through two typical examples about log and test shown in Fig. 5.

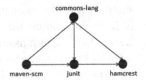

Fig. 5. An example of test projects.

Figure 4(a) reflects that the application project reuses "commons-logging" and "commons-lang" reuses "log4j". In addition, Fig. 4(b) reflects that the application project also reuses "log4j". In fact, two edges in (a) have no connections virtually, which implies software reuse is without hierarchy."commons-logging"offers a simple logger, and "httpclient", a project that is responsible for creating and maintaining a toolset of low level Java clients focused on HTTP and associated protocols, reuses this simple logger rather than "log4j". Moreover, "commons-logging" is also an interface of implementing log, and "mahout", a project that is to build an environment for quickly creating scalable performant machine learning applications in (b) calls implementation through interface to apply to more complex logger, but still needs to include "log4j", which implies software reuse is without transitivity. Thus, we infer that developers prefer to reuse a more comprehensive function of log when they develop most of more complex projects.

Figure 5 reflects the reuse relationships among test projects and their application projects. Project "hamcrest" is a framework of "junit". It also offers the function of more complex test. "commons-lang" as a basic tool kit reused by most projects needs to guarantee its correctness through a large number of complex tests such as the test of the function about string. Thus, it also customizes the function of "hamcrest" based on "junit". While "maven-scm" as a project of high level just reuses "junit" to do a simple and functional test. As for "commons-lang", "maven-scm" reuses its other functions. Thus, we infer that developers usually prefer to reuse a more comprehensive function of test whey they develop most of more common projects.

To some extent, we can infer that whatever they reuse, they all reflect a kind of collaborative development model, which can cut down the time and improve the quality of development. Through these models based on wisdom of developers, we can gain some development experience. Thus, in the aspect of gaining collective intelligence in order to improve this virtual world, software projects in a software world are as important as people in social networks.

The No. 5 in Table 3 is a circular topology structure, which seldom appear in our study. Corresponding to software development, developers ought to avoid reuse software projects that depend on each other as much as possible, especially circular dependency, to reduce the complexity of the system and increase its stability. In spite of this, No. 5 is still needed in some cases as a special pattern in software engineering.

5   Related Works

OSS is highly open and free to share. Thus, they have become one of the most effective development resources in software engineering in recent years. They are suitable for software reuse, which could improve development efficiency by avoiding duplication of efforts and provide better quality. Madanmohan [12] found that more and more software companies seek to speed up the software release to gain a competitive advantage by reusing through investigating many famous software enterprises in United States and India. Brown [13] also found that reusing is given priority to open source resource reuse in commercial software development, whose forms are mainly code lines reuse, algorithm reuse even entire component reuse. Compared with previous works, our work focuses on OSS reuse ecology to analyze the reuse relationships among Java software projects in order to find the groups of projects that are needed effectively and efficiently.

Most works of software ecosystem can be divided into two types. One is that they constructed software ecosystems by technical, business and social dimensions [14]. Moreover, analyzing source codes of projects is a common method to identify the reuse relationships among these projects, such as [15], which need a large amount of storage space and processing time. And community detection algorithms were used to identify communities in GitHub [16], but all of them paid attention to the relationships between developers or between projects of which developers are the mediation. The other is that they focused on the analysis of a certain ecosystem, such as Eclipse [15] and Apache [16].

In our work, we identify the reuse relationships of GitHub-hosted Java software projects by extracting from manager tools and use the community detection and pattern discovery algorithms on the Java reference ecosystem modeled as a network which could be analyzed in-depth from macro and micro perspectives to guide developers on how to reuse software projects accurately, effectively and efficiently better.

6   Conclusion

In this paper, we identify the reuse relationships among GitHub-hosted Java projects by extracting from manager tools and map them to a complex network representing a reference ecosystem regarded as a software world. We use the community detection and pattern discovery algorithms in a specific area to analyze the characteristics and patterns of software reuse in order to reflect collective intelligence about reuse. The results of analyzing the characteristics of the global reference ecosystem of Java are two aspects. Firstly, developers usually prefer to concentrate on a few application-projects and library-projects when reusing. Second, they reuse the projects within the same organization together more often. Especially, there exist some well-known library-projects with cross cutting functionality regarded as the bridges between organizations to be reused by many application-projects in different domains. Another results are from analyzing the patterns. The presence of the frequent patterns suggests that developers' favorite reuse patterns are that a project is reused by other projects or a project reuses other projects, but other two projects have no reuse relationships. Meanwhile, the existence of the other patterns indicates that they select the software projects reused with similar function varies widely depending on different requirements. Meanwhile, these patterns reflect that software reuse is without hierarchy and transitivity, and ought to avoid circular dependency as much as possible.

Our study analyze how OSS are reused on the characteristics and the patterns of reference ecosystems. These analysis can be regarded as collective intelligence and they can be applied to the software development. Our study opens the door for future work in OSS recommendation including the recommendation of the software projects except well-known library-projects with cross cutting functionality and the software projects with similar function, which can guide collaborative software projects development based on collective intelligence.

Acknowledgment.   This work is supported by the National Natural Science Foundation of China (Grant Nos. 61432020, 61472430 and 61502512).

References

1. Jia, Y., Shelhamer, E., Donahue, J., et al.: Caffe: convolutional architecture for fast feature embedding In: Proceedings of the ACM International Conference on Multimedia, pp. 675–678. ACM (2014)
2. Luo, Q., Zhong, D.: Using social network analysis to explain communication characteristics of travel-related electronic word-of-mouth on social networking sites. Tour. Manag. **46**, 274–282 (2015)
3. Dreżewski, R., Sepielak, J., Filipkowski, W.: The application of social network analysis algorithms in a system supporting money laundering detection. Inf. Sci. **295**, 18–32 (2015)
4. Šubelj, L., Bajec, M.: Community structure of complex software systems: analysis and applications. Physica A: Stat. Mech. Appl. **390**(16), 2968–2975 (2011)
5. Ossher, J., Bajracharya, S., Lopes, C.: Automated dependency resolution for open source software. In: Proceedings of 7th Working Conference on Mining Software Repositories, pp. 130–140. IEEE (2010)
6. Gousios, G., Spinellis, D.: GHTorrent: GIThub's data from a firehose. In: Proceedings of the 9th Working Conference on Mining Software Repositoties, pp. 12–21. IEEE (2012)
7. Newman, M.E.J., Girvan, M.: Finding and evaluating community structure in networks. Phys. Rev. E **69**(2), 26113 (2004)
8. Blondel, V.D., Guillaume, J., Lambiotte, R., Lefebvre, E.: Fast unfolding of communities in large networks. J. Stat. Mech. Theor. Exp. **10**, P10008 (2008)
9. Bastian, M., Heymann, S., Jacomy, M.: Gephi: an open source software for exploring and manipulating networks. In: International AAAI Conference on Weblogs and Social Media (2009)
10. Milo, R., Shen-Orr, S., Itzkovitz, S., et al.: Network motifs: simple building blocks of complex networks. Science **298**(5594), 824–827 (2002)
11. Wernicke, S., Rasche, F.: FANMOD: a tool for fast network motif detection. Bioinformatics **22**(9), 1152–1153 (2006)
12. Szyperski, C.: Component Software: Beyond Object-Oriented Programming. Addison-Wesley, Boston (1998)
13. Madanmohan, T.R., De', R.: Open source reuse in commercial firms. IEEE Softw. **21**(6), 62–69 (2004)
14. Brown, A.W., Booch, G.: Reusing open-source software and practices: the impact of open-source on commercial vendors. In: Gacek, C. (ed.) ICSR 2002. LNCS, vol. 2319, pp. 123–136. Springer, Heidelberg (2002)
15. Werner, C., Jansen, S.: A systematic mapping study on software ecosystem from a three-dimensional perspective. Software Ecosystems: Analyzing and Managing Business Networks in the Software Industry, pp. 59–81 (2013)
16. Businge, J., Serebrenik, A., Van den Brand, M.: Survival of eclipse third-party plug-ins. In: Processing of 28th International Conference on Software Maintenance, pp. 368–377 IEEE (2012)
17. Syed S., Jansen, S.: On cluster in open source ecosystem. In: Proceedings of International Workshop on Software Ecosystems, pp. 19–32. Citeseer (2013)

Specific Data Mining Model of Massive Health Data

Cuixia Li[✉], Shuyan Zhang, and Dingbiao Wang

School of Software and Applied Technology, Zhengzhou University,
97 Wenhua Road, Jinshui District, Zhengzhou 450002, Henan Province,
People's Republic of China
qyliying@126.com

Abstract. The mining of massive medicine data is one of the most widely problem in our world. However, it's efficiency and accuracy are still not satisfactory. Many traditional mining algorithms which calculate repeatedly to reduce the dependence between data always ignore the correlation between them. To improve the effect of diagnosis, we extract some special features by conducting a preliminary classification and identification. Then, the specific characteristics of various medical data is mined by correlation mining method. The simulation experimental results demonstrate the validity of the improved algorithm.

Keywords: Medical data · Specific data · Feature extraction · Correlation

1 Introduction

With continuous innovation and development of medical technology, more and more diseases have been overcome. At the same time, all kinds of medical data has also shown an explosive growth. Then how to dig out useful information from massive medical data has become a hot research topic. More and more scholars focus on this field. The main massive medical data mining methods include association rules mining, clustering algorithm based mining and Apriori model based mining [1]. Since it has importance in improving the efficiency of the diagnosis and treatment of disease, clustering based mining research has a broad space for development and much attraction from the scholars [2, 3].

Because massive medical data is usually stored in different structure database, the traditional mining methods for sparse database which has good performance of mining will greatly reduce for populated database [4]. This is due to the traditional algorithm in the particular data mining process needs to repeatedly scan the database [5].

In view of the defects of the traditional algorithm, the specific data mining method based on the feature extraction of massive medical data is proposed. With feature extraction method of massive medical data for preliminary classification and recognition, then the correlation degree of the characteristics of various medical data is calculated, finally the useful medicine information is mined. Simulation results also demonstrate the advantage of the improved algorithm.

© Springer Science+Business Media Singapore 2016
W. Che et al. (Eds.): ICYCSEE 2016, Part I, CCIS 623, pp. 632–640, 2016.
DOI: 10.1007/978-981-10-2053-7_56

2   The Ordinary Mining Method of Massive Medical Data

The method of ordinary data mining in massive medical data is described as follows (abbreviated as OMA) [5, 6]:

(1)   set the massive medical data of the sample set size n, order $I = 1$, set K random initial clustering center is: $Z_j(I)$, $j = 1, 2, \cdots, k$;
(2)   Using the following formula to calculate the distance between each medical data sample and cluster center

$$JC(I) = \sum_{j=1}^{k} \sum_{k=1}^{n_j} \left\| x_k^j - Z_j(I) \right\|^2 \tag{1}$$

(3)   Update the cluster centers by using the following formula

$$Z_j(I+1) = \frac{1}{n_j} \sum_{i=1}^{k} X_i^j, \ j = 1, 2, \cdots, k \tag{2}$$

(4)   Judging the end conditions of algorithm, if it is $\|JC(I) - JC(I-1)\| < \xi$, the operation is ended. On the contrary, we repeat Step 3.

According to the above method, we can mine the characteristics of the rough similar specific medical data and this requires the use of the relevant principles of gray correlation. In the class database of medical data, further extract those data, which has a high degree of relevance, and use these data as the basis for accurate data mining.

Correlation analysis is a mining method according to the gray system theory and develops into a mining method for relating the degree of association data. The core idea is that judge the similarity between the different by using the degree of similarity between the curves. The particular methods are as follows [7]:

(1)   Construct the sequence matrix of medical data. After use clustering classifier for classification and data mining in specific medical data, we need the further association sort analysis. Reference sequence, also known as the mining for the characteristics of specific medical data, can be described with T_0. By using each data in the medical data classification database form a comparison sequence, which can be described with $T_1, T_2, \ldots, T_n$, so the sequence matrix constructed by $n+1$ sequence can be described by the following formula.

$$(T_0, T_1, T_2, \ldots, T_n) = \begin{bmatrix} T_0(1) & T_1(1) & \cdots & T_n(1) \\ | & | & | & | \\ T_0(m) & T_1(m) & \cdots & T_n(m) \end{bmatrix} \tag{3}$$

(2)   Dimensionless Treatment. In order to eliminate the dimension, medical data is processed by the formula (4) to obtain the dimensionless matrix such as (5)

$$T_i'(k) = T_i(k)/T_i(1)$$
$$i = 1, 2, \ldots, n; k = 1, 2, \ldots, m \tag{4}$$

$$(T_0', T_1', T_2', \ldots, T_n') = \begin{bmatrix} T_0'(1) & T_1'(1) & \cdots & T_n'(1) \\ | & | & | & | \\ T_0'(m) & T_1'(m) & \cdots & T_n'(m) \end{bmatrix} \tag{5}$$

(3) Calculate the correlation coefficient between the $T_0'(k)$ and $T_i'(k)$, the formula is as follows.

$$\xi_{0i}(k) = \frac{\min_i \min_k |T_0'(k) - T_i'(k)| + \rho \max_i \max_k |T_0'(k) - T_i'(k)|}{|T_0'(k) - T_i'(k)| + \rho \max_i \max_k |T_0'(k) - T_i'(k)|} \tag{6}$$

Among them, $i = 1, 2, \ldots, n$, $k = 1, 2, \ldots, m$, $\rho \in [0, 1]$, ρ is the discrimination coefficient, which usually set the value as 0.5. Use the following formula to describe the matrix of correlation coefficient.

$$\begin{bmatrix} \xi_{01}(1) & \cdots & \xi_{0n}(1) \\ | & | & | \\ \xi_{01}(m) & \cdots & \xi_{0n}(m) \end{bmatrix} \tag{7}$$

(4) It can be calculated the correlation between other medical data and particular data by using the following formula.

$$r_{01} = \frac{1}{m} \sum_{k=1}^{m} \xi_i(k), i = 1, 2, \ldots, n \tag{8}$$

3 The Method of Specific Data Mining Based on Feature Extraction

During the process of specific data mining in centralized massive medical data, the traditional algorithm has good mining performance for sparse database, but for dense database mining performance will be significantly reduced. Thus, this paper proposes a method of specific data mining based on feature extraction in massive medical data.

3.1 Feature Extraction of Specific Data

The method of achieving specific data mining in massive medical data is based on the characteristics of specific medical data and extracts features. And then classify and recognize specific medical data extracted from massive medical data by using feature fusion.

During the process of going on extracting specific medical data, we need to weight for different medical data, to express the importance of these medical data in massive medical data. Usually we used Boolean method, that is, when a word in particular medical data appear in particular data, weighted 1, otherwise 0. The formulas are as follows [8]:

$$w_{ij} = \log(f_{ij} + 1.0) * \left(1 + \log\frac{1}{N}\sum_{k=1}^{N}\left[\frac{f_{ik}}{n_i}\log\left(\frac{f_{ik}}{n_i}\right)\right]\right) \tag{9}$$

Among them, f_{ij} is the frequency of words appearing in the medical data. n_i is the number of occurrences of the particular feature. f_{ik} is the frequency of words i appearing in the medical data k. N is the number of the medical data.

During the process of the specific data mining, we need to adopt normalized method to process. It can classify the keywords in medical data accurately for describing the importance of different keywords to achieve identifying the keywords in specific medical data. The weight of keywords can be modified as follows adaptively [9].

$$x(L_{ac}) = 1 - \frac{L_{ac}}{\max(L_{ac}) + l} \tag{10}$$

Using the following formula can describe the relationship between the number of medical data and the maximum correlation coefficient.

$$\max(L_{ac}) = \log_2 k \tag{11}$$

Among them, max is used to describe the maximum value of the correlation features and information entropy feature after extract feature in medical data. K is the number of categories of medical data; L is relevant coefficient. The method, which indicate how weight the inter-class and intra-class information entropy of comprehensive feature, can be described with the following formulas.

$$\begin{cases} W_{ik}(d) = \frac{IDF_1}{IDF_{const}} \times a(L_{ac}) \\ IDF_1 = coff_1 + tf_{ik}(d) \times \log\left(\frac{N}{n_k} + 0.01\right) \\ IDF_{const} = coff_{const} + \\ \sqrt{\sum_{i=1}^{n}(tf_{ik}(d))^2 \times \left[\log(\frac{N}{n_k} + 0.01)\right]^2} \end{cases} \tag{12}$$

Among them, $conff$ is the fixed coefficient of IDF_1 and $conff_{onst}$ is the fixed coefficient of IDF_{const}. In the above formula, we can achieve the desired effect of medical data classification through adjusting $conff_1$ and $conff_{const}$ continuously.

3.2   The Achievement of Specific Medical Data Mining

During the specific medical data mining, first of all to assume the database to be mined area. This area is regarded as an image, and the resolution of the image is $p \times q$. In the area W to be mined, assuming that we need to mine Q pieces of same specific medical data, coordinate parameters for each specific medical data is (x_i, y_i). The parameters of each particular medical data are the same, thus there is the same mining radius. Set mining radius to r, then the radius of the mining in mining area is A, corresponding to the center of the each coordinate of medical data (x_i, y_i) is $W = \{x_i, y_i, r\}$. Among them, W_i is mining radius, and target pixel which set to be mined in the mining area (the specific medical data) is (x, y). The calculation formula between the target pixel and the source node can be obtained by the principle Euler formula.

$$h(w_i, k) = \sqrt{(x_i - x)^2 + (y_i - y)^2} \tag{13}$$

Set the probability which the pixel point $A(x, y)$ in the mining area is mined by $K(r_i)$, and the probability formula is as follows.

$$K_{wpg}(x, y, w_i) = \begin{cases} 1 & d(w_i, k) \leq r - r_u \\ u(\frac{-a_1 \theta \beta_1}{\theta_2 \beta_2}) + a_2 & r - r_u < d(w_i k) < r + r_u \\ 0 & else \end{cases} \tag{14}$$

Among them, $r_u (0 < r_u < r)$ is the parameter of communication performance during data mining process, and $\alpha_1, \alpha_2, \beta_1, \beta_2$ is the medical segment detection coefficient data, which is related to the characteristics of medical data, and θ_i is an input parameter variable.

In a particular medical data mining process, the distance between medical data x_i and x_j can be described by the following formula

$$Q = \sum_{j \neq i} H(r - \|x_i - x_j\|) \tag{15}$$

Use the following formula to calculate the maximum difference component in the similarity of the two medical data.

$$\|x_i - x_j\| = \max_{1 \leq k \leq m} |x_{i-(k-1)\tau} - x_{j-(k-1)\tau}| \tag{16}$$

The similarity of medical data is not more than the vector of r, known as the associated vector. Setting the one dimensional medical data sequence to n, the number of vector points in phrase space reconstruction is $N = n - (m - 1)\tau$, and the presence of association in medical data accounted for the proportion of all possible kinds of $N(N - 1)/2$ pairs can be calculated by the following formula.

$$C_m(r) = \frac{2}{N(N - 1)} \sum_{i=1}^{N} \sum_{j=i+1}^{N} H(r - \|x_i - x_j\|) \tag{17}$$

According to the method outlined above, at first, we classify and identify preliminarily by using feature extraction for massive medical data, and then calculate the degree of correlation between the various characteristics of medical data by using relational mining, finally achieve accurate data mining for medical data by using the size of the association.

4 Experimental Results and Analysis

4.1 Lab Setting

In order to verify the effectiveness of the improved algorithm to be used in specific data mining in massive medical data, it is necessary to conduct an experiment. Experimental platform: CPU is Intel Core Duo E7500 2.8 GHz, memory is 4G DDR3, using simulation software Matlab 7.0 build experimental environment. The data in experiment is from a city center hospital, Henan Province. Comparative tests use traditional algorithms. The Simulation parameters constructed during the experiment can be described with the following Table 1, and adopt classified statistical for all medical data by using random medical vocabulary:

Table 1. Simulation parameters setting

Medical vocabulary	The amount of medical data	The amount of medical vocabulary
Anemia	23	343
Anaerobe	9	432
Antibody	12	322
Immunodeficiency	25	257
Monocytogenes	34	254
Cellular proliferation	26	423
Diplobacillus	6	316
Allergic history	15	358
Reduce swelling and ease pain	23	342
Vomit	26	257
Rheumatism	18	241
Cascara sagrada	35	273
Blood transfusion	31	421
Artifical breathing	24	273

4.2 Result Analysis

Through the sample data from form Table 1, we can know that all kinds of classification of medical data are balanced reasonably. The distribution of classified coefficient for each category of medical data can be described by Fig. 1.

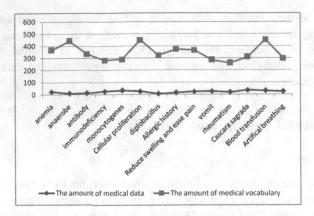

Fig. 1. Distribution of medical vocabulary

As it can be seen from Fig. 1, the classification of medical vocabulary is basically balanced and the inner class variance is small.

Using different algorithms to go on medical data mining specifically, mining results can be obtained by describing with Figs. 2 and 3.

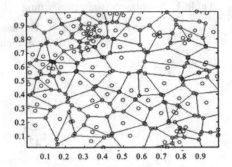

Fig. 2. The mining results of OMA algorithms

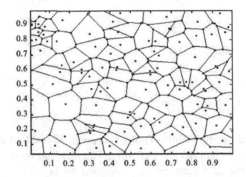

Fig. 3. The mining results of improved algorithms

From comparing the experimental results, we can go on detailed vocabulary Mining Classification by using the improved algorithm for specific data mining in massive medical data. This is due to the improved algorithm can calculate the correlation of medical data, and can extract the features of medical data effectively, thus avoiding the disadvantages of the traditional algorithm to achieve specific data-mining accurately in massive medical data.

During the above experiment, the time of different algorithms can be described in Table 2.

Table 2. Comparison of different algorithms mining time

Algorithm	Time of mining/s
OMA algorithm	25
Improved algorithm	14

According to the result from the front table, we can know that discovering particular data in mass medical data by using the improved algorithm is able to shorten data-mining time effectively and enhance the efficiency of particular medical data mining time. This shows that the improved algorithm can be applied to particular data-mining in large different medical data.

5 Conclusion

After analyzing the weaknesses of traditional algorithm, this paper put forward a particular data mining method which bases on feature extracting in mass medical data. Classify and identify preliminarily by using feature extraction in massive medical data, then calculate the degree of correlation between the various characteristics of medical data by using relational mining, finally achieve accurate data mining for medical data. The simulation results show that the improved algorithm was adopted to shorten data mining time to 11 s. We can spend less time and gain higher accuracy.

References

1. Han, I., Kamber, M.: Data Mining: Concepts and Techniques, pp. 335–389. Morgan Kaufmann Publishers, Berlin (2000)
2. Rosales, R.E., Bharat Rao, R.: Guest editorial: special issue on impacting patient care by mining medical data. Data Min. Knowl. Discov. **20**(3), 325–327 (2010)
3. Wang, Y.-F., Chang, M.-Y., Chiang, R.-D.: Mining medical data: a case study of endometriosis. J. Med. Syst. **37**, 9899 (2013)
4. Ceruto, T., Lapeira, O., Tonch, A., Plant, C., Espin, R., Rosete, A.: Mining medical data to obtain fuzzy predicates. In: Bursa, M., Khuri, S., Elena Renda, M. (eds.) ITBAM 2014. LNCS, vol. 8649, pp. 103–117. Springer, Heidelberg (2014)

5. Chen, H., Fuller, S.S., Friedman, C., Hersh, W.: Knowledge management, data mining, and text mining in medical informatics. In: Chen, H., Fuller, S.S., Friedman, C., Hersh, W. (eds.) Medical Informatics. ISIS, vol. 8, pp. 3–33. Springer, Heidelberg (2005)
6. Panwong, P., Iam-On, N.: Predicting transitional interval of kidney disease stages 3 to 5 using data mining method. In: 2016 Second Asian Conference on Defence Technology (ACDT), pp. 145–150 (2016)
7. Li, H., Guo, C.: Survey of feature representations and similarity measurements in time series data mining. Appl. Res. Comput. **20**(5), 1285–1291 (2013)
8. Gu, B., Sheng, V.S., Tay, K.Y., Romano, W., Li, S.: Incremental support vector learning for ordinal regression. In: IEEE Transactions on Neural Networks and Learning Systems (2015). doi:10.1109/TNNLS.2014.2342533
9. Chu, Y., Wang, Z., Chen, M., Xia, L., Wei, F., Cai, M.: Transfer learning in large-scale short text analysis. In: Zhang, S., Wirsing, M., Zhang, Z. (eds.) KSEM 2015. LNCS, vol. 9403, pp. 499–511. Springer, Heidelberg (2015)

Specular Detection and Removal for a Grayscale Image Based on the Markov Random Field

Fang Yin[1,3(✉)], Tiantian Chen[1], Rui Wu[2], Ziru Fu[1], and Xiaoyang Yu[3]

[1] School of Computer Science and Technology, Harbin University of Science and Technology,
Harbin 150080, China
[2] School of Computer Science and Technology, Harbin Institute of Technology,
Harbin 150001, China
[3] Instrument Science and Technology Postdoctoral Research Station,
Harbin University of Science and Technology, Harbin 150080, China
13936421412@163.com

Abstract. Specular detection and removal has been a hot topic in the field of computer vision. Most of the existing methods are mainly for color images, but grayscale images are widely used. For a single grayscale image with only intensity information, highlight detection and removal becomes a difficult issue. To solve this problem, the single grayscale image highlight detection and removal method based on Markov random field is presented. Each reflection component modeling is estimated by geometric relation of surface normal in diffuse and specular reflection component in the framework of Markov random field. Their maximum a posteriori estimation is calculated under Bayesian formula and highlight area is detected. Finally, image inpainting method based on the BSCB model removes highlights. Experiment reveals that this method can effectively detect grayscale image specular reflection area, improve highlight areas the repair rate.

Keywords: Computer vision · Specular detection · Markov random field

1 Introduction

An opaque object image is formed by reflecting incident light on the surface. Highlight presents the characteristics of the light source mainly, but it can be seen as the surface characteristics of the object in the visual effects. When the general image intensity values below a certain value, it belongs to the scope of diffuse, meanwhile, people's vision feel softer effect. When the reflected light is very strong, the image shows a highlight effect, and then people will have dazzling visual sense. Because of the highlight, surface texture features will weaken or even disappear, the original color of the object is obscured. Highlight results in losing partial areas information and affects the image quality, which handles computer vision image is a big distraction. It often leads to image segmentation, recognition and matching error. To be able to extract accurate object feature information and ensure that the image can be applied in image segmentation, recognition and matching, and other fields, highlights detection and removal technology is essential. Most of the existing highlight detection method is mainly for color image, rarely for the

© Springer Science+Business Media Singapore 2016
W. Che et al. (Eds.): ICYCSEE 2016, Part I, CCIS 623, pp. 641–649, 2016.
DOI: 10.1007/978-981-10-2053-7_57

grayscale image, but the grayscale image is very common in the field of computer vision. For a small amount of information available, grayscale image highlight detection and removal is a difficult problem. To solve this problem, gray image specular detection and removal method based on Markov random field is presented.

2   Related Work

There are many single image highlight detection and repair methods. Current technology is more mature, more widely used method is divided into the following categories: illumination-constrained inpainting, color space conversion method, dichromatic reflection model, etc.

2.1   Illumination-Constrained Inpainting

Compared different color characteristics between specular and diffuse light, illumination-constrained inpainting [1] algorithm gives a method of interaction detection surface color highlight areas. The method is different from the traditional highlight image areas repair methods, it fully uses some useful information included highlight pixels in the guiding the repair process. General inpainting method and illumination constraints are combined through a combination of constrained complementary color process (e.g. pixel value, the illumination chromaticity analysis, light color smoothness, etc.). The algorithm ensures it can overcome the shortcomings that the general inpainting method can not remain surface subtle shading. Compared with the previous detection and removal method of a single highlight image, the algorithm can provide a better the illumination chromaticity analysis to obtain more accurate results. However, this method requires manual intervention. Because algorithm is more complex, a large amount of information is required, so it is a huge time-consuming operation and is not suitable for real-time image processing.

2.2   Color Space Conversion Method

Color space conversion method is based on the RGB color space, studying the highlight areas repair problems from another angle [2, 3]. Because each component in the RGB space not only contains the luminance information but also contains color information, and each component has a certain correlation, it increases the difficulty of repair highlight areas. If a color space can separate chrominance and luminance, then repair problem of highlight areas will become simple. Color space conversion method can achieve the desired effect, which repair highlights area well after histogram equalization for the luminance Y, but the method still exists insufficient, e.g. color space conversion method will make the image texture details lost adjusting luminance Y.

2.3 Dichromatic Reflection Model

Dichromatic reflection model describes essentially [4–6] two reflection processes, specular reflection Component and diffuse reflection Component. When the object surface is relatively smooth, specular reflection Component is stronger than the diffuse reflection Component; on the contrary, the surface is rough, diffuse reflection Component is stronger than the specular reflection Component. There is a link between surface structure and spectrum of the incident light. If the surface structure is smooth, spectral structure does not change, and the color will not change; on the contrary, because the rough surface will lead to changing the spectrum structure, so the color will change. On the basis of the dichromatic reflection model, the each pixel color of object is regarded as a linear combination between the specular reflection component and the diffuse reflection component.

Klinker et al. [7] proposed a single image highlights detection and removal algorithm according to Shafer's dichromatic reflection model [8]. Klinker found a T-shaped formed from distribution diffuse pixels and highlight pixel in the RGB color space. Principal component analysis was used to fitting diffuse and light color vector in the diffuse reflection area and highlight regions. Specular reflection component was removed quickly by the use of these two vectors for projection. However, the highlight pixel clusters usually is distorted because of the surface roughness, geometry [9] et al. Therefore, principal component analysis to estimate the light source colors is usually inaccurate, which largely reduces the versatility of the method.

Although the above methods are able to achieve a good effect of highlight detection and removal, these methods analysis mainly for color image. It can not be applied to grayscale image highlight detection and recovery. The single grayscale image highlight detection and removal method based on Markov random field is presented.

3 Reflection Model

On the basis of the dichromatic reflection model, because normal direction of the diffuse light points is inconsistent and the normal direction of the specular reflection light is unchanged, using the geometric relationship between the diffuse and specular surface normal vector locates the highlight area. Relationships between geometric constraints are shown as Fig. 1, The dichromatic reflection model is given as follow:

$$I = m_d(\vec{n}, \vec{s}, \vec{v}) \int_\lambda f(\lambda)e(\lambda)c_d(\lambda)d\lambda + m_s(\vec{n}, \vec{s}, \vec{v}) \int_\lambda f(\lambda)e(\lambda)c_s(\lambda)d\lambda. \tag{1}$$

Here $m_d(\vec{n}, \vec{s}, \vec{v})$ and $m_s(\vec{n}, \vec{s}, \vec{v})$ represents the weight function of diffuse and specular reflection respectively; $\vec{n}$ is the surface direction, $\vec{s}$ is direction of the light source, $\vec{v}$ is direction of the view; $f(\lambda)$ represents sensor transfer function for the three primary colors; $e(\lambda)$ represents the incident spectral energy distribution function; $c_d(\lambda)$ is diffuse reflectance; $c_s(\lambda)$ is specular reflectance.

Specular reflection is reflected parallelly by the bundle of parallel incident light from the surface to a certain direction, mainly reflecting the characteristics of the source. As shown in Fig. 1, specular reflection of the incident light (i.e. light source), surface normal vector and the reflected light (viewing direction) is in the

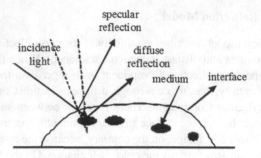

Fig. 1. Geometry of specular reflectance.

same plane, its incidence angle (the angle $\vec{s}$ between $\vec{n}_s$) and the reflection angle (the angle $\vec{v}$ between $\vec{n}_s$) is equal. According to the general definition, specular surface normal vector can be expressed in the form of Formula 1.

$$\vec{n}_s = \frac{\vec{s} + \vec{v}}{\|\vec{s} + \vec{v}\|}. \tag{2}$$

Specular reflection model is defined by the Beckmann distribution as follows:

$$P_s(i,j) = \frac{1}{\sqrt{2\pi}\sigma_s} \exp\left[-\frac{1}{2}\left(\frac{\theta}{\sigma_s}\right)^2\right]. \tag{3}$$

$P_s(i,j)$ is specular reflection distribution, σ_s is the variance of the angular distribution parameter control.

Diffuse reflection is projected by parallel incident light to opaque objects, due to the normal direction of each point is inconsistent, resulting in the phenomenon of reflection light reflected irregularly in all directions. Diffuse reflection light reflecting surface features of the object is shown as Fig. 1. According to Lambert's law, the intensity value of the point (i,j) is usually given by the formula:

$$I(i,j) = \vec{n}_{i,j} \cdot \vec{v}_{i,j}. \tag{4}$$

where $I(i,j)$ is the surface normal vector of the pixel (i,j).

We assume that the observed specular intensities follow a Gaussian distribution. Under these assumptions we can write

$$P_d(i,j) = \frac{1}{\sqrt{2\pi}\sigma_d} \exp\left[-\frac{1}{2}\left(\frac{I(i,j) - \vec{n}_{i,j} \cdot \vec{v}_{i,j}}{\sigma_d}\right)^2\right]. \tag{5}$$

Where $P_d(i,j)$ is the diffuse distribution with variance σ_d^2, the mean value is defined by $\vec{n}_{i,j} \cdot \vec{v}_{i,j}$.

For a given single grayscale highlight image I, maximum posterior probability is estimated by Bayes formula:

$$(P_d(i,j), P_s(i,j)) = \operatorname*{argmax}_{P_d, P_s} I(I_d, I_s | I). \tag{6}$$

Bayesian criterion:

$$P(I_d, I_s | I) \propto P(I | I_d, I_s) P(I_d | I_s) P(I_s) \tag{7}$$

Since the original image without highlight and the specular reflection component are uncorrelated, Eq. (7) can be simplified to:

$$P(I_d, I_s | I) \propto P(I | I_d, I_s) P(I_d) P(I_s) \tag{8}$$

where $P(I | I_d, I_s)$ is the likelihood function, showing before and after the restoration error obedience distribution, $P(I_d)$ and $P(I_s)$, respectively, represent a priori of I_d and I_s).

We assume that likelihood function $P(I | I_d, I_s)$ follow a Gaussian distribution. Under these assumptions we can write

$$P(I | I_d, I_s) = \prod_{i,j} \exp(-|I(i,j) - I_d(i,j) - I_s(i,j)|^2) \tag{9}$$

In Fig. 2 we provide some more detailed analysis of the specular reflection model and diffuse reflection model. Figure 2a shows three pixels: a (a diffuse pixel), b (a specular pixel located between pixels a and c), and c (the highlight's brightest pixel).

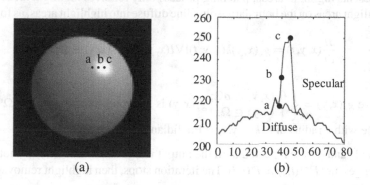

$$(a) \qquad\qquad\qquad (b)$$

Fig. 2. (a) Original image. (b) Specular reflection model and diffuse reflection model.

4   Image Inpainting Method Based on the BSCB Model

Image inpainting of highlight areas can be seen as the original image the recovering process. Most inpainting methods analyze color pictures, so it can not be applied in a single grayscale image. BSCB model is presented by Bertalmio, Sapiro, Caselles and Ballester according to actual image patching process of manual repairing artists, and it is an image inpainting method established based on partial differential equations. Its main idea is based on the manual repair experience, edge information diffusion into the area to be repaired along the isolux line direction, shown in Fig. 3.

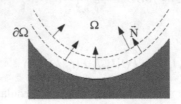

Fig. 3. BSCB model repair process of information diffusion.

We suppose $I_0(i,j)$ is the original highlight grayscale image with size m × n, and establish a set of image sequences $I(i,j,n)$, here $I(i,j,0) \rightarrow I_0(i,j)$, and $\lim_{n\to\infty} I(i,j,n) = I_R(i,j)$, in which $I_R(i,j)$ is the final result of repair.

BSCB model patching process as following two steps:

(1) mathematical model of BSCB repair process can be written as follows:

$$\text{for } I^{n+1}(i,j) = I^n(i,j) + \Delta t I_t^n(i,j) \text{ any } (i,j) \in \Omega. \tag{10}$$

where n represents iteration times, (i,j) is the pixel coordinates point of the grayscale image, Δt represents iterative step, $I_t^n(i,j)$ is image $I^n(i,j)$ without highlight, Ω represents highlight areas, patching process only removes highlight areas.

(2) highlight areas on the boundary isolux line diffuse into highlight areas, as following:

$$\frac{\partial I}{\partial t}(x,y,t) = g_\varepsilon(x,y)k(x,y,t)|\nabla I(x,y,t)|, \quad (x,y) \in \Omega^\varepsilon. \tag{11}$$

where $g_\varepsilon(x,y) = \begin{cases} 0 & (x,y) \in \partial\Omega; \\ 1 & (x,y) \in \Omega. \end{cases}$ $g_\varepsilon(x,y)$ is a smooth function on the Ω^ε. Ω^ε is a circle with a radius of ε. $k(x,y,t)$ is Euclidian curvature of isolux line.

Steps (1) and (2) are repeated until the output image sequence $I^n(i,j)$ without significant changes, i.e. $I^{n+1}(i,j) \approx I^n(i,j)$. The iteration stops, then highlight removal process is complete.

5   Experimental Result

Figure 4a, b, and c shows the result of highlight detection and removal. Figure 5a shows the images in highlight areas relatively steep. However, Fig. 5b is smoother than the Fig. 5a. In order to further quantify the comparison, to verify the validity of this method, as shown in Fig. 6. Figure 6a and b show, respectively, original image of Fig. 3a the first row and image after highlight removal. Compared with Fig. 6a, higher luminance values reduce and luminance value distribution remained largely unchanged in the Fig. 6b. Experiment reveals that the method is better in processing details; it is able to effectively detect small specular reflection area hidden in grayscale images, and only remove the highlight region, so it does not affect the other areas of the image information, retaining the original image information in the maximal degree. Compared with the traditional algorithms, the method is better to prevent image distortion.

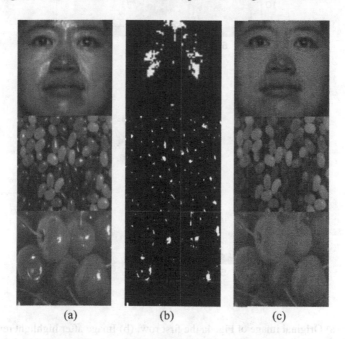

(a)   (b)   (c)

Fig. 4. (a) Original image. (b) Specular image. (c) Image removed specular

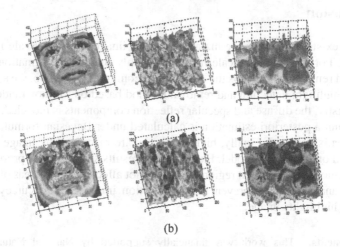

(a)

(b)

Fig. 5. (a) Original three-dimensional diagram. (b) Three-dimensional diagram after removed the highlight

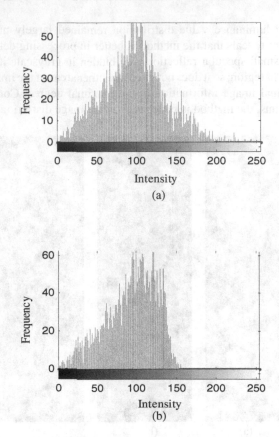

Fig. 6. (a) Original image of Fig. 4a the first row. (b) Image after highlight removal.

6 Conclusion

Most of the existing methods are mainly for color images, but grayscale images are widely used. For a single grayscale image with only intensity information, specular detection and removal has been is the difficulty of computer vision. So, the single gray-scale image highlight detection and removal method based on Markov random field is provided. Firstly, the diffuse and specular reflection components is modeled. Secondly, their maximum a posteriori estimation is calculated under Bayesian formula, and then highlight area is detected. Finally, highlight areas are recovered by image inpainting method based on the BSCB model. Experimental results show that the proposed algo-rithm only remove the highlight region, so it does not affect the other areas of the image information, and it is better to prevent image distortion, improves the accuracy of surface recovery for highlights image.

Acknowledgments. This work was financially supported by National Natural Science Foundation of China (61440025), the research project of science and technology of Heilongjiang provincial education department (12541119).

References

1. Tan, P., Lin, S., Quan, L., Shum, H.Y.: Highlight removal by illumination-constrained inpainting. In: IEEE International Conference on Computer Vision, vol. 1, pp. 164—169. IEEE Press, Nice (2003)
2. Schluns, K., Koschan, A.: Global and local highlight analysis in color images. In: Proceedings of 1st International Conference on Color Graphics Image Processing, pp. 300–304 (2000)
3. Lee, H.-C.: Method for computing the scene-illuminant chromaticity from specular highlights. J. Opt. Soc. Am. A: **3**, 1694–1699 (1986)
4. Tan, P., Quan, L., Lin, S.: Separation of highlight reflections on textured surfaces. In: 2006 IEEE Computer Society Conference on Computer Vision and Pattern Recognition, pp. 1855–1860. IEEE Press, New York (2006)
5. Yoon, K.J., Choi, Y., Kweon, I.S.: Fast separation of reflection components using a specularity-invariant image representation. In: 2006 IEEE International Conference on Image Processing, pp. 973–976. IEEE Press, Atlanta (2006)
6. Yang, Q., Wang, S., Ahuja, N.: Real-time specular highlight removal using bilateral filtering. In: Daniilidis, K., Maragos, P., Paragios, N. (eds.) ECCV 2010, Part IV. LNCS, vol. 6314, pp. 87–100. Springer, Heidelberg (2010)
7. Klinker, G.J., Shafer, S.A., Kanade, T.: The measurement of highlights in color images. Int. J. Comput. Vis. **2**, 7–32 (1990)
8. Shafer, S.A.: Using color to separate reflection components. Color Res. Appl. **10**, 210–218 (1985)
9. Noak, C.L., Shafer, S.A.: Anatomy of a color histogram. In: Proceedings of the IEEE Computer Vision and Pattern Recognition, pp. 599–605. IEEE Press, Champaign (1992)

Teaching Reformation and Practice of ".NET Program Design" Course Oriented CDIO Model

RuiGai Li[✉] and AChuan Wang

College of Information and Computer Engineering, Northeast Forestry University,
Harbin, China
lirg751@163.com

Abstract. ".NET Program Design" is a very important professional elective course in the Department of Computer Science and Technology. It is a practical, comprehensive curriculum with highly specialized applications. This paper analyses the disadvantages of traditional teaching model in the .NET Program Design, and discusses the methods and practices for teaching reform such as introducing CDIO engineering education model to classroom teaching with project cases, arranging teaching content properly, choosing modern teaching methods and reform on the curriculum assessment system. The practice has proved that this model with highly effective in teaching process can promote the training of students abilities in all aspects.

Keywords: .NET Program Design · CDIO engineering education model · Teaching reformation

In recent years, NET platform and Java platform has become the two common platforms for network program development. Accordingly, ".NET Program Design" course has an increasingly important position in the field of computer related areas of undergraduate education activities. It is a highly comprehensive and theoretical course with rich teaching content and strong practice. The learning of this course, not only can cultivate students' programming skills, practical ability and computer thinking, but also can cultivate students' engineering skills and team cooperation ability. However, in the long-term teaching process, teachers paid too much attention to the theory of teaching and ignored the practice activities, which resulted in many serious problems such as students' poor practice ability, weak engineering ability and so on. The emergence of CDIO engineering education concept provided a practical way for the teaching reform of the course.

CDIO engineering education concept was founded by four universities, including Massachusetts Institute of Technology and the Royal Institute of technology in Sweden. It advocated "learning by doing" and "project-based teaching" education model. In the teaching process, it focused on the life cycle of engineering projects which lasted from research & development to operation. In order to guide students to take the active learning, it paid more attention to the organic link between the curriculums, and claimed to learn the engineering course from the view of practice. In CDIO training syllabus, engineering graduates' ability was divided into four levels, including engineering basic knowledge, personal ability, team working skill and engineering capability. It is helpful

© Springer Science+Business Media Singapore 2016
W. Che et al. (Eds.): ICYCSEE 2016, Part I, CCIS 623, pp. 650–656, 2016.
DOI: 10.1007/978-981-10-2053-7_58

for student to reach a predetermined target in the four levels through a comprehensive training [1].

1 Position and Function of the Course

".NET Program Design" is a very important and professional elective course in computer science and technology, information security, software engineering, information management and information system, geographic information system etc. It is the also a core curriculum for software engineering or other related majors, and plays a very important role in achieving the personnel training objectives and enhancing the employability of students.

".NET Program Design" plays a connecting link between the courses in the personnel training system of computer science. ".NET Program Design" has the inheritance relation in the teaching content to the leading courses, such as "advanced language program design", "C++ program design", and provides a basic programming and algorithm realization tool for the practice of follow-up courses, such as "digital image processing", "multimedia technology".

Teaching quality of ".NET Program Design" is directly related to the student's employment rate. NET platform occupies a larger share in the system development market. NET and J2EE, as two present major popular system development platforms, are often compared with each other. However, continued warming of J2EE training market led to a situation that the J2EE programmer is relatively saturated. The NET programmer is relatively scarce, which resulted in that the demand for NET developers always maintained a relatively high level. Employment prospect of NET platform is very broad, so the course construction will play a greater role in promoting the employment rate of students.

2 Problems Existing in the Current Educational Model of ".NET Program Design"

Our teaching team conducted an extensive investigation and study on the teaching situation about ".NET Program Design" in many colleges and universities, and found that the traditional classroom teaching mode had obvious advantages in knowledge impartment, but it was very difficult to accomplish the cultivation of students ability of practice and engineering quality in practice teaching process. The main problems mainly focused on the four following aspects:

(1) Teaching mode is out of date: At present, teacher-centered classroom teaching mode is still very popular in many colleges and universities. Teacher plays a leading role in the teaching process, so that students can only accept knowledge passively. Ignoring the individual differences among students leads to a serious problem that students' interest and sense of participation is very low. Affected by the decrease of class hours, the teaching content is relatively simple, and teaching efficiency is low. At the same time, the traditional classroom teaching environment is relatively

simple. In the classroom, the interaction between teachers and students is less; while it is very difficult for students and teachers to communicate effectively after class.

(2) The teaching content is scattered: In traditional teaching mode, teacher chose and planned teaching content based on the analysis of content structure of the course, divided the knowledge system of course into several sessions, with each session covering one knowledge point, and then taught course knowledge point step by step. In teaching process, students learnt to master the knowledge points step by step, and wrote simple code to verify the validity of the knowledge in practice process and to solve some simple problems. The relation between knowledge points was ignored, so it is very difficult for students to develop a knowledge system of the course and gain a strong practical ability. In addition, it spent too much emphasis on grammar, so students often learnt a bunch of grammatical knowledge, and a little about how to develop practical projects. They were also very confused when they started to design and develop a real project. Meanwhile, it was less even rare to cultivate the student's ability about how to determine the personnel division in a large project, and how to coordinate and manage the manipulation of a comprehensive project, which resulted in a very serious problem that students' team cooperation ability is very weak.

(3) Teaching technology is not reasonable: At present, it is very common for many colleges and universities to use multimedia technology as a presentation tool for a variety of courses. It plays a good display effect in the process of teaching. However, it is not suitable for all courses. The use of multimedia technology should be combined with the specific characteristics of the course. Multimedia technology makes students spend too much attention to the knowledge point that is projected onto the screen, and ignore the overall structure of the course.

(4) The course assessment mode is too simple: Many universities selected "Written Examination + Computer Examination" as the main test mode for ".NET Program Design" course. This mode can't really examine the true ability of students, so it discouraged the learning enthusiasm of students. Therefore, it is necessary to formulate a scientific and reasonable evaluation mechanism. In addition, the daily study and examination was aimed at a single student than at a team, and this resulted in the lack of training in teamwork and communication abilities of students.

3   The Feasibility Analysis of Introducing the Concept of CDIO into the Course Teaching Reform

The concept of CDIO inherited and developed the idea about reform of engineering education that was proposed at 20 years ago in the United States, and what's more, it had been put forward systematically 12 standards for capacity-building, implementation guideline, implementation process and result verification which was very easy to operate in the actual teaching work. Its implementation guide was very comprehensive including many aspects such as training plan, teaching methods, teacher training, student assessment and learning environment. CDIO emphasized that the cognitive process required a double stimulation. That is to say, not only can the engineers understand the abstract knowledge logically, but also to experience the application of the knowledge in the actual environment, in order to obtain a deeper understanding [2].

Tsinghua University introduced CDIO teaching mode to the teaching process of "data structure" and "principle of database system", and achieved an effective teaching effect. CDIO teaching mode strengthened students' multiple abilities such as self-learning, solving practical problem, coordination of communication and team cooperation abilities. Many examples had achieved good teaching results, such as the EIP-CDIO talent training model proposed by Fujian University of Technology and the A-H-CDIO talent training model proposed by Shantou University by the end of 2005. Both domestic and overseas experience had shown that the idea and method of "learning by doing" in CDIO was advanced and feasible, and it was suitable for all aspects of the teaching process of engineering [3]. Because ".NET Program Design" is of obvious characteristics in engineering, it is very suitable to develop the teaching activity with project-driven teaching method. The software development process is similar to the four parts of CDIO Engineering Education, so CDIO engineering education mode is especially suitable for the teaching process of courses in computer-related majors.

4 Practice of Teaching Reform of ".NET Program Design" Based on CDIO Engineering Education Thought

CDIO proposed 12 operable executive standards, and systematically summarized the main engineering education demand. The central idea was to guide relevant staff of engineering education through 12 standards to meet the need of Engineering Education based on the available resources in different conditions [4]. Referring to the 12 standards of CDIO, the author carried out a large number of explorations with a result of many useful experiences in the actual teaching process. The specific implementation process can be summed up in the following aspects:

4.1 Set up a CDIO Environment Background

The development of software project takes the production cycle of software product as the environment of engineering education. The author built a teaching environment based on CDIO engineering education mode, and introduced many actual software projects for teaching activities. Based on the project case, the students were required to complete a specific project development from the four aspects of conception, design, implementation and operation. Therefore, from the beginning the teaching process, students were divided into several learning groups, 3 to 5 persons per group, then each group became a development team with a specific software project. Through the whole teaching period, teachers arranged teaching content according to a software project example, and team members worked according to the role of division of labor, learnt from each and helped each other. It was proved that this method can effectively improve the students' ability of teamwork and cooperation.

4.2   Determine Learning Goals

The goal of this course is to require students to use Visual Studio. NET to write, debug and operate a web application with utility, standard and good readability. It makes the students have an overall conception about conception, design, implementation, and operation of a dynamic network software project from a relatively high level. It attached much importance to practice experience of student rather than a specific grammar in detail. The course goals can meet the requirements about project development of small and medium software enterprises.

4.3   Integrate Various Information Technology Tools

During the classroom teaching period of the course, a variety of information technology means were integrated to create an advanced teaching environment, in order to meet the needs of students for exploratory learning and active learning. Aimed at the key points, difficult points and confusable concepts, teachers recorded video presentations and uploaded it to the network platform, so that the students can use the online learning environment to take self-study and have a better understanding about the content. At the same time, more extracurricular activities were carried out. The students were organized online to discuss and study the latest technologies of .NET platform. It was very helpful for students to deepen the understanding about classroom teaching content, broaden their horizons, and improve their programming abilities.

4.4   Select the Appropriate Teaching Content

During the teaching period, teachers in team group gave full consideration to the differences between students and unique character of every student, and chose the project case of moderate difficulty for every development group according to the specific conditions of students in group. According to the regular pattern of education development and talent cultivation, teachers organized classroom teaching activities and each teaching process scientifically and appropriately. The teaching content was arranged reasonably aimed at the specific issues in the process of the development and combined with the specific implementation process of the project case. At the same time, the choice of teaching content considered the actual needs of the project and requirements of enterprise for development environment and programming ability, so that students can produce a sense of identity to teaching content and teaching methods.

4.5   Reform on Curriculum Assessment Model

Because the traditional simple assessment method can't truly measure the actual ability of students, the assessment method of this course was adjusted to an integrated assessment model which gave priority to a comprehensive practical examination and supplemented by a variety of assessment methods. Curriculum assessment results consist of four parts:

(1) Project performance results: 40 points. It mainly inspects the situation of completion, operation, organization management, project report, and question answering about the project of each development group from the overall point of view.

(2) Final written examination results: 30 points. It is mainly to test students for the situation of grasping basic theory knowledge with an open-book written examination. The paper contains multiple-choice questions, fill-in-the-blank questions, short answer questions, program reading questions, and programming design questions.

(3) Innovation performance results: 20 points. This assessment method lets the student select a specific problem during the implement or the process of a project to study, through accessing to relevant literature, and combining with the current curriculum cutting-edge technology and method to design specific solutions and implementation, then to write an article as the evidence of evaluation.

(4) Class performance results: 10 points. It mainly examines the contribution of students to the discussion forum, attendance, question answering in class and preview before class in ordinary teaching process.

This integrated assessment model examines the students' abilities of multi-aspects, including learning ability, practical ability, problem analysis ability and innovation ability. It can stimulate the learning interest of students and is very useful for improving the comprehensive ability of students.

5   Conclusion

CDIO Engineering Education Model regards the process of concept, design, implementation and operation of an actual project as engineering education background, pays more attention to training of students' practical ability, self-learning ability, innovation ability, team cooperation ability according to the specific needs of the industry for engineering talent. Practice has proved that introducing CDIO Education Mode with project case driven teaching method to teaching process can effectively stimulate student's study enthusiasm, and develop the students' abilities to find and solve problems, control and manage an actual project in cooperation with others.

Acknowledgment. This paper is supported by two projects:

1. 《.NET Program Design》 Key Curriculum Construction Project sponsored by Northeast Forestry University in 2014;

2. Heilongjiang Province Higher Education Teaching Reform Project 《the Research and Practice of Undergraduate Computer Class "Excellent Engineers Training Project"》 (JG2013010103).

References

1. Wang, S.W., Hong, C.: CDIO: the classical model of Engineering Education in MIT – An Unscrambling on the CDIO Syllabus. J. High. Educ. Sci. Technol. 28(4), P116–119 (2009)
2. Ye, D., Wei, F., Han, S.: The study of ASP.NET programming teaching reform under the concept of CDIO Education Thought, Fujian. Computer **4**, 37–38 (2011)
3. Gu, P., Shen, M., Li, S.: From CDIO to EIP-CDIO – A probe into the Mode of Talent Cultivation in Shantou University. High. Eng. Educ. Res. **1**, 12–20 (2008)
4. Wang, G.: Interpretation and thinking of CDIO engineering education model. China High. Educ. Res. **5**, 86–87 (2009)

The Framework Design of Intelligent Checkers

Biao Wang[1], Lijuan Jia[2(✉)], He Quan[1], and Changsong Zheng[1]

[1] School of Automation, Harbin University of Science and Technology,
Harbin 150080, Heilongjiang, China
[2] School of Computer Science and Technology, Harbin University of Science and Technology,
Harbin 150080, Heilongjiang, China
515333669@qq.com

Abstract. As one of the research direction in the field of artificial intelligence, the computer game has a great practical significance. Checkers become a represent of complete knowledge game that because of it has few chess pieces, and moves easily, So it can reflect the thought of the computer game better. First, the recursive algorithm of artificial intelligence can achieve the independent moves and checkmates in draughts game program. Second, the optimization of evaluation function makes the program choose beneficial move ways. Third, the game tree can improve the level of computer in-depth thinking, simplify the search scope, and save computer thinking time by pruning, which improve the level of chess program. In this paper, overall framework of intelligent checkers is built from board representation, game tree, valuation optimization. Based on the accumulation of experience, algorithm upgrade, depth analysis of existing framework, then we give a more optimized system framework and then strengthen the computer game level.

Keywords: Computer game · Checkers · Framework design · Game tree

1 Introduction

The national contest of college students' computer game [1] has been held four sessions since 2011, aimed at regarding high resistance [2] board games as the research carrier. We can simulate our interest in learning program and improve the team cooperation ability by the contest. Checkers as one of the computer game events at the competition, the oldest and most popular one of the puzzles all over the world. The competition abroad has been held since the 1970s, the old age, students and youth who all take part in the competition, The ability is more and more strong and the compete increasingly fierce; From the beginning of this century, the domestic hosts game championship, the domestic has carried out college students' computer game series since recent years, which obtained the response of the college students amateur. There are few of quality strong university in the contest. Computer game reflects comprehensive knowledge level of the student's team.

Our framework of checkers to mainly optimize the combination of recursive algorithm, game tree, ALPHA BETA algorithm and algorithm of mini-max [3]; then we optimize these algorithms in game competition. We divide moves of chess into three

© Springer Science+Business Media Singapore 2016
W. Che et al. (Eds.): ICYCSEE 2016, Part I, CCIS 623, pp. 657–667, 2016.
DOI: 10.1007/978-981-10-2053-7_59

categories of kill, sacrifice and layout. So chess can judge the situation in the corresponding position and the framework is intelligent.

In this paper, the first part of the pieces introduces how chess move, the use of algorithm and the situation of optimization; in the second part, we improve the framework by the experience in competition; finally, we make a summary.

2   Initial Design Checkers Framework

2.1   Understanding of Checkers

The international rules of 10*10 draughts board are the same black and white checkerboard, the board is in the middle of the both side, the right corner of each player is blank grid. As shown in Fig. 1.

Fig. 1. Checkers 100

This paper involves several jargons:

Kill: For a given position, after we win the right of moving the chess, if the chess can jump to eat, the movement of maximum number of killed chess is regarded as optimal killed.

Sacrifice: For a given position, after we win the right of moving the chess and all movements, the movement of killed chess is regarded as sacrifice.

Layout: For a given position, after we win the right of moving the chess and all movements, the movement of being not killed chess is regarded as layout.

If we could kill chess, we would kill the chess in checker rules. And we are based on the movement of maximum number of killed chess. Therefore, so many factors should be considered in the algorithm design. Among them, kill is to find out the several ways to kill most chess and select the optimal way to move [4]. Sacrifice forces the opponent to kill our chess; furthermore, we can kill more chess of opponent to make opponent loss more chesses. This is especially important, and may even turn around the situation. Therefore, we use game tree [5] on sacrifice algorithm method, the depth of exploration can achieve greater harm to the match; layout is launched when we have yet fight. Because the solid foundation plays the significant role in the future offensive, in this way can blow to the opponent. Rather, we are in the troubled [6].

2.2 Board Representations

There are two commonly used expressions of the board. One is two-dimensional array board, the other is bit boards [7]. In which two dimensional array boards sees simple and intuitive, so it's easy to find a potential problem and shorten the time of debugging. And we use the two-dimensional array board frequently. Surely, considering the efficiency issues, bit board will have obvious advantages. This paper adapts the two-dimensional array board.

As shown in Fig. 1, we can use 10*10 integer two-dimensional array to represent the 10*10 squares, where 0 means no pieces, 1 means black chess, 2 means king of black; -1 means white chess; -2 means white king. The corresponding numerical can real-time express the board.

2.3 Way to Generate

When we get the right of move, we choose the optimal movement in all movements. Here introduce the concept of generator moves [8], which generate all reasonable way to move. As shown in Fig. 2.

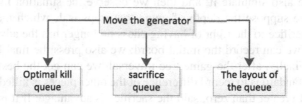

Fig. 2. Move the generator

All reasonable way is divided into three categories: kill, sacrifice and layout. And we take them into the corresponding queue, we can analyze the content of the three queue. Obviously, when all three queues is NULL, which means no chess can go or pieces. According to the rules, we will be jailed for check. And we can see, optimal kill moves in the queue more priority than sacrifice and layout. That is to say, kill queue is not null, sacrifice and layout queen has no means. Sacrifice and layout queue is the same level.

2.4 Game Tree Implementation

Game tree can accurately record the game of two people or a program process. So the game tree algorithm is mainly used in the process of the sacrifice queue, in the sacrifice queue to open game tree. In case of sacrifice, we would simulate it, the opponent will

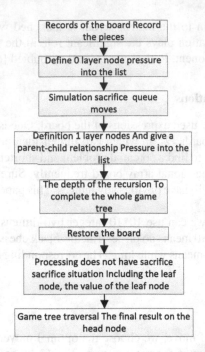

Fig. 3. Depth of 1 game tree creation process

jump to eat, we also simulate it, and then we observe the situation in the following rounds. Now we suppose the depth of game tree steps is 1, which means, from the beginning of sacrifice to the right of having chess no longer has the advantage kill. At the same time, we can record the initial board; we also press the final board into leaf nodes of tree. Finally, after the game tree traversal, we can get the best results. Here, evaluate good standard is to record difference of the other pieces is killed and killed our pieces, which if greater than zero, said the sacrifice is advantage; If it is equal to zero, same said casualties; If less than zero, say we are cheated.

Game tree information need to simulate all processes of optimal kill the queue for ensuring the completeness of the algorithm. All levels use the algorithm of the max value relationship from leaf nodes back to head node. Finally, the best information will be stored in the head nodes. Step depth is assumed to be 1. If step depth is 2 that means: from first one sacrifice to the end of kill each other, we still have the right to play chess (Kill queue is null), and sacrifice queue is not null. Furthermore, we move the chess basing on the best information. After a program test: In the normal computer game within the time allowed, a computer can further step 4; Node numbers is more than 18000.

The program searches the optimal solution with ALPHA - BETA search [9], which reduces some node traversal and saves time. Furthermore, program is able to deal data with a deeper level.

The following steps give a depth of 1 game tree creation process. And the process showed in Fig. 4.

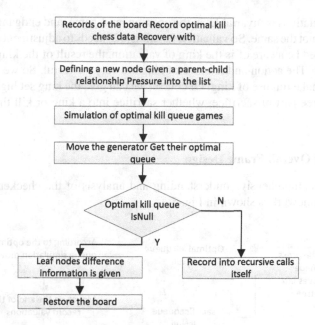

Fig. 4. Recursive procedure

When considering the steps in the depth of 2, just sacrifice queue in the case of in game tree. Actually the layout queue may also have a different way, so sacrifice in step 2 is not the only way. We can found the best optimal movement in step 1 and step 2 sacrifices. That is to say, if step 2 sacrifice situations were bad, we could choose best in step 1; in step 2 we can choose sacrifice and moves in layout. So we can optimize for depth use a few steps at the same use a few game tree. And we find the optimal solution in the game tree.

2.5 Evaluation Function

For computer games, a lot of kind chess in a limited period of time is can't be totally search. Although checkers branch is less, we still can't search totally [10], This valuation function is particularly important.

The effect of value function is given a board to evaluate, determine the general trend. Furthermore, we decide how to make the best judgment to moves. Valuation function mainly used optimal selection and layout of the games.

There are many valuation characteristics. Such as the number of king, ordinary pieces quantity, suppressing the other pieces, formation (Including chain, triangle [11], and so on), distribution of pieces, optimal number of pieces, sacrifice piece count, and so on [11]. In many assessment characteristics, we must focus on the number of the pieces. This program can save pieces in order to win; besides, many features need to integrate together. This involves many features of the weight problem. We can roughly determine the weight of each feature by experience. It's important to note that not all features of

proportional relations is invariable; in the opening, mid-game, and endgame, the weight of valuation is not the same. So valuation function [12] needs to adjustment dynamically.

We still need be aware of is the king of valuation, the result of the king being especially precious. The action and damage of king are very powerful. So we try to protect our king and make full use of king. Of course, we can give the king set high proportion. So the game tree part of sacrifice, whether sacrifice into a king or kill the opponent's king, it works.

2.6 Primary Overall Frame Design

The above to comprehensive understanding and analysis of the checkers, completed application framework is shown in Fig. 5.

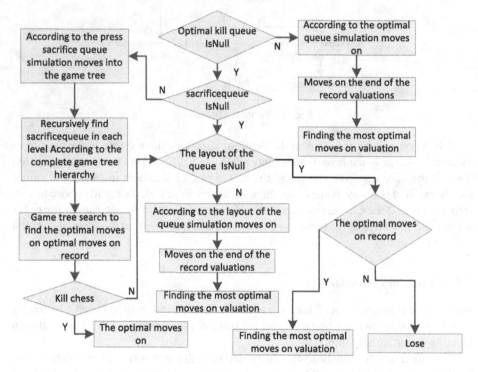

Fig. 5. Primary overall frame design

From Fig. 5 can be seen kill movement and layout way are based on a valuation function to determine. Sacrifice part with game tree, when on the attack can achieve step 4, layer can achieve the higher than 14th floor (With the algorithm optimization and upgrading, you can reach a higher level). Not only introduced in the valuation is optimally killed, but also introduced sacrificed. If the step was 3, while the rest of that, we could prevent continuous sacrifice within each other in three steps. To some extent, we ensure number of chess. So in general, the framework of application level has been very good at chess, we have also made very good result. Of course, as we further study and explore, the framework can also optimize [13] and upgrade for achieving a better game level.

2.7 Analysis on the Running Status of Program Design Under the Primary Framework

The statistics of the program running time in the primary framework are shown in Table 1.

Table 1. The statistics of program running time

Options	Level	Average time (ms)
Attack	1	300
	2	4223
	3	9101
	4	15306
	5	45542
Defense	1	654
	2	5321
	3	12056
	4	32678

Comprehensive program running time and game requirements, the attack is divided into 4 layers, defense, 3 layers. Due to exist many possible to layout, so the defensive time required more than attack. In the current level, this had exceeded the amateur level. The framework of the game in the national has awarded two awards in 2015 computer game competition, the group, third. The game framework is simple, but the attack ability. Of course, this design of framework also exists some problems, after this game, aiming at design; we have made a further optimization.

3   Optimal Design of Checkers Framework

3.1   To Improve Analysis

Framework of game tree works in sacrifice queue and the optimal kill and layout part is weaker; almost no combined with optimal kill and layout, which may ignore some important situation. Furthermore, we can make errors or miss some good opportunity. Sacrifice part moves or not is not determined by the situation, but to judge whether the whole process is beneficial, if we could kill the more chess of opponent. Of course, It is no problem that at point of view in the number of pieces is preferred. However, if more than kill a chess pieces, make the situation is bad for us and formations, this strategy may not be wise.

So come to the conclusion: first, the game tree lists all possible situation; second, the game tree leaf nodes as the final judge are achieved by valuation function.

3.2   Optimize Moves

From the above analysis, if the game tree has all possible situations, so sacrifice and layout can be considered together. Optimization moves generator is shown in Fig. 6.

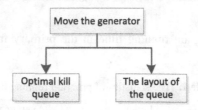

Fig. 6. Optimization of the move generator

The optimized way generator only superior to kill and layout. That is to say, in view of a parent node, his all subsequent nodes are either kill queue all elements or queue all elements layout [14]. On the one hand, this situation conforms to the rules; on the other hand, sacrifice thoughts are included in this situation. In game tree have the thought of sacrifice and coordinate all the circumstances.

3.3   Valuation Function Research

Leaf nodes in game tree are determined by valuation function, so the valuation function is trusted to reflect the actual status of board. If valuation function can't bear the responsibility, making a wrong decision, valuation function would have an impact on the level of chess.

So the valuation functions that need to add more comprehensive content, the details need further consideration. The most important is the ability to grasp the situation of the state overall situation being good for us. On the basis of the analysis more details would have greater significance. If the valuation function not do so, under the framework of the program is not always to beyond the previous frame or even worse, this is also the significance of the existence of the front frame. If the valuation function is not complete and accurate grasping chess, it is better to analysis kill a few simple pieces in sacrifice.

Judge to the details situation involve multiple factors. Emphasis weight is also different in the different stages. Of course, it is important to reservation power of pieces; and according to the actual situation, to some extent, in order to better determine relationship to piece power opponent with me, there is a certain equivalence relation between king and normal.

If the search is a fixed depth, horizon effect will appear [15]. That is to say, when the search is up to a fixed depth, valuation function to evaluate leaf node. If the leaf node in the stage of frequent to change pieces, it is clear that this assessment is not accurate. Because even if the situation now in our favor, but the other party can immediately jump to eat our many pieces; obviously we would get into a passive or failure. There are some methods solve the horizon effect, we can use the valuation function enhancement processing, let the valuation function can analyze a simple potential changes in the piece power. Such as chess can be analyzed in a exchange state, Killed state or be sacrifice state, based on the early primary system framework [16], detection and analysis of the several situation are easy, and the time complexity is not high also.

There are plenty of features for valuation function needs analysis, will be accessible to the king, and so on, for example, only the more features will be more able to reflect the real chessboard, this chess level will be stronger.

3.4   Optimization Design of Overall Framework

Based on the above understanding and analysis of the checkers, optimization program framework is shown in Fig. 7.

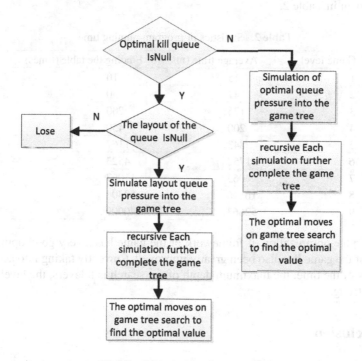

Fig. 7. Optimization framework

Seen from the Fig. 7 game tree is the entire generator moves in a reasonable way, the building of the game tree is similar to the early design process. Under the framework of game tree each layer contains all the possible moves, the final assessment of the leaf nodes are performed by the valuation function. Finally, the optimal solution is chosen.

Obviously, the design framework of coordination in Fig. 7 is better than Fig. 5, besides, the design framework of coordination in Fig. 6 is more perfect. In general, framework in Fig. 6 is more perfect for checkers, piece level also have larger development space and better compatibility [17]. With optimization algorithm and new exploration of the algorithm, chess level will also increase quickly. For the more characteristic parameters involved in the process of moves, artificial neural network and machine learning [18] can be applied to optimize the evaluation function. Over rivals and debug member of accumulating good layout information and moves on strategy, the evaluation function can adjust themselves.

3.5 Analysis of Program Operation Status Under the Optimization Framework

Compared to the framework of basic game, the optimal game framework has a further improvement, the chess of attack and defensive are unified, both of the game tree, the depth of the valuation, get the best plan. Combined with optimization scheme of alpha-beta pruning, permutation table, history heuristic, to reduce the amount of calculation of the game tree, the time complexity is further reduced. Statistics of program running time as shown in Table 2.

Table 2. Statistics of program running time

Game level	Average time (ms)	Finding the table (times)
1	15	10
2	47	50
3	125	200
4	200	450
5	843	1167
6	1752	4523
7	4065	11067
8	16546	52304
9	52663	140009

Visibly, the optimization of framework of the game has a very good optimization; the level of the game has also been greatly improved. Currently taking into account the limitations of the time, the maximum depth of the search is 8 layers, the level of chess has very strong.

4   Conclusion

This team in the checkers framework designed in the fourth session of Chinese college students in computer game series checkers 100, we achieved The second prize of good grades and accumulated a lot of experience, learning a lot of excellent algorithm is benefit for further exploration, and in the late game for the framework has been optimized and improved, so that the level of chess, has a significant improvement. After constant improvement, in the field of checkers 100, we must make greater achievement in the field of checkers.

References

1. Wang, J., Xu, X.: The frontier of artificial intelligence–the national contest of college students' computer game. Comput. Educ., Beijing **7**, 14–18 (2012)
2. Wang, X.: PC Game Programming (Man-Machine Game). Chongqing University Press, Chongqing (2002)
3. Yue, W.: A high-performance checkers engine design. Comput. Prog. Skills Maintenance **32**(02), 63–68 (2014)
4. Yajing, L.: Main technical analysis of six children of a game of chess computer. Comput. Knowl. Technol. **7**(10), 2310–2312 (2011)
5. Russell, S.J.: Artificial intelligence: a modern approach (3). Jianping Yin. Qinghua University Press, Beijing (2013)
6. Sea, K., Haykin, S.: neural networks and machine learning. Shen FuRao. mechanical industry publishing house, Beijing (2011)
7. Yajing, L.: Main technical analysis of six children of a game of chess computer. Comput. Knowl. Technol. **7**(10), 2310–2312 (2011)
8. Zhang, C., Liu, H.: Game tree heuristic search of alpha beta pruning technology research. Comput. Eng. Appl. **16**, 54–56 (2008)
9. Zhang, M., Li, F.: A new game tree search methods. J. Shandong Univ. (Eng.) **6**, 1–7 (2009)
10. Shafiei, N., van Breugel, F.: Towards model checking of computer games with Java PathFinder. IEEE Press **13**, 25–29 (2013)
11. Shu, K., Hu, F.: Chinese chess computer game engine improvement. J. Artif. Intell. **10**, 39–42 (2009)
12. Wang, J., Wang, T.: Chinese chess computer game system evaluation function of adaptive genetic algorithm. J. Northeastern Univ. **10**, 949–953 (2005)
13. Zhou, W., Wang, Y., Ma, Q.: Changes of the situation to seek advantage game. J. Syst. Simul. **17**, 5–9 (2008)
14. Bottcher, N., Martinez, H.P., Serafin, S.: Procedural audio in computer games using motion controllers: an evaluation on the effect and perception. Hindawi Publishing Corp. **11**, 94–97 (2013)
15. Xu, C., Ma, Z., Xu, X., Li, X.: Machine game oriented real-time difference study. J. Comput. Sci. **8**, 219–223 (2010)
16. Yu, W., Yang, X.: Intellectual games interpretation characteristics and development prospects analysis. J. Shandong Sports Institute. **27**, 5–10 (2011)
17. Gao, Y., Chen, S., Liu, X.: Reinforcement learning research review. Acta Automatica. **1**, 87–97 (2004)
18. Wallace, S.A., Russell, I., Markov, Z.: Integrating games and machine learning in the undergraduate computer science classroom. ACM. **8**, 231–236 (2008)

The Software Behavior Trend Prediction
Based on HMM-ACO

Ziying Zhang[1(✉)], Dong Xu[1], and Xin Liu[2]

[1] College of Computer Science and Technology, Harbin Engineering University,
Harbin, Heilongjiang, China
{zhangziying,xudong}@hrbeu.edu.cn
[2] College of Automation, Harbin Engineering University, Harbin, Heilongjiang, China
liuxin@hrbeu.edu.cn

Abstract. For the HMM exists defects in application in the aspect of software behavior prediction, namely, HMM could trap into local optimization because of the problem of B-parameter, which results in the decrease of HMM's precision. This paper builds a new model HMM-ACO through combining Ant Colony Optimization (ACO) algorithm with HMM, with system calls as the data source, improving the prediction accuracy rate of HMM. In order to eliminate the HMM's reflection on observations characteristics, this paper puts forward a new approach to recognize software behavior with hidden states.

Keywords: HMM · ACO · Behavior recognition · Behavior prediction

1 Introduction

The prediction of software behavior has been attracting numerous scholars' attention, and some corresponding prediction tools have been put forward, such as Neural network [1, 2], Bayesian network [3] and Dirichlet model, etc. Although these tools have realized software behavior prediction to some degree, they still have some shortcomings, especially there are larger restrictions on the prediction environment. However, Hidden Markov Model (HMM) [4] avoids above restrictions and relatively successful, but the model is not precisely enough and with poor robustness, these defects could reduce the prediction accuracy rate of the model. These problems seriously hinder HMM's applicability in reality. To address the problem, this paper builds a new model HMM-ACO through combining Ant Colony Optimization (ACO) [5, 6] algorithm with HMM, improving the prediction accuracy rate of HMM, and to further improving its applicability.

2 HMM-ACO

HMM is sensitive to initial parameters, especially B-parameter, which makes model fall into a local optimum in training. Ant Colony Optimization (ACO) Algorithm uses an optimal search strategy which is based on a wide range of groups, and ACO has a better global optimization ability. This paper uses ACO to optimize B-parameter of HMM.

© Springer Science+Business Media Singapore 2016
W. Che et al. (Eds.): ICYCSEE 2016, Part I, CCIS 623, pp. 668–677, 2016.
DOI: 10.1007/978-981-10-2053-7_60

It develops a so-called optimal training model based on HMM (HMM-ACO). Because the hidden states of HMM can reflect essential characteristics of observations and a sequence of hidden states can use less states changes to reflect more changes of an observation sequence, the HMM-ACO aims at recognizing software behavior with hidden states.

The HMM-ACO for software behavior prediction mainly includes the following four parts.

(1) Process ACO to optimize HMM.
(2) Establish a repository (i.e., HMM-ACO training process) for software behavior prediction. The process adopts offline training because it is time-consuming.
(3) Conduct software behavior recognition to judge normal or abnormal status of software behavior. If it is abnormal, a beforehand processing is followed. Otherwise, a behavior prediction is followed.
(4) Conduct software behavior prediction to check the most possible state. If the state is not in the scope of controllable safety, generating an alert to system administrator for actions, thus prevent occurrence of potentially malicious events.

The whole process is shown in Fig. 1 below.

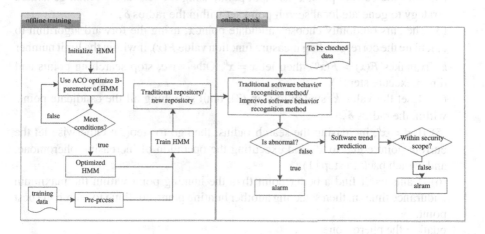

Fig. 1. Software behavior recognition and trend prediction based on HMM-ACO

3 Software Behavior Recognition and Behavior Prediction

System calls are occurring at the resource calling interface between user and kernel spaces. They are several built-in functions in operating system's kernel which provides core services for applications. The system call sequences [7] refers to a chronological order of sequences generated by a running program. Using system calls as the data source of HMM, the calls corresponds to all the HMM observations, and the essential characteristic of a system call corresponds to a hidden state of HMM. Software behavior recognition and behavior prediction can be accomplished based on a proper HMM model.

3.1 HMM-ACO

First of all, establish a continuous search space Ω, the search space represents a collection of HMM parameters, namely $x = (\pi_1, \ldots, \pi_N, a_{11}, \ldots, a_{NN}, b_{11}, \ldots, b_{NM})^T$. x is a point in the search space, and represents a solution of continuous space. Ants search the best point in the space, the process is divided into two steps, searching solution and updating the pheromone. Ants search solutions in the search space, to find the optimal feasible solution. If an ant finds a relatively optimal solution, it will increase the point's pheromone, and attract other ants to do further search. The two operations are as follows.

1. Searching solution
   Ants embark from the nest, and select a hunting point x^s randomly, and set x^s the best point, in the begin, $x^s = x^b$, x^b represents the best point. In the near of x^s, ants continue to search for the optimal solution. Specific steps are as follows:
   (1) Set the ant search radius δ', ants search the best solution in the scope.
   (2) With the current point x^s as input point, using the feasible solution generated strategy to generate local search point set within the radius δ'.
   (3) The ants randomly choose candidate point x, using the forward algorithm to calculate the corresponding measure function value $F(x)$, if within the limit number k, $\exists x$,makes $F(x) > F(x^b)$, then let $x = x^b$; Otherwise, stop searching points and direct execute step(5).
   (4) Reset the value k, switch to step (3), until searching all the candidate points within the radius δ'.
   (5) If $x^s = x^b$, increasing the search radius, then go to step(2); Otherwise let the current point equal to x^b, and executing the operation of increasing pheromone, and switch back to step (1).
   (6) If ants can't find a better point than the hunting point within the maximum endurance times n, then selecting another hunting point, continue to search the best point.
2. Updating the pheromone
   The information communication between ants includes the following three parts:
   (1) In the same area, ant a sends messages to another ant b, the messages contains the hunting location of ant a, the best point x^b of ant a and its corresponding measure function value $F(x^b)$.
   (2) Ant b will save other ants' messages from the same area into its own stack.
   (3) Ant b will randomly select message in the stack and compare the value $F(x^b)$ with its $F'(x^b)$, if $F'(x^b) < F(x^b)$, then ant b will move to the ant which has sent message $F(x^b)$ to ant b and continue to search better point; Otherwise, ant b will discard the poor message and randomly send its message to another ant in the same area.

Ants can find the best point in the search space through the above two operations, then the HMM's parameters has been optimized completely, and get the model HMM-ACO.

3.2 Establishment of Repository

The repository represents the characteristics of normal behaviors. It is the basis of software behavior prediction. On the one hand, it is very difficult to obtain abnormal sequences of system calls. On the other hand, abnormal patterns are always changing in practice. We use the set of normal system calls as short sequences for training data. HMM represents the characteristics of normal behaviors which can be obtained from training data, and Baum Welch algorithm is used to multiple sequences training. The parameters of the trained HMM and other related parameters are preserved in repository.

According to the need of improved software behavior recognition method, the new repository holds the characterized HMM's parameters representing the characteristics of normal behaviors, the standard set of hidden state sequences and the threshold δ used to judge whether the whole sequence is abnormal.

Here lists three parts of the repository.

(1) HMM parameters
(2) HMM is obtained with the set of normal short system call sequences as training data and Baum Welch algorithm. HMM $= (\pi, A, B)$ represents the characteristics of normal behavior. Store these parameters (π, A, B) in the repository as the basic parameters for software behavior recognition.
(3) The standard set of short hidden state sequences
(4) Calculate the set of best short hidden state sequences (remove the repetitive sequences) by HMM and Viterbi algorithm, using the set of training short system call sequences as input data. Use the best short hidden state sequences as the standard set which is stored in the repository as the basis of software behavior recognition.
(5) In order to compare with traditional software behavior recognition method, a threshold δ is taken the same value as threshold in the traditional repository.

3.3 Software Behavior Recognition

Online behavior recognition is to identify the input sequence behavior status: normal or abnormal.

Considering HMM uses fewer hidden states to represent more system call sequences, we put forward an improved software behavior recognition which is based on hidden states. The corresponding steps are described as follows:

(1) Data preprocessing. Slice every system call of input sequences to the set of K-length short system call sequence.
(2) Calculate the set of short hidden state sequences with HMM and Viterbi algorithm, using the short sequences obtained in Step (1) as input data.

(3) Statistical analyzing the numbers of the set of short hidden state sequences obtained in Step (2) not existing in the standard set of short hidden state sequences in repository. Calculate the percentage.

(4) If the percentage is greater than the given threshold value δ, the whole sequence is recognized as an abnormal behavior. At this moment, it is necessary to pass a warning message system administrator in time. Otherwise, it will be recognized as normal behavior.

3.4   Software Trend Prediction

The premise of behavior prediction is that the whole sequence has been recognized to be a normal behavior in the part of software behavior recognition.

Software trend prediction is described as follows:

(1) Slice a K-length system call sequence (O1, O2, …, OK) from the last of the whole sequence.

(2) Calculate the corresponding hidden state sequence (q1, q2, …qK) of (O1, O2, …, OK) using HMM and Viterbi algorithm.

(3) Obtain the next most likely hidden state using qK of (q1, q2, …qK) and the A-parameter of HMM. The detailed process is as follows: the maximum value at qK line of A-parameter of HMM is the next most likely hidden state.

4   Simulation Experiment

4.1   Data Source Selection and Parameter Setting

The experimental data of this paper is downloaded from website http://www.cs.unm.edu/~immsec/data/, the site stores a large set of system call sequences which is collected by professor Forrest in his intrusion detection experiment. These data is saved to different files which is shown in Tables 2 and 3. In the optimization process of B-parameter, relevant parameters of ACO are shown in Table 1 below.

Table 1. Relevant parameters of ACO

Parameter's name	Parameter's value
Pheromone volatilization factor ρ	0.5
The intensity of pheromone Q	100
The number of ant m	15
The parameter of α	1
The parameter of β	5

4.2   The Comparison of HMM and HMM-ACO in Software Behavior Recognition

Firstly, the traditional software behavior recognition mechanism based on HMM, GA-HMM and HMM-ACO is tested, and the recognition results of these three models are compared and analyzed.

Table 2. The maximum mismatch rate of the system call short sequence in the normal process (%)

File location	Bounce	Bounce1	Bounce2	Plus	Queue
HMM	40.9	39.3	4.4	28.7	52.5
GA-HMM	40.9	39.3	2.29	22.9	52.5
HMM-ACO	36.3	37.9	3.29	22.9	51.4

Tables 2 and 3 are the normal process and the abnormal process of the system call short sequences in the traditional behavior recognition mechanism of the maximum mismatch table. By two tables show that: the normal process in short sequences of system calls of maximum matching rate or abnormal process in short sequences of system calls the biggest mismatch rate. HMM-ACOsignificantly better than the performance of the HMM method; compared with the hybrid GA-HMM, HMM-ACO in some files in the maximum mismatch rate is higher than the first, but overall still HMM-ACO effect is better in terms of behavior recognition.

Table 3. The maximum mismatch rate of the system call short sequence in the exception process (%)

File location	sm-280	sm-314	sm10763
HMM	99.2	61.5	91.7
GA-HMM	78.3	61.5	64.6
HMM-ACO	76.3	62.5	63.2

From Figs. 2 and 3: compared with HMM, both sequences of system calls in the normal course of recognition and the recognition sequences of system calls in abnormal process, the rate of HMM-ACO identification are significantly higher than the HMM; compared with the hybrid GA-HMM, HMM-ACO recognition effect is better than GA-HMM except some special files.

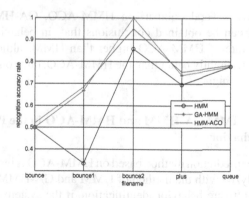

Fig. 2. Comparison of the behavior recognition accuracy of the short sequence of system call in the normal process

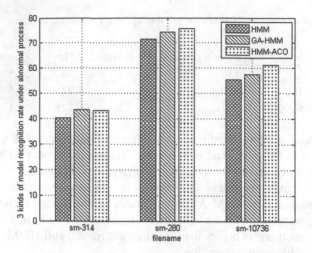

Fig. 3. Comparison of behavior recognition rate of system call short sequence in abnormal process

From Table 4: in activity recognition, the correct rate of HMM-ACO is higher than the HMM, and the error rate is lower; and compared with the hybrid GA-HMM, the accurate rate of HMM-ACO model to higher, but in the wrong rate aspects, GA-HMM slightly accounted for advantage than HMM-ACO.

Table 4. The two models in behavior recognition accuracy and error rate than the table

Model	TA	FA
HMM	0.593	0.283
GA-HMM	0.644	0.217
HMM-ACO	0.665	0.221

Through the above several comparison HMM-ACO, GA-HMM and HMM in behavior recognition can be obtained conclusions that: in behavior recognition, the recognition accurate rate of HMM-ACO higher than HMM, although GA-HMM in terms of the error rate to slightly better than the HMM-ACO, but from the whole, HMM-ACO recognition rate is higher.

4.3   Comparison of HMM,GA-HMM and HMM-ACO in the Prediction of Software Behavior

The software behavior prediction method based on HMM-ACO is tested, and the method is compared and analyzed with the method of HMM and GA-HMM.

In the period of software behavior identification, if the system call sequence to be detected is normal, the next step is to predict the behavior; otherwise, the exception handling. And because after the identification of different model, was judged to be normal to be detection of short sequence number is not the same, cannot be compared, so Figs. 4, 5, and 6 not delete in behavior recognition stage is identified as abnormal detected sequences, but compared with all the normal and abnormal to be sequence detection of prediction accuracy.

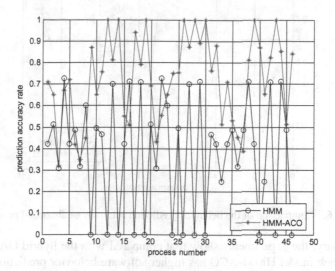

Fig. 4. Comparison of the accuracy of the behavior of the normal process

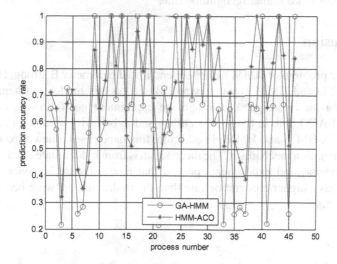

Fig. 5. Comparison of the accuracy of the behavior of the normal process

By Figs. 4, 5 and 6 shows: in software behavior prediction, the accuracy of HMM-ACO higher than the HMM; and compared to the HMM-ACO GA-HMM, although individual processes the prediction accuracy rate not good as GA-HMM, but overall, accuracy rate of HMM-ACO model in forecasting more higher.

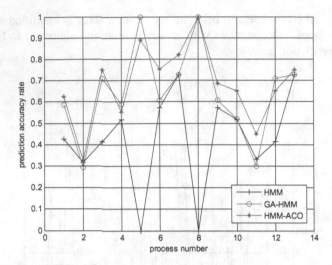

Fig. 6. Comparison of the behavior prediction accuracy of abnormal process

In summary, the experiments show that compared with the hybrid GA-HMM and HMM, the new model HMM-ACO has higher software behavior prediction accuracy; new software behavior recognition mechanism than traditional software behavior recognition mechanism has higher recognition rate.

5   Conclusion

Aiming at the problem of HMM model depend initial parameter B strong, HMM-ACO combined ACO and HMM is put forward. Based on this model a system to predict software behavior is designed. Through the experiment got following conclusions: model HMM-ACO both in terms of accuracy and in behavior prediction accuracy are better than GA-HMM and HMM. In addition, this paper uses a new type of software behavior recognition mechanism. The mechanism is from hidden state level to recognize software behavior, and experiments proved that new software behavior recognition mechanism has a higher recognition rate than the traditional software behavior recognition mechanism.

References

1. Arar, Ö.F., Ayan, K.: Software defect prediction using cost-sensitive neural network. Appl. Soft Comput. **33**, 263–277 (2015)
2. Niantang, L., Yu, W., Yu, L., et al.: Dynamic power management scheme based on software behavior prediction. Comput. Eng. **41**(6), 269–273, 279 (2015)
3. Laradji, I.H., Alshayeb, M., Ghouti, L.: Software defect prediction using ensemble learning on selected features. Inf. Softw. Technol. **58**, 388–402 (2015)
4. Rabiner, L., Juang, B.: An introduction to hidden Markov models. IEEE ASSP Mag. **3**(1), 4–16 (1986)

5. López-Ibáñez, M., Stützle, T., Dorigo, M.: Ant colony optimization: a component-wise overview. **18**(5), 14–37 (2015)
6. Pang, W., Wang, K., Wang, Y., et al.: Clonal selection algorithm for solving permutation optimisation problems: a case study of travelling salesman problem. In: International Conference on Logistics Engineering, Management and Computer Science (LEMCS 2015). Atlantis Press (2015)
7. Forrest, S., Hofmeyr, A., Somayaji, A., et al.: A sense of self for unix processes. In: IEEE Symposium on Proceedings of IEEE Security and Privacy, pp. 120–128 (1996)

Towards Realizing Sign Language-to-Speech Conversion by Combining Deep Learning and Statistical Parametric Speech Synthesis

Xiaochun An, Hongwu Yang[⊠], and Zhenye Gan

College of Physics and Electronic Engineering,
Northwest Normal University, Lanzhou 730070, China
yanghw@nwnu.edu.cn

Abstract. This paper realizes a sign language-to-speech conversion system to solve the communication problem between healthy people and speech disorders. 30 kinds of different static sign languages are firstly recognized by combining the support vector machine (SVM) with a restricted Boltzmann machine (RBM) based regulation and a feedback fine-tuning of the deep model. The text of sign language is then obtained from the recognition results. A context-dependent label is generated from the recognized text of sign language by a text analyzer. Meanwhile, a hidden Markov model (HMM) based Mandarin-Tibetan bilingual speech synthesis system is developed by using speaker adaptive training. The Mandarin speech or Tibetan speech is then naturally synthesized by using context-dependent label generated from the recognized sign language. Tests show that the static sign language recognition rate of the designed system achieves 93.6 %. Subjective evaluation demonstrates that synthesized speech can get 4.0 of the mean opinion score (MOS).

Keywords: Deep learning · Support vector machine · Static sign language recognition · Context-dependent label · Hidden Markov model · Mandarin-Tibetan bilingual speech synthesis

1 Introduction

There are a large number of speech disorders in the world owing to various reasons. It is difficult for speech disorders to communicate with other people except for those who master sign language. This not only makes it difficult for speech disorders to live a normal life in the community, some matters concerned with the speech disorders also have to be assisted by a sign language translator. In recent years, computer vision-based sign language recognition [1] attracted more and more attentions for its natural and convenient way to interact with each other [2,3]. Meanwhile, the computer is able to convert text to natural and fluent speech due to the development of speech synthesis technology. Since the hidden Markov model (HMM)-based statistical parametric speech synthesis [4] can synthesize different languages speech of different speakers by using speaker adaptive transformation [5], this method has become a hot-spot for cross-language

W. Che et al. (Eds.): ICYCSEE 2016, Part I, CCIS 623, pp. 678–690, 2016.
DOI: 10.1007/978-981-10-2053-7_61

speech synthesis [6]. However, existing studies only do research on sign language recognition technologies and speech synthesis problems separately. Therefore the problem that speech disorders have difficulty in communicating with normal people is not considered because of the lacking of study on sign language-to-speech conversion.

The realization of a sign language-to-speech conversion system plays an important role in daily communication between normal people and speech disorders. Although the data glove is used to achieve realtime sign language-to-speech conversion in [7], the operators should wear gloves to locate the tracking data, which gives operators a great complexity and inconvenience. Therefore, it is extremely difficult to apply the system to the real life.

In order to satisfy the needs of communication with speech disorders, the paper realizes a sign language-to-speech conversion system. The static sign languages are recognized with deep learning method [8]. A Mandarin-Tibetan bilingual speech synthesis system is realized using speaker adaptive training. A sign language-to-Mandarin or Tibetan bilingual speech synthesis system is finally achieved through combining speech synthesis system with the sign language recognition.

2 System Framework

The framework of the proposed sign language to speech conversion system is illustrated in Fig. 1. After collecting the sign language input by users, the data obtained from data integration is used to build data cubes, which are the input of the deep model. In order to initialize the weights of the deep model, the weight between two adjacent layers are adjusted through adopting restricted Boltzmann machine (RBM) [9]. Then the essential characteristics of the samples are obtained by using the feedback fine-tuning of the deep model. 30 kinds of static sign language tested by 2 different people are recognized and classified with support vector machine (SVM) [10]. The text of sign language is then obtained from the recognition results. Context-dependent Chinese labels or Tibetan labels, which will be used in HMM-based bilingual speech synthesis, are generated according to the semantic rules of recognized sign language. Mandarin speech or Tibetan speech is finally synthesized from context-dependent labels.

3 Sign Language Recognition

The data obtained by data integration is used to build data cubes that are the input of the deep model. The weights of the entire deep model system are initialized by adopting the method of RBM [11]. The paper firstly samples the hidden layers as reference layers after the conversion from visible layers to hidden layers in order to obtain the status of each node in the hidden layers. Then the reverse transformation is conducted from hidden layers to visible layers. Finally the conversion is successfully made from visible layers to hidden layers. The reconstructed object of visible layers and hidden layers are obtained after above

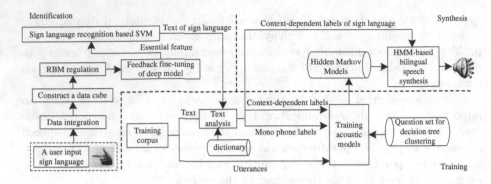

Fig. 1. The framework of sign language to speech conversion system

3 times conversions to achieve the purpose of adjusting the parameters of RBM by reducing the difference between the original object and the reconstructed object. The detailed processing steps are shown as follows:

– For the neural node i of all the hidden layers in Eq. 1, do sampling to get h_{0i}

$$Q(h_{0i} = 1|v_0) \tag{1}$$

The mapping operation is performed between adjacent layers according to the Eq. 2.

$$sigm(-b_i - \sum_j W_{ij}v_{0j}) \tag{2}$$

where, v_0 is a sample in the training sample set of RBM, h_{0i} is the obtained state of each node in the hidden layer after sampling.

– For the neural node j of all the visible layers in Eq. 3, do sampling to get v_{1j}.

$$P(v_{1j} = 1|h_0) \tag{3}$$

Perform the mapping operation between the layers according to the Eq. 4.

$$sigm(-c_j - \sum_i W_{ij}h_{0i}) \tag{4}$$

– For the used neural node i in hidden layers

$$Q(h_{1i} = 1|v_1) \tag{5}$$

Perform the mapping operation between the layers according to the Eq. 6.

$$sigm(-b_i - \sum_j W_{ij}v_{1j}) \tag{6}$$

– Update the offset parameters of linking weights with Eq. 7.

$$W = W - \varepsilon(h_0 v'_0 - Q(h_1 = 1|v_1)v'_1)$$
$$b = b - \varepsilon(h_0 - Q(h_1 = 1|v_1)) \tag{7}$$
$$c = c - \varepsilon(v_0 - v_1)$$

where, W is the link-weight matrix between layers of RBM, B is the bias in hidden layer of RBM, C is the bias in input layer of RBM.

The overall architecture of the deep model is divided into 5 layers. With the level increasing, the dimension of deep representation decreases. In the feedback fine-tuning stage, we perform conversion from down to up by using identification model [12]. Then up to the highest level, conversion of the generated model is obtained from up to down to generate the reconstruction of each level. Constantly adjusting the error between the original representation and reconstructed representation makes it acceptable.

Sign language recognition is a multi-class classification problem. Meanwhile, the performance of SVM is influenced by the kernel parameters [13] and error penalty factor C. The parameter α in RBF controls the number of SVM and generalization ability. The smaller its value is, the more the number of SVM is. Meanwhile, its accuracy becomes poor. In addition, C controls the complexity of the model and the approximate error. C is smaller so that the approximate error of the data become larger, but the model becomes simple. The paper firstly normalizes the feature data. Then we use a grid search to exhaustive parameters of RBF. Through the experiments, C and γ are taken $M = 12$ and $N = 11$ respectively. We search the (C, γ) and obtain its accuracy. Finally when the best parameter is $(8, 0.5)$, the accuracy is up to 100 %. After obtaining the optimal parameters of SVM, we predict the test sets according to the acquired model in the training process. In the end, the predicted results are obtained by adopting the superior classification surface to predict the feature data.

4 Sign Language to Speech Synthesis

4.1 Realization of the Mandarin-Tibetan Bilingual Speech Synthesis

We adopt the speaker adaptive training (SAT) [14] to achieve a Mandarin-Tibetan bilingual speech synthesis system. We select a large Mandarin multi-speaker-based speech corpus and a small Tibetan one-speaker-based speech corpus to train an average mixed-lingual voice model [15] using the speaker adaptive training. Mandarin speech corpus or Tibetan speech corpus is then used to perform the speaker adaptation transformation to obtain a speaker adaptive Mandarin model or Tibetan model for synthesis of the Mandarin Speech or Tibetan speech. The average mixed-lingual model is trained from Mandarin multi-speaker-based corpus and Tibetan one-speaker-based corpus. Mandarin-Tibetan bilingual speech synthesis process is shown in Fig. 2.

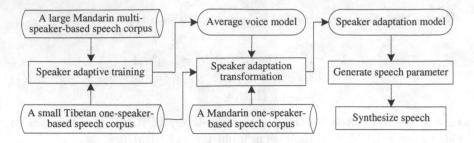

Fig. 2. The process of Mandarin-Tibetan bilingual speech synthesis

4.2 Generating Context-Dependent Labels of Sign Language

A semantic dictionary is designed to obtain semantic of sign language based on the meaning of sign language. Firstly, the semantic of each sign language is expressed in Chinese or Tibetan. Then a Chinese-Tibetan bilingual text analysis is employed to obtain the context information of semantic of each sign language including the initials and finals, syllables, words, prosodic word, prosodic phrase and sentence. Context-dependent labels are generated according to the context information of each sign language and saved to a sign language dictionary for the speech synthesis system.

We select all initials and final of Mandarin and Tibetan, as well as short pause and long silence as the synthesis units. Under the guidance of the grammar dictionaries and grammar rules, sign language text is firstly dealt through the text normalization. The word boundary is obtained from sentences automatically by using the grammatical analysis. The prosodic word boundary and prosodic boundary of the sign language text are predicted by adopting the transformation-based error driven learning algorithm (TBL) [16]. The context-dependent labels of the sign language are finally obtained by above steps as shown in Fig. 3.

5   Experiments

5.1   Experimental Data

In the experiments, we employ 2 persons playing 30 kinds of static sign languages to construct a sample set. 30 deep learning models are finally generated, where the number of samples in each sign language is 1000. The above 30 kinds of static sign languages are shown in Fig. 4.

To evaluate the quality of the converted speech, we use the speaker adaptation transformation to synthesize Mandarin and Tibetan utterances. 60 Mandarin utterances and 60 Tibetan utterances are synthesized according to the recognized sign language. We invite 22 speakers of Tibetan Lhasa dialect to evaluate the quality of synthesized Tibetan speech, as well as 22 Mandarin speakers to evaluate the quality of synthesized Mandarin speech. Meanwhile, an objective evaluation is made on the intelligibility of synthesized speech. We respectively

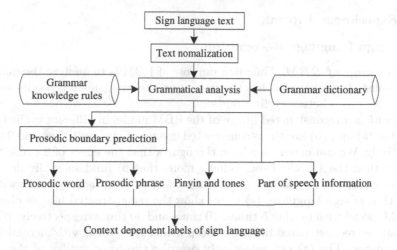

Fig. 3. Context-dependent label generation of sign language

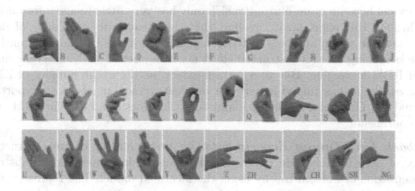

Fig. 4. 30 kinds of static sign languages

calculate the root mean square error (RMSE) of duration and fundamental frequency between the original speech and the synthetic speech from the same speaker.

In addition, we also evaluate the quality of synthesized speech aiming at sign languages. We firstly play the synthesized speech of the recognized sign languages to people who understand the sign language. Then the speech corresponding to the recognized sign languages is selected from 4 given alternative answers according to the heard speech. We finally calculate the average accuracy rate of judgment. Meanwhile, the synthesized speech is played to subjects who understand the sign language. The subjects are asked to select the sign languages that can express the meaning of the synthesized speech. By this way, we can determine whether the synthesized speech can express the meaning of sign languages and can calculate the average accuracy rate of judgment.

5.2   Experimental Results

5.2.1   Sign Language Recognition

The Training of RBM. The main purpose of RBM is to analyze the different cyclic adjusted number that has the impact on the experimental results so that we can pick out a better cyclic value.

Figure 5 is a reconstructed image of the RBM model in different cyclic times, where the (a) and (b) are the reconstructed images of cycling 1 time and 2 times respectively. We can clearly see from the figures that the effect of 2 cyclic times is better than the 1 cyclic time. What's more, the (b) fundamentally describes the basic outline of the entire sign language, while the (a) is very vague without any outline of sign language. (c) to (e) show the reconstructed images obtained by RBM model with cycling 5 times, 10 times and 30 times, respectively. We can find that the reconstructed images of deep model gets worse with increasing the cyclic number. The (c) can reluctantly describe the basic outline of the entire sign language, while the (d) is difficult to be distinguished to a certain extent. As is shown in (e), when cycling 30 times, the reconstructed images by using the deep model become more blurred, which only has very little image information.

Through this experiment, we find that the reconstruction ability of deep model and the cyclic numbers of RBM regulation are not a linear relationship. In the beginning, the reconstructed effect is getting better with the cyclic number of RBM increasing. But to a certain extent, as the cyclic number of RBM increases, the reconstructed images not only get worse, but also lose important information. Therefore, we finally adopt 10 cyclic times to adjust the RBM after the experiments.

Feedback Fine-Tuning Experiments. We firstly aim at the reconstruction ability of deep model to do the tests. The deep model generates the abstract representation of bottom-top by using the identified model. The representation of every layer in the recognition model is reconstructed with the generated model. Then we continue to regulate the cycle and make the reconstructed image closer to the original image.

Figure 6 is the reconstructed images in the different cyclic numbers. The (a) is a deep abstract process the first time the images experience the bottom-top recognition model. Then the deep reconstruction process is finished through the up-down generated model. Experiments show that although the first deep reconstructed image is not very clear, we can see the basic framework of the sign language. The (b) is reconstructed image after cycling 50 times. The (c) shows the reconstructed image cycling 150 times. The (d) is the result of cycling 200 times. We can discover that the reconstructed sign language become clearer and clearer with the fine-tuning number increasing. The error between the reconstructed images and original images is quite low when iterating to 200 times.

Figure 7 illustrates a mutative process of the reconstructed error in the training process of deep model. As can been seen, the error is gradually decreasing with the cyclic number increasing. It drops very fast at the beginning of the cycle, while its margin gets smaller and smaller as the cyclic number becomes

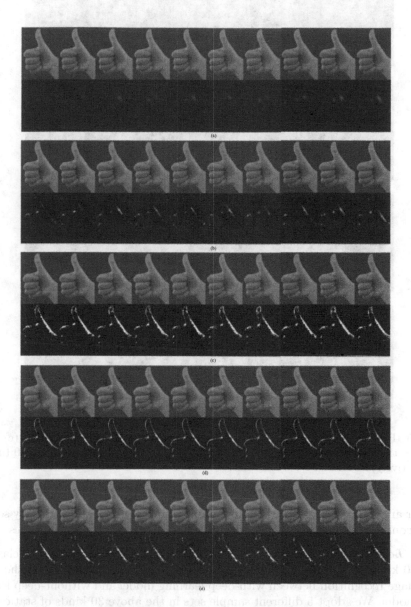

Fig. 5. Reconstructed images of RBM model in different cyclic times (the (a), (b), (c), (d) and (e) are the reconstructed images of cycling 1 time, 2 times, 5 times, 10 times and 30 times, respectively)

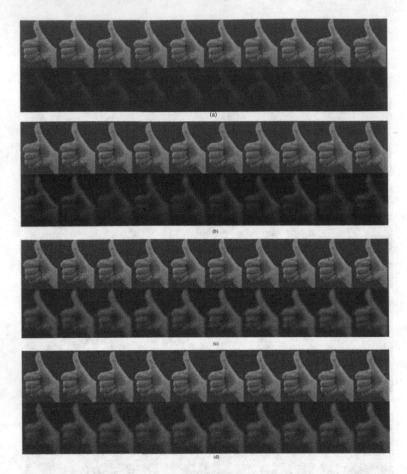

Fig. 6. Reconstructed images in the different cyclic number (The (a), (b), (c) and (d) are the reconstructed hands of cycling 1 time, 50 times, 150 times and 200 times, respectively)

larger and larger. However, it still maintains a downward trend. We finally select to reconstruct the sign language images when the cycle is up to 200 times.

Sign Language Recognition. We choose the multi-class SVM [17] to classify the 30 kinds of static sign languages shown in Fig. 4. Then we compare the sign language recognition between with-deep learning model and without-deep learning model. We adopt 5 different sample sets in the above 30 kinds of static sign languages, which are 1000:1000, 1200:800, 1400:600, 1600:400 and 1800:200 in the training sample sets respectively. We mark them from number 1 to number 5. The precisions of sample set under the different conditions are obtained as shown in Table 1. From the Table 1 we can see that the performance of without-deep learning model is better than the with-deep learning model when the ratio of the training sample sets is 1:1. The with-deep learning model becomes better

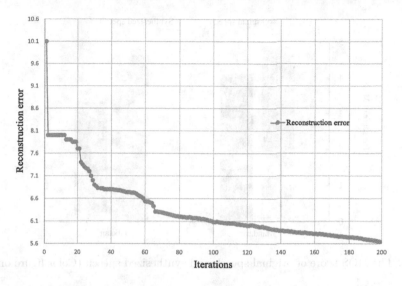

Fig. 7. A mutative process of reconstruction error in the training of deep model

Table 1. The precision of sample sets under the different conditions

Number	1	2	3	4	5
With-deep learning	0.89	0.98	0.95	0.98	0.88
Without-deep learning	0.94	0.96	0.86	0.93	0.81

when the unbalanced training set increases while without-deep learning model becomes a little worse. This illustrates that adopting with-deep learning model is able to obtain the better recognition results. It also explains that with-deep learning model extracts the essential characteristics of the target, so the unbalanced training set can be still classified effectively.

5.2.2 Speech Synthesis

We adopt the mean opinion score (MOS) test to evaluate the synthesized speech. Firstly we randomly play the 60 synthetic Mandarin to the 22 Mandarin subjects who are requested to carefully listen to these utterances and score the naturalness of every utterance by a 5-point score. After conducting MOS evaluation, we ask the subjects to describe the overall intelligibility about the Chinese speech which can express the different recognized sign languages. In the Tibetan MOS evaluation, we use the same method to evaluate the Tibetan speech. We also randomly play the synthesized Tibetan Lhasa dialect utterances to the 22 Tibetan subjects who are requested to carefully listen to these utterances and score the naturalness of every utterance by a 5-point score.

Figure 8 compares the average MOS score of the synthesized Mandarin speech and synthesized Tibetan speech. As can be seen, the average MOS score of the

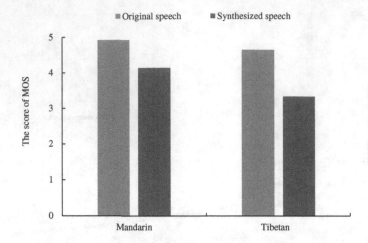

Fig. 8. The MOS score of original speech and synthesized speech (Color figure online)

synthesized Mandarin speech is over 4.0, which is better than the synthesized Tibetan speech. The synthesized Tibetan speech obtains 3.0 scores which has good intelligibility.

Meanwhile, in order to measure the synthesized speech, we also respectively calculate the root mean square error (RMSE) of duration and fundamental frequency between the original speech and the synthetic speech from the same pronunciation people. The results are shown in Tables 2 and 3. As can be seen, the duration and fundamental frequency of the original speech and the synthetic speech are relatively small, where the Mandarin is smaller than the Tibetan.

In addition, we invite 22 subjects who understand sign language to evaluate the quality of synthesized speech for sign languages. The synthesized speech of the recognized sign languages is firstly played to subjects who are requested to listen to the synthesized speech carefully and select the speech from 4 given alternative answers according to the heard speech. We calculate the average accuracy rate of judgment. Then we randomly play 10 kinds of the synthesized

Table 2. RMSE of duration between the original speech and the synthetic speech

Synthesized speech	Mandarin	Tibetan
RMSE (ms)	0.156	0.243

Table 3. RMSE of fundamental frequency between the original speech and the synthetic speech

Synthesized speech	Mandarin	Tibetan
RMSE (Hz)	10.6	15.7

Table 4. Evaluate the quality of synthesized speech for sign language

Evaluated ways	Word	Phrase	Sentence	Sign language
Accuracy (%)	72.49	86.83	96.36	89.87

speech of phrases which can express the meaning of sign languages. The subjects are asked to select the answers corresponding to the sign languages from 3 given alternative answers according to the heard speech. Meanwhile, we calculate the average accuracy rate of judgment. Finally the synthesized speech is played to the subjects to select the sign language that expresses the meaning of synthesized speech. By this way, we can determine whether the sign language can express the synthesized speech and calculate the average accuracy rate of judgment.

We also calculate the average accuracy rate aiming at evaluating the quality of synthesized speech for sign languages as shown in Table 4. We can find that the average accuracy rate of synthesized speech expressing the sign languages have 3 situations under the condition of known sign languages. The syllables are up to 72.49 %, the words account for 86.3 %, while the sentences are 96.36 %. Meanwhile, in the case of playing synthesized speech and asking the subjects to select the recognized sign language, the average accuracy rate of recognizing the sign language counts for 89.87 %.

6   Conclusions

This paper presents a framework for realizing sign language to Mandarin-Tibetan bilingual speech conversion. Firstly, the semantic of each sign language is designed to generate the context-dependent labels in accordance with the recognized sign language. Meanwhile, we use speaker adaptive training techniques to achieve a Mandarin-Tibetan bilingual speech synthesis system. Mandarin or Tibetan speech is then naturally synthesized by using context-dependent label generated from the recognized sign language. Experimental results show that the designed system achieves 93.6 % of recognition rate on static sign language recognition. The system also achieves 4.0 of average MOS score on Mandarin speech and 3.0 of the average MOS score on Tibetan speech. Further work will be focused on more complex sign language characteristic shape and optimize algorithm so that it can improve recognition rate and recognition speed. Then we will use well-designed small Tibetan corpus to improve the speech quality of synthetic Tibetan speech in order to make it much better for speech disorders to live in the world.

Acknowledgements. The research leading to these results was partly funded by the National Natural Science Foundation of China (Grant No. 61263036, 61262055), Gansu Science Fund for Distinguished Young Scholars (Grant No. 1210RJDA007) and Natural Science Foundation of Gansu (Grant No. 1506RJYA126).

References

1. Wachs, J.P., Kölsch, M., Stern, H., Edan, Y.: Vision-based hand-gesture applications. J. Commun. ACM. **54**(2), 60–71 (2011)
2. Ren, Z., Yuan, J., Zhang, Z.: Robust hand gesture recognition based on finger-earth mover's distance with a commodity depth camera. In: Proceedings of the 19th ACM International Conference on Multimedia, pp. 1093–1096. ACM (2011)
3. Doliotis, P., Stefan, A., McMurrough, C., et al.: Comparing gesture recognition accuracy using color and depth information. In: Proceedings of the 4th International Conference on Pervasive Technologies Related to Assistive Environments, p. 20. ACM (2011)
4. Zen, H., Tokuda, K., Black, A.W.: Statistical parametric speech synthesis. J. Speech Commun. **51**(11), 1039–1064 (2009)
5. Yamagishi, J., Kobayashi, T., Nakano, Y., et al.: Analysis of speaker adaptation algorithms for HMM-based speech synthesis and a constrained SMAPLR adaptation algorithm. J IEEE Trans. Audio Speech Lang Process. **17**(1), 66–83 (2009)
6. Bourlard, H., Dines, J., Magimai-Doss, M., Garner, P.N., Imseng, D., Motlicek, P., Valente, F.: Current trends in multilingual speech processing. J. Sadhana **36**(5), 885–915 (2011)
7. Tang Y., Xu J., Fang Z.: A design of the portable gloves which is identified sign language and work on embedded systems. In: Proceedings of the Ninth National Information Acquisition and Processing Conference, pp. 5–8 (2011). (in Chinese)
8. Bengio, Y.: Learning deep architectures for AI. Found. Trends Mach. Learn. J. **2**(1), 1–127 (2009)
9. Salakhutdinov, R., Hinton, G.E.: Deep Boltzmann machines. In: International Conference on Artificial Intelligence and Statistics, pp. 448–455 (2009)
10. Shuping, L., Yu, L., Jun, Y., et al.: Hierarchical static hand gesture recognition by combining finger detection and HOG features. J. Image Graph. **20**(6), 781–788 (2015)
11. Feng, F., Li, R., Wang, X.: Deep correspondence restricted Boltzmann machine for cross-modal retrieval. J. Neurocomput. **154**, 50–60 (2014)
12. Yu, D., Deng, L., Dahl, G.: Roles of pre-training and fine-tuning in context-dependent DBN-HMMs for real-world speech recognition. In: Proceedings of NIPS Workshop on Deep Learning and Unsupervised Feature Learning (2010)
13. Hu, M., Chen, Y., Kwok, J.T.Y.: Building sparse multiple-kernel SVM classifiers. IEEE Trans. Neural Netw. **20**(5), 827–839 (2009)
14. Yang, H., Oura, K., Wang, H., et al.: Using speaker adaptive training to realize Mandarin-Tibetan cross-lingual speech synthesis. J. Multimed. Tools Appl. **74**(22), 9927–9942 (2015)
15. Peng, X., Oura, K., Nankaku, Y., Tokuda, K.: Cross-lingual speaker adaptation for HMM-based speech synthesis considering differences between language-dependent average voices. In: Proceedings of ICSP, pp. 605–608 (2010)
16. Yang, H., Zhu, L.: Predicting Chinese prosodic boundary based on syntactic features. J. Northwest Norm. Univ. (Natural Science) **49**(1), 41–45 (2013)
17. Wang, S.M., Gao, Y., Luo, L.: Human posture recognition based on DAG-SVMS. Adv. Mater. Res. **1042**, 117–120 (2014)

Understand the Customer Preference of Different Market Segments from Online Word of Mouth: Evidence from the China Auto Industry

Xudong Liu[1]([✉]), Lina Zhou[1], and Jinquan Gong[2]

[1] Harbin Institute of Technology, Harbin 150001, China
{cameran, zln}@hit.edu.cn
[2] SUNY University at Buffalo, Buffalo, NY 14260, USA
jinquan.gong@gmail.com

Abstract. With the explosive growth of social media, manufacturers and marketers are increasingly using public opinions in social media for their business decision making. This article aims to understand the customer preference of different market segments from online word of mouth. Using the real data collected from the leading online community of China auto and the ordered Logit model, we find that from the viewpoint of car types, the price is the most important factor for the customer satisfaction. However, from the viewpoint of price range, the price is no more the most important factor for the customer satisfaction. From both viewpoints, the results are similar that different car attributes have different weights on satisfaction. Our findings will make the car manufacturers and marketers come away with a more detailed understanding of the mindset of Chinese consumer groups, and keep a competitive advantage in the fierce China auto market.

Keywords: Online word of mouth · Customer preference · Social media analysis

1 Introduction

Online WOM is a useful tool for customers to reduce perceived risk by searching for information before buying new products [1]. After a purchase has been made, online WOM offers an easy and convenient way for consumers to comment on their acquisitions, complain about their dissatisfaction, share details with friends, or even argue with vendors [2]. Researches have shown that WOM communication is more influential than communications through other sources such as editorial recommendations or advertisements, because it is perceived to provide comparatively reliable information [3]. A survey on Wall Street Journal also showed that 71 % of online U.S. adults use online consumer reviews for their purchases and 42 % of them trust such a source [4].

Online WOM is also useful for marketers and manufacturers. On one hand, online WOM has been regarded as a positive and effective marketing tool through which to

© Springer Science+Business Media Singapore 2016
W. Che et al. (Eds.): ICYCSEE 2016, Part I, CCIS 623, pp. 691–702, 2016.
DOI: 10.1007/978-981-10-2053-7_62

execute business strategies [5]. For example, some businesses regularly post product information online and sponsor promotional events in online communities [6], or even manipulate online reviews strategically to influence purchase decisions [7]. On the other hand, online WOM is useful for the manufacturers and marketers to understand the customers' requirement. For example, manufacturers can use online WOM to decide to create new models or redesign existing models [8]. Manufacturers and marketers can derive the pricing power of product features [9], or estimate the aggregated customer preferences [10] to decide on which product attributes to promote and highlight in online or offline advertisements.

China has overtaken the US to become the largest car market in the world from 2009 [11], however the personal car ownership in China still remains low. According to CAAM [12], the car ownership rate is 105.83 cars per thousand people in the end of year 2014. The rapid growth makes the online automobile communities very popular. For example, Autohome, the leading China online automobile community, reports more than 17 million unique visitors per day [13]. Autohome provides a sophisticated online WOM column for the car buyers to share their experience of the car buying and usage. Using the China auto market as the setting for our study, we hope to address the question: How to infer the customers' preferences of the different car segments from online word of mouth? We believe the findings of our study will make the car manufacturers and marketers come away with a more detailed understanding of the mindset of Chinese consumer groups, and keep a competitive advantage in the fierce China auto market.

The rest of this paper is organized as follows. After briefly discussing the relationship between WOM and customer satisfaction, we propose the research hypotheses and build our empirical models. Using data collected from Autohome, we then present the statistical results and findings. The paper concludes with a review of the theoretical and managerial implications of our results, and a discussion of the limitations of this study and future research directions.

2 Relationship Between Customer Satisfaction and WOM and Hypothesis

Satisfaction is an outcome of purchase and use resulting from the buyer's comparison of the rewards and costs of the purchase in relation to the anticipated consequences [14]. Customer satisfaction with a product or service has emerged as a key driver of WOM [15]. Anderson found the relationship between WOM and customer satisfaction has an asymmetric U-shape that extremely satisfied customers and extremely dissatisfied customers engage in greater word of mouth than the customers experiencing moderate levels of satisfaction [16]. Similar to Anderson, Hu et al. found the WOM have J-shaped distribution that people tend to write reviews only when they are either extremely satisfied or extremely unsatisfied, and people who feel the product is average might not be bothered to write a review [17]. Obviously, the WOM implicates the satisfaction and preference of the customers.

Consumer behavior models typically suggest that (dis)satisfaction results from an evaluation of the rewards and sacrifices associated with the purchase [18]. Customers always need to balance many factors to make a satisfied purchase decision. In general, the price is believed to be the dominant determinant factor of satisfaction, and the car is still a quite expensive product compared with the income of the normal citizen in China, so we propose that price is the most important factor and has a negative impact on satisfaction judgments of the car. Although the price is believed the dominant determinant of satisfaction, it is not the unique factor. Many other factors, such as exterior, interior, space, fuel consumption, etc., also affect the customer satisfaction with a car. For the car manufacturers and marketers, they need to understand the preferences of the car buyers to improve the level of customer satisfaction and to maximize profit.

As the key consequence of customer satisfaction [15], the WOM of a product or service also includes many attributes and the ratings or reviews for these attributes. For example, Autohome provides a 5-star rating system (Likert scale) for the WOM of a car, and the attributes of the WOM include exterior, interior, space, comfort, handling, engine, fuel consumption, and performance price ratio. In general, the ratings or the sentiments of these attributes in WOM are different, and this reflects the preference of the car buyer. For example, a customer may evaluate a car as follow: 5 stars for space and exterior, 4 stars for the fuel consumption and handling, 2 stars for interior and comfort, 5 stars for the performance price ratio. We can conclude that the customer is very satisfied with the space and exterior of the car, but unsatisfied with the interior and comfort of the car. Performance price ratio can be regarded as an overall indicator of customer satisfaction. It refers to a product's ability to deliver performance for its price, and it equals to the performance divided by the price. Products with a higher performance price ratio are more desirable. For example, as mentioned above, the car buyer is very satisfied with the car as a whole, although he is unsatisfied with the interior and comfort of the car, and we can infer the car buyer thinks the price is reasonable for him. Contrarily, perceived price performance inconsistency has a stronger negative effect on satisfaction judgments [19].

According to the size or price of cars, the auto market can be divided into different segments. We propose the car attributes have positive effect on the satisfaction, and in different market segments the car attributes have different weights on the satisfaction. That is to say, we can use this information to infer the aggregate preference of the car buyers in different market segments. For example, for the microcar segment, the fuel consumption may be viewed as the most important factor besides price. And for the compact car segment, the space may be the most important factor besides price.

The paper by Decker et al. [10] is particularly close to our research. In their paper, because the WOM data only include review text and one overall rating of the product, lack numeric ratings for the product attributes, they firstly used natural language processing technique to extract the product attributes and the customer option (pros or cons), and then used Poisson, negative binomial and latent class Poisson regression approaches to estimate the aggregate consumer preferences. And limit to data, they only had about 20000 product reviews covering 4 leading brands in Germany mobile phone market. Another paper by Archak et al. [9] is also very close to our research. Similar to Decker et al. [10], they firstly used text mining to incorporate review text in a consumer choice model by decomposing textual reviews into segments describing different product features, then examined the economic impact of different product attributes and opinions on product sales by using the GMM and Dynamic GMM. Both Decker [10] and Archak [9] didn't consider the consumers' preference of different segments.

In this paper, the numeric ratings for product attributes can be got directly, so we need not to use natural language processing and text mining techniques to extract the product attributes and opinions. And because the dependent variable is ordinal, we use the ordered Logit model to examine the preference of the customers. By analyzing the preference of the customers, the results can help car manufacturers deeply understand the customers' requirements and then make correct decision for product development or improvement, and can also help the marketers decide on which features to promote and highlight in advertisements.

3   Research Methodology

3.1   Data

Data for this study were collected from Autohome [13], which is the leading online destination for automobile consumers in China. The data set includes more than 500,000 customer reviews covering almost all the car models in China auto markets from September 2012 to July 2015. In contrast to reviews of Amazon, the reviews of Autohome include more WOM information. There are 7 ratings for car quality evaluations and 1 overall rating: exterior, interior, space, comfort, handling, engine, fuel consumption and performance-price satisfaction. Besides these numeric ratings, the car model, buy date, buy address, dealer, real fuel consumption, mileage, usage purpose, photos, purchase price, and the text reviews are also provided. In addition, the customer also can add more text reviews whenever they want. Because the text reviews are consistent to the ratings, we need not use nature language processing or text mining technique to analyze them. Figure 1 is a sample of WOM from Autohome.

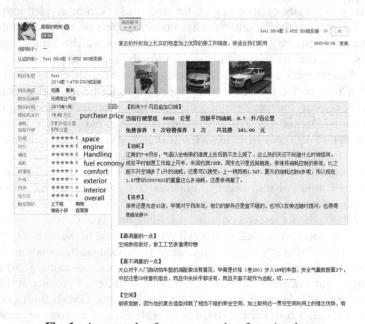

Fig. 1. An example of customer review from Autohome

For the purpose of our analysis, we excluded cars that were: (a) with incorrect price, for example a car with price 99,990,000 RMB which is the default value for purchase price; (b) not belong to the passenger cars. And according to Autohome, the cars are classified into 11 types: microcar (A00), economic car (A0), compact car (A), mid-size car (B), full-size car (C), luxury car (D), Sports car (S), SUV, MPV, Pick-up (P) and Microvan (MV). And from the viewpoint of car price, the cars are classified into 10 ranges. After the data distillation, our final data set consists of 108 car brands, 729 car models, 12273 car sub-models and 502789 car ratings. The key statistical data are shown in Tables 1, 2 and 3.

Table 1. Statistics of the overall data

Attributes	Original data	Distillated data
# Car companies	137	133
# Car brands	110	108
# Car models	741	729
# Car sub-models	12467	12273
# Car ratings	504982	502789
# Unique reviewers	444220	442488

Table 2. Statistics for different car types

Car types		A00	A0	A	B	C	D	S	SUV	MPV	P	MV
Min price[a]		2.08	2.25	3.32	6.05	6.8	30	9.98	3.56	3.49	5	2.3
Max price[a]		23	49.5	59.72	133	196	409.9	264	339	86	77	10
Average price[a]		4.22	7.07	10.67	20.65	44.34	108.95	40.97	19.72	13.33	13.13	4.83
Ratings	#	14613	63444	201758	73985	8545	874	1041	115877	9780	538	13099
	%	2.9 %	12.6 %	40.1 %	14.7 %	1.7 %	0.2 %	0.2 %	23 %	1.9 %	0.1 %	2.6 %

[a]10 thousands RMB.

Table 3. Statistics for different price ranges

Price ranges[a]		2–5	5–10	10–15	15–20	20–25	25–30	30–35	35–50	50–70	>70
Ratings	#	30627	182758	144958	57623	36071	19513	8555	14347	4384	3953
	%	6.1 %	36.3 %	28.8 %	11.5 %	7.2 %	3.8 %	1.7 %	2.9 %	0.9 %	0.8 %

[a]10 thousands RMB.

3.2 Empirical Model

We are interested in understanding the relationship between online WOM and customer satisfaction for different car segments. As mentioned above, besides the price, the word of mouth from Autohome includes 7 ratings for the car quality attributes and 1 overall satisfaction rating considering both performance and price. Inspired by [10], we model the overall rating as a function of the price and the 7 ratings for the car quality attributes. We specify the equation as bellow, and the variables are descripted in Table 4.

$$PPR = \beta_{0i} + \beta_{1i}SPA_{ij} + \beta_{2i}ENG_{ij} + \beta_{3i}DRV_{ij} + \beta_{4i}GAS_{ij} \\ + \beta_{5i}CMF_{ij} + \beta_{6i}INT_{ij} + \beta_{7i}EXT_{ij} + \beta_{8i}PRC_{ij} + \varepsilon_{ij} \tag{1}$$

In above equation, the subscript i means the car types (or price ranges), and the subscript j means the review id of the i-th car type (or price ranges). Because the price has different scale from the other 7 ratings (using the Likert scale), in order to make them the same scale, we use the transformed price to estimate the weights of the car attributes and price. The transformed equation is as follow:

$$TPRC_{ij} = 5*PRC_{ij}/\max(PRC_i) \tag{2}$$

So the final equation is as bellow:

$$PPR = \beta_{0i} + \beta_{1i}SPA_{ij} + \beta_{2i}ENG_{ij} + \beta_{3i}DRV_{ij} + \beta_{4i}GAS_{ij} \\ + \beta_{5i}CMF_{ij} + \beta_{6i}INT_{ij} + \beta_{7i}EXT_{ij} + \beta_{8i}TPRC_{ij} + \varepsilon_{ij} \tag{3}$$

Because the dependent variable is an ordinal variable (Likert scale), we employ the ordered logit model to estimate the coefficients. Obviously, the performance price ratio has a negative relationship with price, so we expect $\beta_8 < 0$.

Our empirical model is based on the conjoint analysis theory [20, 21]. The focus of conjoint analysis is squarely on the measurement of buyer preferences for product attribute levels (including price) and the buyer benefits that may flow from the product attributes [20]. And conjoint analysis has become the most used marketing research method for analyzing consumer trade-offs [21]. Our collected data from Autohome is very suitable for this scenario. And the customer preference in the same market

Table 4. Description of variables

Variable name	Description
i	id of car types (or price ranges)
j	review id of the i-th car type (or price ranges)
PRC	Purchase price
SPA	rating of space satisfaction (Likert scale)
ENG	rating of Engine satisfaction (Likert scale)
DRV	rating of Handling satisfaction (Likert scale)
GAS	rating of fuel consumption satisfaction (Likert scale)
CMF	rating of Comfort satisfaction (Likert scale)
INT	rating of Interior satisfaction (Likert scale)
EXT	rating of Exterior satisfaction (Likert scale)
PPR	rating of overall satisfaction (Likert scale)
TPRC	Transformed Purchase price(Likert scale)

segment can be viewed as the homogenous preference, so we apply the similar model used by Decker et al. [10], that is to say we also model the overall product rating as a function of the relative product attributes.

There are two differences between our model and the model of Decker et al. First, we use the ordered Logit model to analysis the consumers' preference, but Decker et al. used Poisson regression. This is because our WOM data use the Likert scale for the product attributes' rating and the overall rating. The ratings for product attributes used by Decker et al. only has two values (1 and 2) to indicate the pros and cons inferred by the text mining technique. And our WOM data are ordered data, so we apply the ordered Logit model. Second, we didn't consider the brand factor in our model but Decker et al. took it into consideration. The reason is that our aim is to analysis the customer preference of different market segments, not of the concrete brands. Limit to the data, Decker et al. just inferred the customer preference from four leading brands in German mobile phone market.

4 Results and Discussions

4.1 Results from Viewpoint of Car Types

We use the ordered Logit model to analysis the data from the viewpoint of car types, and the results are shown in Table 5. We can make several inferences from the regression coefficients. First, the coefficient on price is negative for all of the car types. And the price is the most important factor for the customer satisfaction, except for the sports car and pick-up segments which only occupy a small market share in total. That is to say, the price is very sensitive for most of the Chinese car buyers. The result also suggests the price has more impact on the family cars (including A0, A, B, etc.) than the luxury cars (including C, D, etc.). So the first hypothesis is confirmed.

Second, as we expected, most of the coefficients on the product attributes are positive and significant. As a whole, the comfort, interior, space and fuel consumption attributes have more weights than other attributes for the Chinese car buyers, However, for the different car types, the car attributes have different weights on the customer satisfaction of the Chinese car buyers. We also use Likelihood ratio test (LRT) to examine our hypothesis, the results (LR chi2(120) = 28323.40, p = 0.000) reject the hypothesis that the coefficients for all the segments are the same and align very well with our expectation. So the second hypothesis is confirmed. Especially, we also find that the A cars and SUV cars are most popular for the Chinese car buyers. And for A cars, the interior, engine, comfort and handling are more important for the Chinese car buyers. The fuel consumption, space, comfort and interior are believed more important for SUV cars.

4.2 Results from Viewpoint of Price Ranges

In general, a car buyer always has an acceptable price range when he or she is going to buy a car, so we believe it is more meaningful to analysis the customer preference from the viewpoint of price ranges. According to Autohome, the price range can be divided

Table 5. Ordered Logit regression results for car types

	All	A00	A0	A	B	C	D	S	SUV	MPV	P	MV
TPRC	-2.60***	-1.75***	-2.67***	-2.48***	-2.15***	-1.59***	-1.23***	-.122	-1.75***	-.824***	-.079	-.564***
	(.0209)	(.0513)	(.0397)	(.0184)	(.0317)	(.0740)	(.1406)	(.1020)	(.0257)	(.0413)	(.1202)	(.0339)
SPA	.390***	.307***	.431***	.217***	.142***	.156**	.251	.206**	.487***	.591***	.256	.545***
	(.0041)	(.0199)	(.0105)	(.0068)	(.0123)	(.0435)	(.1493)	(.0763)	(.0095)	(.0505)	(.1339)	(.0325)
ENG	.278***	.039	.334***	.467***	.332***	.541***	.194	.420***	.171***	.245***	.295*	.176***
	(.0044)	(.0252)	(.0126)	(.0068)	(.0124)	(.0382)	(.1080)	(.1045)	(.0097)	(.0346)	(.1334)	(.0281)
DRV	.290***	.306***	.215***	.362***	.360 ***	.234***	.522***	-.051	.319***	.245***	.306*	.231***
	(.0048)	(.0258)	(.0131)	(.0077)	(.0131)	(.0392)	(.1175)	(.1202)	(.0109)	(.0354)	(.1336)	(.0291)
GAS	.376***	.623***	.472***	.303***	.341***	.315***	.516***	.208**	.489***	.370***	.495***	.502***
	(.0035)	(.0206)	(.0100)	(.0055)	(.0095)	(.0289)	(.0827)	(.0657)	(.0079)	(.0261)	(.1070)	(.0229)
CMF	.412***	.170***	.332***	.398***	.521***	.485***	.344**	.219**	.409***	.384***	.532***	.380***
	(.0046)	(.0276)	(.0136)	(.0073)	(.0121)	(.0383)	(.1081)	(.0797)	(.0103)	(.0333)	(.1372)	(.0279)
EXT	.235***	.418***	.321***	.227***	.219***	.349***	-.161	.261	.300***	.431***	.322	.375***
	(.0050)	(.0256)	(.0135)	(.0079)	(.0148)	(.0482)	(.1534)	(.1543)	(.0110)	(.0346)	(.1380)	(.0287)
INT	.391***	.453***	.534***	.548***	.546***	.678***	.601***	.330***	.369***	.505***	.810***	.410***
	(.0044)	(.0270)	(.0130)	(.0073)	(.0125)	(.0392)	(.1227)	(.0860)	(.0093)	(.0329)	(.1262)	(.0283)
N	502789	14512	63242	201576	73777	8540	874	1041	115835	9776	537	13079
R^2	0.152	0.180	0.200	0.180	0.160	0.165	0.134	0.060	0.169	0.190	0.293	0.189

*$p < 0.05$; **$p < 0.01$; ***$p < 0.001$.

Table 6. Ordered Logit regression results for price ranges

	All	2-5	5-10	10-15	15-20	20-25	25-30	30-35	35-50	50-70	>70
TPRC	-2.60***	-.286***	-.311***	-.620***	.160***	-.128**	-.064	-.095	-.257***	.158*	-.841***
	(.0209)	(.0178)	(.0078)	(.0121)	(.0236)	(.0393)	(.0596)	(.1058)	(.0397)	(.0725)	(.0765)
SPA	.390***	.269***	.396***	.464***	.486***	.296***	.258***	.380***	.312***	.339***	.292
	(.0041)	(.0140)	(.0070)	(.0079)	(.0129)	(.0164)	(.0211)	(.0320)	(.0239)	(.0416)	(.0500)
ENG	.278***	.126***	.314***	.488***	.113***	.148***	.318***	.372***	.415***	.406***	.300***
	(.0044)	(.0177)	(.0075)	(.0081)	(.0133)	(.0174)	(.0233)	(.0384)	(.0279)	(.0481)	(.0565)
DRV	.290***	.275***	.361***	.386***	.307***	.373***	.236***	.213***	.096**	-.010	.342***
	(.0048)	(.0181)	(.0082)	(.0091)	(.0146)	(.0187)	(.0253)	(.0392)	(.0297)	(.0503)	(.0543)
GAS	.376***	.533***	.374***	.312***	.215***	.324***	.342***	.233***	.265***	.270***	.409***
	(.0035)	(.0137)	(.0062)	(.0065)	(.0102)	(.0130)	(.0174)	(.0279)	(.0206)	(.0346)	(.0360)
CMF	.412***	.272***	.464***	.441***	.612***	.598***	.617***	.598***	.587***	.516***	.278***
	(.0046)	(.0193)	(.0081)	(.0088)	(.0134)	(.0170)	(.0225)	(.0337)	(.0268)	(.0449)	(.0462)
EXT	.235***	.387***	.260***	.202***	.228***	.309***	.354***	.207***	.328***	.312***	.368***
	(.0050)	(.0175)	(.0081)	(.0097)	(.0157)	(.0197)	(.0279)	(.0458)	(.0361)	(.0578)	(.0656)
INT	.391***	.423***	.526***	.398***	.322***	.408***	.396***	.598***	.547***	.350***	.395***
	(.0044)	(.0186)	(.0078)	(.0084)	(.0130)	(.0161)	(.0220)	(.0348)	(.0278)	(.0451)	(.0456)
N	502789	30627	182758	144958	57623	36071	19513	8555	14347	4384	3953
R^2	0.152	0.180	0.200	0.180	0.160	0.165	0.134	0.060	0.169	0.190	0.293

*$p < 0.05$; **$p < 0.01$; ***$p < 0.001$.

into 10 intervals as shown in the first line of Table 6, where 2–5 means 20,000 RMB to 50,000 RMB, and so forth. We use the same model to analysis the data, the results are shown in Table 6.

First, most of the coefficients of purchase price are negative, but the price is no more the most important factor of customer satisfaction. Only for the cars with price range 10–15, the price plays the most important role. And it is interesting that for the cars with price range 15–20 and 50–70, the price is positive and significant. The price is not significant for the cars with price range 25–30 and 30–35. So the first hypothesis is partially supported from the viewpoint of price ranges.

Second, for the different price ranges, the car attributes have different weights on the customer satisfaction of the Chinese car buyers. We also use Likelihood ratio test (LRT) to examine our hypothesis, the results (LR chi2(108) = 18478.72, p = 0.000) reject the hypothesis that the coefficients for all the segments are the same and align very well with our expectation. So the second hypothesis is confirmed. Especially, we find that the cars with price range from 50,000 RMB to 150,000 RMB account for the largest market share. And for the cars with price range from 50,000 RMB to 100,000 RMB, the Chinese car buyers pay more attention to the interior, comfort and space. The price, engine, space, and comfort are believed more important for cars with price range from 100,000 RMB to 150,000 RMB.

In sum, from the viewpoint of car types, the price is the most important factor for the customer satisfaction, however, from the viewpoint of price ranges, the price is no more the most important factor for the customer satisfaction. From both viewpoint, it is same that different car attributes have different weights on satisfaction.

5   Conclusions and Management Implications

In this paper, we examine the relationships between online WOM and customer satisfaction in China auto industry. From the leading China auto online community, Autohome, we collected more than 500,000 customer reviews covering almost all the car models in China auto market from September 2012 to July 2015. These data are used to calibrate the ordered Logit model, we find that from the viewpoint of car types, the price is the dominant determinant factor of satisfaction for the Chinese car buyers, however, from the viewpoint of price range, the price is not the most important factor of satisfaction for the Chinese car buyers. From both viewpoint, the price has negative impact on customer satisfaction in the majority of cases. And from both viewpoint, it is same that different car attributes have different weights on satisfaction. We also use Likelihood ratio test (LRT) to examine our hypothesis, the results align very well with our expectation.

Several studies have examined that online WOM is useful for the marketers and manufacturers [5–10], our study contributes to this growing literature. And there are two important contributions of this paper that are very useful for the car marketers and manufacturers in China. First, in common sense, people think that the price is the dominant determinant factor that influence the car buyers' purchase decisions. However in our paper we find that from the perspective of price range, the price is no more the most important factor. This means that if the car manufacturers can deeply understand the customers' requirements and make the correct market positioning, they need not to use the low price strategy to keep or improve their firm competitiveness, and have the chance to make more profit.

Furthermore, our second contribution is to help the car marketers and manufacturers to deeply understand the customers' requirements and to make the correct market positioning. We show them in different market segments, the different car attributes have different weights on satisfaction. Such information can help manufacturers facilitate changes in product design as well as help marketers decide on which product attributes to promote and highlight in online or offline advertisements.

Our research has certain limits. Firstly, our study is based on the assumption that there are no fake reviews, but in fact, because of the free speech nature of online website, it is not possible to avoid the spammers who might be led by commercial or even intentionally manipulative interests. Secondly, limit to data, some factors, e.g. brand image, time to market, cost of maintenance, safety, quality and service, are not considered in this paper. Another issue is self-selection bias. According to [17], the online ratings of products have J-shaped distribution which is because people tend to write reviews only when they are either extremely satisfied or extremely unsatisfied, and people who feel the product is average might not be bothered to write a review. In fact, it is a general phenomenon for the online reviews. However the information of the customers who didn't post review cannot be got, so Hackman two step estimate method can't be used to the correct the select bias.

According to above limits, future work could be devoted to filter the fake reviews to improve the veracity of the results. In order to avoid the self-selection bias, a possible method is to integrate multi-sourced data to calibrate the models. Another direction is to add temporal dimension to the presented work, and it would be interesting to investigate the evolution of the preference of the customers.

Acknowledgments. We thank the anonymous reviewers for their constructive and meaningful advices. This research is partially supported by NKBRPC (2014CB340506) and China Scholarship Council. Any opinions, findings, and conclusions expressed in this material are those of the authors and do not necessarily reflect the views of China Scholarship Council.

References

1. Zhu, F., Zhang, X.: Impact of online consumer reviews on sales: the moderating role of product and consumer characteristics. J. Mark. **74**(2), 133–148 (2010)
2. Lu, Q., Ye, Q., Law, R.: Moderating effects of product heterogeneity between online word-of-mouth and hotel sales. J. Electron. Commer. Res. **15**(1), 1–12 (2014)
3. Bambauer-Sachse, S., Mangold, S.: Brand equity dilution through negative online word-of-mouth communication. J. Retail. Consum. Serv. **18**(1), 38–45 (2011)
4. Spors, K.: How Are We Doing? Wall Street J. R9 (2006)
5. Litvin, S.W., Goldsmith, R.E., Pan, B.: Electronic word-of-mouth in hospitality and tourism management. Tourism Manag. **29**(3), 458–468 (2008)
6. Mayzlin, D.: Promotional chat on the internet. Mark. Sci. **25**(2), 155–163 (2006)

7. Dellarocas, C.: Strategic manipulation of internet opinion forums: implications for consumers and firms. Manag. Sci. **52**(10), 1577–1593 (2006)

8. Feng, J., Papatla, P.: Is online word of mouth higher for new models or redesigns? An investigation of the automobile industry. J. Interact. Mark. **26**(2), 92–101 (2012)

9. Archak, N., Ghose, A., Ipeirotis, P.G.: Deriving the pricing power of product features by mining consumer reviews. Manag. Sci. **57**(8), 1485–1509 (2011)

10. Decker, R., Trusov, M.: Estimating aggregate consumer preferences from online product reviews. Int. J. Res. Mark. **27**(4), 293–307 (2010)

11. Economist.com 23rd Oct 2009. http://www.economist.com/node/14732026. Accessed 31 July 2015

12. China's car ownership rate exceed 100 cars per thousand people for the first time (in Chinese). http://www.caam.org.cn/zhengche/20150227/1505148841.html. Accessed 31 July 2015

13. About Autohome (in Chinese). http://www.autohome.com.cn/about/index.htm. Accessed 31 July 2015

14. Churchill Jr., G.A., Surprenant, C.: An investigation into the determinants of customer satisfaction. J. Mark. Res. **19**(4), 491–504 (1982)

15. Lang, B., Hyde, K.F.: Word of mouth: what we know and what we have yet to learn. J. Consum. Satisfaction Dissatisfaction Complaining Behav. **26**, 3–18 (2013)

16. Anderson, E.W.: Customer satisfaction and word of mouth. J. Serv. Res. **1**(1), 5–17 (1998)

17. Hu, N., Zhang, J., Pavlou, P.A.: Overcoming the J-shaped distribution of product reviews. Commun. ACM **52**(10), 144–147 (2009)

18. Howard, J.A., Sheth, J.N.: The Theory of Buyer Behavior. Wiley, New York (1969)

19. Voss, G.B., Parasuraman, A., Grewal, D.: The roles of price, performance, and expectations in determining satisfaction in service exchanges. J. Mark. **62**(4), 46–61 (1998)

20. Green, P.E., Krieger, A.M.: Segmenting markets with conjoint analysis. J. Mark. **55**(4), 20–31 (1991)

21. Green, P.E., Krieger, A.M., Yoram, W.: Thirty years of conjoint analysis: reflections and prospects. Interfaces **31**(3 supplement), S56–S73 (2001)

Using Gaussian Mixture Model to Fix Errors in SFS Approach Based on Propagation

Wenmin Huang[1], Jiquan Ma[1,2(✉)], and Enbin Zhang[1,2]

[1] College of Computer Science and Technology, Heilongjiang University, Harbin, China
{350387018,277729287}@qq.com, majiquan@hlju.edu.cn
[2] Key Laboratory of Database and Parallel Computing of Heilongjiang Province, Harbin, China

Abstract. A new Gaussian mixture model is used to improve the quality of propagation method for SFS in this paper. The improved algorithm can overcome most difficulties of propagation SFS method including slow convergence, interdependence of propagation nodes and error accumulation. To slow convergence and interdependence of propagation nodes, stable propagation source and integration path are used to make sure that the reconstruction work of each pixel in the image is independent. A Gaussian mixture model based on prior conditions is proposed to fix the error of integration. Good result has been achieved in the experiment for Lambert composite image of the front illumination.

Keywords: Shape from shading · Propagation method · Silhouette · Gaussian mixture model · Surface reconstruction

1 Introduction

Shape from shading plays an important role in surface reconstruction. It attracts lots of researchers' attention for the fact that the algorithm can reconstruct the shape of items only base on one image. The main difficulty of shape from shading problem is that it is an ill-posed problem to infer the height just based on one image's shading. To overcome the difficulty, smooth constraint and integrate constraint are usually introduced. In recent years, most researches on SFS concentrate their work on prior-conditions [1]. Propagation method is a classical SFS method. The basic idea of the method is propagating the height information from initialized point to the whole image. Most representative method of propagation method includes characteristic expansion method, minimum downhill method, contour method and fast marching method [2–5]. But these methods suffer from general difficulties of propagation method. Nowadays probabilistic methods like belief propagation have been used to solve SFS problem and many of them concentrate on fixing the error of basic SFS method [6–8]. This paper introduces a Gaussian mixture model to improve the reconstruction work of propagation method.

© Springer Science+Business Media Singapore 2016
W. Che et al. (Eds.): ICYCSEE 2016, Part I, CCIS 623, pp. 703–712, 2016.
DOI: 10.1007/978-981-10-2053-7_63

2   Basic Model of SFS Problem

The basic model of SFS problem builds on three basic assumptions: 1. the item to be reconstructed has a uniform albedo and Lambert faces; 2. the light source is assumed as infinite parallel light; 3. the picture is taken under orthogonal imaging condition.

Under previous assumptions, near the optical axis, the relationship between gray scale images and normalized reflection intensity can be depicted as Eq. 1.

$$I_N(x, y) = \frac{E(x, y) - E_{min}}{E_{max} - E_{min}} \tag{1}$$

Where $E(x, y)$ is the gray scale intensity, E_{min} is the minimal value of E, E_{max} is the maximal value of E.

Based on the lambert assumption the light reflection equation can be shown as 2.

$$I_N(x, y) = \frac{\vec{N} \cdot \vec{S}}{\|\vec{N}\| \cdot \|\vec{S}\|} \tag{2}$$

If the light source vector is parallel to the optical axis, it can be expressed as a unit vector. Considering $Z(x, y)$ is the depth we will reconstruct from the image, $p(x, y)$ the gradient of Z for x-axis direction and $q(x, y)$ is the gradient of Z for y-axis direction. Equation 3 shows the relationship between image intensity and height gradient.

$$I_N(x, y) = \frac{1}{\sqrt{p^2 + q^2 + 1}} \tag{3}$$

Equation 3 describes that intensity plays a binding effect on height's change and it is impossible to solve the depth z only by intensity constraint. As a result, SFS is an ill-posed problem. To fix the illness, we will introduce a silhouette prior condition based on statistics and a stable propagation method for SFS.

3   Propagation Method for SFS

3.1   A Silhouette Prior Condition Based on Statistics

For most regular objects, silhouette is a very important feature to predict the shape. Stephan Richter finds a relationship between silhouette and object's normal vector by a statistic work [9]. He uses the relationship in his machine learning SFS method. In this article, we try to use it to fix the illness of SFS problem.

Firstly, Stephan Richter defines the relative distance from a pixel to its boundary as Eq. 4:

$$d_{\text{rel}} = \min_{b \in B} \min_{m \in M} \frac{\|p - b\|}{\|m - b\|} \tag{4}$$

Where b is an arbitrary point at the boundary, m is an arbitrary point at the mediate axis, p is the pixel point we will construct. Mediate axis is defined as a set which consists of points with same distance to the boundary.

The angle to the boundary is defined by the gradient of the relative distance as Eq. 5:

$$\beta(p) = -\frac{\nabla d_{rel}(p)}{\|\nabla d_{rel}(p)\|} \tag{5}$$

After a statistic work based on MIT, Blob database, the law can be depicted as Eq. 6:

$$\begin{cases} \varphi = cos^{-1} d_{rel}(p) \\ \theta = \beta(p) \end{cases} \tag{6}$$

Where φ is the polar angle and θ is the azimuth angle of a spherical coordinate system. It shows that the normal vector can be predicted by the pixel's distance to the boundary.

This statistic law is valuable for that we can predict an item's normal by calculating its relative distance to silhouette. Moreover, in most situations, silhouette information is easily to extract. Figure 1 shows the relative distance defined by Eq. 4 of two synthetic images. The statistic law can only apply to convex object.

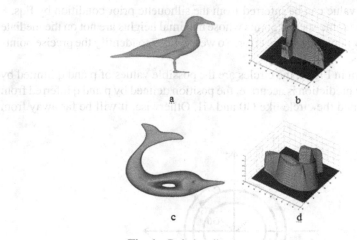

Fig. 1. Relative distance to the boundary

Now, we introduce a new prior condition to SFS problem. In the next section, this prior condition will be used to estimate the height gradient.

3.2 Estimate the Height Gradient

Before we reconstruct the height of object's surface, Eq. 6 is used to recover the height gradient. Firstly, Eq. 6 can be rewritten as Eq. 7.

$$p^2 + q^2 = \left(\frac{1}{I_N}\right)^2 - 1 \tag{7}$$

From Eq. 7, we can conclude p and q distribute on a circle whose radius is decided by I_N. Because polar angle and azimuth angle are coarsely predicted by Eq. 6, we can use Eq. 7 to check the predicted result. The normalized object's normal vector $N(n_x, n_y, n_z)$ can be inferred from φ and θ as Eq. 8.

$$\begin{cases} n_x = sin\varphi cos\theta \\ n_y = sin\varphi sin\theta \\ n_z = cos\varphi \end{cases} \tag{8}$$

Then the gradients p and q prepared for propagation can be calculated by Eq. 9.

$$\begin{cases} p = -\dfrac{n_x}{n_z} \\ q = -\dfrac{n_y}{n_z} \end{cases} \tag{9}$$

Thus gradients' value can be inferred from the silhouette prior condition by Eqs. 8 and 9. It is obvious to some special points whose maximal heights are not on the mediate axis, the statistic law may cause large error. So we use Eq. 7 to identify the precise points from the error points.

As what is shown in Fig. 2, the circles are the possible values of p and q limited by Eq. 7. If our statistic prediction is accurate, the position defined by p and q inferred from Eq. 9 should be around the circle like G0 and G1. Otherwise, it will be far away from the circle like G2.

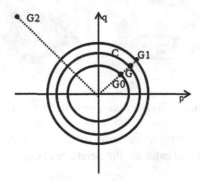

Fig. 2. Ranges for gradient

If the prediction is accurate, we connect the point G1 and original point O. Then the intersection point of OG1 and circle C is chosen as a fixed point to G1. Otherwise a bilinear interpolation method will be used to instead statistic prediction. The ring in the

Fig. 2 depicts the range of error. Width of the ring is defined as γ. Fixed value of p and q can be calculated as Eq. 10: where (p_0, q_0) is the predicted value inferred from Eq. 9, R_0 can be calculated by $\sqrt{p_0^2 + q_0^2}$. In experiment γ is set as 0.2 to obtain a good result.

$$
\begin{cases}
\dfrac{p}{q} = \dfrac{p_0}{q_0} \\
p^2 + q^2 = \left(\dfrac{1}{I_N}\right)^2 - 1
\end{cases}
\quad if \quad
\dfrac{\left| R_0 - \sqrt{\left(\dfrac{1}{I_N}\right)^2 - 1} \right|}{\sqrt{\left(\dfrac{1}{I_N}\right)^2 - 1}} < \gamma
\tag{10}
$$

3.3 Propagation Method for Height

There are three big problems of propagation SFS method: the propagation is basically a integration process and the error will accumulate as its path grows; in many propagation methods, one node can't be reconstructed until its neighbor's height has already been known; at last, when one node received two different propagation messages, it's hard to decide which one is more accurate.

To overcome above difficulties, this paper uses stable propagation source and path which can be inferred from silhouette.

As what is shown in Fig. 3, propagation path is defined as the shortest path from the reconstructing point to the boundary. Obviously, every point in the image has its own reconstructing process. So problem 1 and problem 2 are resolved. Moreover, parallel algorithm can be applied to the reconstruction to improve algorithm's efficiency.

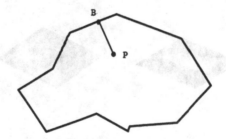

Fig. 3. Propagation for the reconstruction height

In characteristic expansion method, Horn defined the characteristic line as Eq. 11.

$$
(\Delta x, \Delta y) = (R_p \Delta s, R_q \Delta s)
\tag{11}
$$

In the image, change of height can be described as 12

$$
\Delta H(x, y) = p(x_0, y_0)\Delta x + q(x_0, y_0)\Delta y
\tag{12}
$$

For $P(x_k, y_k)$ is a reconstructing point in the image, the integration path can be shown as $S\{(x_i, y_i)\}$ i $\in (1, k)$. $\{(p_i, q_i)\}$ is the gradient of each point, the propagation process can be defined as Eq. 13 [10].

$$H(x_k, y_k) = \sum_{i=1}^{i=k} (p_i R_{p_i} + q_i R_{q_i}) \Delta s \tag{13}$$

Where Δs is propagation distance, R_{p_i} is I_N's partial derivative for p, R_{q_i} is I_N's partial derivative for q.

In practice, we consider the eight points as a neighborhood in propagation. Then we use a vector to judge which point should be transmitted. The vector always points to the destination node which means P in Fig. 3.

The experiment result of propagation method can be depicted as Figs. 4 and 5. Figure 4 is input of four images and Fig. 5 is the propagation result of the input.

Fig. 4. Synthetic grey level image as input

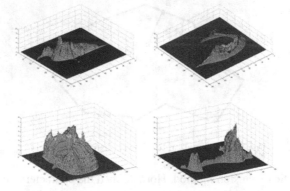

Fig. 5. Propagation algorithm's result

The synthetic image is captured by models downloaded from the online artificial database 3D66 [11].

From Figs. 4 and 5, we can see the error most concentrate on the central area of the image. The distance of integration path has a big effect on the propagation process. But in general, the propagation result can coarsely describe the shape of reconstructing object. In the next section, we will use a probabilistic model to fix the integration error.

4   Using Gaussian Mixture Model to Fix Errors

4.1   A Fix Based on Gaussian Mixture Model

In this section, a few prior conditions are used to fix the propagation error in a Gaussian mixture model. Gaussian mixture model has been widely used in motion detection and machine learning [12–14]. To a convex object, in the boundary the pixel has a low-level intensity and in the central area the intensity is high. By observing, to most convex objects, height may accumulate as d_{rel} increases, but when the d_{rel} arrives at 1 the height reaches its max value. We argue that if there is a relationship between d_{rel} and normal vector, there must be a coarse relationship between d_{rel} and height.

Equation 14 is used to describe the relationship between d_{rel} and height.

$$H(d_{\text{rel}}) = d_{\text{rel}}(2 - d_{\text{rel}})h_k \tag{14}$$

When argument satisfies $d_{rel} = 0$, it means the point is on the boundary and for $d_{rel} = 1$ means the point is on the medial axis. It is obviously that Eq. 14 satisfies the boundary condition of height.

The derivative of $H(d_{rel})$ can be shown as Eq. 15.

$$H'(d_{\text{rel}}) = 2(1 - d_{\text{rel}})h_k \tag{15}$$

We can see the function $H'(d_{rel})$ is a monotonically decreasing function whose range is between 0 and $2h_k$. It satisfies the fact that the height's gradient reaches its max value on the boundary and reaches its min value on the medial axis.

To include more prior conditions to fix the height, we use a Gaussian mixture model as Eq. 16.

$$p(H = h_0) = w_{smooth}\left(e^{-(1-w_1)\cdot\frac{(H - H_0)^2}{2}} + e^{-w_1\cdot\frac{(H - H_1)^2}{2}} \right) \tag{16}$$

In Eq. 16, w_{smooth} is smoothness factor which can be defined as Eq. 17.

$$w_{\text{smooth}}(H) = e^{-\delta} \tag{17}$$

Where δ is defined as Eq. 18 and m is the max value of the gradients in the four neighborhood.

$$\delta = \frac{|H(x+1,y) - H(x,y)| + |H(x,y)| - H(x-1,y) + |H(x,y-1) - H(x,y)| + |H(x,y) - H(x+1,y)|}{4*m}$$

(18)

Factor w_1 can be defined as Eq. 19. It means that we only use the Eq. 14 to fix the central area where errors accumulate. k is a control factor. In experiment the value of k is 0.8.

$$w_1 = k \cdot d_{rel}$$

(19)

At last, we scan the possible values of H to find the most probable value of each pixel.

4.2 Experiment

The experiment implements our algorithm on four synthetic images including pigeon, dolphin, pepper and deer. The reconstruction result of pigeon and dolphin is depicted as Figs. 6 and 7.

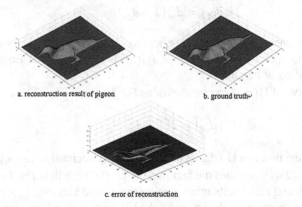

a. reconstruction result of pigeon   b. ground truth

c. error of reconstruction

Fig. 6. Reconstruction result of pigeon

There are two ways to calculate the error: if the ground truth is unknown, MSE of intensity is used to measure the error; if the ground truth is known, MSE of height is used to instead of intensity.

Because of the different scales of each coordinate system, we use a normalized height to measure the error. The mean error and mean square error of propagation method and fixed propagation method are shown in Tables 1 and 2.

Table 1. Error of the propagation method

Synthetic image	Pigeon	Dolphin	Pepper	Deer
Mean error	0.0299	0.0404	0.0276	0.0334
MSE	0.0018	0.0045	0.0013	0.0026

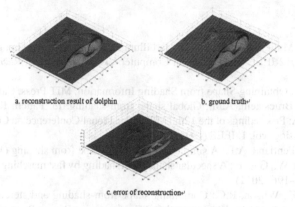

a. reconstruction result of dolphin   b. ground truth

c. error of reconstruction

Fig. 7. Reconstruction result of dolphin

Table 2. Error of the fixed propagation method

Synthetic image	Pigeon	Dolphin	Pepper	Deer
Mean error	0.2164	0.1460	0.2242	0.2006
MSE	0.0717	0.0395	0.0811	0.0778

From Tables 1 and 2, We can see that the average error of fixed propagation method is substantially concentrated in 0.03 and the max MSE is 0.0045 for complex surface shape as image dolphin. For original propagation method the mean error is around 0.2 while the MSE is around 0.07. The error is significantly limited by using our Gaussian mixture model.

From Figs. 6 and 7 we can see the central area has been fixed by our probabilistic method. But in the boundary, the reconstruction result is not so good. One of the possible reason is Eq. 14 is not a suitable prior hypothesis when the reconstructing points are approaching boundary. So, further work should concentrate on study of prior conditions of boundary points.

5   Conclusion

This paper proposes a Gaussian mixture model to fix the error of propagation SFS method. Four synthetic images have been reconstructed using the proposed algorithm. The experiment result shows that the model can fix most flaws of propagation but the model itself may attract error in the boundary. The most drawback of our method is that it can only be applied to convex object. As a result, the future work will concentrate on the prior features of points in the boundary and more appropriate probabilistic model to mix different prior conditions together.

References

1. Barron, J.T., Malik, J.: Shape, albedo, and illumination from a single image of an unknown object. In: 2012 IEEE Conference on Computer Vision and Pattern Recognition (CVPR). IEEE (2012)
2. Horn, B.K.P.: Obtaining Shape from Shading Information. MIT Press, Cambridge (1989)
3. Kimmel, R., Bruckstein, A.M.: Global shape from shading. In: Pattern Recognition, 1994. Conference A: Proceedings of the 12th IAPR International Conference on Computer Vision & Image Processing, vol. 1. IEEE (1994)
4. Bichsel, M., Pentland, A.P.: A simple algorithm for shape from shading (1992)
5. Wang, G., Su, W., Gao, F.: A specular shape from shading by fast marching method. Procedia Eng. **24**, 192–196 (2011)
6. Haines, T.S.F., Wilson, R.C.: Combining shape-from-shading and stereo using Gaussian-Markov random fields. In: 19th International Conference on Pattern Recognition, 2008, ICPR 2008. IEEE (2008)
7. Wilhelmy, J., Krüger, J.: Shape from shading using probability functions and belief propagation. Int. J. Comput. Vis. **84**(3), 269–287 (2009)
8. Potetz, B.: Efficient belief propagation for vision using linear constraint nodes. In: IEEE Conference on Computer Vision and Pattern Recognition, 2007, CVPR 2007. IEEE (2007)
9. Richter, S.R., Roth, S.: A discriminative approach to perspective shape from shading in uncalibrated illumination. Comput. Graph. **53**, 72–81 (2015)
10. Bruckstein, A.M.: On shape from shading. Comput. Vis. Graph. Image Process. **44**(2), 139–154 (1988)
11. http://www.3d66.com/model_1_8.html
12. Stauffer, C., Grimson, W.E.L.: Adaptive background mixture models for real-time tracking. In: IEEE Computer Society Conference on Computer Vision and Pattern Recognition, 1999, vol. 2. IEEE (1999)
13. Huang, Y., et al.: A Gaussian mixture model based classification scheme for myoelectric control of powered upper limb prostheses. IEEE Trans. Biomed. Eng. **52**(11), 1801–1811 (2005)
14. Zivkovic, Z.: Improved adaptive Gaussian mixture model for background subtraction. In: Proceedings of the 17th International Conference on Pattern Recognition, 2004, ICPR 2004, vol. 2. IEEE (2004)

Author Index

Printed in the United States
By Bookmasters